Lecture Notes in Computer Science　12305

More information about this series at http://www.springer.com/series/7412

Yuxin Peng · Qingshan Liu ·
Huchuan Lu · Zhenan Sun ·
Chenglin Liu · Xilin Chen ·
Hongbin Zha · Jian Yang (Eds.)

Pattern Recognition and Computer Vision

Third Chinese Conference, PRCV 2020
Nanjing, China, October 16–18, 2020
Proceedings, Part I

Springer

Editors
Yuxin Peng
Peking University
Beijing, China

Huchuan Lu
Dalian University of Technology
Dalian, China

Chenglin Liu
Chinese Academy of Sciences
Beijing, China

Hongbin Zha
Peking University
Beijing, China

Qingshan Liu
Nanjing University of Information Science
and Technology
Nanjing, China

Zhenan Sun
Chinese Academy of Sciences
Beijing, China

Xilin Chen
Institute of Computing Technology
Chinese Academy of Sciences
Bcijing, China

Jian Yang
Nanjing University of Science
and Technology
Nanjing, China

ISSN 0302-9743   ISSN 1611-3349   (electronic)
Lecture Notes in Computer Science
ISBN 978-3-030-60632-9   ISBN 978-3-030-60633-6   (eBook)
https://doi.org/10.1007/978-3-030-60633-6

LNCS Sublibrary: SL6 – Image Processing, Computer Vision, Pattern Recognition, and Graphics

This Springer imprint is published by the registered company Springer Nature Switzerland AG
The registered company address is: Gewerbestrasse 11, 6330 Cham, Switzerland

Preface

Welcome to the proceedings of the Third Chinese Conference on Pattern Recognition and Computer Vision (PRCV 2020) held in Nanjing, China.

PRCV is the merger of Chinese Conference on Pattern Recognition (CCPR) and Chinese Conference on Computer Vision (CCCV), which are both the most influential Chinese conferences on pattern recognition and computer vision, respectively. Pattern recognition and computer vision are closely interrelated and the two communities are largely overlapping. The goal of merging CCPR and CCCV into PRCV is to further boost the impact of the Chinese community in these two core areas of artificial intelligence and further improve the quality of academic communication. Accordingly, PRCV is co-sponsored by four major academic societies of China: the Chinese Association for Artificial Intelligence (CAAI), the China Computer Federation (CCF), the Chinese Association of Automation (CAA), and the China Society of Image and Graphics (CSIG).

PRCV aims at providing an interactive communication platform for researchers from academia and industry. It promotes not only academic exchange, but also communication between academia and industry. In order to keep at the frontier of academic trends and share the latest research achievements, innovative ideas, and scientific methods in the fields of pattern recognition and computer vision, international and local leading experts and professors are invited to deliver keynote speeches, introducing the latest advances in theories and methods in the fields of pattern recognition and computer vision.

PRCV 2020 was hosted by Nanjing University of Science and Technology and was co-hosted by Nanjing University of Information Science and Technology, Southeast University, and JiangSu Association of Artificial Intelligence. We received 402 full submissions. Each submission was reviewed by at least three reviewers selected from the Program Committee and other qualified researchers. Based on the reviewers' reports, 158 papers were finally accepted for presentation at the conference, including 30 orals, 60 spotlights, and 68 posters. The acceptance rate is 39%. The proceedings of PRCV 2020 are published by Springer.

We are grateful to the keynote speakers, Prof. Nanning Zheng from Xi'an Jiaotong University, China, Prof. Jean Ponce from PSL University, France, Prof. Mubarak Shah from University of Central Florida, USA, and Prof. Dacheng Tao from The University of Sydney, Australia.

We give sincere thanks to the authors of all submitted papers, the Program Committee members and the reviewers, and the Organizing Committee. Without their contributions, this conference would not be a success. Special thanks also go to all of the sponsors and the organizers of the special forums; their support made the conference a success. We are also grateful to Springer for publishing the proceedings and especially to Ms. Celine (Lanlan) Chang of Springer Asia for her efforts in coordinating the publication.

We hope you find the proceedings enjoyable and fruitful.

September 2020

Yuxin Peng
Qingshan Liu
Huchuan Lu
Zhenan Sun
Chenglin Liu
Xilin Chen
Hongbin Zha
Jian Yang

Organization

Steering Committee Chair

Tieniu Tan   Institute of Automation, Chinese Academy of Sciences, China

Steering Committee

Xilin Chen   Institute of Computing Technology, Chinese Academy of Sciences, China

Chenglin Liu   Institute of Automation, Chinese Academy of Sciences, China

Yong Rui   Lenovo, China

Hongbin Zha   Peking University, China

Nanning Zheng   Xi'an Jiaotong University, China

Jie Zhou   Tsinghua University, China

Steering Committee Secretariat

Liang Wang   Institute of Automation, Chinese Academy of Sciences, China

General Chairs

Chenglin Liu   Institute of Automation, Chinese Academy of Sciences, China

Xilin Chen   Institute of Computing Technology, Chinese Academy of Sciences, China

Hongbin Zha   Peking University, China

Jian Yang   Nanjing University of Science and Technology, China

Program Chairs

Yuxin Peng   Peking University, China

Qingshan Liu   Nanjing University of Information Science and Technology, China

Huchuan Lu   Dalian University of Technology, China

Zhenan Sun   Institute of Automation, Chinese Academy of Sciences, China

Organizing Chairs

Xin Geng	Southeast University, China
Jianfeng Lu	Nanjing University of Science and Technology, China
Liang Xiao	Nanjing University of Science and Technology, China
Jinshan Pan	Nanjing University of Science and Technology, China

Publicity Chairs

Zhaoxiang Zhang	Institute of Automation, Chinese Academy of Sciences, China
Jiaying Liu	Peking University, China
Wankou Yang	Southeast University, China
Lianfa Bai	Nanjing University of Science and Technology, China

International Liaison Chairs

Jingyi Yu	ShanghaiTech University, China
Shiguang Shan	Institute of Computing Technology, Chinese Academy of Sciences, China

Local Coordination Chairs

Wei Fang	JiangSu Association of Artificial Intelligence, China
Jinhui Tang	Nanjing University of Science and Technology, China

Publication Chairs

Risheng Liu	Dalian University of Technology, China
Zhen Cui	Nanjing University of Science and Technology, China

Tutorial Chairs

Gang Pan	Zhejiang University, China
Xiaotong Yuan	Nanjing University of Information Science and Technology, China

Workshop Chairs

Xiang Bai	Huazhong University of Science and Technology, China
Shanshan Zhang	Nanjing University of Science and Technology, China

Special Issue Chairs

Jiwen Lu   Tsinghua University, China
Weishi Zheng   Sun Yat-sen University, China

Sponsorship Chairs

Lianwen Jin   South China University of Technology, China
Jinfeng Yang   Civil Aviation University of China, China
Ming-Ming Cheng   Nankai University, China
Chen Gong   Nanjing University of Science and Technology, China

Demo Chairs

Zechao Li   Nanjing University of Science and Technology, China
Jun Li   Nanjing University of Science and Technology, China

Competition Chairs

Wangmeng Zuo   Harbin Institute of Technology, China
Jin Xie   Nanjing University of Science and Technology, China
Wei Jia   Hefei University of Technology, China

PhD Forum Chairs

Tianzhu Zhang   University of Science and Technology of China, China
Guangcan Liu   Nanjing University of Information Science
                                and Technology, China

Web Chair

Zhichao Lian   Nanjing University of Science and Technology, China

Finance Chair

Jianjun Qian   Nanjing University of Science and Technology, China

Registration Chairs

Guangyu Li   Nanjing University of Science and Technology, China
Weili Guo   Nanjing University of Science and Technology, China

Area Chairs

Zhen Cui   Nanjing University of Science and Technology, China
Yuming Fang   Jiangxi University of Finance and Economics, China

Contents – Part I

Computer Vision and Application

Contents – Part II

Contents – Part III

Computer Vision and Application

Medical CT Image Super-Resolution via Cyclic Feature Concentration Network

Xingchen Liu and Juncheng Jia[✉]

School of Computer Science and Technology, Soochow University, Suzhou 215006, China
jiajuncheng@suda.edu.cn

Abstract. Clear and high-resolution medical computed tomography (CT) images are of great significance for early medical diagnosis. In practice, there are many cases of medical CT images that lead to poor image quality which complicates the diagnosis process. This paper proposes a new type of neural network that enables low-resolution images to reconstruct high-resolution images and aids in the early diagnosis of medical CT images. The network mainly consists of a low-level feature extraction module, a cyclic feature concentration block, and a reconstruction module. First, the underlying feature extraction module extracts the effective feature information of the low-resolution image. Then, the cyclic feature concentration block further extracts the feature information. Finally, the reconstruction module reconstructs the high-definition image. It can be seen from the experiment results that the method proposed in this paper achieved the best results on two objective indicators.

Keywords: CT images · Neural network · Feature extraction · Medical diagnosis

1 Introduction

Computed tomography (CT) plays an important role in medical diagnosis and decision-making [1]. It relies on invisible X-rays to image the patient's bones and soft tissues, then reconstructs the received signals into a multilayer CT image sequence to support doctors' clinical decision.

High-resolution medical CT images are of great help for the detection of medical abnormalities. The image resolution of a CT imaging system is limited by the X-ray focus size, detector element spacing, reconstruction algorithm and other factors. In general, there are several methods to improve the resolution of medical CT images: upgrading imaging sensors, improving reconstruction algorithms and enhancing images after reconstruction. The former two methods are hardware-oriented measures that are usually more expensive and may bring higher doses of X-ray radiation to the patient, which adversely affects health. It was pointed out in the literature [2] that X-ray exposure may increase the risk of cancer induction. Therefore, there is a tradeoff between obtaining high-resolution CT images and keeping low CT radiation, which leads to increasing attention of using super-resolution (SR) image processing techniques to enhance the resolution of CT images [3–7]. SR is to obtain a high-resolution CT image from a

© Springer Nature Switzerland AG 2020
Y. Peng et al. (Eds.): PRCV 2020, LNCS 12305, pp. 3–13, 2020.
https://doi.org/10.1007/978-3-030-60633-6_1

low-resolution CT image. In order to improve the reconstruction effect of medical CT images, we propose a new type of cyclic feature concentration network (CFCN). This network mainly extracts effective feature information through a cyclic feature concentration block, which can quickly and accurately recover high-quality images. The main contributions are as follows:

(1) We propose a cyclic feature concentration network that can quickly and accurately recover high-definition medical CT images.
(2) The proposed cyclic feature concentration block contains four feature concentration blocks, which can extract effective feature information step by step.
(3) Compared with other methods, our proposed method achieved the best results in 3 different testsets.

2 Related Work

In order to improve the image resolution and signal-to-noise ratio, many algorithms have been proposed. These methods are roughly divided into two parts: (1) model-based reconstruction methods [8–10], these technologies accurately simulate the degradation of the image process and reconstruct according to the feature specifications of the projection data; (2) learning-based methods [11–13], which learn non-linear mappings from training datasets containing paired low-resolution and high-definition images to recover lost high frequencies detail. In particular, the method based on sparse representation has attracted more and more attention because it shows strong performance in retaining image features, suppressing noise and artifacts. Dong et al. [13] applied adaptive sparse domain selection and adaptive regularization. Excellent results were obtained both in terms of visual perception and peak signal-to-noise ratio (PSNR). Zhang et al. [12] proposed a patch-based technique to enhance the super-resolution of 4D-CT images. These results show that the super-resolution method based on learning can greatly improve the overall image quality, but the results may still lose the nuances of the image and produce a blocky appearance.

With the rapid development of deep learning, it plays an important role in computer vision tasks [14]. The layered features and representations generated from convolutional neural networks are used to enhance the discriminative ability of visual quality. Super-resolution models [15–17]. Deep learning-based methods optimize image quality by learning functions in an end-to-end manner. More importantly, once a super-resolution model based on a Convolutional Neural Network (CNN) is trained, achieving super-resolution is purely feedforward propagation, which requires very low computational overhead.

In the field of medical imaging, deep learning is an emerging method with great potential [18]. Chen et al. [19] proposed a deep densely connected super-resolution network to reconstruct high-resolution brain magnetic resonance (MR) images. Yu et al. [20] developed two CNN-based advanced models with jump connections to promote high-frequency textures, which were then fused with upsampled images to generate super-resolution images. Recently, adversarial learning has also become more and more popular, which has enabled CNN to learn feature representations from complex data distributions and has achieved great success. Adversarial learning is based on generative adversarial networks (GANs). Wolterink et al. [21] proposed an unsupervised conditional GAN to optimize the non-linear mapping, which successfully improved the overall image quality. Mardani et al. [22] used a GAN reconstruction method based on compressed sensing (CS) in the GANCS-based GANCS network to reconstruct high-quality MR images. In addition, in order to ensure data consistency in the learned manifold domain, a least square penalty is applied to the training process. You et al. [23] proposed a semi-supervised deep learning method based on generative adversarial network (GAN), which can accurately recover high-resolution CT images from low-resolution counterparts. Chi et al. [24] proposed a unified deep convolutional neural network (DCNN) framework to simultaneously process noise reduction and super-resolution of CT images and obtained good results.

3 Proposed Method

In this section, a cyclic feature concentration network (CFCN) for medical CT images based on the work of [17] is introduced. First, we introduce the entire network structure, then we introduce the various components of the network in detail.

3.1 Network Structure

The cyclic feature concentration network proposed in this paper is shown in Fig. 1(a). The entire network consists of three parts. The first part is a low-level feature extraction module composed of 2 convolutional blocks. The second part is cyclic feature concentration block composed of 4 feature concentration blocks. The last part uses the sub-pixel convolution algorithm mentioned in [25] as the reconstruction module. We add bicubic interpolation to the reconstruction module, so that the model can quickly converge during the training process and increase the image details that might be ignored previously.

First, downsample the high-definition image to a low-resolution image and store the low-definition image as a tensor input network with a dimension of 3. The image tensor enters two convolutional layers whose output dimension is 64 and accesses the activation function LRelu to extract feature information. Then, the preliminary extracted feature information enters the CBlock. The composition of the CBlock is described in detail in 3.2. The CBlock can effectively extract local long-path and short-path features. Specifically, the proposed CBlock mixes two different types of features together and the last added 1×1 compression unit can reduce the dimension of the feature map, reduce the parameters and ease the training difficulty. We add two loop structures to the 4 CBlock which can effectively transfer features to the long path and shorten the training time. This is also the main structure of the network. The feature information processed by the cyclic feature concentration block generate a residual image through the reconstruction module. In order to obtain the HR image, we perform element-by-element addition on the generated residual image and the up-sampled LR image subjected to bicubic interpolation.

Here, x represents the input (LR image) of the network, y represents the output (HR image) of the network. The first part of the underlying feature extraction module can be expressed as:

$$A_0 = f(f(x)) \tag{1}$$

It consists of two 3×3 convolution blocks, where f represents the processing of the convolution block, A_0 represents the output of the underlying feature extraction module and the second part of the input. The first two feature concentration blocks in the cyclic feature concentration block can be expressed as:

$$A_k = F_k(A_{k-1}), k = 1, 2 \tag{2}$$

When the value of k is 1, it represents the first feature concentration block, and the value of 2 represents the second feature concentration block. where F_k represents the processing of the k-th CBlock, and A_k and A_{k-1} represent the output and input of the k-th

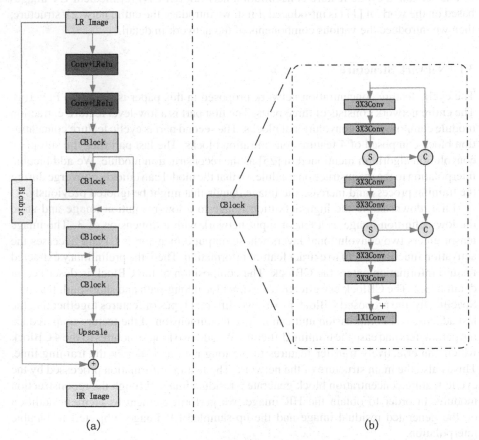

(a)   (b)

Fig. 1. The overall structure of the proposed cyclic feature concentration network

CBlock respectively. Equation 3 represents the third and fourth feature concentration block.

$$A_k = F_k(A_{k-1} + A_{k-2}), k = 3, 4 \tag{3}$$

Therefore, the entire CFCN network can be expressed as:

$$y = T(A_k) + N(x) \tag{4}$$

where T represents the processing of the reconstruction module and N represents the bicubic interpolation operation.

3.2 CBlock Unit

The structure of the CBlock unit is shown in Fig. 1(b). The block is roughly regarded as three parts. The first part and the second part both use two 3×3 convolutional layers, each of which has a leaky rectified linear unit (LReLU) activation function. The third part consists of two 3×3 convolution layers and one 1×1 convolution layer. The 1×1 convolutional layer is used to reduce the dimension and extract the information related to the subsequent network.

The feature map of the first module is extracted through the first part of the short path (2 layers), so they can be regarded as local short path features. Considering that the deep network has stronger expressive ability, we will send some local short-path features to the next module. In this way, the feature maps of the second and third modules naturally become local long-path features. We divide the feature map with the output dimension of 64 into feature maps with the dimensions of 16 and 48. The feature map with the dimension of 48 represents the reserved short path function and the other part of the feature map with the dimension of 16 represents the short path function to be enhanced. The series operation of the short path to be added through the former two parts become the local long path feature. After obtained the long-path and short-path feature maps, we summarize these two types of features to obtain richer and more effective feature information. In summary, the short path of each part is mainly to improve the expression ability of the network and finally add a 1x1 convolutional layer to compress the redundant information in the feature maps.

The first part of the module is composed of two convolution layers. The output of the second convolution layer is divided into two parts. The input of the first part is set to A_{K-1}, we can get:

$$M_1^k = S_1(C_a(A_{k-1})) \tag{5}$$

where A_{K-1} represents the output of the previous CBlock and the input of the current Cblock, C_a means convolution operation, S_1 represents slice operation, M_1^k represents the output of the first part of the k-th CBlock. The second part can be expressed as:

$$M_2^k = S_2(C_a(M_1^k)) \tag{6}$$

where S_2 represents slice operation, M_2^k represents the output of the second part of the k-th CBlock. The connection operations of the first part can be expressed as:

$$N_1^k = C(S(C_a(A_{k-1}), 1/s), A_{k-1}) \tag{7}$$

where C represents concatenation operation and S rep-resents slice operation, $1/s$ represents the part cut from first part and is connected to A_{K-1}. The connection operations of the second part can be expressed as:

$$N_2^k = C(S(C_a(M_1^k), 2/s), N_1^k) \tag{8}$$

where $2/s$ represent the part cut from second part. Through these operations, not only the different feature in-formation can be combined, but also the remaining feature information can be transmitted. Finally, the processing through the 1×1 convolutional layer can be expressed as:

$$A_k = f(C_a(M_2^k) + N_2^k) = f(C_a(S_2(C_a(M_1^k))) + C(S(C_a(M_1^k), 2/s), N_1^k)) \tag{9}$$

where f represents the processing of the 1×1 convolution layer. We omitted the activation function and weight parameters here.

3.3 Reconstruction Module

After the cyclic feature concentration block processing, we added a reconstruction module to process the generated feature map. In this module, a sub-pixel convolution algorithm is used [25]. The input of the entire network is an LR image, which does not need to be processed by traditional interpolation. Obtaining the input image by interpolation will increase the number of parameters of the entire network. By directly input the original low-resolution image to speed up the training process, the final output can be reconstructed by the reconstruction module to the same size image as the ground truth image.

3.4 Loss Function

In this paper, two loss functions are selected to test the direct difference between the reconstructed HD image I_i and the corresponding original HD image $\hat{I}_i$. The first one uses the mean square error (MSE), which is widely used for the loss function of image recovery. The formula is as follows:

$$l_{MSE} = \frac{1}{N} \sum_{i=1}^{N} \left\| I_i - \hat{I}_i \right\|_2^2 \tag{10}$$

The second loss function is the mean absolute error (MAE), which has the following formula:

$$l_{MAE} = \frac{1}{N} \sum_{i=1}^{N} \left\| I_i - \hat{I}_i \right\|_1 \tag{11}$$

According to the experiments, it is found that the model with MSE can improve the performance of training networks with MAE. Therefore, this paper first trains the network with the MAE loss function and then fine-tunes it with the MSE loss function.

4 Experiments

4.1 Datasets

For the experimental dataset, the TCGA-LUAD data from the Cancer Imaging Archive (TCIA) website was selected. This dataset is a CT image of lung cancer. Randomly selected data from dozens of cases were preprocessed into CT images, we compose a training set of 800 images and a validation set of 80 images. Regarding the test sets, three images of different cut planes were selected and three different test sets were randomly formed. In order to compare the effects of reconstruction, two objective indicators commonly used for testing images were selected: peak signal-to-noise ratio (PSNR) [26] and structural similarity index (SSIM) [27]. In order to better compare the reconstruction results, the proposed algorithm is compared with existing algorithms on three different scales (2, 3, and 4).

4.2 Implementation Details

We downsample the original HD images at different scales (2, 3, 4) to get the images needed for training. Most of our convolutional layers use 64 filters, a 3×3 convolution kernel, and a step size of 1. At the end of each CBlock, a 1×1 convolution kernel is used. The batch size is set to 16. The initial learning rate is set to 2 $e-4$ and is accompanied by decay of 0.5 every 2000 epochs. The entire training process is set to 10,000 epochs. The experimental environment is a TensorFlow deep learning framework based on Ubuntu sys-tem. We use CUDA9 + cuDNN7 to accelerate GPU operations. The hardware configuration for training and testing is: Intel Xeon E5-2630 v4 CPU, Nvidia GeForce GTX2080Ti GPU and 32G memory.

4.3 Comparisons with State-of-the-Arts

In order to verify the effectiveness of the method in this paper, SRCNN [28], VDSR [29], IDN [17] algorithms were selected for comparison. Table 1 shows the results of several different algorithms on three test sets. From the two objective indicators of PSNR and SSIM, it can be seen that the method proposed in this paper achieves the best results no matter which scale it is.

The VDSR method is a network composed of 20 layers of convolutions. It uses interpolation to enlarge the LR picture to the desired size and then serves as the input to the network. Each time through a layer of network, the feature map will become smaller, the paper only uses the method of 0 to keep its size unchanged. The DRRN network introduced recursive blocks. Both methods require the original LR image to be inserted to the required size before being applied to the network. This preprocessing step not only increases the computational complexity but also makes the original LR image excessively smooth and blurry, thus losing some details. These methods extract features from the interpolated LR image, thus failing to establish an end-to-end mapping from the original LR to the HR image. The IDN network uses a deconvolution layer in the reconstruction layer to enlarge to the original image size. In general deconvolution, there will be a large number of 0-filled areas, which may adversely affect the result. In view of the above problems, the input image in the method proposed in this paper is a low-resolution picture after downsampling which can reduce the computational complexity. We use the sub-pixel convolution algorithm in the reconstruction module. By combining a single pixel on a multi-channel feature map into a unit on a feature map, each pixel on the feature map is equivalent to a sub-pixel on the new feature map. In this way, reconstruction from low-resolution images to high-resolution images can be achieved. It can be seen from Table 1 that our proposed CFCN has better performance than other methods.

In order to visually compare the reconstruction effects of different algorithms, Figs. 2, 3 and 4 show the reconstruction effects of different algorithms on different scales. For the test image, one image in each of the three test sets is selected. According to the picture display, it can be seen that the reconstruction effect of the SRCNN algorithm is the worst, with considerable blur. Although the VDSR algorithm has some hints, the detail texture is still insufficient. The effect of IDN reconstruction is already clear. Although the method proposed in this paper is similar to the visual perception of IDN results, the proposed method still has the best objective index results. According to the summary, the algorithm in this paper can effectively extract the feature information of the original image and quickly and accurately reconstruct the high-definition image.

Table 1. PSNR and SSIM of different algorithms at 3 scales, bold indicates the best.

	Scale	SRCNN PSNR/SSIM	VDSR PSNR/SSIM	IDN PSNR/SSIM	CFCN PSNR/SSIM
Test1	2	41.43/0.9513	42.89/0.9542	44.62/0.9575	**45.54/0.9690**
	3	38.79/0.9322	40.83/0.9404	42.56/0.9434	**43.55/0.9566**
	4	37.65/0.9266	38.76/0.9281	40.48/0.9310	**41.67/0.9467**
Test2	2	44.17/0.9743	45.39/0.9774	46.84/0.9789	**47.71/0.9857**
	3	42.62/0.9688	43.17/0.9695	44.62/0.9710	**45.57/0.9790**
	4	38.97/0.9598	40.81/0.9609	42.26/0.9624	**43.52/0.9728**
Test3	2	41.84/0.9665	43.09/0.9686	45.90/0.9703	**46.82/0.9793**
	3	38.97/0.9574	40.65/0.9586	43.45/0.9603	**44.50/0.9704**
	4	37.88/0.9445	38.19/0.9469	40.99/0.9486	**42.33/0.9615**

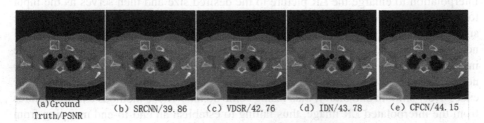

(a) Ground Truth/PSNR   (b) SRCNN/39.86   (c) VDSR/42.76   (d) IDN/43.78   (e) CFCN/44.15

Fig. 2. Reconstruction effect of different algorithms at scale 2

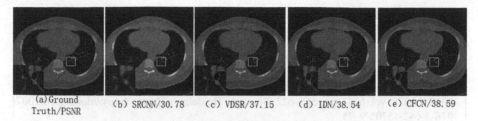

(a) Ground Truth/PSNR   (b) SRCNN/30. 78   (c) VDSR/37. 15   (d) IDN/38. 54   (e) CFCN/38. 59

Fig. 3. Reconstruction effect of different algorithms at scale 3

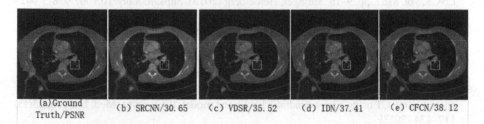

(a) Ground Truth/PSNR   (b) SRCNN/30. 65   (c) VDSR/35. 52   (d) IDN/37. 41   (e) CFCN/38. 12

Fig. 4. Reconstruction effect of different algorithms at scale 4

5   Conclusion

In this paper, a novel cyclic feature concentration network (CFCN) is proposed. The network uses cyclic feature concentration block to extract valid features to reconstruct high-definition images. To further compare the results, we performed experiments on three different scales. Our method is compared with other methods on PSNR and SSIM indicators. The method proposed in this paper is better. It has practical significance for early medical diagnosis of CT images and can assist diagnosis. Future work will further optimize the network, restore high-definition details and perform different optimizations for medical images of different parts.

Acknowledgment. This work is supported in part by grants from the China Postdoctoral Science Foundation No. 2017M611905, Suzhou Industrial Technology Innovation Special Project (Livelihood Technology)—Technology Demonstration Project No. SS201701 No. SYSD2019152, A Project Funded by the Priority Academic Program Development of Jiangsu Higher Education Institutions (PAPD).

References

1. Kak, A.C., Slaney, M., Wang, G.: Principles of computerized tomographic imaging. Med. Phys. **29**(1), 107 (2002)
2. Miglioretti, D.L., Johnson, E., Williams, A., et al.: The use of computed tomography in pediatrics and the associated radiation exposure and estimated cancer risk. JAMA Pediatr. **167**(8), 700–707 (2013)

3. La Rivière, P.J., Bian, J., Vargas, P.A.: Penalized-likelihood sinogram restoration for computed tomography. IEEE Trans. Med. Imaging **25**(8), 1022–1036 (2006)
4. Jiang, M., Wang, G., Skinner, M.W., et al.: Blind deblurring of spiral CT images. IEEE Trans. Med. Imaging **22**(7), 837–845 (2003)
5. Poot, D.H.J., Van Meir, V., Sijbers, J.: General and efficient super-resolution method for multi-slice MRI. In: Jiang, T., Navab, N., Pluim, J.P.W., Viergever, M.A. (eds.) MICCAI 2010. LNCS, vol. 6361, pp. 615–622. Springer, Heidelberg (2010). https://doi.org/10.1007/978-3-642-15705-9_75
6. Shi, W., et al.: Cardiac image super-resolution with global correspondence using multi-atlas patchmatch. In: Mori, K., Sakuma, I., Sato, Y., Barillot, C., Navab, N. (eds.) MICCAI 2013. LNCS, vol. 8151, pp. 9–16. Springer, Heidelberg (2013). https://doi.org/10.1007/978-3-642-40760-4_2
7. Odille, F., Bustin, A., Chen, B., Vuissoz, P.-A., Felblinger, J.: Motion-corrected, super-resolution reconstruction for high-resolution 3D cardiac cine MRI. In: Navab, N., Hornegger, J., Wells, W.M., Frangi, A.F. (eds.) MICCAI 2015. LNCS, vol. 9351, pp. 435–442. Springer, Cham (2015). https://doi.org/10.1007/978-3-319-24574-4_52
8. Zhang, R., Thibault, J.B., Bouman, C.A., et al.: Model-based iterative reconstruction for dual-energy X-ray CT using a joint quadratic likelihood model. IEEE Trans. Med. Imaging **33**(1), 117–134 (2013)
9. Yu, Z., Thibault, J.B., Bouman, C.A., et al.: Fast model-based X-ray CT reconstruction using spatially nonhomogeneous ICD optimization. IEEE Trans. Image Process. **20**(1), 161–175 (2010)
10. Thibault, J.B., Sauer, K.D., Bouman, C.A., et al.: A three-dimensional statistical approach to improved image quality for multislice helical CT. Med. Phys. **34**(11), 4526–4544 (2007)
11. Yang, J., Wright, J., Huang, T.S., et al.: Image super-resolution via sparse representation. IEEE Trans. Image Process. **19**(11), 2861–2873 (2010)
12. Zhang, Y., Wu, G., Yap, P.T., et al.: Reconstruction of super-resolution lung 4D-CT using patch-based sparse representation. In: 2012 IEEE Conference on Computer Vision and Pattern Recognition, pp. 925–931. IEEE (2012)
13. Dong, W., Zhang, L., Shi, G., et al.: Image deblurring and super-resolution by adaptive sparse domain selection and adaptive regularization. IEEE Trans. Image Process. **20**(7), 1838–1857 (2011)
14. LeCun, Y., Bengio, Y., Hinton, G.: Deep learning. Nature **521**(7553), 436–444 (2015)
15. Ledig, C., Theis, L., Huszár, F., et al.: Photo-realistic single image super-resolution using a generative adversarial network. In: Proceedings of the IEEE Conference on Computer Vision and Pattern Recognition, pp. 4681–4690 (2017)
16. Dong, C., Loy, C.C., Tang, X.: Accelerating the super-resolution convolutional neural network. In: Leibe, B., Matas, J., Sebe, N., Welling, M. (eds.) ECCV 2016. LNCS, vol. 9906, pp. 391–407. Springer, Cham (2016). https://doi.org/10.1007/978-3-319-46475-6_25
17. Hui, Z., Wang, X., Gao, X.: Fast and accurate single image super-resolution via information distillation network. In: Proceedings of the IEEE Conference on Computer Vision and Pattern Recognition, pp. 723–731 (2018)
18. Wang, G., Kalra, M., Orton, C.G.: Machine learning will transform radiology significantly within the next 5 years. Med. Phys. **44**(6), 2041–2044 (2017)
19. Chen, Y., Shi, F., Christodoulou, Anthony G., Xie, Y., Zhou, Z., Li, D.: Efficient and accurate MRI super-resolution using a generative adversarial network and 3D multi-level densely connected network. In: Frangi, A.F., Schnabel, J.A., Davatzikos, C., Alberola-López, C., Fichtinger, G. (eds.) MICCAI 2018. LNCS, vol. 11070, pp. 91–99. Springer, Cham (2018). https://doi.org/10.1007/978-3-030-00928-1_11

20. Yu, H., Liu, D., Shi, H., et al.: Computed tomography super-resolution using convolutional neural networks. In: 2017 IEEE International Conference on Image Processing (ICIP), pp. 3944–3948. IEEE (2017)
21. Wolterink, J.M., Leiner, T., Viergever, M.A., et al.: Generative adversarial networks for noise reduction in low-dose CT. IEEE Trans. Med. Imaging 36(12), 2536–2545 (2017)
22. Mardani, M., Gong, E., Cheng, J.Y., et al.: Deep generative adversarial neural networks for compressive sensing MRI. IEEE Trans. Med. Imaging 38(1), 167–179 (2018)
23. You, C., Li, G., Zhang, Y., et al.: CT super-resolution GAN constrained by the identical, residual, and cycle learning ensemble (GAN-CIRCLE). IEEE Trans. Med. Imaging 39(1), 188–203 (2019)
24. Chi, J., Zhang, Y., Yu, X., et al.: Computed tomography (CT) image quality enhancement via a uniform framework integrating noise estimation and super-resolution networks. Sensors 19(15), 3348 (2019)
25. Shi, W., Caballero, J., Huszár, F., et al.: Real-time single image and video super-resolution using an efficient sub-pixel convolutional neural network. In: Proceedings of the IEEE Conference on Computer Vision and Pattern Recognition, pp. 1874–1883 (2016)
26. Huynh-Thu, Q., Ghanbari, M.: Scope of validity of PSNR in image/video quality assessment. Electron. Lett. 44(13), 800–801 (2008)
27. Wang, Z., Bovik, A.C., Sheikh, H.R., et al.: Image quality assessment: from error visibility to structural similarity. IEEE Trans. Image Process. 13(4), 600–612 (2004)
28. Dong, C., Loy, C.C., He, K., et al.: Image super-resolution using deep convolutional networks. IEEE Trans. Pattern Anal. Mach. Intell. 38(2), 295–307 (2015)
29. Kim, J., Lee, J.K., Lee, K.M.: Accurate image super-resolution using very deep convolutional networks. In: Proceedings of the IEEE Conference on Computer Vision and Pattern Recognition, pp. 1646–1654 (2016)

Generative Landmark Guided
Face Inpainting

Yang Yang and Xiaojie Guo[✉]

College of Intelligence and Computing, Tianjin University, Tianjin, China
yangyangcic@tju.edu.cn, xj.max.guo@gmail.com

Abstract. It is challenging to inpaint face images in the wild, due to
the large variation of appearance, such as different poses, expressions and
occlusions. A good inpainting algorithm should guarantee the realism of
output, including the topological structure among eyes, nose and mouth,
as well as the attribute consistency on pose, gender, ethnicity, expression,
etc. This paper studies an effective deep learning based strategy to deal
with these issues, which comprises of a facial landmark predicting subnet
and an image inpainting subnet. Concretely, given partial observation,
the landmark predictor aims to provide the structural information (*e.g.*
topological relationship and expression) of incomplete faces, while the
inpaintor is to generate plausible appearance (*e.g.* gender and ethnicity)
conditioned on the predicted landmarks. Experiments on the CelebA-
HQ and CelebA datasets are conducted to reveal the efficacy of our
design and, to demonstrate its superiority over state-of-the-art alternatives both qualitatively and quantitatively. In addition, we assume that
high-quality completed faces together with their landmarks can be utilized as augmented data to further improve the performance of (any)
landmark predictor, which is corroborated by experimental results on
the 300W and WFLW datasets. The code is available at https://github.
com/YaN9-Y/lafin.

Keywords: Face inpainting · Face alignment · Data augmentation

1 Introduction

Image inpainting (*a.k.a.* image completion) refers to the process of reconstructing
lost or deteriorated regions of images, which can be applied to various tasks such
as image restoration and editing [1,25]. Undoubtedly, one expects the completed
result to be realistic, so that the reconstructed regions can be hardly perceived.
Compared with natural scenes like oceans and lawns, manipulating faces, the
focus of this work, is more challenging. Because the faces have much stronger
topological structure and attribute consistency to preserve. Figure 1 shows two
such examples. Very often, given the observed clues, human beings can easily
infer what the lost parts possibly, although inexactly, look like. As a consequence,

---

This is a student paper.

© Springer Nature Switzerland AG 2020
Y. Peng et al. (Eds.): PRCV 2020, LNCS 12305, pp. 14–26, 2020.
https://doi.org/10.1007/978-3-030-60633-6_2

a slight violation on the topological structure and/or the attribute consistency in the reconstructed face highly likely leads to a significant perceptual flaw.

1.1 Previous Arts

Various image inpainting methods have been developed over the last decades. This part briefly reviews classic and contemporary works closely related to ours.

Fig. 1. Two completion results by our method. For each group, from left to right: corrupted inputs, plus landmarks predicted from the inputs, and final results.

Traditional Inpainting Methods. In this category, diffusion-based and patch-based approaches are two representative branches. Diffusion-based approaches [2, 27] iteratively propagate low-level features around the occluded areas but they are limited to reconstructing structureless and small-size regions. While patch-based methods [1,6] attempt to copy similar blocks from elsewhere to the target regions. On the one hand, their computational cost of calculating the similarity between blocks is expensive. On the other hand, they all hypothesize that the missing part can be found elsewhere, which does not always hold in practice.

Deep Learning-Based Methods. Recently, deep learning based methods have become the mainstream for image inpainting. The context encoder [18], as a pioneer deep-learning method for image completion, introduces an encoder-decoder network trained with an adversarial loss [4]. After that, plenty of follow-ups have been proposed to improve the performance from various aspects. For instance, the scheme by [7] employs both the global and local discriminators to accomplish the task. Another attempt proposed in [29] designs a coarse-to-fine network structure and applies a self-attention layer to connect related features at distant spatial locations. Besides, Yu *et al.* and Liu *et al.* [13,28] upgrade the convolutional layers for making networks adaptive to the masked input. However, most of the above-mentioned methods can barely keep the structure of the original image, and the inpainted result frequently tends to be blurry, especially on large occluded areas. For the sake of maintaining the structure of corrupted images, a number of methods, such as [17,26], try to first predict the edge information for corrupted images, then apply it as a condition to guide the inpainting. These methods work well on small corrupted regions though, when the corruption becomes larger, the performance significantly degrades as it is not easy to predict reasonable edges inside the masked regions, leading to unsatisfactory results.

Deep Face Inpainting Methods. Specific to face completion, the authors of [12] introduces face parsing loss to better preserve the structure. But this work often suffers from the color inconsistency, and lacks of ability in handling faces with large poses. Besides, [9, 28] directly ask users to manually label edges for generating corresponding results. Although providing a flexible way to editing faces, sometimes it is difficult/inconvenient for users to input precise edge information. To relive the requirement from users, Nazeri *et al.* applied a network to predict the edges [17], which however suffers from inaccurate/unreasonable prediction on large holes. Moreover, we argue that, for face completion, both face parsing and edge information are relatively redundant, which may even degenerate the performance when feeding slightly inaccurate information into the inpainting module. Facial landmarks are better to act as the indicator, which are neat, sufficient, and robust to reflect the structure of face, please see Fig. 2 for an example. Many works have successfully applied landmarks to the task of face generation, such as [21, 31] and [30]. It is worth noting that, different from the generation task [31] and [30], in our problem, the landmarks need to be obtained from the corrupted images.

Fig. 2. Illustration of different facial features. From left to right: the input, Canny edges, landmarks, edges by connecting the landmarks, and parsing regions.

1.2   Challenges and Considerations

As stated previously, completing face images in the wild is challenging. A qualified face inpainting algorithm should carefully take into account the following two concerns to guarantee the realism of output: 1) Faces are of strong structure. The topological relationship among facial features including eyebrows, eyes, nose and mouth is always well-organized. The completed faces must satisfy this topology structure primarily; and 2) The attributes of face, such as pose, gender, ethnicity, and expression, should be consistent across the inpainted regions and the observed part. Otherwise, a slight violation on these two factors will result in a significant perceptual flaw.

Why Adopt Landmarks? This work employs facial landmarks as structural guidance, because of their compactness, sufficiency and robustness. One may ask whether the edge or parsing information provide more powerful guidance than the landmarks? If the information is precise, the answer is yes. But, taking the strategy using edges [17] as an example, it is not easy to generate reasonable edges in challenging situations like large-area corrupted faces with large-poses, the redundant and inaccurate information would instead hurt the performance. Alternatively, a set of landmarks (pre-defined fiducial points) always exists, no matter what situation the face is in. Further, the landmarks can be viewed as the discrete/ordered points sampled on the **key** edges/regions of face, which are sufficient to conversely reform the **key** edges/facial regions (face parsing) with redundant information removed. Compared with the edge [17] and parsing [12] information, for one thing, the landmarks are much neater and more robust, please see Fig. 2 for illustration. For another thing, once the landmarks for a face are obtained, the topology structure, pose and expression can be subsequently determined. Moreover, the landmarks are more convenient to control from the editing perspective. These properties support that using landmarks is a better choice for face completion.

1.3 Contributions

This paper presents a deep network, namely Generative **La**ndmark Guided **F**ace **In**paintor (LaFIn for short), which comprises of a facial landmark predicting subnet and an image inpainting subnet, for solving the face inpainting problem. The main contributions can be summarized in the following aspects: 1) As analyzed, facial landmarks are neat, sufficient, and robust to act as the indicator for face inpainting. We construct a module for predicting landmarks on incomplete faces, which reflect the topological structure, pose and expression of the target face; 2) To complete faces, we design an inpainting subnet that employs the predicted landmarks as guidance. For the attribute consistency, the subnet harnesses distant spatial context and connects temporal feature maps; 3) Extensive experiments are conducted to reveal the efficacy of our design and, show its advances over state-of-the-art alternatives both qualitatively and quantitatively. In addition, we can use the completion results to help boosting the performance of data-driven landmark detectors. Our another contribution is: 4) The completion can generate various plausible new faces **conditioned** on the landmarks. Thus, the generated face and the corresponding (ground-truth) landmarks can be employed as the augmented data to relieve the pressure from manual annotation. The effectiveness of this manner is confirmed by experimental results on the WFLW and 300W datasets.

2   Methodology

We build a deep network, denoted as LaFIn. As schematically illustrated in Fig. 3, the network is composed of a subnet for predicting landmarks, and another one for generating new pixels conditioned on the predicted landmarks. In the next subsections, we shall detail the network.

2.1   Landmark Prediction Module

The landmark prediction module $\mathcal{G}_L$ aims to retrieve a set of ($n = 68$ in this work) landmarks from a corrupted face image $I^M := I \circ M$, i.e. $\hat{L} \in \mathbb{R}^{2 \times n} := \mathcal{G}_L(I^M; \theta_L)$, with θ_L the trainable parameters. Technically, $\mathcal{G}_L$ can be accomplished by any landmark detector like [23,24]. Please notice that, for the target inpainting task, what we expect from the landmarks is more about the underlying topology structure and some attributes (pose and expression) than the precise location of each individual landmark. The following may explain the reason: considering the landmarks on face contour for an example, with the corresponding region fixed, shifting them along the contour will not affect the final result much. Consequently, we build a simple yet sufficiently effective $\mathcal{G}_L$. Our $\mathcal{G}_L$ is built upon the MobileNet-V2 model proposed in [19], which focuses on feature extraction. The final landmark prediction is achieved by fully connecting the fused feature maps at different rear stages, as illustrated in Fig. 3. The training loss for $\mathcal{G}_L$ is simply as follows:

$$\mathcal{L}_{lmk} := \|\hat{L} - L_{gt}\|_2^2, \tag{1}$$

where L_{gt} denotes the ground-truth landmarks, and $\|\cdot\|_2$ stands for the ℓ_2 norm.

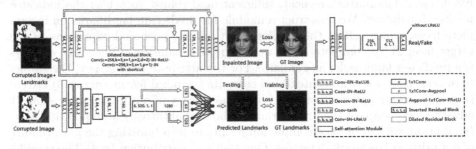

Fig. 3. The architecture of the proposed model. For a corrupted image, the landmarks are first estimated by the landmark prediction module. Then the inpaint module applies the landmarks as prior to inpaint the image. The notations c, k, s and p stand for the channel number, kernel size, stride, and padding, respectively. In addition, for each inverted residual block, it contains a sequence of identical layers repeating t times, and the expansion factor is f.

2.2   Image Inpainting Module

The inpainting network $\mathcal{G}_P$ desires to complete faces by taking corrupted images I^M and their (predicted or ground-truth) landmarks L ($\hat{L}$ or L_{gt}) as input, i.e. $\hat{I} := \mathcal{G}_P(I^M, L; \theta_P)$, with θ_P the network parameters. This subnet comprises of a generator and a discriminator.

Generator. Overall, the generator is based on the U-Net structure. More specifically, the network consists of three gradually down-sampled encoding blocks, followed by 7 residual blocks with dilated convolutions and a long-short term attention block. Then, the decoder processes the feature maps gradually up-sampled to the same size as input. The long-short attention layer [32] is harnessed to connect temporal feature maps, and the stacked dilated blocks are to enlarge the receptive filed so that features in a wider range can be taken into account. Besides, shortcuts are added between corresponding encoder and decoder layers. Moreover, the 1×1 convolution operation is executed before each decoding layer as the channel attention to adjust weights of features from the shortcut and last layer. In such a way, the network can better make use of distant features both spatially and temporally. The structure of the generator can be found in Fig. 3.

Discriminator. In this work, our discriminator is built upon the 70×70 Patch-GAN architecture [8]. To stabilize the training process, we introduce the spectral normalization (SN) [16] into the blocks of the discriminator. Besides, an attention layer is inserted to adaptively treat the features. It is worth to notice that the works like [7] employ local and global discriminators to ensure the global coherency and local consistency. Differently, our discriminator adopts only one judger to accomplish the job, which takes an image and its landmarks as input, *i.e.* $\mathcal{D}(I, L; \theta_D)$ with θ_D the parameters. The reasons are: 1) the generated results are conditioned on the landmarks, already ensuring the global structure; and 2) the attention layer concentrates more on the attribute consistency. The configuration of our discriminator can be found in Fig. 3.

Loss. We use a combination of a per-pixel loss, a perceptual loss, a style loss, a total variation loss and an adversarial loss, for training the inpaintor.

(I) The per-pixel loss is defined as follows:

$$\mathcal{L}_{pixel} := \frac{1}{N_m} \| \hat{I} - I \|_1, \tag{2}$$

where $\| \cdot \|_1$ stands for the ℓ_1 norm. Notice that we use the mask size N_m as the denominator to adjust the penalty. It means that for a small occlusion, the inpainted result should be accurate, while if the corruption is large, the restriction can be relaxed as long as the structure and consistency are rational.

(II) The perceptual loss measures the difference of feature maps extracted from a pre-trained network, which is calculated in the following manner:

$$\mathcal{L}_{perc} := \sum_p \frac{\| \phi_p(\hat{I}) - \phi_p(I) \|_1}{N_p \times H_p \times W_p}, \tag{3}$$

where $\phi_p(\cdot)$ denotes the N_p feature maps with size $H_p \times W_p$ of the p-th layer from the pre-trained network. *relu1_1*, *relu2_1*, *relu3_1*, *relu4_1* and *relu5_1* of the VGG-19 pre-trained on the ImageNet are utilized to calculated the perceptual loss, as well as the style loss described below.

(III) The style loss computes the style distance between two images as follows:

$$\mathcal{L}_{style} := \sum_p \frac{1}{N_p \times N_p} \| \frac{G_p(\hat{I} \circ M) - G_p(I \circ M)}{N_p \times H_p \times W_p} \|_1, \tag{4}$$

where $G_p(x) = \phi_p(x)^T \phi_p(x)$ stands for the Gram Matrix corresponding to $\phi_p(x)$. **(IV)** The total variation loss is utilized to suppress the checkerboard artifact, which is defined as:

$$\mathcal{L}_{tv} := \frac{1}{N_I} \|\nabla \hat{I}\|_1, \tag{5}$$

where N_I is the pixel number of I, and ∇ is the first order derivative, containing ∇_h (horizontal) and ∇_v (vertical).

(V) The adversarial loss adopts the LSGAN proposed in [15], due to its stability during the training process and the advance in visual quality, which is as follows:

$$\mathcal{L}_{adv_G} := \mathbb{E}[(\mathcal{D}(\mathcal{G}_P(I^M, L), L_{gt}) - 1)^2],$$
$$\mathcal{L}_{adv_D} := \mathbb{E}[\mathcal{D}(\hat{I}, L_{gt})^2] + \mathbb{E}[(\mathcal{D}(I, L_{gt}) - 1)^2]. \tag{6}$$

The total loss with respect to the generator yields:

$$\mathcal{L}_{inp} := \mathcal{L}_{pixel} + \lambda_{perc}\mathcal{L}_{perc} + \lambda_{sty}\mathcal{L}_{style}$$
$$+ \lambda_{tv}\mathcal{L}_{tv} + \lambda_{adv}\mathcal{L}_{adv_G}. \tag{7}$$

We use $\lambda_{perc} = 0.1$, $\lambda_{style} = 250$, $\lambda_{tv} = 0.1$ and $\lambda_{adv} = 0.01$ in our experiments. The whole training procedure alternatively minimizes $\mathcal{L}_{inp}$ for the generator $\mathcal{G}_P$ and $\mathcal{L}_{adv_D}$ for the discriminator $\mathcal{D}$ until converged.

2.3   Training Strategy

The generator is desired to complete image via $\hat{I} := \mathcal{G}(I^M)$. For face images, their strong regularity, like the landmarks considered by our design $\hat{I} := \mathcal{G}_P(\mathcal{G}_L(I^M), I^M)$, could benefit model reduction and training procedure, as the space is considerably restricted by the regularity. Intuitively, the training for $\mathcal{G}_P$ and $\mathcal{G}_L$ can be finished jointly. Technically, it is feasible. But, in practice, it is not a good choice. The reasons are as follows: 1) the loss for $\mathcal{G}_L$, say $\mathcal{L}_{lmk}$, computes over a small number of (only 68 in this work) **locations**, which is incompatible with $\mathcal{L}_{inp}$. In other words, the parameter tuning is extremely hard; and 2) even with the well-tuned parameters, the performance of both $\mathcal{G}_L$ and $\mathcal{G}_P$ may be too inaccurate especially at the beginning of training, which consequently leads to low-quality landmark prediction and inpainting results. These two coupled factors very likely drag the training into dilemmas, like bad points of convergence and/or high prices of training. Thus, we decouple the joint model into the landmark prediction and inpainting modules, and train them separately. It is worth to note that we actually have trained the model in a joint way with different carefully-tuned settings, the best shot is still inferior to our separate training. In experiments shown in this work, the landmark prediction model and the inpainting model are trained using 256×256 images and optimized by the Adam optimizer [11] with $\beta_1 = 0$ and $\beta_2 = 0.9$, and the learning rate $= 10^{-4}$. The learning rate of the discriminator is 10^{-5}. We use batch size $= 16$ for the landmark prediction module and batch size $= 4$ for the inpainting model.

3   Experimental Validation on Face Inpainting

In this part, we evaluate the face inpainting performance of our LaFIn on the CelebA-HQ face dataset [10,14]. The masks used for training come from the random mask dataset [13] and additional block masks randomly generated. The competitors include Context Encoder (CE) [18], Generative Face Completion (GFC) [12], Contextual Attention (CA) [29], Geometry Aware Face Completion (GAFC) [20], Pluralistic Image Completion (PIC) [32], and EdgeConnect (EC) [17]. For quantitatively measuring the performance difference among the competitors, we employ PSNR, SSIM [22] and FID [5], as metrics. For PSNR and SSIM, higher values indicate better performance, while for FID, the lower the better. As the ground-truth landmarks are unavailable for the CelebA-HQ dataset, we apply the results by FAN [3] to perform as the reference information for training our landmark predictor.

Result Comparison. Table 1 reports the performance of CA, EC, PIC and our LaFIn with different types and sizes of mask. As can be seen from the numbers in Table 1, EC is superior over PIC and CA in most cases, as it employs the edge information to help inpainting. Overall, our LaFIn outperforms the others by large margins in terms of all PSNR, SSIM and FID, except for the case of center falling behind PIC in terms of FID 6.63 *vs.* 4.98. This comparison verifies that the landmarks are stronger and more robust guidance than the edges for face inpainting.

Table 1. Quantitative comparison on the CelebA-HQ dataset in terms of PSNR, SSIM and FID on random and center masks.

Mask	PSNR			SSIM			FID		
	10–20%	40–50%	Center	10–20%	40–50%	Center	10–20%	40–50%	Center
CA	27.51	20.29	24.13	0.942	0.761	0.864	7.29	38.83	7.39
PIC	30.33	22.58	24.22	0.968	0.838	0.870	2.72	9.24	**4.98**
EC	30.73	23.44	24.82	0.971	0.859	0.874	2.33	8.78	8.26
Ours	**31.48**	**24.22**	**25.92**	**0.975**	**0.883**	**0.905**	**2.05**	**7.12**	6.63

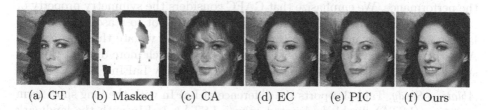

(a) GT   (b) Masked   (c) CA   (d) EC   (e) PIC   (f) Ours

Fig. 4. Visual comparison with state-of-the-art techniques on CelebA-HQ.

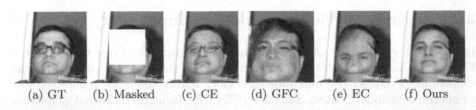

(a) GT   (b) Masked   (c) CE   (d) GFC   (e) EC   (f) Ours

Fig. 5. Visual comparison between the competitors on the CelebA dataset.

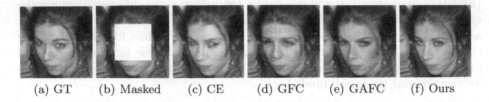

(a) GT   (b) Masked   (c) CE   (d) GFC   (e) GAFC   (f) Ours

Fig. 6. Visual comparison between the competitors on the CelebA dataset.

Further quantitative comparisons with CE, EC and GFC under center masks on CelebA are shown in Table 2. Figure 4 depicts a visual comparison among CA, EC, PIC, and LaFIn, from which, we can see that LaFIn can generate more natural-looking and visually striking results even on the cases with large poses and extreme occlusions. Figure 5 and 6 further provide visual comparisons of CE, GFC, GAFC, EC and LaFIn on two samples from CelebA. Note that GFC utilizes the face parsing information and GAFC uses both the landmark and parsing to guide the inpainting. The results by GFC suffer from the face component shifting problem. GAFC* seems to somewhat mitigate the problem due to the introduction of landmarks, but still inferior to our LaFIn. This comparison tells that the redundancy of the face parsing prior may alternatively hurt the performance. We emphasize that GAFC considers the symmetry property of faces and low rankness of mask in the loss, which are not so reasonable because large poses of faces and random corruptions can easily violate these properties. Also, the editing on parsed regions (+landmarks) is much more difficult than on sparse landmarks. Figure 7 gives several more results by LaFIn.

Table 2. Quantitative comparison on the CelebA dataset in PSNR, SSIM and FID on center masks.

Metric	CE	GFC	EC	Ours
PSNR	25.46	21.04	25.83	**26.25**
SSIM	0.909	0.766	0.899	**0.912**
FID	**1.731**	14.958	3.519	3.512

Ablation Study. Table 3 reports the difference of LaFIn with the long short term attention (LSTA) disabled (denoted as w/o LSTA), LaFIn with the landmark guidance canceled (w/o LMK), and the complete LaFIn. From the numbers

---

* Since neither the code nor implementation details of GAFC is available, when this paper is prepared, we only compare the cases cropped from the GAFC paper.

(a)   (b)   (c)   (d)   (e)   (f)

Fig. 7. (a) and (d) GT images. (b) and (e) Masked images. (c) and (f) Results.

(a)   (b)   (c)   (d)   (e)   (f)

Fig. 8. (a) Ground truth image. (b) the result without landmark information. (c) and (e) present two versions of landmark on the masked image. (d) and (f) give the results conditioned on (c) and (e), respectively.

reported in Table 3, both the LSTA and LMK help the task of face inpainting. Specifically, the LSTA influences more than the landmark indication on the cases with relatively small corruptions. This phenomenon is reasonable because the completed part should pay more attention on the attribute consistency to make the results visually coincident to the observed (large) area. While for the cases with relatively large masks, the attribute consistency is barely violated in the generated result as there is few information given to match. Alternatively the landmark information is more important to ensure the structure well-preserved. The above corroborates the principle of our design, say the LSTA is for the attribute consistency and the landmarks for the main structure. To view the effect of landmark, Fig. 8 shows the inpainting results based on different landmark templates. By varying the templates (mouth), the completed faces accordingly change with much better visual quality than the one without adopting any landmark information. This experiment also informs us that editing faces is viable by manipulating the landmark template. Affirmatively, operating sparse landmarks is more convenient than modifying parsed regions (together with landmarks [20]).

Table 3. Ablation study on different configurations of LaFIn.

Mask		w/o LSTA	w/o LMK	Ours
PSNR	10–20%	31.02	31.10	**31.48**
	40–50%	24.00	23.75	**24.22**
	Center	25.75	24.90	**25.92**
SSIM	10–20%	0.973	0.972	**0.9754**
	40–50%	0.879	0.869	**0.883**
	Center	0.902	0.879	**0.905**
FID	10–20%	2.18	2.26	**2.05**
	40–50%	8.25	7.72	**7.12**
	Center	8.56	7.21	**6.63**

Table 4. Comparison in NME on the WFLW dataset.

LAB	LAB$_{aug}$	LaFIn	LaFIn$_{aug}$
5.66	5.43	6.79	5.92

Table 5. Comparison in NME on the 300W dataset.

Dataset	Common	Challenging	Full
LaFIn	4.69	8.95	5.42
LaFIn$_{aug}$	4.45	8.91	5.21

4 Further Finding on Data Augmentation

Most of data-driven approaches, if not all, require well-labeled data, which is time consuming and labor intensive. Like the original motivation of GANs, it attempts to produce more samples for training networks. Specifically for facial landmark detectors/predictors, one wants to generate diverse plausible faces given the ground-truth landmarks. Intrinsically, this is how our work stands. For an image I, we are able to obtain the augmented data I_{aug} through $I_{aug} := \mathcal{G}_P(L_{gt}, M \circ I)$, where L_{gt} is the landmark of I, and M stands for any mask. By doing so, for the image I, the augmented faces vary with different masks. The discriminator will make sure that the inpainted results match L_{gt}. An example is shown in Fig. 9, from which we can see that the features of I_{aug} are significantly different from those of I with the same landmarks. Consequently, the pair of (I_{aug}, L_{gt}) can be used for training.

To validate the effectiveness of such a data-augmentation manner, we feed the augmented data into both our $\mathcal{G}_L$ and LAB [23] on the WFLW dataset [23]. We notice that LAB is carefully built for the task of facial landmark detection, while the landmark module in LaFIn is much simpler and smaller because as previously explained, in our task, the landmarks can be not that accurate as long as they can provide the main structure of faces. Therefore, our performance in NME (normalized mean error by inter-ocular factor) is inferior to LAB. Nevertheless, as can be viewed from Table 4, the augmentation improves both the performance of LAB and our LaFIn. In addition, we also test LaFIn on the 300W dataset, the numerical results in Table 5 consistently reveal the effectiveness of the augmentation. Notice that no obvious difference in inpainted results is observed using the landmark predictors without and with augmentation, which again verifies that our inpainting module is robust against variation in landmarks, and can produce striking results as long as the structure is reasonably offered.

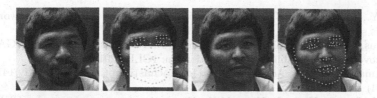

Fig. 9. From left to right: the original image, masked image with GT landmarks, augmented result, and augmented image with GT landmarks.

5   Conclusion

In this study, we have developed a generative network, namely LaFIn, for completing face images. The proposed LaFIn first predicts the landmarks then accomplishes the inpainting conditioned on the predicted landmarks. Our principle is that the landmarks are neat, sufficient, and robust to perform as guidance for providing the structural information to the face inpainting module. For ensuring the attribute consistency, we designed to harness distant spatial context and connect temporal feature maps. Extensive experiments have been conducted to verify our claims, reveal the efficacy of our design and, demonstrate its advances over state-of-the-art alternatives both qualitatively and quantitatively. Furthermore, we proposed to use our LaFIn to augment face-landmark data for relieving manual annotation in the task of landmark detection. The effectiveness of this manner has been experimentally confirmed.

References

1. Barnes, C., Shechtman, E., Finkelstein, A., Goldman, D.B.: PatchMatch: a randomized correspondence algorithm for structural image editing. ACM Trans. Graph. (ToG) **28**(3), 24:1–24:11 (2009)
2. Bertalmio, M., Sapiro, G., Caselles, V., Ballester, C.: Image inpainting. In: 27th Annual Conference on Computer Graphics and Interactive Techniques, pp. 417–424 (2000)
3. Bulat, A., Tzimiropoulos, G.: How far are we from solving the 2D & 3D face alignment problem? (and a dataset of 230,000 3D facial landmarks). In: ICCV, pp. 1021–1030 (2017)
4. Goodfellow, I., et al.: Generative adversarial nets. In: NeurIPS, pp. 2672–2680 (2014)
5. Heusel, M., Ramsauer, H., Unterthiner, T., Nessler, B., Hochreiter, S.: GANs trained by a two time-scale update rule converge to a local Nash equilibrium. In: NeurIPS, pp. 6626–6637 (2017)
6. Huang, J.B., Kang, S.B., Ahuja, N., Kopf, J.: Image completion using planar structure guidance. ACM Trans. Graph. (TOG) **33**(4), 129:1–129:10 (2014)
7. Iizuka, S., Simo-Serra, E., Ishikawa, H.: Globally and locally consistent image completion. ACM Trans. Graph. (ToG) **36**(4), 107 (2017)
8. Isola, P., Zhu, J.Y., Zhou, T., Efros, A.A.: Image-to-image translation with conditional adversarial networks. In: CVPR, pp. 1125–1134 (2017)

9. Jo, Y., Park, J.: SC-FEGAN: face editing generative adversarial network with user's sketch and color. arXiv:1902.06838 (2019)
10. Karras, T., Aila, T., Laine, S., Lehtinen, J.: Progressive growing of GANs for improved quality, stability, and variation. In: ICLR (2018)
11. Kingma, D.P., Ba, J.: Adam: a method for stochastic optimization. arXiv:1412.6980 (2014)
12. Li, Y., Liu, S., Yang, J., Yang, M.H.: Generative face completion. In: CVPR, pp. 3911–3919 (2017)
13. Liu, G., Reda, F.A., Shih, K.J., Wang, T.-C., Tao, A., Catanzaro, B.: Image inpainting for irregular holes using partial convolutions. In: Ferrari, V., Hebert, M., Sminchisescu, C., Weiss, Y. (eds.) ECCV 2018. LNCS, vol. 11215, pp. 89–105. Springer, Cham (2018). https://doi.org/10.1007/978-3-030-01252-6_6
14. Liu, Z., Luo, P., Wang, X., Tang, X.: Deep learning face attributes in the wild. In: ICCV, pp. 3730–3738 (2015)
15. Mao, X., Li, Q., Xie, H., Lau, R.Y., Wang, Z., Paul Smolley, S.: Least squares generative adversarial networks. In: ICCV, pp. 2794–2802 (2017)
16. Miyato, T., Kataoka, T., Koyama, M., Yoshida, Y.: Spectral normalization for generative adversarial networks. arXiv:1802.05957 (2018)
17. Nazeri, K., Ng, E., Joseph, T., Qureshi, F., Ebrahimi, M.: EdgeConnect: generative image inpainting with adversarial edge learning. arXiv:1901.00212 (2019)
18. Pathak, D., Krahenbuhl, P., Donahue, J., Darrell, T., Efros, A.: Context encoders: feature learning by inpainting. In: CVPR, pp. 2536–2544 (2016)
19. Sandler, M., Howard, A., Zhu, M., Zhmoginov, A., Chen, L.C.: MobileNetV2: inverted residuals and linear bottlenecks. In: CVPR, pp. 4510–4520 (2018)
20. Song, L., Cao, J., Song, L., Hu, Y., He, R.: Geometry-aware face completion and editing. In: AAAI, pp. 2506–2513 (2019)
21. Sun, Q., Ma, L., Joon Oh, S., Van Gool, L., Schiele, B., Fritz, M.: Natural and effective obfuscation by head inpainting. In: CVPR, pp. 5050–5059 (2018)
22. Wang, Z., Bovik, A.C., Sheikh, H.R., Simoncelli, E.P., et al.: Image quality assessment: from error visibility to structural similarity. IEEE TIP **13**(4), 600–612 (2004)
23. Wu, W., Qian, C., Yang, S., Wang, Q., Cai, Y., Zhou, Q.: Look at boundary: a boundary-aware face alignment algorithm. In: CVPR, pp. 2129–2138 (2018)
24. Xiao, S., et al.: Recurrent 3D-2D dual learning for large-pose facial landmark detection. In: CVPR, pp. 1633–1642 (2017)
25. Xie, J., Xu, L., Chen, E.: Image denoising and inpainting with deep neural networks. In: NeurIPS, pp. 341–349 (2012)
26. Xiong, W., et al.: Foreground-aware image inpainting. In: CVPR, pp. 5840–5848 (2019)
27. Yamauchi, H., Haber, J., Seidel, H.P.: Image restoration using multiresolution texture synthesis and image inpainting. In: Computer Graphics International, pp. 120–125. IEEE (2003)
28. Yu, J., Lin, Z., Yang, J., Shen, X., Lu, X., Huang, T.S.: Free-form image inpainting with gated convolution. arXiv:1806.03589 (2018)
29. Yu, J., Lin, Z., Yang, J., Shen, X., Lu, X., Huang, T.S.: Generative image inpainting with contextual attention. In: CVPR, pp. 5505–5514 (2018)
30. Zakharov, E., Shysheya, A., Burkov, E., Lempitsky, V.: Few-shot adversarial learning of realistic neural talking head models. arXiv:1905.08233 (2019)
31. Zhang, J., Zeng, X., Pan, Y., Liu, Y., Ding, Y., Fan, C.: FaceSwapNet: landmark guided many-to-many face reenactment. arXiv:1905.11805 (2019)
32. Zheng, C., Cham, T.J., Cai, J.: Pluralistic image completion. In: CVPR, pp. 1438–1447 (2019)

Multi-granularity Multimodal Feature Interaction for Referring Image Segmentation

Zhenxiong Tan[1], Tianrui Hui[2,3], Jinyu Chen[4], and Si Liu[4(✉)]

[1] Beijing Forestry University, Beijing, China
Yuanshi9815@outlook.com
[2] Institute of Information Engineering, Chinese Academy of Sciences, Beijing, China
huitianrui@iie.ac.cn
[3] School of Cyber Security, University of Chinese Academy of Sciences,
Beijing, China
[4] Beihang University, Beijing, China
{chenjinyu,liusi}@buaa.edu.cn

Abstract. Referring image segmentation aims to segment the entity referred by a natural language description. Previous methods tackle this problem by conducting multimodal feature interaction between image and words or sentence only. However, considering only single granularity feature interaction tends to result in incomplete understanding of visual and linguistic information. To overcome this limitation, we propose to conduct multi-granularity multimodal feature interaction by introducing a Word-Granularity Feature Modulation (WGFM) module and a Sentence-Granularity Context Extraction (SGCE) module, which can be complementary in feature alignment and obtain a comprehensive understanding of the input image and referring expression. Extensive experiments show that our method outperforms previous methods and achieves new state-of-the-art performances on four popular datasets, i.e., UNC (+1.45%), UNC+ (+1.63%), G-Ref (+0.47%) and ReferIt (+1.02%).

Keywords: Referring segmentation · Multi-granularity · Multimodal feature interaction

1 Introduction

Given a natural language expression, the goal of referring image segmentation is to segment the object or stuff described by the expression. In this paper, we define the referred object or stuff as the *referent*. As illustrated in Fig. 1, traditional semantic or instance segmentation is limited in a predefined set of categories to understand the image (e.g., "person"). Different from these two tasks, referring segmentation is conditioned on a natural language expression which can specify

---

The first author is a student.

© Springer Nature Switzerland AG 2020
Y. Peng et al. (Eds.): PRCV 2020, LNCS 12305, pp. 27–39, 2020.
https://doi.org/10.1007/978-3-030-60633-6_3

the referent in various and complex linguistic forms (e.g., "man in a gray suit with a red tie"), making this task more challenging. There are many potential applications (e.g., language-guided image editing or object retrieval) based on the referring image segmentation techniques.

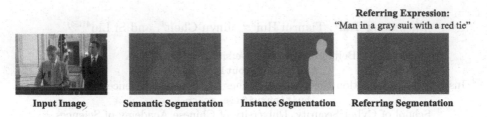

Referring Expression:
"Man in a gray suit with a red tie"

Input Image   Semantic Segmentation   Instance Segmentation   Referring Segmentation

Fig. 1. Examples of typical segmentation tasks. Different from semantic/instance segmentation, referring image segmentation has a much larger category set conditioned on the input natural language expression.

Previous methods treat this task with multimodal feature interaction in single granularity, i.e., word granularity or sentence granularity. Some works only model the interaction between image feature and *words* features by dynamic filters [17] or attentional fusion [18–20], which obtains the partial understanding of the referring expression while lacking global context information. Other works [14,16] only fuse the image feature with *sentence* feature in a direct or recurrent manner where local context of some informative words is omitted in the fused multimodal feature. It is clear that both word-granularity and sentence-granularity multimodal interaction are crucial for more accurate comprehension of visual and linguistic features, yielding finer segmentation results.

Based on the above analysis, we propose a multi-granularity multimodal feature interaction approach which introduces two novel modules, namely Word-Granularity Feature Modulation (WGFM) module and Sentence-Granularity Context Extraction (SGCE) module, which model multimodal interaction in both word granularity and sentence granularity for comprehensive understanding of visual and linguistic information. Concretely, our WGFM module first densely concatenates and fuses image feature with each word feature to enable the image interact with each word respectively. Multimodal features of all the words are further summed and normalized to form an modulation cuboid to modulate the image feature elementwisely, which captures local multimodal context and highlights the image regions well matched with each word. Then, our SGCE module conducts cross-modal attention between the modulated image feature and the sentence feature to extract a global multimodal context vector relevant to the whole sentence for identifying the referent in a holistic manner. Afterwards, the global context vector are adaptively concatenated to each spatial location on the image feature to complement local word-granularity multimodal context with global sentence-granularity multimodal context. Finally, multiple levels of features from CNN backbone are refined by the proposed two modules and fused to exploit both low-level details and high-level semantics, forming a comprehensive multimodal feature representation for mask prediction.

The main contributions of this paper are summarized as follows: (1) We propose a multi-granularity multimodal feature interaction approach which comprehensively understands multimodal information. Extensive experiments on four popular referring image segmentation benchmarks show that our method outperforms all the previous state-of-the-arts, demonstrating its effectiveness. (2) Our approach introduces a Word-Granularity Feature Modulation (WGFM) module which enables the image feature to interact with each word feature to capture local word-granularity multimodal context and highlight image regions highly responsive to each word. (3) Our approach also introduces a Sentence-Granularity Context Extraction (SGCE) module which models the interaction between the image feature and the whole sentence feature to extract global multimodal context information for holistically identifying the referent.

2 Related Work

2.1 Semantic Segmentation

Fully Convolutional Networks (FCN) [1] transforms fully connected layers into convolution layers to enable a classification network to output semantic labels. FCN brings a huge progress to semantic segmentation area. DeepLab [2–4] introduces atrous convolution into FCN model. The atrous convolutions control the resolution at which feature responses are computed and effectively enlarge the field of view of filters. PSPNet [6] and APCNet [5] proposes pyramid pooling module to aggregate multi-scale context. Recent DANet [7] and CFNet [8] utilizes self-attention mechanism [9] to capture long-range contextual information in network and achieve impressing performance. In this paper, we tackle a more challenging task whose segmentation target is specified by natural language.

2.2 Referring Expression Comprehension

Referring expression comprehension aims to localize the object instance described by a natural language referring expression in an image. Many works localize the instance in bounding box level. This task can be modeled to find the most related object to the expression in the image [10,11]. MAttNet [12] computes the matching scores between the referring expression and objects utilizing fine-grained linguistic components including subject, location and relationship. Liao *et al.* [13] reformulate the referring expression comprehension task as a correlation filtering process and speed up the computing process to real time.

To locate the referred object more precisely than bounding box, Hu *et al.* [14] first proposes the referring segmentation problem and generates segmentation mask by directly concatenating and fusing the visual feature from CNN and the linguistic feature from LSTM. Liu *et al.* [15] fuse visual and linguistic features with multimodal LSTM in multiple time steps to obtain better multimodal features. Besides, dynamic filters [17] for each word further enhance feature fusion.

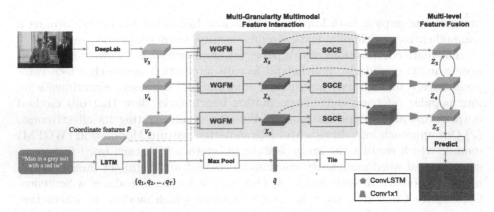

Fig. 2. The overall architecture of our method. Visual and linguistic features are sequentially processed by our WGFM and SGCE modules for multi-granularity feature interaction. Multi-level features are further fused to generate the foreground mask.

To better utilize the visual context information, KWN [18] uses word attention to highlight keywords and aggregate visual context to enhance visual features. For multimodal context extraction, cross-modal self-attention [20] is proposed to capture long-range dependencies between visual and linguistic information based on multi-level feature fusion [16]. Visual and textual co-embedding is also explored in [19] to compute resemblance between the expression and each image region. Adversarial training [21] and cycle-consistency learning [22] are also investigated to promote the segmentation performance. Our method models the feature interaction in both word-granularity and sentence-granularity to tackle the limitation of previous methods.

3 Method

The overall architecture of our method is illustrated in Fig. 2. Our model takes an image and a natural language expression as input and encodes their features by DeepLab [3] and LSTM [23]. Features from two modalities are further processed by our multi-granularity multimodal feature interaction approach to obtain comprehensive multimodal features. Our approach is composed of two modules, namely the Word-Granularity Feature Modulation (WGFM) module and the Sentence-Granularity Context Extraction (SGCE) module. The WGFM module first modulates image feature with each word feature to capture local multimodal context and highlight image regions matched with each word. Then, the modulated image feature and sentence feature are fed into the SGCE module to extract global multimodal context information to refine multimodal context. Finally, multi-level features from DeepLab are fused to predict the mask of the referent. We will detail each part in the following sections.

3.1   Visual and Linguistic Feature Encoding

We adopt DeepLab [3] to encode the image feature. As CNN backbones usually contain multiple levels (e.g., 5 levels in ResNet-101 [24]), we use 3 levels of visual features V_3, V_4, V_5 to exploit both low-level details and high-level semantics. For ease of presentation, we use $V \in \mathbb{R}^{H \times W \times C_v}$ to denote a certain stage of visual feature to elaborate our method. ℓ_2 normalization is performed on the channel dimension of V to get a robust representation. For the input expression with T words, we first embed each word into a word embedding vector and then adopt LSTM [23] to encode each embedding vector into a hidden state vector $q_i \in \mathbb{R}^{C_l}, i = 1, 2, ..., T$ as the linguistic feature $Q = [q_1; q_2; ...; q_T] \in \mathbb{R}^{T \times C_l}$ followed by ℓ_2 normalization as well. In addition, we also form a coordinate feature $P \subset \mathbb{R}^{H \times W \times 8}$ to encode spatial position information into the model following prior works [16, 19, 20].

3.2   Multi-granularity Multimodal Feature Interaction

Our proposed multi-granularity multimodal feature interaction approach consists of two modules named WGFM and SGCE, which conduct multimodal interaction between visual and linguistic features in both word-granularity and sentence-granularity, capturing both local and global multimodal context to comprehensively enhance the multimodal feature for mask prediction.

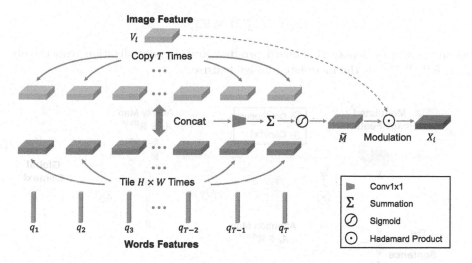

Fig. 3. Illustration of our WGFM module. The tiled words features and copied image feature are densely concatenated, fused and normalized to obtain the modulation cuboid $\tilde{M}$. The image feature V is further modulated by $\tilde{M}$ elementwisely.

Word-Granularity Feature Modulation. As illustrated in Fig. 3, the image feature $V \in \mathbb{R}^{H \times W \times C_v}$ and words features $Q = [q_1; q_2; ...; q_T] \in \mathbb{R}^{T \times C_l}$ are fed into our WGFM for feature modulation. To enable image feature to interact with each word feature, we copy V for T times to get the dense image feature $\tilde{V} \in \mathbb{R}^{T \times H \times W \times C_v}$, and also tile each word feature $q_i \in \mathbb{R}^{C_l}$ for $H \times W$ times to get the dense words features $\tilde{Q} \in \mathbb{R}^{T \times H \times W \times C_l}$. Then, we concatenate two dense features $\tilde{V}$ and $\tilde{Q}$ along channel dimension and fuse them with a 1×1 convolution to obtain the joint multimodal feature $M \in \mathbb{R}^{T \times H \times W \times C_m}$ as follows:

$$M = f_1([\tilde{V}, \tilde{Q}]; \theta_1), \tag{1}$$

where $[,]$ denotes concatenation, $f_1(; \theta_1)$ denotes a 1×1 convolution layer whose parameter is θ_1. Afterwards we conduct summation along the T dimension of M to obtain the aggregated multimodal feature in word-granularity and normalize it to produce the modulation cuboid $\tilde{M} \in \mathbb{R}^{H \times W \times C_m}$ as follows:

$$\tilde{M} = \sigma(\sum_{i=1}^{T} M_i), \tag{2}$$

where $M_i \in \mathbb{R}^{H \times W \times C_m}$ and σ denotes *Sigmoid* function.

Finally, the original image feature V is modulated elementwisely by modulation cuboid $\tilde{M}$ to capture local word-granularity context and highlight image regions which are highly responsive to each word:

$$X = \ell_2(\tilde{M} \odot V), \tag{3}$$

where $\odot$ and ℓ_2 denote Hadamard product and ℓ_2 normalization respectively, $X \in \mathbb{R}^{H \times W \times C_m}$ is the modulated image feature.

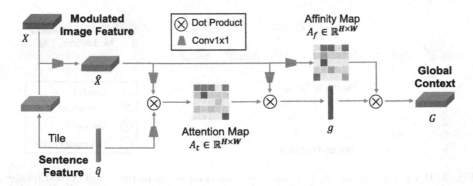

Fig. 4. Illustration of our SGCE module. The global context vector g is extracted from the fusion of the modulated image feature $\hat{X}$ and the sentence feature $\hat{q}$. Then, g is recovered to G using the affinity map from $\hat{X}$.

Sentence-Granularity Context Extraction. In order to extract global multimodal context information relevant to the whole sentence, we conduct max pooling on all the T words features to obtain the sentence feature $\hat{q} \in \mathbb{R}^{C_l}$. As illustrated in Fig. 4, the modulated image feature X is first concatenated and fused with tiled sentence feature to obtain the multimodal feature $\hat{X} \in \mathbb{R}^{H \times W \times C_m}$. Then $\hat{X}$ and sentence feature $\hat{q}$ are transformed by 1×1 convolutions. Cross-modal attention is further performed between these two transformed features to obtain the attention map $A_t \in \mathbb{R}^{H \times W}$ with necessary reshape operations:

$$A'_t = f_2(\hat{X}; \theta_2) \otimes f_3(\hat{q}; \theta_3), \tag{4}$$

$$A_t = Softmax(\frac{A'_t}{\sqrt{C_m}}), \tag{5}$$

where $f_2(; \theta_2)$ and $f_3(; \theta_3)$ denotes convolution layers whose parameters are θ_2 and θ_3, $\otimes$ denotes matrix dot product. Each element of A_t represents the normalized relevance between feature of each spatial location on $\hat{X}$ and feature of the sentence. Then, we utilize A_t to extract global context vector from transformed $\hat{X}$ with necessary reshape operations:

$$g = A_t \otimes f_4(\hat{X}; \theta_4), \tag{6}$$

where $f_4(; \theta_4)$ denotes the convolution layer whose parameter is θ_4, $g \in \mathbb{R}^{C_m}$ is the global multimodal context vector which aggregates multimodal information mostly relevant to the whole sentence.

After extracting global context vector g, we need to adaptively concatenate it back to each spatial location on $\hat{X}$ to construct grid format feature. Here we adopt a straightforward approach to project $\hat{X}$ from C_m to 1 on the channel dimension, and then apply *Sigmoid* function to normalize it, yielding the affinity map $A_f \in \mathbb{R}^{H \times W}$. Finally, we utilize A_f and g to generate global multimodal context $G \in \mathbb{R}^{H \times W \times C_m}$ with necessary reshape and transpose operations:

$$G = A_f \otimes g. \tag{7}$$

3.3   Multi-level Feature Fusion

We follow prior works [16,19,20] to exploit both low-level details and high-level semantics by fusing multi-level features from CNN backbone. For each level of image feature $V_i, i = 3, 4, 5$, we feed it into the proposed MGFI block and obtain global context G_i. Then, G_i is concatenated and fused with modulated image feature X_i and tiled sentence feature to produce single-level multimodal feature Z_i, as shown in Fig. 2. Afterwards we adopt simplified Mutan [25] fusion scheme to fuse multi-level features Z_i, which is formulated as bellow:

$$\hat{Z}_i = Z_i + \sum_{j \in \{3,4,5\} \setminus \{i\}} \sigma(f_q(\hat{q}; \theta_q)) \odot f_{ij}(Z_j; \theta_{ij}), i = 3, 4, 5, \tag{8}$$

where $f(; \theta)$ denotes convolution layers with parameters θ, σ denotes *Sigmoid* function. The fused features $\hat{Z}_i$ are further aggregated through ConvLSTM [26] followed by a 3×3 convolution to generate the foreground mask.

4 Experiments

4.1 Setup

Datasets. We train our model on four popular referring image segmentation datasets including UNC [28], UNC+ [28], ReferIt [30] and G-Ref [29]. UNC and UNC+ are both based on MS-COCO [31]. The difference is that location words are not allowed in UNC+. They contain $19,994$, and $19,992$ images with $142,209$ and $141,564$ referring expressions for $50,000$ objects, respectively. Referents in ReferIt dataset include not only objects but also stuff and this dataset has $19,894$ images with $130,525$ expressions for $96,654$ referents. Though G-Ref is also based on MS-COCO, referring expressions in it are longer than those of UNC and UNC+. G-Ref has $26,711$ images with $104,560$ expressions for $54,822$ objects.

Evaluation Metrics. Following prior works [16,19,20], we adopt overall IoU (Intersection over Union) and Prec@X (Precision at X) as metrics to evaluate our model. Overall IoU calculates total intersection area over total union area of all the test samples. Prec@X measures the percentage of predictions with IoU higher than the X where $X \in \{0.5, 0.6, 0.7, 0.8, 0.9\}$.

Implementation Details. For input images, we resize them to 320×320 and encode them with DeepLab-ResNet101 [3] pretrained on PASCAL-VOC [33] dataset. The channel dimensions of features C_v, C_l and C_m are set to 1000. For model training, we use Adam optimizer [32], and set the initial learning rate and weight decay as $1e-5$ and $5e-4$ respectively. Parameters of the CNN backbone are freezed during training. The output score maps of the network are bilinearly

Table 1. Comparison with state-of-the-art methods on four datasets using *overall IoU* as metric. "n/a" denotes MAttNet does not use the same split as other methods.

Method	UNC			UNC+			G-Ref	ReferIt
	val	testA	testB	val	testA	testB	val	test
Hu *et al.* [14], ECCV2016	–	–	–	–	–	–	28.14	48.03
RMI [15], ICCV2017	45.18	45.69	45.57	29.86	30.48	29.50	34.52	58.73
DMN [17], ECCV2018	49.78	54.83	45.13	38.88	44.22	32.29	36.76	52.81
KWA [18], ECCV2018	–	–	–	–	–	–	36.92	59.09
Qiu *et al.* [21], TMM2019	50.46	51.20	49.27	38.41	39.79	35.97	41.36	60.31
RRN [16], CVPR2018	55.33	57.26	53.95	39.75	42.15	36.11	36.45	63.63
MAttNet [12], CVPR2018	56.51	62.37	51.70	46.67	52.39	40.08	n/a	–
CMSA [20], CVPR2019	58.32	60.61	55.09	43.76	47.60	37.89	39.98	63.80
Chen *et al.* [22], BMVC2019	58.90	61.77	53.81	–	–	–	44.32	–
STEP [19], ICCV2019	60.04	63.46	57.97	48.19	52.33	40.41	46.40	64.13
Ours	**61.38**	**64.29**	**59.42**	**49.54**	**52.85**	**42.04**	**46.87**	**65.15**

upsampled to 320×320 to compute losses with resized ground-truth masks. Standard binary cross-entropy loss is adopted to train the network. In addition, we also adopt Dense-CRF [27] to refine the segmentation masks like the prior works [16,19,20] for fair comparison.

4.2 Comparison with State-of-the-Arts

Table 1 summarizes the comparison results with state-of-the-arts on 4 benchmark datasets. We can observe that our method consistently outperforms previous methods on all the dataset splits, showing the effectiveness of modeling feature interaction in both word-granularity and sentence granularity. On UNC+ dataset which contains no location words, our method boosts the performance by 1.35% and 1.63% IoU on val set and testB set respectively, indicating that our method can well match the expression with the image even without important location information. Previous methods merely obtain marginal improvements on ReferIt test set while our method achieves 1.02% IoU improvement, which also demonstrates that our method can well adapt to different datasets.

Table 2. Ablation studies on the UNC val set. All models use the same backbone CNN (DeepLab-ResNet101) and DenseCRF for postprocessing.

Method	prec@0.5	prec@0.6	prec@0.7	prec@0.8	prec@0.9	IoU (%)
Baseline	47.32	38.22	28.24	16.13	3.88	47.79
+WGFM only	50.55	41.31	31.02	18.74	4.29	49.54
+SGCE only	58.98	50.71	40.49	25.61	6.17	53.58
+WGFM and SGCE	**63.60**	**56.39**	**46.60**	**31.15**	**8.18**	**56.31**
Multi-Level-Baseline	70.72	63.87	54.40	38.95	12.22	60.22
+WGFM only	71.40	65.46	55.96	40.23	13.15	61.22
+SGCE only	70.78	64.28	55.05	39.55	12.81	60.50
+WGFM and SGCE	**72.20**	**66.37**	**57.02**	**41.83**	**14.06**	**61.38**

4.3 Ablation Study

As shown in Table 2, we perform ablation study on UNC val set to evaluate our proposed modules. Our baseline simply concatenates the image feature, tiled sentence feature and coordinate feature for mask prediction. The proposed WGFM module and SGCE module can bring about considerable performance improvements when they are independently inserted into the baseline, indicating the effectiveness of the proposed modules. Introducing SGCE module outperforms baseline by a large margin, which are mainly because that the global multimodal context from sentence granularity can better guide the model to identify the referent holistically. Moreover, combining the two modules together is able to further boost the performance, demonstrating that multi-granularity feature interaction can complement with each other to comprehensively understand multimodal information.

In addition, we also conduct ablation study based on multi-level features where all the models use simplified Mutan [25] fusion for fair comparison. Results in Table 2 show that our proposed modules obtain consistent performance gains comparing with single-level situation, indicating that our method is also effective over the Mutan fusion scheme.

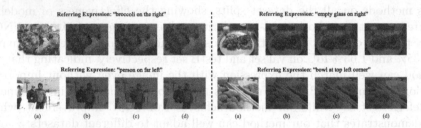

Fig. 5. Qualitative results of our model. (a) Input images. (b) Ground-truth masks. (c) Results generated by the baseline model. (d) Results generated by our full model.

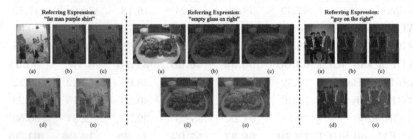

Fig. 6. Visualization of affinity maps and attention maps from our SGCE module. (a) Input images. (b) Ground-truth masks. (c) Output predictions. (d) Attention maps from SGCE. (e) Affinity maps from SGCE.

4.4   Visualization Analysis

Qualitative Results. Figure 5 shows some qualitative examples generated by our full model and the baseline model. From the top-left and the bottom-right examples, we can find that the predictions made by our full model can fully cover the broccoli and the bowl referred by queries, while the masks generated by the baseline model are incomplete or excessive. In the bottom-left example, the baseline model makes unclear judgement between the two men in the left.

The top-right example shows that our full model can also make roughly correct prediction for the transparent glass, showing that the proposed multi-granularity feature interaction approach can well capture the boundary of objects.

Attention and Affinity Maps. To better understand how our model works, we visualize the attention maps and affinity maps from SGCE module based on the *Res4* level of visual feature in Fig. 6. For the left example with query text of *"fat man purple shirt"*, the attention map focuses on the purple T-shirt of the fat man. The affinity map also successfully distinguishes the *"man"* category from others. For the query *"empty glass on right"*, the attention map focuses on the correct glass on the right, and the affinity map distinguishes the area of *"glass"* category from others. The right example is similar to the two ones mentioned above. In general, the value distributions of the attention and affinity maps are consistent with the semantics of the input expressions, demonstrating that our model can achieve a comprehensive understanding of multimodal information.

5   Conclusion

In this paper, we focus on the referring image segmentation task. The limitation of previous methods lies in the absence of jointly considering feature interaction in multiple granularities. Therefore, we propose a multi-granularity multimodal feature interaction approach composed of a Word-Granularity Feature Modulation (WGFM) module and a Sentence-Granularity Context Extraction (SGCE) module to comprehensively understand multimodal information. Our method achieves state-of-the-art performance on four popular datasets. In the future, we plan to explore more structural analysis of multimodal information.

Acknowledgement. This work is supported by National Natural Science Foundation of China under Grant No. 61876177 and Beijing Natural Science Foundation under Grant No. L182013 and No. 4202034.

References

1. Long, J., Shelhamer, E., Darrell, T.: Fully convolutional networks for semantic segmentation. In: CVPR (2015)
2. Chen, L.-C., Papandreou, G., Kokkinos, I., Murphy, K., Yuille, A.L.: Semantic image segmentation with deep convolutional nets and fully connected CRFs. arXiv preprint arXiv:1412.7062 (2014)
3. Chen, L.-C., Papandreou, G., Kokkinos, I., Murphy, K., Yuille, A.L.: DeepLab: semantic image segmentation with deep convolutional nets, atrous convolution, and fully connected CRFs. TPAMI **40**(4), 834–848 (2017)
4. Chen, L.-C., Papandreou, G., Schroff, F., Adam, H.: Rethinking atrous convolution for semantic image segmentation. arXiv preprint arXiv:1706.05587 (2017)
5. He, J., Deng, Z., Zhou, L., Wang, Y., Qiao, Y.: Adaptive pyramid context network for semantic segmentation. In: CVPR (2019)

6. Zhao, H., Shi, J., Qi, X., Wang, X., Jia, J.: Pyramid scene parsing network. In: CVPR (2017)
7. Fu, J., et al.: Dual attention network for scene segmentation. In: CVPR (2019)
8. Zhang, H., Zhang, H., Wang, C., Xie, J.: Co-occurrent features in semantic segmentation. In: CVPR (2019)
9. Wang, X., Girshick, R., Gupta, A., He, K.: Non-local neural networks. In: CVPR (2018)
10. Hu, R., Rohrbach, M., Andreas, J., Darrell, T., Saenko, K.: Modeling relationships in referential expressions with compositional modular networks. In: CVPR (2017)
11. Yang, S., Li, G., Yu, Y.: Cross-modal relationship inference for grounding referring expressions. In: CVPR (2019)
12. Yu, L., et al.: MAttNet: modular attention network for referring expression comprehension. In: CVPR (2018)
13. Liao, Y., et al.: A real-time cross-modality correlation filtering method for referring expression comprehension. arXiv preprint arXiv:1909.07072 (2019)
14. Hu, R., Rohrbach, M., Darrell, T.: Segmentation from natural language expressions. In: Leibe, B., Matas, J., Sebe, N., Welling, M. (eds.) ECCV 2016. LNCS, vol. 9905, pp. 108–124. Springer, Cham (2016). https://doi.org/10.1007/978-3-319-46448-0_7
15. Liu, C., Lin, Z., Shen, X., Yang, J., Lu, X., Yuille, A.: Recurrent multimodal interaction for referring image segmentation. In: ICCV (2017)
16. Li, R., et al.: Referring image segmentation via recurrent refinement networks. In: CVPR (2018)
17. Margffoy-Tuay, E., Pérez, J.C., Botero, E., Arbeláez, P.: Dynamic multimodal instance segmentation guided by natural language queries. In: Ferrari, V., Hebert, M., Sminchisescu, C., Weiss, Y. (eds.) ECCV 2018. LNCS, vol. 11215, pp. 656–672. Springer, Cham (2018). https://doi.org/10.1007/978-3-030-01252-6_39
18. Shi, H., Li, H., Meng, F., Wu, Q.: Key-word-aware network for referring expression image segmentation. In: Ferrari, V., Hebert, M., Sminchisescu, C., Weiss, Y. (eds.) ECCV 2018. LNCS, vol. 11210, pp. 38–54. Springer, Cham (2018). https://doi.org/10.1007/978-3-030-01231-1_3
19. Chen, D.-J., Jia, S., Lo, Y.-C., Chen, H.-T., Liu, T.-L.: See-through-text grouping for referring image segmentation. In: ICCV (2019)
20. Ye, L., Rochan, M., Liu, Z., Wang, Y.: Cross-modal self-attention network for referring image segmentation. In: CVPR (2019)
21. Qiu, S., Zhao, Y., Jiao, J., Wei, Y., Wei, S.: Referring image segmentation by generative adversarial learning. TMM 22(5), 1333–1344 (2019)
22. Chen, Y.-W., Tsai, Y.-H., Wang, T., Lin, Y.-Y., Yang, M.-H.: Referring expression object segmentation with caption-aware consistency. In: BMVC (2019)
23. Hochreiter, S., Schmidhuber, J.: Long short-term memory. Neural Comput. 9(8), 1735–1780 (1997)
24. He, K., Zhang, X., Ren, S., Sun, J.: Deep residual learning for image recognition. In: CVPR (2016)
25. Ben-Younes, H., Cadene, R., Cord, M., Thome, N.: MUTAN: multimodal tucker fusion for visual question answering. In: ICCV (2017)
26. Shi, X., Chen, Z., Wang, H., Yeung, D.-Y., Wong, W.-K., Woo, W.-C.: Convolutional LSTM network: a machine learning approach for precipitation nowcasting. In: NeurIPS (2015)
27. Krahenbuhl, P., Koltun, V.: Efficient inference in fully connected CRFs with Gaussian edge potentials. In: NeurIPS (2011)

28. Yu, L., Poirson, P., Yang, S., Berg, A.C., Berg, T.L.: Modeling context in referring expressions. In: Leibe, B., Matas, J., Sebe, N., Welling, M. (eds.) ECCV 2016. LNCS, vol. 9906, pp. 69–85. Springer, Cham (2016). https://doi.org/10.1007/978-3-319-46475-6_5

29. Mao, J., Huang, J., Toshev, A., Camburu O., Yuille, A.L., Murphy, K.: Generation and comprehension of unambiguous object descriptions. In: CVPR (2016)

30. Kazemzadeh, S., Ordonez, V., Matten, M., Berg, T.: ReferItGame: referring to objects in photographs of natural scenes. In: EMNLP (2014)

31. Lin, T.-Y., et al.: Microsoft COCO: common objects in context. In: Fleet, D., Pajdla, T., Schiele, B., Tuytelaars, T. (eds.) ECCV 2014. LNCS, vol. 8693, pp. 740–755. Springer, Cham (2014). https://doi.org/10.1007/978-3-319-10602-1_48

32. Kingma, D.P., Ba, J.: Adam: a method for stochastic optimization. arXiv preprint arXiv:1412.6980 (2014)

33. Everingham, M., Van Gool, L., Williams, C.K.I., Winn, J., Zisserman, A.: The Pascal visual object classes (VOC) challenge. Int. J. Comput. Vis. **88**, 303–338 (2010). https://doi.org/10.1007/s11263-009-0275-4

Hyperspectral Image Denoising Based on Graph-Structured Low Rank and Non-local Constraint

Haitao Chen and Haitao Yin[✉]

College of Automation and College of Artificial Intelligence, Nanjing University of Posts and Telecommunications, Nanjing 210023, China
haitaoyin@njupt.edu.cn

Abstract. Hyperspectral image (HSI) denoising is an effective image processing technology to improve the quality of HSI. By exploiting the spectral-spatial characteristics of HSI, this paper proposes a HSI denoising method based on the Graph-structured Low Rank model and Non-Local constraint (GLRNL). Graph-structured low rank model can reveal the spectral similarity in HSI and preserve the distribution characteristics of spectral bands. The graph structure is used to ensure that the denoised HSI has similar spectral characteristics as the real HSI, and reduce the spectral distortions. Moreover, the non-local constraint in GLRNL adopts the spatial-spectral self-similarity and provides an effective side prior information for low rank calculation. Comparing with some traditional low rank and sparse model based HSI denoising methods, the proposed method achieves the competitive results on different simulated and real HSI data with various types of noise.

Keywords: Hyperspectral image denoising · Low rank · Non-local constraint · Graph structure

1 Introduction

With the development of hyperspectral sensor technology, the bands number of hyperspectral images (HSIs) ranges from dozens to hundreds. The observed regions can be described with rich spectral information. For example, the reflective optics system imaging spectrometer (ROSIS) and the Airborne Visible/Infrared Imaging Spectrometer (AVIRIS) can provide the HSIs with 115 bands and 224 bands, respectively. The rich spectral characteristics in HSI is helpful for environmental monitoring, agriculture and forestry, weather prediction and military scouting [1]. Due to the limitation of technique and weather conditions, the HSIs are corrupted by various noise during the acquisition stage, which reduces the image quality and affects the practical applications. HSI denoising is an effect tool to suppress noise and improve image quality, which is critical for further applications.

This research was supported by the National Natural Science Foundation of China under grant No. 61971237 and No. 61501255.

HSI denoising is a research hotspot in remote sensing and its applications. The traditional denoising approaches include the wavelet transform [2], non-local means [3], principal component analysis [4] and total variation [5]. The traditional methods study the spectral correlation of HSI incompletely, and maybe result in some artifacts and distortions. Currently, sparse and low rank models are two powerful tools to represent image, and are widely used to image restoration. The definition of sparse representation is that image can be represented as the product of overcomplete dictionary and sparse coefficients. Qian et al. [6] introduce the nonlocal similarity and spectral-spatial structure of HSI into sparse representation. Xing et al. [7] report a HSI denoising method based on three-dimensional spatial-spectral blocks and a Bayesian dictionary learning. Different from the sparse element prior information in classical sparse model, the low rank model measures the sparsity of matrix rank aiming to improve the description ability on the spatial structure. Zhang et al. [8] present a low rank based denoising method, and define the spectral similarity of HSI as the low rank property of matrix. Xue et al. [9] design a joint spectral-spatial low-rank regularization based spectral similarity and nonlocal spatial low rank dictionary. To preserve multidimensional structure of HSI, Xie et al. [10] develop a tensor sparse and low rank model based HSI denoising method. Zhang et al. [11] combine the low rank tensor decomposition and spatial–spectral total variation regularization by nonlocal similar patches to restore the clean HSI. The existing low rank approaches commonly rely on the spatial and spectral similarity of HSI, modelled as rank-minimization problem. However, the spectral distribution characteristic is also an important information of HSI, which is rarely considered in the sparse and low-rank based denoising procedures.

In order to exploit the spectral distribution characteristics of HSI, we propose a novel low rank based denoising method with spectral graph constraint, termed as Graph-structured Low Rank model and Non-Local constraint (GLRNL). In GLRNL, we create a graph to connect the spectral channels of HSI and preserve the spectral distribution of denoised HSI as natural HSI, and consequently reduce the spectral distortions further. In addition, we propose a non-local regularization, deriving from the spatial similarity of HSI, as a side prior information to boost the robustness of low rank optimization problem. Finally, the weighted nuclear norm is used to construct the low-rank minimization problem instead of the traditional one.

The organization of this paper is as follows. Section 2 briefly reviews the related works. The proposed method is introduced in Sect. 3. Experimental results and comparisons are given in Sect. 4. Section 5 concludes this paper.

2 Related Works

Commonly, sparse and low rank model jointly consider the sparsity and low rank of data, and receives wide attentions. Robust principal component analysis (RPCA) [12] is the representative one, which can reveal the low dimensional structure of high dimensional data. RPCA decomposes the input data as a low rank part and a sparse part, which has various applications, such as image restoration [13], pansharpening [14], face recognition

[15], and saliency detection [16]. Formally, let $\mathbf{X}$ be the observed data, and RPCA decomposes $\mathbf{X}$ as

$$\mathbf{X} = \mathbf{H} + \mathbf{Q} \tag{1}$$

where $\mathbf{H}$ is the low rank matrix and $\mathbf{Q}$ is the sparse matrix. $\mathbf{H}$ and $\mathbf{Q}$ can be solved via the following optimization problem

$$\min_{\mathbf{H},\mathbf{Q}} \text{rank}(\mathbf{H}) + \lambda \|\mathbf{Q}\|_0 \ s.t. \ \mathbf{X} = \mathbf{H} + \mathbf{Q} \tag{2}$$

where $\lambda > 0$ is a balanced parameter, $\text{rank}(\mathbf{H})$ is the rank of $\mathbf{H}$ and $\|\mathbf{Q}\|_0$ computes the number of nonzero entries. Generally, Problem (2) is solved by the following relaxed version

$$\min_{\mathbf{H},\mathbf{Q}} \|\mathbf{H}\|_* + \lambda \|\mathbf{Q}\|_1 \ s.t. \ \mathbf{X} = \mathbf{H} + \mathbf{Q} \tag{3}$$

where $\|\mathbf{H}\|_* = \sum_i \sigma_i(\mathbf{H})$ is the nuclear norm, $\sigma_i(\mathbf{H})$ is the i th singular value of $\mathbf{H}$, and $\|\cdot\|_1$ denotes the ℓ_1-norm.

In Problem (3), the low rank component is solved by singular value threshold (SVT) algorithm. However, the traditional nuclear norm ignores the importance of singular values and SVT processes each singular value equally. In fact, larger singular value contains more image information than smaller one. Intuitively, adaptive SVT can improve the accuracy and robustness of low rank matrix reconstruction. Hence, Gu et al. [13] propose a weighted nuclear norm

$$\|\mathbf{H}\|_{\omega,*} = \sum_i \omega_i \sigma_i(\mathbf{H}) \tag{4}$$

where $\omega = [\omega_1, \omega_2, \ldots \omega_n]$ and $\omega_i > 0$ is non-negative weight assigned to $\sigma_i(\mathbf{H})$. The weight ω_i is set as

$$\omega_i^{(t+1)} = c/(\sigma_i(\mathbf{H}^{(t)}) + \varepsilon) \tag{5}$$

where $\mathbf{H}^{(t)}$ is the low rank matrix at the t-th iteration and $\omega_i^{(t+1)}$ is the related weight at the $(t + 1)$-th iteration, c is a positive constant and ε is a small constant used to avoid division by zero. In addition, Yoo et al. [17] propose another adaptive weighted nuclear norm with the weight as $\omega_i = c/(\alpha\sigma_i + \varepsilon)$, where α controls ratio of the singular values. Jia et al. [18] exploit discriminative weighted nuclear norm to denoise low-dose X-ray computed tomography (CT) image, and the weight is defined as $\omega_i = \beta E/(\sigma_i(\mathbf{H}) + \varepsilon)$, where β is a positive constant and E is the local entropy of an image block.

3   Proposed Method

3.1   Denoising Model

During HSI acquisition, HSI is corrupted by different types of noise, such as Gaussian noise, deadline noise, stripe noise and impulse noise. Except for Gaussian noise, the deadline noise, impulse noise and stripe noise are generally sparse distribution. Therefore, the noisy HSI is formulated as

$$\mathbf{Y} = \mathbf{L} + \mathbf{S} + \mathbf{N} \tag{6}$$

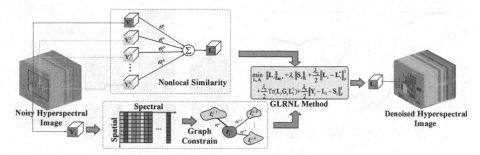

Fig. 1. Flowchart of GLRNL.

where $\mathbf{Y} \in \mathbb{R}^{m \times n \times b}$ is noisy HSI, $\mathbf{L} \in \mathbb{R}^{m \times n \times b}$ is target clean one, $\mathbf{S} \in \mathbb{R}^{m \times n \times b}$ is sparse noise (including deadlines, impulse noise and stripes), and $\mathbf{N} \in \mathbb{R}^{m \times n \times b}$ is the Gaussian noise, $m \times n$ denotes the spatial size of HSI and b is the HSI spectral number.

In this paper, we divide the HSI into 3-D patches and restore each patch sequentially, which can preserve the local details and provide an effective low rank definition [18]. Let $\mathbf{Y}_i \in \mathbb{R}^{p \times b}$ be the i-th 3-D patch ($\sqrt{p} \times \sqrt{p} \times b$) of noisy image $\mathbf{Y}$, reshaped as a $p \times b$ matrix. Then, $\mathbf{Y}_i$ can be represented as

$$\mathbf{Y}_i = \mathbf{L}_i + \mathbf{S}_i + \mathbf{N}_i \tag{7}$$

where $\mathbf{L}_i, \mathbf{S}_i$ and $\mathbf{N}_i$ correspond to the i-th patch of $\mathbf{L}, \mathbf{S}$ and $\mathbf{N}$, respectively. The proposed denoising model is

$$\min_{\mathbf{L}_i, \mathbf{S}_i} \|\mathbf{L}_i\|_{\omega,*} + \lambda_1 \|\mathbf{S}_i\|_1 + \frac{\lambda_2}{2} \|\mathbf{L}_i - \mathbf{L}_i^*\|_F^2 + \frac{\lambda_3}{2} \mathrm{Tr}(\mathbf{L}_i \mathbf{G}_i \mathbf{L}_i^T) + \frac{\lambda_4}{2} \|\mathbf{Y}_i - \mathbf{L}_i - \mathbf{S}_i\|_F^2 \tag{8}$$

where $\|\mathbf{L}_i\|_{\omega,*}$ is the weighted nuclear norm of $\mathbf{L}_i$ and its weight vector ω is defined as [13], λ_i ($i = 1, 2, 3, 4$) are the tradeoff parameters. In Problem (8), $\mathbf{L}_i^*$ and $\mathbf{G}_i$ are the non-local prior information of $\mathbf{L}_i$ and the graph Laplacian matrix respectively, and will be defined next. Figure 1 is the flowchart of proposed method.

HSI has lots of 3-D similar patterns, called as non-local similarity. Exploiting these intrinsic structures can improve the denoising performance. In Problem (8), $\|\mathbf{L}_i - \mathbf{L}_i^*\|_F^2$ is a special non-local constrain to guide the calculation of $\mathbf{L}_i$. $\mathbf{L}_i^*$ is the weighted average of similar patches of $\mathbf{Y}_i$, i.e. $\mathbf{L}_i^* = \sum_{k=1}^{K} \theta_i^k \mathbf{Y}_i^k$, and $\mathbf{Y}_i^k$ is the k-th similar patch of $\mathbf{Y}_i$. The weight θ_i^k is defined as

$$\theta_i^k = \exp\left(-\frac{\|\mathbf{Y}_i - \mathbf{Y}_i^k\|_F^2}{2\epsilon^2}\right), \; \theta_i^k = \frac{\theta_i^k}{\sum_{k=1}^{K} \theta_i^k} \tag{9}$$

The rich spectral information is a specific characteristic of HSI, acquired through the consecutive spectral ranges. To explain bands distribution, we employ correlation coefficient to describe band correlation between reference band (band 1) and other bands,

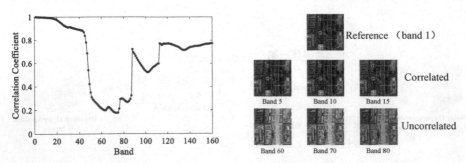

Fig. 2. Band correlation between reference band (band 1) (left) and each band, and the corrected bands and uncorrected bands in Washington DC Mall (right).

as shown in Fig. 2 (left). In addition, some correlated bands and uncorrelated bands are also shown in Fig. 2 (right). It can be seen that the spectral channels in HSI have high correlation and special bands distribution characteristics. Therefore, we model the bands distribution characteristics and related correlation as a graph constrain $\mathrm{Tr}(\mathbf{L}_i\mathbf{G}_i\mathbf{L}_i^T)$. In this graph, the spectral channels $\mathbf{L}_i = \{l_i^j\}_{j=1}^b$ are used as nodes. The weight between the nodes l_i^j and l_i^k is defined as

$$d_i^{j,k} = \exp\left(-\frac{\left\|y_i^j - y_i^k\right\|_2^2}{c\rho^2}\right) \tag{10}$$

where y_i^j and y_i^k are the j-th and k-th spectral channels of $\mathbf{Y}_i$. Then, the graph embedding for $\mathbf{L}_i$ is computed as

$$\frac{1}{2}\sum_{j,k=1}^b d_i^{j,k}\left\|l_i^j - l_i^k\right\|_2^2 = \mathrm{Tr}(\mathbf{L}_i\mathbf{G}_i\mathbf{L}_i^T) \tag{11}$$

where $\mathbf{G}_i = \mathbf{C}_i - \mathbf{A}_i$ is Laplacian matrix, $\mathbf{A}_i = \{d_i^{j,k}\}$, $\mathbf{C}_i$ is the diagonal matrix given by $\mathbf{C}_i^{jj} = \sum_{k=1}^b d_i^{j,k}$.

3.2 Solver

To solve Problem (8), we introduce an auxiliary matrix $\mathbf{Z}_i = \mathbf{L}_i$, i.e.,

$$\min_{\mathbf{L}_i,\mathbf{S}_i,\mathbf{Z}_i} \|\mathbf{L}_i\|_{\omega,*} + \lambda_1\|\mathbf{S}_i\|_1 + \frac{\lambda_2}{2}\|\mathbf{L}_i - \mathbf{L}_i^*\|_F^2 + \frac{\lambda_3}{2}\mathrm{Tr}(\mathbf{Z}_i\mathbf{G}_i\mathbf{Z}_i^T) + \frac{\lambda_4}{2}\|\mathbf{Y}_i - \mathbf{L}_i - \mathbf{S}_i\|_F^2$$

$$s.t.\ \mathbf{Z}_i = \mathbf{L}_i \tag{12}$$

Then, the alternating direction method of multipliers (ADMM) algorithm is used to seek the solution of Problem (12). The related augmented Lagrange function is

$$\mathcal{L}(\mathbf{L}_i, \mathbf{S}_i, \mathbf{Z}_i, \mathbf{\Lambda}_i) = \|\mathbf{L}_i\|_{\omega,*} + \lambda_1\|\mathbf{S}_i\|_1 + \frac{\lambda_2}{2}\|\mathbf{L}_i - \mathbf{L}_i^*\|_F^2 + \frac{\lambda_3}{2}\mathrm{Tr}(\mathbf{Z}_i\mathbf{G}_i\mathbf{Z}_i^T)$$

$$+ \frac{\lambda_4}{2}\|\mathbf{Y}_i - \mathbf{L}_i - \mathbf{S}_i\|_F^2 + \frac{\mu}{2}\|\mathbf{L}_i - \mathbf{Z}_i\|_F^2 + \langle \mathbf{\Lambda}_i, \mathbf{Z}_i - \mathbf{L}_i \rangle \quad (13)$$

where μ is the penalty parameter, $\mathbf{\Lambda}_i$ is the Lagrange multiplier. The minimum optimization problem on $\mathcal{L}(\mathbf{L}_i, \mathbf{S}_i, \mathbf{Z}_i, \mathbf{\Lambda}_i)$ is solved by following subproblems alternatively.

1) $\mathbf{L}_i^{(t+1)}$ subproblem:

$$\min_{\mathbf{L}_i} \|\mathbf{L}_i\|_{\omega,*} + \|\mathbf{L}_i - \mathbf{D}\|_F^2 \quad (14)$$

where $\mathbf{D} = [\lambda_2 \mathbf{L}_i^* + \lambda_4(\mathbf{Y}_i - \mathbf{S}_i^{(t)}) + \mathbf{\Lambda}_i^{(t)} + \beta \mathbf{Z}_i^{(t)}]/(\mu + \lambda_2 + \lambda_4)$. Problem (14) can be solved by weighted SVT operation [13], i.e.,

$$\mathbf{L}_i^{(t+1)} = \mathbf{U}\mathcal{S}_\omega(\mathbf{\Sigma})\mathbf{V}^T \quad (15)$$

In Eq. (15), $\mathbf{D} = \mathbf{U}\mathbf{\Sigma}\mathbf{V}^T$ is the singular value decomposition of matrix $\mathbf{D}$, $\mathcal{S}_\omega(\mathbf{\Sigma})$ is the soft-thresholding operator with weight vector ω defined as:

$$\mathcal{S}_\omega(\mathbf{\Sigma})_i = \max(\eta_i - \omega_i, 0) \quad (16)$$

where $\mathbf{\Sigma} = \text{diag}(\eta_i)$ and η_i is the i-th singular value of $\mathbf{D}$.

2) $\mathbf{S}_i^{(t+1)}$ subproblem:

$$\min_{\mathbf{S}_i} \lambda_1 \|\mathbf{S}_i\|_1 + \frac{\lambda_4}{2}\left\|\mathbf{S}_i - (\mathbf{Y}_i - \mathbf{L}_i^{(t+1)})\right\|_F^2 \quad (17)$$

$\mathbf{S}_i^{(t+1)}$ is solved as $\mathbf{S}_i^{(t+1)} = \mathcal{S}_{\lambda_1/\lambda_4}\left[\mathbf{S}_i - (\mathbf{Y}_i - \mathbf{L}_i^{(t+1)})\right]$, and $\mathcal{S}$ is the soft-thresholding operator.

3) $\mathbf{Z}_i^{(t+1)}$ subproblem:

$$\min_{\mathbf{Z}_i} \frac{\lambda_3}{2}\text{Tr}(\mathbf{Z}_i\mathbf{G}_i\mathbf{Z}_i^T) + \frac{\mu}{2}\left\|\mathbf{Z}_i - (\mathbf{L}_i^{(t+1)} - \frac{\mathbf{\Lambda}_i^{(t)}}{\beta})\right\|_F^2 \quad (18)$$

The closed-form solution of Problem (18) is $\mathbf{Z}_i^{(t+1)} = (\mu\mathbf{L}_i^{(t+1)} - \mathbf{\Lambda}_i^{(t)})(\lambda_3\mathbf{G}_i + \mu\mathbf{I})^{-1}$.

4) $\mathbf{\Lambda}_i^{(t+1)}$ updating:

$$\mathbf{\Lambda}_i^{(t+1)} = \mathbf{\Lambda}_i^{(t)} + \mu(\mathbf{Z}_i^{(t+1)} - \mathbf{L}_i^{(t+1)}) \quad (19)$$

Moreover, we employ $\left\|\mathbf{Z}_i^{(t+1)} - \mathbf{L}_i^{(t+1)}\right\|_\infty \leq \varepsilon$ as the stopping criteria to alternatively iterate the above four subproblems and set $\varepsilon = 10^{-6}$ in the following experiments.

4 Experiments and Analyses

In this section, we present the experiments on the simulated and real noisy data to evaluate the performances of proposed method. Five methods are employed for performance comparisons, including the LRRSDS [19], NAILRMA [20], LLRGTV [21], TDL [22]

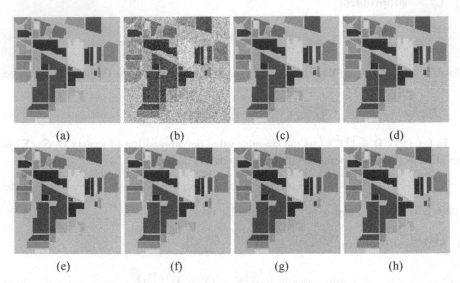

Fig. 3. Denoised results of India data on case 1 (band 52). (a) Clean image, (b) Noisy image, (c) NAILRMA, (d) LRRSDS, (e) LLRGTV, (f) TDL, (g) FSSLRL, (h) GLRNL.

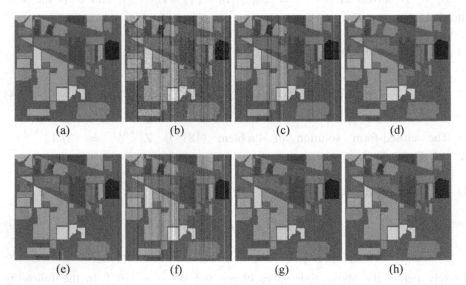

Fig. 4. Denoised results of India data on case 3 (band 193). (a) Clean image, (b) Noisy image, (c) NAILRMA, (d) LRRSDS, (e) LLRGTV, (f) TDL, (g) FSSLRL, (h) GLRNL.

and FSSLRL [23]. Four indexes including mean peak signal-to-noise ratio (MPSNR), mean structure similarity (MSSIM) [24], Erreur Relative Globale Adimensionnelle de Synthèse (ERGAS) [25], and spectral angle mapper (SAM) [26] are applied to measure the quality of denoised images. Better denoised results are reflected by higher MSSIM and MPSNR values. As for the ERGAS and SAM indices, the lower values indicate better quality.

4.1 Simulated Data Experiments

In this type of experiment, the Indian Pines data[1] is selected as the test data, which contains 145×145 spatial pixels and 224 spectral bands. To simulate the noisy HSI data, the pixel values of each band are normalized within the range of [0, 1] and then the Gaussian noise, deadline noise and impulse noise are added into the clean image. In the following experiments, we study four cases.

Case 1: Zero-mean Gaussian noise with noise level $\sigma = 0.1$ is added to all bands.

Case 2: Zero-mean Gaussian noise is added to all bands and the noise level σ is randomly sampled from 0.05 to 0.15.

Case 3: All bands were corrupted by Gaussian noise with noise level $\sigma = 0.05$. Besides, 40 bands are randomly chosen to add the stripe noise. The number of stripes in each band is randomly set from 20 to 40.

Case 4: Gaussian noise is added as case 2. In addition, impulse noise with a percentage of 20% is added to randomly chosen 20 bands. Moreover, the deadlines with width of 1–3 are added to 20 bands, in which 10 bands are selected from the impulse noise bands and other 10 bands are randomly sampled from the rest bands.

Figure 3 and Fig. 4 show the denoised images of different methods on India (band 52 and band 193). By visual comparisons, it can be seen that our method suppresses noise and preserves local details effectively. NAILRMA, TDL and LRRGTV can remove Gaussian noise, but are not effect on removing stripe noise. Table 1 reports the quantitative evaluation results of different denoising methods on different cases for simulated data. It can be seen that our method is competitive to the tested methods, and even exceeds the others on the simulated experiments.

4.2 Real Data Experiments

In this type of experiment, we evaluate the performance of proposed method on real HYDICE Urban[2] data and Earth Observing-1(EO-1) Hyperion[3] data. In the HYDICE Urban data, bands 104–108, 139–151, and 207–210 are removed, which seriously polluted by the atmosphere and water absorption. The used HYDICE Urban data size is $307 \times 307 \times 189$. The size of original EO-1 data is $1000 \times 400 \times 242$ and we crop the size of $400 \times 200 \times 166$ region (removing the water absorption bands) as the test data. The values of each spectral band are normalized as [0, 1].

---

[1] https://engineering.purdue.edu/~biehl/MultiSpec/aviris_documentation.html.

[2] http://www.tec.army.mil/hypercube.

[3] http://www.lmars.whu.edu.cn/prof_web/zhanghongyan/resource/noise_EOI.zip.

Table 1. Quantitative comparison of different methods on simulated Data

Noise	Indexes	LRRSDS	NAIRLMA	LRRGTV	TDL	FSSLRL	GLRNL
Case 1	MPSNR	32.0451	37.6678	35.9941	36.0291	32.2973	**37.9867**
	MSSIM	0.8162	0.9414	0.9546	**0.9675**	0.9364	0.9544
	ERGAS	61.6091	31.0338	39.7550	38.0987	59.8459	**30.0145**
	SAM	0.0482	0.0214	0.0229	0.0213	0.0295	**0.0204**
Case 2	MPSNR	32.4269	37.0387	36.1255	34.0587	32.4669	**37.6155**
	MSSIM	0.8278	0.9485	0.9521	0.8864	0.9402	**0.9565**
	ERGAS	59.5647	**29.5657**	39.2604	49.0392	58.5699	31.5057
	SAM	0.0465	0.0211	0.0227	0.0356	0.0303	**0.0204**
Case 3	MPSNR	36.5264	39.7728	37.3241	38.9212	38.9453	**40.7228**
	MSSIM	0.9291	0.9500	0.9767	0.9234	0.9746	**0.9832**
	ERGAS	37.7464	39.9302	36.0208	57.2289	28.2845	**24.5283**
	SAM	0.0291	0.0318	0.0197	0.0483	**0.0180**	**0.0180**
Case 4	MPSNR	37.5458	38.4603	37.4991	27.4736	38.5096	**39.7964**
	MSSIM	0.9594	0.9786	**0.9824**	0.6143	0.9637	0.9765
	ERGAS	45.4213	53.9836	48.2853	212.3922	50.9225	**30.0859**
	SAM	0.0320	0.0455	0.0270	0.1837	0.0394	**0.0221**

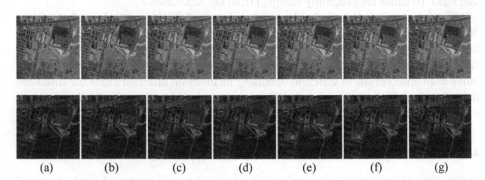

(a)   (b)   (c)   (d)   (e)   (f)   (g)

Fig. 5. Denoised results of real HYDICE Urban data bands 103(upper) and 188(bottom). (a) Original image, (b) NAILRMA, (c) LRRSDS, (d) LRRGTV, (e) TDL, (f) FSSLRL, (g) GLRNL.

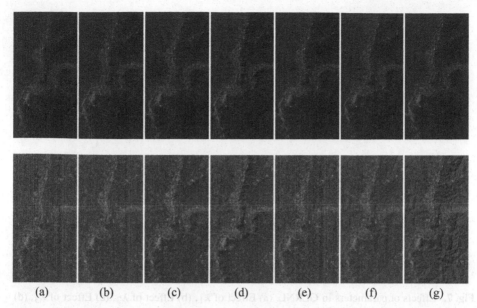

(a)   (b)   (c)   (d)   (e)   (f)   (g)

Fig. 6. Denoised results of real EO-1 data bands 2(upper) and 132(bottom). (a) Original image, (b) NAILRMA, (c) LRRSDS, (d) LRRGTV, (e) TDL, (f) FSSLRL, (g) GLRNL.

Figure 5 and Fig. 6 show the denoised results of HYDICE Urban data (bands 103 and 188) and EO-1 data (bands 2 and 132), respectively. By the visual comparison, it can be seen that TDL results in some artifacts, as shown in Figs. 5(e) and 6(e). NAILRMA, LRRSDS, LRRGTV, and FSSLRL exhibit better visual effect, but still have some stripes noise. Overall, our proposed method can remove different types noise, preserve local spatial details, and provides a better visual effect. The visual comparisons demonstrate that the non-local and graph-structured constraints in our method is effective in preserving spatial and spectral information.

4.3 Effects of Parameters

In the proposed method, the parameters λ_1, λ_2, λ_3, λ_4 and patch size p are the major parameters to influence the performance of proposed method. To study the effects of these parameters, we test the proposed method with different parameters settings. Taking the effect of λ_1 as example, λ_1 is changed as [0.05, 0.1, 0.15, 0.2, 0.25, 0.3, 0.35, 0.4] and the other parameters are fixed. The proposed method is implemented on the simulated noisy Indian data with different λ_1. To describe concisely, the obtained indexes values are normalized within [0, 1] by $(x - \min(x))/(\max(x) - \min(x))$. Figure 7(a) shows the trends of performance with λ_1 changes. Then, the effects of λ_2, λ_3, λ_4 and patch size p are assessed using the same implementation as λ_1, as shown in Fig. 7(b)–(e). Based on the trends in Fig. 7, λ_1, λ_2, λ_3, λ_4 and patch size p are set as 0.35, 2×10^{-4}, 2×10^{-4} and 20, respectively.

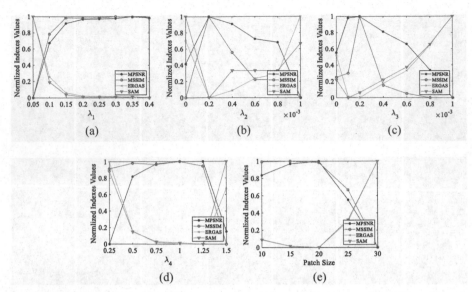

Fig. 7. Effects of parameters in GLRNL. (a) Effect of λ_1, (b) Effect of λ_2, (c) Effect of λ_3, (d) Effect of λ_4, (e) Effect of patch size.

5 Conclusions

In this paper, we proposed a HSI denoising algorithm based on graph structured low rank and non-local constraint. The non-local constraint can effectively maintain the structure of HSI and improve the robustness of algorithm, and the graph structure can preserve the distribution characteristics of HSI spectral bands, which can reduce the spectral information distortion. The experimental results on simulated data and real noisy data show the effectiveness of our method. In the future, we will study intelligent parameters selection and optimization methods to improve the result further.

References

1. Rasti, B., Scheunders, P., Ghamisi, P., et al.: Noise reduction in hyperspectral imagery: overview and application. Remote Sens. **10**(3), 482–510 (2018)
2. Atkinson, I., Kamalabadi, F., Jones, D.: Wavelet-based hyperspectral image estimation. In: 2003 IEEE International Geoscience and Remote Sensing Symposium (IGARSS), pp. 743–745 (2003)
3. Qian, Y., Shen, Y., Ye, M., et al.: 3-D nonlocal means filter with noise estimation for hyperspectral imagery denoising. In: 2012 IEEE International Geoscience and Remote Sensing Symposium (IGARSS), pp. 1345–1348 (2012)
4. Chen, G., Qian, S.: Denoising of hyperspectral imagery using principal component analysis and wavelet shrinkage. IEEE Trans. Geosci. Remote Sens. **49**(3), 973–980 (2011)
5. Yuan, Q., Zhang, L., Shen, H.: Hyperspectral image denoising employing a spectral-spatial adaptive total variation model. IEEE Trans. Geosci. Remote Sens. **50**(10), 3660–3677 (2012)

6. Qian, Y., Ye, M.: Hyperspectral imagery restoration using nonlocal spectral-spatial structured sparse representation with noise estimation. IEEE J. Sel. Topics Appl. Earth Observ. Remote Sens. **6**(2), 499–515 (2013)
7. Xing, Z., Zhou, M., Castrodad, A., et al.: Dictionary learning for noisy and incomplete hyperspectral images. SIAM J. Imag. Sci. **5**(1), 33–56 (2012)
8. Zhang, H., He, W., Zhang, L., et al.: Hyperspectral image restoration using low-rank matrix recovery. IEEE Trans. Geosci. Remote Sens. **52**(8), 4729–4743 (2014)
9. Xue, J., Zhao, Y., Liao, W., et al.: Joint spatial and spectral low-rank regularization for hyperspectral image denoising. IEEE Trans. Geosci. Remote Sens. **56**(4), 1940–1958 (2018)
10. Xie, Q., Zhao, Q., Meng, D., et al.: Multispectral images denoising by intrinsic tensor sparsity regularization. In: Proceedings of the IEEE Conference on Computer Vision and Pattern Recognition (CVPR), pp. 1692–1700 (2016)
11. Zhang, H., Liu, L., He, W., et al.: Hyperspectral image denoising with total variation regularization and nonlocal low-rank tensor decomposition. IEEE Trans. Geosci. Remote Sens. **58**(5), 3071–3084 (2020)
12. Candès, E.J., Li, X., Ma, Y., et al.: Robust principal component analysis. J. ACM **58**(3), 1–37 (2011)
13. Gu, S., Zhang, L., Zuo, W., et al.: Weighted nuclear norm minimization with application to image denoising. In: Proceedings of the IEEE Conference on Computer Vision and Pattern Recognition (CVPR), pp. 2862–2869 (2014)
14. Yang, S., Zhang, K., Wang, M.: Learning low-rank decomposition for pan-sharpening with spatial-spectral offsets. IEEE Trans. Neural Netw. Learn. Syst. **29**(8), 3647–3657 (2018)
15. Wright, J., Yang, A.Y., Ganesh, A., et al.: Robust face recognition via sparse representation. IEEE Trans. Pattern Anal. Mach. Intell. **31**(2), 210–227 (2009)
16. Qu, Y., Guo, R., Wang, W., et al.: Hyperspectral anomaly detection through spectral unmixing and dictionary based low rank decomposition. IEEE Trans. Geosci. Remote Sens. **56**(8), 4391–4405 (2018)
17. Yoo, S., Wang, Z., Seo, J.: Adaptive weighted nuclear norm minimization for removing speckle noise from optical coherence tomography images. In: 41st Annual International Conference of the IEEE Engineering in Medicine and Biology Society (EMBC), pp. 2687–2690 (2019)
18. Jia, L., Zhang, Q., Shang, Y., et al.: Denoising for low-dose CT image by discriminative weighted nuclear norm minimization. IEEE Access **6**, 46179–46193 (2018)
19. Sun, L., Jeon, B., Zheng, Y., et al.: Hyperspectral image restoration using low-rank representation on spectral difference image. IEEE Geosci. Remote Sens. Lett. **14**(7), 1151–1155 (2017)
20. He, W., Zhang, H., Zhang, L., et al.: Hyperspectral image denoising via noise-adjusted iterative low-rank matrix approximation. IEEE J. Sel. Topics Appl. Earth Observ. Remote Sens. **8**(6), 3050–3061 (2015)
21. He, W., Zhang, H., Shen, H., et al.: Hyperspectral image denoising using local low-rank matrix recovery and global spatial-spectral total variation. IEEE J. Sel. Topics Appl. Earth Observ. Remote Sens. **11**(3), 713–729 (2018)
22. Peng, Y., Meng, D., Xu, Z., et al.: Decomposable nonlocal tensor dictionary learning for multispectral image denoising. In: Proceedings of the IEEE Conference on Computer Vision and Pattern Recognition (CVPR), pp. 2949–2956 (2014)
23. Sun, L., Jeon, B., Soomro, B.N., et al.: Fast superpixel based subspace low rank learning method for hyperspectral denoising. IEEE Access **6**, 12031–12043 (2018)
24. Wang, Z., Bovik, A.C., Sheikh, H.R., et al.: Image quality assessment: from error visibility to structural similarity. IEEE Trans. Image Process. **13**(4), 600–612 (2004)

25. Wald, L., Ranchin, T., Mangolini, M.: Fusion of satellite images of different spatial resolutions: assessing the quality of resulting images. Photogramm. Eng. Remote Sens. **63**(6), 691–699 (1997)
26. Yuhas, R.H., Goetz, A.F., Boardman, J.W.: Discrimination among semi-arid landscape endmembers using the spectral angle mapper (SAM) algorithm. In: Proceedings of the Summaries 3rd Annual JPL Airborne Geoscience Workshop, pp. 147–149 (1992)

Automatic Tooth Segmentation and 3D Reconstruction from Panoramic and Lateral Radiographs

Mochen Yu[1], Yuke Guo[2], Diya Sun[1], Yuru Pei[1(✉)], and Tianmin Xu[3]

[1] Key Laboratory of Machine Perception (MOE), Department of Machine Intelligence, Peking University, Beijing, China
peiyuru@cis.pku.edu.cn
[2] Luoyang Institute of Science and Technology, Luoyang, China
[3] School of Stomatology, Peking University, Beijing, China

Abstract. The panoramic and lateral radiographs are commonly used in orthodontic dentistry to acquire patient-specific tooth morphology for diagnosing and treatment planning. Considering the variational dentition configurations and image blurs caused by device-specific artefacts and structure overlapping, the robust tooth segmentation and 3D reconstruction from radiographs remain a challenging issue. We propose a deformable exemplar-based conditional random fields (CRF) model for tooth segmentation and 3D shape estimation from the panoramic and lateral radiographs. The shared tooth foreground in the lateral and the panoramic radiographs are utilized for consistent labeling. The 3D deformable exemplars are introduced to provide a regularization of tooth contours to improve the tooth parsing in noisy and ambiguous radiographs. We introduce an alternating optimization scheme to solve the discrete superpixel labels and the continuous deformation of the 3D exemplars simultaneously. Extensive experiments on clinically obtained radiographs demonstrate that the proposed approach is effective and efficient for both tooth segmentation and 3D shape estimation from the panoramic and lateral radiographs.

Keywords: Tooth segmentation · 3D reconstruction · Deformable exemplar-based CRF

1   Introduction

Malocclusion introduces risks of tooth decay, excessive pressures on the temporomandibular joint, and aesthetic problems. There is an increasing demand for orthodontic treatments considering the high prevalence of malocclusion. Panoramic radiograph (*pan*) and lateral radiograph (*lat*), with low radiation exposure compared with 3D computed tomography, are routines in clinical orthodontics, providing insights into dental geometries and inner decays economically and efficiently. Automatic radiograph parsing and tooth segmentation

© Springer Nature Switzerland AG 2020
Y. Peng et al. (Eds.): PRCV 2020, LNCS 12305, pp. 53–64, 2020.
https://doi.org/10.1007/978-3-030-60633-6_5

is a cornerstone for qualitative measurements of tooth shapes and movements for orthodontic treatment planning and evaluations. However, image blurring, common in radiographs due to structure overlapping, shifting from the focal trough, and slight subject's movements during the capturing process pose a major problem for the automatic tooth segmentation and 3D estimation.

The automatic segmentation of 2D radiographs has been studied in past decades. A large variety approaches have been proposed for the automatic tooth segmentation from radiographs, including thresholding [28], active contours and the level sets [14,29], the region growing [13], the probabilistic model [11], the morphological edge detection [22], and the deep learning [7,10,19]. An evaluation and survey on dental radiograph segmentation are referred to [23,30]. In order to handle the blurry radiographs, an adaptive power law transformation is used to reduce the contrast variations between teeth and alveolar bones by the local singularities of the enhanced image [16]. Thresholding on its own is limited to get tooth contour and often serves as initializations of dental segmentation. The level set method realizes automatic gums lesion detection in periapical dental X-ray images [15]. However, the active contour and level set-based segmentation highly rely on interactive initialization and prone to error accumulations. The deep learning-based technique, such as the U-Net [24] and the Mask-RCNN [10], handles automatic contour sketching from the *lat* and the *pan*, where a large amount of annotated images are required to learn network parameters.

The *pan* encodes the complete dental arches with clear tooth contours except for root tips, while having no constant horizontal and vertical magnifications. The *lat* provides frontal tooth orientations and spatial relationships of dental arches while being blurry due to the overlapping of symmetric craniofacial structures. The *pan* and *lat* are complementary in clinical orthodontics for providing insight into 3D geometries. The co-segmentation in computer vision community develops with the availability of images with common contents, utilizing the consistent foreground modeling [18,25]. Whereas, it is limited to identify the tooth boundaries considering blurry radiographs with structure overlapping and focal-losing. The statistical model utilizes shape priors for 2D image parsing and 3D shape estimation from radiographs by shape registration. The tooth contours can be efficiently solved in a reduced subspace, though the linear models are limited to parse the dental radiographs with shapes not covered by the subspace [27,32]. The statistical 3D model is commonly used for 3D shape estimation from radiographs, in which the projected intensity images or silhouettes are required to be consistent with the 2D radiographs [4–6,34]. The thin-plate-spline deformation can further refine the statistically-reconstructed surface [33]. Kang et al. [12] proposed to optimize pose and shape by formulating the coupled problems as a maximum penalized likelihood estimation for solving shape parameters in a subspace. However, the above 3D shape estimation breaks up the radiograph parsing and 3D shape reconstruction, where the radiograph parsing is either predefined [20] or using partial and incomplete contours by edge filters [4]. To the best of our knowledge, there is no framework addressing the simultaneous segmentation and 3D complete tooth estimation from the *pan* and *lat*.

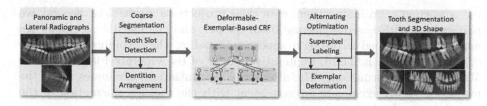

Fig. 1. Flowchart of our system. Given paired panoramic radiograph (*pan*) and lateral radiograph (*lat*), we utilize the integral projection and rigid transformation for coarse segmentation and exemplar arrangement. Then the 3D deformable exemplars-based CRF is used to infer tooth segmentation and 3D tooth reconstruction from *pan* and *lat* simultaneously, which are solved by the alternating optimization method. Finally, the system outputs the 3D tooth and semantic segmentations of the *pan* and *lat*.

In this paper, we propose a deformable exemplar-based conditional random fields (CRF) model for simultaneous tooth segmentation and 3D tooth estimation from the paired *pan* and *lat* (Fig. 1). The proposed framework is built upon multi-label CRF with labeling regularization using non-rigidly deformed tooth exemplars. We initialize tooth partition using horizontal and vertical integral projections. Given paired *pan* and *lat*, we utilize the shared tooth foreground for consistent labeling between radiographs. To improve the tooth segmentation in noisy and ambiguous radiographs, we utilize the shape priors from the 3D deformable exemplars, where the tooth contours on the *pan* and *lat* are required to be consistent with projections of 3D exemplars. We introduce an alternating optimization scheme to solve the discrete superpixel labels and the continuous deformation of the 3D exemplars simultaneously.

In our evaluations, we show 3D tooth estimation and radiograph segmentation with semantic labeling from the *pan* and *lat*. Our results outperform compared tooth segmentation methods from radiographs and obtain accurate semantic labeling of an individual tooth. The main contributions of this paper are as follows:

- We propose the 3D deformable exemplars-based CRF model for tooth segmentation and 3D reconstruction from the *pan* and *lat* simultaneously.
- We present an alternating optimization scheme to solve the discrete semantic labels of radiographs and continuous 3D deformations for the patient-specific 3D teeth.

2 Method

We address the simultaneous tooth segmentation and 3D reconstruction from the input paired *pan* and *lat* under an exemplar-based CRF framework, as shown in Fig. 1. The 3D exemplars are introduced to refine teeth contours, especially to the confusing contours due to image blurring. First, we conduct coarse segmentation by the integral projection to locate teeth slots. The 3D exemplars are arranged in the consistent coordinate system with their projections inside slots. Second, we perform the CRF-based segmentation of the *pan* and *lat*. The 3D exemplars are non-rigidly deformed according to superpixel labeling while avoiding the collision of neighboring teeth. Third, we iteratively infer of the superpixel labels and the 3D exemplar deformations. The alternating optimization terminates when the projection of 3D exemplars are consistent with the tooth segmentation of the paired *pan* and *lat*.

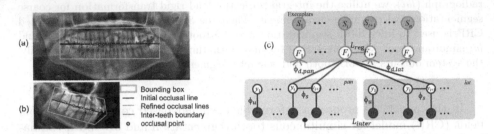

Fig. 2. Coarse segmentation of (a) the *pan* and (b) the *lat* images. (c) 3D exemplar-based CRF.

2.1 Coarse Segmentation and Dentition Arrangement

We perform the coarse segmentation to identify the tooth slots using horizontal integral projection [11] to find the initial occlusal line of the *pan* and *lat* images, where the region of interests are plotted as polygons (Fig. 2(a, b)). Each tooth is segmented by a vertical integral projection when given the initial occlusal line. The initial tooth boundary is orthogonal to the occlusion line, and the orientation of slot boundaries are tuned to minimize the integral values. The partial occlusal line is modified accordingly to the slot boundaries. The piecewise occlusal line between the upper and lower teeth is orthogonal to the local slot boundaries. The occlusal points are located at the center of piecewise occlusal lines, as shown in Fig. 2(a, b).

The 3D tooth exemplars are arranged heuristically according to the coarse segmentation of radiographs. We follow the consistent coordinate system in [20], where the *pan* is wrapped with the surface determined by the rotation trajectory and rays originating from the X-ray source. The *lat* radiograph is captured when the X-ray source is on one side of the head. We conduct a rigid transformation of the 3D exemplars to project exemplars inside teeth slots in the consistent coordinate system.

2.2 Deformable Exemplar-Based CRF

Given the input *pan* and *lat* pair (I_{pan}, I_{lat}), the goal is to infer labels Y_{pan} and Y_{lat} of the radiographs regarding the individual tooth, as well as deformations F

of tooth exemplars S. Without loss of generality, we decompose each image into superpixels. The graph $G = (\mathcal{N}, \mathcal{E})$ is built upon the superpixel set U_{pan} and U_{lat} of the input pan and lat, together with the exemplar deformation F (see Fig. 2(c)). $\mathcal{N}$ denotes the nodes with respect to superpixels and deformation variables. $\mathcal{E}$ denotes the edges connecting adjacent superpixels, and those between superpixels of pan and lat for consistent segmentation with joint foregrounds. There are also edges between the superpixel variables and the deformable exemplars for the regularization of teeth contours.

Intra-image Labeling Consistency. We address the intra-image labeling consistency using the unary potential derived from the superpixel appearance and the relationship potential for smooth labeling of neighboring superpixels in local patches. We introduce the exemplar-based regularization utilizing silhouettes of 3D deformable exemplars to avoid boundary confusion in the blurry region by exploiting well-defined exemplar topologies. The loss function of the intra-image labeling consistency is defined as follows:

$$L_{intra} = \sum_{\kappa \in \{pan, lat\}} \left\{ \sum_{u_i \in U_\kappa} [-\ln \phi_u(y_i|u_i) + \sum_{u_j \in Neib(u_i)} \phi_s(u_i, u_j)] + \phi_{d,\kappa} \right\}.$$

(1)

The unary term ϕ_u is defined considering the labeling consistencies with the superpixel appearances. $y_i \in Y$ denotes the label of superpixel u_i. The potential of local superpixel $u_i \in U_\kappa$ is computed based on the output of the support vector machine (SVM) classifiers as in [21], and $\phi_u(y_i|u_i) = (1 + \exp(af(u_i) + b))^{-1}$. f denotes the output of the linear SVM classifier for the superpixel u_i with predefined parameters a and b. The one-vs-rest SVM with the chi-square kernel is used to build the classifier. Here we use a combinatorial feature descriptor of superpixels. The intensity descriptor is defined by a 10-bin intensity histogram of pixels inside a superpixel. The contextual descriptor is defined as the chi-square distance to 2-ring surrounding superpixels. We also build a 30-bin SIFT histogram upon a dictionary obtained by the K-means clustering.

The pairwise potential ϕ_s for the smoothing labeling of neighboring superpixels is defined by the global boundary probabilities (gbp) [2].

$$\phi_s(u_i, u_j) = (1 - h(u_i, u_j)) [y_i \neq y_j] + h(u_i, u_j) [y_i = y_j],$$

(2)

where $[\cdot]$ denotes an indicator function. $h(\cdot)$ is a sum function of gbp values on the shared boundary. In case the neighboring superpixels have different labels, the sum of gbp $h(u_i, u_j)$ gets a large value along the shared boundary.

The 3D exemplar-based regularization is introduced to improve the contour definition and handle the confusing contours due to image blurring. The exemplar-based regularization term ϕ_d is defined by the chamfer distance between the contours ν of the projected exemplars and the 2D dental contours c defined by the superpixel labeling Y. The projection function regarding to the pan image is determined by the device-specific ray orientation curves [20]. We follow the linear pin-hole model to define the projection function of the *lat* using given X-ray device parameters.

$$\phi_{d,\kappa} = \sum_{i=1}^{n_\kappa} d_{cd}\left(\nu_i^\kappa(S_i, F_i), c_i(Y_\kappa, U_\kappa)\right), \kappa \in \{pan, lat\}. \tag{3}$$

d_{cd} denotes the chamfer distance. n_κ denotes tooth number of radiograph κ. $\nu_i(S_i, F_i)$ denotes the projected tooth contour of the i-th tooth exemplar S_i with deformation parameters F_i. The tooth contour in radiographs is obtained by aggregating the superpixels with the same label. $c_i(Y_\kappa, U_\kappa)$ denotes the tooth contours determined by label Y_κ assigned to superpixel set U_κ.

Inter-image Correlation. Under the observation that there exist joint foreground teeth in the input paired radiographs, we introduce the inter-image correlation constraint for consistent labeling. The loss function of the inter-image correlation is defined by the foreground difference between corresponding tooth in the radiograph pair.

$$L_{inter} = \sum_{i=1}^{n} \alpha_i \chi^2 \left(q_{pan,i}, q_{lat,i}\right), \tag{4}$$

where $q_{pan,i}$ and $q_{lat,i}$ denote the histogram of superpixels inside the i-th tooth regarding the input paired *pan* and *lat*. The chi-squared distance χ^2 is used to measure the intensity histogram differences. The weight α_i is set at 1 when there is corresponding tooth i on the *pan* and *lat* images, and 0 otherwise. The tooth number n is set to 28 in our experiments.

Exemplar Regularization. The exemplar regularization term is used to preserve the topology of the 3D deformable exemplars and to avoid intersections between neighboring teeth.

$$L_{reg} = \sum_{i=1}^{n} \left\{ \sum_{j=1}^{n_{b,i}} \sum_{v_k \in \mathcal{N}eib_b(v_j)} \|\delta(v_j, v_k)\|^2 + \sum_{S_j \in \mathcal{N}eib_s(S_i)} \sum_{x \in S_j} \xi(d_s(x, S_i)) \right\}. \tag{5}$$

We utilize the embedded deformation [31] defined on a simplified base mesh. The smoothness constraint is imposed on nodes of the base meshes. The mesh smoothness of the 3D exemplar is preserved by minimizing the transformation difference between local transformation of base node v_j and that derived by its neighbors as in [31]. The transformation difference $\delta(v_j, v_k) = R_k(v_j - v_k) + v_k + t_k - (v_j + t_j)$. R and t denote a 3×3 rotation matrix and a 3D translation vector respectively. $n_{b,i}$ denotes the node number of the i-th tooth's base mesh. The second term is to penalize the intersections of neighboring teeth. We consider the signed distance d_s between a vertex and a neighboring 3D exemplar. In case the vertex is outside the exemplar surface, the distance is positive. ξ is a product of the indicator function and a large penalty constant, and $\xi(z) = [z < 0] \cdot A$. The constant A is set to 10^6 in our experiments.

Loss. The overall function of tooth segmentation and 3D reconstruction from *pan* and *lat* is defined as follows:

$$L = \gamma_{intra} L_{intra} + \gamma_{inter} L_{inter} + \gamma_{reg} L_{reg}. \tag{6}$$

The constant coefficients γ_{intra}, γ_{inter}, and γ_{reg} are used to balance the intra-image labeling consistency, the inter-image correlation, and the exemplar regularization, and set to 1, 1, and 2 respectively in our experiments.

Alternating Optimization. The optimization of the objective function (Eq. 6) is not trivial considering the discrete superpixel labels and the continuous deformation variables. We present an alternating optimization scheme to solve the superpixel labeling together with the exemplar deformations. Here we decompose the original optimization problem into two subproblems, i.e., the 2D superpixel labeling and 3D exemplar deformation. First, the initially arranged exemplars are used to compute the reference contours in ϕ_d. The distance transformation for the chamfer distance is imposed on the exemplar contours. The deformation variables are fixed, and the 3D exemplar regularization term L_{reg} is removed. We only consider the inference of the discrete labels in the first subproblem. Second, the continuous deformation variables F are to be solved given the labeling in the first phase. The unary and pairwise potentials in the intra-image consistency term L_{intra} and the inter-image correlation term L_{inter} are removed. The distance transformation for the chamfer distance is imposed on the 2D contour determined by superpixel labeling. We solve the nonrigid exemplar deformation in the second subproblem. The two problems at the i-th iteration are:

$$Y^{(i)} = \arg\min L(Y^{(i-1)}, F^{(i-1)}), \text{ and} \tag{7}$$

$$F^{(i)} = \arg\min L(Y^{(i)}, F^{(i-1)}). \tag{8}$$

The QPBO algorithm [26] is employed to solve Eq. 7. Considering the regularization term of the intersection penalty defined by the step function, the second subproblem is non-smooth. The pattern search algorithm [3] is employed to solve Eq. 8. We perform this alternating optimization until the deformed exemplars are consistent with the tooth segmentation on the *pan* and *lat* images.

3 Experiments

The experimental dataset includes 20 real *pan* and *lat* radiographs obtained from orthodontic patients. The *pan* and *lat* are captured by Orthopantomograph OP 100 device of Instrumantarium Imaging Incorporation with the resolution of 1050×500 and 720×900, respectively. There are also 20 simulated radiograph pairs obtained from cone-beam CT (CBCT) images using volume rendering. The simple linear iterative clustering (SLIC) method [1] is employed for the superpixel decomposition of the *pan* and *lat* pairs.

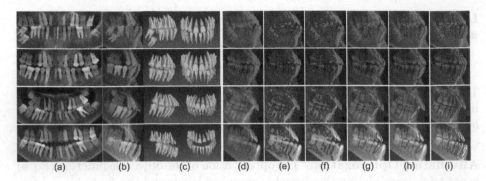

Fig. 3. Tooth segmentation of (a) the *pan* and (b) *lat* images. (c) 3D tooth reconstruction in two viewpoints. Tooth segmentation comparisons on (d) *lat* images using (e) Otsu thresholding [28], (f) CRF [9], (g) AAM [8], (h) EASM [17], and (i) ours.

The tooth segmentation is evaluated using the Jaccard similarity coefficient (JSC), $e^J = \frac{2|U \cap U'|}{|U|+|U'|}$ when given the superpixel set U labeled as an individual tooth by the automatic segmentation method and the ground truth U'. The 3D shape is evaluated using the mean surface deviation (MSD) between the reconstructed teeth surfaces and the manually-labeled from the CBCT images. We compare our method with the mainstream techniques in radiograph-based tooth detection and segmentation, including the Otsu thresholding [28], the AAM [8], the extended ASM (EASM) [17], and the CRF [9] methods.

3.1 Qualitative Evaluation

Qualitative results on radiograph segmentation and 3D tooth estimation are shown in Fig. 3(a–c). Table 1 shows the JSC of the tooth segmentation obtained by the proposed method, the Otsu thresholding [28], the CRF [9], the AAM [8], and the EASM [17] methods. We perform the qualitative evaluations on four types of teeth, including the incisor, the canine, the premolar, and the molar, on the *pan* and *lat* images. Figure 3(d–i) and Fig. 4 show the comparison on tooth segmentation. As we can see, it is hard to obtain complete and regular tooth contours using thresholding considering image blurring in the *pan* images because of surrounding alveolar bones and the focal losing, especially in the root regions. The CRF-based method imposes the pairwise regularization on the image segmentation and helps to identify the smooth tooth contours. However, the resulted tooth contours by the CRF without using the shape priors are prone to be erroneous, especially in the root regions. The thresholding and the CRF-based methods are limited to separate tooth contours confronted with the intercuspation, where the upper and the lower dental arches are overlapping. The AAM and the EASM-based method locates the image contours using a statistical 2D shape model, in which the segmentation performance may deteriorate when the linear subspace does not cover the shapes of the testing images. The proposed method employs the 3D shape priors, in which the deformable 3D exemplars facilitate the accurate contour location even in the tooth root region. The silhouettes of the 3D exemplars act as a regularization for smooth and complete teeth contours. Figure 3(i) and Fig. 4(f) show the overlapping projected contours of output 3D teeth. Note that the proposed method is feasible to handle the overlapping of the upper and lower dental arches in the intercuspation state facilitated by 3D exemplars.

Table 1. The JSC of tooth segmentation from the *pan* and *lat* images using compared methods.

	Incisor		Canine	Premolar		Molar
	pan	*lat*	*pan*	*pan*	*lat*	*pan*
Otsu [28]	0.78 ± 0.05	0.73 ± 0.02	0.76 ± 0.02	0.74 ± 0.03	0.65 ± 0.05	0.56 ± 0.06
AAM [8]	0.80 ± 0.10	0.79 ± 0.08	0.69 ± 0.08	0.70 ± 0.07	0.68 ± 0.08	0.77 ± 0.08
EASM [17]	0.83 ± 0.03	0.80 ± 0.06	0.67 ± 0.05	0.76 ± 0.04	0.67 ± 0.08	0.80 ± 0.08
CRF [9]	0.81 ± 0.08	0.72 ± 0.07	0.80 ± 0.05	0.73 ± 0.05	0.70 ± 0.08	0.73 ± 0.09
Ours	$\mathbf{0.90 \pm 0.03}$	$\mathbf{0.88 \pm 0.03}$	$\mathbf{0.91 \pm 0.03}$	$\mathbf{0.91 \pm 0.02}$	$\mathbf{0.89 \pm 0.05}$	$\mathbf{0.87 \pm 0.06}$

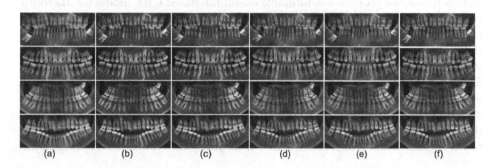

(a)   (b)   (c)   (d)   (e)   (f)

Fig. 4. Tooth segmentation of *pan* images. (a) Input. (b) Otsu thresholding [28]. (c) CRF [9]. (d) AAM [8]. (e) EASM [17]. (f) Ours.

We measure the MSDs of four types of teeth, as shown in Table 2. The MSDs of the proposed method are 0.56 mm, 1.43 mm, 1.68 mm, and 1.83 mm, regarding the incisor, the canine, the premolar, and the molar, respectively. As we can see, the error of the molar is larger than the other teeth. We think the reason is that it is difficult to infer the 3D shape due to the structure overlapping of the multi-root teeth from radiographs.

Parameter Analysis. The *pan* and *lat* are represented by superpixels, which greatly reduces the computational complexities. We report image segmentation accuracies with different numbers of superpixels, as shown in Table 2. We observe

Table 2. The MSDs and segmentation accuracies (JSC) using different feature channels (intensity-*in*, SIFT-*si*, Context-*co*) and different numbers of superpixels.

	MSD (mm)	Feature Channel			Superpixel Num. ($\times 10^3$)		
		fea_{in}	$fea_{in,si}$	$fea_{in,si,co}$	1.25	2.5	3.75
Incisor	0.56 ± 0.09	0.88 ± 0.04	0.89 ± 0.02	$\mathbf{0.90 \pm 0.03}$	0.85 ± 0.02	0.89 ± 0.04	$\mathbf{0.90 \pm 0.03}$
Canine	1.43 ± 0.34	0.87 ± 0.02	0.88 ± 0.03	$\mathbf{0.91 \pm 0.03}$	0.85 ± 0.03	$\mathbf{0.91 \pm 0.02}$	$\mathbf{0.91 \pm 0.03}$
Premolar	1.68 ± 0.39	0.83 ± 0.03	0.86 ± 0.04	$\mathbf{0.91 \pm 0.02}$	0.85 ± 0.05	0.90 ± 0.03	$\mathbf{0.91 \pm 0.02}$
Molar	1.83 ± 0.42	0.82 ± 0.02	$\mathbf{0.87 \pm 0.04}$	0.87 ± 0.06	0.84 ± 0.04	0.86 ± 0.04	$\mathbf{0.87 \pm 0.06}$

the performance improvements of the tooth segmentation with increasing numbers of superpixels. This agrees with common sense that the superpixels with small granularity are capable of capturing details in tooth boundaries. In our experiments, we use a combinatorial feature descriptor of superpixels. Table 2 shows the tooth segmentation performance with different feature channels. As we can see, the combinational feature using the intensity, the SIFT, and the contextual features produces the best results.

4   Conclusion

In this paper, we propose a deformable exemplar-based CRF model for 3D tooth reconstruction and segmentation from the input paired *pan* and *lat*. We utilize the shared tooth foreground in the *pan* and *lat* for consistent labeling of the individual tooth. The 3D deformable exemplars provide tooth shape priors, facilitating tooth segmentation from blurry radiographs due to device-specific artefacts and structure overlapping. We present an alternating optimization scheme to solve the discrete superpixel labels and the continuous deformation of the 3D exemplars simultaneously. Experiments on clinically-obtained radiographs demonstrate that the proposed approach is effective and efficient in conducting tooth segmentation and reconstruction from noisy and ambiguous radiographs.

Acknowledgments. This work was supported by NSFC 61876008.

References

1. Achanta, R., Shaji, A., Smith, K., Lucchi, A., Fua, P., Süsstrunk, S.: SLIC superpixels compared to state-of-the-art superpixel methods. IEEE Trans. Pattern Anal. Mach. Intell. **34**(11), 2274–2282 (2012)
2. Arbelaez, P., Maire, M., Fowlkes, C., Malik, J.: Contour detection and hierarchical image segmentation. IEEE Trans. Pattern Anal. Mach. Intell. **33**(5), 898–916 (2011)
3. Audet, C., Dennis Jr., J.E.: Analysis of generalized pattern searches. SIAM J. Optim. **13**(3), 889–903 (2002)
4. Baka, N., et al.: 2D–3D shape reconstruction of the distal femur from stereo X-ray imaging using statistical shape models. Med. Image Anal. **15**(6), 840 (2011)

5. Benameur, S., Mignotte, M., Labelle, H., De Guise, J.: A hierarchical statistical modeling approach for the unsupervised 3-D biplanar reconstruction of the scoliotic spine. IEEE Trans. Biomed. Eng. **52**(12), 2041–2057 (2005)
6. Benameur, S., Mignotte, M., Parent, S., Labelle, H., Skalli, W., de Guise, J.: 3D/2D registration and segmentation of scoliotic vertebrae using statistical models. Comput. Med. Imaging Graph. **27**(5), 321–337 (2003)
7. Chin, C., Lin, J., Wei, C., Hsu, M.: Dentition labeling and root canal recognition using ganand rule-based system. In: 2019 International Conference on Technologies and Applications of Artificial Intelligence (TAAI), pp. 1–6 (2019)
8. Edwards, G.J., Taylor, C.J., Cootes, T.F.: Interpreting face images using active appearance models. In: Proceedings of the Automatic Face and Gesture Recognition 1998, pp. 300–305 (1998)
9. Fulkerson, B., Vedaldi, A., Soatto, S.: Class segmentation and object localization with superpixel neighborhoods. In: IEEE International Conference on Computer Vision, pp. 670–677. IEEE (2009)
10. Jader, G., Fontineli, J., Ruiz, M., Abdalla, K., Pithon, M.M., Oliveira, L.: Deep instance segmentation of teeth in panoramic X-ray images. In: 2018 31st SIB-GRAPI Conference on Graphics, Patterns and Images (SIBGRAPI), pp. 400–407 (2018)
11. Jain, A.K., Chen, H.: Matching of dental X-ray images for human identification. Pattern Recogn. **37**(7), 1519–1532 (2004)
12. Kang, X., Yau, W.P., Taylor, R.H.: Simultaneous pose estimation and patient-specific model reconstruction from single image using maximum penalized likelihood estimation (MPLE). Pattern Recogn. **57**, 61–69 (2016)
13. Lai, Y.H., Lin, P.L.: Effective segmentation for dental X-ray images using texture-based fuzzy inference system. In: Blanc-Talon, J., Bourennane, S., Philips, W., Popescu, D., Scheunders, P. (eds.) ACIVS 2008. LNCS, vol. 5259, pp. 936–947. Springer, Heidelberg (2008). https://doi.org/10.1007/978-3-540-88458-3_85
14. Li, S., Fevens, T., Krzyżak, A., Li, S.: An automatic variational level set segmentation framework for computer aided dental X-rays analysis in clinical environments. Comput. Med. Imaging Graph. **30**(2), 65–74 (2006)
15. Lin, P.L., Huang, P.Y., Huang, P.W.: An automatic lesion detection method for dental X-ray images by segmentation using variational level set. In: International Conference on Machine Learning and Cybernetics (ICMLC), vol. 5, pp. 1821–1825. IEEE (2012)
16. Lin, P., Huang, P., Huang, P., Hsu, H., Chen, C.: Teeth segmentation of dental periapical radiographs based on local singularity analysis. Comput. Methods Programs Biomed. **113**(2), 433–445 (2014)
17. Milborrow, S., Nicolls, F.: Locating facial features with an extended active shape model. In: Forsyth, D., Torr, P., Zisserman, A. (eds.) ECCV 2008. LNCS, vol. 5305, pp. 504–513. Springer, Heidelberg (2008). https://doi.org/10.1007/978-3-540-88693-8_37
18. Mukherjee, L., Singh, V., Dyer, C.R.: Half-integrality based algorithms for cosegmentation of images. In: IEEE Conference on Computer Vision and Pattern Recognition, pp. 2028–2035. IEEE (2009)
19. Pei, Y., Liu, B., Zha, H., Han, B., Xu, T.: Anatomical structure sketcher for cephalograms by bimodal deep learning. In: BMVC 2013 (2013)
20. Pei, Y., et al.: Personalized tooth shape estimation from radiograph and cast. IEEE Trans. Biomed. Eng. **59**(9), 2400–2411 (2012)

21. Platt, J., et al.: Probabilistic outputs for support vector machines and comparisons to regularized likelihood methods. In: Advances in Large Margin Classifiers, vol. 10, no. 3, pp. 61–74 (1999)
22. Pushparaj, V., Gurunathan, U., Arumugam, B.: An effective dental shape extraction algorithm using contour information and matching by Mahalanobis distance. J. Digit. Imaging **26**(2), 259–268 (2013)
23. Rad, A.E., Mohd Rahim, M.S., Rehman, A., Altameem, A., Saba, T.: Evaluation of current dental radiographs segmentation approaches in computer-aided applications. IETE Tech. Rev. **30**(3), 210–222 (2013)
24. Ronneberger, O., Fischer, P., Brox, T.: Dental X-ray image segmentation using a U-shaped deep convolutional network. In: ISBI (2015)
25. Rother, C., Minka, T., Blake, A., Kolmogorov, V.: Cosegmentation of image pairs by histogram matching-incorporating a global constraint into MRFs. In: IEEE Computer Society Conference on Computer Vision and Pattern Recognition, vol. 1, pp. 993–1000. IEEE (2006)
26. Rother, C., Kolmogorov, V., Lempitsky, V., Szummer, M.: Optimizing binary MRFs via extended roof duality. In: IEEE Conference on Computer Vision and Pattern Recognition, pp. 1–8. IEEE (2007)
27. Saad, A., El-Bialy, A., Kandil, A., Sayed, A.: Automatic cephalometric analysis using active appearance model and simulated annealing. Int. J. Graphics Vision Image Process. **6**, 51–67 (2006). Special Issue on Image Retrieval and Representation
28. Said, E.H., Nassar, D.E.M., Fahmy, G., Ammar, H.H.: Teeth segmentation in digitized dental X-ray films using mathematical morphology. IEEE Trans. Inf. Forensics Secur. **1**(2), 178–189 (2006)
29. Shah, S., Abaza, A., Ross, A., Ammar, H.: Automatic tooth segmentation using active contour without edges. In: Biometrics Symposium: Special Session on Research at the Biometric Consortium Conference, pp. 1–6. IEEE (2006)
30. Silva, G., Oliveira, L., Pithon, M.: Automatic segmenting teeth in X-ray images: trends, a novel data set, benchmarking and future perspectives. Expert Syst. Appl. **107**, 15–31 (2018)
31. Sumner, R., Schmid, J., Pauly, M.: Embedded deformation for shape manipulation. ACM Trans. Graph. (TOG) **26**(3), 80 (2007)
32. Yue, W., Yin, D., Li, C., Wang, G., Xu, T.: Automated 2-D cephalometric analysis on X-ray images by a model-based approach. IEEE Trans. Biomed. Eng. **53**(8), 1615–1623 (2006)
33. Zheng, G., et al.: Accurate and robust reconstruction of a surface model of the proximal femur from sparse-point data and a dense-point distribution model for surgical navigation. IEEE Trans. Biomed. Eng. **54**(12), 2109–2122 (2007)
34. Zheng, G.: Effective incorporating spatial information in a mutual information based 3D–2D registration of a CT volume to X-ray images. Comput. Med. Imaging Graph. **34**(7), 553–562 (2010)

Blind Super-Resolution
with Kernel-Aware Feature Refinement

Ziwei Wu[1]($\boxtimes$), Yao Lu[1], Gongping Li[1], Shunzhou Wang[1], Xuebo Wang[1],
and Zijian Wang[2]

[1] Beijing Key Laboratory of Intelligent Information Technology,
Beijing Institute of Technology, Beijing, China
{wzw_cs,vis_yl,gongping_li,shunzhouwang,wangxuebo}@bit.edu.cn
[2] China Central Television, Beijing, China
wangzijian@cctv.com

Abstract. Numerous learning-based super-resolution (SR) methods
that assume the blur kernel is known in advance or hand-crafted perform
excellently on the synthesized images. While these methods fail to produce satisfactory results when applied to real-world images. Therefore,
several blind SR methods are proposed, where the blur kernel is unavailable. However, they only consider the Gaussian blur kernel, which is
not enough for real-world images. To address this problem, we propose
to synthesize training data with real blur kernels estimated from real
sensor images. Further, we find that models trained with the data synthesized with the estimated real blur kernels still cannot perform well
enough without awareness of the blur kernel. Thus, we first propose a
new kernel predictor (KP) to predict the blur kernel from the input
low-resolution image. Then we design the kernel-aware feature refinement (KAFR) module to utilize the predicted blur kernel to facilitate the
restoration of the high-resolution image. Extensive experiments on synthesized and real-world images show that the proposed method achieves
superior performance against the state-of-the-art (SOTA) methods both
on effectiveness and efficiency.

Keywords: Blind super-resolution · Kernel estimation · Blur kernel
utilization

1 Introduction

Single image super-resolution (SISR) aims at recovering a high-resolution (HR)
image with more content and textures from the corresponding low-resolution
(LR) image. As a low-level computer vision task, SISR has a wide range of
applications [8,15].

A general degradation model that describes the relationship between HR
image x and LR image y is given by

$$y = (x \otimes k) \downarrow_s + n_\sigma,　\qquad (1)$$

Z. Wu—This is a student paper.

© Springer Nature Switzerland AG 2020
Y. Peng et al. (Eds.): PRCV 2020, LNCS 12305, pp. 65–78, 2020.
https://doi.org/10.1007/978-3-030-60633-6_6

where $x \otimes k$ denotes the convolution operation between HR image x and blur kernel k, $\downarrow_s$ denotes the down-sampling operation with scale factor s, and n is an additive white Gaussian noise (AWGN) with noise level σ. Recently, the deep learning based super-resolution (SR) methods have achieved remarkable performance not only on the peak signal-to-noise ratio (PSNR) [17,26,27], but also on visual quality [11,16]. Most SR methods have a simple assumption that the blur kernel in Eq. (1) is known in advance or hand-crafted, e.g. bicubic kernel. However, in real-world scenarios, the situation is much more complicated. The corruption of the HR image can be caused by multiple factors, such as the camera shake or compression artifacts [4,28]. Therefore, these methods that assume the blur kernel is known in advance or hand-crafted fails to produce satisfactory results [4]. To solve this problem, numerous blind SR methods [6,18,24] based on deep learning are proposed, where the blur kernel is unknown. Although they have achieved impressive performance, they only consider the Gaussian kernels, which is not enough for real-world images as analyzed above. Moreover, empirical and theoretical studies have demonstrated that the performance of SISR deteriorated seriously when the blur kernel deviates from the real one [5].

Furthermore, it has also been proven in [6] that kernel mismatch will lead to over-smoothing or over-sharpening SR results. Thus, to accurately estimate the blur kernel is crucial for the blind SR problem. Although there is a lot of researches focus on kernel estimation [13,21], they cannot be applied to blind SR problem directly. They estimate the blur kernel from the blurred image while we need to predict the blur kernel from the downsampled version of the blurred image, named the LR image. Here we focus on kernel estimation in blind SR problem. Gu et al. [6] do not predict the entire blur kernel but the kernel vector projected with the Principal Components Analysis (PCA) technique. They use an iterative kernel correction scheme to iteratively correct the predicted kernel vector. However, this strategy is computationally expensive. Zhang et al. [24] use a grid search strategy to roughly determine the degradation parameters, but this strategy only works with a regular blur such as the Gaussian blur or the uniform motion blur.

In this paper, we focus on blind SR methods towards real-world image super-resolution. Firstly, we need to synthesize LR images that close to real-world images. One strategy is to generate LR images with a generative adversarial network (GAN) [2,4]. However, the GAN-based method [4] is unstable and has a chance to generate failure cases. Another strategy is to perform convolution operation on the HR image with real blur kernels and downsample them to get LR images. Thus, inspired by [28], we estimate the real blur kernels from real sensor images, and then synthesize the training data with the estimated kernels.

Secondly, it is non-trivial to predict the blur kernels from the input LR images in blind SR problem [5,6]. Hence, we propose an end-to-end kernel predictor (KP) in the blind SR problem, which can extract the blur kernels from the LR images effectively and accurately. Besides, we also investigate how to utilize the blur kernel predicted by our kernel predictor to facilitate the reconstruction of the HR image. In [24], they first project the blur kernel to a low dimension kernel vector with a learned PCA projection matrix. Then they stretch the kernel vector into degradation maps and concatenate the degradation maps with the LR image and feed it to the network. However, this strategy is not the optimal choice [6]. Similarly, Gu et al. [6] concatenate the image features with the degradation maps and utilize kernel information with spatial feature transform (SFT) layers [20]. Nevertheless, estimating real blur kernels [13] on the real-world images is time-consuming. We cannot get sufficient of real kernels to learn the PCA projection matrix. In addition, it is easier to learn a proper PCA projection matrix for homogeneous blur kernels, e.g. Gaussian blur kernels, than the real blur kernels due to the diversity of the real blur kernels. Therefore, we design the kernel-aware feature refinement (KAFR) module to perform kernel utilization. In the KAFR module, the predicted kernel is transformed into a group of channel-wise parameters to refine the image features kernel-orientedly. Then the refined image features are used to produce a modulation tensor to further refine the image features. By introducing the KAFR module into our network, we achieve superior performance on multiple real degradations.

We summarize our contributions as follows: (1) We provide a new data synthesizing strategy for the blind SR problem, which can produce training data that is closed to the real-world low-resolution images. (2) We propose an end-to-end kernel predictor for kernel estimation in the blind SR problem, which can extract the blur kernels from the input LR images effectively and accurately. (3) We propose a new strategy to utilize the estimated blur kernel to facilitate the restoration of the HR image. We use the predicted blur kernel to refine the image features with the well-designed KAFR module.

2   Related Work

Kernel Estimation. Most kernel estimation algorithms introduce the image prior into the maximum a posterior (MAP) framework and formulate the task of estimating the blur kernel as the standard optimization problem. Pan et al. [13] introduce an L_0-regularization term to enforce sparsity on the dark channel of latent images and effectively estimate the blur kernel from the natural image by solving the optimization problem. Besides, Xu et al. [21] use edge information extracted by a deep network to construct objective function and develop an optimization scheme to estimate the blur kernel. Different from the methods mentioned above, we need to predict the blur kernel from the LR image which is a downsampled version of the blurred image. Thus, we propose an end-to-end trainable kernel predictor in blind SR problem using to effectively and accurately estimate the blur kernels from the LR images.

Blind Super-Resolution. Blind SR assumes that the blur kernel is unknown and unavailable. Existing blind SR methods can be classified into two categories depending on whether it is supervised or not. The unsupervised methods [18, 22, 25] break the limitation of training by HR−LR image pairs. Shocher et al. [18] exploit the internal recurrence of a natural image and train an image-specific model for every test image. However, this approach is computationally expensive. Another two methods [22,25] are inspired by image-to-image translation [31] and propose the Cycle-in-Cycle architecture to map the LR image to the HR image space. As for supervised methods, Zhang et al. [24] use a grid search strategy to roughly determine the degradation parameters (i.e. blur kernel and noise level). And then they use the dimensionality stretching strategy to stretch the degradation parameters into degradation maps and concatenate them with the LR image as the input of the network. To make a step forward, Gu et al. [6] use an iterative kernel correction method to iteratively correct the kernel vector and propose a new SR network with SFT layers [20] to utilize the kernel information. However, they only consider the Gaussian kernels, which is not enough for real-world images. Different from the previous works, we synthesize the training data with the real blur kernels and propose a new method to utilize the predicted blur kernel to facilitate the reconstruction of the HR image.

3   Method

3.1   Methodology

In blind super-resolution problem, researchers try to recover the HR image from the LR image without any information of the blur kernel in Eq. (1). As shown in [6], the kernel mismatch will deteriorate the SR performance. Thus, it is essential to accurately estimate the blur kernel from the input LR image. Kernel estimation is a common task in blind deblurring [13,14,21]. A general blur model can be given by

$$b = l \otimes k + n, \tag{2}$$

where b, l, k, n denotes the blur image, latent or sharp image, blur kernel and an AWGN, respectively. And they get the latent image and the blur kernel by solving the following minimization problem

$$\min_{l,k} \|b - l \otimes k\|_2^2 + \mu_1 R_l(l) + \mu_2 R_k(k), \tag{3}$$

where the $\mu_1 R_l(l)$ and $\mu_2 R_k(k)$ are the regulation terms. To make a step forward, they estimate the latent image and the blur kernel by alternately solving the following two minimization problems with an initial blur kernel

$$l_{i+1} = arg \min_{l_{i+1}} \|b - l_{i+1} \otimes k_i\|_2^2 + \mu_1 R_l(l_{i+1}), \tag{4}$$

and

$$k_{i+1} = arg \min_{k_{i+1}} \|b - l_{i+1} \otimes k_{i+1}\|_2^2 + \mu_2 R_k(k_{i+1}). \tag{5}$$

However, the kernel estimation in Eq. (3) is fundamentally inapplicable for blind SR. Because they predict the blur kernel from the blur image while we need to estimate the blur kernel from the LR image which is a downsampled version of the blur image. Moreover, we do not need to restore a sharp LR image either.

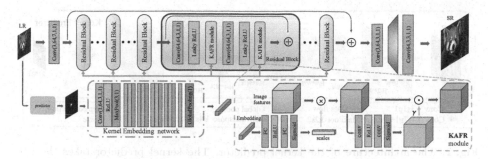

Fig. 1. Overall architecture of the proposed method. Given an LR image, our system first predicts the blur kernel from it. Then the predicted blur kernel is fed into the kernel embedding network to learn a kernel embedding for all KAFR modules. In the KAFR module, the kernel embedding is used to produce a group of channel-wise parameters to refine the image features extracted by the convolution layers. Then the refined image features are used to generate a modulation tensor to further refine the image features. Zoom in for the best view.

Besides, this iterative scheme is very time-consuming. So we propose an end-to-end trainable kernel predictor to estimate the blur kernel from the LR image. We can get the kernel predictor by solving the following problem with a data-driven method

$$\theta_{kp} = arg \min_{\theta_{kp}} \| \Gamma(x) - K \|_2^2 + \lambda * \Re(\Gamma(x)), \tag{6}$$

where $\Gamma(\cdot)$ denotes the kernel predictor, θ_{kp} denotes the parameters of the kernel predictor, K is the ground truth kernel, $\Re(\cdot)$ is regulation terms, λ is the weight of the regulation terms. In experiments we set $\lambda = 0.01$.

After predicting the blur kernel, the next problem is how to use it to facilitate the restoration of the HR image. There is a dimensionality stretching strategy [6,24] for kernel utilization which needs adequate blur kernels to learn a PCA projection matrix. However, this strategy does not work for us because we do not have enough blue kernels. We use the state-of-the-art blind deblurring method [14] to estimate the real blur kernel from a real sensor image. It costs about 5 min to produce one blur kernel on an Intel Core i7-8700k CPU. We spend 10 days on computing more than 2000 blur kernels but it still not enough to learn a proper PCA projection matrix as used in [6] (30000 Gaussian blur kernels). In addition, it is much easier to learn a PCA projection matrix for homogeneous blur kernels (Gaussian blur kernel) than the real blur kernels because of the diversity of the real blur kernels. So we design the KAFR module to utilize the predicted blur kernel to refine the image features.

3.2 Proposed Method

The overall architecture of our kernel-aware super-resolution (KASR) framework is illustrated in Fig. 1. We first estimate the blur kernel form the input LR image and then feed it into the kernel embedding extraction network to learn a

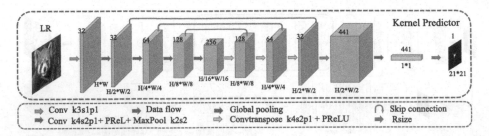

Fig. 2. The architecture of the kernel predictor. The kernel predictor takes the LR image as input and extracts the blur kernel from it.

kernel embedding for all the KAFR modules. Meanwhile, the LR image is also fed into the backbone network to extract image features which will be refined in the KAFR module.

Kernel Predictor. The overall architecture of our kernel predictor is shown in Fig. 2. The kernel predictor takes the LR as input and extracts the blur kernel from it. For the regulation terms in (6), we take four factors into consideration [2].

$$\Re(\kappa) = \omega_1 L_{sum} + \omega_2 L_{mass\_center} + \omega_3 L_{boundary} + \omega_4 L_{sparsity}, \quad (7)$$

where $\omega_1 = 0.5, \omega_2 = 1, \omega_3 = 0.5, \omega_4 = 1$ and κ is the predicted blur kernel.

- $L_{sum} = |\sum_{i,j} \kappa_{i,j} - 1|$, the sum of the predicted blur kernel should equal to 1 otherwise the energy of the kernel will overflow.
- $L_{mass\_center} = \left\| \frac{\sum_{i,j} K_{i,j} \cdot (i,j)}{\sum_{i,j} K_{i,j}} - \frac{\sum_{i,j} \kappa_{i,j} \cdot (i,j)}{\sum_{i,j} \kappa_{i,j}} \right\|_2$, the mass center of the predicted kernel and the ground truth kernel should be the same.
- $L_{boundary} \sum_{i,j} |\kappa_{i,j} * m_{i,j}|$, values close to the boundary of the kernel should be zero, $m_{i,j}$ is a penalty mask that values exponentially grow with distance from the center of κ.
- $L_{sparsity} = \sum_{i,j} |\kappa_{i,j}|^{\frac{1}{p}}, p > 1$, penalize small values to encourage sparsity to prevent the network from over-smoothing kernels.

We use an encoder-decoder architecture for its large receptive field because the input image could have an arbitrary shape. Except for the first one, convolution layers in the kernel predictor are followed by a PReLU layer with an initial parameter of 0.1 and a max-pooling layer with kernel size of 2 and stride of 2. The deconvolution layers in the kernel predictor are followed by a PReLU layer with an initial parameter of 0.1.

Kernel Aware Feature Refinement Module. The architecture of our KAFR module is shown in Fig. 1. The predicted blur kernel is fed into the kernel embedding extraction network to produce a kernel embedding for all the KAFR modules. Then the kernel embedding is used to refine the image features. There are

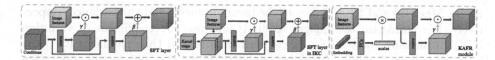

Fig. 3. Illustration of the SFT layer [20], SFT in IKC [6] and our KAFR module. $\odot$, $\oplus$, and $\otimes$ denote the element-wise multiplication, element-wise addition and channel-wise multiplication, respectively. Zoom in for the best view.

also other methods that transform the subsidiary information into the features. SFT [20] introduces the segmentation prior to generate a pair of modulation parameters to modulate the image features to produce results with more realistic textures. Similarly, IKC [6] first stretches the blur kernel into degradation maps, then concatenates the degradation maps with the image features to produce the modulation parameters. Differently, we use a learned kernel embedding to generate a group of channel-wise parameters to refine the image features. Then we use a modified SFT [20] layer to further refine the image features. We only use the element-wise multiplication and reduce the element-wise addition in the SFT layer. An illustration of the SFT layer [20], SFT in IKC [6], and our KAFR module are provided in Fig. 3. Our KAFR module can be formulated by

$$KAFR(F_{img}, E_\kappa) = \Psi(\Phi(E_\kappa) \bullet F_{img}) * (\Phi(E_\kappa) \bullet F_{img}), \tag{8}$$

where F_{img} and E_κ denote the image features and the learned kernel embedding, Ψ and Φ denote the convolution layers and fully connected layers in KAFR module, $*$ and $\bullet$ denote the element-wise and channel-wise multiplications, respectively.

Our KAFR module is somewhat similar to the attention mechanism [29,30], but there is a fundamental difference between them. The attention mechanism tries to enhance the features with the internal information while our KAFR module uses the subsidiary external information to refine the image features.

4 Experiment

4.1 Dataset and Implementation Details

Dataset. We use the SOTA blind deblurring method [13] to estimate the real blur kernels from the real sensor images from DPED [9] captured by iPhone3GS. We randomly crop the image to 512×512 patches and use the code from [13] to estimate the blur kernels from the patches. We find that patches which are lack of textures might result in the failure of the kernel estimation algorithm. Thus, we set restrictions for the cropped patches p as follows

$$\delta_1 \le |\frac{\sqrt{Var(p)}}{Mean(p)}| \le \delta_2, \tag{9}$$

where $Mean(p)$ and $Var(p)$ denote the mean intensity and the variance, respectively. We normalize the images to [0, 1] and use the coefficient of variation (CV) to measure the patches. We experimentally find that patches with a too large or too small value of CV are not suitable for the blur kernel estimation algorithm in [13]. Patches with large CV values are mostly cropped from the junction of a bright area and a dark area (e.g. sky and buildings) and patches with small CV values are mostly cropped from the area which is lack of textures, such as the floor or the sky. Both of these two cases will result in failure of the kernel estimation, and we name these failure kernels as bad kernels. An illustration of this is provided in Fig. 4. We experimentally set $\delta_1 = 0.2, \delta_2 = 0.3$ and $0.2 \leq Mean(p) \leq 0.5$ (avoiding to crop patches from sky or other highlight area) and spend more than 10 days on collecting 2353 blur kernels. Despite the effort we make to avoid failure cases, there are some bad kernels. We remove most of the bad kernels and remain a few of them to achieve robustness. Finally, we have 2000 real blur kernels for training and 65 kernels for testing.

| 0.002 | 0.634 | 0.268 | 0.244 |

Fig. 4. Illustration of the kernel estimation results on different patches. The top-left corner is the blur kernel predicted from the patch and the numbers under the patches are the corresponding CV (coefficient of variation) values. Zoom in for the best view.

Our HR images are from DIV2K [1] and Flickr2K [19]. Then the training set has 3450 images of 2K high-quality. We randomly select a batch of blur kernels from the 2000 training blur kernels for each batch of HR images. We first blur the HR images with the selected blur kernels each to each then downsample them with bicubic interpolation to get the training LR images. We test the proposed method on the popular benchmarks, Set5 [3], Set14 [23], BSD100 [12], Urban100 [7] and the validation set of the DIV2K [1].

Implementation Details. We conduct the experiments on scale factor 4, and both the training and the testing are accomplished on RGB channels. When applying to real-world images, we use an AGWN with kernel width $\sigma = 15$. We divide our training procedure into two stages. First, we remove the kernel predictor from the KASR network as our backbone network which is named KASR-η. We separately train the KASR-η and the kernel predictor. Specifically, we use the real kernel to pretrain the KASR-η. Then, when the kernel predictor is finished, we assemble these two components and fix the kernel predictor to fine-tune our final model KASR. We use the mean square error (MSE) between the HR image and the SR result to train the KASR-η. We use 16 residual blocks where every block contains 2 KAFR modules. We use Adam [10] with $\beta_1 = 0.9, \beta_2 = 0.999$ to optimize our model. The learning rate is initialized as 1×10^{-4} and decreases 0.2 times for every 100000 iterations. We implement our method with Pytorch framework and train it on a Tesla V100 GPU.

4.2   Comparison with the SOTA

We compare the proposed method with the SOTA non-blind methods RDN [27] and RCAN [26]. Also, we make comparisons with the SOTA blind methods IKC [6] and ZSSR [18]. To further demonstrate the effectiveness of the proposed method, we use the training data synthesized with the estimated real blur kernels to retrain RDN [27] without the awareness of the blur kernel and name it RDNMD. We also retrain the SFTMD in [6], which use the dimentionality stretching method to utilize the kernel information. The quantitative results, peak signal to noise ratio (PSNR), and structural similarity (SSIM) are shown in Table 1 and a visual example is provided in Fig. 5. We can see those non-blind methods RDN and RCAN fail both in visual and quantitative results when applied to multiple degradations. And comparing with the SOTA blind method IKC, we can see that only considering the Gaussian blur kernel is not enough for the real-world degradations. Although we use the LR images synthesized with the estimated real blur kernels to retrain the RDN, it still cannot perform well enough which indicates the estimation and utilization of the blur kernel is indispensable for the blind SR problem. And the comparison between KP (our kernel predictor)+SFTMD and KASR shows that the dimensionality stretching strategy is not optimal when there are not adequate blur kernels.

Notice that we have 65 test kernels, which means we need to test 65 times on every test dataset. ZSSR [18] trains an image-specific model for every test image. And IKC [6] uses an iterative kernel correction scheme to iteratively correct the predicted kernel vector. Both of them are very time-consuming. It takes more than 2 weeks for ZSSR and more than 11 days for IKC to finish all the tests on an NVIDIA RTX2080Ti GPU, while our method can make it in 3 h.

We also test the blind methods on the real-world images randomly downloaded from the internet. A visual comparison can be found in Fig. 6. Comparing with RDNMD and KP+SFTMD [6], our method achieves high contrast. And comparing with IKC [6] and ZSSR [18], the proposed method provides results with fewer artifacts. Also, the proposed method produces results with more visually pleasing textures according to the second row.

4.3   Ablation Studies

Ablation for Kernel Predictor. We use mean square error (MSE) and mean absolute error (MAE) to evaluate our kernel predictor. We investigate the effect of the regulation terms. The quantitative results can be found in Table 2 and a visual comparison is provided in Fig. 7. As one can see, although the kernel

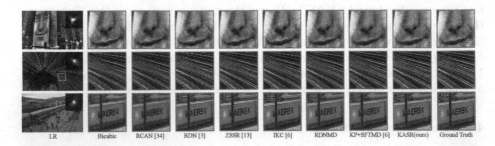

LR   Bicubic   RCAN [34]   RDN [3]   ZSSR [13]   IKC [6]   RDNMD   KP+SFTMD [6]   KASR(ours)   Ground Truth

Fig. 5. Comparison of the SR performance on different test kernels. The top-right corner of the image is the test blur kernel. The first image is "$img\_076.png$" from the Urban100 [7]. The left two are "$0828.png$" and "$0850.png$" from the validation set of the DIV2K [1]. Zoom in for the best view.

Table 1. Quantitative comparison of the SOTA methods and the proposed KASR on Set5, Set14, BSD100, Urban100, and the validation set of the DIV2K dataset. Notice that method marked by "$\diamond$" is trained with the Gaussian kernels, and methods marked by "$\star$" are trained with the real blur kernels. KP denotes our kernel predictor. The results are the mean value on 65 test kernels and formatted to PSNR/SSIM. Red color indicates the best results, and blue indicates the second best.

Method	Set5 [3]	Set14 [23]	BSD100 [12]	Urban100 [7]	DIV2K_val [1]
RCAN [26]	24.35/0.647	22.85/0.584	23.31/0.568	20.73/0.551	24.84/0.642
RDN [27]	26.97/0.818	24.04/0.707	24.57/0.689	21.61/0.701	26.66/0.786
Bicubic	26.24/0.773	23.94/0.668	24.37/0.648	21.45/0.634	26.36/0.750
ZSSR [18]	24.62/0.708	23.34/0.644	23.68/0.628	21.30/0.637	26.53/0.765
IKC [6]$^\diamond$	28.02/0.828	25.59/0.714	24.76/0.683	22.79/0.708	26.62/0.784
RDNMD*	28.39/0.843	25.34/0.728	25.45/0.702	23.47/0.743	27.94/0.804
KP+SFTMD [6]*	27.76/0.825	25.04/0.712	25.18/0.685	23.10/0.720	27.59/0.798
KASR* (ours)	28.74/0.852	25.64/0.738	25.60/0.708	23.58/0.751	28.13/0.809

(a) LR   (b) ZSSR [13]   (c) IKC [6]   (d) RDNMD   (e) KP+SFTMD [6]   (f) KASR (ours)

Fig. 6. Visual comparison of SR performance on real-world images. Zoom in for the best view.

predictor trained without regulation terms achieves better performance in terms of MSE and MAE, kernel predictor trained with regulation terms provides more visually favorable results.

Fig. 7. Visual comparison of the estimated blur kernels. (a), (d) the ground truth kernels. (b), (e) results of kernel predictor trained without regulation terms. (c), (f) results of kernel predictor trained with regulation terms.

Table 2. Quantitative comparison of the kernel predictor and its degradation version on Set5, BSD100 and the validation set of the DIV2K dataset. The results are formatted to MSE $\times 10^{-5}$/MAE $\times 10^{-3}$.

Setting	Set5 [3]	BSD100 [12]	DIV2K_val [1]
MSE only	3.16/2.21	2.54/1.29	5.03/4.61
MSE + $\lambda\Re(\kappa)$	6.69/2.59	6.33/1.87	7.33/3.80

Ablation for KAFR Module. We also investigate the architecture of the KAFR module. We compare the KASR-η with other architecture settings. We first remove the γ tensor branch in the KAFR module and name it KASR-α. Namely, we reduce the element-wise refinement and retain the channel-wise one only. Then we append an SFT layer behind the channel-wise refinement and name it KASR-sft. Besides, we make a comparison with the SFTMD [6]. All the models above-mentioned are trained with the real blur kernels. The comparison results are in Table 3. We can see that the dimensionality stretching strategy (SFTMD) is not optimal when there are not adequate blur kernels. And appending an SFT layer behind the channel-wise refinement, denoted by KASR-sft, is also not the best choice. And comparing KASR-α and KASR-η, we find that only conduct channel-wise refinement is not enough.

Table 3. Quantitative comparison of the KASR-η and other architectures on Set5, Set14, BSD100, Urban100 and the validation set of the DIV2K dataset. The results are formatted to PSNR/SSIM. Red color indicates the best results, and blue indicates the second best.

Method	Set5 [3]	Set14 [23]	BSD100 [12]	Urban100 [7]	DIV2K_val [1]
KASR-α	29.28/0.856	25.92/0.742	25.83/0.713	24.20/0.762	28.53/0.813
KASR-sft	29.24/0.855	25.87/0.741	25.83/0.714	24.26/0.764	28.56/0.814
SFTMD [6]	29.11/0.854	25.85/0.741	25.79/0.713	24.23/0.765	28.49/ 0.814
KASR-η	29.45/0.861	26.07/0.745	25.93/0.717	24.61/0.776	28.75/0.819

5 Conclusion

We propose a new blind SR method, named KASR, which towards real-world image super-resolution. In particular, we synthesize training data with real blur kernels estimated from real sensor images. In the KASR network, we first predict the blur kernel from the LR image and then utilize the predicted blur kernel by KAFR modules to facilitate the restoration of the HR image. Extensive experiments show that our KASR-η perform excellently with a reliable blur kernel. And our kernel predictor leaves much to be desired. In the future, we will explore how to estimate the blur kernel more precisely.

Acknowledgement. This work is supported by the National Natural Science Foundation of China (No. 61273273), by the National Key Research and Development Plan (No. 2017YFC0112001), and by China Central Television (JG2018-0247).

References

1. Agustsson, E., Timofte, R.: NTIRE 2017 challenge on single image super-resolution: dataset and study. In: CVPR Workshop (2017)
2. Bell-Kligler, S., Shocher, A., Irani, M.: Blind super-resolution kernel estimation using an internal-GAN. In: NeurIPS, pp. 284–293 (2019)
3. Bevilacqua, M., Roumy, A., Guillemot, C., Alberi-Morel, M.L.: Low-complexity single-image super-resolution based on nonnegative neighbor embedding. In: BMVC (2012)
4. Bulat, A., Yang, J., Tzimiropoulos, G.: To learn image super-resolution, use a GAN to learn how to do image degradation first. In: Ferrari, V., Hebert, M., Sminchisescu, C., Weiss, Y. (eds.) ECCV 2018. LNCS, vol. 11210, pp. 187–202. Springer, Cham (2018). https://doi.org/10.1007/978-3-030-01231-1_12
5. Efrat, N., Glasner, D., Apartsin, A., Nadler, B., Levin, A.: Accurate blur models vs. image priors in single image super-resolution. In: ICCV (2013)
6. Gu, J., Lu, H., Zuo, W., Dong, C.: Blind super-resolution with iterative kernel correction. In: CVPR (2019)
7. Huang, J.B., Singh, A., Ahuja, N.: Single image super-resolution from transformed self-exemplars. In: CVPR (2015)

8. Huang, Y., Shao, L., Frangi, A.F.: Simultaneous super-resolution and cross-modality synthesis of 3D medical images using weakly-supervised joint convolutional sparse coding. In: CVPR (2017)
9. Ignatov, A., Kobyshev, N., Timofte, R., Vanhoey, K., Van Gool, L.: DSLR-quality photos on mobile devices with deep convolutional networks. In: ICCV (2017)
10. Kingma, D.P., Ba, J.: Adam: a method for stochastic optimization. arXiv preprint arXiv:1412.6980 (2014)
11. Ledig, C., et al.: Photo-realistic single image super-resolution using a generative adversarial network (2017)
12. Martin, D., Fowlkes, C., Tal, D., Malik, J., et al.: A database of human segmented natural images and its application to evaluating segmentation algorithms and measuring ecological statistics. In: ICCV (2001)
13. Pan, J., Sun, D., Pfister, H., Yang, M.H.: Blind image deblurring using dark channel prior. In: CVPR (2016)
14. Pan, L., Hartley, R., Liu, M., Dai, Y.: Phase-only image based kernel estimation for single image blind deblurring. In: CVPR (2019)
15. Rasti, P., Uiboupin, T., Escalera, S., Anbarjafari, G.: Convolutional neural network super resolution for face recognition in surveillance monitoring. In: Perales, F.J.J., Kittler, J. (eds.) AMDO 2016. LNCS, vol. 9756, pp. 175–184. Springer, Cham (2016). https://doi.org/10.1007/978-3-319-41778-3_18
16. Sajjadi, M.S., Scholkopf, B., Hirsch, M.: EnhanceNet: single image super-resolution through automated texture synthesis. In: ICCV (2017)
17. Shi, W., et al.: Real-time single image and video super-resolution using an efficient sub-pixel convolutional neural network. In: CVPR (2016)
18. Shocher, A., Cohen, N., Irani, M.: "Zero-shot" super-resolution using deep internal learning. In: CVPR (2018)
19. Timofte, R., Agustsson, E., Van Gool, L., Yang, M.H., Zhang, L.: NTIRE 2017 challenge on single image super-resolution: methods and results. In: CVPR Workshop (2017)
20. Wang, X., Yu, K., Dong, C., Change Loy, C.: Recovering realistic texture in image super-resolution by deep spatial feature transform. In: CVPR (2018)
21. Xu, X., Pan, J., Zhang, Y.J., Yang, M.H.: Motion blur kernel estimation via deep learning. IEEE TIP **27**(1), 194–205 (2017)
22. Yuan, Y., Liu, S., Zhang, J., Zhang, Y., Dong, C., Lin, L.: Unsupervised image super-resolution using cycle-in-cycle generative adversarial networks. In: CVPR Workshop (2018)
23. Zeyde, R., Elad, M., Protter, M.: On single image scale-up using sparse-representations. In: Boissonnat, J.-D., et al. (eds.) Curves and Surfaces 2010. LNCS, vol. 6920, pp. 711–730. Springer, Heidelberg (2010). https://doi.org/10.1007/978-3-642-27413-8_47
24. Zhang, K., Zuo, W., Zhang, L.: Learning a single convolutional super-resolution network for multiple degradations. In: CVPR (2018)
25. Zhang, Y., Liu, S., Dong, C., Zhang, X., Yuan, Y.: Multiple cycle-in-cycle generative adversarial networks for unsupervised image super-resolution. IEEE TIP **29**, 1101–1112 (2019)
26. Zhang, Y., Li, K., Li, K., Wang, L., Zhong, B., Fu, Y.: Image super-resolution using very deep residual channel attention networks. In: Ferrari, V., Hebert, M., Sminchisescu, C., Weiss, Y. (eds.) ECCV 2018. LNCS, vol. 11211, pp. 294–310. Springer, Cham (2018). https://doi.org/10.1007/978-3-030-01234-2_18
27. Zhang, Y., Tian, Y., Kong, Y., Zhong, B., Fu, Y.: Residual dense network for image super-resolution. In: CVPR (2018)

28. Zhou, R., Susstrunk, S.: Kernel modeling super-resolution on real low-resolution images. In: ICCV (2019)
29. Zhou, T., Wang, S., Zhou, Y., Yao, Y., Li, J., Shao, L.: Motion-attentive transition for zero-shot video object segmentation. In: AAAI (2020)
30. Zhou, T., Wang, W., Qi, S., Ling, H., Shen, J.: Cascaded human-object interaction recognition. In: CVPR (2020)
31. Zhu, J.Y., Park, T., Isola, P., Efros, A.A.: Unpaired image-to-image translation using cycle-consistent adversarial networks. In: ICCV (2017)

Semi-supervised Learning to Remove Fences from a Single Image

Wei Shang, Pengfei Zhu, and Dongwei Ren[✉]

Tianjin Key Lab of Machine Learning, College of Intelligence and Computing,
Tianjin University, Tianjin, China
{shangwei,zhupengfei}@tju.edu.cn, rendongweihit@gmail.com

Abstract. When taking photos in the outdoors, e.g., playgrounds, gardens and zoos, fences are often inevitable to interfere with the perception of the background scenes. Thus it is an interesting topic on removing fences from a single image, which is dubbed image de-fencing. Generally, image de-fencing can be tackled within the two-stage framework, i.e., first detecting fences and then restoring background image. In existing methods, detecting fences mask cannot guarantee accuracy due to the diverse patterns of fences. Meanwhile, state-of-the-art supervised learning-based restoration methods usually fail in well handling inaccurate fences mask, yielding artifacts in restored background image. In this paper, we propose a semi-supervised framework for removing fences from a single image, where a simple yet effective recurrent network is proposed for supervised learning to detect fences mask and unsupervised learning is employed for robust restoration of background image. Specifically, we propose a recurrent network for fences mask detection, where convolutional LSTM is adopted for progressive detection. Instead of strict fences mask detection using standard ℓ_1 loss, we adopt an asymmetric loss to make detected fences mask be more inclusive of true fences. Then motivated by the success of deep image prior (DIP) for several image restoration tasks, we propose to employ unsupervised learning of three DIP networks for modeling background image, fences layer and fences mask, respectively. Moreover, we propose to adopt Laplacian smoothness loss function for refining the detected fences mask, making the restoration of background image be more robust to detection errors of fences mask. Experimental results validate the effectiveness of our semi-supervised image de-fencing approach.

Keywords: Image restoration · Image de-fencing · Inpainting · Deep image prior · Semi-supervised learning

1 Introduction

Fences are very common in our daily life, which can be seen everywhere especially in playgrounds, gardens, zoos, etc. While fences form an isolation zone

---

The first author is a second year master student.

© Springer Nature Switzerland AG 2020
Y. Peng et al. (Eds.): PRCV 2020, LNCS 12305, pp. 79–90, 2020.
https://doi.org/10.1007/978-3-030-60633-6_7

Fig. 1. An example of de-fencing using our proposed semi-supervised approach.

and provide security, they inevitably bring adverse perception interference when taking photos in these scenes. Removing fences from captured image, also named as image de-fencing, is thus an interesting research topic, by which non-occluded background scene images would be recovered with visually plausible quality. In existing methods, image de-fencing is basically formulated as a two-phase restoration problem, i.e., first detecting fences and then restoring background image [2]. For example, fences should be first detected to obtain a mask, and then existing inpainting methods can be adopted to predict a non-occluded image with plausible visual quality.

To date, considerable research works on image de-fencing have been proposed in the literature, which can be roughly grouped into two categories, i.e., single-image de-fencing and multi-image de-fencing. Typical multi-image de-fencing methods includes [4,8,11], etc. In [8], a semi-automated de-fencing algorithm with convolutional neural networks for detecting fence pixels using a video of the dynamic scene. Du et al. [4] proposed to employ the segmentation approach based on convolutional neural networks, and then proposed a fast recovery algorithm with the help of optical flow. Benefiting from temporal or multi-view information, these multi-image de-fencing methods would perform better in removing fences and restoring background image. However, these methods usually require multiple images captured in specific mode or setting, and it is not trivial to collect feasible multiple images for average users, restricting the practical application of multi-image de-fencing.

In contrast, single image de-fencing is more practical but more challenging problem [3–5,9,10,12,14], which is currently rarely studied. Farid et al. [5] proposed a semi-automated approach, where interactive information from users are deployed to detect fences. In [3], a two-stage deep learning approach was proposed, where two conditional GANs are individually adopted for predicting fences mask and generating the inpainted background image, respectively. These fully supervised single image-defencing methods [3,4] extremely rely on the accuracy of detected fences mask. However, fence mask detection errors are usually inevitable in the first stage due to the diverse patterns of fences. And image inpainting methods in the second stage is not able to well handle detection errors, yielding severe visual artifacts.

Image inpainting refers to the process of filling missing or restoring damaged areas in an image. Generally, image inpainting methods can be broadly divided into two categories: classical exemplar-based [1] and deep CNN-based methods [16,20–22]. Early CNNs-based methods [21] are proposed for handling images with small holes. Subsequently, Shift-Net [22] are developed to modify the encoder-decoder architecture for better utilizing the surrounding textures from known region. Recently, LBAM [20] was proposed to introduce a module with learnable bidirectional attentional maps that is incorporated into the U-Net architecture for filling in irregular holes with fine-detailed textures and sharp structures. Moreover, DIP [18] showed that the structure of a generator network is sufficient to capture a great deal of low-level image statistics prior for many restoration tasks including image inpainting.

In this paper, we propose a semi-supervised single image de-fencing approach, where fences mask detection is accomplished using supervised learning a recurrent network and unsupervised restoration method is proposed for robust fences removal. The overview of our proposed semi-supervised single-image de-fencing approach is illustrated in Fig. 2. In the fences mask prediction phase, we propose a simple recurrent network, where LSTM is adopted for progressive detection. It is expected that our recurrent network can detect fences mask gradually in different stages, which is effective and suitable for this task. In supervised training, instead of strict fences mask detection using standard ℓ_1 loss, we adopt an asymmetric loss [7] to make detected fences mask be inclusive of more true fences. That is, over-estimated fences can be filled in the following unsupervised learning phase, while under-estimated fences are more difficult to handle and yield remaining fences artifacts in restored background image. Then in the unsupervised learning phase for robust restoration, motivated by the success of deep image prior (DIP [18]) for several image restoration tasks, we employ three DIP networks for modeling background image, fences layer and fences mask, respectively. To improve the robustness to detection errors of fences mask, we propose to adopt Laplacian smoothness function [13] as the loss for learning DIP networks to refine the detected fences mask. Also, the asymmetric loss can be employed in the second robust restoration phase to facilitate the refinement of fences mask.

Our semi-supervised approach has two merits: (i) By adopting asymmetric loss for supervised training, the predicted fences mask is often inclusive of true fences, which can guarantee the over-estimated fences will be removed more clear. (ii) By adopting Laplacian smoothness loss and asymmetric loss for unsupervised training, the fences mask can also be refined during restoring background image. We summarize our contributions from three aspects:

- A semi-supervised learning framework for single image de-fencing is proposed for robust fences removal from a single image, which contains a supervised fences mask detection network and an unsupervised restoration network.
- We propose a recurrent network for fences mask detection, where LSTM is adopted for progressive detection. The recurrent network can be trained using supervised learning to detect fences mask gradually in different stages
- We propose an unsupervised learning of DIP networks for robust restoration of background image, where mask detection errors can be refined by the

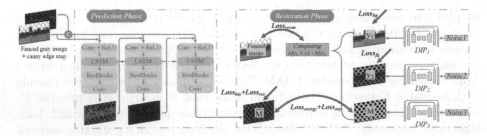

Fig. 2. The overview of our semi-supervised single image de-fencing approach. In the prediction phase, a recurrent network is proposed for predicting fences mask $\widetilde{\mathbf{M}}$ In the restoration phase, two DIPs jointly decompose an input image into fences layer $\mathbf{y}_2$ and background image layer $\mathbf{y}_1$, respectively. Another DIP network is adopted to obtain the refined fences mask $\mathbf{M}$. The details of training process can be found in Sec. 2.1 for supervised learning and Sec. 2.2 for unsupervised learning.

Laplacian smoothness loss and asymmetric loss, making the restoration of background image be more robust to fences mask detection errors.

2 Proposed Method

In this section, we present the semi-supervised learning framework, as shown in Fig. 2. Overall, the semi-supervised learning approach follows the two-stage framework, i.e., (i) supervised learning to detect fences mask, and (ii) unsupervised learning for robust restoration of background image.

2.1 Supervised Learning to Detect Fences Mask

The task of the fences mask detection can be viewed as an image-to-image translation problem in which we convert a fenced image to a fences mask. Denote the ground-truth fences mask as $\mathbf{M}_{gt}$, $\mathbf{I}_{fen}$ as the input image, and the detection mask as $\widetilde{\mathbf{M}}$. It is very difficult to estimate fences mask directly. And thus we decompose this task into several states, i.e., a progressive detection network. For stage t, the current fences mask can be estimated using

$$\widetilde{\mathbf{M}}^t = f_{\text{detect}}(\mathbf{I}_{fen}, \widetilde{\mathbf{M}}^{t-1}),\tag{1}$$

where f_{detect} is a simple mask detection network. We note that, instead of taking color fenced image as $\mathbf{I}_{fen}$, it is suggested that fences would be better predicted in gray space along with its edge maps [3]. In this work, the gray fenced image and its edge map by canny operation are concatenated to act as $\mathbf{I}_{fen}$. As for the network architecture, we propose to adopt a simple recurrent network, where ConvLSTM is adopted to propagate the deep features across stages.

Recurrent Network. As shown in the prediction phase of Fig. 2, the parameters of different stages are shared. The prediction network f_{detect} consists of four parts: (i) The input layer f_{in} which contains a convolutional layer takes the concatenation of fenced image $\mathbf{I}_{fen}$ and $\widetilde{\mathbf{M}}^{t-1}$ from stage $t-1$ as input, (ii) recurrent layer $f_{\text{recurrent}}$ contains ConvLSTM [17], (iii) three residual blocks (*ResBlocks*) f_{res} for extracting deep features, and (iv) output layer f_{out} with a convolutional layer for generating detected fences mask. For stage t, the inference of f_{detect} can be formulated as,

$$\mathbf{z}^t = f_{\text{in}}(\widetilde{\mathbf{M}}^{t-1}, \mathbf{I}_{fen}),$$
$$\mathbf{h}^t = f_{\text{recurrent}}(\mathbf{h}^{t-1}, \mathbf{z}^t), \qquad (2)$$
$$\widetilde{\mathbf{M}}^t = f_{\text{out}}(f_{\text{res}}(\mathbf{h}^t)),$$

where f_{in}, f_{res} and f_{out} are stage-invariant, i.e., network parameters are reused across different stages. In the following, we present the implementation details of key modules in f_{detect}. We note that all the filters are with size 3×3, padding 1×1 and stride 1.

Supervised Learning. We design task-oriented loss function to induce the prediction phase. Considering that fences mask is binary, and thus we adopt a binary loss [18] to force the output as close to be binary as possible. The loss function is formulated as,

$$\mathcal{L}_{bin} = \left(\sum_i |\widetilde{\mathbf{M}}(i) - 0.5| \right)^{-1}. \qquad (3)$$

And given the ground-truth fences mask, the standard ℓ_1 loss is,

$$\mathcal{L}_{rec} = \|\widetilde{\mathbf{M}} - \mathbf{M}_{gt}\|_1. \qquad (4)$$

The strict ℓ_1 loss is expected to obtain exactly accurate detection. However, it is difficult to detect some masks very well especially when fences in the testing dataset are different from the training dataset. For de-fencing, under-estimated fences would not be well handled, while over-estimated regions can be filled in restoration phase. Consequently, we need to impose more penalty to under-estimation error, making the model can generalize well to more cases. Thus, we propose an asymmetric loss [7],

$$\mathcal{L}_{asy} = \sum_i \left| \alpha - \mathbb{I}_{(\widetilde{\mathbf{M}}(i) - \mathbf{M}_{gt}(i)) < 0} \right| \cdot |\widetilde{\mathbf{M}}(i) - \mathbf{M}_{gt}(i)|, \qquad (5)$$

where $\mathbb{I}_e = 1$ for $e < 0$ and 0 otherwise. By setting $0 < \alpha < 0.5$, we can force $\widetilde{\mathbf{M}}$ to be inclusive of $\mathbf{M}_{gt}$. To sum up, the loss for training detection network f_{detect} is

$$\mathcal{L} = a \cdot \mathcal{L}_{bin} + b \cdot \mathcal{L}_{asy}, \qquad (6)$$

where a and b are hyper-parameters, $a = 0.1$ and $b = 1$ are set in our training process. In [4], the authors provide a training dataset containing 545 pairs of fence images and ground-truth fences, based on which our detection network can then be trained. We resize the training images into 908×512 and crop patches with size 256×256. During training, the batch size is set as 4. The ADAM algorithm is adopted to train the model with an initial learning rate 1×10^{-4}, and ends after 100 epochs. When reaching 30, 50 and 80 epochs, the learning rate is decayed by multiplying 0.2.

2.2 Unsupervised Learning for Robust Restoration

Three DIP Networks for De-fencing. In unsupervised learning, the composition of fenced image $\mathbf{I}$ can be formulated as,

$$\mathbf{I}(i) = \mathbf{M}(i)\mathbf{y}_1(i) + (1 - \mathbf{M}(i))\mathbf{y}_2(i), \tag{7}$$

where $\mathbf{y}_1$ is fences layer and $\mathbf{y}_2$ is background image layer, and a binary mask $\mathbf{M}$ indicates the location of fences. Motivated by DIP [18] and Double DIP [6], we can employ three DIP networks to generate $\mathbf{y}_1$, $\mathbf{y}_2$ and $\mathbf{M}$, respectively. The fences mask $\mathbf{M}$ should be forced to be close to the estimated $\widetilde{\mathbf{M}}$ from the detection phase. And then by forcing the composed image to be close to the input fenced image, three DIP networks can be optimized in an unsupervised learning manner. As for the network architecture, three DIP networks adopt the default architecture in [18].

Fenced image   Fecens predicted by ℓ_1 loss   Fecens predicted by asy loss

Fig. 3. Examples of fences mask detection using ℓ_1 loss and asymmetric loss respectively.

Unsupervised Learning. The training phase of restoration is approximately divided two stages: (i) In the first stage, the fences mask is fixed as $\widetilde{\mathbf{M}}$ from detection phase to generate coarse restoration results. (ii) In the second stage, fences mask can be refined along with generating finer restoration results. By fixing fences mask as $\widetilde{\mathbf{M}}$, the DIP networks for $\mathbf{y}_1$ and $\mathbf{y}_2$ can be trained using reconstruction loss [6],

$$\begin{aligned}
\mathcal{L}_{bg} &= \|(\mathbf{I} - \mathbf{y}_1) \odot (1 - \widetilde{\mathbf{M}})\|^2, \\
\mathcal{L}_{fg} &= \|(\mathbf{I} - \mathbf{y}_2) \odot \widetilde{\mathbf{M}}\|^2,
\end{aligned} \tag{8}$$

where $\mathbf{I}$ is fenced image, and $\odot$ represents the entry-wise product. So the total loss in the first stage to generate coarse results can be formulated as,

$$\mathcal{L}_1 = \mathcal{L}_{fg} + \mathcal{L}_{bg}. \tag{9}$$

Then in the second stage, we propose several loss functions to generate finer results and refine the fences mask. We first adopt the reconstruction loss based on the composition model Eq. (7),

$$\mathcal{L}_{recon} = \|\mathbf{I} - \mathbf{M} \odot \mathbf{y}_2 - (1 - \mathbf{M}) \odot \mathbf{y}_1\|^2, \tag{10}$$

where $\mathbf{M}$ is the mask regenerated by the mask DIP network, which can be refined during learning process. It is a natural solution that the binary loss Eq. (3) can be adopted for training mask DIP network. Also we can adapt the asymmetric loss Eq. (5) to unsupervised learning,

$$\mathcal{L}_{uasy} = \sum_i \left| \alpha - \mathbb{I}_{(\mathbf{M}(i) - \widetilde{\mathbf{M}}(i)) < 0} \right| \cdot |\mathbf{M}(i) - \widetilde{\mathbf{M}}(i)|, \tag{11}$$

where the ground-truth mask is substituted by $\widetilde{\mathbf{M}}$ estimated using recurrent network.

However, it is not enough because the capability of detection network is finite especially when the mask detected is incomplete. Hence, we propose to employ the Laplacian smoothness function [13] as loss function to make it more robust in the mask regeneration phase. The Laplacian smoothness function is the energy function for image matting proposed by Levin et al. [13]. By assuming local smoothness of fences and background information, the Laplacian smoothness function can be adopted to refine $\mathbf{M}$ by minimizing the loss

$$\mathcal{L}_{smooth} = \mathbf{M}^T \mathbf{L} \mathbf{M} + \lambda (\mathbf{M} - \widetilde{\mathbf{M}})^T \mathbf{Q} (\mathbf{M} - \widetilde{\mathbf{M}}), \tag{12}$$

where $\mathbf{Q} \in \mathbb{R}^{N \times N}$ is a diagonal matrix whose nonzero elements indicate the presence of fences. We note that fences mask $\widetilde{\mathbf{M}} \in \mathbb{R}^N$ is reshaped to be a 1D vector in Laplacian smoothness loss, while it is a 2D image in other parts. This should not cause ambiguity according to context. The matrix $\mathbf{L}$ is a special Laplacian-like matrix given by

$$\mathbf{L}_{ij} = \sum_{n|(i,j) \in p_n} \left(\delta_{ij} - w_{ij}^n \right), \tag{13}$$

where

$$w_{ij}^n = \frac{1}{|p_n|} \left(1 + (\mathbf{I}_i - \boldsymbol{\mu}_n)^T \left(\boldsymbol{\Sigma}_n + \frac{\varepsilon}{|p_n|} \mathbf{I} \right)^{-1} (\mathbf{I}_j - \boldsymbol{\mu}_n) \right). \tag{14}$$

In the above formula, i, j are pixels in the patch p_n, centered on the pixel n; $\boldsymbol{\mu}_n \in \mathbb{R}^{3 \times 1}, \boldsymbol{\Sigma}_n \in \mathbb{R}^{3 \times 3}$ are the mean and covariance of the patch whose number of pixels is $|p_n|$; $\mathbf{I} \in \mathbb{R}^{3 \times 3}$ is the identity matrix, and ε provides an additional control over the smoothness. Then this function can be converted to be "tensor-friendly",

$$\mathbf{M}^T \mathbf{L} \mathbf{M} = \sum_{n=1}^{N} \sum_{i=1}^{9} \sum_{j=1}^{9} w_{ij}^n \left(\mathbf{M}_i - \mathbf{M}_j \right)^2, \tag{15}$$

where we sum over all overlapping patches around N pixels in $\mathbf{M}$, as well as over all possible combinations of pixel pairs i, j in a given 3×3 patch, where the total number of combinations is $(3^2) \times (3^2) = 81$. The weights w_{ij}^n are given in (14). Finally, the final Laplacian smoothness loss function is formulated as,

$$\mathcal{L}_{smooth} = \sum_{b=1}^{B} \sum_{n=1}^{N} \sum_{k=1}^{K} \mathbf{W} \odot \left(\mathbf{M}_I - \mathbf{M}_J \right)^2$$
$$+ \lambda \sum_{b=1}^{B} \sum_{n=1}^{N} \left(\mathbf{M} - \widetilde{\mathbf{M}} \right)^2, \tag{16}$$

where $k \in [1 \dots K = 81]$ indexes the pixel pairs i, j in a 3×3 patch, and $\mathbf{W} \in \mathbb{R}^{B \times N \times 81}$ is the matrix of weights. $\mathbf{M}_I, \mathbf{M}_J \in \mathbb{R}^{B \times N \times 81}$ are repetitions of $\mathbf{M}$; the first represents $\mathbf{M}_I$ in index $i \rightarrow (1, \dots, 1, 2, \dots, 2, \dots, 9, \dots, 9) \in \mathbb{R}^{81}$, and the second represents $\mathbf{M}_J$ in index $j \rightarrow (1, 2, \dots, 9, 1, 2, \dots, 9, \dots, 1, 2, \dots 9) \in \mathbb{R}^{81}$. $\widetilde{\mathbf{M}} \in \mathbb{R}^{B \times N}$ is the hint image. We verify its effectiveness in Sect. 3 by experiments.

And finally the total loss in the second stage can be formulated as,

$$\mathcal{L}_2 = a \cdot \mathcal{L}_{smooth} + b \cdot \mathcal{L}_{uasy} + c \cdot \mathcal{L}_{bin}, \tag{17}$$

where $b = 0.5$, a and c vary with the number of iterations. Iterations are 2,000 and 4,000 in the first and second stages respectively. The optimization process is done using ADAM optimizer with a learning rate 0.001, and then we can obtain visually plausible results by this coarse-to-fine unsupervised learning for robust restoration.

3 Experimental Results

In this section, we evaluate the performance of our proposed semi-supervised de-fencing approach. First, ablation studies are conducted to analyze the effectiveness of asymmetric loss and Laplacian smoothness loss for fences mask detection and robust restoration. Then, we try to compare with other methods. However, recent de-fencing methods [3,4] do not release their source codes and executable programs. Therefore, we compare with state-of-the-art inpainting methods for restoring background image given the estimated fences mask by our method.

Our method is implemented using Pytorch [15]. Supervised learning and unsupervised learning models are trained on a PC equipped with an NVIDIA TITAN V GPU.

Preliminary Mask $\widetilde{\mathbf{M}}$ Refined Mask w/o $\mathcal{L}_{smooth}$ Refined Mask w/ $\mathcal{L}_{smooth}$

Fenced image   Background image w/o $\mathcal{L}_{smooth}$ Background image w/ $\mathcal{L}_{smooth}$

Fig. 4. The effectiveness of Laplacian smoothness loss for the refinement of fences mask in unsupervised learning phase.

3.1 Ablation Study

Currently, there is only one dataset for image de-fencing [4], in which only the pairs of fenced image and fences mask are provided. Thus it is not feasible to evaluate the models using quantitative metrics of restored background images. So in the ablation study, we only show several cases to verify the effectiveness of asymmetric loss in supervised learning phase and Laplacian smoothness loss in unsupervised learning phase.

Fenced image   LBAM[20]   SN[22]   Ours

Fig. 5. Visual comparison on testing image from [4].

- **Asymmetric Loss.** In supervised learning phase, two recurrent models are trained using traditional ℓ_1 loss Eq. (4) and the proposed asymmetric loss Eq. (5), respectively. As shown in Fig. 3, we show an example of fences mask detection using these two models. One can see that the fences masks detected by ℓ_1 loss is worse than those by asymmetric loss. In contrast, our asymmetric loss can provide better detection results. We note that the minor disconnections will lead to visible artifacts in restored background image. The proposed asymmetric loss Eq. (5) can lead to a over-estimated fences mask.

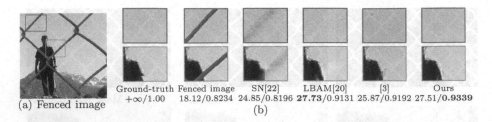

(a) Fenced image

Ground-truth Fenced image   SN[22]   LBAM[20]   [3]   Ours
+∞/1.00   18.12/0.8234 24.85/0.8196 **27.73**/0.9131 25.87/0.9192 27.51/**0.9339**
(b)

Fig. 6. Qualitative and quantitative comparison on the image from [3].

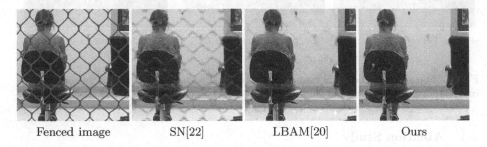

Fenced image   SN[22]   LBAM[20]   Ours

Fig. 7. Qualitative comparison on an image taken by our mobilephone.

- **Laplacian Smoothness Loss.** In unsupervised learning phase, Laplacian smoothness loss Eq. (16) is adopted to refine the estimated fences mask. As shown in Fig. 4, the detected fences mask $\widetilde{M}$ can be refined during unsupervised learning to obtain M. Without the Laplacian smoothness loss, $\widetilde{M}$ can still be refined by binary loss in (17). But the result by adding Laplacian smoothness loss can make the fences mask more accurate. Correspondingly, the restored background image with Laplacian smoothness loss are visually plausible, while the result without Laplacian smoothness loss suffers from artifacts especially in *tree* regions. By incorporating Laplacian smoothness loss, the unsupervised learning is more robust to fences mask detection error.

3.2 Comparison with State-of-the-Art Methods

Since the source codes of de-fencing methods are not available, we compare the proposed method with several state-of-the-art inpainting methods including Shift-Net(SN) [22] and LBAM [20]. The fences mask is estimated by our fences mask detection network. We also note that existing inpainting models are trained using central mask or random mask that are quite different from fences. Thus, to make a fair comparison, SN and LBAM are fine-tuned on Places dataset [23] on which the fences masks from [4] are added to synthesize training pairs of fenced image and clean background image. As compared in Fig. 5, LBAM and SN can roughly remove fences. But in the close-ups, artifacts at fences regions are still remained. Our method can generate much more visually plausible background image, where the fences regions are filled with consistent contents with background. Furthermore, we borrow one fenced image from [3], whose ground-truth background image is available. As shown in Fig. 6, we report the restored background images, where $-/-$ indicates PSNR/SSIM [19]. The results of SN and LBAM are obtained by our fine-tuned models, while the result by [3] is provided in their paper. One can see that our method can achieve comparable quantitative metrics. Although PSNR by our method is inferior to LABM, the visual effect is much better, leading to higher SSIM value. Finally, we compare our method with SN and LBAM on an image taken by our mobilephone. Our method achieves visually favorable result that is significantly superior to SN and LBAM.

4 Conclusion

In this paper, we proposed a semi-supervised learning framework for single image de-fencing. In the supervised learning phase, a recurrent network is proposed to predict fences mask gradually. The asymmetric loss can be adopted for over-estimating fences mask. In the unsupervised learning phase, three DIP networks are adopted to model background image, fences layer and fences mask, respectively. By incorporating Laplacian smoothness loss, the estimated fences mask can also be refined for robust restoration of background image. Experimental results have validated the effectiveness of our semi-supervised learning approach for single image de-fencing.

References

1. Barnes, C., Shechtman, E., Finkelstein, A., Goldman, D.B.: PatchMatch: a randomized correspondence algorithm for structural image editing. ACM Trans. Graph. (ToG) **28**, 24 (2009)
2. Criminisi, A., Pérez, P., Toyama, K.: Region filling and object removal by exemplar-based image inpainting. IEEE Trans. Image Process. **13**(9), 1200–1212 (2004)
3. Divyanshu, G., Shorya, J., Utkarsh, T., Pratik, C., Wang, L.: Fully automated image de-fencing using conditional generative adversarial networks. arXiv preprint arXiv:1908.06837 (2019)
4. Du, C., Kang, B., Xu, Z., Dai, J., Nguyen, T.: Accurate and efficient video de-fencing using convolutional neural networks and temporal information. In: 2018 IEEE International Conference on Multimedia and Expo (ICME), pp. 1–6. IEEE (2018)
5. Farid, M.S., Mahmood, A., Grangetto, M.: Image de-fencing framework with hybrid inpainting algorithm. SIViP **10**(7), 1193–1201 (2016). https://doi.org/10.1007/s11760-016-0876-7
6. Gandelsman, Y., Shocher, A., Irani, M.: "Double-DIP": unsupervised image decomposition via coupled deep-image-priors, June 2019
7. Guo, S., Yan, Z., Zhang, K., Zuo, W., Zhang, L.: Toward convolutional blind denoising of real photographs. In: 2019 IEEE Conference on Computer Vision and Pattern Recognition (CVPR) (2019)

8. Jonna, S., Nakka, K.K., Sahay, R.R.: My camera can see through fences: a deep learning approach for image de-fencing. In: 2015 3rd IAPR Asian Conference on Pattern Recognition (ACPR), pp. 261–265. IEEE (2015)
9. Jonna, S., Nakka, K.K., Sahay, R.R.: Deep learning based fence segmentation and removal from an image using a video sequence. In: Hua, G., Jégou, H. (eds.) ECCV 2016. LNCS, vol. 9915, pp. 836–851. Springer, Cham (2016). https://doi.org/10. 1007/978-3-319-49409-8_68
10. Jonna, S., Voleti, V.S., Sahay, R.R., Kankanhalli, M.S.: A multimodal approach for image de-fencing and depth inpainting. In: 2015 Eighth International Conference on Advances in Pattern Recognition (ICAPR), pp. 1–6. IEEE (2015)
11. Khasare, V.S., Sahay, R.R., Kankanhalli, M.S.: Seeing through the fence: image de-fencing using a video sequence. In: 2013 IEEE International Conference on Image Processing, pp. 1351–1355. IEEE (2013)
12. Kumar, V., Mukherjee, J., Mandal, S.K.D.: Image defencing via signal demixing. In: Proceedings of the Tenth Indian Conference on Computer Vision, Graphics and Image Processing, p. 11. ACM (2016)
13. Levin, A., Lischinski, D., Weiss, Y.: A closed-form solution to natural image matting. IEEE Trans. Pattern Anal. Mach. Intell. **30**(2), 228–242 (2007)
14. Liu, Y., Belkina, T., Hays, J., Lublinerman, R.: Image defencing. In: Proceedings of CVPR (2008)
15. Paszke, A., et al.: PyTorch: an imperative style, high-performance deep learning library. In: Advances in Neural Information Processing Systems, pp. 8024–8035 (2019)
16. Pathak, D., Krahenbuhl, P., Donahue, J., Darrell, T., Efros, A.A.: Context encoders: feature learning by inpainting. In: Proceedings of the IEEE Conference on Computer Vision and Pattern Recognition, pp. 2536–2544 (2016)
17. Shi, X., Chen, Z., Wang, H., Yeung, D.Y., Wong, W.K., Woo, W.c.: Convolutional LSTM network: a machine learning approach for precipitation nowcasting. In: Advances in Neural Information Processing Systems, pp. 802–810 (2015)
18. Ulyanov, D., Vedaldi, A., Lempitsky, V.: Deep image prior. In: Proceedings of the IEEE Conference on Computer Vision and Pattern Recognition, pp. 9446–9454 (2018)
19. Wang, Z., Bovik, A.C., Sheikh, H.R., Simoncelli, E.P.: Image quality assessment: from error visibility to structural similarity. IEEE Trans. Image Process. **13**(4), 600–612 (2004)
20. Xie, C., et al.: Image inpainting with learnable bidirectional attention maps. In: Proceedings of the IEEE International Conference on Computer Vision, pp. 8858–8867 (2019)
21. Xie, J., Xu, L., Chen, E.: Image denoising and inpainting with deep neural networks. In: Advances in Neural Information Processing Systems, pp. 341–349 (2012)
22. Yan, Z., Li, X., Li, M., Zuo, W., Shan, S.: Shift-Net: image inpainting via deep feature rearrangement. In: Ferrari, V., Hebert, M., Sminchisescu, C., Weiss, Y. (eds.) Computer Vision – ECCV 2018. LNCS, vol. 11218, pp. 3–19. Springer, Cham (2018). https://doi.org/10.1007/978-3-030-01264-9_1
23. Zhou, B., Lapedriza, A., Khosla, A., Oliva, A., Torralba, A.: Places: a 10 million image database for scene recognition. IEEE Trans. Pattern Anal. Mach. Intell. **40**, 1452–1464 (2017)

Confidence-Aware Adversarial Learning for Self-supervised Semantic Matching

Shuaiyi Huang, Qiuyue Wang, and Xuming He[✉]

ShanghaiTech University, Shanghai, China
{huangshy1,wangqy2,hexm}@shanghaitech.edu.cn

Abstract. In this paper, we aim to address the challenging task of semantic matching where matching ambiguity is difficult to resolve even with learned deep features. We tackle this problem by taking into account the confidence in predictions and develop a novel refinement strategy to correct partial matching errors. Specifically, we introduce a Confidence-Aware Semantic Matching Network (CAMNet) which instantiates two key ideas of our approach. First, we propose to estimate a dense confidence map for a matching prediction through self-supervised learning. Second, based on the estimated confidence, we refine initial predictions by propagating reliable matching to the rest of locations on the image plane. In addition, we develop a new hybrid loss in which we integrate a semantic alignment loss with a confidence loss, and an adversarial loss that measures the quality of semantic correspondence. We are the first that exploit confidence during refinement to improve semantic matching accuracy and develop an end-to-end self-supervised adversarial learning procedure for the entire matching network. We evaluate our method on two public benchmarks, on which we achieve top performance over the prior state of the art. We will release our source code at https://github.com/ShuaiyiHuang/CAMNet.

Keywords: Semantic correspondence · Confidence · Refinement · Self-supervised adversarial learning

1   Introduction

Dense correspondence estimation has been a core building block for a variety of computer vision tasks [9, 26]. Recently, traditional instance-level correspondence has been extended to the problem setting of semantic-level matching, which aims to align different object instances of the same category [20]. Semantic matching has attracted growing attention [6, 19, 24] and demonstrated practical value in real-world applications [27]. However, an effective strategy remains elusive largely due to the presence of background clutters, severe intra-class variation and viewpoint change, as well as difficulty in data annotation (Fig. 1).

This work was supported in part by the Shanghai NSF Grant No.18ZR1425100.

© Springer Nature Switzerland AG 2020
Y. Peng et al. (Eds.): PRCV 2020, LNCS 12305, pp. 91–103, 2020.
https://doi.org/10.1007/978-3-030-60633-6_8

Fig. 1. Motivation. From left to right are the source image, the target image, and the estimated confidence map. The point in the target image (yellow circle) was initially matched to the wrong position in the source image (background, red rhombus). After a step of correction through the confidence map, it was matched to a more correct position (foreground, blue diamond). (Color figure online)

To tackle those challenges, recent research efforts have adopted the learning-based representation, which have significantly improved semantic matching quality compared with traditional methods [6,15,19,21,23]. Early learning methods in semantic matching require strong supervision in the form of ground truth correspondences, which is difficult to obtain for a large set of real images [5]. As a result, recent trend has focused on weakly-supervised methods [10,24] that only employ matching image pair, but they are mostly limited to small-scale datasets. A more promising strategy is to leverage self-supervised learning where image pairs and ground-truth correspondences are generated synthetically using random transformations [8,19,23]. This can significantly reduce annotation cost and makes it possible to utilize large-scale single image datasets along with their existing labels (e.g. masks) for additional constraints [19].

Despite the increasingly large training datasets, existing deep network based semantic matching typically rely on a deterministic neural networks which do not incorporate uncertainty in their correspondence estimation. Due to the implicit ambiguity in feature matching, such simplistic approaches lack the capacity to measure the quality of prediction results and to properly refine their initial predictions. While some prior work attempt to improve initial estimation through an iterative refinement process [8,23], they often suffer from error propagation when initial predictions are of low quality.

In this paper, we propose to incorporate uncertainty reasoning into self-supervised semantic matching in order to estimate and correct low-quality correspondence results. Toward this goal, we first introduce a pixel-wise confidence estimation mechanism to determine whether the initial prediction for each location on image plane is reliable or not. We then develop a confidence-aware refinement procedure to update initial predictions. We exploit the self-supervision setting to generate dense ground truth labels for confidence estimation and introduce an additional adversarial loss to improve overall prediction consistency. A key advantage of our approach is the tight integration of the model design and self-supervised learning strategy, which allows us to effectively train this confidence-aware semantic matching network.

Specifically, we present a novel Confidence-Aware Semantic Matching Network (CAMNet) that estimates matching confidences and base on which refines the correspondence prediction. Our network consists of four parts. First, a Base Correspondence Network produces initial correspondence predictions. Then, a Confidence Estimation Network is developed to measure pixel-wise confidence of initial predictions. Moreover, taking as input initial predictions and estimated confidence maps, we design a Confidence-aware Refinement Network to produce refined correspondences by propagating reliable predictions. We train our model with an alignment loss, a confidence loss and an adversarial loss in an end-to-end fashion. As we develop all the training signals for correspondence, confidence, and adversarial learning from synthetically warped image pairs, we refer to this novel learning scheme as confidence-aware adversarial learning for self-supervised semantic matching.

We extensively evaluate our method on two standard benchmarks, including PF-Willow [5] and PF-PASCAL [5]. Our experimental results demonstrate the strong performance of our model over the prior state-of-the-art approaches. Our main contribution are threefold:

- We propose a confidence-aware refinement strategy for semantic matching and are the first to consider generating confidence ground truth with the self-supervised setting.
- We introduce adversarial training to mitigate the distortion problem in self-supervised semantic matching task.
- Our self-supervised learning strategy achieves top performance in two standard benchmarks.

2   Related Work

2.1   Semantic Correspondence

Early semantic matching approaches usually leverage ground truth correspondences for strong supervision [5,15]. Collecting such annotations under large intra-class appearance and shape variations is label-intensive and may be objective. Consequently, recent work has focused on weakly-supervised [10,24] or self-supervised setting. Self supervision has attracted growing attention as it further saves human effort and enables to leverage large-scale single image datasets along with their existing labels for additional constraints. Rocco et al. [23] first propose to synthetically generate the image pair and ground truth correspondences from an image itself. Seo et al. [8] extend this idea with an offset-aware correlation kernel to filter out distractions. Junghyup et al. [19] further utilize images annotated with binary foreground masks and subject to synthetic geometric deformations for this task. However, few methods consider the uncertainty in model prediction or incorporate an estimated prediction quality in a refinement process in order to correct partial errors, especially in challenging cases.

2.2   Confidence Estimation

Confidence estimation has been widely explored and showing promising improvement in stereo matching and depth completion [4,18]. Recently, Sangryul et al. first introduce confidence in weakly-supervised semantic matching [13]. However, confidence is learned without any available supervision and is used for regularization in loss function instead of reasoning during prediction. In contrast, we are the first to consider generating confidence ground truth as supervision, and directly utilize confidence in model design to guide refined correspondence prediction in self-supervised semantic matching.

2.3   Generative Adversarial Networks

Generative adversarial networks (GANs) have gained increasing attention [1,11, 22]. Inspired from the success of GANs, we introduce adversarial learning to enforce the global consistency of warped images, which has never been explored before in semantic matching. We utilize a discriminator to distinguish real data from fake data, and simultaneously to guide the generator to produce higher quality correspondences. As the real data for training GANs are from synthetically warped image pairs, we call our learning scheme as self-supervised adversarial training.

3   Model Architecture

3.1   Overview

Semantic alignment aims to establish dense correspondences between a source image I_s and a target image I_t. A typical CNN-based method computes a correlation map between the convolution features of two images, from which a dense flow field is predicted as final output.

Our objective is to augment the correspondence prediction with a confidence estimation mechanism so that the model is capable of propagating reliable information during inference and producing a consistent flow prediction from the correlation map. To achieve this, we adopt a refinement strategy: given an initial dense correspondence between two feature maps, we first estimate the probability of being correctly predicted for each feature location. The estimated confidence map is then fed into a refinement process to guide information propagation and prediction of final correspondence.

In this section, we introduce a novel network dubbed Confidence-aware Semantic Matching Network (CAMNet) to implement our strategy. Our network comprises three main submodules, including a Base Correspondence Network (Sect. 3.2), a Confidence Estimation Network (Sect. 3.3), and a Confidence-aware Refinement Network (Sect. 3.4). The entire network is learned in a jointly manner. Below we will describe these submodules of CAMNet in details, and defer the discussion of model learning to Sect. 4. An overview of our model architecture is shown in Fig. 2.

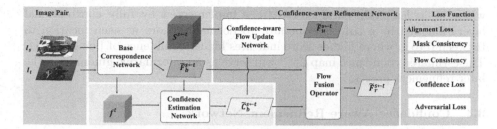

Fig. 2. Overview of the confidence-aware semantic matching network.

3.2 Base Correspondence Network

The first module of our CAMNet is the Base Correspondence Network, aiming to produce an initial semantic flow given the input image pair. To this end, we adopt a differentiable semantic correspondence network as our module design, which consists of a feature extractor, a correlation computation operator and a base flow prediction module[1].

Concretely, the input image pair $\mathbf{I}_s$ and $\mathbf{I}_t$ are first passed through a CNN-based feature extractor, which extract features maps $f^s \in \mathbb{R}^{d \times h_s \times w_s}$ and $f^t \in \mathbb{R}^{d \times h_t \times w_t}$ respectively. We adopt ResNet-101 [7] with additional adaptation layers to parameterize feature extractors as in [19]. Subsequently, we compute a 3D correlation map $S^{s \leftarrow t} \in \mathbb{R}^{(h_s \times w_s) \times h_t \times w_t}$ from L2 normalized feature map f^s and f^t, which describes pairwise similarity for any two locations between $\mathbf{I}_s$ and $\mathbf{I}_t$. Given 3D correlation map $S^{s \leftarrow t}$, we then apply a flow prediction module to obtain initial semantic flow predictions $\tilde{\mathcal{F}}_b^{s \leftarrow t} \in \mathbb{R}^{2 \times h_t \times w_t}$. While any differentiable flow prediction networks can be adopted here, we choose the kernel soft argmax layer proposed in [19] due to its superior performance. The initial semantic flow $\tilde{\mathcal{F}}_b^{s \leftarrow t}$ will be refined in the later process.

3.3 Confidence Estimation Network

To correct partial errors in the initial prediction, we introduce a confidence estimation mechanism to determine what information is trustworthy for guiding the refinement. To this end, we design a confidence estimation network to generate a prediction confidence map from the input image features and predicted flows. By taking the input and output into consideration, the network aims to detect inconsistent flow configurations, which may not be easily discovered by checking output predictions alone.

Concretely, given the target feature map f^t and the initial flow $\tilde{\mathcal{F}}_b^{s \leftarrow t}$, the confidence estimation network $\mathbb{C}$ estimates a base confidence map $\tilde{\mathcal{C}}_b^{s \leftarrow t} \in \mathbb{R}^{1 \times h_t \times w_t}$ as follows:

$$\tilde{\mathcal{C}}_b^{s \leftarrow t} = \mathbb{C}(f^t, \tilde{\mathcal{F}}_b^{s \leftarrow t}; \theta_{\mathbb{C}}) \tag{1}$$

---

[1] We note that any differentiable semantic correspondence network can be used here.

where $\theta_{\mathbb{C}}$ is learnable parameter. $\mathbb{C}(\cdot)$ is implemented by fully convolutional layers followed by a softmax operator to normalize score at each location to between 0 to 1, where higher score indicates higher confidence of being correct. The estimated confidence map $\tilde{\mathcal{C}}_b^{s \leftarrow t}$ will serve as an informative guidance in the later refinement process.

3.4   Confidence-Aware Refinement Network

Our third module, Confidence-aware Refinement Network, aims at refining the initial flow under the guidance of estimated confidence map. It consists of a confidence-aware flow update and a flow fusion operator described as follows.

Confidence-Aware Flow Update. Different from the initial flow prediction network (cf. Sect. 3.2), our confidence-aware flow update network takes both the confidence map and the correlation map as input, and aims to propagate reliable matching information for flow update.

Specifically, we feed the confidence map $\tilde{\mathcal{C}}_b^{s \leftarrow t}$ for the initial flow and the correlation map $S^{s \leftarrow t}$ into the confidence-aware flow update network $\mathbb{U}(\cdot)$ and compute an updated flow $\tilde{\mathcal{F}}_u^{s \leftarrow t} \in \mathbb{R}^{2 \times h_t \times w_t}$ as follows:

$$\tilde{\mathcal{F}}_u^{s \leftarrow t} = \mathbb{U}(\tilde{\mathcal{C}}_b^{s \leftarrow t}, S^{s \leftarrow t}; \theta_{\mathbb{U}}) \tag{2}$$

where $\theta_{\mathbb{U}}$ is learned parameter.

Flow Fusion. Finally, we feed the base flow, updated flow and the base confidence map into a flow fusion operator to obtain the final refined flow $\tilde{\mathcal{F}}_r^{s \leftarrow t} \in \mathbb{R}^{2 \times h_t \times w_t}$, which enable us to explicitly suppress initial predictions with low-confidence. The confidence maps act as gates that control which pixel position initial predictions $\tilde{\mathcal{F}}_b^{s \leftarrow t}$ will be replaced by the updated flow $\tilde{\mathcal{F}}_u^{s \leftarrow t}$. We choose the following flow fusion operator to generate high quality refined flows:

$$\tilde{\mathcal{F}}_r^{s \leftarrow t} = \tilde{\mathcal{F}}_b^{s \leftarrow t} \odot \tilde{\mathcal{C}}_b^{s \leftarrow t} + \tilde{\mathcal{F}}_u^{s \leftarrow t} \odot (1 - \tilde{\mathcal{C}}_b^{s \leftarrow t}). \tag{3}$$

where $\odot$ denotes the point-wise product of confidence maps and flows.

4   Confidence-Aware Adversarial Learning

We utilize a self-supervised learning strategy for network training by exploiting an image dataset with foreground segmentation masks. Our method first transforms each image and its corresponding foreground mask with random transformations to generate training image pairs as in [19]. During training, we make predictions from target to source and source to target in two directions for each input image pair. Based on this setup, we develop a multi-task loss that includes three components: an alignment loss that enforces the bidirectional matching consistency, a confidence loss that supervises the confidence network, and an adversarial loss on the transformed images within foreground region. We will describe these loss terms in detail below and omit multiply every loss with foreground masks for notation brevity.

4.1   Alignment Loss

We first adopt the self-supervised training loss proposed in SFNet [19] to encourage correspondences to be established within foreground masks and measure consistency between flow estimations in two directions, which has the following form:

$$\mathcal{L}_a(\tilde{\mathcal{F}}^{s\leftarrow t}, \tilde{\mathcal{F}}^{t\leftarrow s}) = \lambda \cdot \mathcal{L}^{mask}(\tilde{\mathcal{F}}^{s\leftarrow t}, \tilde{\mathcal{F}}^{t\leftarrow s}, M^t, M^s) + \mathcal{L}^{flow}(\tilde{\mathcal{F}}^{s\leftarrow t}, \tilde{\mathcal{F}}^{t\leftarrow s}) \quad (4)$$

where λ is the weight parameter for balancing mask and flow consistency loss, M^t and M^s are the binary ground truth foreground masks for the target and the source image respectively. Our alignment loss is defined on the initial flow and the refined flow jointly:

$$\mathcal{L}_{align} = \gamma \cdot \mathcal{L}_a(\tilde{\mathcal{F}}_b^{s\leftarrow t}, \tilde{\mathcal{F}}_b^{t\leftarrow s}) + \mathcal{L}_a(\tilde{\mathcal{F}}_r^{s\leftarrow t}, \tilde{\mathcal{F}}_r^{t\leftarrow s}) \quad (5)$$

where γ is used to balance loss for base flow and refined flow.

4.2   Confidence Loss

As we have ground truth flow in self-supervised learning, the ground truth base confidence map $\mathcal{C}_b^{s\leftarrow t}$ can be obtained by thresholding the error between the predicted base flow $\tilde{\mathcal{F}}_b^{s\leftarrow t}$ and the ground truth flow $\mathcal{F}^{s\leftarrow t}$. This enables us to directly supervise confidence estimation. The confidence loss measures the quality between estimated confidence map and the ground truth confidence map as follows:

$$\mathcal{L}_c(\tilde{\mathcal{C}}^{s\leftarrow t}, \tilde{\mathcal{C}}^{t\leftarrow s}, \mathcal{C}^{s\leftarrow t}, \mathcal{C}^{t\leftarrow s}) = \mathbb{CE}(\tilde{\mathcal{C}}^{s\leftarrow t}, \mathcal{C}^{s\leftarrow t}) + \mathbb{CE}(\tilde{\mathcal{C}}^{t\leftarrow s}, \mathcal{C}^{t\leftarrow s}) \quad (6)$$

where $\mathbb{CE}$ is the cross entropy loss.

In addition to the confidence maps $\tilde{\mathcal{C}}_b^{t\leftarrow s}$ and $\tilde{\mathcal{C}}_b^{t\leftarrow s}$, we also estimate refined confidence maps $\tilde{\mathcal{C}}_r^{t\leftarrow s}$ and $\tilde{\mathcal{C}}_r^{s\leftarrow t}$ for the refined flow in order to further regularize confidence estimation network. Hence our final confidence loss is defined as

$$\mathcal{L}_{confi} = \beta \cdot \mathcal{L}_c(\tilde{\mathcal{C}}_b^{s\leftarrow t}, \tilde{\mathcal{C}}_b^{t\leftarrow s}, \mathcal{C}_b^{s\leftarrow t}, \mathcal{C}_b^{t\leftarrow s}) + \mathcal{L}_c(\tilde{\mathcal{C}}_r^{s\leftarrow t}, \tilde{\mathcal{C}}_r^{t\leftarrow s}, \mathcal{C}_r^{s\leftarrow t}, \mathcal{C}_r^{t\leftarrow s}) \quad (7)$$

where β is the weight parameter balancing the two loss terms.

4.3   Adversarial Loss

We introduce an adversarial loss for enforcing global consistency in flow prediction. To this end, we first build an image generator $\mathbf{G}$ by integrating our model (described in Sect. 3) with a warping operator $\mathcal{W}$. Considering matching from target to source, $\mathbf{G}$ generates synthetically warped source images $\tilde{\mathbf{I}}_t$ given an image pair $(\mathbf{I}_s, \mathbf{I}_t)$ as follows:

$$\tilde{\mathbf{I}}_t = \mathcal{W}\left(\mathbf{I}_s; \tilde{\mathcal{F}}_r^{s\leftarrow t}\right) = \mathbf{G}(\mathbf{I}_s, \mathbf{I}_t) \quad (8)$$

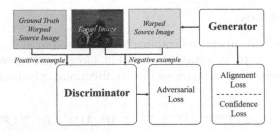

Fig. 3. Overview of self-supervised adversarial training for semantic match-ing. The negative example and positive example composed of the target image and the warped source image are important for adversarial training. The discriminator and the generator are trained alternately.

We define a Discriminator $\mathbf{D}$ to distinguish warped images from the ground-truth flows and the generated flows. The input of the discriminator contains either source images warped by the ground truth flow (denoted as $\mathbf{I}_t^*$) or by the predicted flow, i.e. $\tilde{\mathbf{I}}_t$, and both of them are concatenated with the corresponding target image. From the input pair, the discriminator should learn whether the flow is correct s.t. the warped source image align with target image. $\mathbf{D}$ outputs a one channel prediction map with values between 0 to 1 where higher value indicates the more $\mathbf{D}$ believes that the input pair is real data.

In self-supervision, we apply random transformation in the form of ground truth flow $\mathcal{F}^{s \leftarrow t}$ to the single source image $\mathbf{I}_s$ to form a training image pair $(\mathbf{I}_s, \mathbf{I}_t)$. Thus, the source image warped by ground truth flow $\mathbf{I}_t^*$ is essentially the target image $\mathbf{I}_t$ itself in the self-supervised setting:

$$\mathbf{I}_t^* = \mathbf{I}_t = \mathcal{W}\left(\mathbf{I}_s; \mathcal{F}^{s \leftarrow t}\right) \tag{9}$$

Consequently, we call our framework as self-supervised adversarial learning as real input pair for training GANs is developed from self-supervision (i.e., no real annotations are needed).

Adversarial loss is used for mitigating distortion problem in semantic match-ing. The discriminator and the generator is trained alternatively as shown in Fig. 3. After training is done, the generator (without warping operator) is used as a correspondence estimator and the discriminator can be removed.

Training the Generator. The generator $\mathbf{G}$ tries to generate realistic images to fool the discriminator, and our loss term aims to minimize the following Least Square GAN loss bidirectionally

$$\mathcal{L}_{adv} = (\mathbf{D}(\tilde{\mathbf{I}}_s, \mathbf{I}_s) - 1)^2 + (\mathbf{D}(\tilde{\mathbf{I}}_t, \mathbf{I}_t) - 1)^2 \tag{10}$$

which enforces the generator to produce reasonable warped images.

Our overall loss for the generator consists of the alignment loss, the confidence loss and the adversarial loss terms with μ_1, μ_2 as weight parameters as follows,

$$\mathcal{L}_G = \mu_1 \cdot \mathcal{L}_{align} + \mu_2 \cdot \mathcal{L}_{confi} + \mathcal{L}_{adv} \tag{11}$$

Training the Discriminator. The discriminator is used to distinguish predicted flow-image pair from real flow-image pair. Considering both matching directions, the discriminator loss for the real data pair $\mathcal{L}_{real}$ and predicted data pair $\mathcal{L}_{fake}$ is as follows

$$\mathcal{L}_{real} = (\mathbf{D}(\mathbf{I}_s^*, \mathbf{I}_s) - 1)^2 + (\mathbf{D}(\mathbf{I}_t^*, \mathbf{I}_t) - 1)^2 \tag{12}$$

$$\mathcal{L}_{fake} = (\mathbf{D}(\tilde{\mathbf{I}}_s, \mathbf{I}_s) - 0)^2 + (\mathbf{D}(\tilde{\mathbf{I}}_t, \mathbf{I}_t) - 0)^2 \tag{13}$$

where $\mathbf{I}_t^*$ is the same as $\mathbf{I}_t$ in self-supervision as shown in Eq. 9. Similarly, $\mathbf{I}_s^*$ equals to $\mathbf{I}_s$. Consequently, the overall loss for the discriminator is defined as:

$$\mathcal{L}_D = \mathcal{L}_{real} + \mathcal{L}_{fake} \tag{14}$$

5   Experiments

We conduct experiments on standard datasets PF-PASCAL [5] and PF-WILLOW [5] to evaluate our method for semantic matching. We first present the implementation details in Sect. 5.1. Then, we show quantitative and qualitative results of the two datasets in Sect. 5.3 and Sect. 5.2. Ablation study is given in Sect. 5.4.

5.1   Implementation Details

Images are resized into the size of 320×320. Following SFNet [19], our model is trained on the PASCAL VOC 2012 segmentation dataset [3] which includes a foreground mask for each image and PF-PASCAL [5] validation split is used for model selection. The confidence estimation network is implemented with 3×3 convolutional filters sequentially followed by BN and ReLU. We multiply correlation map by estimated confidence map as the input of the confidence-aware refinement network. The confidence-aware refinement network is implemented with architecture similar to Dense-Net and the last layer of which is of two channels. We adopt PatchGAN [12] as our discriminator. To validate generalization ability of our method, we test our trained model on PF-WILLOW dataset [5] test split without finetuning. We set $\lambda = 0.188$, $\gamma = \beta = 0.4$, $\mu_1 = 288.0$, $\mu_2 = 18.0$.

5.2   PF-WILLOW Benchmark
Dataset and Evaluation Metric. The PF-WILLOW dataset [5] contains 900 image pairs selected from 100 images with four categories. We report the PCK scores [28] ($\alpha = 0.05, 0.10, 0.15$) w.r.t bounding box size.

Fig. 4. (a) Alignment examples on PF-WILLOW from our model. (b) Qualitative comparisons on PF-PASCAL. We show the ground truth and predicted keypoints in squares and dots respectively, with their distance in target images depicting the matching error.

Table 1. Evaluation results on PF-WILLOW [5] test split.

Method	Supervision	PCK($\alpha = 0.05$)	PCK($\alpha = 0.10$)	PCK($\alpha = 0.15$)
HOG+PF-LOM [5]	–	0.284	0.568	0.682
DCTM [17]	Strong	0.381	0.610	0.721
UCN-ST [2]	Strong	0.241	0.540	0.665
CAT-FCSS [15]	Strong	0.362	0.546	0.692
SCNet [6]	Strong	0.386	0.704	0.853
WeakAlign [24]	Weak	0.382	0.712	0.858
RTN [14]	Weak	0.413	0.719	0.862
NCNet [25]	Weak	0.440	0.727	0.854
CNNGeo [23]	Self	–	0.560	–
A2Net [8]	Self	–	0.680	–
SFNet [19]	Self (+mask)	0.459	0.735	0.855
Ours	Self (+mask)	**0.464**	**0.746**	**0.863**

Experimental Results. Table 1 compares our CAMNet with other SOTA approaches. The PCK values ($\alpha = 0.05, 0.10, 0.15$) of our method are 46.4%, 74.6% and 86.3% respectively, outperforming the previously published best self-supervised method SFNet by 0.5%, 1.1%, and 0.8% respectively. Figure 4 shows that our method can effectively handle background clutters and viewpoint changes.

5.3   PF-PASCAL Benchmark

Dataset and Evaluation Metric. There are 1351 image pairs on PF-PASCAL [5] benchmark. The key point annotations are only used for evaluation. In line with previous works, we report PCK [28] ($\alpha = 0.05, 0.10, 0.15$) w.r.t image size.

Experimental Results. Table 2 shows the detailed comparison. Our method achieves the best results on $\alpha = 0.05, 0.1, 0.15$ in self-supervised setting, which outperforms the previous best self-supervised method SFNet [19] by 1.3%, 1.6%

Table 2. Evaluation results on PF-PASCAL [5] test split.

Method	Supervision	PCK($\alpha = 0.05$)	PCK($\alpha = 0.10$)	PCK($\alpha = 0.15$)
ProposalFlow [5]	–	0.314	0.625	0.795
DCTM [17]	Strong	0.342	0.696	0.802
SCNet [6]	Strong	0.362	0.722	0.820
WeakAlign [24]	Weak	0.460	0.758	0.884
RTN [14]	Weak	0.552	0.759	0.852
NC-Net [25]	Weak	0.543	0.789	0.860
SAM-Net [16]	Weak	**0.601**	0.802	0.869
CNNGeo [23]	Self	–	0.600	–
A2Net [8]	Self	–	0.680	
SFNet [19]	Self (+mask)	0.536	0.819	0.906
Ours	Self (+mask)	0.549	**0.835**	**0.910**

and 0.4% respectively, demonstrating our method's effectiveness. Figure 4 shows qualitative comparision between our model and SFNet [19].

5.4 Ablation Study

We select SFNet [19] as our baseline and report PCK($\alpha = 0.1$) on PF-PASCAL [5] and PF-WILLOW [5] test split to analyse the effectiveness of our proposed individual modules. As shown in Table 3, PCK results increase steadily when sequentially adding our proposed modules, leading to 1.1% and 1.6% improvement on the PF-WILLOW and PF-Pascal respectively in the end. Although adding GANs does not show significant improvements in numbers, it indeed mitigates distortion problems and provides better visual quality, and does not add any additional computational complexity during inference.

Table 3. Ablation study on PF-PASCAL [5] and PF-WILLOW [5].

Method	PCK (PF-WILLOW)	PCK (PF-Pascal)
SFNet [19]	0.735	0.819
Baseline + Refinement	0.740	0.825
Baseline + Confidence-aware Refinement	0.745	0.833
Baseline + Confidence-aware Refinement + GAN	**0.746**	**0.835**

6   Conclusion

In this paper, we proposed an effective deep network CAMNet for semantic matching. First, we developed a confidence-aware refinement procedure by directly utilizing confidence in model design to guide refined correspondence prediction. In addition, we introduce adversarial training to mitigate distortion in semantic alignment. Finally, we design a novel end-to-end framework with self-supervision to enable effective confidence and adversarial learning. Experimental results on two standard benchmarks confirmed the effectiveness of our method.

References

1. Bao, J., Chen, D., Wen, F., Li, H., Hua, G.: CVAE-GAN: fine-grained image generation through asymmetric training. In: ICCV (2017)
2. Choy, C.B., Gwak, J., Savarese, S., Chandraker, M.: Universal correspondence network. In: NeurIPS (2016)
3. Everingham, M., Van Gool, L., Williams, C.K., Winn, J., Zisserman, A.: The pascal visual object classes (VOC) challenge. Int. J. Comput. Vis. **88**(2), 303–338 (2010)
4. Gidaris, S., Komodakis, N.: Detect, replace, refine: deep structured prediction for pixel wise labeling. In: CVPR (2017)
5. Ham, B., Cho, M., Schmid, C., Ponce, J.: Proposal flow: semantic correspondences from object proposals. IEEE Trans. Pattern Anal. Mach. Intell. **40**(7), 1711–1725 (2017)
6. Han, K., et al.: Scnet: learning semantic correspondence. In: CVPR (2017)
7. He, K., Zhang, X., Ren, S., Sun, J.: Deep residual learning for image recognition. In: CVPR (2016)
8. Seo, P.H., Lee, J., Jung, D., Han, B., Cho, M.: Attentive semantic alignment with offset-aware correlation kernels. In: Ferrari, V., Hebert, M., Sminchisescu, C., Weiss, Y. (eds.) ECCV 2018. LNCS, vol. 11208, pp. 367–383. Springer, Cham (2018). https://doi.org/10.1007/978-3-030-01225-0_22
9. Horn, B.K., Schunck, B.G.: Determining optical flow. Artif. Intell. **17**(1–3), 185–203 (1981)
10. Huang, S., Wang, Q., Zhang, S., Yan, S., He, X.: Dynamic context correspondence network for semantic alignment. In: ICCV (2019)
11. Iizuka, S., Simo-Serra, E., Ishikawa, H.: Globally and locally consistent image completion. ACM Trans. Graph. **36**(4), 107 (2017)
12. Isola, P., Zhu, J.Y., Zhou, T., Efros, A.A.: Image-to-image translation with conditional adversarial networks. In: CVPR (2017)
13. Jeon, S., Min, D., Kim, S., Sohn, K.: Joint learning of semantic alignment and object landmark detection. In: ICCV (2019)
14. Kim, S., Lin, S., Jeon, S.R., Min, D., Sohn, K.: Recurrent transformer networks for semantic correspondence. In: NeurIPS (2018)
15. Kim, S., Min, D., Ham, B., Lin, S., Sohn, K.: Fcss: fully convolutional self-similarity for dense semantic correspondence. IEEE Trans. Pattern Anal. Mach. Intell. **41**(3), 581–595 (2019)
16. Kim, S., Min, D., Jeong, S., Kim, S., Jeon, S., Sohn, K.: Semantic attribute matching networks. In: CVPR (2019)
17. Kim, S., Min, D., Lin, S., Sohn, K.: Dctm: discrete-continuous transformation matching for semantic flow. In: CVPR (2017)
18. Kim, S., Min, D., Kim, S., Sohn, K.: Unified confidence estimation networks for robust stereo matching. IEEE Trans. Image Process. **28**(3), 1299–1313 (2018)
19. Lee, J., Kim, D., Ponce, J., Ham, B.: Sfnet: learning object-aware semantic correspondence. In: CVPR (2019)

20. Liu, C., Yuen, J., Torralba, A.: Sift flow: dense correspondence across scenes and its applications. IEEE Trans. Pattern Anal. Mach. Intell. **33**(5), 978–994 (2010)
21. Novotny, D., Larlus, D., Vedaldi, A.: Anchornet: a weakly supervised network to learn geometry-sensitive features for semantic matching. In: CVPR (2017)
22. Perarnau, G., Van De Weijer, J., Raducanu, B., Álvarez, J.M.: Invertible conditional GANs for image editing (2016). arXiv preprint arXiv:1611.06355
23. Rocco, I., Arandjelović, R., Sivic, J.: Convolutional neural network architecture for geometric matching. In: CVPR (2017)
24. Rocco, I., Arandjelović, R., Sivic, J.: End-to-end weakly-supervised semantic alignment. In: CVPR (2018)
25. Rocco, I., Cimpoi, M., Arandjelović, R., Torii, A., Pajdla, T., Sivic, J.: Neighbourhood consensus networks. In: NeurIPS (2018)
26. Scharstein, D., Szeliski, R.: A taxonomy and evaluation of dense two-frame stereo correspondence algorithms. Int. J. Comput. Vis. **47**(1 3), 7 42 (2002)
27. Taniai, T., Sinha, S.N., Sato, Y.: Joint recovery of dense correspondence and cosegmentation in two images. In: CVPR (2016)
28. Yang, Y., Ramanan, D.: Articulated human detection with flexible mixtures of parts. IEEE Trans. Pattern Anal. Mach. Intell. **35**(12), 2878–2890 (2012)

Multi-scale Dense Object Detection in Remote Sensing Imagery Based on Keypoints

Qingxiang Guo[1], Yingjian Liu[1,2(✉)], Haoyu Yin[1], Yue Li[1], and Chaohui Li[1]

[1] Department of Computer Science and Technology, Ocean University of China, Qingdao 266100, China
[2] Department of Computer Science, University of North Carolina at Charlotte, Charlotte, NC 28223, USA
liuyj@ouc.edu.cn

Abstract. Object detection is a popular topic of computer vision and has attracted increasing attention in the field of remote sensing. However, object detection in remote sensing imagery is a very challenging task. Many existing detection methods depend on densely tiled anchor boxes, which require redundant computation resources and careful fine-tuning. In recent years, some anchor-free detectors have been proposed and show excellent performance. In this paper, we propose a novel anchor-free method for remote sensing object detection, named as multi-scale dense object detector based on keypoints (*MDKD*). Our method detects object in remote sensing imagery as a single keypoint, which needs neither predefined anchor boxes nor fallible pairing operation. For remote sensing scenario, we dedicate to improve the recall of detection results by allowing more keypoint samplings. In addition, we take advantage of balanced L1 loss to predict bounding boxes more accurately. Multi-scale test improves the performance further more. Our proposed *MDKD* method is verified on two widely used benchmarks, i.e. NWPU VHR-10 and DOTA datasets. The experimental results indicate that our keypoint based detector is effective on remote sensing imagery and outperforms several popular anchor-based methods.

Keywords: Remote sensing · Object detection · CNN · Anchor free · Keypoint detection

1 Introduction

Object detection in remote sensing imagery has been a challenging task for decades. It aims at locating objects of interests in imagery captured by satellites or other air crafts, and recognizing their categories. Due to the rapid development of remote sensing technologies, researchers can access very-high-resolution (VHR) aerial images more easily than ever before. Object detection in remote sensing images has become more and more attractive because of high demands

© Springer Nature Switzerland AG 2020
Y. Peng et al. (Eds.): PRCV 2020, LNCS 12305, pp. 104–116, 2020.
https://doi.org/10.1007/978-3-030-60633-6_9

in various applications, e.g. urban planning, target monitoring, etc. Compared with natural sensing imagery, object detection in remote sensing imagery faces three main challenges: complex background pattern, densely packed objects, and various appearance of objects, as shown in Fig. 1. Traditional detection methods in remote sensing have made a series of advances. However, their performance can hardly satisfy the increasing demands.

Fig. 1. Main challenges for object detection in remote sensing imagery, i.e. complex background (left), dense layout (middle) and variations of sizes (right).

In recent years, a lot of CNN-based methods have been proposed and played an important role in object detection [1–5]. However, most existing high-accuracy methods for remote sensing detection are based on powerful backbone and a mass of predefined anchor boxes. The former induces complex computation and huge memory footprint. And the latter requires careful fine-tuning of hyper-parameters related to anchor boxes, e.g. scales and aspect ratios. Improper hyper-parameters can result in bad performance and redundant computation. Moreover, misalignment between anchor box and feature map can generate improper visual information, especially for tiny objects.

After CornerNet [6], two CenterNet [7,8] are popular keypoint based detectors. Both of them show excellent performance on natural sensing detection task. We propose a keypoint based detector for remote sensing imagery, named as multi-scale dense object detector based on keypoints (*MDKD*). CenterNet proposed in [7] is chosen as *MDKD*'s base model because of its straightforward detection head and powerful expansibility. For 2D object detection task, CenterNet predicts one center point for each object and regresses width and height of bounding box directly, as shown in Fig. 2. *MDKD* makes a series of modification to be more suitable for object detection in remote sensing imagery. For efficiency consideration, *MDKD* adopts DLA-34 [9] instead of widely-used Hourglass-104 [10] as its backbone network. The contributions of this paper can be summarized as follows:

(1) *MDKD* detects objects in remote sensing imagery by a standard framework of keypoint estimation, without any predefined anchor box or keypoint pairing operation [6,8].

(2) We pick more samples on center point map to improve detection performance especially for densely packed objects. And we utilize balanced L1 loss [11] in our *MDKD* for better bounding box prediction.
(3) *MDKD* is evaluated on two widely used remote sensing datasets, i.e. NWPU VHR-10 [12] and DOTA [13]. Experimental results indicate that it outperforms the baseline models, including some popular anchor-based methods. In addition, multi-scale testing improves accuracy more notably than it does in natural sensing detection task.

Fig. 2. Detect objects as points.

2   Methods

Because of the dense layout and various shape of objects in remote sensing imagery, anchor-based detectors need plenty of anchor boxes. It will result in complex computation and elaborate fine-tuning of hyper-parameters related to anchor boxes. We propose a novel anchor-free method, i.e. *MDKD*, for object detection in remote sensing imagery. *MDKD* transfers CenterNet [7], a keypoint based detector, to remote sensing detection task via some straightforward but effective modifications. Figure 3 illustrates the detection framework of *MDKD*.

2.1   Detect Densely Packed Objects as Points

MDKD solves the problem of object detection via a standard framework of keypoint estimation. It estimates a unique center point for each object and regress other properties(e.g. width and height) from the center point directly.

The model requires a $W \times H \times 3$ image as input. Following backbone and up-sampling layers, there are three similar detection heads, as shown in Fig. 3. Each detection head will finally output a map with spatial resolution of $\lfloor \frac{W}{4} \rfloor \times \lfloor \frac{H}{4} \rfloor$.

To be specific, the first heatmap $\hat{Y} \in [0,1]^{\frac{W}{4} \times \frac{H}{4} \times C}$ indicates how likely there is a center point of an instance belonging to the corresponding category c. And it is trained by a modified version of focal loss [5]:

$$L_{hm} = -\frac{1}{N} \sum_{xyc} \begin{cases} (1 - \hat{Y}_{xyc})^\alpha \log(\hat{Y}_{xyc}), & \text{if } Y_{xyc} = 1 \\ (1 - Y_{xyc})^\beta (\hat{Y}_{xyc})^\alpha \log(1 - \hat{Y}_{xyc}), & \text{otherwise} \end{cases} \tag{1}$$

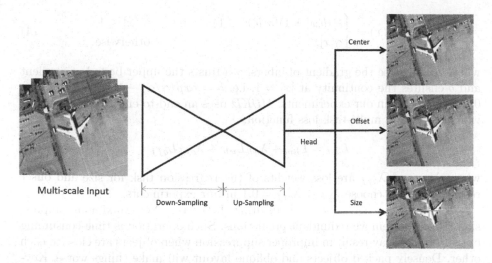

Fig. 3. Detection framework of *MDKD*.

where α and β are hyper-parameters for down-weighting losses of easy negative samples. We choose $\alpha = 2$ and $\beta = 4$ in our experiments. N is the total number of keypoints.

Because of the down-sampling and rounding operation, local offsets are generated when mapping center point from original image to heatmap. In order to compensate the offset and obtain more accurate location, *MDKD* predicts another map $\hat{O} \in \mathbb{R}^{\frac{W}{4} \times \frac{H}{4} \times 2}$ to represent the offsets along two axis. It is trained as a standard regression problem:

$$L_{off} = \frac{1}{N} \sum_P L1(\hat{O}_P - O_P) \qquad (2)$$

where $L1$ denotes L1 loss, only computing for all **positive** samples. It is worthy to consider offset prediction as a category-agnostic task to save computation.

The third map $\hat{S} \in \mathbb{R}^{\frac{W}{4} \times \frac{H}{4} \times 2}$ predicts the size (i.e. width and height) of bounding box, whose center point falls in the corresponding location. It is also considered as a category-agnostic regression task:

$$L_{wh} = \frac{1}{N} \sum_P BL1(\hat{S}_P - S_P) \qquad (3)$$

where $BL1$ denotes balanced L1 loss. During training, we notice that the loss contribution of size prediction task occupies about 12% of the loss of keypoint estimation task in most cases. However, due to the huge variations of objects' size in remote sensing imagery, accurate prediction for bounding box is very crucial and challenging. To balance the training processes of keypoint detection task and size regression task, we replace the original regression loss with balanced L1 loss [11]:

$$L_{bl1}(x) = \begin{cases} \frac{\alpha}{b}(b|x|+1)ln(b|x|+1) - \alpha|x|, & \text{if } |x| \leq 1 \\ \gamma|x|, & \text{otherwise} \end{cases} \tag{4}$$

where α will raise the gradient of inliers, γ adjusts the upper bound of gradient and b ensures the continuity at $|x| = 1$, i.e. $b = exp\,(\gamma/\alpha) - 1$. We choose $\alpha = 0.5$ and $\gamma = 1.5$ in our experiments. MDKD uses an end-to-end training scheme, which requires a multi-task loss function:

$$L_{det} = L_{hm} + \lambda_{wh}L_{wh} + \lambda_{off}L_{off} \tag{5}$$

where λ_{wh} and λ_{off} are loss weights of the regression task for size and offset respectively. We choose $\lambda_{wh} = \lambda_{off} = 0.1$ in our experiments.

Many previous detection methods utilize IoU-based Non-Maximum Suppression (NMS) to remove redundant predictions. Such operation is time-consuming operation and may result in improper suppression when objects are close to each other. Densely packed objects and oblique layout will make things worse. Keypoint based detector only picks at most K (e.g. 100) peaks with highest values across all categories on the heatmap as detection results. This can be implemented via max pooling easily. However, there may be too many densely packed objects in remote sensing imagery. According to the statistics [13], there can be about 2000 instances within a single image in DOTA dataset. Although we can split original images into smaller patches, the default value K=100 makes it difficult to detect all objects properly. Therefore, we assign parameter K a larger value in order to obtain more detection results within a single input image. This straightforward method brings remarkable improvement especially for those categories with tiny size and dense distribution.

2.2 Network Architecture for Multi-scale Testing

Considering efficiency, MDKD adopts relatively lightweight Deep Layer Aggregation (DLA) network rather than Hourglass-104 network as its backbone network. For example, our MDKD equipped with DLA-34 only runs about 198 ms per image and needs 1323 MB memory. Under the same circumstances, the model equipped with Hourglass-104 runs about 344 ms per image for testing and needs 2267 MB memory. By adding more up-sampling layers and skip-connections, the original DLA network is improved further more. Figure 4 illustrates the architecture of modified DLA network. Regular convolutions in down-sampling layers are also replaced by deformable convolutions [14]. As a result, receptive field can fit the shape of objects better and extract more powerful features, as shown in Fig. 5.

This improved architecture has the capacity of detecting keypoints properly even if they are massive and densely packed. However, many keypoints are discarded in post-processing of original CenterNet [7]. In this paper, more objects can be detected via sampling more keypoints as previously mentioned in Sect. 2.1. However, the huge variants of objects' size in remote sensing imagery is hard to handle because only the last feature map is used to conduct detection.

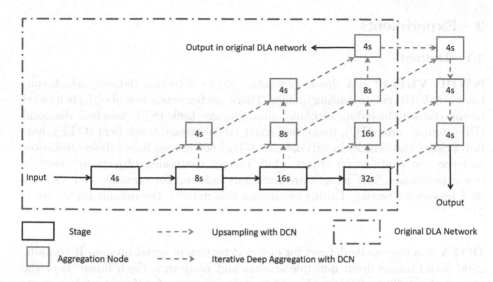

Fig. 4. Architecture of modified DLA network. The number within box denotes downsampling stride.

A simple but effective multi-scale testing is used to solve this problem. Given an input image, it is resized to different sizes first. Each image will be processed by the network independently. Then, some result sets for the same image with different scales are obtained. Intuitively, smaller objects can be detected more easily in high-resolution scales and vice versa. Finally, those results are merged via soft-NMS [15] to get the final detection results. It is notable that remote sensing detection benefits more from multi-scale testing compared with natural sensing detection, e.g. 5.7% mAP vs. 2.4% mAP [7] on DOTA validation set and COCO [16] dataset, respectively.

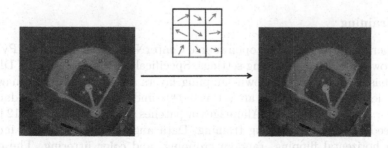

Fig. 5. Deformable convolution.

3 Experiments

3.1 Datasets

NWPU VHR-10 is a classic geospatial object detection dataset, which contains 800 VHR remote sensing images. There are ten categories of objects labeled in the dataset, including airplane, ship, storage tank (ST), baseball diamond (BD), tennis court (TC), basketball court (BC), ground track field (GTF), harbor, bridge, and vehicle. In this dataset, 650 of 800 images have labeled instances and the rest contains no object. Only images containing objects are used in our experiments. 70% images are randomly selected for training and the rest 30% images for testing. During testing on this dataset, the default input size is 512×512.

DOTA is a large-scale dataset for object detection in aerial images. It contains 2806 aerial images from multiple sensors and platforms. Each image is of the size of about 4000×4000 pixels and contains objects of different scales, orientations and shapes. In DOTA datasets, there are 188,282 instances belonging to 15 categories, i.e., plane, baseball diamond (BD), bridge, ground field track (GTF), small vehicle (SV), large vehicle (LV), ship, tennis court (TC), basketball court (BC), storage tank (ST), soccerball field (SBF), roundabout (RA), harbor, swimming pool (SP) and helicopter (HC). Those images are randomly divided into training set, validation set and testing set in a ratio of 3:1:2. All images are split into a series of 1024×1024 patches with an overlap of 200 pixels. 15,749 training samples are obtained in total. Our *MDKD* is evaluated on DOTA with the official development kit. We compute mAP of validation set locally in ablation studies. While we compute mAP of testing set by submitting results to online evaluation platform[1] for comparison with other methods. For this more challenging dataset, we keep patch's original resolution, i.e. 1024×1024, for testing as default.

3.2 Training

Our experiments are based on open source CenterNet[2] implemented by PyTorch and follow some default training settings. Specifically, we used modified DLA-34 as our base network whose down-sampling layers are pretrained on ImageNet. Weight decay and momentum are set as 0.0001 and 0.9, respectively. Adam [17] optimizer is used for training. All images or patches are resized to 512×512 before being feed into network during training. Data augmentation strategy includes random horizental flipping, random cropping, and color jittering. The experiments are conducted on a machine with two GeForce RTX 2080 GPUs and 16G memory in total. The model is trained for 140 epochs with a batch-size of 24(~92k iterations for DOTA and ~2.5k iterations for NWPU VHR-10). The

---

[1] http://captain.whu.edu.cn/dotaweb/evaluation.html
[2] https://github.com/xingyizhou/CenterNet

initial learning rate is 1.25e-4 and we drop learning rate by 0.1 at the 90-th and the 120-th epoch, respectively.

3.3 Inference

Given an input image, the model will predict three feature maps with respect to center point location, local offset and the size of bounding box. We pick at most top K peaks on the first map and collect their corresponding offsets and sizes as outputs. Then, those outputs are decoded to bounding boxes. During multi-scale testing, soft-NMS [15] is applied for merging results of different scales. We emprically set the scale list to be [0.5, 0.75, 1, 1.25, 1.5]. Mean Average Percision(mAP) metric is used to evaluate our methods following [13].

3.4 Ablation Study

Backbone Network. Deep Layer Aggregation (DLA) [9] is proposed to aggregate feature information from different levels within a neural network via iterative deep aggregation (IDA) and hierarchical deep aggregation (HDA). In [7], DLA is enhanced by introducing deformable convolutions and adding more skip connections from bottom layers. For efficiency, the model with relatively small scale, i.e. DLA-34, is used in *MDKD*. Results of using different base networks are shown in Table 1. Here we use DLA-v0 to denote the original DLA network following [7]. It is obvious that modified DLA network is able to provide better detection results on both NWPU VHR-10 and DOTA validation datasets.

Table 1. Results with different networks (%)

	DLA-v0	DLA-34
DOTA-val	61.04	64.63
NWPU VHR-10	87.8	90.0

Regression Loss. As illustrated in [7], L1 loss outperforms smooth L1 loss by 2.4% mAP on MS COCO dataset. However, this gap is relatively small in remote sensing datasets. We use different losses to train the bounding box regression and the results are shown in Table 2. Since the imbalance between keypoint prediction task and size regression task has been relieved, balanced L1 loss brings higher accuracy than its two counterparts.

Number of Peak Samplings. [3]Keypoint based detector picks at most top-K (pairs of) peaks on the heatmap as detection results and most works [6–8] set

---

[3] This part is mainly designed for DOTA dataset, since the maximum count of objects within single image is less than 100 in NWPU VHR-10 dataset.

Table 2. Results with different regression loss.(%)

	L1 loss	Smooth L1 loss	Balanced L1 loss
DOTA-val	64.09	63.94	64.63
NWPU VHR-10	87.7	87.9	90.0

$K = 100$ or smaller as default. However, it is not enough for object detection in remote sensing imagery because of the dense layout of objects. In our experiments, we increase the value of K in order to make the model detecting more objects. Note that results with very low confidence have no effect on mAP computation. In Table 3, categories whose instances are usually distributed densely within a single imagery benifit more from larger K, such as small-vehicle, large-vehicle, ship and storage-tank. As for harbor, we found that there can be more than one center points were detected for the same instance. Therefore, sampling more peaks is more likely to find the correct one. We choose $K = 200$ as default for trade-off between mAP and post-processing time.

Table 3. Results on DOTA validation set with different K (%).

K	plane	BD	bridge	GTF	SV	LV	ship	TC	BC	ST	SBF	RA	harbor	SP	HC	mAP
100	90.01	67.4	44.32	62.17	**57.22**	63.13	**61.72**	90.64	60.48	**77.42**	49.11	66.57	**67.93**	73.09	38.3	64.63
150	90.04	67.75	44.08	62.55	**62.43**	**64.73**	70.02	90.64	60.22	**80.88**	49.06	67.4	**71.26**	73.16	37.72	66.13
200	90.04	68.22	44.09	62.76	**65.44**	65.21	77.05	90.64	60.32	**81.20**	49.17	67.45	**72.48**	73.06	37.85	66.99
300	90.05	67.88	44.18	62.05	**66.44**	67.15	78.14	90.64	60.28	**81.10**	48.59	67.85	**73.22**	73.09	37.72	67.22

3.5   Comparison

Our *MDKD* is compared with other popular methods on both DOTA testing dataset and NWPU VHR-10 dataset. The results are shown in Table 4 and Table 5. It can be seen that our method outperforms several previous detection method, including some popular anchor-based methods. Note that Faster R-CNN [1] and R-FCN [2] use relatively heavy ResNet-101 [18] as their backbone while our *MDKD* uses modified DLA-34. It demonstrates that keypoint based detectors have the capability of good performance on object detection in remote sensing imagery. For run-time speed, *MDKD* can run at a speed of 296 ms per image with default input size, i.e. 512 512, on RTX 2080. Single-scale testing can run at a speed of 61 ms per image, with relatively lower mAP. Finally, some visualized detection results including bounding boxes and keypoints can be seen in Fig. 6.

Table 4. Comparison with other methods on DOTA testing set. *SDKD* denotes single-scale testing (%)

	YOLOv2 [4]	R-FCN [2]	FR-H [1]	SSD [3]	CenterNet [7]	SDKD	MDKD
backbone	GoogleNet [19]	ResNet101	ResNet101	Inceptionv2 [20]	DLA-34	DLA-34	DLA-34
plane	76.9	81.01	80.32	57.85	88.33	88.02	**88.88**
BD	33.87	58.96	77.55	32.79	69.64	72.41	**81.88**
bridge	22.73	31.64	32.86	16.14	47.24	47.14	**50.96**
GTF	34.88	58.97	**68.13**	18.67	55.29	59.86	64.57
SV	38.73	49.77	**53.66**	0.05	41.7	49.45	50.59
LV	32.02	45.04	52.49	36.93	61.24	62.02	**65.79**
ship	52.37	49.29	50.04	24.74	65.75	73.13	**74.47**
TC	61.65	68.99	90.41	81.16	90.23	90.46	**90.55**
BC	48.54	52.07	75.05	25.1	72.79	74.11	**81.43**
ST	33.91	67.42	59.59	47.47	75.69	77.74	**83.29**
SBF	29.27	41.83	**57**	11.22	37.93	38.72	54.40
RA	36.83	**51.44**	49.81	31.53	33.11	30.92	37.61
harbor	36.44	45.15	61.69	14.12	66.85	71.18	**75.75**
SP	38.26	53.3	56.46	9.09	67.61	**73.16**	72.96
HC	11.61	33.89	41.85	0	**59.14**	56.52	58.87
mAP	39.2	52.58	60.46	29.86	62.16	64.32	**68.80**

Table 5. Comparison with other methods on NWPU VHR-10. *SDKD* denotes single-scale testing (%).

	Densely [21]	COPD [22]	DAPNet [23]	RICNN [12]	D-R-FCN [24]	Faster [1]	SDKD	MDKD
plane	99.1	62.3	99.9	88.4	87.3	90.9	**100.0**	99.9
ship	**91.8**	68.9	80.8	77.3	81.4	80.2	68.8	70.1
ST	84.2	63.7	78.9	85.3	63.6	77.9	98.6	**99.1**
BD	94.6	83.3	90.9	88.1	90.4	90.9	**99.3**	99.1
TC	92.6	32.1	**98.7**	40.8	81.6	90.3	92.3	94.6
BC	85.0	36.3	89.6	58.5	74.1	81.8	93.0	**95.9**
GTF	98.4	85.3	90.3	86.7	90.3	88.6	99.5	**100.0**
harbor	**93.8**	55.3	85.6	68.6	75.3	80.3	89.2	92.6
bridge	**91.4**	14.8	79.4	61.5	71.4	70.4	88.9	88.7
vehicle	83.6	44.0	80.3	71.1	75.5	78.8	86.5	**93.0**
mAP	91.5	54.6	87.5	72.6	79.1	83.0	91.6	**93.3**

Fig. 6. Visualization of object detection results and center point estimation results.

4   Conclusion

In this paper, we propose a novel method, namely *MDKD*, for remote sensing detection based on keypoint estimation. This method is anchor-free and represents object as single center point. It allows more keypoint samplings within single image to detect densely packed objects. Balanced L1 loss is used to regress variant sizes of bounding boxes better. And multi-scale testing can improve the detection performance notably. Our *MDKD* equipped with relatively lightweight DLA network outperforms several popular detectors on both NWPU VHR-10 and challenging DOTA datasets.

Acknowledgement. This work was supported partially by the National Natural Science Foundation of China under Grant No. 61572448, No. 61673357, and by the Key R&D Program of Shandong Province, China under Grant No. 2018GSF120015. This work was partially done when Y. Liu visited the University of North Carolina at Charlotte with a scholarship from the China Scholarship Council.

References

1. Ren, S., He, K., Girshick, R., Sun, J.: Faster r-cnn: towards real-time object detection with region proposal networks. In: 2017 IEEE Transactions on Pattern Analysis and Machine Intelligence (TPAMI), vol. 39, pp. 1137–1149 (2017)
2. Dai, J., Li, Y., He, K., Sun, J.: R-fcn: object detection via region-based fully convolutional networks. In: NeurIPS 2016, Barcelona, pp. 379–387 (2016)
3. Liu, W., et al.: SSD: single shot multibox detector. In: Leibe, B., Matas, J., Sebe, N., Welling, M. (eds.) ECCV 2016. LNCS, vol. 9905, pp. 21–37. Springer, Cham (2016). https://doi.org/10.1007/978-3-319-46448-0_2

4. Redmon, J., Farhadi, A.: YOLO9000: better, faster, stronger. In: 2017 IEEE Conference on Computer Vision and Pattern Recognition (CVPR), Puerto Rico, pp. 6517–6525 (2017)
5. Lin, T., Goyal, P., Girshick, R., He, K., Dollár, P.: Focal loss for dense object detection. In: 2017 IEEE International Conference on Computer Vision (ICCV), Venice, pp. 2999–3007 (2017)
6. Law, H., Deng, J.: CornerNet: detecting objects as paired keypoints. In: Ferrari, V., Hebert, M., Sminchisescu, C., Weiss, Y. (eds.) Computer Vision – ECCV 2018. LNCS, vol. 11218, pp. 765–781. Springer, Cham (2018). https://doi.org/10.1007/978-3-030-01264-9_45
7. Zhou, X., Wang, D., Krähenbühl, P.: Objects as points (2019). CoRR abs/1904.07850
8. Duan, K., Bai, S., Xie, L., Qi, H., Huang, Q., Tian, Q.: Centernet: keypoint triplets for object detection. In: 2019 IEEE International Conference on Computer Vision (ICCV), Seoul, pp. 6568–6577 (2019)
9. Yu, F., Wang, D., Shelhamer, E., Darrell, T.: Deep layer aggregation. In: 2018 IEEE Conference on Computer Vision and Pattern Recognition (CVPR), Salt Lake City, pp. 2403–2412 (2018)
10. Newell, A., Yang, K., Deng, J.: Stacked hourglass networks for human pose estimation. In: Leibe, B., Matas, J., Sebe, N., Welling, M. (eds.) ECCV 2016. LNCS, vol. 9912, pp. 483–499. Springer, Cham (2016). https://doi.org/10.1007/978-3-319-46484-8_29
11. Pang, J., Chen, K., Shi, J., Feng, H., Ouyang, W., Lin, D.: Libra r-cnn: towards balanced learning for object detection. In: 2019 IEEE Conference on Computer Vision and Pattern Recognition(CVPR), Long Beach, pp. 821–830 (2019)
12. Cheng, G., Zhou, P., Han, J.: Learning rotation-invariant convolutional neural networks for object detection in VHR optical remote sensing images. IEEE Trans. Geosci. Rem. Sens. **54**, 7405–7415 (2016)
13. Xia, G., et al.: DOTA: a large-scale dataset for object detection in aerial images. In: 2018 IEEE Conference on Computer Vision and Pattern Recognition (CVPR), Salt Lake City, pp. 3974–3983 (2018)
14. Dai, J., et al.: Deformable convolutional networks. In: 2017 IEEE International Conference on Computer Vision (ICCV), Venice, pp. 764–773 (2017)
15. Bodla, N., Singh, B., Chellappa, R., Davis, L.S.: Soft-NMS - improving object detection with one line of code. In: 2017 IEEE International Conference on Computer Vision (ICCV), Venice, pp. 5562–5570 (2017)
16. Lin, T.-Y., et al.: Microsoft COCO: common objects in context. In: Fleet, D., Pajdla, T., Schiele, B., Tuytelaars, T. (eds.) ECCV 2014. LNCS, vol. 8693, pp. 740–755. Springer, Cham (2014). https://doi.org/10.1007/978-3-319-10602-1_48
17. Kingma, D., Ba, J.: Adam: a method for stochastic optimization. In: 3rd International Conference on Learning Representations (ICLR), San Diego (2015)
18. He, K., Zhang, X., Ren, S., Sun, J.: Deep residual learning for image recognition. In: 2016 IEEE Conference on Computer Vision and Pattern Recognition (CVPR), Las Vegas, pp. 770–778 (2016)
19. Szegedy, C., et al.: Going deeper with convolutions. In: 2015 IEEE Conference on Computer Vision and Pattern Recognition (CVPR), Boston, pp. 1–9 (2015)
20. Ioffe, S., Szegedy, C.: Batch normalization: accelerating deep network training by reducing internal. In: 2015 International Conference on Machine Learning (ICML), Lille, pp. 448–456 (2015)
21. Tayara, H., Chong, K.: Object detection in very high-resolution aerial images using one-stage densely connected feature pyramid network. Sensors **18**, 3341 (2018)

22. Cheng, G., Han, J., Zhou, P., Li, K.: Multi-class geospatial object detection and geographic image classification based on collection of part detectors. ISPRS J. Photogramm. Remote Sens. **98**, 119–132 (2014)

23. Cheng, L., Liu, X., Li, L., Jiao, L., Tang, X.: Deep adaptive proposal network for object detection in optical remote sensing images (2018). CoRR abs/1807.07327

24. Xu, Z., Xu, X., Wang, L., Yang, R., Pu, F.: Deformable convNet with aspect ratio constrained NMS for object detection in remote sensing imagery. Remote Sens. **9**, 1312 (2017)

Ship Detection in SAR Images Based on Region Growing and Multi-scale Saliency

Qi Hu[1,2]([⊠]), Shaohai Hu[1,2], and Shuaiqi Liu[3]

[1] Institute of Information Science, Beijing Jiaotong University, Beijing 100044, China
qihu_hbu@163.com, shhu@bjtu.edu.cn
[2] Beijing Key Laboratory of Advanced Information Science and Network Technology, Beijing 100044, China
[3] College of Electronic and Information Engineering, Hebei University, Baoding 071002, China

Abstract. Synthetic aperture radar (SAR) is one of the most widely used remote sensing monitoring methods for large-scale marine activities. Due to the influence of speckle noise, sea clutter and complex environment, ship detection of SAR images is still a challenging task. Based on the multi-layer selective cognition characteristics of the human visual system, we propose a ship target detection algorithm based on region growing and multi-scale saliency. First, the layered rough-fine land-sea segmentation is used to remove the effect of land scattering. Second, the non-subsampled Laplacian pyramid (NSLP) filter is applied to decompose the image at different scales. Then, the saliency region of the transformed coefficients is extracted by spectral residual (SR). And the constant false alarm rate (CFAR) algorithm is used to further filter the false alarm and extract target more accurately. Finally, saliency sub-images of different scales are fused to get the final detection results. Experimental results show that the algorithm not only effectively suppresses the influence of land and sea clutter, but also can improve the detection rate.

Keywords: SAR image · Target detection · Region growing · Saliency

1 Introduction

In the past few decades, various platforms have collected a large amount of SAR data and completed many tasks such as fishery monitoring, warship reconnaissance, marine waste monitoring, oil spill detection, and immigration control [1, 2]. As an important marine application, ship detection has been extensively studied. Many classic ship detection

---

The first author is a student.
This research was funded by the Fundamental Research Funds for the Central Universities No. 2020YJS033; Natural Science Foundation of China under grant 61401308 and 61572063; Natural Science Foundation of Hebei Province under grant F2016201142, F2020201025 and F2018210148; Science research project of Hebei Province under grant BJ2020004; Opening Foundation of Machine vision Engineering Research Center of Hebei Province under grant 2018HBMV02.

© Springer Nature Switzerland AG 2020
Y. Peng et al. (Eds.): PRCV 2020, LNCS 12305, pp. 117–128, 2020.
https://doi.org/10.1007/978-3-030-60633-6_10

systems have been designed under the framework of automatic target recognition (ATR) in SAR images. As the first step of SAR-ATR, target detection has a crucial influence on the recognition accuracy and speed of SAR-ATR system.

In traditional ship detection methods, two steps are usually adopted. First, land-sea segmentation based on texture and shape features is implemented to eliminate land interference and improve target search efficiency; then, feature extraction and machine learning methods are used to detect ship targets. For remote sensing images, the purpose of land-sea segmentation is to accurately separate sea and land, and the segmentation results will provide useful information for coastline extraction and ship detection. The boundary accuracy of the land-sea segmentation results directly affects the ship detection [3, 4], and the segmentation error of the land area will increase the complexity of the ship detection process.

As the most commonly used detection method at present, the CFAR algorithm can provide a detection threshold with a constant false alarm probability in a uniform background, which can better avoid the influence of background noise, clutter and interference changes. A large number of statistical models are used to model the coherent speckle of different scenes in SAR images, mainly including Gaussian distribution, Gamma, lognormal, K and G distributions [5]. However, the complex parameter estimation algorithm based on statistical models of above distribution greatly increases the computational burden of the CFAR detector. In order to improve the detection performance of the CFAR algorithm under non-uniform background conditions, Novak et al. proposed many improved versions of CFAR, but these algorithms lack universality. In practical applications, the main challenge lies in the detection and identification of ships under complex backgrounds. CFAR cannot achieve satisfactory detection results in complex environments and large-scale SAR images.

With the development of deep learning methods, there are many target detection algorithms using deep neural network model, which improve the shortcomings of traditional learning methods to a certain extent. The commonly used network models include automatic encoder, Boltzmann machine and convolution neural network. In particular, convolutional neural networks have emerged, such as Alex Network (AlexNet), VGG, Google Network (GoogleNet) and Residual Network (Res-net), as well as many target detection models based on them, including SSD, Yolov1 and Faster-RCNN, etc. These methods have gradually become the mainstream in the field of SAR ship detection.

In recent years, people have noticed that the ability of human visual attention systems to detect targets from complex scenes in optical images is very fast and reliable. Many excellent computational visual attention models have been proposed to simulate the structure of human visual systems. Itti et al. [6] proposed a biologically-inspired visual attention model, which measures the saliency of an image through center-to-surround contrast. From the perspective of information theory, Bruce et al. [7] proposed an attention model based on information maximization, which uses Shannon's self-information measure to calculate the saliency of image regions. This model requires a large number of image block samples to measure the SI of the image area. It is not easy to obtain a large number of SAR image samples in the detection stage.

The above definitions of saliency are based on local contrast. At the same time, some other algorithms focus on saliency construction based on global contrast. From the perspective of the spectral domain, Hou et al. [8] proposed a spectral residual method based on Fourier transform. They analyzed the log spectrum of the input image and extracted the SR of the spectral domain image to construct a saliency map. Yu et al. [9] studied the relationship between the sign of discrete cosine transform coefficients and visual saliency, and proposed a pulse cosine transform (PCT) visual attention model. Recently, Cheng et al. [10] proposed a saliency detection algorithm based on global area contrast, which can generate a high-quality saliency map. However, when calculating the saliency value of an area, it is necessary to compare the area with all other areas in the entire image, which is quite time-consuming.

Although the existing ship detection methods based on saliency have better detection results than traditional methods, there are still some problems that need to be further solved. First of all, most detectors are good for detecting images with a single background and prominent objects. However, it is not robust to speckle noise or fluctuations in cluttered scenes. This may increase false alarms or even detection losses. Second, the saliency map can represent the area of interest, but the estimation of the target position is too rough, and the accurate contour information for target positioning cannot be provided. It is not enough to rely on the saliency map to infer the accurate ship target. Therefore, it is expected that both the fuzzy selection mechanism and the accurate contour information can be taken into account to obtain more reliable detection results. Based on the land-sea mask and region growing, we perform a layered land-sea segmentation method to eliminate land interference and improve target search efficiency. Moreover, the non-subsampled Laplacian filter is used to decompose the image at different scales, and the transformed image is processed by saliency detection and CFAR detection, so that it can also obtain accurate detection results under complex background.

2 Related Work

2.1 Land-Sea Segmentation

Land-sea segmentation algorithms can generally be divided into two types: edge-based segmentation and region-based segmentation. The edge-based image segmentation algorithm uses edges to detect the contours of an area. Edge-based segmentation uses discontinuities in the image, while region-based segmentation algorithms take advantage of the similarity in the image. Region-based segmentation method reduces computational requirements by processing regions instead of pixels, helps the optimization process converge to global solutions more efficiently, and alleviates the problem of noisy images by using region statistics instead of individual pixel values. There are many methods based on regions, such as region growing or merging.

The region growing method is a simple and effective segmentation method. This method can segment regions with the same characteristics in the image, and retain clear boundary contour information and segmentation results. Region growing is the process of selecting pixels with similar characteristics and combining them into regions. Lu et al. [11] used the region growing method to perform functional magnetic resonance activation detection. Region-based segmentation has a good effect when segmenting water and land. The result of edge-based segmentation is not very sure, because every object in the scene has edges, and it is a difficult task to distinguish which edge belongs to which object. In the case of region-based segmentation, the image is segmented using pixel similarity instead of difference. Therefore, it is more likely to segment objects clearly.

2.2 Spectral Residual

Inspired by the human visual attention mechanism, some researchers began to develop more intelligent biological models for ship target detection. Since ships are prominent objects in the vast near-shore areas, the human eye can effortlessly focus on these areas in order to quickly understand the scene. Therefore, the development of biologically inspired target area automatic selection calculation models has proved to be a promising application method. Yu et al. [12] used the aforementioned PCT visual attention model to detect ships in SAR images. To ensure the detection rate of weak targets in SAR images, Liu et al. [13] improved Itti's model and proposed a visual attention detection algorithm for SAR images based on singular value decomposition (SVD). Based on the prior knowledge of ships in water instead of land, Hou et al. [14] proposed a SAR image ship detection method based on visual attention model to improve the ship detection performance. Bi et al. [15] used the PCT model to select ships at the stage of the candidate area they focused on. In addition, Wang et al. [16] proposed a saliency detector based on pattern recursion for target detection in a non-uniform SAR background. These saliency-based ship detectors use a selective search strategy, which is more efficient than sliding window-based methods. Furthermore, instead of simply relying on local contrast, they incorporated meaningful high-level cues into their models to make them more effective.

Usually, a saliency region detection algorithm with good real-time performance adopts the global search method, which also reduces the search cost of local traversal. Hou et al. [8] proposed a global visual saliency region extraction method based on image frequency domain calculation processing, namely spectral residual (SR). It is implemented by fast Fourier transform and has the advantages of simple and fast calculation. In this paper, we present an effective SAR image ship detection method, which makes full use of the advantages of the spectral residual method to obtain better detection results.

3   The Proposed SAR Image Ship Detection Method

In order to eliminate land interference in SAR images of complex background and large scene and improve target search efficiency, this paper performs land-sea segmentation based on land-sea mask and region growing. Fully considering the fuzzy selection mechanism and accurate contour information, saliency region extraction and CFAR detection are performed on multi-scale images to obtain accurate detection results. The proposed SAR image ship target detection flow chart is shown in Fig. 1.

3.1   Layered Rough-Fine Land-Sea Segmentation

In order to realize the land-sea segmentation of SAR images, we first look for the largest connected region in SAR image. Furthermore, the smooth and robust land-sea

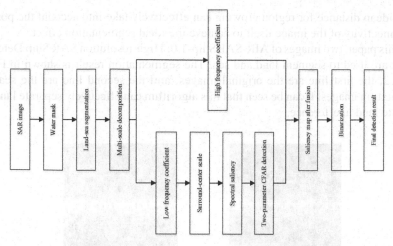

Fig. 1. SAR image ship target detection flow chart

segmentation results are obtained by morphological operation, and the land-sea mask is obtained to roughly estimate the land-sea segmentation. This land-sea mask is a binary image, where 1 represents land pixels and 0 represents ocean pixels.

Then we use the region growing algorithm to segment the image accurately. Wu et al. [17] selected seed points for region growing based on two-dimensional OTSU. This method introduces the gray information and spatial information of the image. Suppose the gray image is I. The pixel value $I(i,j)$ and the average gray value $G(i,j)$ of the neighborhood form a binary group. This binary group contains the gray information of pixel s and its neighborhood information t. In areas where gray values are continuous, the difference between s and t is not large. In the marginal area, the gap between the two is larger. The two-dimensional OTSU obtains an appropriate set of binary thresholds (S, T) by calculation. If t_r represents the variance between classes, then (S, T) satisfies:

$$t_r(S, T) = \text{MAX}\{t_r(s, t)\} \tag{1}$$

Suppose the image is divided into two categories:

$$I(i,j) = \begin{cases} 1, & s \leq S \ and \ t \leq T \\ 0, & \text{others} \end{cases} \tag{2}$$

We use the segmented foreground (point with a pixel value of 1) as a candidate for seed point selection, and then use the adjacent pixels of the seed. If the difference between its intensity value and the region average value is below a certain threshold, a pixel will be added in the region. The Euclidean distance is used to calculate the difference between the intensity value and the region average value:

$$d_{i,j} = \|I(i,j) - G(i,j)\|_2^2 \tag{3}$$

Each time a new pixel is added to the segmented area, and the intensity average of the area is updated. The use of two-dimensional OTSU to select seed points and the use of Euclidean distance for region growing can effectively take into account the position and connectivity of the image itself to achieve the ideal segmentation effect.

In this paper, two images of AIR-SARShip-1.0: High-resolution SAR Ship Detection Dataset are used to segment land and sea. The segmentation result is shown in Fig. 2. In Fig. 2, the first line are the original images, and the second line are the sea land segmentation images. It can be seen that this algorithm can effectively separate land and eliminate land interference.

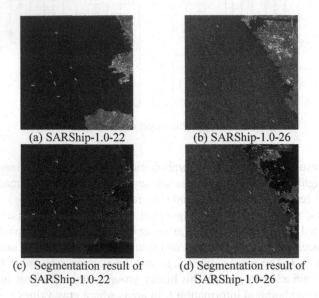

(a) SARShip-1.0-22   (b) SARShip-1.0-26

(c) Segmentation result of   (d) Segmentation result of
       SARShip-1.0-22   SARShip-1.0-26

Fig. 2. Land-sea segmentation of SAR images

3.2 SAR Image Multi-scale Decomposition

The image pyramid is a kind of multi-scale expression of an image, which is mainly used for image segmentation. It is a simple but effective structure to explain images with multiple resolutions [18]. To describe the high-frequency information, we use Laplacian pyramid (LP) for multi-scale decomposition. After the upper layer image is upsampled and Gaussian convolved, the predicted image is subtracted from each layer image of the Gaussian pyramid. The resulting series of difference images are the Laplace pyramid decomposition images. Its mathematical definition is

$$L_i = GP_i - UP(GP_{i+1}) * g_{5 \times 5} \tag{4}$$

where GP_i represents the image of layer i of the Gaussian pyramid, which has a total of N layers; the operation $UP(\bullet)$ is to perform upsampling, that is, to map the pixel at (x, y) in the original image to the position $(2x + 1, 2y + 1)$ in the target image; $g_{5 \times 5}$ is Gaussian kernel with 5×5 size. The Laplacian pyramid obtained according to the formula has

N-1 layers, and the image of the i-th layer is marked as L_i. The pyramid selected in this paper consists of $LP_1, LP_2, \ldots, LP_{N-1}, GP_N$. The hybrid pyramid with the top layer of Gaussian pyramid can retain the general outline information of the image, and the detailed information is provided by the images of each layer of the Laplacian pyramid.

At some scales, when small targets are fused into the SAR image background, large areas of clutter are still clearly visible. Therefore, choosing the appropriate scale has become a key step in SAR image target detection. We use the surround scale and the center scale to exclude some large clutter and small clutter that are obviously not the target of interest [19].

3.3 Saliency Detection

Studying the spectral residual calculation model shows that, compared to the log amplitude spectrum, the phase spectrum is the part that really contributes to the saliency map. Based on this, the classical spectral residual model is improved [20]. The new visual saliency calculation model is

$$P_M(f) = \text{FFT}[I(x)] / |\{\text{FFT}[I(x)]\}| \tag{5}$$

$$R_M(f) = BP(f) \cdot P_M(f) \tag{6}$$

$$S_M(x) = \text{FFT}^{-1}[R_M(f)] \tag{7}$$

where $BP(f)$ refers to a band-pass filter, which is a Gaussian filter with center frequency f_0 and cut-off frequency Δf in this model. $P_M(f)$ is the phase spectrum of the original image. $R_M(f)$ represents the spectrum residual. $S_M(x)$ is the saliency map. The improved saliency calculation model of spectrum residual mainly includes two steps: normalization of the original spectrum and band-pass filtering in the frequency domain.

Compared with the previous spectral residual method, the improved method does not require logarithmic amplitude spectrum calculation, low-pass filtering, exponential calculation in the frequency domain and Gaussian filtering in the spatial domain, which greatly improves the calculation speed. The improved method contains two control parameters f_0 and Δf, which can be adjusted according to the target characteristics, so the calculation model has certain adaptability.

Generally, the gray value of the target in the SAR image is higher than the gray value of the background pixels of its neighboring area. Therefore, the potential area of the ship target is usually included in the obtained saliency region [21]. The method of region extraction can greatly reduce the amount of data in the subsequent local traversal detection, and improve the practical performance of the detection algorithm to a certain extent.

After extracting the saliency regions using the spectral residual method, the image will become blurred and the target contour information is not obvious enough. Therefore, we use the classic two-parameter CFAR detection algorithm [22] to further remove false alarms and extract more accurate target contour information.

4   Experimental Results and Analysis

In order to verify the performance of the proposed algorithm, we use AIR-SARShip-1.0: High-resolution SAR Ship Detection Dataset for testing [23]. The dataset is derived from the GF-3 satellite. It is a public sample dataset of SAR ship targets for wide-range scenes. It contains 31 SAR images, and image resolutions include 1 m and 3 m. The image size is mostly 3000 × 3000. Scene types include ports, islands and reefs, sea surfaces of different levels of sea state, etc. At present, this dataset is mainly used to support applications such as ship target detection in complex scenarios. The test images are as shown in Fig. 2.

We compare the proposed SAR image ship target detection algorithm based on region growing and multi-scale saliency with the classic detection algorithm. The comparison algorithms include the two-parameter CFAR algorithm [22], CFAR ship detection based on lognormal mixture models (LNM-CFAR) [24], semiparametric clutter estimation for ship detection (SCE-CFAR) [25], visual attention-based target detection (VAB) [26] and ship target detection method based on saliency detection (SD) [27]. Because VAB and SD not only highlight the ship target, but also highlight the land, so we also segment the land and sea. First, the above algorithms are used to detect the ships in Fig. 2(a) and Fig. 2(b), and the detection results are shown in Fig. 3 and Fig. 4, respectively.

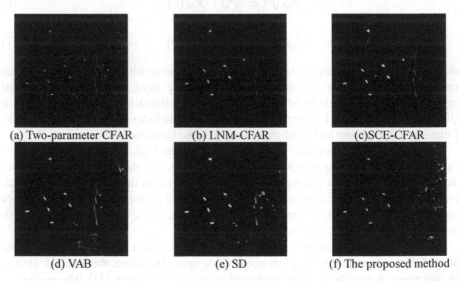

(a) Two-parameter CFAR   (b) LNM-CFAR   (c)SCE-CFAR

(d) VAB   (e) SD   (f) The proposed method

Fig. 3. SAR image ship target detection results of Fig. 2(a)

It can be seen from Fig. 3 and Fig. 4 that there are still a lot of sea clutter in the image after the detection using the two-parameter CFAR algorithm. LNM-CFAR algorithm can suppress most of the clutter, but the effect is still poor. SCE-CFAR algorithm can effectively suppress the clutter, and remove the land impact. But at the same time, the nearshore ship target is also removed. Both the VAB algorithm and the SD algorithm effectively suppress the influence of sea clutter and remove most of the land, but the

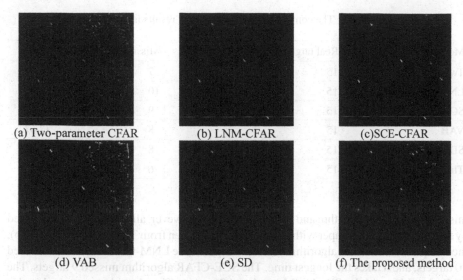

Fig. 4. SAR image ship target detection results of Fig. 2(b).

remaining land clutter still affects the detection of nearshore ship targets. The proposed algorithm not only eliminates the influence of the land area, but also retains the obvious target in the visual cognition. It smoothly suppresses most of the clutter, and the detection effect is the best.

In order to evaluate the effectiveness of the algorithm more objectively, the objective indicators of the algorithm detection results in Fig. 3 and Fig. 4 are shown in Table 1 and Table 2. We ensure that the algorithms are executed under the same conditions. Each algorithm is run 5 times, and the average value of the 5 running times is taken. The false alarm rate of CFAR is set to 10^{-3}. The size of background window is 55×55 and the size of protection window is 33×33.

Table 1. The comparison of experimental results in Fig. 3.

Methods	Real targets	Detected targets	Missing targets	Time/s
Two-parameter CFAR	14	12	2	943.32
LNM-CFAR	14	10	4	3378.53
SCE-CFAR	14	11	3	1239.92
VAB	14	12	2	37.58
SD	14	12	2	16.73
The proposed method	14	14	0	822.82

It can be seen from Table 1 that for Fig. 2(a), two targets are missed by the two parameter CFAR algorithm. Four targets are missed by the LNM-CFAR algorithm with the longest time. Three targets are missed by the SCE-CFAR algorithm. Two targets are

Table 2. The comparison of experimental results in Fig. 4.

Methods	Real targets	Detected targets	Missing targets	Time/s
Two-parameter CFAR	15	8	7	951.17
LNM-CFAR	15	5	10	3394.85
SCE-CFAR	15	6	9	1282.11
VAB	15	7	8	37.67
SD	15	7	8	16.95
The proposed method	15	9	6	835.28

missed by the VAB algorithm and the SD algorithm. However, all real targets are detected by the algorithm in this paper with less time. It can be seen from Table 2 that for Fig. 2(b), the two parameter CFAR algorithm missed 7 targets. The LNM-CFAR algorithm missed 10 targets, and it takes the longest time. The SCE-CFAR algorithm missed 9 targets. The VAB algorithm and the SD algorithm missed 8 targets. The algorithm proposed in this paper missed 6 targets, which achieved a high detection rate and took a short time. Compared with the above detection algorithms, CFAR based detection algorithms have a higher detection rate, but usually takes more time. The saliency based algorithm takes less time, but the detection effect of the target will be affected. In this paper, land-sea segmentation is performed based on region growing, and the advantages of CFAR and multi-scale saliency are both used. The proposed algorithm can effectively suppress the influence of clutter and has higher detection efficiency.

In order to fully illustrate the performance of the algorithm proposed in this paper, we use 31 images of the dataset for experiments and calculate the average precision (AP). The experimental results are shown in Table 3. It can be seen from Table 3 that the detection rate of the proposed method is the highest, which shows that the algorithm in this paper has strong robustness and is suitable for most SAR images.

Table 3. Average precision of dataset.

Methods	Two-parameter CFAR	LNM-CFAR	SCE-CFAR	VAB	SD	Proposed method
AP(%)	82.6	72.2	74.8	77.2	79.4	88.5

5   Conclusion

In view of the shortcomings of traditional target detection methods in high-resolution SAR images with complex scene, this paper proposes a ship target detection algorithm based on region growing and multi-scale saliency. Land-sea mask and region growing algorithms are used to segment the image into sea and land to remove the effects of land scattering. Then the image is decomposed by non-subsampling Laplace filter at different scales, and the spectral residual method and CFAR algorithm are utilized to seeks saliency maps and extracts targets more accurately. Experimental results show that the algorithm can not only effectively suppress the influence of land and sea clutter, but also improve the detection efficiency and extract targets more accurately.

References

1. Arii, M., Koiwa, M., Aoki, Y.: Applicability of SAR to marine debris surveillance after the great East Japan Earthquake. IEEE J. Sel. Top. Appl. Earth Observations Remote Sens. **7**(5), 1729–1744 (2014)
2. Cheng, Y., Liu, B., Li, X., Nunziata, F., Xu, Q., Ding, X.: Monitoring of oil spill trajectories with COSMO-SkyMed X-Band SAR images and model simulation. IEEE J. Sel. Top. Appl. Earth Observations Remote Sens. **7**(7), 2895–2901 (2014)
3. Liu, G., Zhang, Y., Zheng, X., Sun, X., Fu, K., Wang, H.: A new method on inshore ship detection in high-resolution satellite images using shape and context information. IEEE Geosci. Remote Sens. Lett. **11**(3), 617–621 (2014)
4. Zhu, C., Zhou, H., Wang, R., Guo, J.: A novel hierarchical method of ship detection from spaceborne optical image based on shape and texture features. IEEE Trans. Geosci. Remote Sens. **48**(9), 3446–3456 (2010)
5. Gao, G., Shi, G.: CFAR ship detection in nonhomogeneous sea clutter using polarimetric SAR data based on the notch filter. IEEE Trans. Geosci. Remote Sens. **55**(8), 4811–4824 (2017)
6. Itti, L., Koch, C., Niebur, E.: A model of saliency-based visual attention for rapid scene analysis. IEEE Trans. Pattern Anal. Mach. Intell. **20**(11), 1254–1259 (1998)
7. Bruce, N., Tsotsos, J.: Saliency based on information maximization. In: 18th International Conference on Neural Information Processing Systems, pp. 155–162. Springer-Verlag (2005)
8. Hou, X., Zhang, L.: Saliency detection: a spectral residual approach. In: 2007 IEEE Computer Society Conference on Computer Vision and Pattern Recognition, Minneapolis, Minnesota, USA, pp. 1–8. IEEE (2007)
9. Yu, Y., Wang, B., Zhang, L.: Pulse discrete cosine transform for saliency-based visual attention. In: 2009 IEEE 8th International Conference on Development and Learning, Shanghai, China, pp. 1–6. IEEE (2009)
10. Cheng, M., Mitra, N., Huang, X., Torr, P., Hu, S.: Global contrast based salient region detection. IEEE Trans. Pattern Anal. Mach. Intell. **37**(3), 569–582 (2015)
11. Lu, Y., Jiang, T., Zang, Y.: Region growing method for the analysis of functional MRI data. NeuroImage **20**(1), 455–465 (2003)
12. Yu, Y., Ding, Z., Wang, B., Zhang, L.: Visual attention-based ship detection in SAR images. Adv. Neural Netw. Res. Appl. **67**, 283–292 (2010)
13. Liu, S., Cao, Z., Li, J.: A SVD-based visual attention detection algorithm of SAR image. In: Zhang, B., Mu, J., Wang, W., Liang, Q., Pi, Y. (eds.) The Proceedings of the Second International Conference on Communications, Signal Processing, and Systems. LNEE, vol. 246, pp. 479–486. Springer, Cham (2014). https://doi.org/10.1007/978-3-319-00536-2_55
14. Hou, B., Yang, W., Wang, S., Hou, X.: SAR image ship detection based on visual attention model. In: IEEE International Geoscience and Remote Sensing Symposium, pp. 2003–2006 (2013)
15. Bi, F., Zhu, B., Gao, L., Bian, M.: A visual search inspired computational model for ship detection in optical satellite images. IEEE Geosci. Remote Sens. Lett. **9**(4), 749–753 (2012)
16. Wang, H., Xu, F., Chen, S.: Saliency detector for SAR images based on pattern recurrence. IEEE J. Sel. Top. Appl. Earth Observations Remote Sens. **9**(7), 2891–2900 (2016)

17. Wu, H., Zhou, Yu., Zhou, Y., Chen, X., Xiang, L., Li, Z.: Image segmentation using region growing based on 2D OTSU to selected seed points. J. Atmos. Environ. Opt. **8**(6), 448–453 (2013). (in Chinese)
18. Yan, C., Liu, C.: A ship target detection method of SAR image based on saliency detection. J. Univ. Chin. Acad. Sci. **36**(03), 401–409 (2019). (in Chinese)
19. Wang, Z., Du, L., Zhang, P., Li, L., Wang, F., Xu, S.: Visual attention-based target detection and discrimination for high-resolution SAR images in complex scenes. IEEE Trans. Geosci. Remote Sens. **56**, 1–18 (2017)
20. Xiong, W., Xu, Y., Cui, Y., Li, Y.: Geometric feature extraction of ship in high-resolution synthetic aperture radar images. Acta Photonica Sinica **47**(01), 55–64 (2018). (in Chinese)
21. Itti, L.: Models of bottom-up and top-down visual attention. California Institute of Technology, Computer Science Department Pasadena, CA, United States (2000)
22. Novak, L., Owirka, G., Brower, W.: The automatic target-recognition system in SAIP. Lincoln Lab. J. **10**(2), 187–202 (1997)
23. Sun, X., Wang, A., Zhi, R., Sun, Y.: AIR-SARShip-1.0: high-resolution SAR ship detection dataset. J. Radars **8**(6), 852–862 (2019). (in Chinese)
24. Cui, Y., Yang, J., Yamaguchi, Y.: CFAR ship detection in SAR images based on lognormal mixture models. In: 3rd International Asia-Pacific Conference on Synthetic Aperture Radar, Seoul, South Korea, pp. 1–3. IEEE (2011)
25. Cui, Y., Yang, J., Yamaguchi, Y., Singh, G., Park, S., Kobayashi, H.: On semiparametric clutter estimation for ship detection in synthetic aperture radar images. IEEE Trans. Geosci. Remote Sens. **51**(5), 3170–3180 (2013)
26. Wang, Z., Du, L., Zhang, P., Li, L., Wang, F., Xu, S.: Visual attention-based target detection and discrimination for high-resolution SAR images in complex scenes. IEEE Trans. Geosci. Remote Sens. **56**(4), 1855–1872 (2017)
27. Yan, C., Liu, C.: A ship target detection method of SAR image based on saliency detection. J. Univ. Chin. Acad. Sci. **36**(3), 401–409 (2019). (in Chinese)

Hybrid Dilated Convolution Network Using Attentive Kernels for Real-Time Semantic Segmentation

Jiankai He⬤, Bin Jiang(✉)⬤, Chao Yang⬤, and Wenxuan Tu⬤

College of Computer Science and Electronic Engineering, Hunan University,
Changsha, China
{hejiankai,jiangbin,yangchaoedu,twx}@hnu.edu.cn

Abstract. Though current semantic segmentation methods achieve high accuracy, most of them suffer from low speed, massive memory usage, and high computation complexity. To avoid these problems, we propose a light-weight network called Hybrid Dilated Convolution Network (HDCNet). HDCNet mainly consists of the Hybrid Scale-Aligned Block (HSAB) and the Attentive Depthwise Separable Block (ADSB). The HSAB adopts multiple small kernel convolutions with small-scale dilation rates to extract local information and applies several large kernel convolutions with large-scale dilation rates to encode global information, respectively. We further explore the best option to match the kernel size to the dilation scale. The ADSB is designed to decrease redundant parameters and enhance the critical information by depthwise separable convolution and mixed convolution kernels. In this way, ADSB and HSAB jointly encode multi-scale context information to improve model performance. Thereafter, we combine integrated local information with global information to generate final prediction results. Extensive experiments on Cityscape dataset have demonstrated that the proposed method reaches a better trade-off between accuracy and efficiency compared with other start-of-the-art methods. In particular, HDCNet obtains 72.82% MIoU with only 2.02M and 16.8 GFLOPs.

Keywords: Semantic segmentation · Model efficiency · Multi-scale contexts

1 Introduction

Semantic segmentation technology plays a key role in medical imaging, satellite remote sensing, and self-driving [1]. In the real world, the core challenge is how

---

This work is partially supported by the National Natural Science Foundation of China under Grant No. 61702176.
The first author is a graduate student.

© Springer Nature Switzerland AG 2020
Y. Peng et al. (Eds.): PRCV 2020, LNCS 12305, pp. 129–141, 2020.
https://doi.org/10.1007/978-3-030-60633-6_11

to keep both high accuracy and real-time speed under resource-constrained environments. Therefore, it is essential to design a semantic segmentation method that could achieve a trade-off between accuracy and efficiency.

Previous methods often adopt complex structures to improve accuracy. Chen et al. [2] propose the Atrous Spatial Pyramid Pooling (ASPP) module to use diverse dilation convolutions to capture contextual information. Gao Huang et al. [3] propose the dense block, which connects each layer to the corresponding layer in a feed-forward structure. These methods have achieved significant improvement in accuracy but suffer from heavy overheads and expensive computation. Due to the limitation in memory and computation resources, these methods are not feasible to be deployed on edge devices.

To solve the above problems, some works have tried to design lightweight models to achieve a real-time speed. Sandler et al. [4] propose depthwise separable convolution, which replaces a full convolutional filter with a factorized one that divides a convolution into two sub-convolutions. Mehta et al. [5] present the EESP unit, which adopts group point-wise and depth-wise dilated separable convolutions to efficiently learn multi-scale context representations. However, some problems such as the detailed information loss in these models lead to accuracy performance decrease, resulting in the final MIoU score notably drops to 63% or even lower.

Recently, some works force on the balance between accuracy and speed. For instance, Li et al. [6] propose the DFANet that aggregates discriminative features through sub-network and sub-stage cascade respectively. Though these state-of-the-art methods could generate accurate results and maintain a real-time speed, most of them are still limited in terms of memory footprint or computational complexity.

In this paper, we propose an efficient and effective semantic segmentation network called Hybrid Dilated Convolution Network (HDCNet). It aims to reach a better trade-off between accuracy and efficiency. The architecture of HDCNet is illustrated in Fig. 1. HDCNet consists of Attentive Spatial Pyramid Module (ASPM) and Global Context-Aware Module (GCAM). Specifically, ASPM consists of the Hybrid Scale-Aligned Block (HSAB) and the Attentive Depthwise Separable Block (ADSB). In order to improve the model efficiency, we utilize group point-wise and depth-wise dilated separable convolutions to reduce parameters and computational complexity. In pursuit of better accuracy, we firstly mix up multiple kernel sizes in depth-wise dilated separable convolutions and propose the best option to match the kernel size to the dilation scale. Then we design a lightweight attention-based on depthwise separable convolution to enhance the critical information representation. Finally, we add two efficient long-range shortcut connection between the backbone and GCAM, which adds critical information to the output information of ASPM.

The key contributions of this paper are three-fold:

1. A light-weight network called HDCNet is proposed for semantic segmentation task. Compared with other models, our method achieves better trade-off in terms of speed, accuracy, and memory.

2. An effective block HSAB is proposed, which mix up multiple kernel sizes in depth-wise dilated separable convolutions, and explore the best option to match the kernel size to dilation scale. It improves the ability of convolution kernel to grasp information, expands the receptive field, so as to increases the model accuracy.
3. An efficient block ADSB is designed, which adopts a multi-layer mixed depth-wise convolution connection method to enhance the information representations. The attention mechanism effectively improves the network accuracy while the computation burden increases slightly.

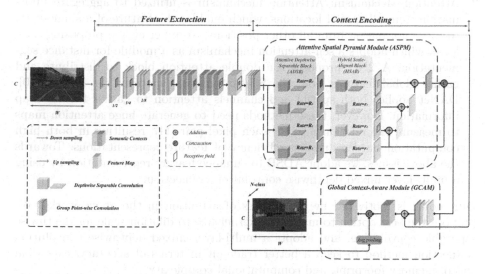

Fig. 1. An overview of Hybrid Dilated Convolution Network (HDCNet). The dotted red box and blue box represent the Attentive Spatial Pyramid Module (ASPM) and the Global Context-Aware Module (GCAM), respectively. (Color figure online)

The remainder of this paper is organized as follows. Section 2 reviews related works on semantic segmentation and some lightweight methods. Section 3 presents the model that we designed in detail. Section 4 carries out experiments and discusses the results. Section 5 draws a conclusion.

2   Related Work

In this section, we will review the related works of semantic segmentation. It can be mainly divided into the following two types.

1) Multi-Scale Feature Fusion: Multi-scale is a common method to makes full use of the context information [7]. To improve the accuracy, Zhao et al. [8], propose the pyramid pooling module, different sizes of information are generated by pyramid pooling. Chen et al. propose to integrate multi-scale contexts

by proposing Atrous Spatial Pyramid Pooling (ASPP) module with diverse dilation rates. Despite their success, most of these methods lead to a high computation burden. The latter category tries to increase efficiency. Sandler et al. [4] introduce depthwise separable convolutions as an efficient replacement for traditional convolution layers. Though these approaches are effective, the low accuracy makes these models not competent for practical application. Different from the aforementioned works, the proposed HDCNet enhances the model accuracy and efficiency by designing lightweight yet powerful HSAB, which utilizes the best option to match the kernel size to dilation scale to learn representations from large effective receptive fields.

2) Attention Mechanism: Attention mechanism is utilized to aggregate information from all spatial locations, which enables the feature of a single-pixel to fuse information from all other positions. Wang et al. [9]. proposing non-local, which adopts a self-attention mechanism as a module for instance segmentation. A^2Net proposes the double attention block to distribute and gather global features from the Spatio-temporal space of the images [10]. DANet applies both spatial and channels attention to gather feature map information. However, these methods need to generate huge attention maps to measure the relationships for each pixel-pair [11], resulting in both high computation burden and less efficiency of feature representations. Towards the above issues, we adopt ADSB to enhance feature representation by using a multi-layer mixed depthwise convolution connection.

Our model is motivated by the success of attention in the above works. We propose the best option to match the kernel size to dilation scale for depthwise separable convolution, and adopts a multi-layer mixed depthwise convolution connection method to keep a better trade-off in terms of accuracy, execution speed, memory footprint, and computational complexity.

3   HDCNet

3.1   Overview

In this section, we introduce a lightweight architecture called HDCNet for semantic segmentation, which mainly consists of the Hybrid Scale-Aligned Block (HSAB) and the Attentive Depthwise Separable Block (ADSB). HDCNet aims to achieve an overall trade-off in terms of accuracy and efficiency.

As showed in Fig. 1, given an input, we firstly feed it into our backbone network to obtain the semantic features. The encoder Output Stride (OS) is reasonably set to 8 for high-resolution datasets to save memory resources.

Next, we perform the Attentive Spatial Pyramid Module. ASPM consists of the Hybrid Scale-Aligned Block (HSAB) and the Attentive Depthwise Separable Block (ADSB). As shown in the red box of Fig. 1. For the first step, we feed semantic features outputted from the tail of the backbone stage7 into ADSB. ADSB is designed as a single-layer depthwise separable convolution. This design has two advantages. On the one hand, it can enhance the information representations by increasing the depth of the network. On the other hand, depthwise separable convolution performs convolutional kernels for each channel separately, thus reducing the number of parameters and computational cost. For the second step, we feed multi-scale feature maps into HSAB. HSAB utilizes the best match of the kernel size to dilation scale to increase the effective receptive field, which considerably boosts the model accuracy.

Finally, we perform the Global Context-Aware Module (GCAM). As shown in the blue box of Fig. 1. We add two efficient long-range shortcut connection between the stage7 and the outputted from ASPM. It is used to supply key information and expand the dimensions of output feature maps from ASPM efficiently respectively. In the following sections, we will elaborate on the design of ASPM and GCAM in detail. Especially the HSAB and the ADSB in ASPM.

3.2   Attentive Spatial Pyramid Module (ASPM)

As depicted in the red box of Fig. 1, ASPM consists of the HSAB and the ADSB. First of all, given low-dimensional feature maps (generated by Stage7) as an input $F_S = \left\{ f_{S_1}^{H \times W}, f_{S_2}^{H \times W}, \cdots, f_{S_{C_i}}^{H \times W} \right\}$, we firstly transform low-dimensional F_S into high-dimensional $F_S' = \left\{ f_{S_1}'^{H \times W}, f_{S_2}'^{H \times W}, \cdots, f_{S_{C_0}}'^{H \times W} \right\}$, which can reduce the parameters without affecting the accuracy. Secondly, we put F_S' into ADSB as $F_D = \left\{ f_{D_1}^{H \times W}, f_{D_2}^{H \times W}, \cdots, f_{D_0}^{H \times W} \right\}$. Finally, we feed F_D into HSAB to produce $F_M = \left\{ f_{M_1}^{H \times W}, f_{M_2}^{H \times W}, \cdots, f_{M_{C_0}}^{H \times W} \right\}$. In this way, ADSB and HSAB jointly encode multi-scale context information to improve model performance.

$$F_S' = \delta \left(\rho \omega^{1 \times 1} \times \delta \left(d\omega^{1 \times 1} \times F_S + b \right) + b \right) \tag{1}$$

where $\delta(\cdot)$ is the operations of both Batch Normalization (BN) [12] and Parametric Rectified Linear Unit (PReLU) function [13]. $\rho\omega^{1 \times 1}$ and $d\omega^{1 \times 1}$ are represent 3×3 depth-wise convolution and a 1×1 point-wise convolution, respectively. b is the bias vector. We summarize all notations in Table 1.

Table 1. Basic notations for the proposed method.

Notations	Meaning
$H \times W$	The size of feature maps
C_i	The number of input channels
C_0	The number of output channels
F_S	The low-dimensional shallow feature maps
F_S'	The high-dimensional shallow feature maps
F_D	The ADSB output feature map
F_M	The HSAB output feature map

Hybrid Scale-Aligned Block (HSAB). As discussed in Sect. 2, the depthwise separable convolution is beneficial to model efficiency improvement, but suffers from low accuracy. Inspired by the theory of multiple kernel sizes and dilated convolution [14], Hybrid Scale-Aligned Block is proposed to improve the model accuracy by capture multi-scale contexts and enlarge the field of view.

1)Mixed Kernel Strategy: Recent research results have shown larger kernel sizes can potentially improve model accuracy and efficiency. Larger kernels tend to capture high-resolution patterns with more details at the cost of more parameters and computations. Small kernels tend to capture low-resolution patterns with fewer parameters. To capture multi-scale contexts, we mix up different kernel sizes in a single convolution. The convolution kernels are $n_1 \times n_1, n_2 \times n_2, n_3 \times n_3, n_4 \times n_4$. According to Tan observation [14], the convolution accuracy first goes up from 3×3 to 9×9, but then drops down quickly when the kernel size is larger than 9×9, suggesting very large kernel sizes can hurt both accuracy and efficiency. Therefore, n_1, n_2, n_3, and n_4 are generally taken as $\{3, 5, 7, 9\}$.

2) Dilation Strategy: To enlarge the field of view, we need to increase the convolution dilation rates. Because large kernels can capture high-resolution patterns with more details at the cost of more parameters and computations, we adopt two small kernel convolutions with small-scale dilation rates to extract local information and apply two large kernel convolutions with large-scale dilation rates to encode global information, respectively. Besides, the size of an effective receptive field is smaller [15]. To address the above issue, we set a four-group depthwise separable convolution dilation rate as $\{1, 2, 4, 8\}$, which can make the HSAB receptive field relatively large than the feature map. This strategy namely Relatively Large Strategy (RLS) as shown in Fig. 2, the max effective receptive field:

$$R_M = W \times n_{M_4} \tag{2}$$

$$n_{M_i} = (n_i - 1) \times (r_i - 1) + n_i \tag{3}$$

Our experiments show that RLS can achieve the best results.

3) Other Possible Strategies: We have several other strategies as shown in Fig. 2. The four convolutions in HSAB are respectively represented by M_1, M_2, M_3 and M_4, which dilation rates are r_1, r_2, r_3, r_4 respectively.

(1) Relatively Small Strategy (RSS): A strategy that receptive field relatively small than feature map, the purpose is to make full of the HSAB effective receptive field. We let $r_4 = 7$, the max effective receptive field:

$$R_M = n_{M_4} \times n_{M_4} \tag{4}$$

(2) Smaller-Scale Strategy (SSS): Recent studies show that dilated convolution has reasonable performance for small kernels, but the accuracy drops quickly for large kernels [14], so we let $r_4 = 1$, the max effective receptive field:

$$R_M = n_{M_3} \times n_{M_3} \tag{5}$$

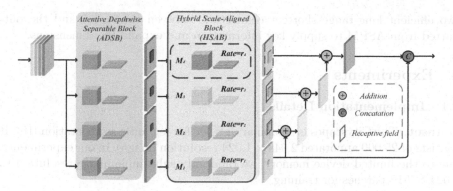

Fig. 2. Architecture of the proposed ASPM. M_i: The ith convolution of HSAB. r_i: The dilation rate of M_i.

(3) Larger-Scale Strategy (LSS): In order to capture more global information, we further expand the HSAB receptive field. We let $r_4 = 10$, the max effective receptive field:

$$R_M = W \times n_{M_4} \tag{6}$$

Attentive Depthwise Separable Block (ADSB). To enhance the critical information representation, we adopt a multi-layer mixed depthwise convolution connection method to enhance the information representations show in Fig. 1. ADSB is a flexible convolutional with several design choices.

1) Convolution Kernel Size N: Previous studies have shown that 4 groups of kernels $\{3 \times 3, 5 \times 5, 7 \times 7, 9 \times 9\}$ can further benefit the model efficiency and accuracy. So we set the ADSB convolution kernel as a fixed size $\{3 \times 3, 5 \times 5, 7 \times 7, 9 \times 9\}$.

2) Dilation Rate R: Dilated convolution can expand the receptive field without extra parameters and computations. However, the stacking of dilated convolutions will lead to the gridding effect, which would hurt the accuracy. In order to avoid the gridding effect, we set the dilation rates as a fixed size $r_i = \{1, 1, 1, 1\}$.

3) Convolutional Layers L: Deepening the network is another effective way to improve the accuracy of the model. We call the added convolution layers L, convolution layers of ADSB in the ith experiment are $L = i - 1$, $i = 1, 2, 3 \cdots$. Our experiment shows that when $L = 1$, the model achieves better accuracy.

3.3 Global Context-Aware Module (GCAM)

As depicted in the blue box of Fig. 1, the feature map F_M generated by ASPM has integrated rich multi-scale information. However, F_M will lose some detail information after the downsampling operation. To tackle this dilemma, we add two efficient long-range shortcut connection between the stage7 and the out-putted from ASPM to supply key information and expand the dimensions.

4   Experiments

4.1   Implementation Details

Dataset. The Cityscapes is a popular dataset for semantic segmentation [16]. It consists of 25,000 annotated 2,048 × 1,024 resolution images. In our experiments, due to the limited device memory, we randomly subsample image resolution to 1,024 × 512 patches for training.

Training Details. In our experiments, we implement all experiments in Pytorch 3.5, which is based on the Windows 10 system. The experimental hardware environment mainly consists of Intel (R) Core (TM) i7-8700k processor and NVIDIA TITAN X GPU cards. All ablation results are evaluated on Cityscapes val set.

4.2   Modified MobileNet V2

Ablation for Backbone Choices. To show the advantages of FCN-Mobilenet V2 as the backbone, we selected 4 FCN-based models as baselines.

As shown in Table 2, FCN-Shufflenet shows efficiency in terms of memory. However, the low accuracy makes FCN-Shufflenet not competent for practical application. FCN-DenseNet121 is 0.86% more accurate than FCN-Mobilenet V2, but incurs about 4 times parameters. We observe that FCN-MobileNet V3 achieves a better trade-off between accuracy and efficiency than accuracy-oriented. Nevertheless, FCN-MobileNet V2 outperforms FCN-MobileNet V3 in terms of accuracy by 3.52% MIoU, which increases the negligible competitional cost. To sum up, all the above comparisons demonstrate that employing MobileNet V2 as a feature extractor could more efficiently bring a great benefit for semantic segmentation.

Table 2. Ablation studies on backbone choices. #par: the number of parameters. #ns: network size.

Method	#Par (M)	#NS (MB)	MIoU (%)
FCN-Shufflenet V2	0.61	2.48	50.57
FCN-ResNet18	11.30	44.19	66.82
FCN-DenseNet121	7.09	28.11	68.41
FCN-MobileNet V3-Large	1.29	5.05	64.03
FCN-MobileNet V2	**1.93**	**7.68**	**67.55**

Table 3. Comparison between different strategies, "RSS" represents Relatively Small Strategy, "SSS" represents Smaller-Scale Strategy, "LSS" refers to Larger-Scale Strategy and "RLS" refers to Relatively Large Strategy.

Dilated strategy	Strategy type	#Par (M)	MIoU (%)
–	–	2.02	71.39
√	RSS	2.02	71.21
√	SSS	2.02	71.37
√	LSS	2.02	70.06
√	RLS	**2.02**	**72.82**

4.3   Comparisons of Different Strategies in HSAB

We conduct some experiments based on the FCN-Mobilenet V2, ADSB, and GCAM. As presented in Table 3, RSS gets even worse results than the baseline, SSS accuracy is equivalent to the baseline, but it is lower than RLS. There is a great gap between LSS and RSS. Our point of view is that the larger dilation rates lead to the larger receptive field, which improves the performance of the model. When the receptive field is too large, the correlation between the information is reduced, which leads to a decrease in the accuracy of the model. Eventually, the accuracy of the model quickly declines due to the loss of effective information. Obviously, HDCNet of these strategies have the same amount of the network parameter, but only RLS brings optimal detection performance improvement.

4.4   Ablation for Attentive Depthwise Separable Block

We conduct experiments to investigate the relationship between the number of attention layers in ASPM and the detection performance by taking FCN-MobileNetv2, RLS, and GCAM as the baseline. As shown in Table 4, we add our attention layer to the depthwise separable convolution of ASPM. We set the number of layers as L, starting from $L = 0$. With the L increase, model accuracy first goes up, but then drops down quickly, suggesting only increasing the number of attention layers may not bring better results. We attribute the superiority of $L = 1$ as two aspects: a) Single-layer depthwise separable convolution can avoid the gridding effect and reducing computational cost. b) Single-layer depthwise separable convolution enhances the information representations and achieves higher accuracy.

4.5   Ablation Studies on the Semantic Segmentation Dateset

In order to show the effect of our modules, we conducted ablation experiments on our modules. As shown in Table 5.

Table 4. Detection results on the Cityscapes val set.

Layer number	#Par (M)	#NS (MB)	MIoU (%)
0	2.01	7.85	72.01
1	**2.02**	**7.89**	**72.82**
2	2.03	7.93	71.75
3	2.05	8.17	72.00
4	2.06	8.21	71.38

Table 5. Comparison of with/without our module on the Cityscapes val set.

Baseline	ASPM			GCAM	#Par(M)	MIoU(%)
	HSAB		ADSB			
	Dilated strategy	Kernel strategy				
√					1.98	69.16
√		√	√	√	2.02	71.39
√	√		√	√	2.00	70.06
√	√	√		√	2.01	72.01
√	√	√	√		2.02	71.90
√	√	√	√	√	2.02	72.82

4.6 Evaluation on Cityscapes Dataset

In this section, we compare HDCNet with other state-of-the-art methods on Cityscapes test set. As shown in Table 6, performance is measured in terms of class-wise mean intersection over union (MIoU). "-" indicates that previous works have not published the corresponding value. Execution Speed (ES) is measured on NVIDIA TITAN X GPU cards.

Table 6 displays the results of HDCNet compared with high-accuracy methods. We can observe that FCN8s [17], DeepLab V2 [2], and PSPNet [8] take much time for model inference due to the limited performance and heavy structure of itself. In contrast, HDCNet achieves significant progress in both speed and computation cost when dealing with a high-resolution image, and obtains a similar or slightly better accuracy compared with most of them. Without any post-processing operations (e.g., Dense CRF [2]), the proposed HDCNet reaches a better trade-off among the overall performance.

As shown in Table 6, we compare HDCNet with fast models, and come to the following conclusions. Previous lightweight methods such as SegNet [18] achieve a fast speed, while failing to provide an accurate scene description for semantic segmentation. Besides, ENet [19] has excellent performance on model efficiency. However, these methods sacrificing too much model accuracy.

Recent lightweight methods (e.g., ERFNet [20], DFANet [6], ICNet [21]) have achieved a good trade-off between accuracy and speed. However, deploying these

Table 6. Object detection comparison between our methods and state-of-the-art detectors on Cityscapes val set.

Method	Resolution	FLOPs (G)	ES (FPS)	#Params (M)	MIoU (%)
PSPNet [8]	713 × 713	412.2	0.78	250.8	78.4
FCN8s [17]	1024 × 512	136.2	2	134.5	65.3
DeepLab V2 [2]	1024 × 512	457.8	0.25	44.0	70.4
DenseASPP121 [23]	1024 × 512	155.8	-	28.6	76.2
SegNet [18]	640 × 360	286.0	14.6	29.5	56.1
ENet [19]	640 × 360	3.8	135.4	0.4	58.3
ERFNet [20]	1024 × 512	53.5	41.7	2.1	68.0
ICNet [21]	1024 × 512	-	46.3	26.5	69.5
Skip-ShuffleNet [22]	1024 × 512	6.2	-	1.0	58.3
Skip-MobileNet [22]	1024 × 512	15.4	45	3.4	62.4
DFANet [6]	1024 × 512	1.7	160	7.8	70.3
CIFReNet [15]	1024 × 512	16.5	34.5	1.9	70.9
MixEESPNet(Ours)	640 × 360	7.44	66.7		
	713 × 713	16.6	30.3	2.02	72.82
	1024 × 512	16.8	31.3		

models on edge devices is difficult due to heavy memory footprint or computational complexity. Differently, HDCNet takes full advantage of depthwise separable convolution to relieve the resource burden, which achieves 72.82% MIoU and yields a real-time speed of 31.3 FPS on a 1024 × 512 resolution image, makes a good performance improvement on both accuracy and efficiency.

5   Conclusion

In this paper, we propose a lightweight Hybrid Dilated Convolution Network (HDCNet). Specifically, our method consists of two core components: Attentive Spatial Pyramid Module (ASPM) and Global Context-Aware Module (GCAM). We utilize ASPM to extract information, GCAM to supply information. Specifically, ASPM consists of the Hybrid Scale-Aligned Block (HSAB) and the Attentive Depthwise Separable Block (ADSB). The ADSB adopts a multi-layer mixed depthwise convolution connection method to enhance the information representations. The HSAB matches the kernel size to dilation scale in depthwise separable convolutions to improves the ability of the model to grasp information. Experiments show that the proposed HDCNet not only achieves high accuracy results but also relieves the burden of model efficiency, which makes it of great potentiality for deployment on resource-constrained devices.

In future works, we will make further discussions about two aspects: a) design a more robust model with better performance. b) Further, improve the segmentation effect of image boundary.

References

1. Hong, W.X., Wang, Z.Z., Yang, M., Yuan, J.S.: Conditional generative adversarial network for structured domain adaptation. In: CVPR, pp. 1335–1344 (2018)
2. Chen, L.C., Papandreou, G., Kokkinos, I., Murphy, K., Yuille, A.L.: DeepLab: semantic image segmentation with deep convolutional nets, atrous convolution, and fully connected CRFs. IEEE TPAMI **40**, 834–848 (2018)
3. Huang, G., Liu, Z., Van Der Maaten, L., Weinberger, K.Q.: Densely connected convolutional networks. In: CVPR, pp. 2261–2269 (2017)
4. Sandler, M., Howard, A.G., Zhu M.L., Zhmoginov, A., Chen, L.C.: MobileNetV2: inverted residuals and linear bottlenecks. In: CVPR, pp. 4510–4520 (2018)
5. Mehta, S., Rastegari, M., Shapiro, L.G., Hajishirzi, H.: ESPNetv2: a light-weight, power efficient, and general purpose convolutional neural network. In: CVPR, pp. 9190–9200 (2019)
6. Li, H.C., Xiong, P.F., Fan, H.Q., Sun, J: Deep feature aggregation for real-time semantic segmentation. In: CVPR, pp. 9522–9531 (2019)
7. Ding, H.H., Jiang, X.D., Shuai, B., Liu, A.Q., Wang, G.: Context contrasted feature and gated multi-scale aggregation for scene segmentation. In: CVPR, pp. 2393–2402 (2018)
8. Zhao, H., Shi, J., Qi, X., Wang, X., Jia, J.Y.: Pyramid scene parsing network. In: CVPR, pp. 6230–6239 (2017)
9. Wang, X.L., Girshick, R.B., Gupta, A., He, K.M.: Non-local neural networks. In: CVPR, pp. 7794–7803 (2018)
10. Chen, Y.P., Kalantidis, Y., Li, J.S., Yan, S.C., Feng, J.S.: A2-Nets: double attention networks. In: NeurIPS, pp. 350–359 (2018)
11. Fu, J., Liu, J., Tian, H.J., Li, Y., Bao, Y.J., Fang, Z.W., Lu, H.Q.: Dual attention network for scene segmentation. In: CVPR, pp. 3146–3154 (2019)
12. Ioffe, S., Szegedy, C.: Batch normalization: accelerating deep network training by reducing internal covariate shift. In: ICML, pp. 448–456 (2015)
13. He, K.M., Zhang, X.Y., Ren, S.Q., Sun, J.: Delving deep into rectifiers: surpassing human-level performance on imagenet classification. In: ICCV, pp. 1026–1034 (2015)
14. He, K.M., Zhang, X.Y., Ren, S.Q., Sun, J.: Delving deep into rectifiers: surpassing human-level performance on imagenet classification. In: ICCV, pp. 1026–1034 (2015)
15. Jiang, B., Tu, W.X., Yang, C., Yuan, J.: Context-integrated and feature-refined network for lightweight object parsing. In: IEEE, pp. 5079–5093 (2020)
16. Cordts, M., et al.: The cityscapes dataset for semantic urban scene understanding. In: CVPR, pp. 3213–3223 (2016)
17. Shelhamer, E., Long, J., Darrell, T.: Fully convolutional networks for semantic segmentation. IEEE Trans. Pattern Anal. Mach. Intell. **39**(4), 640–651 (2017)
18. de Oliveira Jr., L.A., et al.: SegNetRes-CRF: a deep convolutional encoder-decoder architecture for semantic image segmentation. In: IJCNN, pp. 1–6 (2018)
19. Paszke, A., Chaurasia, A., Kim, S., Culurciello, E.: Enet: a deep neural network architecture for real-time semantic segmentation. In: arXiv preprint arXiv:1606.02147 (2016)
20. Romera, E., Alvarez, J.M., Bergasa, L.M., Arroyo, R.: Erfnet: efficient residual factorized convnet for real-time semantic segmentation. IEEE TITS **19**, 263–272 (2018)

21. Zhao, H.S., Qi, X.J., Shen, X.Y., Shi, J.P., Jia, J.Y.: ICNet for real-time semantic segmentation on high-resolution images. In: arXiv preprint arXiv:1704.08545v2 (2018)
22. Siam, M., Gamal, M., Abdel-Razek, M., Yogamani, S.K., Jägersand, M.: RTSeg: real-time semantic segmentation comparative study. In: ICIP, pp. 1603–1607 (2018)
23. Yang, M.K., Yu, K., Zhang, C., Li, Z.W., Yang, K.Y.: DenseASPP for semantic segmentation in street scenes. In: CVPR, pp. 3684–3692 (2018)

Damage Sensitive and Original Restoration Driven Thanka Mural Inpainting

Nianyi Wang[1]([⊠]), Weilan Wang[1], Wenjin Hu[1], Aaron Fenster[2], and Shuo Li[2]

[1] Northwest Minzu University, Lanzhou, Gansu, China
livingsailor@gmail.com
[2] University of Western Ontario, London, ON, Canada

Abstract. Thangka murals are an important part of the cultural heritage of Tibet, but many precious murals were damaged during the Tibetan history. Three reasons cause existing methods to fail to provide a feasible solution for Thanka mural restoration: 1) damaged Thanka murals contain multiple large irregular broken areas; 2) damaged Thanka murals should be repaired with the original content instead of imaginary content; and 3) there is no large Thanka dataset for training. We propose a damage sensitive and original restoration driven (DSORD) Thanka inpainting method to resolve this problem. The proposed method consists of two parts. In the first part, instead of using existing arbitrary mask sets, we propose a novel mask generation method to simulate real damage of the Thanka murals, both masked Thanka and the mask generated by our method are inputted into a partial convolution neural network for training, which makes our model familiar with a variety of irregular simulated damages; and in the second part, we propose a 2-phase original-restoration-driven learning method to guide the model to restore the original content of the Thanka mural. Experiments on both simulated and real damage demonstrated that our DSORD approach performed well on a small dataset (N = 3000), generated more realistic content, and restored better the damaged Thanka murals.

Keywords: Image inpainting · Thanka restoration · Mask generation

1 Introduction

As a kind of Tibetan encyclopedia (Wang et al. 2012), Thanka images contain a variety of colors, complex structures, and extremely fine textures along with profound religious and cultural meanings, making Thanka murals different from natural images. During Tibet's long history, many precious Thangka murals were damaged. Thanka inpainting plays a significant role in the protection of Tibetan cultural heritage in modern times. Traditional non-digital manual repair methods not only require professional Thanka painting skills but also may result in unrecoverable inpainting results or even permanent damage. This makes digital inpainting of Thanka murals urgent (Fig. 1).

Although image inpainting methods have progressed, there is no available solution for damaged Thanka murals due to three main reasons.

© Springer Nature Switzerland AG 2020
Y. Peng et al. (Eds.): PRCV 2020, LNCS 12305, pp. 142–154, 2020.
https://doi.org/10.1007/978-3-030-60633-6_12

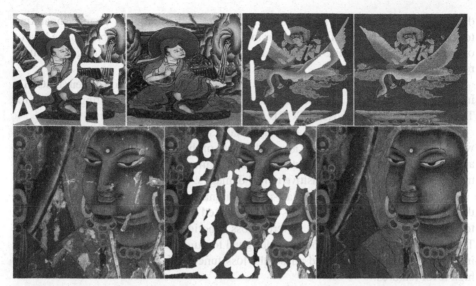

Fig. 1. Our method achieves satisfactory inpainting results on both simulated Thanka mural damage (row 1) and real Thanka mural damage (row 2) in one-step without any post processing. The size of images is 512 × 512.

First, damaged Thanka murals contain multiple irregular broken areas, which make both traditional non-learning and standard convolution-based learning methods hard to inpaint. (1) Traditional methods are mainly based on image statistics or redundancy, and they inpaint target area with texture details. However, they cannot capture the global structure and image semantics, which makes them unable to deal well with large complex holes (Iizuka et al. 2017). Furthermore, inpainting large areas will be very time consuming for them. (2) Most learning-based methods also fail to inpaint irregular areas, because they perform standard convolutions, in which non-hole areas (valid pixels) and hole areas (invalid pixels) are treated indiscriminately (Liu et al. 2018). This inherent characteristic of the standard convolution inevitably prevents convolution neural network (CNN) based methods from generating accurate inpainting results (Yu et al. 2019a).

Second, there are not enough intact Thanka images for training, which keeps generative adversarial network (GAN) based methods from achieving good performance. In the past, digital protection of Thanka murals has been paid little attention (Wang et al. 2012), resulting in few digital Thanka images. GAN is another important method for inpainting; however, it is hard to train GANs well on small datasets (Goodfellow et al. 2014). In our research, we have a small dataset that contains about 3000 Thanka images. Among the GAN-based methods proposed in the past 2 years, we tested Generative Multi-column Convolutional Neural Networks (GMCNN) (Wang et al. 2018) and Gated Convolution (Yu et al. 2019a) to train with our Thanka dataset. Both methods failed to generate satisfactory results.

Third, the objective of Thanka inpainting is to restore a Thanka mural to a state as similar to the original content. However, the objective of existing learning-based methods is to inpaint a hole with a reasonable content. Inpainting with an original content is different from inpainting with a reasonable content, which requires Thanka inpainting to restore a damaged mural to as similar to the original mural as possible. To the best of our knowledge, the existing learning-based methods seldom focus on original content restoration.

In this paper, we propose a damage sensitive and original restoration driven (DSORD) inpainting method to resolve the multiple irregular broken areas inpainting problem for Thanka murals. The proposed DSORD Thanka inpainting consists of two parts.

In the first part, we propose a damage shape sensitive mask generation method, in which mask shapes are simulated as real damages of the Thanka murals. Compared with arbitrary masks, our masks simulate real Thanka damage better, which makes the model trained by our masks more sensitive to real damages.

In the second part, we propose a 2-phase original restoration driven learning. Unlike the existing learning-based methods in which fixed loss functions are usually adopted to train the network, we adopt different loss functions for different training phases. The loss function of the first phase is designed to guide the model to reconstruct pixel information, and in the second phase, the loss function is designed to concentrate on high level features.

In order to avoid the disadvantages of standard convolutions in CNN-based methods and the unsatisfactory result of GAN-based methods on a small dataset, we adopt partial convolution-based (Liu et al. 2018) Unet encoder-decoder architecture. Our model is trained to learn not only image features but also mask features.

Experiments on a Thanka dataset show that our end-to-end method deals well by inpainting multiple holes in one-step and in a short time (300 ms for multiple holes in a 512×512 image) and without any post-processing. Comparisons with both simulated damage and real damage demonstrated the performance of our method.

In summary, the main contributions of this work are:

1. We propose a damage sensitive mask generation method for Thanka murals. The model trained by our masks fills better irregular damaged areas of Thanka murals.
2. We propose a 2-phase original restoration driven learning method, which contributes to generation of more realistic image content.
3. For the first time, we introduce a learning-based method to inpaint traditional Thanka murals. The proposed method performs well on a small dataset and can generate satisfactory inpainting results.

2   Related Work

2.1   Image Inpainting Method

Traditional Methods. There are two main types of traditional inpainting methods: geometry-based methods and patch-based methods (Wang et al. 2019). Geometry-based methods use partial differential equation to propagate local structure from the exterior to the interior of a hole. Patch-based methods are built on texture synthesis and statistical properties of the nearest image pixel neighbors. Of all traditional methods, PatchMatch

(Barnes et al. 2009) is one of the best. PatchMatch fills holes faster than the other traditional methods, but when similar texture cannot be found in the image, the inpainting result will be very bad.

For all traditional methods, there are three major problems. First, they cannot capture the global structure and image semantics. Second, they fail to fill complex holes. Third, filling a large hole is very time consuming.

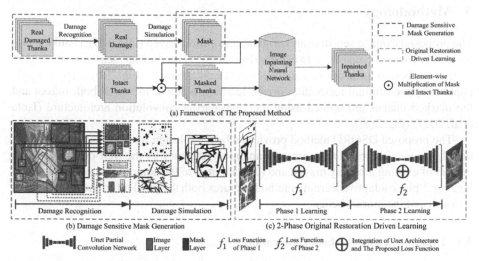

Fig. 2. Framework of the proposed damage sensitive and original restoration driven (DSORD) Thanka mural inpainting method.

Learning-based Methods. The vast majority of learning-based methods are based on either a CNN or a GAN, or on both together. Early methods only dealt with rectangular areas (Yu et al., 2019b). Context Encoders (Pathak et al. 2016) was the first attempt to fill asymmetric areas. Other new strategies such as Dilated convolution (Iizuka et al. 2017), coarse-to-fine architecture (Karras et al. 2017), attention mechanism (Yu et al. 2018), and edge-guided (Nazeri et al. 2019) methods were proposed to obtain better image features. Partial convolution was proposed to replace standard convolution, which improves greatly irregular area inpainting performance (Liu et al. 2018).

However, all these methods have limitations. GANs are hard to train with a small dataset, and CNN-based methods cannot deal well with irregular areas inpainting because of the use of standard convolutions. Though the existing partial convolution-based methods can process irregular areas well, they all use arbitrary masks, and they don't pay attention to the relationship between mask and inpainting performance.

Mask Research. Mask set preparation is a key step for the inpainting task because different mask shapes and sizes determine different inpainting modes. However, most existing inpainting methods use arbitrary masks. More recently, researchers are starting to pay attention to mask design. Publications have reported on generating masks by occlusion and dis-occlusion of video frames (Liu et al. 2018), by generating three types of masks with different hyper-parameters (Chang et al. 2019) and by designing two types of masks to simulate real world inputs (Xiong et al. 2019). However, neither the relationship between mask and inpainting performance was discussed, nor the quantitative analysis was provided in these studies.

3 Methodology

The proposed damage sensitive and original restoration driven (DSORD) Thanka mural inpainting method consists of two parts (Fig. 2(a)): (1) a damage sensitive mask generation method to simulate real damages of Thanka murals (Fig. 2(b)), and (2) a 2-phase adaptive original-restoration-driven learning method, in which both masks and the masked images are trained using the Unet partial convolution architecture (Isola et al. 2017) (Fig. 2(c)).

The proposed DSORD method provides a better solution for Thanka mural restoration due to two factors: (1) we improve inpainting accuracy by the well-designed masks instead of existing arbitrary masks, and (2) compared to the commonly used fixed learning, our 2-phase adaptive learning method captures both the pixel information and visual semantics of the murals better.

3.1 Damage Sensitive Mask Generation

Thanka murals usually contain 2 types of damage: (a) long cracks, creases and scratches. In most cases they have long and narrow shapes with some bends and irregular curvature; and (b) damaged irregular spots or block areas. These damages appear randomly on damaged Thanka murals. Different from other learning-based methods that use arbitrary masks, we adopt a dynamic algorithm to generate masks for training. Specifically, we use random lines and elliptical curves of different lengths, widths, angles and locations to simulate the first type of damage, and use random solid circles and polygons of different sizes, and locations to simulate the second type of damage. For each shape, its length, width, angle, size and number of shapes are randomly and independently generated by the algorithm. Each mask image has the same size as the Thanka image. Figure 2(b) shows a simulation of damage to a Thanka mural using the proposed mask. Figure 3 shows some mask samples, in which more shapes means a more serious damage.

Fig. 3. Mask examples generated by our damage sensitive mask generation method. More shapes mean heavier damages.

Summary of Advantages: Compared with existing arbitrary masks, our masks simulate real Thanka damage better, which makes the model trained by our masks more sensitive to real damages.

3.2 Original Restoration Driven Learning

The design of the loss function depends on the objective of the training. In this work, the objective of Thanka mural inpainting is to restore the original content of the damaged areas. Our approach to achieve this goal is to train the neural network in 2 independent phases, where different loss functions are used in each training phase. Let L_{phase1}, L_{phase2} be the total loss of phase 1 and phase 2 respectively. In each training phase, the total loss consists of 3 different loss parts: pixel reconstruction loss, L_R, perceptual loss, L_p, and style loss, L_s. L_{phase1} and L_{phase2} are defined as:

$$L_{phase1} = \alpha_1 L_R + \beta_1 L_p + \lambda_1 L_s \tag{1}$$

$$L_{phase2} = \alpha_2 L_R + \beta_2 L_p + \lambda_2 L_s \tag{2}$$

where α_1, β_1, λ_1, α_2, β_2, and λ_2 are the coefficients of the different loss components.

First, we define pixel reconstruction loss, L_R, perceptual loss, L_p, and style loss, L_s, as follows:

Pixel Reconstruction Loss L_R: Given the masked input image, I_{in}, ground truth image, I_{gt}, output image, I_{out}, and mask image, M, (M is a binary matrix with 0 for the area of the hole and 1 for non-hole area), we denote pixel reconstruction loss of the hole area, non-hole area, and total area as L_{hole}, $L_{non-hole}$ and L_R respectively, which are defined as:

$$L_{hole} = \frac{1}{N} \left\| (I_{gt} - I_{out}) \odot (1 - M) \right\|_1 \tag{3}$$

$$L_{non-hole} = \frac{1}{N} \left\| (I_{gt} - I_{out}) \odot M \right\|_1 \tag{4}$$

$$L_R = 6L_{hole} + L_{non-hole} \tag{5}$$

where $\odot$ is an element-wise multiplication, and N is the number of elements of the image input ($N = H*W*C$, H, W, and C are the image height, width and channel size respectively).

Perceptual Loss L_p: Since L_R loss cannot capture high level visual features, we introduce a perceptual loss L_p, which is defined on the VGG16 network (Gatys *et al.* 2015) and is pretrained by ImageNet (Russakovsky *et al.* 2015). We define a composite image I_{comp} as:

$$I_{comp} = M \odot I_{gt} + (1 - M) \odot I_{out} \tag{6}$$

Then L_p is defined as:

$$L_p = \sum_{i=1}^{P} \frac{\left\| \Psi^i(I_{gt}) - \Psi^i(I_{out}) \right\|_1}{N_i} + \sum_{i=1}^{P} \frac{\left\| \Psi^i(I_{gt}) - \Psi^i(I_{comp}) \right\|_1}{N_i} \tag{7}$$

where $\psi^i(\bullet)$ denotes the feature map of the i-th pooling layer, N_i denotes the number of elements in $\psi^i(I_{gt})$. As shown in Eq. (7), the perceptual loss L_p consists of 2 parts, the first part computes the difference between the output image I_{out} and the ground truth image I_{gt}, the second part computes the difference between the composite image I_{comp} and the ground truth image I_{gt}. Three pooling layers of the pre-trained VGG-16, namely *pool-1*, *pool-2*, and *pool-3* are used to extract the image features.

Style Loss L_s: We further introduce a style loss, L_s, that was defined in [Liu *et al.* 2018], which computes the feature maps on the pooling layers of VGG-16. L_s also consists of 2 parts:

$$L_s^{I_{out}} = \sum_{i=1}^{P} \frac{1}{C_i C_i} \left\| K_i((\Psi^i(I_{gt}))^T (\Psi^i(I_{gt})) - (\Psi^i(I_{out}))^T (\Psi^i(I_{out}))) \right\|_1 \tag{8}$$

$$L_s^{I_{comp}} = \sum_{i=1}^{P} \frac{1}{C_i C_i} \left\| K_i((\Psi^i(I_{gt}))^T (\Psi^i(I_{gt})) - (\Psi^i(I_{comp}))^T (\Psi^i(I_{comp}))) \right\|_1 \tag{9}$$

$$L_s = L_s^{I_{out}} + L_s^{I_{comp}} \tag{10}$$

where C_i is the channel size of the i-th layer, and K_i is the normalization factor $1/C_i*H_i*W_i$ for the i-th layer (H_i and W_i denote height and width).

2-Phase Learning: For the first phase, we focus on pixel reconstruction. We assigned a very small value to the coefficient λ_1 of L_s, which constraints the neural network from losing pixel-level detail information. For the second phase of the training, the coefficient λ_2 of L_s is assigned a large value to guide the network to learn more about the high level features. In our experiments, we set $\lambda_1 = 1$, $\lambda_2 = 120$. Analogous to (Liu *et al.* 2018), we set $\alpha_1 = \alpha_2 = 1$, $\beta_1 = \beta_2 = 0.05$.

Summary of Advantages: Different from existing fixed learning, our 2-phase leaning provides a novel learning strategy for the neural network in which the model is trained with different loss functions in two independent phases. Each phase is assigned a different objective. This approach enables the deep neural network model to capture both pixel information and visual semantics more precisely than the existing fixed learning methods.

4   Experiments

4.1   Dataset and Experiment Implementation

We collected about 3000 undamaged Thanka images as our dataset. Each image was resized to 512×512. The dataset was randomly split into a training set (80%), a validation set (10%), and a test set (10%). We adopted encoder-decoder Unet architecture

(Ronneberger *et al.* 2015). The standard convolutions were replaced by partial convolutions. The design of the network layer and the partial convolutions are similar to (Liu *et al.* 2018). We trained our model on both our mask and the mask provided in the (Liu *et al.* 2018). Two GAN-based methods GMCNN (Wang *et al.* 2018) and Gated Convolution (Yu *et al.* 2019a) were also trained for comparison.

4.2 Comparison and Analysis

(A) Performance on Simulated Damage

Qualitative Evaluation: We denote 3 inpainting methods from (Yu *et al.* 2019a), (Wang *et al.* 2018) and (Liu *et al.* 2018) as Gated Conv, GMCNN, and PConv respectively. As shown in Fig. 4, the Gated Conv method tends to generate unrealistic image content (see 4(b)), while the GMCNN method failed to deal with large irregular holes (see 4(c)). The PConv method generated reasonable content, but the results were always full of obvious artifacts (see 4(d)). Compared with other methods, our results provide more realistic content (see 4(e)).

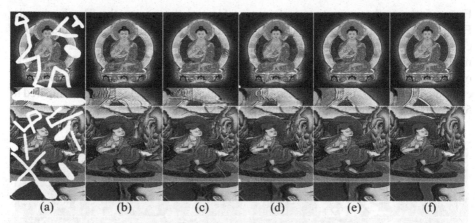

(a)   (b)   (c)   (d)   (e)   (f)

Fig. 4. Comparison on simulated damage shows that our results (e) are more realistic and closer to the original murals (f) than Gated Conv (b), GMCNN (c) and PConv (d). (a) is the damaged Thanka mural.

Quantitative Evaluation: We define four simulated damage levels Level1, Level2, Level3, to Level4. Damage degree: Level1 <Level2 <Level3 <Level4. We evaluated inpainting performance on 1391 Thanka images. The peak signal-to-noise ratio (PSNR) and structure similarity (SSIM) metrics were used as evaluation indexes. As shown in Table 1, except for the slightest damage Level1, both PSNR and SSIM of our method outperformed the other methods.

Table 1. Results generated on 1391 test images. Except for the slightest damage Level1, both PSNR and SSIM metrics generated by our method outperformed the other three methods.

	Damage	Gated Conv	GMCNN	PConv	Proposed
PSNR	Level1	**33.7307**	33.0126	28.8150	33.1154
	Level2	24.6559	23.8553	23.8825	**25.6004**
	Level3	24.7559	23.5947	23.9366	**25.6517**
	Level4	20.1791	18.9029	20.2521	**21.2284**
SSIM	Level1	**0.9821**	0.9802	0.9680	0.9767
	Level2	0.8889	0.8734	0.8733	**0.8928**
	Level3	0.8947	0.8799	0.8787	**0.8975**
	Level4	0.7247	0.6860	0.7071	**0.7418**

(B) Performance on Real Damage

Figure 5 shows the inpainting performance for real damage to the Thanka murals. Figure 5(a) shows the real damaged Thanka murals, 5(b) shows the inpainting process; 5(c), (d), (e) and (f) show inpainting results of the Gated Conv, GMCNN, PConv methods, and our method respectively. In (c), it is easy to find the unrealistic blocks on Buddha's face and the mismatched patches at the bottom left. Completely distorted color appears in 5(d). In 5(e), there are many artifactual lines and textures. Compared to 5(c), 5(d) and 5(e), our results shown in 5(f) achieved the best restoration.

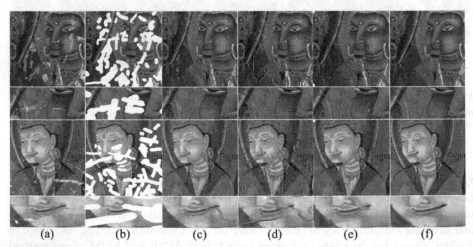

    (a)   (b)   (c)   (d)   (e)   (f)

Fig. 5. Comparison on real damaged Thanka murals shows that our results (f) achieve better visual performance on image structure and texture than Gated Conv (c), GMCNN (d) and PConv (e). (a) shows the damaged murals. (b) shows the damage processing. Since these are real damaged murals, there is no ground truth.

(C) Mask Comparison

The mask performance was tested with damage Level2, Level3, and Level4. Figure 6 shows the comparison of PSNR and SSIM of the 2 models trained by an arbitrary mask and our mask. The results show that the model trained by our mask performed better. Figure 7 shows the visual comparison of the different masks. The first row is the original Thanka mural, the second and the third rows are the cropped resulting images of the 2 models that were trained by the arbitrary mask and our mask. The final row is the ground truth image. We can easily find that there are more unrealistic textures and obvious artifacts in the second row (trained by the arbitrary mask). The results in the third row (trained by our mask) are more realistic and closer to the ground truth (the last row).

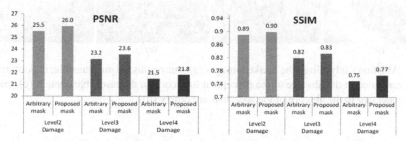

Fig. 6. Quantitative comparison shows that the model trained with our mask outperformed the model trained by arbitrary mask under different damage levels (tested on 1391 images).

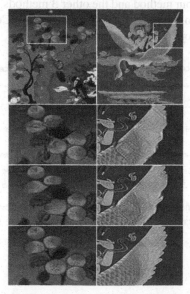

Fig. 7. Visual comparison of the two masks. The results of the model trained by our mask (row 3) are more realistic and closer to the ground truth (row 4) than the results of the model trained by the arbitrary mask (row 2).

4.3 Further Discussion

Ablation Study. Figure 8 shows a visual comparison of the ablation study. In Fig. 8, (a) is the masked input, (b) is the result generated by our loss function; (c) is the result generated in the absence of the perceptual loss L_p, and (d) is the result generated in the absence of the style loss L_s. We can see that (c) contains unreasonable content while (d) is very blurry. Compared to (c) and (d), our results (b) are more plausible.

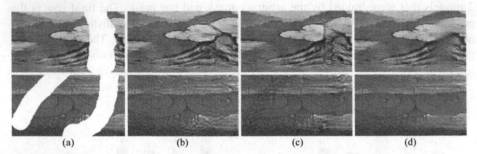

| (a) | (b) | (c) | (d) |

Fig. 8. Ablation study. Given the masked images (a), our results (b) are plausible; in contrast, the results generated in the absence of the perceptual loss contain unreasonable content (c), and the results generated in the absence of the style loss are blurry (d).

Robustness. We used ten levels of simulated damages (from the slightest Level1 to the heaviest Level10) to test the robustness of our method. We set the PSNR and SSIM values generated by our method as the baseline, and then computed the differences of PSNR and SSIM between our method and the other 3 methods. Figure 9 shows that with the increase in the degree of damage, PSNR and SSIM of PConv, GMCNN and Gated Conv decrease faster than our method.

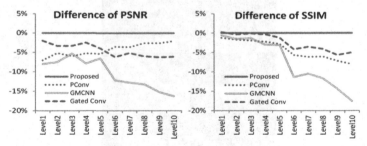

Fig. 9. Comparison of robustness. With the increase of damage degree, PSNR and SSIM of PConv, GMCNN and Gated Conv decrease faster than our method (ours are set as baseline).

5 Conclusion

We proposed a damage sensitive and original restoration driven (DSORD) Thanka mural inpainting method. Our end-to-end method inpaints multiple broken areas in a short time

(300 ms for multiple holes in a 512 × 512 image) and without any post-processing. Experimental results on simulated damage and real damage show that our method is able to restore damaged Thanka murals. Compared with other methods, our results are more realistic and closer to the original Thanka mural paintings.

Although our method was only applied and verified in Thanka mural restoration, it is designed for a category of real scenes that face the following common problems: 1) small dataset; 2) large irregular holes; 3) restoring original content rather than imaginary content. The method can be applied to the following fields: a) medical image inpainting; b) other mural restoration; c) damaged old photograph restoration; d) ancient calligraphic handwriting restoration, etc.

Acknowledgements. This work is jointly supported by NSFC (Grant No. 61862057), the Program for Innovative Research Team of SEAC ([2018]98), and the Fundamental Research Funds for the Central Universities (No. 31920200066).

References

Iizuka, S., et al.: Globally and locally consistent image completion. ACM Trans. Graph. (ToG) **36**(4), 107 (2017)

Barnes, C., et al.: PatchMatch: a randomized correspondence algorithm for structural image editing. ACM Trans. Graph. (ToG). **28**(3), 24 (2009). ACM

Liu, G., et al.: Image inpainting for irregular holes using partial convolutions. In: Proceedings of the European Conference on Computer Vision (ECCV) (2018)

Yu, J., et al.: Free-form image inpainting with gated convolution. In: Proceedings of the IEEE International Conference on Computer Vision (2019)

Wang, W., et al.: Research outline and progress of digital protection on thangka. Adv. Top. Multimedia Res. 67 (2012)

Goodfellow, I., et al.: Generative adversarial nets. In: NIPS'14: Proceedings of the 27th International Conference on Neural Information Processing Systems, vol. 2, pp. 2672–2680, December 2014

Wang, Y., et al.: Image inpainting via generative multi-column convolutional neural networks. In: Advances in Neural Information Processing Systems, pp. 331–340 (2018)

Wang, H., et al.: Inpainting of Dunhuang murals by sparsely modeling the texture similarity and structure continuity. J. Comput. Cult. Heritage (JOCCH) **12**(3), 17 (2019)

Yu, T., et al.: End-to-end partial convolutions neural networks for Dunhuang grottoes wall-painting restoration. In: Proceedings of the IEEE International Conference on Computer Vision Workshops (2019)

Pathak, D., et al.: Context encoders: feature learning by inpainting. In: Proceedings of the IEEE conference on computer vision and pattern recognition (2016)

Karras, T., et al.: Progressive growing of GANs for improved quality, stability, and variation. arXiv preprint arXiv: 1710.10196 (2017)

Yu, J., et al.: Generative image inpainting with contextual attention. In: Proceedings of the IEEE Conference on Computer Vision and Pattern Recognition (2018)

Nazeri, K., et al.: EdgeConnect: generative image inpainting with adversarial edge learning. arXiv preprint arXiv: 1901.00212 (2019)

Zhang, W., et al.: A precise-mask-based method for enhanced image inpainting. Math. Probl. Eng. **2016**, 1–5 (2016)

Richard, M.M.O.B., et al.: Fast digital image inpainting. In: Appeared in the Proceedings of the International Conference on Visualization, Imaging and Image Processing. (VIIP 2001), Marbella, Spain (2001)

Chang, Y.L., et al.: Free-form video inpainting with 3D gated convolution and temporal patchGAN. In: Proceedings of the IEEE International Conference on Computer Vision (2019)

Isola, P., et al.: Image-to-image translation with conditional adversarial networks. In: Proceedings of the IEEE conference on computer vision and pattern recognition (2017)

Gatys, L.A., et al.: A neural algorithm of artistic style. arXiv preprint arXiv: 1508.06576 (2015)

Russakovsky, O.: Imagenet large scale visual recognition challenge. Int. J. Comput. Vis. 115(3), 211–252 (2015)

Ronneberger, O., Fischer, P., Brox, T.: U-Net: convolutional networks for biomedical image segmentation. In: Navab, N., Hornegger, J., Wells, W.M., Frangi, A.F. (eds.) MICCAI 2015. LNCS, vol. 9351, pp. 234–241. Springer, Cham (2015). https://doi.org/10.1007/978-3-319-24574-4_28

Xiong, W., et al.: Foreground-aware image inpainting. In: Proceedings of the IEEE Conference on Computer Vision and Pattern Recognition, pp. 5840–5848 (2019)

SEGM: A Novel Semantic Evidential Grid Map by Fusing Multiple Sensors

Junhui Li[1,2(✉)], Heping Li[3], and Hui Zeng[1,2]

[1] School of Automation and Electrical Engineering, University of Science
and Technology Beijing, Beijing 100083, China
812899049@qq.com, hzeng@163.com
[2] Beijing Engineering Research Center of Industrial Spectrum Imaging, Beijing 100083, China
[3] Institute of Automation, Chinese Academy of Sciences, Beijing 100190, China
heping.li@ia.ac.cn

Abstract. A map that can fully express environment is of great significance for intelligent robots. Traditional grid maps such as occupancy grid map can only express the simple "free or occupied". So, it is difficult to meet the demand of tasks in a complex environment. This paper proposes a framework called Semantic Evidential Grid Map (SEGM) for generating a semantic evidential grid map based on Dempster-Shafer theory of evidence, which can combine laser scanner and stereo camera to make a map more accurate in expressing the environment without losing any details. As a result, this work allows a better handling of uncertainty information and provides more detailed semantic representations. It will be very helpful for intelligent robots or autonomous vehicles to execute tasks such as localization, perception and navigation in a complex environment. The experimental results on the KITTI CITY dataset show that the proposed method can provide a more detailed and accurate map representation than the traditional methods.

Keywords: Robot vision · Semantic evidential grid map · Theory of evidence · Multi-sensors

1 Introduction

In the new coronavirus epidemic that outbroke at the end of 2019, intelligent robots have played a significant role, such as helping to diagnose diseases, measure temperature by face recognition, patrol and disinfect all the time. A map that can accurately and fully represent the surrounding environment is essential for these intelligent systems to finish the above-mentioned tasks successfully and efficiently.

The uncertainty caused by noise from sensor data has a great impact on the map expression. There are serval methods to deal with the uncertainty such as Bayesian, Fuzzy and Evidential Theory. The traditional map commonly used for robot navigation is occupancy grid map [1, 2], which expresses the probability of obstacles in each

---

J.Li—Student

Y. Peng et al. (Eds.): PRCV 2020, LNCS 12305, pp. 155–166, 2020.
https://doi.org/10.1007/978-3-030-60633-6_13

grid by using Bayesian framework. This kind of map has some shortcomings such as the inadequate representation of environment, slow updating speed, and insensitivity to dynamic obstacles. Therefore, the evidential approach based on Dempster-Shafer theory (DST) [3] was proposed. It can handle both noisy and conflicting information directly and help fuse various sensor information effectively.

There are numerous works on building evidential grid maps using different types of sensors, such as laser scanner [4], radar [5] and stereo camera [6]. And there are some approaches using a variety of different sensors together to improve the accuracy of the map [7, 8]. Particularly, with the continuous upgrade of robot tasks and the urgent need for driverless cars in the complex urban environment, maps are required with the ability that can express higher-level semantics in the environment such as roads, sidewalks, traffic signals and even lane lines. Moras [9] made use of DST in the context of mobile perception in dynamic environments and Kurdej et al. [10] proposed to merge a prior map of semantic contexts into an evidential grid map which further improved the expressive ability of the map. Although the method in [7] used multi-sensor including laser and stereo camera, the camera is only used as a supplement to detect obstacles and it has been not fused with laser further. The method in [10] only used a laser to distinguish "Occupied" and "Free" and got semantic information from a prior semantic map.

This paper proposes a framework for constructing a semantic evidential grid map. It uses an improved method of evidence theory to fuse stereo camera and laser scanner. During this process, a novel mass function and fusion rule according to the sensor characteristics is used to express the environment more completely and accurately. What's more, semantic information is directly obtained by segmenting the images from stereo camera, which means we can obtain a semantic evidential grid map using a stereo camera.

This paper is organized as follows. Section 2 outlines the details of our proposed approach. Section 3 describes the experimental procedure to validate our methodology and discusses experimental results. Section 4 draws conclusions.

2 Approach

As shown in Fig. 1, this paper proposes a framework of generating a semantic evidential grid map from two sensors' data. Firstly, we establish laser scanner and stereo camera model to map sensor data into the evidential grid map and generate two evidential grid maps, one for the laser sensor $(M_{L,t})$ and the other for the stereo camera $(M_{S,t})$, and then fuse them to get a temporary grid map at time t $(M_{T,t})$. Subsequently, we fuse the temporary map and global map from 0 to t − 1 to express the entire world. Finally, a global map which can represent the environment where vehicles can drive is output.

2.1 Dempster-Shafer Theory of Evidence

DST is a method to model uncertainty of information similar to Bayesian without strong constraints. However, it also has its limitations that the assumption about evidence independence is not easy to achieve in reality. But due to the independence of different sensors, it can be applied to sensor data fusion reliably. In DST, there is an exclusive set

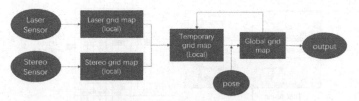

Fig. 1. Overview of the proposed framework. We assume that the laser and the stereo camera data are aligned and the corresponding pose is known in advance.

$\Omega = \{\omega_1, \omega_2, \ldots, \omega_n\}$, which can express different states of the environment in the real world, being called the frame of discernment (FOD).

The most important difference with the theory of probability is that a mass of evidence is attributed not only to a single subset but also to any subset of the Ω, including the empty set (ϕ) and the whole set (Ω). Beliefs about some pieces of evidence are modeled by the attribution of mass to the corresponding set. This mass assignment m^Ω, known as a mass function or basic belief assignment (BBA), is defined as a mapping:

$$m : 2^\Omega \to [0, 1]$$
$$\sum_{A \subseteq \Omega} m(A) = 1 \tag{1}$$
$$m(\phi) = 0$$

Multiple BBAs can be combined using the fusion rules under the premise of being independent of each other. Assuming that m_1 and m_2 are the mass functions defined in the same FOD, the rule of evidence fusion marked as "$\oplus$" is defined as follows:

$$m_1 \oplus m_2(A) = \frac{1}{1 - K} \sum_{B \cap C = A} m_1(B) \cdot m_2(C) \tag{2}$$

Where K is the normalization constant:

$$K = \sum_{B \cap C = \phi} m_1(B) \cdot m_2(C) \tag{3}$$

Note that the premise of applying this formula is that two sets of mass functions are defined in the same FOD.

2.2 Evidential Mapping

The process of generating an evidential grid map at time t can be visualized in Fig. 2, which consists of three main steps: firstly, we will model the scanner and stereo binocular camera, including defining their FODs and corresponding BBAs. Then, we will generate two local grid maps $M_{L,t}$ and $M_{S,t}$. Secondly, after unifying the coordinate of $M_{L,t}$ and $M_{S,t}$, a local temporary fusion map at time t $M_{T,t}$ will be obtained by combining them with the formula (2), and finally, the $M_{T,t}$ will be adjusted according to the position and orientation of the robot. The formula (2) will be used again to get a global map $M_{G,t}$ from time 0 to t.

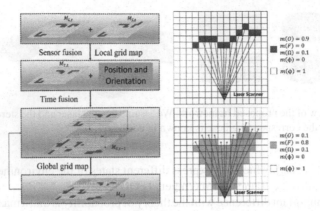

Fig. 2. Left is the visualization of generating a global evidential grid map at time t. **Right** is an example of defining the BBA of laser scanner.

2.3 Sensor Model of Mapping Data into Grid Map

The model of laser scanner

The laser sensor can provide more reliable information about whether there are obstacles in a specific location or not. At time t, the laser sensor can provide a 2D laser scanning result obtained by a rapid rotation of a laser beam which consists of the position of obstacles. Cells the laser beam crossed are considered as Free (F), while the laser beam impacted are considered as Occupied (O). Laser scanner can only distinguish between these two states, so the FOD is defined as $\Omega^L = \{F, O, \Omega, \phi\}$, where F represents free, O represents occupied, and Ω and ϕ represent complete set and empty set, respectively. We will use the method proposed in [11] with a little modification to define the BBA of each cell in $M_{L,t}$, which increases the O evidence on the impacted cells and the F evidence on the crossed cells:

If the cell is impacted, we increase the O evidence as same as [11]:

$$m_{L,t}(O) = \lambda \qquad m_{L,t}(F) = 0$$
$$m_{L,t}(\Omega) = 1 - \lambda \qquad m_{L,t}(\phi) = 0 \qquad (4)$$

If the cell is crossed, we increase not only F evidence but also O evidence because the laser scanner cannot detect the obstacle sometimes:

$$m_{L,t}(F) = \lambda_F \qquad m_{L,t}(O) = \lambda_O$$
$$m_{L,t}(\Omega) = 1 - \lambda_F - \lambda_O \qquad m_{L,t}(\phi) = 0 \qquad (5)$$

λ is determined by the confidence of laser scanner. An example is shown in Fig. 2.

The model of Stereo Camera

We use the binocular stereo camera as the second sensor in the system because it can record more details of the environment, such as obstacles that are difficult to detect by a 2D laser scanner due to its height and angle limitations. However, the information obtained from images is not very reliable because of some factors like illumination and scale. Therefore, we must pay more attentions when defining the BBA to ensure the accuracy of map representation. We need three steps to establish $M_{s,t}$.

Disparity Estimation

To map the images captured by the stereo camera into the 2D aerial view map, we have to know the depth information of pixels. For binocular stereo cameras, calculating a disparity map between two images is a general approach. Commonly used methods such as BM, SGBM and GC algorithm provided by OpenCV function library are easy to implement but have slow calculation speed, low accuracy, and difficulty in matching weak texture regions which adversely affect the mapping accuracy of the grid map. Therefore, we adopt the disparity estimation method of proportional belief propagation based on ELAS mentioned in [12], whose accuracy and speed have been greatly improved compared with above methods.

Image Semantic Segmentation

In order to fully express details of the environment, the information obtained from images should not only have the difference of free and occupied, but also include high-level semantic information. With the development of deep learning, semantic segmentation technology has made great progress. We use the method proposed in [13] to perform semantic segmentation and obtain semantic classifications such as road, pedestrian, vehicle, tree, grass, sky and so on as shown in Fig. 3.

Defining the BBA

According to the result of semantic segmentation, the FOD of each cell in $M_{S,t}$ is defined as $\Omega^S = \{Obstacle, Road, Person, Car, Grass / Sidewalk, \Omega, \phi\}$, where *Obstacle* represents other fixed obstacles except the above mentioned in Ω^S.

Firstly, we map the semantic information on the image to the grid map. For every pixel on the image, we have known its coordinates on the image (u, v), disparity information d and semantic tags *type* now. According to the method proposed in [14] based on the geometric structure of binocular stereo camera (see Fig. 3), the projection relationship of a pixel on the image plane (u, v) to the grid map plane $P_D(x, y)$ is as follows:

$$\begin{cases} x = \frac{b}{2} + \frac{(u-u_0)b}{d} \\ y = \frac{fb}{d} \end{cases} \tag{6}$$

Where d is the corresponding disparity, b is the length of baseline of stereo camera, f is the focal length, (u_0, v_0) is the principal point of the image.

Sequentially, the BBA of each cell in the grid map $M_{S,t}$ should be defined. Unlike laser scanner, we must consider more factors to express the map. As shown in Fig. 4, the resolution is the size of a cell (m). A pixel mapped to a cell obviously has a different weight according to its different position like green and red point. The closer a point is to the center of the cell, the higher its weight. Therefore, we use a two-dimensional Gaussian distribution to describe the weight as follows:

$$weight = e^{-\frac{(x-x_{cell})^2+(y-y_{cell})^2}{2\sigma^2}} \tag{7}$$

Where (x, y) is the coordinates of projection point and (x_{cell}, y_{cell}) is the center coordinates of the cell.

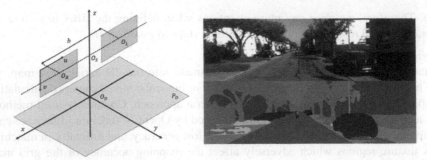

Fig. 3. Geometric structure of binocular stereo camera coordinate system. The line connecting the optical centers of left and right camera is called baseline, and its middle point is recorded as origin O_S. The center of mapping plane O_D is the projection of O_S.

At the same time, for the reason that the accuracy of pixels captured by camera decreases as the distance from the camera's optical center increases, we design a reliability factor which will drop after the distance exceeds the reliability threshold to attenuate the influence of the remote points as shown in Fig. 4 with definition as follows:

$$factor = 1 - \frac{1}{1 + e^{-(distance-D)}} \tag{8}$$

Where D equals to $40b$, which represents the reliable range of the camera.

Now, we can get the BBA definition of each grid in $M_{S,t}$. Take semantic classification as *Car* as an example:

$$m_{s,t}(\{Car\}) = \lambda * factor * weight \quad m_{s,t}(B) = 0$$
$$m_{s,t}(\Omega) = 1 - \lambda * factor * weight \quad m_{s,t}(\phi) = 0 \tag{9}$$

Where B is all the remaining subset of Ω^S except Car, Ω and ϕ.

What's more, there is still a problem in generating $M_{S,t}$ that numerous pixels with different semantic may be projected into the same cell which brings us a great challenge. To solve it, we use the DST to combine the BBA of multiple pixels to deal with the conflict.

$$m_{s,t} = m_{S,t}^{pixel1} \oplus m_{S,t}^{pixel2} \oplus \cdots \oplus m_{S,t}^{pixeln} \tag{10}$$

2.4 Fusion of Sensor Map

Maps obtained by different sensors have their own advantages and disadvantages, so we combine $M_{L,t}$ and $M_{S,t}$ with DST to improve the quality of the map. But there is a premise of using DST that the FOD should be consistent. Accordingly, Ω^L should be expanded to Ω^S by the method proposed in [10]:

$$T(\{F\}) = \{Road, Grass/Sidewalk\}$$
$$T(\{O\}) = \{Obstacle, Person, Car\} \tag{11}$$

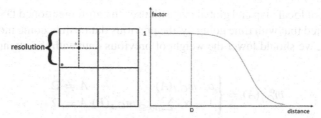

Fig. 4. Left is an example of the distribution of projection points. Right is the trust factor curve.

And the BBA will be also expended:

$$m_{L,t}^{\Omega^S}(B) = m_{L,t}^{\Omega^L}(A)/card\,(T(A)) \tag{12}$$

Where B is the single element in $T(A)$.

In order to fully retain the advantages of sensors and reduce the negative effects caused by shortcomings of sensors, the similarity between BBAs [15] are calculated by:

$$sim = \sqrt{(M_1 - M_2)^T D(M_1 - M_2)/2} \tag{13}$$

Where $M_i = [m_i(A_1), m_i(A_2), \cdots, m_i(A_n)]^T$, n is the number of factors in FOD, D is an n × n matrix with $D_{i,j} = |A_i \cap A_j|/|A_i \cup A_j|$.

Similarity reflects the degree of mutual support of the two BBAs. According to the characteristics of two sensors, BBAs are modified with similarity as follows:
If the cell is observed by laser scanner:

$$m_{L,t}^{sim}(A) = \begin{cases} m_{L,t}(O) & A = O \\ sim * m_{L,t}(F) & A = F \\ 1 - \sum_{B \neq \Omega} m_{L,t}^{sim}(A) & A = \Omega \end{cases} \tag{14}$$

If the cell is not observed by laser scanner:

$$m_{S,t}^{sim}(A) = \begin{cases} sim * m_{c,t}(A) & A \neq \Omega \\ 1 - \sum_{B \neq \Omega} m_{c,t}^{sim}(B) & A = \Omega \end{cases} \tag{15}$$

Then the local temporary fusion grid map $M_{T,t}$ will be obtained with the combination of $M_{L,t}$ and $M_{S,t}$ using DST.

$$m_{T,t} = m_{L,t}^{sim} \oplus m_{S,t}^{sim} \tag{16}$$

2.5 Global Map

In order to build a perception of the overall environment, we need to fuse the local maps at each moment. In this paper, we focus on the generation and expression of maps. So, we assume that the position and orientation in the world coordinate system of sensors at every timestamp have been already known in advance.

The fusion of local map and global map still uses the aforementioned DST. However, it should be noted that with time going by, the previous data will become more and more unreliable, i.e., we should lower the weight of previous data with the attenuation factor α:

$$m_{G,t}^{\alpha}(A) = \begin{cases} \alpha \cdot m_{G,t}(A) & A \neq \Omega \\ 1 - \alpha \cdot \sum_{B \neq \Omega} m_{G,t}(B) & A = \Omega \end{cases} \quad (17)$$

Finally, a global map $M_{G,t}$ from time 0 to t can be obtained:

$$m_{G,t} = m_{G,t-1}^{\alpha} \oplus m_{T,t} \quad (18)$$

For the storage of the map and use for robots subsequently, we traverse the map and select the subset who has the largest evidence value as the cell's state, and then represent it in different colors. In particular, if a cell' state is Ω, it means that this grid is in conflict which probably caused by dynamic obstacles. Therefore, we set these cells to the unknown state to ensure the safety of robots when performing tasks.

3 Experiments

We test our map construction framework on the KITTI CITY dataset. The laser scanner data is simulated by selecting points in a plane from the 3D Lidar data points with the sensor height as the limit. The camera's position and orientation at every time are estimated by ORBSLAM [16].

3.1 Multi-sensor Grid Map Results

The fusion grid map at a certain time is shown in Fig. 5. The resolution of the map is set to 0.2×0.2 m, which means one cell represents 0.04 m^2 in the real world. Different colors correspond to different states that fixed obstacle cells are represented as blue, road cells for driving are gray, cars are purple, pedestrian crossings or grass are light green, pedestrians are red, and unknown areas are black.

As shown in Fig. 5, the laser grid map provides a reliable description of range. However, it cannot detect cars parked on the side of road due to height restrictions as indicated by the green circle. In contrast, the stereo camera can capture this information very well. But large noise in images may impact the accuracy of grid map. The fusion map using traditional method combines two grid maps, including both pros and cons, such as the car in the green circle and noises in the yellow circle. Our method tends to believe in the more reliable evidence when fusing BBAs, i.e., the range of environment detected by the laser scanner and the low obstacles detected by the stereo camera are more reliable.

3.2 Global Map

The color meaning of a cell in the global map is consistent with fusion map. The KITTI CITY dataset[1] consists of 4545 pairs of binocular images and corresponding laser data. Figure 6 shows the global grid map result.

---

[1] http://www.cvlibs.net/datasets/kitti.

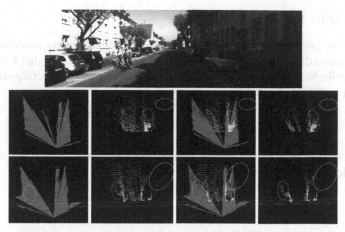

Fig. 5. An example of local grid maps of sensors. **Top** is the picture taken by camera. **Middle** and **Bottom** are laser grid map, stereo grid map, fusion grid map with traditional method and our method of two local areas of street in the top picture. (Color figure online)

As a comparison, we also show the maps generated by the traditional way [7]. As we can see obviously, the global grid map generated by our method is more accurate and robust. In the laser gird map, our method is more sensitive to obstacles that are difficult to detect as shown in red circles. In the stereo grid map, noise points are greatly reduced and the detection of obstacles is more accurate in our map. For example, we can easily see the interval of each car while cars are almost connected in the traditional stereo grid map as shown in blue circles. The fusion grid map combines the advantages of two sensor grid maps. Compared with the traditional map, it provides a more robust estimation of environment. And the more human-friendly interpretation of environment is extremely helpful for robots to make more humane decisions.

Fig. 6. Global grid map result. The left is the entire map. The right is detailed maps in the red box with original BBA (top) and our approach (bottom). From left to right, it's the laser, stereo, and fusion grid map. (Color figure online)

3.3 Quantitative Evaluation

To evaluate an evidential grid map, Entropy and Specificity were proposed in [17]. Entropy indicates the consistency of BBA's distribution in a cell. A higher Entropy indicates the distribution is more inconsistent, i.e., this cell contains conflicting information. Entropy can be calculated as follow:

$$E = -\sum_{A \subseteq \Omega} m(A) \cdot \ln(pl(A)) \tag{19}$$

Where pl(A) is the plausibility express of mass function which represents the highest confidence level of evidence for subset A.

$$pl(A) = \sum_{A \subseteq B \neq \phi} m(B) \ \forall A \in \Omega \tag{20}$$

Specificity indicates the credibility of the mass function distribution. A higher specificity corresponds to a more credible mass distribution. The calculation formula is as follows:

$$S = \sum_{A \subseteq \Omega, A \neq \phi} \frac{m(A)}{card(A)} \tag{21}$$

Where $card(A)$ means the number of factors in subset A.

As shown in Fig. 7. The average Entropy of fusion grid map (blue line is the traditional method and purple line is ours) is higher than the other two grid maps, because a lot of conflicts occur during the fusion of evidence. But the modification of fusion rules helps reduce the inconsistency greatly. For the Specificity, the laser grid map shows the best trust and the stereo grid map shows the worst one, which corresponds to their characteristics. The fusion grid map using traditional method also shows a bad trust, because it inherits both advantages and disadvantages. Our method greatly improved the credibility with taking essences and removing dross.

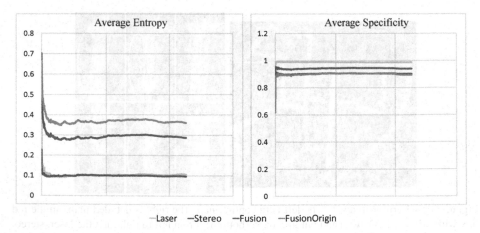

Fig. 7. Average Entropy and Specificity for every frame.

Therefore, we can get a conclusion that the evidential grid map by using the framework proposed in this paper can be more accurate and complete in describing environment. And the semantic information can extremely help robots perform more complex decisions and tasks.

4 Conclusion

This paper proposes a framework of generating a semantic evidential grid map based on DST which makes full use of laser scanner and binocular stereo camera. The experimental results on KITTI CITY dataset prove that it can provide a more accurate map which can describe a more complex environment for a robot executing tasks. The possible future research directions will focus on the detection of dynamic obstacles by fully making use of those conflicting information. However, the proposed method takes an average of 0.46 s for each set of data. This is not enough for real-time applications. So, how to reduce computation complexity is another direction for us in the future.

Acknowledgment. This work is supported by the National Natural Science Foundation of China (Grant No. 61973029).

References

1. Nuss, D., Reuter, S., Thom, M., et al.: A random finite set approach for dynamic occupancy grid maps with real-time application [J]. Int. J. Robot. Res. **37**(8), 841–866 (2018)
2. Singh, R., Nagla, K.S.: Sonar sensor model for the precision measurement to generate robust occupancy grid map [J]. MAPAN **34**(2), 239–257 (2019)
3. Shafer, G.: A Mathematical Theory of Evidence [M]. Princeton university press, New Jersey (1976)
4. Moras, J., Dezert, J., Pannetier, B.: Grid occupancy estimation for environment perception based on belief functions and PCR6 [C]. In: Signal Processing, Sensor/Information Fusion, and Target Recognition XXIV. International Society for Optics and Photonics, vol. 9474, p. 94740P (2015)
5. Mullane, J., Adams, M.D., Wijesoma, W.S.: Evidential versus Bayesian estimation for radar map building [C]. In: 2006 9th International Conference on Control, Automation, Robotics and Vision, pp. 1–8. IEEE (2006)
6. Yu, C., Cherfaoui, V., Bonnifait, P.: Evidential occupancy grid mapping with stereo-vision [C]. In: 2015 IEEE Intelligent Vehicles Symposium (IV), pp. 712–717. IEEE (2015)
7. Valente, M., Joly, C., de La Fortelle, A.: Fusing laser scanner and stereo camera in evidential grid maps [C]. In: 2018 15th International Conference on Control, Automation, Robotics and Vision (ICARCV), pp. 990–997. IEEE (2018)
8. Valente, M., Joly, C., de La Fortelle, A.: Evidential SLAM fusing 2D laser scanner and stereo camera [J]. Unmanned Syst. **7**(03), 149–159 (2019)
9. Moras, J., Cherfaoui, V., Bonnifait, P.: Moving objects detection by conflict analysis in evidential grids [C]. In: 2011 IEEE Intelligent Vehicles Symposium (IV), pp. 1122–1127. IEEE (2011)
10. Kurdej, M., Moras, J., Cherfaoui, V., et al.: Map-aided evidential grids for driving scene understanding [J]. IEEE Intell. Trans. Syst. Mag. **7**(1), 30–41 (2015)

11. Trehard, G., Pollard, E., Bradai, B., et al.: On line mapping and global positioning for autonomous driving in urban environment based on evidential SLAM [C]. In: 2015 IEEE Intelligent Vehicles Symposium (IV), pp. 814–819 IEEE (2015)

12. Ran, Z.: Research on fast and low power binocular parallax estimation method [D]. University of Science and Technology Beijing (2020)

13. Zhu, Y., Sapra, K., Reda, F.A., et al.: Improving semantic segmentation via video propagation and label relaxation [C]. In: Proceedings of the IEEE Conference on Computer Vision and Pattern Recognition, pp. 8856–8865 (2019)

14. Perrollaz, M., Yoder, J.D., Spalanzani, A., et al.: Using the disparity space to compute occupancy grids from stereo-vision [C]. In: 2010 IEEE/RSJ International Conference on Intelligent Robots and Systems, pp. 2721–2726. IEEE (2010)

15. Jousselme, A.L., Grenier, D., Bossé, É.: A new distance between two bodies of evidence [J]. Inform. Fusion 2(2), 91–101 (2001)

16. Mur-Artal, R., Tardós, J.D.: Orb-slam2: an open-source slam system for monocular, stereo, and rgb-d cameras [J]. IEEE Trans. Robot. 33(5), 1255–1262 (2017)

17. Yager, R.R.: Entropy and specificity in a mathematical theory of evidence [J]. Int. J. Gen Syst 9(4), 249–260 (1983)

FF-GAN: Feature Fusion GAN for Monocular Depth Estimation

Ruiming Jia[1](✉) ⓘ, Tong Li[1] ⓘ, and Fei Yuan[2] ⓘ

[1] School of Information Science and Technology, North China University of Technology, Beijing, China
jiaruiming@ncut.edu.cn
[2] Digital Content Technology and Media Service Research Center, Institute of Automation, Chinese Academy of Sciences, Beijing, China

Abstract. Since the results of CNN methods for monocular depth estimation generally suffer the problem of visual dissatisfaction, we propose Feature Fusion GAN (FF-GAN) to address this issue. First, an end-to-end network based on encoder-decoder structure is proposed as the generator of FF-GAN, which can exploit the information of different scales. The encoder of our generator fuse features in different levels with a feature fusion module. The component which can obtain the information of multi-scale receptive field is the main part of the decoder of our generator. Second, in order to match the generator, the discriminator of FF-GAN is designed to efficiently learn the information of different scales by applying pyramid structure. Experiments on public datasets demonstrate the effectiveness of our generator and discriminator. Compared with the CNN methods, the results predicted by FF-GAN are significantly improved in terms of texture loss and edge blur while ensuring accuracy, and the visual effect is better.

Keywords: Conditional Generative Adversarial Network · Encoder-decoder · Monocular depth estimation · Receptive field

1 Introduction

As a key step of scene reconstruction and scene understanding tasks such as 3d object recognition, segmentation, depth estimation is an important research direction of computer vision. Compared with the depth estimation from stereo images and video sequences [1, 2], the progress of depth estimation from a single image is relatively slow for it's an ill-posed problem. Therefore, monocular depth estimation has certain research significance and challenges.

Traditional methods of depth estimation are mainly based on binocular vision system. They are easily affected by the quality of the disparity map. Recently, there are more

This work is supported by Science and Technology Application Innovation Research Project of Zhejiang Provincial Public Security Department (2019TJYYCX007) and North China University of Technology Student Science and Technology Activities Project Funding (218051360020XN114/004).

© Springer Nature Switzerland AG 2020
Y. Peng et al. (Eds.): PRCV 2020, LNCS 12305, pp. 167–179, 2020.
https://doi.org/10.1007/978-3-030-60633-6_14

and more works based on deep learning [3–7], which are generally implemented by Convolutional Neural Network (CNN). The CNN method has higher accuracy compared with traditional methods. However, it is hard to learn the rich texture information and its results generally have the problems of edge blur and unclear texture due to the target loss. Meanwhile, Conditional Generation Adversarial Network (CGAN) [8], which is widely applied for image-to image translation task, can generate rich and clear target images. Thence, this paper proposes a CGAN method for monocular depth estimation: Feature Fusion Generation Adversarial Network (FF-GAN), the specific contributions are as follows:

1) We propose Feature Fusion Encoder-Decoder Network (FFEDNet) as the generator of FF-GAN. Feature Fusion Module (FFM) is proposed for the encoder of FFEDNet, which improves the use of multi-scale features. Multi-Scale Res-Block (MSRB) is the main component of the decoder. It can obtain information of multiple size receptive field.
2) The discriminator of FF-GAN is proposed named Pyramid Matching Network (PMNet), which can fuse the features from different levels in discriminator.
3) Experiments on NYUD v2 (NYU Depth v2) [9] and KITTI [10] datasets show that FF-GAN can predict depth images with richer detailed texture information and sharper edges while ensuring accuracy.

2 Related Work

CNN Methods of Depth Estimation. The CNN methods perform well in monocular depth estimation. Eigen et al. [11] directly predicted dense depth maps from single RGB image in an end-to-end manner by a CNN model. Liao et al. [12] improve the accuracy by Conditional Random Field (CRF). Laina et al. [4] constructed an encoder-decoder network, which adopted ResNet [13] as encoder and up-projection blocks as the main component of the decoder. What's more, the method in [14] combined information of semantic, surface normal, optical flow to build the network in a multi-task learning manner. Although these methods increase the accuracy in common metrics, the estimated depth maps are also more blurred.

Generative Adversarial Network. Generative Adversarial Networks (GAN) [15] is widely used in computer vision tasks. The generator of GAN uses random noise as input, and it can produce a large variety of generated samples which are difficult to monitor. Therefore, Mirza et al. [8] proposed Conditional Generative Adversarial Network (CGAN), its generated samples are constrained by adding conditions in GAN. Classic CGAN networks include Pix2pix [16], CycleGAN [17], et al. To further improving the accuracy of the network model, the attention mechanism had been introduced in the network in [18]. The image generated by GAN has more detailed texture, sharper edge and better human visual effect than images predicted by CNN.

3   Method

CNN methods based on encoder-decoder structure are widely used for monocular depth estimation, but their predicted results have the problems of edge blur and texture missing, as shown in the blue and black rectangular boxes respectively in Fig. 1. Therefore, we adopt a CGAN method, its generator is FFEDNet which is improved from the traditional encoder-decoder network, and its discriminator is PMNet.

Input   Eigen [11]   Ground truth

Fig. 1.   Visual examples of CNN method by Eigen [11].

The overall framework of FF-GAN is shown in Fig. 2, RGB images are the input of FFEDNet. x_{fake} are the output of FFEDNet. x_{real} are the target images. x_{fake} and RGB images form "false" image pairs, x_{real} and RGB images form "true" image pairs. "True and false" image pairs are input of PMNet. PMNet learns to distinguish "true" and "false" image pairs while FFEDNet attempts to generate depth maps close to x_{real}. FFEDNet and PMNet formed a confrontation, and they constantly improved their abilities during the training.

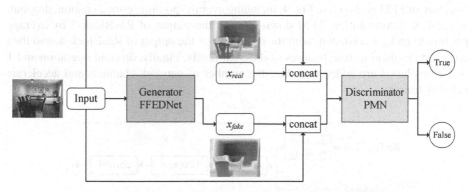

Fig. 2.   Overall framework of our FF-GAN.

3.1   The Generator of FF-GAN: FFEDNet

FFEDNet is a CNN model that can realize monocular depth estimation from end to end. It consists of three parts: encoder, FFB and decoder as shown in Fig. 3, the backbone network of the encoder is ResNet50 [13]. FFM can fuse the feature maps which are outputs of ResBlock-3, ResBlock-4. The decoder contains four cascaded MSRB, a 3 × 3 convolution, up-sampling by bilinear interpolation.

Encoder. We use ResNet50 and a 1 × 1 convolution to encode the low-dimensional RGB image with a resolution of 304 × 228 to high-dimensional feature maps with a resolution of 10 × 8. And the outputs of ResBlock-3 and ResBlock-4 in ResNet50 are the input of FFM.

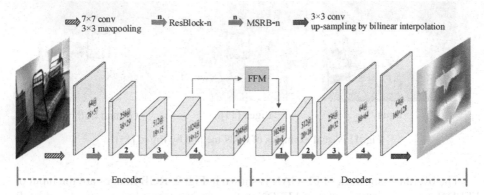

Fig. 3. Network architecture of our FFEDNet.

FFM. We propose FFM to fuse and further utilize the features from different levels. For the reason that encoder-decoder networks generally use the output of encoder's last layer as decoder's input, which lacks the fusion of features at different levels. The network structure of FFM is shown in Fig. 4, including average pooling, concatenation, dropout, and a 1 × 1 convolution. FFM down-samples the output of ResBlock-3 by average pooling to make it consistent with the resolution of the output of ResBlock-4, and then concatenates them to fuse features of different levels. Finally, dropout operation and 1 × 1 convolution are added to reduce the number of network channels and accelerate network training.

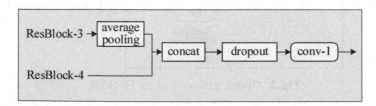

Fig. 4. Network structure of FFM.

Decoder. The decoder of FFEDNet mainly contains four MSRB as shown in Fig. 3. MSRB can obtain the information of multi-scale receptive field during the decoding process. The resolution of MSRB's output is twice as MSRB's input, and the number of MSRB's output channels is half of MSRB's input. After four MSRB, we use a 3 × 3 convolution to reduce the number of channels to 1. Finally, the binary interpolation is added for the consistent resolution with the input image. As shown in Fig. 5, MSRB contains the Residual up-projection unit and the Multi-Scale unit.

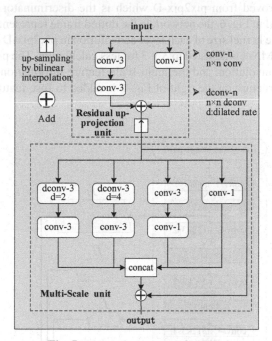

Fig. 5. Network structure of MSRB.

Residual Up-projection Unit. We construct a Residual up-projection unit inspired by the Up-Projection Blocks in [4]. It can decrease the number of channels and improve the resolution of feature maps. As shown in Fig. 5, it adopts the residual structure, the main part uses two 3 × 3 convolutions, and skip connection part uses 1 × 1 convolution. Finally, bilinear interpolation samples the resolution of the feature maps to double.

Multi-Scale unit. Multi-Scale unit is added after Residual up-projection unit for obtaining the information of different receptive fields. Large size of receptive field can obtain global structural information, and small size of receptive field can obtain local detail information. Multi-Scale unit can obtain more information by concatenating convolutions with different kernel sizes. In Fig. 5, the input through 3 × 3 dilated convolution [19] with dilation of 2, 3 × 3 standard convolution; 3 × 3 dilated convolution with dilation of 4, 1 × 1 standard convolution; 3 × 3 standard convolution, 1 × 1 standard convolution; 1 × 1 standard convolution. And their outputs are concatenated. Finally, the input of Multi-Scale unit is added to the concatenation result.

3.2 The Discriminator of FF-GAN: PMNet

The generator of FF-GAN increases the use of multi-scale features, the discriminator of FF-GAN is aimed at doing the same thing. The feature pyramid structure is often used for target detection to achieve more accuracy [20]. And we propose Pyramid Matching Network (PMNet) as discriminator of FF-GAN inspired by it.

PMNet is improved from pix2pix-D which is the discriminator of pix2pix [16]. PMNet is illustrated in Fig. 6, the network in the dotted frame represents pix2pix-D. The difference is that the kernel size of each layer convolution in pix2pix-D is 4 × 4 while the kernel size in the PMNet is adjusted to 3 × 3 for less calculation. The pix2pix-D consists of five layers of convolution, and the multi-scale feature information is not used. The feature pyramid structure on the right of Fig. 6 is added to fuse feature from different layers.

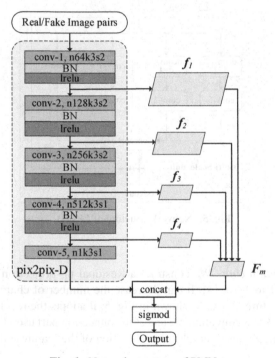

Fig. 6. Network structure of PMNet.

In Fig. 6, "conv-1, n64k3s2" means the first layer of convolution, with filters' number of 64, kernel size of 3 × 3, and step size of 2. f_1–f_4 represents the feature map of different levels. For using multi-scale feature information, we unify f_1–f_4 to the same spatial resolution, and obtain the average response F_m as multi-feature fusion information, which is defined as formula (1):

$$F_m = \frac{1}{n} \sum_{i=1}^{n} S(f_i), \tag{1}$$

Where $S(\cdot)$ represents the down-sampling function. We adopt the nearest neighbor interpolation method to accelerate the network training. By applying the multi-feature information F_m, the network can synthesize the feature information at all levels and increase the robustness.

3.3 Loss Function

FF-GAN takes the RGB image x as input, the target image y as the constraint condition during training. According to the definition of CGAN loss function:

$$L_{cGAN}(G, D) = E_{x,y}[\log D(x, y)] + E_{x,z}[\log(1 - D(x, G(x, z)))], \tag{2}$$

FFEDNet learns how to minimize L_{cGAN}. And PMNet learns how to maximize L_{cGAN}. Isola et al. [16] pointed that adding additional losses such as L_1 to GAN loss functions can improve the accuracy. Therefore, the final loss function of CGAN is:

$$G' = \arg\min_G \max_D C_g L_{cGAN} + C_c L_1, \tag{3}$$

Where C_g and C_c are the weight of L_{cGAN} and L_1 respectively, $C_g = 1$ and $C_c = 100$.

4 Experiments

4.1 Evaluation Metrics

As for previous work [4, 6], the validity of our method is evaluated by the metrics of mean absolute relative error (*Rel*), mean log10 error (*Log10*), root mean squared error (*Rms*) and threshold accuracy ($\delta < 1.25^i$, $i = 1, 2, 3$). The expressions are as follows:

$$Rel = \frac{1}{|N|} \sum_{i=1}^{N} |y_i - \hat{y}_i|/\hat{y}, \tag{4}$$

$$Log10 = \frac{1}{|N|} \sum_{i=1}^{N} |\lg y_i - \lg \hat{y}_i|, \tag{5}$$

$$Rms = \sqrt{\frac{1}{|N|} \sum_{i=1}^{N} \|y_i - \hat{y}_i\|^2}, \tag{6}$$

$$\delta = \max(\frac{y_i}{\hat{y}_i}, \frac{\hat{y}_i}{y_i}) < thr, \tag{7}$$

Where i means the pixel index, N is the total number of all pixels in a depth map, and $y_i, \hat{y}_i$ represent the true depth value at pixel i and the depth value predicted by the network respectively.

4.2 Comparison with Other Methods on NYUD v2

The NYUD v2 dataset contains 120,000 of RGB-depth images with a resolution of 640 × 480, which includes 249 indoor scenes for training and 215 indoor scenes for testing. We adopt the data preprocessing method in [21]. The comparison with other algorithms is shown in Table 1. The last row represents our CGAN method FF-GAN. FFEDNet represents our CNN method, which is trained only by FFEDNet. And its loss function is L1. FFEDNet obtains the best performance in most metrics. FFEDNet is higher than Liao [12] method 14.2% in $\delta < 1.25$, and it is lower than Liu [6] method 34.6% in RMS. Besides, FF-GAN is also better than other algorithms in most metrics, but lower than FFEDNet.

Table 1. Comparison of different depth estimation methods on NYUD v2.

Method	Error (The lower, the better)			Accuracy (The higher, the better)		
	Abs rel	Avg $\log_{10}$	RMS	$\delta < 1.25$	$\delta < 1.25^2$	$\delta < 1.25^3$
Zhuo [5]	0.305	0.122	1.04	0.525	0.838	0.962
Eigen [11]	0.215	−	0.907	0.611	0.887	0.971
Liu [6]	0.230	0.095	0.824	0.614	0.883	0.975
Liao [12]	0.179	0.067	0.665	0.703	0.931	**0.985**
FFEDNet	**0.145**	**0.063**	**0.539**	**0.803**	**0.948**	0.984
FF-GAN	0.153	0.066	0.554	0.788	0.941	0.984

4.3 Visual Comparisons on NYUD v2

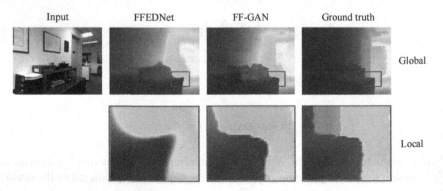

Fig. 7. Visual Comparison of FFEDNet and FF-GAN.

This section compares the visual effect of depth maps predicted by FFEDNet and FF-GAN. As can be seen from the global images in Fig. 7, the depth map predicted by

FF-GAN is more similar to the target map and the texture of objects on the desktop is richer than FFEDNet. By observing the comparison of the local images, we can find that the local edge of the depth maps of FFEDNet is vague, while FF-GAN is clear and correct. Obviously, the common metrics can't reflect the definition of the depth maps, so we especially use Perceptual Index (*PI*) [22] to evaluate the definition of results. *PI* is defined as formula (8):

$$PI = \frac{10 - Ma + NIQE}{2}, \tag{8}$$

Where *Ma* and *NIQE* refer to the metrics in [23] and [24] respectively. *PI* can evaluate the definition of the depth map, and the lower the *PI*, the clearer the edge of the depth map. The image definition comparison between FFEDNet and FF-GAN on NYUD v2 is shown in Table 2. FF-GAN is 18.9% lower than FFEDNet. Compared with FFEDNet, the depth map predicted by FF-GAN is significantly improved in terms of texture loss and edge blur, and the visual effect is better and clearer.

Table 2. Image definition comparison of FF-GAN and FFEDNet.

Method	PI (The lower, the better)
FFEDNet	12.068
FF-GAN	**9.788**

4.4 Experiments of Different Generators and Discriminators Combinations on NYUD v2

We make experiments with different combinations of generators and discriminators. The results on NYUD v2 data set are shown in Table 3. Pix2pix-G refers to the generator of pix2pix, pix2pix-D refers to the discriminator of pix2pix. As can be seen from Table 3, pix2pix-G+PMNet is superior to pix2pix (pix2pix-G+pix2pix-D) in most metrics, which proves the effectiveness of PMNet. FF-GAN (FFEDNet+PMNet) is about 20.7% higher than pix2pix and about 21.0% higher than pix2pix-G+PMNet in $\delta < 1.25$, which proves the effectiveness of FFEDNet.

The corresponding visual comparison of Table 3 is shown in Fig. 8. By observing the table edge in the first row, the chair legs in the second row and the chair in the third row, it can be obviously found that the edge of the depth maps predicted by FF-GAN is more accurate and clearer. By visual comparison, it can be concluded that FF-GAN's generator and discriminator have certain validity, and the performance of FF-GAN is better than pix2pix.

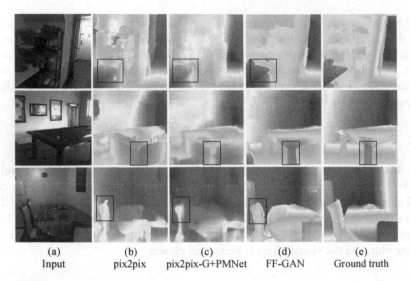

| (a) | (b) | (c) | (d) | (e) |
| Input | pix2pix | pix2pix-G+PMNet | FF-GAN | Ground truth |

Fig. 8. Visual comparison of different generator and discriminator combinations.

Table 3. Experimental comparison of different generator and discriminator combinations.

Generator	Discriminator	Error (The lower, the better)			Accuracy (The higher, the better)		
		Abs rel	Avg $\log_{10}$	RMS	$\delta < 1.25$	$\delta < 1.25^2$	$\delta < 1.25^3$
pix2pix-G	pix2pix-D	0.210	0.094	0.756	0.653	0.882	0.962
pix2pix-G	PMNet	0.207	0.091	0.744	0.651	0.890	0.966
FFEDNet	PMNet	**0.153**	**0.066**	**0.554**	**0.788**	**0.941**	**0.981**

4.5 Comparison with Other Methods on KITTI

KITTI is a dataset about outdoor scenes. We use the method of [21] to preprocess the data, and finally have 46402 images for training and 3200 images for test. The comparison of different methods on KITTI is shown in Table 4. FF-GAN is much higher than methods by Eigen [11], Ma [21], and Godard [3] in all metrics, and it is higher than Yang [7] 1.4% in $\delta < 1.25$. Moreover, it can be seen from Table 4 that the performance of FF-GAN on KITTI is better than FFEDNet.

Visual results of FF-GAN on KITTI are shown in Fig. 9, we can find that the results of FF-GAN are highly similar to the target images. In summary, FF-GAN has achieved good performance on KITTI in both metrics and visual effect. It demonstrates that FF-GAN has certain validity for depth estimation on indoor and outdoor scene data sets.

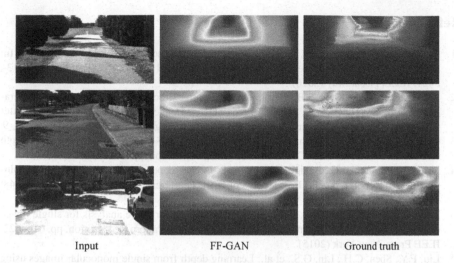

Input   FF-GAN   Ground truth

Fig. 9. Visual examples of FF-GAN on KITTI.

Table 4. Comparison of different depth estimation methods on KITTI.

Method	Error (The lower, the better)			Accuracy (The higher, the better)		
	Abs rel	Avg $\log_{10}$	RMS	$\delta < 1.25$	$\delta < 1.25^2$	$\delta < 1.25^3$
Eigen [11]	0.190	–	7.156	0.692	0.899	0.967
Ma [21]	0.208	–	6.266	0.591	0.900	0.962
Godard [3]	0.115	–	5.164	0.858	0.946	0.970
Yang [7]	**0.097**	–	**4.442**	**0.888**	0.958	0.980
FFEDNet	0.101	0.045	5.392	0.870	0.969	**0.992**
FF-GAN	0.102	**0.044**	5.149	0.878	**0.971**	**0.992**

5 Conclusion

This paper proposes FF-GAN which can predict the depth map from a single RGB image. The generator and discriminator of FF-GAN can fuse features of different levels to make better use of the multi-scale information. FFEDNet combines features by adding FFM in encoder and building decoder with MSRB. PMNet applies pyramid structure to the discriminator of Pix2pix [16] to fuse feature. Experiments show that both FFEDNet and PMNet perform well, and compared with the CNN methods, the predicted depth maps by FF-GAN have richer texture and clearer edge.

References

1. Ha, H., Im, S., Park, J., et al.: High-quality depth from uncalibrated small motion clip. In: Proceedings of IEEE Conference on Computer Vision and Pattern Recognition, pp. 5413–5421. IEEE, Piscataway (2016)
2. Karsch, K., Liu, C., Kang, S.B.: Depth transfer: depth extraction from videos using nonparametric sampling. In: Hassner, T., Liu, C. (eds.) Dense Image Correspondences for Computer Vision, pp. 173–205. Springer, Cham (2016). https://doi.org/10.1007/978-3-319-23048-1_9
3. Godard, C., Aodha, O., Firman, M., et al.: Digging into self-supervised monocular depth estimation. In: International Conference on Computer Vision, pp. 3827–3837 (2019)
4. Laina, I., Rupprecht, C., Belagiannis, V., et al.: Deeper depth prediction with fully convolutional residual networks. In: Proceedings of IEEE Conference on 3D vision, pp. 239–248. IEEE, Piscataway (2016)
5. Zhuo, W., Salzmann, M., He, X.M., et al.: Indoor scene structure analysis for single image depth estimation. In: Conference on Computer Vision and Pattern Recognition, pp. 614–622. IEEE Press, New York (2015)
6. Liu, F.Y., Shen, C.H., Lin, G.S., et al.: Learning depth from single monocular images using deep convolutional neural field. IEEE Trans. Pattern Anal. Mach. Intell. **38**(10), 2024–2039 (2016)
7. Yang, N., Wang, R., Stückler, J., Cremers, D.: Deep virtual stereo odometry: leveraging deep depth prediction for monocular direct sparse odometry. In: Ferrari, V., Hebert, M., Sminchisescu, C., Weiss, Y. (eds.) ECCV 2018. LNCS, vol. 11212, pp. 835–852. Springer, Cham (2018). https://doi.org/10.1007/978-3-030-01237-3_50
8. Mirza, M., Osindero, S.: Conditional generative adversarial nets. arXiv: 1411.1784v1. https://arxiv.org/abs/1411.1784. Accessed 20 April 2020
9. Silberman, N., Hoiem, D., Kohli, P., Fergus, R.: Indoor segmentation and support inference from RGBD images. In: Fitzgibbon, A., Lazebnik, S., Perona, P., Sato, Y., Schmid, C. (eds.) ECCV 2012. LNCS, vol. 7576, pp. 746–760. Springer, Heidelberg (2012). https://doi.org/10.1007/978-3-642-33715-4_54
10. Geiger, A., Lenz, P., Urtasun, R.: Are we ready for autonomous driving the KITTI vision benchmark suite. In: Conference on Computer Vision and Pattern Recognition, pp. 3354–3361. IEEE Press, New York (2012)
11. Eigen, D., Puhrsch, C., Fergus, R.: Depth map prediction from a single image using a multi-scale deep network. In: Proceedings of International Conference on Neural Information Processing Systems, pp. 2366–2374 (2014)
12. Liao, B., Li, H.W.: Image depth estimation model based on atrous convolutional neural network. J. Comput. Appl. **39**(1), 267–274 (2019)
13. He, K., Zhang, X., Ren, S., et al.: Deep residual learning for image recognition. In: Proceedings of IEEE Conference on Computer Vision and Pattern Recognition, pp. 770–778. IEEE, Piscataway (2016)
14. Ranjan, A., Jampani, V., Balles, L., et al.: Competitive collaboration: joint unsupervised learning of depth, camera motion, optical flow and motion segmentation. In: Proceedings of IEEE Conference on Computer Vision and Pattern Recognition, pp. 12240–12249 (2019)
15. Goodfellow, I., Pouget-Abadie, J., Mirza, M., et al.: Generative adversarial nets. In: International Conference on Neural Information Processing Systems, pp. 2672–2680 (2014)
16. Isola, P., Zhu, J.Y., Zhou, T.H., et al.: Image-to-image translation with conditional adversarial networks. In: Conference on Computer Vision and Pattern Recognition, pp. 5967–5976 (2017)
17. Zhu, J.Y., Park, T., Isola, P., et al.: Unpaired image-to-image translation using cycle-consistent adversarial networks. In: International Conference on Computer Vision, pp. 2242–2251 (2017)

18. Mejjati, Y.A., Richardt, C., Tompkin, J., et al.: Unsupervised attention-guided image-to-image translation. In: The 32nd Annual Conference on Neural Information Processing Systems, pp. 3693–3703 (2018)

19. Chen, L., Papandreou, G., Schroff, F., et al.: Rethinking atrous convolution for semantic image segmentation. arXiv Preprint: 1706.05587 (2017)

20. Liu, W., et al.: SSD: single shot multibox detector. In: Leibe, B., Matas, J., Sebe, N., Welling, M. (eds.) ECCV 2016. LNCS, vol. 9905, pp. 21–37. Springer, Cham (2016). https://doi.org/10.1007/978-3-319-46448-0_2

21. Ma, F., Karaman, S.: Sparse-to-dense: depth prediction from sparse depth samples and a single image. In: International Conference on Robotics and Automation, pp. 1–8. IEEE Press, New York (2018)

22. Blau, Y., Mechrez, R., Timofte, R., Michaeli, T., Zelnik-Manor, L.: The 2018 PIRM challenge on perceptual image super-resolution. In: Leal-Taixé, L., Roth, S. (eds.) ECCV 2018. LNCS, vol. 11133, pp. 334–355. Springer, Cham (2019). https://doi.org/10.1007/978-3-030-11021-5_21

23. Ma, C., Yang, C., Yang, X., et al.: Learning a no-reference quality metric for single-image super-resolution. Comput. Vis. Image Underst. 158(1), 1–16 (2017)

24. Mittal, A., Soundararajan, R., Bovik, A.C., et al.: Making a completely blind image quality analyzer. IEEE Sign. Process. Lett. 20(3), 209–212 (2013)

GCVNet: Geometry Constrained Voting Network to Estimate 3D Pose for Fine-Grained Object Categories

Yaohang Han[1], Huijun Di[1(✉)], Hanfeng Zheng[1], Jianyong Qi[2], and Jianwei Gong[2]

[1] Beijing Laboratory of Intelligent Information Technology, School of Computer Science and Technology, Beijing Institute of Technology, Beijing, China
{yaohanghan,ajon,zhf_cs}@bit.edu.cn
[2] Intelligent Vehicle Research Center, School of Mechanical Engineering, Beijing Institute of Technology, Beijing, China
{qijianyong,gongjianwei}@bit.edu.cn

Abstract. As a fundamental AI problem, monocular 3D pose estimation has received much attention. This paper addresses the challenge of estimating full perspective model parameters, including object pose and camera intrinsics, from a single 2D image of fine-grained object categories. To tackle this highly ill-posed problem, we propose a Geometry Constrained Voting Network (GCVNet). It is a unified end-to-end network consisting of four synergic task-specific subnetworks: 1) Fine-grained classification subnetwork, offering fine-grained 3D shape priors. 2) Voting subnetwork, generating 2D measurements. 3) Segmentation subnetwork, providing a foreground mask for voting. 4) PnP subnetwork, estimating the perspective parameters via explicit geometric reasoning, as well as constraining the classification subnetwork to provide proper 3D priors and the voting subnetwork to generate a group of geometric consistent 2D measurements, rather than independent voting for each 2D measurement in the literature. Experiments on challenging datasets demonstrate the superior performance of GCVNet.

Keywords: Pose estimation · Geometric reasoning · Differentiable PnP

1 Introduction

Monocular 3D pose estimation aims to recovery 3D orientation and translation relative to a canonical frame of the specific object from a single 2D image. It is a long-standing fundamental AI problem [7,14,16,22,24], and is currently a hot topic with remarkable progress by using deep neural networks [4,12,20,21,28,32]. Accurate pose estimations are necessary in various practical applications such as

This is a student paper. Special thanks to Megvii Inc. for providing training resources for the paper.

robotic manipulation [29], augmented reality [23], autonomous driving [2], and etc.

This paper addresses a problem of estimating full perspective model parameters, including object pose and camera intrinsics, from a single 2D image of fine-grained object categories. By handling the full perspective model, we do not require pre-calibration of the camera. Thus we can handle any input images without known camera intrinsics. This is in contrast to the previous works which assume known camera intrinsics [10,15,18,20,21,27,32]. Such an assumption would limit their applications. Some works don't assume the known camera intrinsics but use the approximate/simplified camera projection model to reduce the difficulty [19,31]. However, an inadequate camera model may limit their performance. Some other works even totally bypass the camera intrinsics by casting the pose estimation into a classification problem [4,12,26,28]. Nevertheless, the discretization could produce a coarse result.

By considering fine-grained object categories, we deal with the situation of the high intra-class variations presenting in the practical applications. The situation is different from the case in the widely-used benchmark datasets for pose estimation [1,9,32] where only one instance of each category is considered. We also expect that the accurately estimated pose could be exploited to improve the performance of fine-grained object categorization [25].

The addressed problem is highly ill-posed. The challenges lie in two aspects: 1) Inferring 3D information from a single 2D image is essentially ill-posed. While estimating full perspective model parameters will amplify the illness further, due to the high degree of freedom. Therefore, the estimation will heavily rely on the proper prior. 2) The objects from fine-grained categories are usually visually similar, but with high intra-class geometric variations. This leads to the difficulty that the fine-grained categories could not be classified correctly from 2D images, and the wrong sub-category label will provide wrong prior for pose estimation.

To tackle the above-discussed challenges, we propose a GCVNet, as shown in Fig. 1. It is a unified end-to-end network consisting of four synergic task-specific subnetworks: 1) Fine-grained classification subnetwork, offering fine-grained 3D shape models to fit the input 2D images better and to enable a tighter geometric constraint. 2) Voting subnetwork, generating 2D measurements with an effective context aggregation scheme. 3) Segmentation subnetwork, providing a foreground mask for voting. 4) PnP (Perspective-n-Point) subnetwork, estimating the full perspective model parameters via explicit geometric reasoning, as well as constraining the classification subnetwork to provide proper 3D priors and the voting subnetwork to generate a group of geometric consistent 2D measurements.

Compared with two popular paradigms for pose estimation, GCVNet has the following two advantages:

- **The end-to-end methods** [15,17,30–32] based on direct pose regression may generate the results which are geometric non-plausible or with limited accuracy, due to the lack of explicit geometric reasoning in the networks. In the experiment, we compare GCVNet with the work of [30], which also addresses the same problem as us, and show the superior performance of GCVNet.

- **The two-step methods** [6,10,18,19,21,27] regresses the 2D keypoints first, and then solve the PnP for pose estimation. The separation of the two steps could lead to the situation that each 2D point is generated independently, which may not be geometric consistent/optimal and would limit the accuracy of the pose estimated later via the PnP algorithm. While in GCVNet, we integrate these two steps into one single unified network, which allows end-to-end training to propagate the geometric constraint from PnP subnetwork to voting subnetwork to generate a group of geometric consistent 2D points. Experiments show the improved performance of GCVNet over the two-step methods.

In summary, the contributions of our work are:

- A novel idea to make PnP algorithm differentiable, which allows the PnP algorithm to be a part of the network, and thus allows explicit geometric reasoning in the network, yielding geometric plausible and more accurate pose estimation.
- A unified geometry constrained voting network to estimate full perspective model parameters from a single 2D image, with end-to-end training to boost the classification and voting by propagating geometric constraints from PnP subnetwork, improving the accuracy of pose estimation in further.
- State-of-the-art 3D rotation estimation performance and comparable result on other pose estimation metrics on the benchmark datasets StanfordCars3D and CompCars3D.

2   Proposed Method

2.1   Overview

Given an RGB image, we address the 3D pose estimation problem via solving the full perspective model parameters, which consists of the camera intrinsics and the 6DoF pose, including a 3D rotation $\mathbf{R} \in SO(3)$ and a translation $\mathbf{t} \in \mathbb{R}^3$.

The overview of our proposed Geometric Constrained Voting Network is shown in Fig. 1. The GCVNet mainly consists of 6 parts: (a) Encoder, (b) Decoder, (c) Classification Subnetwork, (d) Segmentation Subnetwork, (e) Voting Subnetwork and (f) PnP Subnetwork. The main pipeline of GCVNet has three steps. Firstly, we employ the encoder/decoder backbone to extract features from an image, then regress the pixel-wise 2D offset field and segmentation to vote the position of 2D control points by the voting subnetwork. Secondly, we classify the image to get the fine-grained 3D model for 3D control points by classification subnetwork. Finally, we use 2D-3D correspondence to predict the full perspective model parameters by PnP subnetwork. The proposed voting and PnP subnetwork will be introduced in detail in the rest of this section.

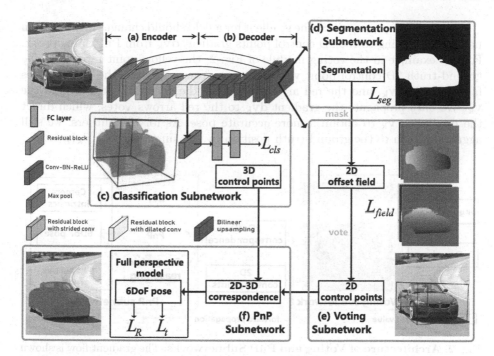

Fig. 1. Overview of our proposed GCVNet. Each 2D image is firstly sent to (a) **Encoder** for feature extraction. The extracted features are processed by the (c) **Classification Subnetwork** to generate 3D control points and fed to (b) **Decoder** for obtaining the high-resolution features. After that, the high-resolution features are fed to (d) **Segmentation Subnetwork** to get a foreground mask and (e) **Voting Subnetwork** to generate the 2D offset filed. In voting subnetwork, the foreground mask of 2D offset field votes to the 2D control points. Finally, (f) **PnP Subnetwork** receives the 2D-3D control points to estimate the full perspective model parameters, including camera intrinsics and 6DoF pose.

2.2 Voting Subnetwork

The voting subnetwork only has a voting module, as shown in Fig. 2. To vote the 2D control points and back propagate the gradient, we adopt the context aggregation voting scheme in the voting module. Unlike the PVNet [20], to calculate the gradient of a vote independently, we use the 2D offset-field with length rather than the unit-vector field. Specifically, the value of 2D offset-field $\mathbf{F_k}(\mathbf{p_i})$ is defined as the offset from a pixel $\mathbf{p_i}$ in the field to the given 2D control point $\mathbf{v_k}$:

$$\mathbf{F_k}(\mathbf{p_i}) = \mathbf{v_k} - \mathbf{p_i} \tag{1}$$

Voting Module. Our voting module is employed in both forward and backward propagation. During forward-propagation, each pixel $\mathbf{F_k}(\mathbf{p_i})$ on the foreground mask $\mathcal{M}$ votes a 2D hypotheses point independently. We use the average of those 2D hypotheses points to calculate the estimated 2D control points $\hat{\mathbf{v}}_\mathbf{k}$.

During back-propagation, the gradient for a field should change the prediction position according to the 2D control points gradient $\mathbf{dv_k}$ from PnP subnetwork. For the example in the voting module of Fig. 2, the yellow point represents the ground-truth 2D control point $\mathbf{v_k}$, the green arrows represent the outlier-votes far away from $\mathbf{v_k}$, and the red arrows represent the accurate votes around the $\mathbf{v_k}$. We *only* propagate the gradient $\mathbf{dv_k}$ to the red arrows/votes, which direct the prediction $\hat{\mathbf{v}}_k$ to obtain a more accurate pose. As for the outliers, we still aggregate them to the ground-truth position by the supervision of $\mathbf{F_k(p_i)}$.

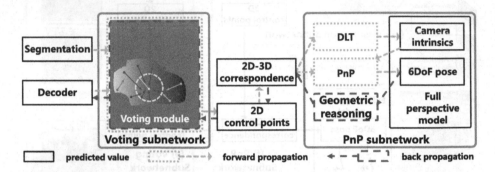

Fig. 2. Architecture of Voting and PnP Subnetworks. The gradient flow is shown among the different modules in the two subnetworks. In the voting subnetwork, we adopt the context aggregation voting scheme that *only* back-propagate the 2D control points gradient to the red vectors/votes in the offset field. In PnP subnetwork, there is no gradient flow through DLT (Direct Linear Transformation) and PnP (Perspective-n-Point) modules. By explicit reasoning the gradient, we use geometric reasoning module to backpropagate the gradient from 6DoF pose to the 2D control points.

2.3   PnP Subnetwork

Our proposed PnP subnetwork, including DLT, PnP, and Geometric reasoning modules, is shown in Fig. 2. The DLT module could roughly estimate the camera intrinsics and 6DoF pose from 2D-3D points as the initial value. The PnP module with the iterative BA then refines the 6DoF pose as the final results. To achieve the end-to-end estimation, we design the geometric reasoning module that back propagates the 6DoF pose error to the 2D control points. To make the rest of this section easy to follow, we summarize the symbols in Table 1.

DLT Module. The DLT module provides the camera intrinsics by solving the over-determined equations. Given n pairs of correspondence 3D control points and the estimated 2D points, we build n equations on the full perspective model parameters as follows:

$$\mathbf{v_i} = 1/\mathbf{s_i}[\mathbf{KT}_\theta(\mathbf{P_i})] \tag{2}$$

where $\mathbf{T}_\theta$ is the rigid transform based 6DoF parameters θ, $\mathbf{s_i}$ is the scale factor.

PnP Module. Given the camera intrinsics, the PnP algorithm could solve the BA-refined 6DoF pose by minimizing the total re-projection error:

$$E(\boldsymbol{\theta}) = \sum_{i=1}^{n} \mathbf{e_i}(\boldsymbol{\theta})^{\mathbf{T}} \mathbf{e_i}(\boldsymbol{\theta}) \tag{3}$$

where the i-th re-projection error is $\mathbf{e_i}(\boldsymbol{\theta}) = 1/s_i[\mathbf{KT}_\theta(\mathbf{P_i})] - \mathbf{v_i}$.

Geometric Reasoning Module. To achieve the end-to-end 6DoF pose estimation, we have to make the PnP algorithm differentiable. However, it is hard work because the PnP algorithm is not a common function but a set of over-determined equations. Inspired by the BA iterations, we want to calculate the gradient of $\boldsymbol{\theta}$ around the extreme point of re-projection error. With the condition of extreme point, we build the bridge of $\boldsymbol{\Delta}\mathbf{v_i}$ and $\boldsymbol{\Delta}\boldsymbol{\theta}$ in Eq. 6. Intuitively, our calculated gradient will give the one more step direction of $\mathbf{v_i}$ after BA and help to learn the $\mathbf{v_i}$ that produces more accurate pose. Our explicit geometric reasoning to the problem has three steps in total:

Firstly, given the $\boldsymbol{\Delta}\mathbf{v_i}$, the i-th disturbed re-projection error is:

$$\acute{\mathbf{e}}_i(\boldsymbol{\theta}) = \mathbf{e_i}(\boldsymbol{\theta}) - \boldsymbol{\Delta}\mathbf{v_i} \tag{4}$$

Secondly, given the $\boldsymbol{\Delta}\boldsymbol{\theta}$, the total disturbed re-projection error is:

$$\begin{aligned}\acute{E}(\boldsymbol{\theta} + \boldsymbol{\Delta}\boldsymbol{\theta}) &= \sum_{i=1}^{n} \acute{\mathbf{e}}_i(\boldsymbol{\theta} + \boldsymbol{\Delta}\boldsymbol{\theta})^{\mathbf{T}} \acute{\mathbf{e}}_i(\boldsymbol{\theta} + \boldsymbol{\Delta}\boldsymbol{\theta}) \\ &\approx \sum_{i=1}^{n} [\acute{\mathbf{e}}_i(\boldsymbol{\theta}) + \mathbf{J_i}\boldsymbol{\Delta}\boldsymbol{\theta}]^{\mathbf{T}} [\acute{\mathbf{e}}_i(\boldsymbol{\theta}) + \mathbf{J_i}\boldsymbol{\Delta}\boldsymbol{\theta}]\end{aligned} \tag{5}$$

where the first order expansion is used as an approximation. $\mathbf{J_i} = \frac{\partial \mathbf{e_i}(\boldsymbol{\theta})}{\partial \boldsymbol{\theta}} \in \mathbb{R}^{2\times 6}$ is the Jacobi matrix.

Table 1. Symbol description in Sect. 2.3 PnP subnetwork.

Symbol	Meaning
$\mathbf{P_i} \in \mathbb{R}^{3\times 1}$	the i-th 3D control point in the world coordinate
$\mathbf{v_i} \in \mathbb{R}^{2\times 1}$	the i-th 2D control point in the image coordinate
$\boldsymbol{\theta} \in \mathbb{R}^{6\times 1}$	the 6DoF parameters including 3D $\mathbf{R}$ and 3D translation $\mathbf{t}$
$\mathbf{K}$	the camera intrinsics in the form of $\left[\begin{smallmatrix} f_x & 0 & c_x \\ & f_y & c_y \\ & & 1 \end{smallmatrix}\right]$
$\boldsymbol{\Delta}\boldsymbol{\theta}/\boldsymbol{\Delta}\mathbf{v_i}$	the small increments for the $\boldsymbol{\theta}$/the $\mathbf{v_i}$
$\mathbf{e_i}(\boldsymbol{\theta})/E(\boldsymbol{\theta})$	the i-th/total re-projection error using the 6DoF parameters $\boldsymbol{\theta}$
$\acute{\mathbf{e}}_i / \acute{E}$	the i-th/total re-projection error disturbed by the small increments $\boldsymbol{\Delta}$

Finally, the total re-projection error in Eq. 5 is at the extreme point after BA. So we use the constraint of the extreme point to bridge the $\mathbf{\Delta\theta}$ and $\mathbf{\Delta v_i}$:

$$\frac{\partial \acute{E}(\theta + \Delta\theta)}{\partial \Delta\theta} = 2\mathbf{H}\Delta\theta + 2\sum_{i=1}^{n}\mathbf{J_i^T}\acute{e}_i(\theta) = 0 \qquad (6)$$

where $\mathbf{H} = \sum_{i=1}^{n}\mathbf{J_i^T J_i}$ is the Hessian matrix. When the $\mathbf{\Delta\theta}$ and $\mathbf{\Delta v_i}$ tend to zeros, the constraint of re-projection error after BA will promise a proper direction for the gradient. The simplified solved gradient for back-propagation is:

$$\frac{\mathbf{d}\theta}{\mathbf{dv_i}} = \mathbf{H}^{-1}\sum_{i=1}^{n}\mathbf{J_i^T} \qquad (7)$$

2.4   Implementation

Network Architecture. Our feature extraction backbone has two parts: the encoder part uses a ResNet50 [11], and the decoder part has a few of skip connection from the different levels of the encoder part as [20]. Using the feature from ResNet50, our classification subnetwork consists of two-layer Multi-Layer Perceptron (MLP) to get the 3D model among the candidates. Assuming GCVNet takes the $H \times W \times 3$ image as input, each image has K 2D control points, and there are C classes of objects. Our backbone will output two tensors: the decoder $H \times W \times (2 + K \times 2)$ tensor for segmentation and offset-field estimation, and a $(C + 1)$ tensor for classification.

Figure 2 shows the architecture of voting and PnP subnetwork. We design them by geometric reasoning without trainable parameters. Based on the decoder of the backbone, our geometric reasoning module and voting module in Fig. 2 is implemented by custom extended function in PyTorch framework, and the forward propagation PnP module to obtain 6DoF pose is implemented by OpenCV.

Multi-Task Loss. We employ the Smooth $\mathcal{L}_1$ loss proposed in [5] as the loss of offset-field L_{field}, the loss of rotation L_R and the loss of translation L_T. Specifically, the L_{field} can be described as:

$$L_{field} = \sum_{k=1}^{n}\sum_{p_i \in \mathcal{M}}\mathcal{L}_1(\Delta F_k(p_i)|_x) + \mathcal{L}_1(\Delta F_k(p_i)|_y) \qquad (8)$$

where $\Delta F_k(p_i) = \hat{F}_k(p_i) - F_k(p_i)$. We employ the Cross-Entropy loss as the loss of segmentation L_{seg}, and the loss of classification L_{cls}. Above all, we supervise the learning of the tasks jointly with a multi-task loss:

$$L = \lambda_{field}L_{field} + \lambda_{seg}L_{seg} + \lambda_{cls}L_{cls} + \lambda_t L_t + \lambda_R L_R \qquad (9)$$

where the different λ are the hyper-parameters to control the composition of the losses L respectively.

3   Experiments

3.1   Datasets and Evaluation Metric

We evaluate the GCVNet on the two datasets: the CompCars3D [13] has $5,696$ images including 113 sub-classes of cars, and the StanfordCars3D [33] has $16,185$ images including 196 sub-classes of cars. For each image, there is only one car with the annotation provided by [30]. Each annotation includes a fine-grained 3D mesh model and the full perspective model parameters to align 3D and 2D. We adopt the well-established metrics in [6,30] as follows:

Rotation. We use the geodesic distance to evaluate the difference of rotation matrix R, and e_R gives the minimal angular distance:

$$e_R = \frac{\|log(R_{gt}^T R_{pred})\|_F}{\sqrt{2}} \tag{10}$$

We report the median of e_R, named *MedErr$_R$*, and the percentage of recalled examples $ACC_{R \leq \frac{\pi}{6}}$ with the threshold of $R \leq \frac{\pi}{6}$.

Translation. To make the distance metric scale-invariant, we use the relative translation distance e_t normalized by the ground truth. We report the median of e_t, named *MedErr$_t$*.

$$e_t = \frac{\|t_{gt} - t_{pred}\|_2}{\|t_{gt}\|_2} \tag{11}$$

Pose. We use the average normalized distance of all transformed model points in 3D space to evaluate the accuracy of 6DoF rigid transformation. We report the median of $e_{R,t}$, named *MedErr$_{R,t}$*.

$$e_{R,t} = \underset{X \in \mathcal{M}}{avg} \frac{d_{bbox}}{d_{img}} \cdot \frac{\|T_{gt}(X) - T_{pred}(X)\|_2}{\|t_{gt}\|_2} \tag{12}$$

where $T(X) = \mathbf{R}X + \mathbf{t}$ is the 3D transformation function. We use $X \in \mathcal{M}$ to represent the 3D point X of the ground truth 3D model $\mathcal{M}$. As the normalization term, d_{bbox} is the 2D bounding box diagonal of the given object in the image and d_{img} is the image diagonal. The ratio of d_{bbox} and d_{img} provides an unbiased metric for 3D pose evaluation without camera intrinsics [6].

Focal Length. We use the relative focal length error e_f normalized by the ground truth. We report the median of e_f, named *MedErr$_f$*.

$$e_f = \frac{f_{gt} - f_{pred}}{f_{gt}} \tag{13}$$

Projection. We use the average normalized distance of all transformed model points in 3D space to evaluate the accuracy of 6DoF rigid transformation. normalized by the ground truth. We report the $MedErr_P$ and $ACC_{P \leq 0.1}$.

$$e_P = \underset{X \in \mathcal{M}}{avg} \frac{\|\boldsymbol{\pi}_{gt}(\boldsymbol{X}) - \boldsymbol{\pi}_{pred}(\boldsymbol{X})\|_2}{d_{bbox}} \tag{14}$$

where $\boldsymbol{\pi}(\boldsymbol{X}) = \mathbf{K}(\mathbf{RX} + \mathbf{t})$ is the full perspective model in Eq. 2.

3.2   Training Details

As [6], we also resize and pad images to a spatial resolution of 512×512 maintaining the aspect ratio. All available samples is used by the origin train-test split for training and evaluation. In this way, we train our model with a batch size of 32 on the platform of 8 NVIDIA GeForce 1080 Ti GPUs. The encoder part of our backbone use the ResNet-50 pre-trained weight on ImageNet [3], and the decoder part is initalized by Kaiming method [8]. During training, the Adam optimizer is employed. We set the initial learning rate as 0.001 and halve it every 20 epochs. We also apply online data augmentation including random cropping, mirroring, rotation and color jittering. To balance individual loss terms, we set $\lambda_{field} = 1.0$, $\lambda_{seg} = \lambda_{cls} = \lambda_t = 10.0$, $\lambda_R = 100.0$ in Eq. 9 according to the loss scale.

3.3   Experiment Results

Comparing with the SOTA. Table 2 shows our evaluation results on the StanfordCars3D and CompCars3D Datasets. We compare our model with the End-to-End (E2E) method [30] and the Two-Step method [6] which is the state-of-the-art results to our best known. As shown in Table 2, our end-to-end field-based model is superior on the rotation metric, which is an essential metric for full perspective model estimation. Without the iterative refinement of focal length on [6], our end-to-end GCVNet model achieves comparable results on other metrics. Also, benefit from our classification subnetwork, our method doesn't need the additional ground-truth label to achieve 2D-3D alignment. In Fig. 3, we show the 2D-3D alignment as the qualitative results .

Comparing with the Baseline. In Table 2, **Ours-TwoStep** methods are the self-ablation of **Ours-GCVNet** on the two datasets. By setting $\lambda_t = \lambda_R = 0.0$ in Eq. 9, **Ours-TwoStep** methods eliminate the influence of the end-to-end 6DoF pose supervision. Table 2 reveals that the GCVNet performs better than our two-step method without the end-to-end supervision on the full perspective model metrics. Compared with our two-step method to estimate the 2D points individually, the GCVNet shows the potential of the overall optimization by end-to-end pose estimation.

Table 2. Quantitative comparisons on the CompCars3D and StanfordCars3D datasets. The $\sqrt{}$ in **E2E** column indicates the method using end-to-end pose estimation. The others are the two-step methods. Red, green, **blue** color indicates the first, second and third place results.

Dataset	Method	E2E	Rotation		Translation	Pose	Focal	Projection	
			$MedErr_R$	$ACC_{R\leq\frac{\pi}{6}}$	$MedErr_t$	$MedErr_{R,t}$	$MedErr_f$	$MedErr_P$	$ACC_{P\leq0.1}$
Comp	$3DField$ [30]	$\sqrt{}$	5.24	97.6	0.330	0.235	**0.323**	0.0785	73.7
	GP^2C-LF [6]	–	5.23	**97.9**	0.261	0.186	0.297	0.0421	95.1
	GP^2C-BB [6]	–	**4.87**	98.1	0.255	0.184	0.295	0.0387	95.7
	Ours-TwoStep	–	4.37	98.1	0.322	0.190	0.379	0.0454	90.2
	Ours-GCVNet	$\sqrt{}$	3.99	98.4	**0.318**	**0.189**	0.376	**0.0431**	90.5
Stanford	$3DField$ [30]	$\sqrt{}$	5.43	**98.0**	0.233	0.180	**0.234**	0.0746	76.4
	GP^2C-LF [6]	–	5.38	98.3	0.193	**0.151**	0.201	**0.0372**	96.2
	GP^2C-BB [6]	–	**5.24**	98.3	0.192	0.147	0.207	0.0325	96.5
	Our-TwoStep	–	5.09	97.5	0.229	0.152	0.252	0.0378	93.6
	Our-GCVNet	$\sqrt{}$	4.92	97.5	**0.220**	0.146	0.243	0.0365	**94.6**

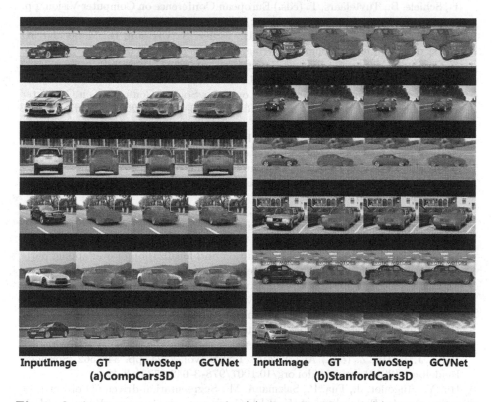

InputImage GT TwoStep GCVNet   InputImage GT TwoStep GCVNet
(a)CompCars3D   (b)StanfordCars3D

Fig. 3. Qualitative comparisons on the (a) CompCars3D and (b) StanfordCars3D datasets. We compare the full perspective parameters estimation results of our GCVNet with the TwoStep method mentioned in Table 2 and show the improved performance of GCVNet by 2D-3D alignment. **Best viewed in digital zoom.**

4 Conclusion

We propose the GCVNet to estimate full perspective model parameters from a single 2D image with the end-to-end supervision of the 6DoF pose. To achieve the end-to-end pose estimation, we propose the explicit geometric reasoning in the PnP subnetwork, as well as the context aggregation voting scheme in the voting subnetwork. Experiments show the superior performance of GCVNet and the extra gain from the end-to-end 6DoF pose supervision. We still have some limitations, like the classification accuracy, which will be followed in future work.

References

1. Brachmann, E., Krull, A., Michel, F., Gumhold, S., Shotton, J., Rother, C.: Learning 6D object pose estimation using 3D object coordinates. In: Fleet, D., Pajdla, T., Schiele, B., Tuytelaars, T. (eds.) European Conference on Computer Vision, pp. 536–551. Springer, Cham (2014). https://doi.org/10.1007/978-3-319-10605-2_35
2. Chen, X., Kundu, K., Zhang, Z., Ma, H., Fidler, S., Urtasun, R.: Monocular 3D object detection for autonomous driving. In: Proceedings of the IEEE Conference on Computer Vision and Pattern Recognition, pp. 2147–2156 (2016)
3. Deng, J., Dong, W., Socher, R., Li, L.J., Li, F.F.: ImageNet: a large-scale hierarchical image database. In: 2009 IEEE Computer Society Conference on Computer Vision and Pattern Recognition (CVPR 2009), 20–25 June 2009, Miami, Florida, USA (2009)
4. Elhoseiny, M., El-Gaaly, T., Bakry, A., Elgammal, A.: A comparative analysis and study of multiview cnn models for joint object categorization and pose estimation. In: International Conference on Machine Learning, pp. 888–897 (2016)
5. Girshick, R.: Fast R-CNN. In: 2015 IEEE International Conference on Computer Vision (ICCV) (2016)
6. Grabner, A., Roth, P.M., Lepetit, V.: GP2C: geometric projection parameter consensus for joint 3d pose and focal length estimation in the wild. In: The IEEE International Conference on Computer Vision (ICCV), October 2019
7. Grimson, W., Lozano-Perez, T.: Recognition and localization of overlapping parts from sparse data in two and three dimensions. In: Proceedings. 1985 IEEE International Conference on Robotics and Automation, vol. 2, pp. 61–66. IEEE (1985)
8. He, K., Zhang, X., Ren, S., Jian, S.: Delving deep into rectifiers: surpassing human-level performance on imagenet classification (2015)
9. Hinterstoisser, S., et al.: Model based training, detection and pose estimation of texture-less 3d objects in heavily cluttered scenes. In: Lee, K.M., Matsushita, Y., Rehg, J.M., Hu, Z. (eds.) ACCV 2012. LNCS, vol. 7724, pp. 548–562. Springer, Heidelberg (2013). https://doi.org/10.1007/978-3-642-37331-2_42
10. Hu, Y., Hugonot, J., Fua, P., Salzmann, M.: Segmentation-driven 6D object pose estimation. In: Proceedings of the IEEE Conference on Computer Vision and Pattern Recognition, pp. 3385–3394 (2019)
11. Jian, S.: Deep residual learning for image recognition. In: IEEE Conference on Computer Vision & Pattern Recognition (2016)
12. Kehl, W., Manhardt, F., Tombari, F., Ilic, S., Navab, N.: SSD-6D: making RGB-based 3D detection and 6d pose estimation great again. In: Proceedings of the IEEE International Conference on Computer Vision, pp. 1521–1529 (2017)

13. Krause, J., Stark, M., Deng, J., Fei-Fei, L.: 3D object representations for fine-grained categorization. In: Proceedings of the IEEE International Conference on Computer Vision Workshops, pp. 554–561 (2013)
14. Lai, K., Bo, L., Ren, X., Fox, D.: A scalable tree-based approach for joint object and pose recognition. In: Twenty-Fifth AAAI Conference on Artificial Intelligence (2011)
15. Li, Y., Wang, G., Ji, X., Xiang, Y., Fox, D.: DeepIM: deep iterative matching for 6D pose estimation. In: Proceedings of the European Conference on Computer Vision (ECCV), pp. 683–698 (2018)
16. Lowe, D.G.: Three-dimensional object recognition from single two-dimensional images. Artif. Intell. **31**(3), 355–395 (1987)
17. Mahendran, S., Ali, H., Vidal, R.: 3D pose regression using convolutional neural networks. In: Proceedings of the IEEE International Conference on Computer Vision, pp. 2174–2182 (2017)
18. Oberweger, M., Rad, M., Lepetit, V.: Making deep heatmaps robust to partial occlusions for 3D object pose estimation. In: Ferrari, V., Hebert, M., Sminchisescu, C., Weiss, Y. (eds.) ECCV 2018. LNCS, vol. 11219, pp. 125–141. Springer, Cham (2018). https://doi.org/10.1007/978-3-030-01267-0_8
19. Pavlakos, G., Zhou, X., Chan, A., Derpanis, K.G., Daniilidis, K.: 6-DoF object pose from semantic keypoints. In: 2017 IEEE International Conference on Robotics and Automation (ICRA), pp. 2011–2018. IEEE (2017)
20. Peng, S., Liu, Y., Huang, Q., Zhou, X., Bao, H.: PVNet: pixel-wise voting network for 6DoF pose estimation. In: Proceedings of the IEEE Conference on Computer Vision and Pattern Recognition, pp. 4561–4570 (2019)
21. Rad, M., Lepetit, V.: Bb8: a scalable, accurate, robust to partial occlusion method for predicting the 3D poses of challenging objects without using depth. In: Proceedings of the IEEE International Conference on Computer Vision, pp. 3828–3836 (2017)
22. Schneiderman, H., Kanade, T.: A statistical approach to 3D object detection applied to faces and cars. Carnegie Mellon University, The Robotics Institute (2000)
23. Shahrokni, A., Vacchetti, L., Lepetit, V., Fua, P.: Polyhedral object detection and pose estimation for augmented reality applications. In: Proceedings of Computer Animation 2002 (CA 2002), pp. 65–69. IEEE (2002)
24. Shimshoni, I., Ponce, J.: Finite-resolution aspect graphs of polyhedral objects. IEEE Trans. Pattern Anal. Mach. Intell. **19**(4), 315–327 (1997)
25. Sochor, J., Herout, A., Havel, J.: BoxCars: 3D boxes as CNN input for improved fine-grained vehicle recognition. In: Proceedings of the IEEE Conference on Computer Vision and Pattern Recognition, pp. 3006–3015 (2016)
26. Su, H., Qi, C.R., Li, Y., Guibas, L.J.: Render for CNN: viewpoint estimation in images using CNNs trained with rendered 3D model views. In: Proceedings of the IEEE International Conference on Computer Vision, pp. 2686–2694 (2015)
27. Tekin, B., Sinha, S.N., Fua, P.: Real-time seamless single shot 6D object pose prediction. In: Proceedings of the IEEE Conference on Computer Vision and Pattern Recognition, pp. 292–301 (2018)
28. Tulsiani, S., Malik, J.: Viewpoints and keypoints. In: Proceedings of the IEEE Conference on Computer Vision and Pattern Recognition, pp. 1510–1519 (2015)
29. Varley, J., DeChant, C., Richardson, A., Ruales, J., Allen, P.: Shape completion enabled robotic grasping. In: 2017 IEEE/RSJ International Conference on Intelligent Robots and Systems (IROS), pp. 2442–2447. IEEE (2017)

30. Wang, Y., et al.: 3D pose estimation for fine-grained object categories. In: Leal-Taixé, L., Roth, S. (eds.) ECCV 2018. LNCS, vol. 11129, pp. 619–632. Springer, Cham (2019). https://doi.org/10.1007/978-3-030-11009-3_38

31. Wu, J., et al.: Single image 3D interpreter network. In: Leibe, B., Matas, J., Sebe, N., Welling, M. (eds.) ECCV 2016. LNCS, vol. 9910, pp. 365–382. Springer, Cham (2016). https://doi.org/10.1007/978-3-319-46466-4_22

32. Xiang, Y., Schmidt, T., Narayanan, V., Fox, D.: PoseCNN: a convolutional neural network for 6D object pose estimation in cluttered scenes. arXiv preprint arXiv:1711.00199 (2017)

33. Yang, L., Luo, P., Change Loy, C., Tang, X.: A large-scale car dataset for fine-grained categorization and verification. In: Proceedings of the IEEE Conference on Computer Vision and Pattern Recognition, pp. 3973–3981 (2015)

Estimated Exposure Guided Reconstruction Model for Low-Light Image Enhancement

Xiaona Liu[1], Yang Zhao[1,2(✉)], Yuan Chen[1], Wei Jia[1], Ronggang Wang[2,3], and Xiaoping Liu[1]

[1] School of Computer Science and Information Engineering,
Hefei University of Technology, Hefei 230009, China
`yzhao@hfut.edu.cn`
[2] Peng Cheng Laboratory, Shenzhen 518000, China
[3] School of Electronic and Computer Engineering, Shenzhen Graduate School,
Peking University, Shenzhen 518055, China

Abstract. Low-light image enhancement is still a challenging task nowadays. On one hand, sensitive methods tend to reproduce light images with severe noise and color deviation. On the other hand, insensitive methods can recover clear and natural results but with much lower brightness. Hence, this paper analyzes several basic mathematical models and then proposes a low-light image enhancement network (LLIENet) based on a basic mathematical model, which contains several modules. First, a MaskNet is proposed to estimate the global illumination prior. Second, BaseNet and CoefficientNet are used to decompose the low-light image into a lightened base image and a subtle coefficient map. Finally, a RefineNet is added to further refine high-frequency details and suppress noise and color deviation. Extensive experiments are evaluated to demonstrate the superiority of the proposed method over several state-of-the-art methods.

Keywords: Low-light image enhancement · Weakly illuminated image

1 Introduction

With the popularity of smartphones, people can take photos anytime. However, images captured under low-light often have low visibility and contrast, and contain severe noise. Such images not only reduce visual quality but also go against

The first author is a student.
This work is supported by the grants of the National Natural Science Foundation of China (Nos. 61972129, 61877016), and the grant of the Key Research and Development Program in Anhui Province (No. 1804a09020036).

Electronic supplementary material The online version of this chapter (https://doi.org/10.1007/978-3-030-60633-6_16) contains supplementary material, which is available to authorized users.

Y. Peng et al. (Eds.): PRCV 2020, LNCS 12305, pp. 193–205, 2020.
https://doi.org/10.1007/978-3-030-60633-6_16

many computer vision tasks, such as target detection, object tracking and so on. Therefore, low-light image enhancement (LLIE) is a fundamental task in the field of image processing.

In the past decades, various LLIE methods have been proposed, which can be roughly divided into four categories: histogram-equalization-based methods [1,3, 15], Retinex-based methods [9,16], dehaze-model-based methods [6,12,23], and deep neural network (DNN)-based methods [13,18]. Histogram equalization is the most basic LLIE technique, but it is easy to lose details in bright areas due to overexposure. Retinex theory is commonly used in traditional LLIE methods, which decomposes the observed image into illumination and reflection layers. However, this kind of method tends to produce noisy results. In dehaze-based models, the inverted low-light image is dehazed and then inverted back to a lighter image. This model works but lacks a certain theoretical basis. Recently, DNN-based methods have achieved impressive results as in many other low-level vision tasks. Unfortunately, these state-of-the-art (SOTA) DNN-based methods still suffer from various artifacts, such as noise, halos around edges, blurry details and color deviation.

In this paper, the architecture of the proposed network is designed based on a basic mathematical model. First, the input image is decomposed into a base image and a coefficient map via the BaseNet and the CoefficientNet, respectively. Owing to the large receptive fields, the U-Net-based BaseNet can effectively enhance the brightness to a uniform level. But many details of high-frequency (HF) texture and colors are missing. A CNN-based CoefficientNet is thus applied to preserve spatial details and adaptively adjust the brightness of each pixel. Since shallow CNN cannot capture global information which is vital in LLIE task, a FCN-based MaskNet is added at the beginning to estimate the global exposure/brightness levels and then this estimated global exposure map is input as a global guidance. At last, the HF residuals and colors are further improved via another shallow CNN, named RefineNet. This RefineNet adjusts the HF details and color deviation according to the original dark image guided by the coefficient map.

The contributions of this work can be summarized as follows:

- An end-to-end network is designed from a basic mathematical formula. Instead of providing labels for each sub-network, we merely design the structure and the output of each module according to their different physical meanings. Experimental results demonstrate that these modules themselves can learn corresponding functions by means of reasonable design.
- We introduce a MaskNet to estimate global brightness information as an important guidance for other modules. This global guidance can effectively improve the subsequent decomposition of the BaseNet and the CoefficientNet. The last RefineNet can further improve the details and colors by considering the original input and HF coefficient map simultaneously.

2   Related Work

At present, a variety of related LLIE methods have been proposed, which are briefly introduced in this section.

Histogram-equalization-based methods [1,3,15] redistribute the gray histogram of the low-light image to expand the range of the gray values, that can improve the image contrast and some dark details. Early Retinex-based methods, e.g., single-scale Retinex [9] and multi-scale Retinex [16], decompose the observed image into illumination layer and reflection layer. Brightened reflection is then obtained by removing the illumination component. However, these methods tend to produce halo and many other artifacts for non-uniform illumination. Wang et al. [20] thus proposed a naturalness preserved enhancement (NPE) model for non-uniform illumination images. Recently, Guo et al. [8] presented a LLIE method based on structure prior named low-light illumination map estimation (LIME). In [17], Ren et al. introduced a camera response model for low-light enhancement (LECAM). Although these traditional methods can effectively brighten images, they cannot handle severe noise contained in low-light images well.

Dehaze-based LLIE methods [6,12,23] are proposed based on the observation that the inverted low-light images are similar to the haze images. In these methods, the inverted low-light images are firstly processed by dehazing methods, and then the dehazed images are converted back to enhanced images. However, this kind of method lacks a convincing theoretical basis and may produce unrealistic results.

With the rapid development of deep learning, DNN has been widely used in low-level vision fields, such as image super-resolution [11] and denoising [22]. Lore et al. [13] introduced an autoencoder to denoise and enhance natural low-light images simultaneously. Liang et al. [18] proposed MSR-Net to directly learn the mapping between low-light images and bright images. Because aligned and paired training data are difficult to be captured, these methods are trained on synthetic datasets which are different from real low-light images. Chen et al. [4] proposed a paired LLIE raw data training set and applied a fully convolutional network to enhance quite dark images by means of raw data. However, this method is designed for extremely low-light condition and may cause unnatural artifacts for weakly illuminated sRGB images. Gharbi et al. [7] introduced deep bilateral learning for real-time image enhancement. Wang et al. [19] proposed a low-light image enhancement network based on image-to-illumination mapping. Motivated by Retinex model, it becomes a trend to learn image decomposition via deep network. Wei et al. [5] presented a RetinexNet to decompose low-light images into illumination and reflection, and the reflection is denoised by additional Block-matching and 3D filtering (BM3D). Zhang et al. [24] were also influenced by the Retinex theory and proposed the KinD network to decompose the images. The KinD achieved faster processing speed with better enhanced results. Unfortunately, these SOTA DNN-based methods still often suffer from color deviation and noise.

3   The Proposed Method

3.1   The Design of Network Architecture

In this paper, we design the proposed network from the basic mathematical model. Suppose the low-light input image is X, we can obtain brightness-enhanced image Y from X via several most basic operations, such as addition, multiplication, and power operation, as follows,

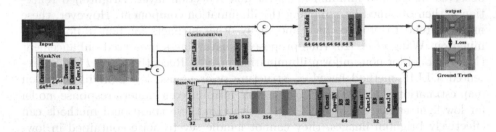

Fig. 1. The architecture of LLIENet, which includes four modules, i.e., MaskNet, BaseNet, CoefficientNet, and RefineNet.

$$Y = X + R, \tag{1}$$
$$Y = X \otimes C,$$
$$Y = (X)^{\gamma},$$

where R denotes a residual map that added to the X, C is a coefficient map and $\otimes$ denotes pixel-wise multiplication, and γ represents a power map which consists of the corresponding power value of each pixel. By comparing the sensitivity, the power operation is the most sensitive to the changes of weights. Hence, its brightness enhancement result is the most significant. Unfortunately, this sensitivity also leads to significant changes of contrast and colors. On the contrary, the addition operation is the least sensitive one, but it can well preserve high-frequency (HF) components and avoid color deviation by means of an added residual map, which is commonly adopted in many low-level vision tasks that need fine details, such as super-resolution [11] and denoising [22]. The multiplication operation balances the benefits of the other two operations, which is more sensitive than addition, and more stable than power function. Hence, multiplication-based models have attracted much attention in traditional low-light enhancement methods, e.g., Retinex-based methods [8,9,16]. However, multiplication is still more sensitive than addition, this kind of model is thus easier to cause color deviation and cannot handle HF details well. Hence, motivated by the merits and demerits of these basic formulas and Retinex model, the basic model of the proposed method is defined as,

$$Y = B \otimes C + R, \tag{2}$$

where B denotes a base layer which contains the basic information of input X and its brightness has been adjusted to a uniform level; C is a coefficient map which is used to adaptively adjust the brightness of each pixel; and R represents the residual map to refine the pixel-level details and colors.

Then we design proper DNN-based module for each component in Eq. (2). To obtain base layer B, commonly used U-Net is adopted as the backbone of our BaseNet, which has very large receptive fields but may lead to the loss of pixel-level details. To calculate C which contains individual coefficient for each pixel, a CNN-based CoefficientNet is utilized to preserve spatial resolution of feature maps. However, the receptive field of shallow CNN is much smaller, thus this CoefficientNet lacks the global brightness information which is crucial to the LLIE task. Hence, a FCN-based MaskNet is designed to estimate the global exposure/brightness levels and then the estimated global exposure map is also input to the CNN as a global prior. At last, an additional CNN, named RefineNet, is utilized to adjust the HF details and improve colors. The low-light X is input to the RefineNet to perceive original information. Furthermore, this module is also guided by the coefficient map C, which characterizes the adjustment degree of each pixel. Note that C is input to RefineNet rather than the base map B, because C contains more HF components than the base layer. As a result, the architecture of the proposed network becomes,

$$Y = F_B(X, M) \otimes F_C(X, M) + F_R(X, F_C(X, M)), \qquad (3)$$

where mask map M is computed by $M = F_M(X)$, and F_B, F_C, F_R, F_M denote the BaseNet, CoefficientNet, RefineNet and MaskNet, respectively. Based on Eq. (3), the architecture of the proposed low-light image enhancement network (LLIENet) is illustrated in Fig. 1.

Note that although we design the network by means of various modules as in Eq. (2), the network is directly end-to-end trained by low-light input images and the corresponding brightened images. That means, there is no ground truth or middle-loss of these M, B, C, R to supervise these modules. Instead of providing fixed labels for each sub-network, we merely design the structure and output of each sub-network according to their different physical meanings. Then, specific features or decomposition of these modules can be naturally learned by the network itself during the end-to-end training stage. For example, in the MaskNet which tends to estimate the global brightness information, we adopt encoder structure to discard most of the details and only focus on the basic global difference information. The BaseNet is applied to obtain base map with a uniform brightness level, and thus commonly used U-Net with large receptive field is adopted to reproduce 3-channel base map B. To obtain the coefficient of each pixel, the CoefficientNet adopts multiple convolutional layers to generate a 1-channel coefficient map C. The RefineNet has a similar structure with CoefficientNet but outputs 3-channel residual maps to adjust the reconstructed image. These different M, B, C, and R maps learned via the network are visualized in Fig. 2. It can be found that although we do not strictly constrain the output of these sub-networks, they can still effectively learn features that correspond to the physical model due to the properly designed network architecture. In the following, structure and implementation of each sub-network are introduced in detail.

3.2 Details of Each Module

MaskNet. As mentioned before, FCN is selected as the backbone of the MaskNet to perceive global brightness and produce a coarse exposure/brightness map. As shown in Fig. 1, there are five 3×3 convolutional layers (conv.) activated by ReLU. The outputs of the first four conv. are downsampled by max-pooling. The feature maps are then enlarged through deconvolution to the same size as the input. Finally, a 1×1 conv. is used to obtain the 1-channel global exposure map. As shown in Fig. 2(b), although these global exposure maps are autonomously learned through the network without artificial labels, the learned maps effectively characterize the global brightness distribution of images.

(a)   (b)   (c)   (d)   (e)   (f)

Fig. 2. (a) Input low-light image. (b) global exposure maps estimated by MaskNet. (c) base maps reproduced by BaseNet. (d) coefficient maps generated by CoefficientNet. (e) results obtained before RefineNet. (f) final results obtained after RefineNet.

BaseNet. To enlarge receptive fields, commonly used U-Net is adopted as the backbone of BaseNet. The estimated global exposure map M and the input image X are concatenated as the input. As shown in Fig. 1, the encoder contains total ten 3×3 conv., ten Leaky ReLU activation, and ten batch normalization layers. Max-pooling layers are added after every two conv. In the decoder block, skip-connections are used to concatenate the features of the encoder. Each concatenation is followed by a 1×1 conv. and two residual blocks (RB). Each RB contains two 3×3 conv., a ReLU unit follows the first conv., and a skip connection. To avoid checkerboard artifacts during upsampling, we use bilinear interpolation and a 3×3 conv. instead of deconvolution layer. The result is finally constrained to the range of [0,1] by means of the sigmoid function. As shown in Fig. 2(c), the BaseNet can enhance the brightness of different images to a uniform level. However, these base maps have significant unnatural artifacts and are partly missing in color and detail.

CoefficientNet. In response to the problems in BaseNet, we adopted simple CNN structure to preserve the HF details of each pixels. The estimated global exposure map M is also input into this module. As shown in Fig. 1, there are total seven 3×3 conv., and the first six conv. are followed by Leaky ReLU activations, and a sigmoid function is added at last layer to constraint the value range. As shown in Fig. 2(d), detailed information of the input image can be obtained via this CoefficientNet.

RefineNet. As mentioned before, the coefficient map C and the original input X are concatenated as input to restore HF details and color information of the image. This sub-network shares the same structure with the CoefficientNet but has a 3-channel output. As shown in Fig. 2(f), the refined output can reproduce natural color and suppress noise simultaneously.

3.3 Loss Function

Pixel-wise MAE and MSE losses are often used in image reconstruction problem, but they often lead to significant halo artifacts in LLIE task. Hence, we adopt SSIM loss and perceptual loss to avoid the halos caused by directly pixel-wise constraints. The perceptual loss [10] computes the differences in latent feature space rather than that of spatial pixels, and SSIM loss [21] focuses on the structural similarity instead of pixel-wise distance. The final loss is defined as:

$$\ell = \ell_{ssim}(Y, \widetilde{Y}) + \frac{\alpha}{N_l} \sum_{i=1}^{N_l} \ell_{mse}(\psi_i(Y), \psi_i(\widetilde{Y})). \tag{4}$$

where $\ell_{ssim}(Y, \widetilde{Y})$ denotes the SSIM loss between estimated result Y and corresponding ground truth $\widetilde{Y}$, the weight α is experimentally set as 0.03, and $\psi_i(\cdot)$ calculates the feature maps in different layers of pretrained VGG network. In our experiment, the ReLU 2_2 layer and the ReLU 5_4 layer of the VGG-19 network are used.

4   Experiment

4.1   Datasets

Although the problem of low-light enhancement has been studied for a long time, most methods are trained on synthetic data [13,18] due to the difficulty of capturing fully paired low-light datasets. Fortunately, Cai et al. [2] provided a large-scale multi-exposure image dataset, which contains 589 elaborately selected high-resolution multi-exposure sequences with 4,413 images. We used their reference images as our labels and selected 453 pairs of relatively aligned low-light images from corresponding mult-exposure sequences. The resolution of the image was kept within 2000×2000 by cropping and resizing. Finally, 5370 image pairs are collected as our training set and 30 image pairs are selected for the evaluation. Then these training images are cropped to 256×256 patches for training.

4.2   Implementation Details

In our experiment, the batch size is set to 8. The data is normalized to the range of $[0, 1]$. The proposed network is trained 300 epochs. We use the ADAM optimizer with the initial learning rate 10^{-4}, and the learning rate is reduced to 10^{-5} after 200 iterations. All experiments were performed on a Nvidia GTX 1080Ti GPU using the PyTorch framework[1].

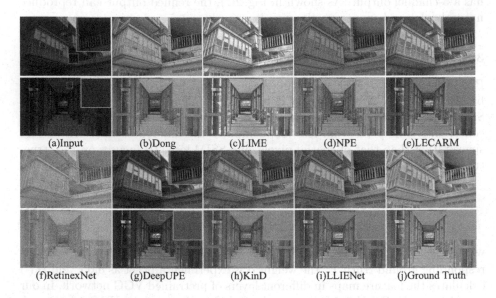

(a)Input	(b)Dong	(c)LIME	(d)NPE	(e)LECARM
(f)RetinexNet	(g)DeepUPE	(h)KinD	(i)LLIENet	(j)Ground Truth

Fig. 3. Results enhanced by different methods on our dataset.

Table 1. Average PSNR, SSIM and NIQE values of various low-light enhancement methods on our dataset.

Method	Dong	LIME	NPE	LECARM	RetinexNet	DeepUPE	KinD	LLIENet
PSNR	17.0464	16.6344	16.9652	18.2213	15.8807	15.5578	18.7276	**19.7253**
SSIM	0.7158	0.7743	0.7626	0.8086	0.6470	0.6889	0.8162	**0.8567**
NIQE	3.4029	2.9801	**2.6875**	2.7958	3.4405	2.7422	2.6991	2.7696

---

[1] Our code is available at https://github.com/xiaonaa/LLIE-code.

4.3   The Results

The proposed method is evaluated on three datasets: MEF [14], NPE-data and its extension (NPE-ex1, NPE-ex2 and NPE-ex3) [20] and our dataset. Full reference image quality assessments (IQA) of PSNR and SSIM are merely used in our dataset because other datasets lack ground truth images. A blind IQA NIQE is thus adopted for all testing sets. We compare the proposed LLIENet with several SOTA methods, including four traditional methods of Dong [6], LIME [8], NPE [20], LECARM [17], and three DNN-based methods of RetinexNet [5], DeepUPE [19] and KinD [24].

Some low-light enhancement results are illustrated in Fig. 3[2]. We can obtain the following findings. First, results of these traditional methods contain severe noise. NPE tends to reproduce unrealistic images, and the results of LECARM are often darker than that of others. Second, RetinexNet may reproduce unnatural color deviation and noise. Third, DeepUPE can recover clear results, but these results are much darker than others. Fourth, the results reconstructed by KinD contain obvious noise, especially in the flat area. Finally, the proposed LLIENet can reproduce brighter images and effectively suppress the HF noise and color deviation simultaneously.

Related IQA values are listed in Table 1. The proposed LLIENet achieves the highest PSNR and SSIM values. That means the proposed method can obtain the least distortion by comparing with ground truth. By comparing the blind IQA values of NIQE, the LLIENet performs similar to DeepUPE but worse than NPE and KinD. Note that although NIQE is a commonly used blind IQA that reflects the degree of similarity between the test image and natural scene statistic model, it still cannot accurately measure the visual quality for LLIE task. For example, although NPE achieves the highest NIQE values, the results reconstructed by NPE are more unreal and unnatural than other methods.

Figure 4 illustrates the results of these methods on MEF and NPE datasets[3]. Similar findings can be observed, the proposed LLIENet can reconstruct bright and natural images and suppress noise and color deviation while retaining the detailed information of the images. Related NIQE results are listed in Table 2. We still use the NIQE metric due to the lack of an accurate IQA method for LLIE task. The proposed LLIENet obtains satisfactory NIQE results but performs lower than DeepUPE. However, by comparing their visual quality, the proposed method can recover much brighter images than the DeepUPE and provide clearer details in dark area.

---

[2] More results can be found in the supplementary file.
[3] More subjective comparisons can also be found in the supplementary file.

Fig. 4. Results enhanced by different methods on the MEF and NPE datasets.

Table 2. Average NIQE values of various low-light enhancement methods on the MEF and NPE datasets.

	Dong	LIME	NPE	LECARM	RetinexNet	DeepUPE	KinD	LLIENet
MEF	4.5074	3.8171	3.5399	3.2033	4.9269	**3.1223**	3.3752	3.1993
NPE-data	4.0734	3.9050	3.6372	3.7734	4.1253	**3.4816**	3.7117	3.5267
NPE-ex1	4.6026	3.7533	3.3081	3.4166	4.5251	**3.2253**	3.3581	3.2339
NPE-ex2	3.5840	3.0157	2.8102	2.9256	3.6958	**2.7641**	2.7839	2.8327
NPE-ex3	4.4681	3.8350	3.2820	3.4030	4.5706	**3.2061**	3.4313	3.3588

4.4   The Ablation Study

The Effect of MaskNet. As shown in Fig. 5(a), the result produced by the network without MaskNet causes color deviation and uneven grayscale, such as partial whitening of the sky and wall. That demonstrates the global prior is useful to correctly learn the overall color and brightness. As listed in Table 3, the MaskNet is also beneficial to reduce the final distortion.

Fig. 5. The results in the ablation study.

Table 3. Average PSNR and SSIM values in the ablation study.

	w/o MaskNet	w/o CoefficientNet	w/o RefineNet	LLIENet
PSNR	19.1329	19.1076	19.3244	**19.5127**
SSIM	0.8458	0.8366	0.8415	**0.8469**

The Effect of CoefficientNet. As shown in Fig. 5(b), some details are blurred without CoefficientNet, which can adjust each pixel individually. By comparing the objective results in Table 3, the network with CoefficientNet performs better than that without CoefficientNet.

The Effect of RefineNet. As illustrated in Fig. 5(c), the network with RefineNet can refine the color of images. By comparing the texture on tire marked with blue rectangles, the RefineNet can also improve the HF details. Moreover, RefineNet can further increase the PSNR values as listed in Table 3.

5   Conclusion

This paper proposed an estimated exposure guided low-light enhancement network. The architecture is designed based on a basic mathematical model, which consists of four modules: a MaskNet to estimate the global illumination prior, a BaseNet to reproduce a lightened base image, a CoefficientNet to fine-tune the brightness of each pixel, and a RefineNet to further enhance HF detail and color of the image. Experimental results demonstrate that the proposed method can significantly brighten low-light images and suppress noise, halos, and color deviation simultaneously.

References

1. Arici, T., Dikbas, S., Altunbasak, Y.: A histogram modification framework and its application for image contrast enhancement. IEEE Trans. Image Process. **18**(9), 1921–1935 (2009)
2. Cai, J., Gu, S., Zhang, L.: Learning a deep single image contrast enhancer from multi-exposure images. IEEE Trans. Image Process. **27**(4), 2049–2062 (2018)
3. Celik, T., Tjahjadi, T.: Contextual and variational contrast enhancement. IEEE Trans. Image Process. **20**(12), 3431–3441 (2011)
4. Chen, C., Chen, Q., Xu, J., Koltun, V.: Learning to see in the dark. In: 2018 IEEE Conference on Computer Vision and Pattern Recognition (CVPR) (2018)
5. Wei, C., Wang, W., Yang, W., Liu, J.: Deep retinex decomposition for low-light enhancement. In: British Machine Vision Conference. British Machine Vision Association (2018)
6. Dong, X., et al.: Fast efficient algorithm for enhancement of low lighting video. In: 2011 IEEE International Conference on Multimedia and Expo, pp. 1–6. IEEE (2011)
7. Gharbi, M., Chen, J., Barron, J.T., Hasinoff, S.W., Durand, F.: Deep bilateral learning for real-time image enhancement. ACM Trans. Graph. (TOG) **36**(4), 1–12 (2017)
8. Guo, X., Li, Y., Ling, H.: LIME: low-light image enhancement via illumination map estimation. IEEE Trans. Image Process. **26**(2), 982–993 (2016)
9. Jobson, D.J., Rahman, Z.u., Woodell, G.A.: Properties and performance of a center/surround retinex. IEEE Trans. Image Process. **6**(3), 451–462 (1997)
10. Johnson, J., Alahi, A., Fei-Fei, L.: Perceptual losses for real-time style transfer and super-resolution. In: Leibe, B., Matas, J., Sebe, N., Welling, M. (eds.) ECCV 2016. LNCS, vol. 9906, pp. 694–711. Springer, Cham (2016). https://doi.org/10.1007/978-3-319-46475-6_43
11. Kim, J., Lee, J.K., Lee, K.M.: Accurate image super-resolution using very deep convolutional networks. In: 2016 IEEE Conference on Computer Vision and Pattern Recognition (CVPR) (2016)
12. Li, L., Wang, R., Wang, W., Gao, W.: A low-light image enhancement method for both denoising and contrast enlarging. In: 2015 IEEE International Conference on Image Processing (ICIP), pp. 3730–3734. IEEE (2015)
13. Lore, K.G., Akintayo, A., Sarkar, S.: LLNet: a deep autoencoder approach to natural low-light image enhancement. Pattern Recogn. **61**, 650–662 (2017)
14. Ma, K., Zeng, K., Wang, Z.: Perceptual quality assessment for multi-exposure image fusion. IEEE Trans. Image Process. **24**(11), 3345–3356 (2015)

15. Pizer, S.M., Amburn, E.P., Austin, J.D., Cromartie, R., Zuiderveld, K.: Adaptive histogram equalization and its variations. Comput. Vis. Graph. Image Process. **39**(3), 355–368 (1987)
16. Rahman, Z.u., Jobson, D.J., Woodell, G.A.: Multi-scale retinex for color image enhancement. In: Proceedings of 3rd IEEE International Conference on Image Processing, vol. 3, pp. 1003–1006. IEEE (1996)
17. Ren, Y., Ying, Z., Li, T.H., Li, G.: LECARM: low-light image enhancement using the camera response model. IEEE Trans. Circuits Syst. Video Technol. **29**(4), 968–981 (2018)
18. Shen, L., Yue, Z., Feng, F., Chen, Q., Liu, S., Ma, J.: MSR-Net: low-light image enhancement using deep convolutional network. arXiv preprint arXiv:1711.02488 (2017)
19. Wang, R., Zhang, Q., Fu, C.W., Shen, X., Zheng, W.S., Jia, J.: Underexposed photo enhancement using deep illumination estimation. In: The IEEE Conference on Computer Vision and Pattern Recognition (CVPR), June 2019
20. Wang, S., Zheng, J., Hu, H.M., Li, B.: Naturalness preserved enhancement algorithm for non-uniform illumination images. IEEE Trans. Image Process. **22**(9), 3538–3548 (2013)
21. Wang, Z., Bovik, A.C., Sheikh, H.R., Simoncelli, E.P.: Image quality assessment: from error visibility to structural similarity. IEEE Trans. Image Process. **13**(4), 600–612 (2004)
22. Zhang, K., Zuo, W., Chen, Y., Meng, D., Zhang, L.: Beyond a gaussian denoiser: residual learning of deep CNN for image denoising. IEEE Trans. Image Process. **26**(7), 3142–3155 (2017)
23. Zhang, X., Shen, P., Luo, L., Zhang, L., Song, J.: Enhancement and noise reduction of very low light level images. In: Proceedings of the 21st International Conference on Pattern Recognition (ICPR 2012), pp. 2034–2037. IEEE (2012)
24. Zhang, Y., Zhang, J., Guo, X.: Kindling the darkness: a practical low-light image enhancer. arXiv preprint arXiv:1905.04161 (2019)

A Novel CNN Architecture for Real-Time Point Cloud Recognition in Road Environment

Duyao Fan[1], Yazhou Yao[1], Yunfei Cai[1(✉)], Xiangbo Shu[1], Pu Huang[2], and Wankou Yang[3]

[1] Nanjing University of Science and Technology, Nanjing 210094, China
cyf@njust.edu.cn
[2] Nanjing Audit University, Nanjing 211815, China
[3] Southeast University, Nanjing 210096, China

Abstract. To ameliorate the problems of disorder, sparseness, and floating occur for 3D LiDAR point cloud in the road environment, we propose a novel deep CNN architecture for real-time point cloud features extraction. Specifically, we first code the 3D position of point cloud by the index of vertical and horizontal directions. In this way, the 3D point cloud can be converted into a multi-channel point feature map. Then, through multi-level features extraction and fusion of the point feature map, the semantic segmentation of the point cloud scene is finally realized. Comprehensive experiments and ablation studies on public available point cloud datasets demonstrate the superiority of our approach. More importantly, our approach has been successfully applied to the perception of the real-world self-driving system. The source code has been made public available at: https://github.com/Lab1028-19/A-Novel-CNN.

Keywords: Point cloud recognition · LiDAR · Self-driving

With the extensive penetration of big data [1–10], the innovation of sensing technology [11–15], and the gradual maturity of deep learning [16–27], highly intelligent unmanned platforms have been valued and developed rapidly. Environment awareness and understanding are the focus in the current research field [28–33]. Since LiDAR has a long detection range, high precision, and strong anti-interference ability, it is widely used in self-driving system. As the essence of the point cloud is a simple sampling of the surface in a real 3D environment, it is very sparse compared to the scale of the scene that can be detected with only one-sided information, making high-level semantic perception tasks difficult. To conclude, there are three key issues in road environment point cloud recognition: 1) point cloud is disordered; 2) point cloud is sparse, and 3) point cloud is floating [34–36].

This work was supported by the National Natural Science Foundation of China (No. 61976116, 61773117), Fundamental Research Funds for the Central Universities (No. 30920021135), and the Primary Research & Development Plan of Jiangsu Province - Industry Prospects and Common Key Technologies (No. BE2017157).

© Springer Nature Switzerland AG 2020
Y. Peng et al. (Eds.): PRCV 2020, LNCS 12305, pp. 206–218, 2020.
https://doi.org/10.1007/978-3-030-60633-6_17

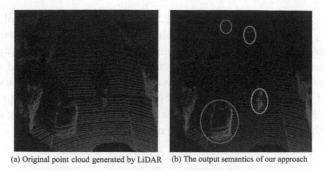

(a) Original point cloud generated by LiDAR   (b) The output semantics of our approach

Fig. 1. The original point cloud generated by LiDAR and the output semantics produced through our approach.

To solve these problems, existing methods [37–40] focused on using traditional methods, while works in [41–43, 49–51] focused on utilizing machine learning methods. The problem is that accuracy is still far from satisfied. Charles *et al.* [44] proposed a new way for point cloud semantics segmentation. It implements an end-to-end point cloud deep neural network that can learn features directly from point clouds. This method has achieved good results in a three-dimensional target recognition, segmentation, and point-wise semantic segmentation. Nevertheless, this method fails to capture local features in space, which greatly limits its application in real-world complex scenarios.

As shown in Fig. 1, we focus on the semantic segmentation task of the point cloud in the road environment. We propose a new network for detecting obstacles and semantic segmentation in an unmanned environment. We first index the point cloud by the angles in both vertical and horizontal directions in spherical coordinates in 3D space. Then the point cloud is transformed into compact two-dimensional multichannel point feature maps as the input of our designed convolution neural networks. Specifically, our proposed network utilizes a single-point micro-perception to perform point cloud local feature extraction for improving the robustness of data missing holes. In addition, our network also uses the jump connection structure to complement features between nets in different levels, then extract and merge multi-level features of point clouds for improving the feature expression caused by sparsity in the point cloud. Finally, the local and global features are combined for up-sampling inverse coding to implement point cloud semantic segmentation. Extensive experiments on publicly available point cloud datasets Apollo [45] and KITTI [46] have proved the validity and applicability of the proposed network in accuracy and real-time performance.

1 The Proposed Approach

The overall structure of our semantic segmentation network is shown in Fig. 2. It is based on the basic network, with deconvolution of the highest layer $(m/16, n/24, c)$ heat map extracted by maxpool_5 feature. At the same time, the feature maps of the maxpool_1, maxpool_2, maxpool_3, and maxpool_4 layers are extracted to perform deconvolution with the same number of feature channels but kernel sizes and step sizes are different. We make all the up-sampled feature maps have the same resolution (m, n) as the input feature map, and merge them on the feature channel, get the point feature of $(m, n, 328)$, then use the PMP module to get the one-hot logits map of (m, n, c), finally we perform softmax on each point to get the semantic segmentation result. In the following subsections, we will give the details of our proposed network.

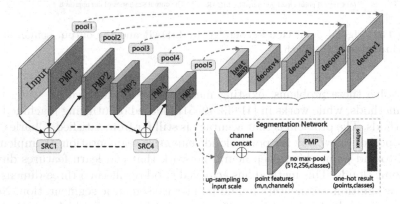

Fig. 2. The overall structure of our semantic segmentation network.

1.1 Processing for Point Cloud

For point cloud recognition tasks in unmanned platform scenes, it is important to extract multi-level spatial features of point clouds. In this work, the original scattered sparse point cloud is indexed by a vertical angle (elevation angle) and horizontal angle (azimuth angle) in space spherical coordinates.

Original Data Analysis: The model of LiDAR used for KITTI data acquisition in urban road environments is Velodyne HDL-64E. In this work, the KITTI-road part is used to the scene point cloud's semantics segmentation in the road environment. This dataset labels the passable areas of the image in pixel level through using the radar and camera to jointly calibrate the parameters. The point cloud is projected onto the front view annotation, filters out point cloud that can't be projected onto the image.

Multi-channel Point Feature Map: Generally speaking, it is necessary to preprocess the disordered point cloud data to obtain an ordered multi-channel point feature map. Our process is as follows:

If it is a multi-line LiDAR point cloud data transmitting in real-time, the decoding can be directly transmitted back according to the UDP data packet, thereby obtaining a point cloud feature map of m rows and n columns, where

Table 1. The features of unit subspace.

Channel	Features	Remarks
0	Number	Number of point clouds
1	Vector	Average value
2	Projection	Average value
3	Intensity	Average value
4	Depth	Average value for X
5	Transverse	Average value for Y
6	Height	Average value for Z

m represents a laser beam, n is a horizontal sampling point divided by a fixed angular resolution, and each point includes both 3D vector distance and reflected intensity values.

If the data contains three-dimensional rectangular coordinates (x, y, z) and intensity value, we first read the multi-line LiDAR vertical scanning angle range, vertical angular resolution, horizontal angular resolution, and other related parameters. Then we convert the 3D Cartesian Coordinates (x, y, z) of the data points into space spherical coordinates (γ, θ, ϕ) through:

$$\begin{cases} \Theta = \cos^{-1} \dfrac{z}{\sqrt{x^2 + y^2 + z^2} + \varepsilon} \\ \varphi = \tan^{-1} \dfrac{y}{x + \varepsilon} \end{cases} \tag{1}$$

where ε is any infinitesimal small amount. Thirdly, we divide the angular unit subspace according to the vertical angular resolution $\Delta\theta$ and the horizontal angular resolution $\Delta\phi$, and obtain the spatial compact point feature of the resolution m × n by:

$$\begin{cases} \mathrm{m} = \Theta/\Delta\Theta - \min_{M} \mathrm{m}_i \\ \mathrm{n} = \varphi/\Delta\varphi - \min_{N} \mathrm{n}_i \end{cases} \tag{2}$$

where $\min_{M} \mathrm{m}_i$ represents the minimum value of the single sample set M, and $\min_{n} \mathrm{n}_i$ represents the minimum value of the single sample set N. As shown in Table 1, we finally construct k features, arrange the point clouds in order, and construct a multi-channel point cloud feature map of size m × n × k.

1.2   The Proposed Semantic Segmentation Network

In this subsection, we present the details of our proposed point cloud semantic segmentation network.

Multichannel Point Feature Map Network: Based on the characteristics of the scene point cloud, a deep convolutional neural network with special structures such as jump connection and point micro-perception is designed to extract feature points for a multi-channel point feature map. Figure 3(a) is a schematic diagram of a multi-channel point feature map convolutional neural network (MFPM-CNN), the upper part of the network is the basic network structure of point cloud classification; the lower part is the network structure for point cloud scene semantic segmentation task.

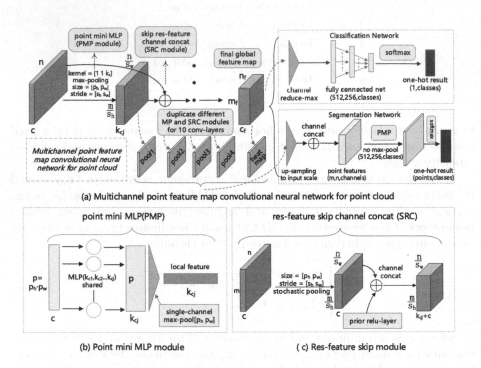

(a) Multichannel point feature map convolutional neural network for point cloud

(b) Point mini MLP module

(c) Res-feature skip module

Fig. 3. The detailed structure of our semantic segmentation network.

Point Mini MLP Module: The multi-channel point feature map generated by the point cloud has some information missing. In addition, the point cloud depth distance measurement error, the horizontal, and vertical angle calculation accuracy cause a line-like missing cavity. The detection areas like the glass mirror and the water surface are missing from the block due to the small energy of the laser radar echo. In the far top area, data loss will occur due to the detection range being too far. To this end, as shown in Fig. 3(b), we propose the point micro perception module (PMP module) as the convolution kernel, which reduces the influence of missing holes in feature extraction to a certain extent and enhances the anti-noise ability of the model. The size of input multichannel point feature map is [m n c]. We use the sliding window, whose size is [ph pw], to extract the features of local region, feature transformation of small-scale shallow-point micro-perception for all features on the longitudinal channel of each point in the window (the c-dimension features of all the initial channels of each point are transformed into k_{cj}-dimensional features by parameter-shared MLP (k_{c1}, $k_{c2}...k_{cj}$), and use ReLU as an activation function). Finally, the maximum pooling of the k_{cj} dimension feature is calculated for the ph × pw points in the window, and k_{cj} dimension local association feature representing the window region is obtained.

There are two characteristics of the features extracted by PMP module. 1) Since the activation function ReLU is used in PMP, the output of each layer is non-negative. The c dimension feature maximum pooling masks the value of 0 in the slid window, but the classical CNN convolution kernel is sensitive to missing points with value of 0:

$$
\left\{
\begin{aligned}
\text{Input} &= \begin{bmatrix} f_{11} & \cdots & f_{1w} \\ \vdots & \ddots & \vdots \\ f_{h1} & \cdots & f_{hw} \end{bmatrix} \\
\text{Kernel} &= \begin{bmatrix} k_{11} & \cdots & k_{1y} \\ \vdots & \ddots & \vdots \\ k_{x1} & \cdots & k_{xy} \end{bmatrix} \\
\text{Output} = \text{Input} \times \text{Kernel} &= \begin{bmatrix} \sum_{i=1}^{x}\sum_{j=1}^{y} f_{11}k_{ij} & \cdots & \sum_{i=1}^{x}\sum_{j=1}^{y} f_{1w}k_{ij} \\ \vdots & \ddots & \vdots \\ \sum_{i=1}^{x}\sum_{j=1}^{y} f_{h1}k_{ij} & \cdots & \sum_{i=1}^{x}\sum_{j=1}^{y} f_{hw}k_{ij} \end{bmatrix}
\end{aligned}
\right.
\tag{3}
$$

PMP can be seen as [1 1] convolution on different channels of the same point multiple times. Compared with the traditional convolution kernel for [h w], the convolution kernel's parameters decrease. 2) Due to the information, the missing point has no change to the corresponding convolution kernel parameter output, the number of effective training times of the convolution kernel parameters under the same training times is not uniform. However, the training pace of parameters is independent of the missing information points in PMP, which improves the training rate of data to model parameters.

Res-feature Skip Module: As shown in Fig. 3(c), the residual feature jump connection module (SRC [47] module) is used to solve the lack of detail caused by the pooling of the feature extraction receptive field. Due to the lack of detail affecting the accuracy of the maximum pooling layer, we leverage the SRC module with stochastic pooling:

$$
\left\{
\begin{aligned}
p_i &= \frac{a_i}{\sum a_k} k \in R_j \\
a_l &: s_j = a_l \text{ where } l \sim P(p_1, p_2, ..., p_{|R_j|})
\end{aligned}
\right.
\tag{4}
$$

where a_i is the value of the i-th point in the convolution area j, p_i is the normalized probability in the area, and l is the final pooled position index. Combining the [m_2 n_2 c_1] dimensional shallow feature with the [m_2 n_2 c_2] deep feature on the c-dimensional channel can obtain multi-channel merge features of [m_2 n_2 $c_1 + c_2$] dimension. The merged features have high-level salient features, and shallow detail features with random compliments, which enrich the expression of point cloud features and enhance the robustness of the model.

Basic Network for Feature Extraction: The semantic segmentation model is oriented to a frame of point cloud collected by LiDAR in the road scene to obtain semantic discrimination of roads, pedestrians, vehicles, etc. For each point, we use convolutional neural networks to extract multi-level features from pre-processed multi-channel point feature maps. We utilize the deconvolution to merge the layers from the upper layer to the lower layer for obtaining the multi-channel point feature map with the same input size. The parameters-shared by a fully connected layer is used to transform the multi-channel features of each point, get a model for semantic segmentation.

According to Fig. 3(a), we design the specific network structure for basic feature extraction. The point cloud is preprocessed into a 7-channel feature map with a resolution of m × n as input. Supplement shallow features to subsequent connection layer through SRC module, which achieves the jump connection of the adjacent maximum pooling layer. Each point micro-perceiver included in the PMP has two layers with shared parameters, the higher the dimension of the network, the higher the abstract semantics that can be expressed. PMP gradually reduces the feature map size through pooling, expands the receptive field of significant feature extraction, and obtains different levels of feature expression. The data is transmitted from the shallow layer to the deep layer except for the jump connection to the SRC module. The size of network layer is represented by the matrix which stores network layer node. The module description indicates the different structures involved in the corresponding network layer, the step and size of convolutional are [1, 1] without special instructions in PMP, and neuron nonlinear activation function is ReLU.

2 Experiment

2.1 Datasets and Evaluation Metrics

We compare our approach with other methods on two publicly available point cloud datasets Apollo [45] and KITTI [46]. Specifically, we compare different models on a local deep learning platform. We divide the original training data into a training and validation set by k-fold cross-validation. We evaluate the generalization ability for M optional models by k trials. Finally, we select the pixel accuracy and mean intersection over union as the evaluation metrics:

$$\begin{cases} \mathrm{PA} = \dfrac{\sum_{i=0}^{k} p_{ii}}{\sum_{i=0}^{k} \sum_{j=0}^{k} p_{ij}} \\[3ex] \mathrm{MIoU} = \dfrac{1}{k+1} \sum_{i=0}^{k} \dfrac{p_{ii}}{\sum_{j=0}^{k} p_{ij} + \sum_{j=0}^{k} p_{ji} - p_{ii}} \end{cases} \tag{5}$$

where the dataset is labeled with $k+1$ categories (includes background), p_{ij} represents the number of pixels that are predicted to be class j and p_{ii} is the

Table 2. The comparison of storage space and information loss in the preprocessed KITTI-road point cloud dataset.

Method	Range	Raw storage	Spatial resolution	Final storage	Change storage	Amount of information	Change in information
3D-Voxel	$40 \times 40 \times 10\,\mathrm{m}^3$	100810×4	$0.2^3\,\mathrm{m}^3$	2×10^6	19.84	19329.65×4	19.17%
Horizontal grid	$40 \times 40 \times 10\,\mathrm{m}^3$	100810×4	$0.2^3\,\mathrm{m}^3$	40000×4	0.39	12241.62×4	12.14%
Multiple views	$40 \times 40 \times 10\,\mathrm{m}^3$	100810×4	$0.03\,\mathrm{m}\|0.015\,\mathrm{m}$	$3 \times 640 \times 640$	3.05	≤ 173215.54	42.96%
Ours	**Unlimited**	$\mathbf{121105 \times 4}$	**angle**	$\mathbf{64 \times 2000 \times 4}$	**1.06**	$\mathbf{97805.27 \times 4}$	**80.76%**

number of pixels that are correctly predicted to be class j. When the model is being trained, the data is flipped horizontally with a probability of 0.5, and we add a slight interference to the normal distribution into the current training batch data according to:

$$\text{noise} \in N_{(0,\sigma^2)} \text{ and } |\text{noise}| < \text{clip.} \qquad (6)$$

We leverage the data enhancement technology to improve the generalization ability of our learned model.

2.2 Comparison of Storage Space and Information Loss

As shown in Table 2, we utilize the KITTI-road point cloud dataset to evaluate the storage space and information loss. To be specific, the data space is limited to the method, the storage is represented by the matrix size of the storage for point cloud, and the amount of information is represented by the number of effective point cloud features.

We use the 289-frame point cloud data of the KITTI-road data set for testing. The average point cloud data per frame is about 121,105 points, the range is limited to $40 \times 40 \times 10$ m^3, and the average point cloud data per frame is about 100,810 points. Each point contains four features: x, y, z, and intensity. The average results in Table 2 are obtained after noise reduction. The 3D voxel grid divides the 3D space of $40 \times 40 \times 10$ m^3 according to the mainstream 0.2 m side cube. Since the point cloud is sparse, empty voxels occupy most of the storage space, and there are fewer overlapping points in the foreground voxel grid. The horizontal grid divides the projection of the three-dimensional space of $40 \times 40 \times 10$ m^3 in the XoY horizontal plane by 0.2 m side. The storage space is reduced a lot but the loss of information is increasing. The point cloud for Multi-views is projected onto XoY, XoZ, and YoZ planes in $40 \times 40 \times 10$ m^3 space. As the sum of the information for three-view projection cannot be accurately calculated, three projection images were generated by 640×640 resolution in the test. As shown in Table 2, our method is obviously superior to the above common preprocessing methods in the comprehensive consideration of storage space and information loss.

2.3   Analysis on PMP and SRC Module

To demonstrate the rationality and necessity of the designed network structure, we conduct experiments on analyzing the performance of point micro-perception (PMP) and feature jump connection (SRC). In the experiment, we use the data from ApolloSpace [45] to train 1000 rounds after deleting the corresponding special structure. We then compare the test results of the new and old networks on the verification set, and discuss the role of the proposed special module.

1) We remove the shallow PMP module and replace the point convolution kernel with the common 2×2 and 3×3 convolution kernels. The other parts are consistent, and the experimental results are shown in Fig. 4. From Fig. 4, we can notice that the average category accuracy decreases obviously and there is spike vibration in steady state. Besides, the recall rate and accuracy are slightly reduced. The explanation is that when the convolution kernel extracts the small target feature, the object feature is diluted by the error. Our proposed PMP module can mask missing data holes and thus is more sensitive to the extraction of detailed features.

2) We remove the SRC module and keep others consistent. The experimental results are present in Fig. 4. From Fig. 4, we can find that the trends of the curve variations are similar, but the overall accuracy drops significantly. The reason is that the feature information obtained in semantic abstraction is degraded, and the SRC module is responsible for transmitting the shallow features to the deep layer for fusion judgment and improving the semantic segmentation precision.

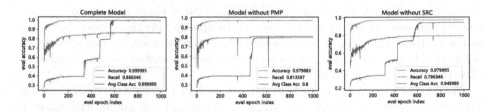

Fig. 4. The analysis on the proposed point micro-perception (PMP) and feature jump connection (SRC) module.

2.4   Semantic Segmentation for Road Drivable Area

To further validate the effectiveness of our approach, we perform experiments on semantic segmentation for road drivable area. Specifically, each frame of data contains 19,320 points, and the dataset for each comparison model is identical. According to the k-fold cross-validation method (k=5), the dataset is scrambled and divided into 5 groups (the obstacle detection is about 1496 frames per group, and the road segmentation is about 57 frames per group). One of the groups was selected as the validation set without repetition, and the other four groups were used as the training set. Table 3 shows the variant models designed for different application scenarios: HD models and fast models.

We use KITTI-road point cloud data for the simulation experiments. We project the point cloud onto the image with the road label, determine the category of each point, and filter out the point cloud outside the field of the front camera. In this experiment, we train and test 1000 rounds of data. The model tends to be stable after 500 rounds. The experimental results are shown in Table 4. It can be seen from Table 4 that the accuracy, recall, and accuracy of our method (HD) have a significant improvement. Our accuracy of the model is 97.11%, the increase is about 2%, and the recall is 94.38% improving 2%, the precision is 96.25% and the increase is about 2%. In addition, the accuracy of the fast model is similar to that of the [48] and [44], but the speed is improved.

Table 3. The detailed parameters for our proposed HD and fast model.

Type	Layers of network	Receptive field	Scale reduces	Channels of up sampling
HD	21c+9dc	[14 18]	[1 1]	344
Fast	11c+5dc	[12 12]	[1/4 1/6]	368

Table 4. HD model and fast Results on road semantic segmentation.

Method	Ours (HD)	Ours (Fast)	PointNet [44]	L05+deconv [48]
Running time	22.32 ms+3.3 ms	**9.12 ms+3.3 ms**	16.24 ms+2.1 ms	13.01 ms+3.1 ms
Accuracy	**97.11**	95.87	94.91	95.92
Recall	**94.38**	93.17	92.91	92.85
Precision	**96.25**	95.13	94.53	94.69
F1-Score (max)	**93.71**	92.84	93.09	91.69

3   Conclusion

In this work, we presented a novel CNN architecture for detecting obstacles and semantic segmentation for LIDAR point cloud data in an unmanned environment. Extensive experiments and ablation studies on two publicly available datasets demonstrate the superiority of our approach in real-time performance and the detection accuracy. Our proposed model has also been successfully applied to the driverless car and shows excellent real-time environment perception performance in the real road test.

References

1. Yao, Y., et al.: Towards automatic construction of diverse, high-quality image dataset. IEEE Trans. Knowl. Data Eng. **32**(6), 1199–1211 (2020)
2. Lu, J., et al.: HSI road: a hyper spectral image dataset for road segmentation, vol. 1–6 (2020)

3. Hua, X., et al.: A new web-supervised method for image dataset constructions. Neurocomputing **236**, 23–31 (2017)
4. Yao, Y., et al.: Exploiting web images for dataset construction: a domain robust approach. IEEE Trans. Multimed. **19**(8), 1771–1784 (2017)
5. Zhang, J., et al.: Extracting visual knowledge from the internet: making sense of image data, vol. 862–873 (2016)
6. Shen, F., et al.: Automatic image dataset construction with multiple textual meta-data, vol. 1–6 (2016)
7. Yao, Y., et al.: A domain robust approach for image dataset construction. In: ACM International conference on Multimedia, pp. 212–216 (2016)
8. Yao, Y., et al.: Bridging the web data and fine-grained visual recognition via alleviating label noise and domain mismatch. In: ACM International Conference on Multimedia (2020)
9. Sun, Z., et al.: CRSSC: salvage reusable samples from noisy data for robust learning. In: ACM International Conference on Multimedia (2020)
10. Zhang, C., et al.: Data-driven meta-set based fine-grained visual recognition. In: ACM International Conference on Multimedia (2020)
11. Liu, H., et al.: Road segmentation with image-LiDAR data fusion in deep neural network. Multimed. Tools Appl. (2019)
12. Han, X., et al.: Deep representation learning for road detection using siamese network. Multimed. Tools Appl. (2019)
13. Zhou, T., et al.: Motion-attentive transition for zero-shot video object segmentation. In: AAAI Conference on Artificial Intelligence (2020)
14. Luo, H., et al.: SegEQA: video segmentation based visual attention for embodied question answering. In: IEEE Conference on Computer Vision, pp. 9667–9676 (2019)
15. Wang, W., et al.: Target-aware adaptive tracking for unsupervised video object segmentation. In: The DAVIS Challenge on Video Object Segmentation on CVPR Workshop (2020)
16. Kirschner, U.: Urban transdisciplinary co-study in a cooperative multicultural working project. In: Luo, Y. (ed.) CDVE 2018. LNCS, vol. 11151, pp. 145–152. Springer, Cham (2018). https://doi.org/10.1007/978-3-030-00560-3_20
17. Yao, Y., et al.: Exploiting web images for multi-output classification: from category to subcategories. IEEE Trans. Neural Netw. Learn. Syst. **31**(7), 2348–2360 (2020)
18. Xu, M., et al.: Deep learning for person reidentification using support vector machines. Adv. Multimed. (2017)
19. Gu, Y., et al.: Clustering-driven unsupervised deep hashing for image retrieval. Neurocomputing **368**, 114–123 (2019)
20. Wang, W., et al.: Set and rebase: determining the semantic graph connectivity for unsupervised cross modal hashing. In: International Joint Conference on Artificial Intelligence, pp. 853–859 (2020)
21. Hu, B., et al.: PyRetri: a PyTorch-based library for unsupervised image retrieval by deep convolutional neural networks. arXiv (2020)
22. Zhang, C., et al.: Web-supervised network with softly update-drop training for fine-grained visual classification. In: AAAI Conference on Artificial Intelligence, pp. 12781–12788 (2020)
23. Yao, Y., et al.: Extracting privileged information for enhancing classifier learning. IEEE Trans. Image Process. **28**(1), 436–450 (2019)
24. Xie, G., et al.: Attentive region embedding network for zero-shot learning. In: IEEE Conference on Computer Vision and Pattern Recognition, pp. 9384–9393 (2019)

25. Shu, X., et al.: Hierarchical long short-term concurrent memory for human interaction recognition. IEEE TPAMI (2019)
26. Xie, G.-S., et al.: SRSC: selective, robust, and supervised constrained feature representation for image classification. IEEE Trans. Neural Netw. Learn. Syst. (2019)
27. Shu, X., et al.: Personalized Age Progression with Bi-level Aging Dictionary Learning. IEEE Trans. Pattern Anal. Mach. Intell. (2018)
28. Yao, Y., et al.: Extracting multiple visual senses for web learning. IEEE Trans. Multimed. **21**(1), 184–196 (2019)
29. Zhang, J., et al.: Extracting privileged information from untagged corpora for classifier learning. In: International Joint Conference on Artificial Intelligence, pp. 1085–1091 (2018)
30. Zhang, C., et al.: Web-supervised network for fine-grained visual classification, vol. 1–6 (2020)
31. Chen, T., et al.: Classification constrained discriminator for domain adaptive semantic segmentation, vol. 1–6 (2020)
32. Yang, W., et al.: Exploiting textual and visual features for image categorization. Pattern Recogn. Lett. **117**, 140–145 (2019)
33. Huang, P., et al.: Collaborative Representation Based Local Discriminant Projection for Feature Extraction. Digit. Signal Proc. **76**, 84–93 (2018)
34. Zhou, S.Y., et al.: Study on method of road detection in vehicle detection and tracking system. Electron. Des. Eng. **20**(2), 157–162 (2014)
35. Liu, Y., et al.: Unstructured road-detection algorithm based on multiple models and optimization. Gongcheng Sheji Xuebao **20**(2), 157–162 (2013)
36. Gang, J.: Point cloud hole filling method based on SVM and space projection. Comput. Eng. **35**(22), 269–271 (2009)
37. Bai, M., et al.: Road detection method based on graph model. Pattern Recog. Artif. Intell. **27**, 655–62 (2014)
38. Wijesoma, W.S., et al.: Road-boundary detection and tracking using ladar sensing. IEEE Trans. Robot. Autom. **20**(3), 456–464 (2004)
39. Guo, Q., et al.: Unstructured road detection based on two-dimensional entropy and contour features. J. Comput. Appl. (7), 56 (2013)
40. Zhu, X., et al.: A real-time road boundary detection algorithm based on driverless cars. Electrical, Electronics and Computer Engineering (2015)
41. Gong, J.W., et al.: Unstructured road recognition using self-supervised multilayer perceptron online learning algorithm. Trans. Beijing Inst. Technol. **34**(3), 261–266 (2014)
42. Zhou, S.Y., et al.: Road detection using support vector machine based on online learning and evaluation. In: IEEE Intelligent Vehicles Symposium, pp. 256–261 (2010)
43. Wang, X.B., et al.: Unstructured road detection based on support vector machine. Sci. Technol. Eng. **11**, 9106–9109 (2011)
44. Charles, R.Q., et al.: Pointnet: deep learning on point sets for 3D classification and segmentation. In: IEEE Conference On Computer Vision and Pattern Recognition, pp. 652–660 (2017)
45. Huang, X., et al.: The apolloscape open dataset for autonomous driving and its application. IEEE TPAMI (2019)
46. Geiger, A., et al.: Are we ready for autonomous driving? the kitti vision benchmark suite. In: IEEE Conference on Computer Vision and Pattern Recognition (2012)
47. He, K., et al.: Deep residual learning for image recognition. In: IEEE Conference on Computer Vision and Pattern Recognition, pp. 770–778 (2016)

48. Velas, M., et al.: CNN for very fast ground segmentation in velodyne lidar data, vol. 97–103 (2018)
49. Sun, Z., et al.: Dynamically visual disambiguation of keyword-based image search. In: International Joint Conference on Artificial Intelligence, pp. 996–1002 (2019)
50. Yang, W., et al.: Discovering and Distinguishing Multiple Visual Senses for Polysemous Words. In: AAAI Conference on Artificial Intelligence, pp. 523–530 (2018)
51. Ding, L., et al.: Approximate kernel selection via matrix approximation. IEEE Trans. Neural Netw. Learn. Syst. **PP**(99), 1–11 (2020)

Lightweight Image Super-resolution with Local Attention Enhancement

Yunchu Yang[1], Xiumei Wang[1,2](✉), Xinbo Gao[1], and Zheng Hui[1]

[1] VIPS Lab, School of Electronic Engineering, Xidian University, Xi'an, China
{yc_yang,zheng_hui}@aliyun.com, wangxm@xidian.edu.cn,
xbgao@mail.xidian.edu.cn
[2] The University of Sydney, Sydney, Australia

Abstract. In recent years, methods based on convolutional neural network (CNN) have been the mainstream in single image super-resolution (SISR). Although these methods have achieved excellent performance, the massive amount of model parameters and heavy computation limit their application. On the other hand, channel attention (CA) mechanism, which can enhance network performance, has also been widely used in SR task recently. However, the channel attention mechanism is introduced from high-level vision tasks to the SR task. The original design of this mechanism doesn't consider the specificity of the SR task. To address these issues, we propose a lightweight expansion and distillation residual network (EDRN) for image super-resolution. Specifically, through the diverse use of different feature channels and different convolution kernel sizes, our network can effectively reduce the amount of parameters while achieving superior performance. To further explore the potential of channel-wise attention in the SR task, we develop a novel plug-and-play local channel attention enhancement strategy (LCAES) to make the network better use the characteristics of local features of the image. Furthermore, comprehensive quantitative and qualitative evaluations demonstrate that the proposed method performs favorably against state-of-the-art SR algorithms in terms of visual quality, reconstruction accuracy, and parameter amount.

Keywords: Image super-resolution · Lightweight expansion and distillation residual network · Local channel attention enhancement strategy

1 Introduction

Single image super-resolution (SR) is a low-level vision task that aims to reconstruct a high-resolution (HR) image from its low-resolution (LR) observation.

This work was supported in part by the National Natural Science Foundation of China under Grant 61972305, 61871308, in part by the Natural Science Basic Research Plan in Shaanxi Province of China 2019JM-090, 2019JM-426.
The first author Yunchu Yang is a M.D. candidate.

© Springer Nature Switzerland AG 2020
Y. Peng et al. (Eds.): PRCV 2020, LNCS 12305, pp. 219–231, 2020.
https://doi.org/10.1007/978-3-030-60633-6_18

The problem of SR is ill-posed since there exist multiple possible HR solutions for any LR input. To address this inverse problem, using deep neural networks to learn the mapping of LR to HR to constrain the solution space has become very popular in recent years.

The first work of using a convolutional neural network (CNN) for SR was SRCNN [4], a three-layer CNN which established a direct relationship between bicubic upsampling LR and HR. Kim et al. [10] proposed VDSR, a deep convolutional neural network with 20 convolution layers and incorporated a global skip connection to enforce residual learning. Lim et al. [15] made a significant breakthrough in terms of SR performance by building a very wide and deep network and removed the batch normalization [9] adopted in previous SRResNet [14]. Recently, Zhang et al. [25] further increased the depth of network and proposed RCAN. They introduced residual in residual (RIR) structure to stabilize network training and channel attention (CA) mechanism [5] to make the network focus on more informative features.

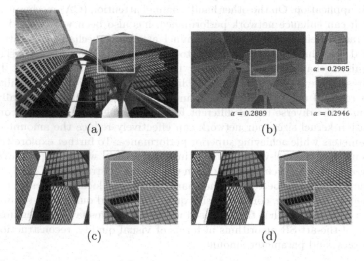

Fig. 1. The comparison of traditional reconstruction strategy and local channel attention enhancement strategy (LCAES) with scale 2 on "img 062" from Urban100. (a) Input test image. (b) Feature map of a channel before calculating attention (α indicates value of channel attention). Right: Entire image using traditional reconstruction strategy. Left: Two adjacent image patches using LCAES. (c) Results of traditional reconstruction strategy. (d) Results of LCAES.

Research on lightweight SR networks also draws much attention due to their small storage space, low memory consumption, and fast inference speed. Hui et al. [8] proposed an information distillation network (IDN) that explicitly divides the previously extracted features into two parts and uses different operations for different divisions. They further introduced a more efficient network

IMDN [7] and designed a more suitable attention mechanism called contrast-aware attention (CCA) for SR task. Ahn *et al.* [1] designed an architecture that implements a cascading mechanism upon a residual network. However, whether the very deep network or a lightweight network, most of the convolution layers in their models always keep the same large width (64) with the same kernel size (3×3), which takes many parameters. And stacking the same setting of convolution layers in a block lacks diverse processing of features.

On the other hand, after reviewing recent SR networks that apply channel-wise attention (including CA and CCA), we find there exists a mismatch between the training and testing phase, which fails to full use of attention mechanism. Most super-resolution networks consistently use the same training settings that crop the small training patch pairs from the LR and HR datasets to constitute a training dataset. While in the network testing phase, the entire LR image is sent into the network, which is generally larger than the size of the training LR patch. This setting is based on the properties of convolutional neural networks, and a large number of training image patch pairs have already captured sufficient variability of natural images. Training on small patch pairs could generate favorable results while reducing training time and resource consumption. However, things change when the network adopts channel-wise attention mechanism. Under the channel-wise attention mechanism, the first step is to calculate the channel-wise statistic $z \in R^C$ of the input feature $X = [x_1, \ldots, x_c, \ldots, x_C]$ which has C feature maps with size $H \times W$. The calculation formula of the c-th element of z in CA is

$$z_c = \frac{1}{HW} \sum_{(i,j) \in x_c} x_c^{i,j},$$ (1)

and for CCA, the formula is

$$z_c = \sqrt{\frac{1}{HW} \sum_{(i,j) \in x_c} \left(x_c^{i,j} - \frac{1}{HW} \sum_{(i,j) \in x_c} x_c^{i,j} \right)^2 + \frac{1}{HW} \sum_{(i,j) \in x_c} x_c^{i,j}}.$$ (2)

Notice that both CA and CCA include global average pooling operation, this operation transforms the size of the input feature map into 1×1 regardless of the content and size of the input. In the training phase, the attention layer extracts the global information of the small image patches, while global information of the whole input image is extracted in the testing phase. As shown in Fig. 1(b), when the input is different image patches cropped from different positions on a test image, its attention value of a channel will change with the content of image patches, this value is also different from the channel attention value when the input is the entire image. When the input image is quite large, the global information extracted on the whole image will cover the local characteristic information, resulting in a decrease in the discrimination ability and adaptability of the network. However, the improving effects of using CA and CCA in the network mask this phenomenon.

To address the above issues, we propose a lightweight expansion and distillation residual network (EDRN) with expansion and distillation residual blocks (EDRB) which reduce the number of parameters. This is achieved through a considerate design of feature channels. Specifically, we reduce the number of channels of the 3 × 3 kernel size convolution layer and make up degradation by using 1 × 1 kernel size convolution layer to expand and compress the channel. This design helps our network achieves high performance on the SR task with a moderate number of parameters. To make full use of the potential of channel-wise attention, we introduce local channel attention enhancement strategy (LCAES). The LCAES is a novel plug-and-play reconstruction strategy that crops the input image into small patches and reconstructs them separately. It could discover the local feature characteristic and improve the performance of the SR network with channel-wise attention without retraining. Figure 1 shows an example of using LCAES.

The contributions of this paper can be summarized as follows:

- We propose a lightweight expansion and distillation residual network (EDRN) with expansion and distillation residual block (EDRB). Our network diversifies the processing of features to full use of the limited parameters.
- We propose a novel reconstruction strategy named local channel attention enhancement strategy (LCAES) to unleash the potential of channel-wise attention and further improve the performance of our network.
- We experimentally prove that our model is competitive in terms of both the amount of parameters and the effect of reconstruction. And our LCAES can also improve the performance of other SR networks.

2 Proposed Method

2.1 Network Structure

As shown in Fig. 2, our EDRN mainly consists of a shallow feature extraction, a deep feature processor, and a upscale module. The shallow feature extraction is implemented by one 3 × 3 convolution. The deep feature processor stacks multiple expansion and distillation residual blocks followed by a 1 × 1 convolution to reduce the number of channels of the concatenated features. An activation layer and a 3 × 3 convolution layer were also added at the end of deep feature extraction. A residual skip connection adds the output of shallow feature extraction to the output of deep feature extraction. The three convolution layers mentioned above all have 64 output channels. The upscale module includes one 3 × 3 convolution with $3 \times s^2$ output channels and a sub-pixel convolution [18], where s denotes the upscale factor.

Denote the input LR image as $\mathbf{I}^{LR}$, the corresponding target HR image as $\mathbf{I}^{HR}$ and our EDRN as $H_{EDRN}(\cdot)$. The super-resolved image $\mathbf{I}^{SR}$ can be generated by

$$\mathbf{I}^{SR} = H_{EDRN}\left(\mathbf{I}^{LR}\right). \tag{3}$$

We use L1 loss function in network training for better fidelity of recovered image. Given a training set $\left\{I_i^{LR}, I_i^{HR}\right\}_{i=1}^{N}$ that has N LR-HR pairs, the loss function we minimize can be expressed by

$$\mathcal{L}(\Theta) = \frac{1}{N} \sum_{i=1}^{N} \left\| H_{EDRN}\left(I_i^{LR}\right) - I_i^{HR} \right\|_1, \tag{4}$$

where Θ denotes the updateable parameter set of our network and $\|\cdot\|_1$ is l_1 norm.

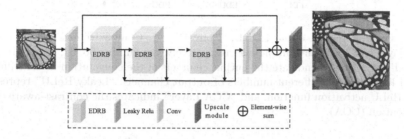

Fig. 2. The architecture of expansion and distillation residual network (EDRN).

2.2 Expansion and Distillation Residual Block

The design of EDRB is illustrated in Fig. 3. EDRB first adopts a 3×3 convolution layer followed by a Leaky ReLU activation function to extract input features. Then, we employ channel split operation on the features to separate the features into distilled features and rough features. For the rough features, we use several expansion and distillation unit (EDU) to further distill and separate the features. Our EDU consists of two 1×1 convolution layer, a 3×3 convolution layers, and a Leaky ReLU activation function. The first 1×1 convolution layer with a large number of output channels is followed by the Leaky ReLU activation function. Its role is to expand the input feature channels to generates rich feature maps for subsequent operations. Then compression on channels achieved by the second 1×1 convolution layer is used to reduce the number of feature channels to the original state before expansion. A 3×3 convolution layer is followed to further process features. For the output of each EDU, we use channel split operation to separate the features into distilled features and rough features. The following stage is concatenating distilled features from each step. Then the CCA layer and a 1×1 convolution layer will process the concatenated features in series. And the input feature of EDRB will be added to output features of the last 1×1 convolution layer.

2.3 Local Channel Attention Enhancement Strategy

Figure 4 illustrates the training and our novel testing strategy of the SR task. During the training phase, the SR network learns a mapping from the LR patch

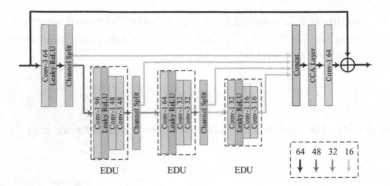

Fig. 3. The architecture of our EDRB, "Conv-1" and "Conv-3" represent 1×1 convolution layer and 3×3 convolution layer respectively, the number 96, 64, 48 and 16 after "Conv-1" or "Conv-3" indicate the output channels of the convolution layer. Different colored lines indicate different numbers of output channels. "Leaky ReLU" represents Leaky ReLU activation function, and "CCA Layer" indicates the contrast-aware channel attention (CCA).

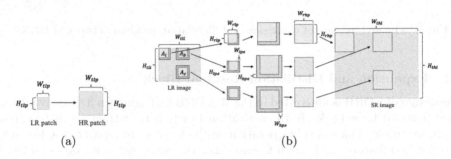

Fig. 4. The diagrammatic sketch of (a) Training situation. (b) Local channel attention enhancement strategy (LCAES).

to the HR patch, as illustrated in Fig. 4(a). In the traditional testing phase, the SR network performs all processing on the whole input LR image and reconstruct the SR image. Our local channel attention enhancement strategy (LCAES) is a novel reconstruction strategy used in the testing phase, which can be plug-and-play on any exiting SR models without retraining. As illustrated in Fig. 4(b), given a test input image which size is $H_{tli} \times W_{tli}$, we first crop the input LR image into three types of LR image patches: The first type of LR image patches, which is named as A_c, is cropped from the four corner areas of the input LR image. The size of this type of image patches is $(H_{rlp} + H_{lpe}) \times (W_{rlp} + W_{lpe})$. They are cropped from the corner to the inside area of the input LR image. The second type of LR image patches A_b are cropped from the image area which is adjacent to one of the four boundaries of the image LR image, their size is $(H_{rlp} + H_{lpe}) \times (W_{rlp} + 2 \times W_{lpe})$. The last type of LR image patches, named as A_i, are cropped from the area which is far from the input image's boundaries and corners, the size of them is $(H_{rlp} + 2 \times H_{lpe}) \times (W_{rlp} + 2 \times W_{lpe})$. There will be some overlapping pixels between different image patches cropped from adjacent parts of the input image. We set H_{rlp} and W_{rlp} close to the training LR patch's size to make the testing process close to the training environment. The role of extra edge H_{lpe} and W_{lpe} is to avoid boundary effects.

After cropping the input LR image into LR image patches, each patch will be fed into the SR network to generate corresponding HR patches. Then we remove the upscaled extra edge (H_{lpe} and W_{hpe} represents the length and height of the upscaled extra edge) and keep the rest $H_{rhp} \times W_{rhp}$ core part. Finally, we put these core parts of HR patches into the corresponding position to synthesize the final HR image. The corners and borders parts of HR image will be reconstructed firstly, then the core part of the HR patches generated from A_i will fill the rest part of HR image and cover the overlapping with reconstructed borders and corners parts.

3   Experiment

3.1   Settings

Following [7,15,25], we train our models on DIV2K dataset [21], which contains 800 high-quality RGB images. To evaluate the performance of the proposed network, we use five widely used benchmark datasets: Set5 [2], Set14 [23], BSD100 [16], Urban100 [6], and Manga109 [17]. We conduct experiments with Bicubic (BI) degradation models. We evaluate our SR results' fidelity with two metrics, including peak signal-to-noise ratio (PSNR) and structure similarity index (SSIM) [22] on luminance channel of transformed YCbCr space. To obtain DIV2K training patch pairs, the HR image patches with size of 192×192 are randomly cropped from HR images, and the LR image patches are cropped from the corresponding LR images which downsampled by HR images with the scaling factors. The mini-batch size is set to 16. We augment the training data with random horizontal flips and 90 rotations. Our model is trained by ADAM optimizer [12] by setting $\beta_1 = 0.9$, $\beta_2 = 0.999$, and $\epsilon = 10^{-8}$. The initial learning rate is set to 2×10^{-4} and halved at every 2×10^5 iterations of back-propagation. We use PyTorch framework to implement our models with a NVIDA TITAN Xp GPU (12G memory). When using LCAES in testing, for convenience, we set $H_{rlp} = W_{rlp}$ and $H_{lpe} = W_{lpe}$. Considering achieving a balanced effect on each dataset and a proper calculation consumption, we set $H_{rlp} = 92$, $H_{lpe} = 12$ for scale factor 2, $H_{rlp} = 62$, $H_{lpe} = 12$ for scale factor 3 and $H_{rlp} = 46$, $H_{lpe} = 8$ for scale factor 4.

3.2   Benchmark Results

Table 1 shows quantitative comparisons with bicubic interpolation and 12 state-of-the-art methods: SRCNN [4], FSRCNN [3], VDSR [10], DRCN [11], LapSRN [13], DRRN [19], MemNet [20], IDN [8], EDSR-baseline [15],

Table 1. Average PSNR/SSIM for scale factor ×2, ×3 and ×4 on datasets Set5, Set14, BSD100, Urban100, and Manga109. Best and second best results are **highlighted** and underlined. (EDRN+L denotes EDRN combined with LCAES)

Method	Scale	FLOPs	Params	Set5 PSNR/SSIM	Set14 PSNR/SSIM	BSD100 PSNR/SSIM	Urban100 PSNR/SSIM	Manga109 PSNR/SSIM
Bicubic	×2	-	-	33.66/0.9299	30.24/0.8688	29.56/0.8431	26.88/0.8403	30.80/0.9339
SRCNN [4]		6.85G	8K	36.66/0.9542	32.45/0.9067	31.36/0.8879	29.50/0.8946	35.60/0.9663
FSRCNN [3]		5.58G	13K	37.00/0.9558	32.63/0.9088	31.53/0.8920	29.88/0.9020	36.67/0.9710
VDSR [10]		571.56G	666K	37.53/0.9587	33.03/0.9124	31.90/0.8960	30.76/0.9140	37.22/0.9750
DRCN [11]		10747.35G	1,774K	37.63/0.9588	33.04/0.9118	31.85/0.8942	30.75/0.9133	37.55/0.9732
LapSRN [13]		96.57G	251K	37.52/0.9591	32.99/0.9124	31.80/0.8952	30.41/0.9103	37.27/0.9740
DRRN [19]		6330.10G	298K	37.74/0.9591	33.23/0.9136	32.05/0.8973	31.23/0.9188	37.88/0.9749
MemNet [20]		2508.07G	678K	37.78/0.9597	33.28/0.9142	32.08/0.8978	31.31/0.9195	37.72/0.9740
IDN [8]		130.83G	553K	37.83/0.9600	33.30/0.9148	32.08/0.8985	31.27/0.9196	38.01/0.9749
EDSR-baseline [15]		295.05G	1,370K	37.99/0.9604	33.57/0.9175	32.16/0.8994	31.98/0.9272	38.54/0.9769
SRMDNF [24]		324.37G	1,511K	37.79/0.9601	33.32/0.9159	32.05/0.8985	31.33/0.9204	38.07/0.9761
CARN [1]		208.01G	1,592K	37.76/0.9590	33.52/0.9166	32.09/0.8978	31.92/0.9256	38.36/0.9765
IMDN [7]		148.60G	694K	38.00/0.9605	33.63/0.9177	<u>32.19</u>/0.8996	32.17/0.9283	<u>38.88</u>/**0.9774**
EDRN (Ours)		128.89G	602K	<u>38.02</u>/<u>0.9606</u>	<u>33.64</u>/<u>0.9180</u>	32.18/<u>0.8997</u>	<u>32.20</u>/<u>0.9284</u>	**38.88**/<u>0.9773</u>
EDRN+L (Ours)		193.95G	602K	**38.03**/**0.9606**	**33.65**/**0.9182**	**32.20**/**0.8999**	**32.27**/**0.9295**	38.87/0.9773
Bicubic	×3	-	-	30.39/0.8682	27.55/0.7742	27.21/0.7385	24.46/0.7349	26.95/0.8556
SRCNN [4]		6.85G	8K	32.75/0.9090	29.30/0.8215	28.41/0.7863	26.24/0.7989	30.48/0.9117
FSRCNN [3]		4.58G	13K	33.18/0.9140	29.37/0.8240	28.53/0.7910	26.43/0.8080	31.10/0.9210
VDSR [10]		571.56G	666K	33.66/0.9213	29.77/0.8314	28.82/0.7976	27.14/0.8279	32.01/0.9340
DRCN [11]		16747.35G	1,774K	33.82/0.9226	29.76/0.8311	28.80/0.7963	27.15/0.8276	32.24/0.9343
LapSRN [13]		214.52G	502K	33.81/0.9220	29.79/0.8325	28.82/0.7980	27.07/0.8275	32.21/0.9350
DRRN [19]		6330.10G	298K	34.03/0.9244	29.96/0.8349	28.95/0.8004	27.53/0.8378	32.71/0.9379
MemNet [20]		2508.07G	678K	34.09/0.9248	30.00/0.8350	28.96/0.8001	27.56/0.8376	32.51/0.9369
IDN [8]		67.02G	553K	34.11/0.9253	29.99/0.8354	28.95/0.8013	27.42/0.8359	32.71/0.9381
EDSR-baseline [15]		149.69G	1,555K	34.37/0.9270	30.28/0.8417	29.09/0.8052	28.15/0.8527	33.45/0.9439
SRMDNF [24]		145.93G	1,528K	34.12/0.9254	30.04/0.8382	28.97/0.8025	27.57/0.8398	33.00/0.9403
CARN [1]		110.98G	1,592K	34.29/0.9255	30.29/0.8407	29.06/0.8034	28.06/0.8493	33.50/0.9440
IMDN [7]		66.92G	703K	34.36/0.9270	<u>30.32</u>/<u>0.8417</u>	<u>29.09</u>/0.8046	<u>28.17</u>/<u>0.8519</u>	33.61/0.9445
EDRN (Ours)		58.15G	611K	<u>34.41</u>/<u>0.9272</u>	30.32/0.8413	29.08/<u>0.8047</u>	28.15/0.8518	<u>33.64</u>/<u>0.9446</u>
EDRN+L (Ours)		104.96G	611K	**34.41**/**0.9272**	**30.36**/**0.8427**	**29.09**/**0.8050**	**28.20**/**0.8533**	**33.64**/**0.9447**
Bicubic	×4	-	-	28.42/0.8104	26.00/0.7027	25.96/0.6675	23.14/0.6577	24.89/0.7866
SRCNN [4]		6.85G	8K	30.48/0.8628	27.50/0.7513	26.90/0.7101	24.52/0.7221	27.58/0.8555
FSRCNN [3]		4.20G	13K	30.72/0.8660	27.61/0.7550	26.98/0.7150	24.62/0.7280	27.90/0.8610
VDSR [10]		571.56G	666K	31.35/0.8838	28.01/0.7674	27.29/0.7251	25.18/0.7524	28.83/0.8870
DRCN [11]		16747.35G	1,774K	31.53/0.8854	28.02/0.7670	27.23/0.7233	25.14/0.7510	28.93/0.8854
LapSRN [13]		120.57G	502K	31.54/0.8852	28.09/0.7700	27.32/0.7275	25.21/0.7562	29.09/0.8900
DRRN [19]		6330.10G	298K	31.68/0.8888	28.21/0.7720	27.38/0.7284	25.44/0.7638	29.45/0.8946
MemNet [20]		2508.07G	678K	31.74/0.8893	28.26/0.7723	27.40/0.7281	25.50/0.7630	29.42/0.8942
IDN [8]		44.62G	553K	31.82/0.8903	28.25/0.7730	27.41/0.7297	25.41/0.7632	29.41/0.8942
EDSR-baseline [15]		106.57G	1,518K	32.09/0.8938	28.58/0.7813	27.57/<u>0.7357</u>	26.04/ <u>0.7849</u>	30.35/0.9067
SRMDNF [24]		83.32G	1,552K	31.96/0.8925	28.35/0.7787	27.49/0.7337	25.68/0.7731	30.09/0.9024
CARN [1]		84.82G	1,592K	32.13/0.8937	**28.60**/0.7806	**27.58**/0.7349	<u>26.07</u>/0.7837	30.47/**0.9084**
IMDN [7]		38.26G	715K	**32.21**/<u>0.8948</u>	28.58/0.7811	27.56/0.7353	26.04/0.7838	30.45/0.9075
EDRN (Ours)		33.34G	623K	32.19/0.8946	<u>28.60</u>/<u>0.7814</u>	27.54/0.7353	26.06/0.7846	<u>30.47</u>/0.9076
EDRN+L (Ours)		56.12G	623K	<u>32.19</u>/**0.8948**	28.60/**0.7816**	<u>27.57</u>/**0.7358**	**26.10**/**0.7863**	30.49/<u>0.9079</u>

SRMDNF [24], CARN [1] and IMDN [7]. In addition to calculating SSIM, PSNR, and parameters, We further investigate the floating point operations per second (FLOPs), which can reflect computing consumption to some extent on each model. We assume the HR image size to be 720p (1280 × 720) to calculate FLOPs. EDRN+L denotes EDRN combined with LCAES. It can be found out that our EDRN performs favorably against other compared approaches on most datasets, and the LCAES can further enhance its performance at the cost of extra FLOPs. Figure 5 shows visual comparisons on ×2, ×3, and ×4 between our approach and others. It shows that our EDRN could generate better visual results in most cases, our LCAES could help our EDRN further focus on local details of the image and generate richer, realistic textures.

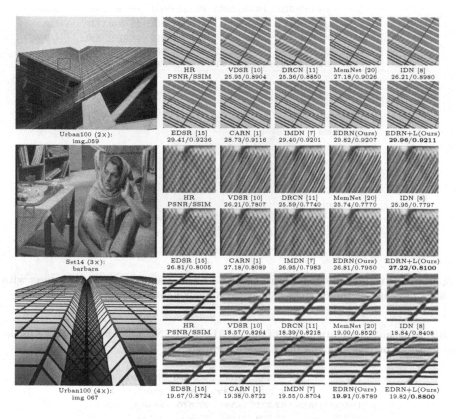

Fig. 5. Visual comparisons of SR methods on Set14 and Urban100 datasets. (EDSR denotes EDSR-baseline, EDRN+L denotes EDRN combined with LCAES)

3.3   Model Analysis

In this subsection, we investigate the effectiveness of EDRB and LCAES.

Ablation Studies of EDU and LCAES: To validate the effectiveness of the EDU, we replace the EDU in our EDRB with a module consisting of a leaky ReLU active layer in the middle of two 3×3 convolution layers to build a basic network. The quantitative results of the basic model correspond to the first row in Table 2. From Table 2, we can infer that a constructing network with EDU can help the network achieve better performance with fewer parameters, and LCAES leads to performance improvement without increasing parameters or retraining network.

Table 2. Investigations of EDU and LCAES.

Scale	EDU	LCAES	Params	Set5	Set14	BSD100	Urban100	Manga109
				PSNR/SSIM	PSNR/SSIM	PSNR/SSIM	PSNR/SSIM	PSNR/SSIM
×4	✗	✗	729K	32.17/0.8943	28.58/0.7811	27.54/0.7351	26.03/0.7837	30.42/0.9068
	✗	✓	729K	32.18/0.8944	28.59/0.7814	27.56/<u>0.7356</u>	<u>26.07</u>/<u>0.7856</u>	30.44/0.9072
	✓	✗	623K	<u>32.19</u>/0.8946	**28.60**/<u>0.7814</u>	27.54/0.7353	26.06/0.7846	<u>30.47</u>/ <u>0.9076</u>
	✓	✓	623K	**32.19/0.8948**	<u>28.60</u>/**0.7816**	**27.57/0.7358**	**26.10/0.7863**	**30.49/0.9079**

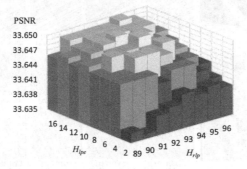

Fig. 6. PSNR values obtained for various H_{rlp} and H_{lpe} settings for Set14 dataset with scale 2. The highest three values are colored. (Color figure online)

Table 3. Quantitative evaluation of our LCAES on PSNR (The highest values of the two methods on 5 benchmark datasets are highlighted).

Scale	Model	Set5	Set14	BSD100	Urban100	Manga109
×4	IMDN	32.210	**28.582**	27.558	26.042	**30.451**
	IMDN+LCAES	**32.221**	28.580	**27.570**	**26.053**	30.436
	RCAN	**32.639**	28.851	27.748	26.748	31.200
	RCAN+LCAES	32.623	**28.859**	**27.767**	**26.812**	**31.208**

Further Investigation of LCAES: First we investigate the settings of parameters H_{rlp}, W_{rlp}, H_{lpe} and W_{lpe} in LCAES. For convenience, we set $H_{rlp} = W_{rlp}$ and $H_{lpe} = W_{lpe}$. Figure 6 illustrates the PSNR results produced by different selecting of H_{rlp} and W_{rlp} on Set14 dataset with upscale factor 2. When training our network with the upscale factor 2, the HR patch size is 192, and the corresponding LR patch size is 96. We find that when H_{rlp} is set close to the training LR patch size would always result in a good performance. The reason for this result is that a H_{rlp} close to the training LR patch size could make the testing phase close to the training environment, which could help the network better play the mapping relationship it learned from training. Another finding is that a large H_{lpe} is also very helpful for improving the performance. We think this is due to a large H_{lpe} can better mitigate the border effect. However, too large H_{lpe} will cause too much extra computation. Thus we choose a moderate value for H_{lpe}.

To further prove the superiority and applicability of LCAES, we perform LCAES on two other different SR networks with different channel-wise attention. We choose RCAN [25], a very deep network with channel attention, and IMDN [7], a lightweight network with contrast-aware channel attention. We plug LCAES into the network directly without retraining. The comparison results of whether to use LCAES on two networks are shown in Table 3. From results, we can find that using LCAES improves the fidelity of reconstructed images on most datasets for different networks. And this phenomenon is more evident on large datasets with large image sizes and complex content such as Urban100 and B100.

4   Conclusion

In this paper, we propose an expansion and distillation residual network for lightweight and accurate single image super-resolution. We expand and compress feature channels of different convolutional layers by using expansion and distillation residual block containing expansion and distillation units to improve SR performance while saving the amount of parameters. Additionally, we discover the fact that channel attention mechanism has not reached its potential in SR and design a local channel attention enhancement strategy to realize channel attention mechanism's potential in SR. Extensive experiments on benchmark datasets illustrate the effectiveness of our EDRN and LCAES in image super-resolution.

References

1. Ahn, N., Kang, B., Sohn, K.-A.: Fast, accurate, and lightweight super-resolution with cascading residual network. In: Ferrari, V., Hebert, M., Sminchisescu, C., Weiss, Y. (eds.) ECCV 2018. LNCS, vol. 11214, pp. 256–272. Springer, Cham (2018). https://doi.org/10.1007/978-3-030-01249-6_16
2. Bevilacqua, M., Roumy, A., Guillemot, C., Alberi-Morel, M.-L.: Low-complexity single-image super-resolution based on nonnegative neighbor embedding. In: Proceedings of the British Machine Vision Conference (BMVC), pp. 135.1–135.10 (2012)
3. Dong, C., Loy, C.C., Tang, X.: Accelerating the super-resolution convolutional neural network. In: Leibe, B., Matas, J., Sebe, N., Welling, M. (eds.) ECCV 2016. LNCS, vol. 9906, pp. 391–407. Springer, Cham (2016). https://doi.org/10.1007/978-3-319-46475-6_25

4. Dong, C., Loy, C.C., He, K., Tang, X.: Image super-resolution using deep convolutional networks. IEEE Trans. Pattern Anal. Mach. Intell. **38**(2), 295–307 (2016)
5. Hu, J., Shen, L., Sun, G.: Squeeze-and-excitation networks. In: The IEEE Conference on Computer Vision and Pattern Recognition (CVPR), pp. 7132–7141 (2018)
6. Huang, J.B., Singh, A., Ahuja, N.: Single image super-resolution from transformed self-exemplars. In: The IEEE Conference on Computer Vision and Pattern Recognition (CVPR), pp. 5197–5206 (2015)
7. Hui, Z., Gao, X., Yang, Y., Wang, X.: Lightweight image super-resolution with information multi-distillation network. In: ACM Multimedia Conference on Multimedia Conference (ACMMM), pp. 2024–2032 (2019)
8. Hui, Z., Wang, X., Gao, X.: Fast and accurate single image super-resolution via information distillation network. In: The IEEE Conference on Computer Vision and Pattern Recognition (CVPR), pp. 723–731 (2018)
9. Ioffe, S., Szegedy, C.: Batch normalization: accelerating deep network training by reducing internal covariate shift. In: International Conference on Machine Learning (ICML), pp. 448–456 (2015)
10. Kim, J., Kwon Lee, J., Mu Lee, K.: Accurate image super-resolution using very deep convolutional networks. In: The IEEE Conference on Computer Vision and Pattern Recognition (CVPR), pp. 1646–1654 (2016)
11. Kim, J., Kwon Lee, J., Mu Lee, K.: Deeply-recursive convolutional network for image super-resolution. In: The IEEE Conference on Computer Vision and Pattern Recognition (CVPR), pp. 1637–1645 (2016)
12. Kingma, D.P., Ba, J.: Adam: A method for stochastic optimization. arXiv preprint arXiv:1412.6980 (2014)
13. Lai, W.S., Huang, J.B., Ahuja, N., Yang, M.H.: Deep Laplacian pyramid networks for fast and accurate super-resolution. In: The IEEE Conference on Computer Vision and Pattern Recognition (CVPR), pp. 5835–5843 (2017)
14. Ledig, C., et al.: Photo-realistic single image super-resolution using a generative adversarial network. In: The IEEE Conference on Computer Vision and Pattern Recognition (CVPR), pp. 105–114 (2017)
15. Lim, B., Son, S., Kim, H., Nah, S., Lee, K.M.: Enhanced deep residual networks for single image super-resolution. In: The IEEE Conference on Computer Vision and Pattern Recognition Workshops (CVPRW), pp. 1132–1140 (2017)
16. Martin, D., Fowlkes, C., Tal, D., Malik, J.: A database of human segmented natural images and its application to evaluating segmentation algorithms and measuring ecological statistics. In: The IEEE International Conference on Computer Vision (ICCV), pp. 416–423 (2001)
17. Matsui, Y., Ito, K., Aramaki, Y., Fujimoto, A., Ogawa, T., Yamasaki, T., Aizawa, K.: Sketch-based manga retrieval using manga109 dataset. Multimed. Tools Appl. **76**(20), 21811–21838 (2017)
18. Shi, W., et al.: Real-time single image and video super-resolution using an efficient sub-pixel convolutional neural network. In: The IEEE Conference on Computer Vision and Pattern Recognition (CVPR), pp. 1874–1883 (2016)
19. Tai, Y., Yang, J., Liu, X.: Image super-resolution via deep recursive residual network. In: The IEEE Conference on Computer Vision and Pattern Recognition (CVPR), pp. 2790–2798 (2017)
20. Tai, Y., Yang, J., Liu, X., Xu, C.: MemNet: a persistent memory network for image restoration. In: The IEEE International Conference on Computer Vision (ICCV), pp. 4549–4557 (2017)

21. Timofte, R., Agustsson, E., Van Gool, L., Yang, M.H., Zhang, L.: NTIRE 2017 challenge on single image super-resolution: methods and results. In: The IEEE Conference on Computer Vision and Pattern Recognition Workshops (CVPRW), pp. 1110–1121 (2017)
22. Wang, Z., Bovik, A.C., Sheikh, H.R., Simoncelli, E.P., et al.: Image quality assessment: from error visibility to structural similarity. IEEE Trans. Image Process. (TIP) 13(4), 600–612 (2004)
23. Zeyde, R., Elad, M., Protter, M.: On single image scale-up using sparse-representations. In: International Conference on Curves and Surfaces, pp. 711–730 (2012)
24. Zhang, K., Zuo, W., Zhang, L.: Learning a single convolutional super-resolution network for multiple degradations. In: The IEEE Conference on Computer Vision and Pattern Recognition (CVPR), pp. 3262–3271 (2018)
25. Zhang, Y., Li, K., Li, K., Wang, L., Zhong, B., Fu, Y.: Image super-resolution using very deep residual channel attention networks. In: Ferrari, V., Hebert, M., Sminchisescu, C., Weiss, Y. (eds.) ECCV 2018. LNCS, vol. 11211, pp. 294–310. Springer, Cham (2018). https://doi.org/10.1007/978-3-030-01234-2_18

Hyperspectral Image Restoration
for Non-additive Noise

Fanlong Zhang[1], Heyou Chang[2(✉)], Zhangjing Yang[1], Tianming Zhan[1],
and Guowei Yang[1]

[1] Nanjing Audit University, Nanjing 211815, Jiangsu, China
{csfzhang,yzj,ztm}@nau.edu.cn, ygw_ustb@163.com
[2] Nanjing XiaoZhuang University, Nanjing 210024, Jiangsu, China
cv_hychang@126.com

Abstract. This work is about a curious phenomenon. Among hyperspectral image (HSI) restoration methods, most of them suppose the observed HSI is the superposition of clean component and noise component. However, in real world applications, the components are often non-additive. The non-additivity admits the uniqueness of attribute. Especially each pixel of image belongs to either clean component or noise component. To separate components robustly, this work proposes a general HSI restoration framework. The novelty of this work lies in exploiting non-additivity of components by a binary mask matrix. Two special cases are discussed. To solve the proposed models, efficient alternating minimization algorithms are developed. Compared with gradient-type algorithms, our algorithms are easy to implement as there is no need to tune optimization parameters like step sizes. The experimental results over simulated and real datasets verified the performance of the proposed method.

Keywords: Hyperspectral Image (HSI) · Low rank · Non-additive assumption · Alternating minimization

1 Introduction

Hyperspectral image (HSI) has distinguishable spectral signatures due to the different minerals characteristics, and thus has being widely applied [1]. The observed HSI is degraded because of various sources of noise [2], which severely limits the performance of subsequent analysis tasks, such as land cover classification [3, 4], target/anomaly detection [5], and unmixing [6]. Therefore, HSI restoration, also called restoration or noise reduction, is the first and a crucial phase of HSI processing, necessitating research for more effective and economic restoration methods. This domain has gained popularity, attracted many researchers [7].

This work was partly supported by the National Natural Science Foundation of China (Grant Nos. 61603192, U1831127, 61806098, and 61876213), Jiangsu Government Scholarship for Overseas Studies under Grant No. JS-2018-068.

© Springer Nature Switzerland AG 2020
Y. Peng et al. (Eds.): PRCV 2020, LNCS 12305, pp. 232–243, 2020.
https://doi.org/10.1007/978-3-030-60633-6_19

The typical two-dimensional image restoration methods, such as block-matching three-dimensional filtering (BM3D) [8], total variation (TV) [9], and the nonlocal-based algorithm [10], can be directly applied to the HSI in a band-by-band manner, since each spectral band can be treated as an independent grey-level image. The merit is the numerous restoration algorithms in image processing can be employed for HSI and processing can be parallel. However, the correlation among the spectral bands is ignored. As a result, restoration approaches considering the spatial-spectral structure of HSI have been developed under different frameworks [11–16].

Given an observed HSI $H \in R^{n \times m}$, with n being the number of bands, m being the number of pixels. The widely used assumption is that the H is decomposed as a combination of true unknown signal X and additive noise S: $H = X + S$.

This decomposition is impossible to solve without more information, since the number of unknowns is twice as many as the given observations in H. The key information is low-rankness of clean HSI X. Specifically, a clean HSI X can be regarded as a low-rank matrix because of the redundancy in the spectral bands. By using this information, many restoration methods are proposed and we can call them methods based on low-rank matrix approximation. In nature, they find and exploit low-dimensional structures in high-dimensional data, aiming to remove the noise in the polluted image to recover the clean image.

All methods above mentioned are based on the same assumption, which is additivity between decomposition terms. However, in real world applications, the decomposition terms are often non-additive. To separate image components robustly in such a situation, this paper employs a binary mask matrix which shows the location of each component, and proposes a novel HSI restoration framework. In [17, 18], the authors also borrowed a binary matrix to mark the attribution of each pixel, and apply into image restoration effectively. The difference is that we only impose the mask matrix on the clean HSI and sparse noise, and our model is general. The main novelties of this work can be summarized as follows.

1) The non-additivity between clean component and sparse component is explored by a binary matrix.
2) A general restoration model is established. The proposed model allows different kinds of assumption on the observation model.
3) An efficient alternating minimization algorithm is developed to solve the proposed model. In algorithm, both sub-problems have a closed-form solution, and thus our algorithm is easy to implement.

The rest of this paper is organized as follows. Section 2 details our motivation and restoration model. Section 3 presents an alternating minimization algorithm. Section 4 compares the performance on simulated and real datasets. Section 5 draws conclusions for this work.

2 Motivation and Model

First, we establish some notations. The numbers of bands and pixels in each band are denoted by n and m, where $m = h \times w$, h and w are the height (rows) and width (columns) of each band. Most of current restoration works are under the additive assumption:

$$H = X + S, \tag{1}$$

where the observation data H is decomposed as sum of two components X and S. In general, the component X represents the clean data (unpolluted data), latent true data, or recovered data, while the component S represents the noise data. The assumption means there are both clean component and noise component in H. Specifically, for any element $H_{i,j}$ of H, under non-additive assumption one portion of $H_{i,j}$ is from $X_{i,j}$, and the remaining portion from $S_{i,j}$. In one word, the attribute of $H_{i,j}$ is un-unique. But in real world applications, the components are often non-additive [15]. For example, an image may consist of a foreground object overlaid on a background, where each pixel belongs to either the foreground or the background. To represent the non-additivity, we aim to learn a binary mask $W_{i,j}$ to mark the attribute of $H_{i,j}$: when $W_{i,j} = 1$, the $H_{i,j}$ is from $S_{i,j}$, while $W_{i,j} = 0$, $H_{i,j}$ is from $X_{i,j}$. Thus, the non-additive model of (1) can be denoted by:

$$H = (1 - W) \circ X + W \circ S, \tag{2}$$

where $\circ$ denotes Hadamard product, $W \in \{0, 1\}^{m \times n}$ is mask matrix with binary entries marking the location of each component. In [28], it was shown that capturing the spectral redundancy by low-rank modeling is more appropriate than full-rank modeling for HSI restoration. As a result, the clean HSI image X can be assumed to be low-rank and factorized by ACM^T. So (2) can be modeled by:

$$H = (1 - W) \circ ACM^T + W \circ S. \tag{3}$$

To solve the components, we give a general optimization problem:

$$\min_{W \in \{0,1\}^{m \times n}, C, S} \left\| H - (1 - W) \circ ACM^T - W \circ S \right\|_F^2 + \lambda f(C) + \gamma g(S). \tag{4}$$

Furthermore, since $W \circ H = W \circ (1 - W) \circ X + W \circ W \circ S = W \circ S$, we can rewrite (4) as following:

$$\min_{W \in \{0,1\}^{m \times n}, C} \left\| H - (1 - W) \circ ACM^T - W \circ H \right\|_F^2 + \lambda f(C) + \gamma g(W), \tag{5}$$

which is equivalent to:

$$\min_{W \in \{0,1\}^{m \times n}, C} l(C, W) := \left\| (1 - W) \circ \left(H - ACM^T \right) \right\|_F^2 + \lambda f(C) + \gamma g(W). \tag{6}$$

Considering the special structure of HSI data, constructing a dictionary in the spectral domain to sparsely represent images is more consistent with the imaging mode and physical meaning of HSIs. One can expand the noisy HSI by scan lines into a pixel matrix, which is used as a training set so that a spectral dictionary can be obtained. The noisy image can be reconstructed with the dictionary and the corresponding sparse codes [19]. Specifically, in (3), the matrix $A \in R^{n \times k}$ represents the spectral dictionary, k is the number of dictionary atoms, $C \in R^{k \times m}$ denotes the sparse codes, M is an identity matrix of size m × m. The dictionary A may be estimated from the data using the HySime algorithm [20] or singular value decomposition (SVD) of H in the case the noise is independent and identically distributed (i.i.d.). As a result, we can obtain a special restoration model as following, which we denote Mask Sparse restoration model for HSI (MaskSpaHSI):

$$\min_{W \in \{0,1\}^{m \times n}, C} \|(1 - W) \circ (H - AC)\|_F^2 + \lambda \|C\|_1 + \gamma \|W\|_1. \tag{7}$$

To exploit the high correlation of the spectral information in HSI, low-rankness of clean HSI image X can be assumed to be the low-rank component of observation, which means letting both A and M be identity matrices. As a result we can obtain a special restoration model, which we denote Mask Low-rank restoration model for HSI (MaskLrHSI):

$$\min_{W \in \{0,1\}^{m \times n}, C} \|(1 - W) \circ (H - C)\|_F^2 + \lambda \|C\|_* + \gamma \|W\|_1. \tag{8}$$

3 Algorithm

3.1 General Framework of Alternative Minimization for (6)

The optimization problem (6) is a joint optimization problem. One approach to solving (6) is by concatenating W and C as a single variable and then directly applying standard iterative algorithms like gradient descent (or its variants); see [21, 22] for an overview. However, as the variables within the problem are naturally parted into two blocks W and C, it can be easier to update just one block variable at a time, without updating all variables simultaneously. The alternating minimization is suitable for such a purpose. Alternating minimization (also known as block-coordinate minimization) sequentially optimizes over one block variable while fixing the other. Specifically, we update variables W and C alternately as follows. In step K, given W^k and C^k, we pursue W^{k+1} and C^{k+1} as follows.

(i) Pursue C^{k+1} by solving

$$\min_C \left\| (1 - W) \circ \left(H - ACM^T \right) \right\|_F^2 + \lambda f(C). \tag{9}$$

Before that, we denote the quadratic term by function $g(\cdot)$ defined by:

$$g(U) = \|U\|_F^2, U = \left(1 - W^k \right) \circ \left(H - ACM^T \right). \tag{10}$$

Now we need to find the derivative of matrix function (10). By the chain rule,

$$\frac{\partial g(U)}{\partial C_{i,j}} = Tr \left[\left(\frac{\partial g(U)}{\partial U} \right)^T \left(\frac{\partial U}{\partial C_{i,j}} \right) \right]. \tag{11}$$

Thus, we have:

$$\frac{\partial g(C)}{\partial C_{i,j}} = Tr\left[2\left(1 - W^k\right)^T \circ \left(H - ACM^T\right)^T \left(\left(W^k - 1\right) \circ A_{(:,i)}M^T_{(:,j)}\right)\right]. \quad (12)$$

Denote the derivative matrix of function $g(\cdot)$ at point C^k by G_{C^k}. Now we linearize the quadratic term in (9) at C^k [23, 24],

$$\left\|\left(1 - W^k\right) \circ \left(H - ACM^T\right)\right\|^2_F \approx \left\|\left(1 - W^k\right) \circ \left(H - AC^k M^T\right)\right\|^2_F + trace\left[G^T_{C^k}\left(C - C^k\right)\right]$$

$$+ \frac{\eta}{2}\left\|C - C^k\right\|^2_F = \frac{\eta}{2}\left\|C - C^k + \frac{1}{\eta}G_{C^k}\right\|^2_F + const, \quad (13)$$

where the symbol *const* means the term without containing variable C. We now turn to the problem of solving a minimization problem of following form:

$$C^{k+1} = \arg\min_C \frac{\eta}{2}\left\|C - C^k + \frac{1}{\eta}G_{C^k}\right\|^2_F + \lambda f(C). \quad (14)$$

Note that the objective function in (14) is strictly convex when $f(\cdot)$ is convex, and so has a unique minimum.

(ii) Pursue W^{k+1} by solving

$$\min_W \left\|(1 - W) \circ \left(H - ACM^T\right)\right\|^2_F + \gamma g(W). \quad (15)$$

Denote $R = H - ACM^T$, one can rewritten (15) as:

$$\min_W \sum_{i,j}\left(1 - W_{i,j}\right)^2 R^2_{i,j} + \gamma g(W). \quad (16)$$

Given a tolerance ε, we use the relative error as the stopping criterion defined by

$$\frac{\left\|l\left(C^{k+1}, W^{k+1}\right) - l\left(C^k, W^k\right)\right\|_F}{\left\|l\left(C^k, W^k\right)\right\|} < \varepsilon. \quad (17)$$

The algorithm is summarized in Algorithm 1:

Algorithm 1. General Framework of alternative minimization for (6)

1: Input observation data $H \in R^{n \times m}$, parameters $A \in R^{n \times p}, M \in R^{m \times q}, \lambda, \gamma, \eta$, iteration step $k = 0$, stopping tolerance ε, and initialize the $C^0 \in R^{p \times q}$ by a random matrix.

2: Computing C^{k+1} by (14);

3: Computing W^{k+1} by (16);

4: Checking stopping criterion: if (17) is satisfied, go to 5; otherwise denote $k = k + 1$ and go to step 2;

5: Output: Optima C^{k+1}, W^{k+1}.

3.2 Algorithm for MaskSpaHSI

Now, we give special cases of the Algorithm 1 for MaskSpaHSI. In MaskSpaHSI, $f(\cdot)$ $= g(\cdot) = \|\cdot\|_1$, $M = I_{m \times m}$, the dictionary A is estimated by the HySime. In this case, optimization (14) has a close-form solution:

$$C^{k+1} = S_{\frac{\lambda}{\eta}}\left(C^k - \frac{1}{\eta}G_{C^k}\right), \tag{18}$$

where $S_\lambda(\cdot)$ is soft threshold operator defined by:

$$S_\lambda(x) = \max(x - \lambda, 0) + \min(x + \lambda, 0). \tag{19}$$

For (16), each element of W^{k+1} can be optimized by

$$\min_{W_{i,j}}(1 - W_{i,j})^2 R_{i,j}^2 + \gamma|W_{i,j}|. \tag{20}$$

which also has a close-form solution:

$$W_{i,j}^{k+1} = \arg\min_{W_{i,j}} \frac{1}{2}(1 - W_{i,j})^2 + \frac{\gamma}{2R_{i,j}^2}|W_{i,j}| = S_{\frac{\gamma}{2R_{i,j}^2}}(1). \tag{21}$$

The algorithm is summarized in Algorithm 2:

Algorithm 2. Alternative minimization for MaskSpaHSI

1: Input observation data $H \in R^{n \times m}$, parameters $A \in R^{n \times p}$, $M = I_{m \times m}$, λ, γ, η, iteration step $k = 0$, stopping tolerance ε, and initialize the $C^0 \in R^{p \times m}$ by a random matrix.

2: Computing $C^{k+1} = S_{\frac{\lambda}{\eta}}(C^k - \frac{1}{\eta}G_{C^k})$;

3: Computing $W_{i,j}^{k+1} = S_{\frac{\gamma}{2R_{i,j}^2}}(1)$;

4: Checking stopping criterion: if (17) is satisfied, go to 5; otherwise denote $k = k + 1$ and go to step 2;

5: Output: Optima C^{k+1}, W^{k+1}.

3.3 Algorithm for MaskLrHSI

Now, we give the algorithm for MaskLrHSI. In MaskSpaHSI, $f(\cdot) = \|\cdot\|_*$, $g(\cdot) = \|\cdot\|_1$, $A = I_{n \times n}$, $M = I_{m \times m}$. In this case, optimization (14) has a close-form solution:

$$C^{k+1} = D_{\frac{\lambda}{\eta}}\left(C^k - \frac{1}{\eta}G_{C^k}\right), \tag{22}$$

where $D_\lambda(\cdot)$ is singular value threshold operator defined, for a given matrix with its SVD $B = U\Sigma V^T$, by:

$$D_\lambda(B) = US_\lambda(\Sigma)V^T. \tag{23}$$

The algorithm is summarized in Algorithm 3:

Algorithm 3. Alternative minimization for MaskLrHSI

1: Input observation data $H \in R^{n \times m}$, parameters $A = I_{n \times n}$, $M = I_{m \times m}$, λ, γ, η, iteration step $k = 0$, stopping tolerance ε, and initialize the $C^0 \in R^{n \times m}$ by a random matrix.

2: Computing $C^{k+1} = D_{\frac{\lambda}{\eta}}(C^k - \frac{1}{\eta}G_{C^k})$;

3: Computing $W_{i,j}^{k+1} = S_{\frac{\gamma}{2R_{i,j}^2}}(1)$;

4: Checking stopping criterion: if (17) is satisfied, go to 5; otherwise denote $k = k + 1$ and go to step 2;

5: Output: Optima C^{k+1}, W^{k+1}.

3.4 Algorithm Convergence

Figure 1 shows the convergence with different η. The iteration number vary significantly to varying η. This is because the parameter η directly decide how accurate the approximation term. At most cases the objective function becomes stable after 20 iterations. Generally speaking, the variation of objective function value is less than 10^{-5} when the number of iterations is over 30. However, the empirical performance of alternating minimization is not sufficiently substantiated by solid convergence guarantees. In fact, although the idea of alternately updating the variables is quite straightforward, the convergence properties for alternating minimization are far more complicated [29].

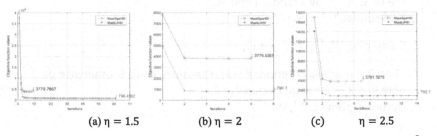

$$(a)\, \eta = 1.5 \qquad (b)\, \eta = 2 \qquad (c) \qquad \eta = 2.5$$

Fig. 1. Objective function values versus iterations, with stopping tolerance $\varepsilon = 10^{-5}$.

4   Experiments

The proposed algorithm is compared with the state-of-the-art algorithms, including fast hyperspectral denoising algorithm (FastHyDe) [12], subspacebased nonlocal low-rank and sparse factorization algorithm (SNLRSF) [13], non i.i.d. mixture of Gaussian algorithm based on low-rank matrix factorization (NMoG-LRMF) [25], block-matching and 4D filtering algorithm (BM4D) [26], HSI restoration algorithm on the basis of low-rank matrix recovery (LRMR) [15], and spatio-spectral total variation algorithm (SSTV) [27].

The parameters for these compared methods were manually adjusted according to their default strategies. All the bands of the HSI datasets are normalized into [0, 1]. The mean peak signal-to-noise ratio (MPSNR) and mean structural similarity index (MSSIM) are employed for quantitative comparison: $\text{MPSNR} = \frac{1}{n} \sum_{i=1}^{n} \text{PSNR}_i$, $\text{MSSIM} = \frac{1}{n} \sum_{i=1}^{n} \text{SSIM}_i$.

4.1 Experiments on Simulated Data

Five HSI datasets are employed in this experiment: Washington DC Mall sub-scene of size $256 \times 256 \times 191$, Pavia Centre sub-scene of size $200 \times 200 \times 103$, Urban of size $307 \times 307 \times 162$, Cuprite of size $250 \times 250 \times 224$, and Japser Ridge of size $100 \times 100 \times 224$. Some bands of these data are heavily contaminated by noise and cannot be regarded as clean ground truth, and thus have to be deleted. The color composite representations of original HSIs are shown in Fig. 2.

(a)   (b)   (c)   (d)   (e)

Fig. 2. The color composite representations of HSIs. (a) Japser Ridge, (b) Cuprite, (c) Urban, (d) Pavia, (e) Washington DC Mall.

To simulate the clean HSI, HSIs are firstly restored via SVD. Then these restored HSIs are considered as clean data in this simulated experiment. To simulate noisy HSIs, we polluted the clean HSI images by adding sparse noise with magnitude 1 for 10% of pixels and Gaussion noise with intensity of 0.02, simultaneously.

Comparisons in terms of MPSNR, MSSIM and time at different scenes and noise levels are presented in Tables 1, 2, and 3. Dealing with different levels of noise under different scenes, proposed MaskSpaHSI or MaskLrHSI yields the best MPSNR performance in most cases. At the same time, MaskSpaHSI and MaskLrHSI run faster than LRMR and SNLRSF although slower than FastHyDe, since the latter is a non-iterative algorithm.

Some restoration results are shown in Fig. 3. Our method can obtain competitive recovery performance in comparison with others. High quality spectral signature is of critical importance to material identification. The quality of reconstructed spectra from different restoration methods can also be inferred from Fig. 4. Clearly, MaskSpaHSI and MaskLrHSI can recover spectral signatures better than others in many cases.

4.2 Experiments on Real Data

Two real HSI datasets are adopted to further demonstrate the restoration performance of MaskSpaHSI and MaskLrHSI. One is the Urban dataset of size $307 \times 307 \times 210$, whose bands are polluted due to dense water vapor and atmospheric effects. Another dataset is

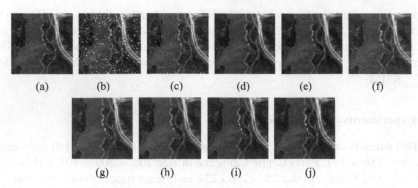

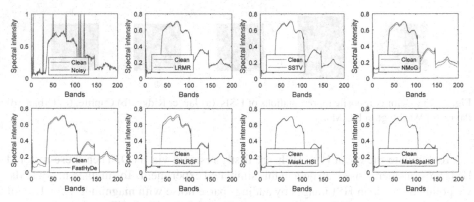

Fig. 3. Recovered band 10 of Japser Ridge under Gaussion noise with intensity of 0.02: (a) clean image, (b) noisy image, recovered images by (c) BM4D, (d) LRMR, (e) SSTV, (f) NMoG, (g) FastHyDe, (h) SNLRSF, (i) MaskSpaHSI, (j) MaskLrHSI.

Fig. 4. Restored spectral signatures of Japser Ridge.

Table 1. MPSNRs of different algorithms on five datasets

Datasets	LRMR	FastHyDe	SNLRSF	MaskSpaHSI	MaskLrHSI
Japser Ridge	28.8021	25.3243	26.8587	30.5436	30.4912
Cuprite	22.6607	25.3847	29.9589	31.1913	31.0105
Urban	20.9258	25.1778	28.5993	33.5955	33.2579
Pavia Centre	15.9882	25.1717	27.9715	32.0908	31.6525
Washington DC Mall	23.1411	25.1447	28.8598	31.1090	30.8626

the Indian Pines dataset of size $145 \times 145 \times 220$, some of whose bands are polluted by a mixture of Gaussian noise, impulse noise, and water absorption. Restoration results from different methods are given in Figs. 5, 6. It is easy to observe that our proposed method achieves competitive visual quality compared to the others.

Table 2. SSMs of different algorithms on five datasets

Datasets	LRMR	FastHyDe	SNLRSF	MaskSpaHSI	MaskLrHSI
Japser Ridge	0.8219	0.7853	0.8084	0.8712	0.8732
Cuprite	0.8193	0.9024	0.9090	0.9428	0.9472
Urban	0.8157	0.7766	0.8799	0.9609	0.9646
Pavia Centre	0.7638	0.7590	0.8299	0.9249	0.9315
Washington DC Mall	0.8297	0.7687	0.8891	0.9316	0.9320

Table 3. Times (s) of different algorithms on five datasets

Datasets	LRMR	FastHyDe	SNLRSF	MaskSpaHSI	MaskLrHSI
Japser Ridge	8.65	0.11	11.34	29.15	25.23
Cuprite	56.48	0.55	53.52	11.72	8.91
Urban	110.21	1.06	102.87	6.73	5.54
Pavia Centre	50.89	0.66	64.00	3.57	2.55
Washington DC Mall	78.46	0.77	74.81	5.62	4.86

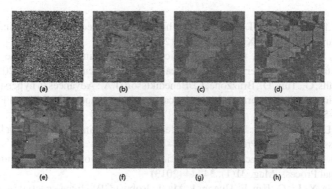

Fig. 5. Restoration results in Indian Pines. (a) Observation; (b) LRMR; (c) SSTV; (d) NMoG; (e) FastHyDe; (f) SNLRSF; (g) MaskLrHSI, (h) MaskSpaHSI.

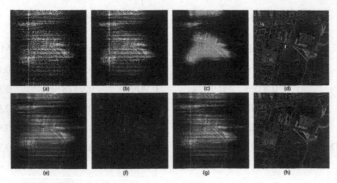

Fig. 6. Restoration results in Urban. (a) Observation; (b) LRMR; (c) SSTV; (d) NMoG; (e) FastHyDe; (f) SNLRSF; (g) MaskLrHSI, (h) MaskSpaHSI.

5 Conclusion

This paper proposed a general HSI restoration framework based on non-additive and low-rank assumptions, by employing a binary mask matrix. Two special cases, MaskSpaHSI and MaskLrHSI, and corresponding alternating minimization algorithms are discussed. MaskSpaHSI and MaskLrHSI can extend to more general forms.

References

1. Gewali, U.B., Monteiro, S.T., Saber, E.: Machine learning based hyperspectral image analysis: a survey. arXiv preprint arXiv:1802.08701 (2018)
2. Rasti, B., Scheunders, P., Ghamisi, P., Licciardi, G., Chanussot, J.: Noise reduction in hyperspectral imagery: overview and application. Remote Sens. **10**(3), 482 (2018)
3. Camps-Valls, G., Tuia, D., Bruzzone, L., Benediktsson, J.A.: Advances in hyperspectral image classification: earth monitoring with statistical learning methods. IEEE Signal Process. Mag. **31**(1), 45–54 (2014)
4. Wang, Z., Liu, J., Xue, J.H.: Joint sparse model-based discriminative K-SVD for hyperspectral image classification. Sig. Process. **133**, 144–155 (2017)
5. Nasrabadi, N.M.: Hyperspectral target detection: an overview of current and future challenges. IEEE Signal Process. Mag. **31**(1), 34–44 (2013)
6. Mei, X., Ma, Y., Li, C., Fan, F., Huang, J., Ma, J.: Robust GBM hyperspectral image unmixing with superpixel segmentation based low rank and sparse representation. Neurocomputing **275**, 2783–2797 (2018)
7. Ghamisi, P., et al.: Advances in hyperspectral image and signal processing: a comprehensive overview of the state of the art. IEEE Geosci. Remote Sens. Mag. **5**(4), 37–78 (2017)
8. Dabov, K., Foi, A., Katkovnik, V., Egiazarian, K.: Image denoising by sparse 3-D transform-domain collaborative filtering. IEEE Trans. Image Process. **16**(8), 2080–2095 (2007)
9. Rudin, L.I., Osher, S., Fatemi, E.: Nonlinear total variation based noise removal algorithms. Physica D **60**(1–4), 259–268 (1992)
10. Buades, A., Coll, B., Morel, J.-M.: A non-local algorithm for image denoising. In: 2005 IEEE Computer Society Conference on Computer Vision and Pattern Recognition (CVPR 2005), vol. 2, pp. 60–65. IEEE (2005)

11. Kong, Z., Yang, X.: Color image and multispectral image denoising using block diagonal representation. IEEE Trans. Image Process. **29**(2), 4247–4259 (2019)
12. Zhuang, L., Bioucas-Dias, J.M.: Fast hyperspectral image denoising and inpainting based on low-rank and sparse representations. IEEE J. Sel. Top. Appl. Earth Obs. Remote Sens. **11**(3), 730–742 (2018)
13. Cao, C., Jie, Yu., Zhou, C., Kai, H., Xiao, F., Gao, X.: Hyperspectral image denoising via subspace-based nonlocal low-rank and sparse factorization. IEEE J. Sel. Top. Appl. Earth Obs. Remote Sens. **12**(3), 973–988 (2019)
14. Zhang, X., et al.: Hyperspectral unmixing via low-rank representation with space consistency constraint and spectral library pruning. Remote Sens. **10**(2), 339 (2018)
15. Zhang, H., He, W., Zhang, L., Shen, H., Yuan, Q.: Hyperspectral image restoration using low-rank matrix recovery. IEEE Trans. Geosci. Remote Sens. **52**(8), 4729–4743 (2014)
16. Fan, H., Li, J., Yuan, Q., Liu, X., Ng, M.: Hyperspectral image denoising with bilinear low rank matrix factorization. Sig. Process. **163**, 132–152 (2019)
17. Jin, M., Chen, Y.: Robust image recovery via mask matrix. In: Cui, Z., Pan, J., Zhang, S., Xiao, L., Yang, J. (eds.) IScIDE 2019. LNCS, vol. 11935, pp. 349–361. Springer, Cham (2019). https://doi.org/10.1007/978-3-030-36189-1_29
18. Minaee, S., Wang, Y.: An ADMM approach to masked signal decomposition using subspace representation. IEEE Trans. Image Process. **28**(7), 3192–3204 (2019)
19. Song, X., Lingda, W., Hao, H., Wanpeng, X.: Hyperspectral image denoising based on spectral dictionary learning and sparse coding. Electronics **8**(1), 86 (2019)
20. Bioucas-Dias, J.M., Nascimento, J.M.P.: Hyperspectral subspace identification. IEEE Trans. Geosci. Remote Sens. **46**(8), 2435–2445 (2008)
21. Jain, P., Kar, P.: Non-convex optimization for machine learning. Found. Trends Mach. Learn. **10**(3–4), 142–336 (2017)
22. Chi, Y., Lu, Y.M., Chen, Y.: Nonconvex optimization meets low-rank matrix factorization: an overview. IEEE Trans. Signal Process. **67**(20), 5239–5269 (2019)
23. Liu, R., Lin, Z., Su, Z.: Linearized alternating direction method with parallel splitting and adaptive penalty for separable convex programs in machine learning. In: Asian Conference on Machine Learning, pp. 116–132 (2013)
24. Lin, Z., Liu, R., Su, Z.: Linearized alternating direction method with adaptive penalty for low-rank representation. In: Advances in Neural Information Processing Systems, pp. 612–620 (2011)
25. Chen, Y., Cao, X., Zhao, Q., Meng, D., Zongben, X.: Denoising hyperspectral image with non-iid noise structure. IEEE Trans. Cybern. **48**(3), 1054–1066 (2018)
26. Maggioni, M., Katkovnik, V., Egiazarian, K., Foi, A.: Nonlocal transform-domain filter for volumetric data denoising and reconstruction. IEEE Trans. Image Process. **22**(1), 119–133 (2013)
27. Aggarwal, H.K., Majumdar, A.: Hyperspectral image denoising using spatio-spectral total variation. IEEE Geosci. Remote Sens. Lett. **13**(3), 442–446 (2016)
28. Rasti, B.: Sparse hyperspectral image modeling and restoration, Ph.D. thesis, Department of Electrical and Computer Engineering, University of Iceland, Reykjavik, Iceland (2014)
29. Li, Q., Zhu, Z., Tang, G.: Alternating minimizations converge to second-order optimal solutions. In: International Conference on Machine Learning, pp. 3935–3943 (2019)

A SARS-CoV-2 Microscopic Image Dataset with Ground Truth Images and Visual Features

Chen Li[1(✉)], Jiawei Zhang[1], Frank Kulwa[1], Shouliang Qi[1], and Ziyu Qi[2]

[1] Microscopic Image and Medical Image Analysis Group, MBIE College,
Northeastern University, Shenyang 110169, China
lichen201096@hotmail.com, 1971087@stu.neu.edu.cn, frank.kulwa@gmail.com,
qisl@bmie.neu.edu.cn
[2] School of Life Science and Technology, University of Electronic Science and
Technology of China, Chengdu 611731, China
qi.ziyu@outlook.com

Abstract. SARS-CoV-2 has characteristics of wide contagion and quick propagation velocity. To analyse the visual information of it, we build a SARS-CoV-2 Microscopic Image Dataset (SC2-MID) with 48 electron microscopic images and also prepare their ground truth images. Furthermore, we extract multiple classical features and novel deep learning features to describe the visual information of SARS-CoV-2. Finally, it is proved that the visual features of the SARS-CoV-2 images which are observed under the electron microscopic can be extracted and analysed.

Keywords: SARS-CoV-2 · Image dataset · Visual features · Ground truth image · Classical feature extraction · Deep learning feature extraction

1 Introduction

It is reported that the severe acute respiratory syndrome coronavirus 2 (SARS-CoV-2) breaks out since the end of December in 2019 [8]. More than 12,690,000 people have been infected around the world till July 12th, 2020 [17], which becomes a global malignant epidemic.

A novel coronavirus was detected for the first time in the laboratory on January 7th, 2020, and the whole genome sequence of the virus was obtained. The first 15 positive cases of novel coronavirus were detected by nucleic acid detection and the virus was separated from one positive patient and observed under an electron microscope. The detection of pathogenic nucleic acid was completed

J. Zhang—Cofirst author. This work is supported by "National Natural Science Foundation of China" (No. 61806047), the "Fundamental Research Funds for the Central Universities" (Nos. N2019003 and N2019005), and the China Scholarship Council (No. 2017GXZ026396).

© Springer Nature Switzerland AG 2020
Y. Peng et al. (Eds.): PRCV 2020, LNCS 12305, pp. 244–255, 2020.
https://doi.org/10.1007/978-3-030-60633-6_20

on January 10th, 2020. The first novel coronavirus with electron microscopic images in China was successfully separated from the Center for Disease Control and Prevention (CDC) on January 24th, 2020. Novel coronavirus nucleic acids were detected in 33 samples from China's CDC on January 26th, 2020, and the virus was successfully separated from positive environmental samples. SARS-CoV-2 outbreak was announced as a public health emergency of international concern (PHEIC) by the World Health Organization (WHO) on January 30th, 2020. It was officially named as COVD-19 (corona virus disease 2019) by WHO on February 11th, 2020. The International Committee on Taxonomy of Viruses (ICTV) has named the virus as Severe Acute Respiratory Syndrome Coronavirus 2 (SARS-CoV-2) [6].

Coronavirus is a kind of plus-strand RNA virus, which can infect many mammals including human beings and can cause cold and some other serious diseases such as Middle East Respiratory Syndrome (MERS) and Severe Acute Respiratory Syndrome (SARS) [4]. The SARS-CoV-2 is a novel coronavirus that has never been found in human bodies. The SARS-CoV-2 is highly infectious and mainly transmitted through close contact and respiratory droplets. Besides, the severe patients may die [15]. The main symptom of human infection is respiratory disease, accompanied by fever, cough and may cause viral pneumonia [1]. There is no effective medicine developed up to now. Because of the rapid mutation of RNA coronavirus and lack of efficient medicine, there are mainly two methods to confirm the case, the first is detecting the positive nucleic acid of coronavirus by using RT-PCR, the other one is the high homology between the results of the virus gene sequence and the known SARS-CoV-2 [3]. Both of the methods above need professional medical equipments and personnel. So the process is time consuming and expensive. Microscopic image analysis provides a new method for rapid coronavirus screening [10–14], thus, several visual features are extracted in the experiments below.

2   SARS-CoV-2 Microscopic Image Dataset (SC2-MID)

2.1   SARS-CoV-2 Visual Properties and Ground Truth Image Preparation

SARS-CoV-2 is a type of β coronavirus, which has envelope with spinous process. The shape of the virus particle is circular or oval, which seems as a solar corona. The tubular inclusions can be detected in the cell which is infected with coronavirus. The spinous process of different coronaviruses has significant differences. The SARS-CoV-2 has a diameter of 60–140 nm. It also has the largest genome as a RNA virus. The mircoscopic image of SARS-CoV-2 is shown in Fig. 1.

In order to obtain the accurate image and extract the visual features of SARS-CoV-2, a dataset is constructed and one of the example images with its Ground Truth (GT) image are shown in Fig. 2. Geometric features and texture features of SARS-CoV-2 are extracted by combining both the original image and GT image.

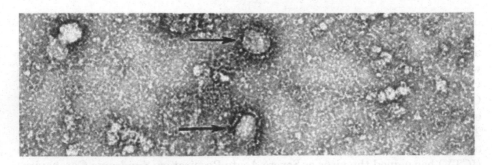

Fig. 1. The electron microscopic image of SARS-CoV-2 [9].

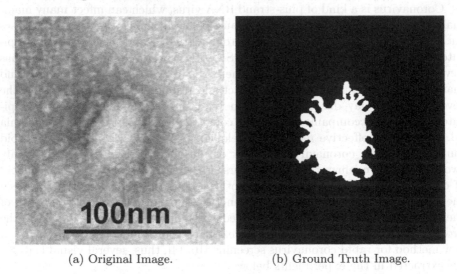

(a) Original Image.   (b) Ground Truth Image.

Fig. 2. An example of original image and ground truth image.

Based on the visual properties of SARS-CoV-2 images, we prepare the GT images according to the rules as follows:

a The GT images in the dataset are generated in pixel level. The foreground of GT images is the SARS-CoV-2 and the pixel value is "1" which represents white. The pixel value of the background is zero ("0") which represents black.

b Then all part of SARS-CoV-2 is drawn from the observation by naked eyes.

c When the spinous region of SARS-CoV-2 is not clear, the GT image is drawn following the value of the particle region. If the SARS-CoV-2 particle is bright part, then the bright region around the particle will be drawn as spinous part of SARS-CoV-2, and vice versa.

2.2 Dataset Construction

Most of the SARS-CoV-2 images are kept by National Health Commission at present. One of the images with scale is shown in Fig. 2(a) and the corresponding GT image is shown in Fig. 2(b). The scale of Fig. 2(a) is labelled in the image precisely and can be used to calculate the real size of the SARS-CoV-2.

The SARS-CoV-2 microscopic image dataset (SC2-MID) we built has 17 electron microscopic images which are separated to 48 single SARS-CoV-2 images. We are willing to share the dataset with the researchers for sharing but not for commercial purposes. If you want to obtain the dataset, please contact our data manager Jiawei Zhang (1971087@stu.neu.edu.cn). The horizontal axes of images are roughly the same length with the scale length in Fig. 2(a) which has a length of 100 nm. The other images are resized and cut based on the scale of Fig. 2(a). The image names with their data sources are shown in Table 1.

The database of SARS-CoV-2 is built which contains 48 electron microscopic images. The corresponding GT images are also shown in Fig. 3.

3 SARS-CoV-2 Visual Feature Extraction

Visual feature extraction is one of the important parts of computer vision. We extract several shape features including basic geometric feature and Hu invariant moment. We also extract several texture features including Histogram of Oriented Gradient (HOG) and Gray-Level Co-occurrence Matrix (GLCM). And we use deep learning such as VGG-16, Xception and DenseNet121 to get the feature map of images. All the features above can be extracted from the dataset we constructed. An example of feature extraction is shown as follows.

3.1 Shape Feature Extraction

Shape feature extraction is one of the most important research topics in describing the true nature of images. The original shape of an object can be precisely stored by using shape features, which is significant for computer vision and image recognition.

Basic Geometric Features: Perimeter refers to the boundary length of the object in the image. The perimeter of the image is composed of several discrete pixel points, which is calculated as the sum of pixel points of the target edge.

The area of an object in an image is usually represented by calculating the sum of pixel points of the object. The GT image is used to calculate the area of SARS-CoV-2 in the original image. The SARS-CoV-2 in GT image is shown as white which is represented by 1, and the background is shown as black which is represented by 0. Scan the GT image and sum the number of pixel points that the value is 1. The process is defined as below:

$$A = \sum_{x=1}^{N} \sum_{y=1}^{M} f(x, y), \tag{1}$$

Table 1. The sources of SARS-CoV-2 dataset.

ImageName	Source
IMG-001 - IMG-004	https://www.infectiousdiseaseadvisor.com/home/topics/gi-illness/covid-19-symptoms-may-need-to-be-extended-to-include-gi-symptoms/
IMG-005 - IMG-008	https://wired.jp/2020/03/08/what-is-a-coronavirus/
IMG-009 - IMG-014	https://www.genengnews.com/news/sars-cov-2-insists-on-making-a-name-for-itself/
IMG-015 - IMG-018	https://www.h-brs.de/en/information-on-coronavirus
IMG-019 - IMG-022	https://www.medicalnewstoday.com/articles/why-does-sars-cov-2-spread-so-easily
IMG-023 - IMG-028	https://www.upwr.edu.pl/news/51004/coronavirus-sars-cov-2-messages.html
IMG-029 - IMG-032	https://www.infectiousdiseaseadvisor.com/home/topics/gi-illness/covid-19-symptoms-may-need-to-be-extended-to-include-gi-symptoms/
IMG-033 - IMG-035	https://www.dzif.de/en/sars-cov-2-dzif-scientists-and-development-vaccines
IMG-036 - IMG-038	https://www.charite.de/en/clinical-center/themes-hospital/faqs-on-sars-cov-2/:
IMG-039 - IMG-041	https://news.harvard.edu/gazette/story/2020/03/in-creating-a-coronavirus-vaccine-researchers-prepare-for-future/
IMG-042	https://newsbash.ru/society/health/17116-kak-vygljadit-koronavirus-pod-mikroskopom-rossijskie-uchenye-sdelali-foto.html
IMG-043	https://www.rbc.ru/rbcfreenews/5e735ff09a7947be392f2bec
IMG-044	http://www.ellegirl.ru/articles/foto-dnya-kak-vyglyadit-koronavirus/
IMG-045 - IMG-046	https://mp.weixin.qq.com/s/zO8rW8W2TgzN2o6JEKcLnQ
IMG-047 - IMG-048	https://br.sputniknews.com/asia-oceania/2020012415043397-china-publica-foto-do-coronavirus-visto-por-microscopio-eletronico/

The area of SARS-CoV-2 is represented by the sum of the pixels that $f(x, y) = 1$.

The major axis and the minor axis are the longest length and the shortest length while linking two random points of an oval. They are usually represented as the major axis and minor axis of the smallest oval which can contain all of the objects in an irregular figure. The major axis and minor axis are defined as below:

$$l_a = max \sqrt{(i_m - i_n)^2 + (j_m - n_n)^2}, \tag{2}$$

where l_a is the length of the major axis, i_m, i_n, j_m, j_n are boundary points of the connected region in four directions.

$$s_a = \frac{p}{t} - l_a, \tag{3}$$

where s_a is the length of minor axis, p is the perimeter, t is the ovality factor defined by s_a/l_a.

From Fig. 2, we segment the image scale and match it to pixels. The pixels of part of the scale are black and the rest pixels of the image are white. So it is easy to calculate the length of scale and calculate the proportion of the image size to true size. And then the proportion can be used to calculate the values of the shape features. The length of scale consists of 206 pixels which means 100 nm for real size, so the proportion is about 0.4854 nm/pixels.

The shape features of SARS-CoV-2 are extracted by combining the GT image and original image, and the true size of these features are calculated by using the proportion above and the values are shown in Table 2. The eccentricity is the ratio of the major axis to the minor axis.

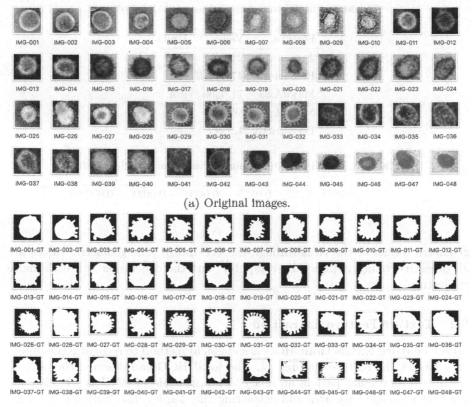

(a) Original images.

(b) Ground truth images.

Fig. 3. The database of gray images and GT images of SARS-CoV-2.

Hu Invariant Moment: Hu invariant moment creates seven invariant moment functions by using the normalized second and third order center distance. The central moment is defined as bellow:

$$\mu_{pq} = \sum_{x=1}^{M} \sum_{y=1}^{N} (x - x_0)^p (y - y_0)^q f(x, y), \tag{4}$$

where (x_0, y_0) is the center of gravity coordinates of the image.

HU is used to describe the properties of images. The Hu invariant moment is widely used because of its high stability while changing the geometric characteristics of the images, which is invariant to translation, rotation and scale transformation. Hu invariant moment is generally used to identify large objects in an image, which can describe the shape of objects well and recognize them quickly. The seven values of Hu invariant moment for Fig. 2 are shown in Table 3.

The geometric values are roughly consistent with the morphology of the virus when observed by the naked eyes. It is much more accurate to obtain the

Table 2. The shape feature values of SARS-CoV-2.

Feature	Value	Unit
Area	2808.98	nm^2
Perimeter	477.4689	nm
MajorAxis	71.5500	nm
MinorAxis	56.8100	nm
Eccentricity	1.2595	–

Table 3. The Hu invariant monent values of SARS-CoV-2.

0.9584	0.1854	0.0270	0.0125	−0.0004	−0.0021	−0.0002

geometric values of the virus through the GT image. The geometric values are all objective values obtained by computer, but the conclusion is a little more subjective.

3.2 Texture Feature Extraction

HOG Feature Extraction: HOG is a classical method to recognize the object and extract the texture features [5]. The local HOG can describe the texture features of particular part in an image. The principle of HOG is selecting the gradient of image edge area and extracting the density distribution coefficient. The first step is dividing the global image into several sub-images based on pixels. Then calculate the oriented gradient values and save them into a matrix. Finally, integrate the matrix based on initial image.

Because of the rich information and high diversity of image, the dimensions of extracted HOG feature are different. So it is necessary to normalize the HOG feature into a 36 dimensional vector, which has 4 blocks in a region with 9 dimensions per block. We extract the HOG feature of SARS-CoV-2 by combining the GT and original images. The normalized histogram with 36 dimensional vectors is shown in Fig. 4.

The vector graph of HOG feature is shown in Fig. 5. The red arrows describe the change of oriented gradient precisely. The vector graph shows good ability in describing the HOG feature of the part of SARS-CoV-2.

GLCM Feature Extraction: GLCM is one of the general methods to describe the texture features of images. The principle is measuring the spatial information between two pixels to describe the texture features. The texture feature refers to the gray-scale relationship between two pixels. Find the corresponding relationship between the pixel and the pixels in eight directions. The GLCM is to combine the co-occurrence matrix between every two pixels in the image.

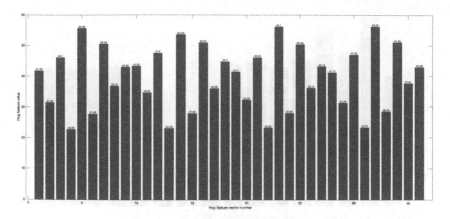

Fig. 4. HOG histogram of SARS-CoV-2.

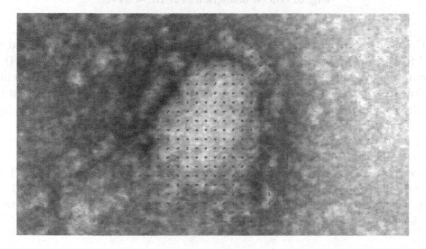

Fig. 5. The vector graph of HOG feature for SARS-CoV-2.

There are four kinds of eigenvalues in GLCM, which are contrast, homogeneity, correlation and energy. Every feature above has 4 dimensional vectors. The normalized histogram with 16 dimensional vectors is shown in Fig. 6.

The edge feature of the image can be well extracted without losing much local details, and the feature of low sensitivity for local geometry and optical transformation can be acquired by HOG and GLCM feature extraction, which can express the texture feature of SARS-CoV-2 effectively.

3.3 Deep Learning Feature Extraction

Convolution neural networks (CNN) have the ability to learn deep features which are substituted to hand crafted features such as corners, edges, blobs and ridges. These features are very robust and invariant to image translational changes.

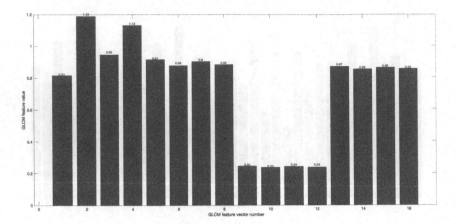

Fig. 6. GLCM histogram of SARS-CoV-2.

CNN uses multiple convolution layers to progressively extract high and low level features from raw input images. They use convolution matrix (kernel) for blurring, sharpening, embossing and edge detection. The lower layers identify general features such as corners and edges, while deep layers extract features specific to the organisms. Example of the CNN is shown in Fig. 7, which shows deep layers of VGG-16 model. With such high extraction power of CNNs, we can use them to extract deep learning features which can be used for detection and classification of SARS-CoV-2. Additionally, the scarcity of SARS-CoV-2 dataset on training CNN can be overcome by the use of transfer learning. Thus, we extract deep learning features from SARS-CoV-2 images using VGG-16 [16], Xception [2] and DenseNet [7] which have been pre-trained on ImageNet dataset. The feature maps extracted from different layers are shown in Fig. 8.

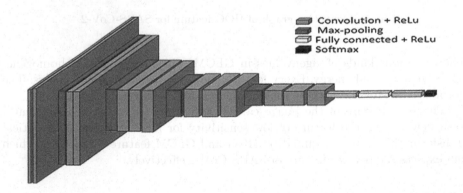

Fig. 7. VGG-16 network for extracting deep learning features.

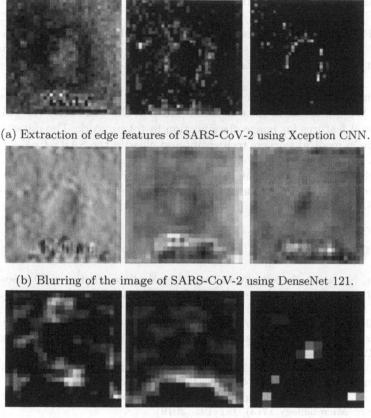

(a) Extraction of edge features of SARS-CoV-2 using Xception CNN.

(b) Blurring of the image of SARS-CoV-2 using DenseNet 121.

(c) Corner and edge feature learning by VGG-16.

Fig. 8. Extracted feature maps of dataset using CNNs.

3.4 Analysis

The SARS-CoV-2 has strong texture heterogeneity by combining the HOG feature and GLCM feature. The shape features and texture features can be extracted from the images in SC2-MID which can be combined and used as eigenvectors in image classification. The combination of texture and geometric features can describe the surface features of the virus precisely, which contains much more abundant information that can help for better classification accuracy in image classification.

Moreover, we can leverage strong feature learning power of deep learning networks (CNN) by extracting deep learning features from SARS-CoV-2, which are robust, invariant to translation changes, do not need image pre-processing and can reduce the need of hand crafted features in segmentation, detection and classification of SARS-CoV-2.

4   Conclusion and Future Work

The SARS-CoV-2 has some recognizable visual information which can be represented by visual features, such as texture and shape features, providing a possibility to describe the morphological property of SARS-CoV-2 for medical workers.

We may get more electron microscopic images in the future to enlarge our dataset. We will establish the evaluation index of the SC2-MID. We will extract more visual features for image classification in future work. The dataset can be used to help medical workers to identify and classify the SARS-CoV-2.

Acknowledgements. We thank B.E. Jiawei Zhang, due to his great work is considered as important as the first author in this paper. We also thank the websites which provide SARS-CoV-2 images.

References

1. Chen, N., et al.: Epidemiological and clinical characteristics of 99 cases of 2019 novel coronavirus pneumonia in Wuhan, China: a descriptive study. Lancet **395**(10223), 507–513 (2020)
2. Chollet, F.: Xception: deep learning with depthwise separable convolutions. In: Proceedings of the IEEE Conference on Computer Vision and Pattern Recognition, pp. 1251–1258 (2017)
3. Chu, D., et al.: Molecular diagnosis of a novel coronavirus (2019-nCoV) causing an outbreak of pneumonia. Clinical Chemistry, January 2020
4. Cui, J., Li, F., Shi, Z.: Origin and evolution of pathogenic coronaviruses. Nature reviews. Microbiology **17**(3), 181–192 (2019)
5. Dalal, N., Triggs, B.: Histograms of oriented gradients for human detection. In: Proceedings of ICPR 2005, pp. 886–893 (2005)
6. Gorbalenya, A., et al.: Severe acute respiratory syndrome-related coronavirus: the species and its viruses - a statement of the coronavirus study group. bioRxiv (2020)
7. Huang, G., Liu, Z., Van Der Maaten, L., Weinberger, K.Q.: Densely connected convolutional networks. In: Proceedings of the IEEE Conference on Computer Vision and Pattern Recognition, pp. 4700–4708 (2017)
8. Hui, D., et al.: The continuing 2019-nCoV epidemic threat of novel coronaviruses to global health - the latest 2019 novel coronavirus outbreak in Wuhan, China. Int. J. Infect. Dis. **91**, 264–266 (2020)
9. Image, B.: The electron microscopic image of SARS-CoV-2. https://baike.baidu.com/item/2019
10. Kulwa, F., et al.: A State-of-the-art survey for microorganism image segmentation methods and future potential. IEEE Access **7**(1), 100243–100269 (2019)
11. Li, C.: Content-based Microscopic Image Analysis. Logos Verlag Berlin GmbH, Gubener Street 47, Berlin, Germany (2016)
12. Li, C., Kulwa, F., Zhang, J., Li, Z., Xu, H., Zhao, X.: A review of clustering methods in microorganism image analysis. In: Pietka, E., Badura, P., Kawa, J., Wieclawek, W. (eds.) Information Technology in Biomedicine. AISC, vol. 1186, pp. 13–25. Springer, Cham (2021). https://doi.org/10.1007/978-3-030-49666-1_2

13. Li, C., Wang, K., Xu, N.: A survey for the applications of content-based microscopic image analysis in microorganism classification domains. Artif. Intell. Rev. **51**(4), 577–646 (2019)

14. Li, C., et al.: A brief review for content-based microorganism image analysis using classical and deep neural networks. In: Pietka, E., Badura, P., Kawa, J., Wieclawek, W. (eds.) ITIB 2018. AISC, vol. 762, pp. 3–14. Springer, Cham (2019). https://doi.org/10.1007/978-3-319-91211-0_1

15. Malik, Y., et al.: Emerging novel coronavirus (2019-nCoV)-current scenario, evolutionary perspective based on genome analysis and recent developments. Vet. Q. **40**(1), 68–76 (2020)

16. Simonyan, K., Zisserman, A.: Very deep convolutional networks for large-scale image recognition. arXiv preprint arXiv:1409.1556 (2014)

17. University, J.H.: Coronavirus COVID-19 global cases by the center for systems science and engineering (CSSE) at johns Hopkins University (JHU). https://coronavirus.jhu.edu/map.html

Image Super-Resolution via Deep Feature Recalibration Network

Jingwei Xin[1], Xinrui Jiang[2], Nannan Wang[2(✉)], Jie Li[1], and Xinbo Gao[1,3(✉)]

[1] School of Electronic Engineering, State Key Laboratory of Integrated Services Networks, Xidian University, Xi'an 710071, China
jwxintt@gmail.com
[2] School of Telecommunications Engineering, State Key Laboratory of Integrated Services Networks, Xidian University, Xi'an 710071, China
nnwang@xidian.edu.cn
[3] Chongqing Key Laboratory of Image Cognition, Chongqing University of Posts and Telecommunications, Chongqing 400065, China
gaoxb@cqupt.edu.cn

Abstract. Recent years have witnessed remarkable progress in convolutional neural network (CNN) based image super-solution (SR) methods. Existing methods tend to deepen the network by means of residual skip connections to achieve better performance. However, these methods are still hard to be applied in real-world applications due to the requirement of its heavy computation. In this paper, we propose a Deep Feature Recalibration Network (DFRN), which strives for efficiency yet effective networks. We divide the process of network nonlinear mapping into two steps: information integration and feature enhancement, and proposed two types of block models: Multi-Scale Information Integration Block (MSIIB) and Feature Recalibration Block (FRB). MSIIB integrates the representation of the input data in the network with different size of receptive fields. FRB enhances the information via obtaining the attention along two different dimensions (channel and plane space of feature maps) respectively. By combining MSIIB and FRB, we provide a more efficient and time-saving method for SISR. Experiments show that the proposed DFRN method outperforms state-of-the-art methods in terms of both objective evaluation metrics (PSNR, SSIM, and running speed) and subjective perception on the generated images.

Keywords: Single image super resolution · Information integration · Feature recalibration · Computational complexity · Time-saving.

Supported in part by the National Key Research and Development Program of China under Grant 2018AAA0103202, in part by the National Natural Science Foundation of China under Grant 61922066, Grant 61876142, Grant 61671339, Grant 61772402, Grant U1605252, and Grant 61976166, in part by the National High-Level Talents Special Support Program of China under Grant CS31117200001, in part by the Fundamental Research Funds for the Central Universities under Grant JB190117, in part by the Xidian University Intellifusion Joint Innovation Laboratory of Artificial Intelligence, and in part by the Innovation Fund of Xidian University.

© Springer Nature Switzerland AG 2020
Y. Peng et al. (Eds.): PRCV 2020, LNCS 12305, pp. 256–267, 2020.
https://doi.org/10.1007/978-3-030-60633-6_21

1   Introduction

Single image super-resolution (SISR) is an important class of image processing techniques and has gained increasing attentions for decades, which aims to reconstruct a high-resolution (HR) image I_{HR} from a low-resolution (LR) image I_{LR}. It guarantees the restoration of the low frequency information in the frequency band and predicts the high frequency information above the cutoff frequency, which enjoys a wide range of real-world applications, such as medical imaging [18], satellite imaging [21], security and surveillance [28], and other high-level computer vision tasks [3–5].

Traditional SISR approaches could be broadly classified into three categories, *i.e.*, interpolation-based methods, model-based optimization methods and deep learning-based methods. Interpolation-based methods include bilinear interpolation, bicubic interpolation and other advanced variants [26]. Since these methods require a small number of computations, they are easy to implement and extremely fast. However, these models often tend to generate blurred images with unpleasant artifacts, especially in areas with much highfrequency information. By exploiting powerful image priors (*e.g.*, the non-local self-similarity prior [22], sparsity prior [24], and so on), model-based optimization methods are flexible to reconstruct relative high-quality HR images, but they need impose more restrictions on the selection of training samples are extremely sensitive to the noise in images. Morever, their performance degrade drastically when the image resolution is magnified at a higher multiple.

Recently, deep learning based methods have been a promising approach to solve SR problem [2]. These methods have been demonstrated state-of-the-art performance by learning a mapping from LR to HR image patches. From then, researchers had carried out its study with different perspectives and obtained plentiful achievements, and various model structure and learning strategies are used in SR networks. (e.g., residual learning [11], recursive learning [6,19,20], skip connection [27] and so on). These CNN-based methods can often achieve satisfactory results, but the increasing model size and the computational complexity severely restrict their applications in the real world.

To address these issues, we propose a Deep Feature Recalibration Network as shown in Fig. 1, which constructs an efficient network structure based on block learning, each block we named feature extract block (FEB) and contains two sub-blocks, Multi-Scale Information Integration Block (MSIIB) and Feature Recalibration Block (FRB). Among them, MSIIB can extract multi-scale features with different reception fields in a parallel structure, while the serial connection mode of FEB will further improve the sufficiency and completeness of final features for image reconstruction. FRB can combine the importance of each feature channel with the importance of each element distribution in the flat space of the feature map, further screen the useful features and suppress redundant information which extracted by each MSIIB. Our contributions are three-fold as follows.

Multi-Scale Information Integration Structure. We propose a Multi-Scale Information Integration Block (MSIIB) structure, which combines multi-branch network with layer sharing.

Feature Recalibration Strategy. We propose the feature recalibration strategy, which extends the attention from channels to the entire feature spaces (both channel and plane space).

Reconstruction Performance. The experimental results demonstrate that our network is best in comparison to state-of-the-art SR methods in terms of both objective metrics (*i.e.* PSNR/SSIM, time consuming) and subjective perception.

2   Related Work

Recently, due to the outstanding learning ability, deep learning methods have been demonstrated with high superiority over other classical example-based methods in solving the problem of SISR. Dong et al. proposed the pioneering work named Super-Resolution Convolution Neural Network(SRCNN) [2], which has established an nonlinear LR-HR mapping and achieves notable improvements over the previous work. Shi *et al.* [17] proposed the Efficient Sub-Pixel Convolutional neural Network (ESPCN) to increase the resolution at the end of the network by sub-pixel convolution layer.

Inspired by VGG-net used for ImageNet classification, Kim *et al.* proposed a Very Deep Convolutional Network (VDSR) [11] and a Deeply-Recursive Convolutional Network(DRCN) [12] for SISR task. VDSR focuses on the training efficiency, which has a long and slender structure that can use a bigger learning rate to accelerate network training. DRCN employs a weight-share method which could get better performance with less parameters. For larger upscaling factor (*e.g.* ×4, ×8 and more), Lai *et al.* [13] proposed a Laplacian Pyramid Super-Resolution Network (LapSRN), which can progressively reconstruct the sub-band residuals of high-resolution images and generates multi-scale prediction only on one network. To decrease the amount of the network parameters as well as get high performance, Tai *et al.* proposed Deep Recursive Residual Network(DRRN) [19], which introduces a very deep model up to 52 convolutional layers with residual learning and recursive module. The authors further proposed Memory Network (MemNet) [20] based on the idea of weights sharing. This network could adaptively combine the multi-scale features by the memory block. In virtue of the concept of information distillation, Hui *et al.* [10] proposed Information Distillation Network (IDN) that employs distillation blocks to gradually extract abundant and efficient features for the reconstruction of HR images. Han *et al.* [7] explore new structures for SR based on this compact RNN view, leading to a dual-state design.

Inspired by the success of very deep models, Lim et al. [15] proposed a very wide network EDSR and a very deep network MDSR which has more than 160 layers. Li et al. [14] proposed a novel multi-scale residual block to make better use of multi-scale information. Due to the difficulty of learning a mapping between LR images and HR images, iterative up and downsampling method is proposed in [8] to revise the sampling results, which can better capture more contextual information. To overcome the gradient vanishing problem, residual channel attention network is adopted in [27], which proposes long and short skip connections in the residual structure to obtain deep residual network.

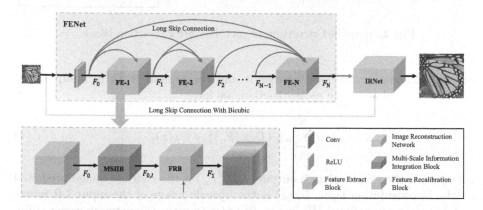

Fig. 1. Pipeline of our proposed DFRN model.

3   Proposed Method

3.1   Basic Network Architecture

Our network structure is shown in Fig. 1, which consists of two parts: feature extraction network (FENet) and image reconstruction network (IRNet). The FENet consists of multiple stacked blocks as shown in Fig. 2. Each block could be divided into a MSIIB and a FRB, which is used to extract the information for HR image reconstruction from the input feature maps.

$$F_0 = f_{conv}(x) \tag{1}$$

where $f_{conv}()$ denotes the feature extracted function by a convolution layer from the input LR image x. Then, F_0 is sent to the cascaded Feature Extract Block.

$$F_n = f_{conv}(F_0, F_1, F_2, ..., F_{n-1}, F_{n,I,R}) \tag{2}$$

$$F_{n,I,R} = F_n^R(F_n^I(F_{n-1})) \tag{3}$$

F_n is the output of Feature Extract Block, and this $f_{conv}()$ is used for the feature fusion. n is the sequence number of the current block, $F_{n,I,R}$ is the output of Feature Recalibration Block, F_n^I and F_n^R is the function of MSIIB and FRB.

The IRNet includes a transposed convolution layer and a convolution layer. The output marked blue denotes the outputs of each FE-Block, which would be

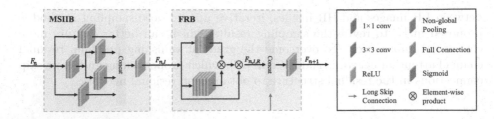

Fig. 2. Simplified structures of each Feature Extraction Block.

utilized to compute the loss between it and the ground truth in a supervised manner. For online image reconstruction, we have

$$I = \sum_{n=1}^{N} \omega_n \left\{ D_{rec}(F_n) \right\} + Up(x), \tag{4}$$

where ω_n is the weight of each output, $D_{rec}(F_n)$ is the residual image predicted by the Deconvolution layer, $Up(x)$ is the upscale image of the input LR iamge and I is the reconstructed HR image. In order to make the network easier train, we employ a global dense connection manner in our network.

3.2   Multi-scale Information Integration Block

Skip connections result in excellent performance (Res-Skip or Dense-Skip). The main reason is that it makes it possible to deepen the network. On the other hand, since feature maps of one layer correlates other features maps from adjacent layers, *i.e.* feature maps of adjacent layers are similar, dense connection is time-consuming and it may be not neccesary to preserve all these connections. This is the basic motivation of our proposed Multi-Scale Information Integration Block (MSIIB). The structure is shown in Fig. 2.

For each Multi-Scale Information Integration Block, we have

$$F_{n,I} = F_n^I(F_{n-1}) \tag{5}$$

$$F_{n,I} = f_{conv}(R_1, R_2, R_3, R_4) \tag{6}$$

$$R_i = w_n^{ri}(F_{n-1}) \tag{7}$$

Where $F_{n,I}$ is the output of MSIIB. Four set of data are sent into the Cancat-layer, correspond to R_1, R_2, R_3, R_4, and $w_n^{ri}(F_n)$ is the weight set of ith branches in MSIIB.

MSIIB adopts a parallel structure. To ensure the diversity of feature information, each branch remain independent to a certain extent. The first branch contain a 1×1 convolution layer and two 3×3 convolution layer. Second and third branches obtain deeper feature information with lower computational complexity, which share the front end of the network. The fourth branch is utilized to retention the input feature information and it also shorten the length of the block by a 1×1 convolution layer. All the branches are able to extract depth features with different receptive fields. Besides, for the virtue of the parallel structure, the output of the four branches are less correlated, which make the richness of information can be guaranteed. In Sect. 4.2, we prove that our MSIIB can get more advanced results than the popular network structures.

3.3 Feature Recalibration Block

In order to boost the representation ability of a network, existing works have shown the benefits of enhancing spatial encoding. Considering the different contribution of each feature map to the result, SENet enhanced the information for different channels. However, these methods give equal treatment to the data in each feature map. Thus, in this paper, considering the importance of each feature map channel and the image plane space at the same time, we proposed Feature Recalibration Block, which could not only obtain the importance of each feature map but also extract the main contour and high frequency region of the image in each feature map.

The structure of FRB is shown in Fig. 2, we introduce two different dimensions respectively to obtain the importance of each feature map channel and the amplitude of each pixel in the image plane space, which could get the weight coefficients corresponding to each pixel in all feature maps.

$$
\begin{aligned}
F_{n,I,R} &= F_n^R(F_{n,I}) \\
&= I_{C,A} * I_{S,A} * F_{n,I}
\end{aligned}
\tag{8}
$$

$I_{C,A}$ and $I_{S,A}$ is the attention information of feature map channels and image plane space.

For the attention of channel space: 1) We utilize non-global pooling instead of global because it could get a set of vectors instead of a real number from one feature map. The vectors retain more pixel and structure information than a real number, which can provide more contribution to the establishment of the relationship between channels. 2) In the experiment, we found that the SENet only has weak gain for SR task. Therefore, we extend it more complete with few parameters and could achieve better performance.

For the attention of image plane space: 1) We used a 1×1 convolution layer to buffer the mapping from the image depth feature of the plane space attention. 2) We choose the dilated convolution as the attention way, which can expand the field of perception over the network, making the gradient easier back propagation and having a low computational cost. Through the experimental validation, we set $d1$, $d2$ and $d3$ are 2,3 and 2 respectively.

3.4 Loss Function

Inspired by LapSRN [13], we send the output of each block to the image reconstruction network. Let x be the input LR image and θ be the set of network parameters to be optimized. Our goal is learning a nonlinear mapping function to generate a high resolution image $x \hat{=} f(\theta, x)$, which is close to the ground truth HR image y. We use the same scale with HR images at the corresponding level as supervision. All of the predictions can be used to compute the final output via weighted averaging. The overall loss function is defined as:

$$L(\theta) = \frac{\alpha}{M} \sum_{m=1}^{M} \left(\left\| y^m - \sum_{n=1}^{N} \omega_n \hat{x}_n^m \right\|_1 + \sum_{n=1}^{N} \| y^m - \hat{x}_n^m \|_1 \right), \qquad (9)$$

where M is the number of training patches, α denotes the learning rate and ω is the optimal weight to compute the final output.

3.5   Experimental Settings

There exist diverse single image super-resolution datasets, the most widely used are the 291 dataset and the DIV2K dataset. Due to the limitation of the computing capacity of the equipment in the real world, we only study all state-of-the-art models that have less than 2M parameters in this work. These methods almost all use 291 dataset as the training set, so this work also uses 291 dataset as the training set. Which dataset consists of the 91 images from Yang et al.[23] and 200 images from the Berkeley Segmentation Dataset(BSD) [16]. And then, we utilize four widely used benchmark datasets, Set5 [1], Set14 [25], BSD100 [16] and Urban100 [9] which are composed of 5, 14, 100 and 100 images respectively.

Each convolutional layer in our network consists of 64 feature maps. We optimize our network by Adam with back propagation. The momentum parameter is set to 0.5, weight decay is set to 2×10^{-4}, and the initial learning rate is set to 5×10^{-4} and being divided a half every 20 epochs. For assessing the quality of SR results, we employ two objective image quality assessment metrics: Peak Signal to Noise Ratio (PSNR) and structural similarity (SSIM). All metrics are performed on the Y-channel (YCbCr color space) of center-cropped, removal of a s-pixel wide strip from each $\times s$ upscaling image border.

4   Experimental Results and Analysis

4.1   Comparison with State-of-the-Art Methods

Performance. To verify the superiority of the proposed DFRN model, almost all state-of-the-art SR methods which trained on 291 dataset are compared in this paper: SRCNN [2], VDSR [11], LapSRN [13], DRRN [19], MemNet [20], IDN [10], DSRN [7].

Table 1 list the PSNR and SSIM of DFRN and state-of-the-arts. It can be seen that our DFRN achieves the best performance among all compared methods on four benchmark datasets. For a better view, we present super-reconstructed images including both general and structured ones for subjective assessment as shown in Fig. 3 and Fig. 4 respectively. Since LapSRN did not provide $\times 3$ upscaling results, we reproduce the corresponding results through downscaling its $\times 4$ upscaling results as [13] does.

Table 1. Benchmark results of state-of-the-art SR methods: Average PSNR/SSIM for ×2, ×3 and ×4 upscaling.

Methods	Scale	Set5		Set14		B100		Urban100	
		PSRN	SSIM	PSRN	SSIM	PSRN	SSIM	PSRN	SSIM
Bicubic	×2	33.66	0.930	30.24	0.869	29.56	0.843	26.88	0.840
SRCNN	×2	36.66	0.954	32.45	0.907	31.36	0.888	29.50	0.895
VDSR	×2	37.53	0.959	33.05	0.913	31.90	0.896	30.77	0.914
LapSRN	×2	37.52	0.959	33.08	0.913	31.80	0.895	30.41	0.910
DRRN	×2	37.74	0.959	33.23	0.914	32.05	0.897	31.23	0.919
MemNet	×2	37.78	0.960	33.28	0.914	32.08	0.898	31.31	0.920
IDN	×2	37.83	0.960	33.30	0.915	32.08	0.899	31.27	0.920
DSRN	×2	37.66	0.959	33.15	0.913	32.10	0.897	30.97	0.916
DFRN	×2	**37.92**	**0.960**	**33.42**	**0.915**	**32.13**	**0.899**	**31.52**	**0.922**
Bicubic	×3	30.39	0.868	27.55	0.774	27.21	0.739	24.46	0.735
SRCNN	×3	32.75	0.909	29.30	0.822	28.41	0.786	26.24	0.799
VDSR	×3	33.66	0.921	29.77	0.831	28.82	0.798	27.14	0.828
LapSRN	×3	33.82	0.923	29.79	0.832	28.82	0.797	27.07	0.827
DRRN	×3	34.03	0.924	29.96	0.835	28.95	0.800	27.53	0.838
MemNet	×3	34.09	0.925	30.00	0.835	28.96	0.800	27.56	0.838
IDN	×3	34.11	0.925	29.99	0.835	28.95	0.801	27.42	0.836
DSRN	×3	33.88	0.922	**30.26**	0.837	28.81	0.797	27.16	0.828
DFRN	×3	**34.23**	**0.926**	30.03	**0.837**	**28.99**	**0.802**	**27.61**	**0.840**
Bicubic	×4	28.42	0.810	26.00	0.703	25.96	0.668	23.14	0.658
SRCNN	×4	30.48	0.863	27.50	0.751	26.90	0.710	24.52	0.722
VDSR	×4	31.35	0.884	28.01	0.767	27.29	0.725	25.18	0.752
LapSRN	×4	31.54	0.886	28.19	0.772	27.32	0.728	25.21	0.755
DRRN	×4	31.68	0.889	28.21	0.772	27.38	0.728	25.44	0.764
MemNet	×4	31.74	0.889	28.26	0.772	27.40	0.728	25.50	0.763
IDN	×4	31.82	0.890	28.25	0.773	27.41	0.730	25.41	0.763
DSRN	×4	31.40	0.883	28.07	0.770	27.25	0.724	25.08	0.747
DFRN	×4	**31.93**	**0.891**	**28.31**	**0.774**	**27.44**	**0.730**	**25.59**	**0.769**

Time Consuming. According to Table 1, VDSR, LapSRN, DRRN, MemNet, IDN and DSRN achieve the good performance except our proposed DFRN method. Considering that DSRN does not public the network framework code, we only compare the remaining six methods with proposed DFRN. Here we use the public codes of the compared algorithms to evaluate the runtime on the machine with Nvidia TITAN X Pascal GPU. Alignment with [10]. Table 2 shows the average execution time on four benchmark datasets, the proposed DFRN is almost faster than all the compared methods.

Table 2. Time-execution comparison on the 4 benchmark datasets with scale factors ×2, ×3 and ×4.

DataSet	Scale	VDSR	DRCN	LapSRN	DRRN	MemNet	IDN	DFRN
Set5	×2	0.054	0.735	0.032	4.343	5.715	0.016	**0.011**
	×3	0.062	0.748	0.049	4.380	5.761	0.011	**0.009**
	×4	0.054	0.735	0.040	4.450	5.728	0.009	**0.009**
Set14	×2	0.113	1.579	0.035	8.540	12.031	0.025	**0.011**
	×3	0.122	1.569	0.061	8.298	11.543	0.014	**0.010**
	×4	0.112	1.526	0.040	8.540	11.956	0.010	**0.010**
BSD100	×2	0.071	0.983	0.018	4.430	5.875	0.015	**0.009**
	×3	0.071	0.996	0.037	4.430	5.897	0.009	**0.009**
	×4	0.071	0.984	0.023	4.373	4.373	**0.007**	0.009
Urban100	×2	0.451	5.010	0.082	26.699	35.871	0.062	**0.014**
	×3	0.514	5.054	0.122	26.693	35.803	0.034	**0.013**
	×4	0.448	5.048	0.100	26.702	37.404	0.022	**0.012**

Table 3. Different block structure comparisons: Inception_Block, Residual_Block, Dense_Block and our proposed MSIIB.

DataSet	Set5	Set14	BSD100	Urban100
Inception_Block	31.56	28.12	27.31	25.23
Residual_Block	32.64	28.07	27.28	25.19
Dense_Block	31.59	28.07	27.27	25.21
MSIIB	**31.73**	**28.23**	**27.37**	**25.34**

4.2 Model Analysis

In this subsection, we conduct ablation study on our network structure and would provide comparisons with prevalent models to demonstrate the superiority of the propsoed network for the SR task.

Multi-scale Information Integration Structure. A comparison of MSIIB with Dense Block, Residual Block and Inception Block is done in the following. For fair comparison, the test network repalces each FEB by these three classical block structures. And we set these classical blocks have similar parameters and computation complexity, and are trained on the 291 dataset for ×4 upscaling. Then, we test these models on four standard datasets as shown in Table 3. The performance shows that our MSIIB achieve the best performance.

Feature Recalibration Strategy. The following experiment gives the effect of FRB by the ablation study, we consider three situations: (1) Basic Net (MSIIB);

Table 4. Comparisons of MSIIB, MSIIB+CA and MSIIB+FR.

DataSet	Set5	Set14	BSD100	Urban100
Basic Net	31.73	28.23	27.37	25.34
With CA	31.80	28.26	27.40	25.42
With FR	**31.93**	**28.31**	**27.44**	**25.59**

Fig. 3. Subjective quality assessment for ×2 upscaling on the general image: *ppt*3 from Set14. Texts are more distinct in Ours (DFRN)

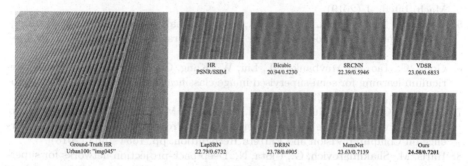

Fig. 4. Subjective quality assessment for ×4 upscaling on the structured image: *img*045 from Urban100. Structured lines in Ours (DFRN) are reconstructed better than in other methods.

(2) Basic with channel attention (CA); (3) Basic with feature recalibration (FR). As shown in Table 4. From the comparison we can found that the method of channel attention can get a certain gain effect, and the method of deep attention can achieve the best performance.

5   Conclusions

In this paper, we proposed an end-to-end deep feature recalibration network (DFRN) for fast image super-resolution. In this network, the information integration method is combined with the feature enhancement method to improve its ability of nonlinear mapping, and achieves an accurate prediction for high frequency information, which can lead to an efficient high resolution image. The experimental results have demonstrated that our method outperforms the existing method by a large margin on benchmarked images. We believe this compact network could be more widely applicable in practice, and is readily used to other machine vision problems such as denoising, compression artifact removal and even recognition.

References

1. Bevilacqua, M., Roumy, A., Guillemot, C., Alberi-Morel, M.L.: Low-complexity single-image super-resolution based on nonnegative neighbor embedding. BMVA Press (2012)
2. Dong, C., Loy, C.C., He, K., Tang, X.: Image super-resolution using deep convolutional networks. IEEE Trans. Pattern Anal. Mach. Intell. **38**(2), 295–307 (2015)
3. Gong, C., Liu, T., Yang, J., Tao, D.: Large-margin label-calibrated support vector machines for positive and unlabeled learning. IEEE Trans. Neural Networks **30**(11), 3471–3483 (2019)
4. Gong, C., Shi, H., Liu, T., Zhang, C., Yang, J., Tao, D.: Loss decomposition and centroid estimation for positive and unlabeled learning. IEEE Trans. Pattern Anal. Mach. Intell., 1 (2019)
5. Gong, C., Tao, D., Liu, W., Liu, L., Yang, J.: Label propagation via teaching-to-learn and learning-to-teach. IEEE Trans. Neural Networks **28**(6), 1452–1465 (2017)
6. Gong, C., Tao, D., Maybank, S.J., Liu, W., Kang, G., Yang, J.: Multi-modal curriculum learning for semi-supervised image classification. IEEE Trans. Image Process. **25**(7), 3249–3260 (2016)
7. Han, W., Chang, S., Liu, D., Yu, M., Witbrock, M., Huang, T.S.: Image super-resolution via dual-state recurrent networks. In: Proceedings of the IEEE Conference on Computer Vision and Pattern Recognition, pp. 1654–1663 (2018)
8. Haris, M., Shakhnarovich, G., Ukita, N.: Deep back-projection networks for super-resolution. In: Proceedings of the IEEE Conference on Computer Vision and Pattern Recognition, pp. 1664–1673 (2018)
9. Huang, J.B., Singh, A., Ahuja, N.: Single image super-resolution from transformed self-exemplars. In: Proceedings of the IEEE Conference on Computer Vision and Pattern Recognition, pp. 5197–5206 (2015)
10. Hui, Z., Wang, X., Gao, X.: Fast and accurate single image super-resolution via information distillation network. In: Proceedings of the IEEE Conference on Computer Vision and Pattern Recognition, pp. 723–731 (2018)
11. Jiwon, K., Jung Kwon, L., Kyoung Mu, L.: Accurate image super-resolution using very deep convolutional networks. In: Proceedings of the IEEE Conference on Computer Vision and Pattern Recognition, pp. 1646–1654 (2016)
12. Kim, J., Kwon Lee, J., Mu Lee, K.: Deeply-recursive convolutional network for image super-resolution. In: Proceedings of the IEEE Conference on Computer Vision and Pattern Recognition, pp. 1637–1645 (2016)
13. Lai, W.S., Huang, J.B., Ahuja, N., Yang, M.H.: Deep Laplacian pyramid networks for fast and accurate super-resolution. In: Proceedings of the IEEE Conference on Computer Vision and Pattern Recognition, pp. 624–632 (2017)

14. Li, J., Fang, F., Mei, K., Zhang, G.: Multi-scale residual network for image super-resolution. In: Ferrari, V., Hebert, M., Sminchisescu, C., Weiss, Y. (eds.) ECCV 2018. LNCS, vol. 11212, pp. 527–542. Springer, Cham (2018). https://doi.org/10.1007/978-3-030-01237-3_32

15. Lim, B., Son, S., Kim, H., Nah, S., Mu Lee, K.: Enhanced deep residual networks for single image super-resolution. In: Proceedings of the IEEE Conference on Computer Vision and Pattern Recognition Workshops, pp. 136–144 (2017)

16. Martin, D., Fowlkes, C., Tal, D., Malik, J., et al.: A database of human segmented natural images and its application to evaluating segmentation algorithms and measuring ecological statistics. ICCV Vancouver (2001)

17. Shi, W., et al.: Real-time single image and video super-resolution using an efficient sub-pixel convolutional neural network. In: Proceedings of the IEEE Conference on Computer Vision and Pattern Recognition, pp. 1874–1883 (2016)

18. Shi, W., et al.: Cardiac image super-resolution with global correspondence using multi-atlas patchmatch. In: Mori, K., Sakuma, I., Sato, Y., Barillot, C., Navab, N. (eds.) MICCAI 2013. LNCS, vol. 8151, pp. 9–16. Springer, Heidelberg (2013). https://doi.org/10.1007/978-3-642-40760-4_2

19. Tai, Y., Yang, J., Liu, X.: Image super-resolution via deep recursive residual network. In: Proceedings of the IEEE Conference on Computer Vision and Pattern Recognition, pp. 3147–3155 (2017)

20. Tai, Y., Yang, J., Liu, X., Xu, C.: MemNet: a persistent memory network for image restoration. In: Proceedings of the IEEE International Conference on Computer Vision, pp. 4539–4547 (2017)

21. Thornton, M.W., Atkinson, P.M., Holland, D.: Sub-pixel mapping of rural land cover objects from fine spatial resolution satellite sensor imagery using super-resolution pixel-swapping. Int. J. Remote Sens. **27**(3), 473–491 (2006)

22. Timofte, R., De, V., Van, L.: A+: adjusted anchored neighborhood regression for fast super-resolution. In: Asian Conference on Computer Vision, pp. 111–126 (2014)

23. Yang, J., Wright, J., Huang, T.S., Ma, Y.: Image super-resolution via sparse representation. IEEE Trans. Image Process. **19**(11), 2861–2873 (2010)

24. Yang, S., Wang, M., Chen, Y., Sun, Y.: Single-image super-resolution reconstruction via learned geometric dictionaries and clustered sparse coding. IEEE Trans. Image Process. **21**(9), 4016–4028 (2012)

25. Zeyde, R., Elad, M., Protter, M.: On single image scale-up using sparse-representations. In: Boissonnat, J.-D., et al. (eds.) Curves and Surfaces 2010. LNCS, vol. 6920, pp. 711–730. Springer, Heidelberg (2012). https://doi.org/10.1007/978-3-642-27413-8_47

26. Zhang, L., Wu, X.: An edge-guided image interpolation algorithm via directional filtering and data fusion. IEEE Trans. Image Process. **15**(8), 2226–2238 (2006)

27. Zhang, Y., Li, K., Li, K., Wang, L., Zhong, B., Fu, Y.: Image super-resolution using very deep residual channel attention networks. In: Ferrari, V., Hebert, M., Sminchisescu, C., Weiss, Y. (eds.) ECCV 2018. LNCS, vol. 11211, pp. 294–310. Springer, Cham (2018). https://doi.org/10.1007/978-3-030-01234-2_18

28. Zou, W.W., Yuen, P.C.: Very low resolution face recognition problem. IEEE Trans. Image Process. **21**(1), 327–340 (2011)

Semantic Ground Plane Constraint in Visual SLAM for Indoor Scenes

Rong Wang[1]([✉]), Wenzhong Zha[1], Xiangrui Meng[1], Fanle Meng[1], Yuhang Wu[1], Jianjun Ge[1], and Dongbing Gu[2]

[1] Key Lab of Cognition and Intelligence Technology, Information Science Academy of China Electronics Technology Group Corporation, Beijing, China
rwang127@foxmail.com
[2] School of Computer Science and Electronic Engineering, University of Essex, Colchester, UK

Abstract. Typical visual simultaneous localization and mapping (SLAM) systems rely on purely geometric elements. Recently semantic SLAM has become popular due to its capability to obtain high-level understanding of the environment. However, most semantic SLAM systems merely aim to produce semantic maps rather than exploiting semantic constraints to boost the accuracy performance. In this paper, we propose a new RGB-D SLAM system for the use in indoor environments. Given that a ground plane represents a significant part of most indoor scenes, our new system exploit a semantic ground plane constraint inferred from a deep neural network along with geometric constraints on the state estimation to improve the performance. More specifically, we introduce a semantically meaningful geometric primitive, namely, a global ground plane to the optimization process of visual SLAM. The global ground model is able to provide an extra constraint on the state estimation through which both frame-to-keyframe and frame-to-model tracking processes are combined, and low-texture ground plane is effectively utilized. Experimental results for the TUM dataset and a real scene have demonstrated the improved performance brought by the semantic constraint in our method.

Keywords: RGB-D SLAM · Semantic segmentation · Semantic constraint · Semantic map

1 Introduction

Simultaneous localization and mapping (SLAM) is a fundamental and widely used technique in areas including robotics, augmented reality, and autonomous driving. Visual SLAM attracts more and more attentions since cameras are an information rich sensor and they are easily available at small size and low cost nowadays. Among all kinds of cameras, RGB-D cameras that can directly provide dense depth information are typically suitable for indoor scenes.

© Springer Nature Switzerland AG 2020
Y. Peng et al. (Eds.): PRCV 2020, LNCS 12305, pp. 268–279, 2020.
https://doi.org/10.1007/978-3-030-60633-6_22

Currently typical visual SLAM systems could not convey high-level cognitive information for advanced tasks, such as motion planning, grasping, manipulation, etc. In recent years, with the significant progress in deep learning, impressive performance in high-level cognitive tasks, such as semantic segmentation, has been achieved. Potentially, semantic SLAM, which integrates semantic information with geometric SLAM, is able to improve the scene understanding of surroundings and accomplish the advanced tasks.

Some existing semantic SLAM systems are often devoted to producing semantic maps where semantic information is not effectively utilized. Some progresses have been made to incorporate semantic information into various stages of SLAM. However, they are either established on object levels which are lack of semantic understanding of the whole scene or optimized in the pose-graph framework which cannot efficiently utilize semantic information for each frame.

In this paper, we introduce a global ground plane as a semantic constraint in the optimization process of RGB-D SLAM to improve its accuracy, and the resultant system is also able to produce a dense semantic 3D map. The inspiration comes from the fact that a ground plane is commonly seen in indoor scenes, especially from the viewpoints of ground mobile robots. The global ground plane could constrain both the points lying on it and the camera poses where it is observed. Based on this semantic ground plane constraint, our method can not only combine both frame-to-keyframe and frame-to-model tracking processes, but also effectively utilize the information provided by low-texture ground plane. Potentially this kind of semantic constraint could be extended to any recognizable semantic parts of the scenes, which have prior 3D geometric models. The main contributions of this paper include:

- A complete semantic SLAM system is proposed, which can improve the accuracy of pose estimation and produce a dense semantic 3D map;
- We take advantage of semantic segmentation in the process of SLAM, where a global ground plane is established as a semantic constraint to refine both tracking and mapping modules;
- The improved performance of our method with respect to the accuracy and robustness has been demonstrated on the public TUM RGB-D dataset [1] and in a real scene.

The rest of this paper is organized as follows. After this introduction, some related works about semantic SLAM and plane-based SLAM are reviewed. Then our proposed method is given in detail, including how to utilize semantic information in SLAM and produce a semantic map. Subsequently, the experimental section provides quantitative and qualitative results of our method to demonstrate its effectiveness and accuracy. Finally, a brief conclusion and discussion are drawn.

2   Related Work

In this section, we provide a review of recent related works under the topics of semantic SLAM and plane-based SLAM.

2.1   Semantic SLAM

Semantic SLAM has become more and more popular due to the capability to produce high-level semantic attributes. Most semantic SLAM systems are dedicated to generate semantic maps where semantic information is not fully explored [2–4]. How to use semantic information to boost SLAM is an open issue. Semantic information was used in feature matching to improve the robustness [5,6]. Semantic label probability was incorporated into the odometry optimization function to obtain medium-term tracking [7]. DS-SLAM [8] relied on semantic labels to reduce the impact of dynamic objects. Detected objects have also been used in data association to incorporate high-level information in SLAM [9,10]. [11] and [12] demonstrated the semantic objects can help to recover camera poses under large viewpoint changes. Furthermore, attributes of detected objects can be jointly optimized in conjunction with other geometric elements in the scene [13–16]. Semantically-defined objects can be incorporated into the process of SLAM [17–19].

2.2   Plane-Based SLAM

Since planes are typical structures in indoor environments, plane-based SLAM attracts a lot of attention. Plane could be treated as a feature representation and used just like point and line elements [20,21]. Planes also provide extra structural relations that could constrain geometric model of the scene [22,23]. Moreover, it was common to introduce plane parameters into graph optimization where they were jointly optimized along with camera poses and map points [24,25]. CPA-SLAM [26] included a global plane model into a direct RGB-D SLAM with an EM framework. It was also required to jointly correct the camera poses and plane parameters in the global scale within graph optimization. The concept of semantic ground plane in our method is similar to that in CPA-SLAM, but our use of ground plane is conceptually straightforward, more compatible with the framework of SLAM algorithm and able to contain higher cognitive information.

3   System Introduction

In this section, we will present the implementation of our proposed method in detail. First, the framework of our method is presented. Then the method to build semantic ground plane is described. Next, the detail of applying the established semantic constraint to both tracking and mapping modules in SLAM is given. Finally, we present the way to build dense semantic 3D map.

3.1   System Overview

Since ORB-SLAM [27] acquires satisfactory performance in most practical situations, our system is built upon it. Semantic segmentation is incorporated into this feature-based SLAM system in order to introduce the semantic constraint

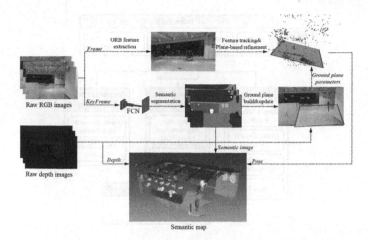

Fig. 1. The overview of our proposed system. Semantic segmentation of keyframe image is used to establish a global ground plane. Then the global ground plane is used to refine each pose estimation and 3D point computation. Dense semantic map is generated based on the camera poses, depth images and semantic segmentation results.

and meanwhile produce semantic map. We model the environment with a global ground plane which is used as semantic constraint since the ground is a typical indoor feature and carries high-level knowledge. The ground plane is incrementally constructed from the result of the semantic segmentation of keyframes. We then develop a novel formulation to integrate the global ground plane constraint into the processes of pose estimation and 3D point computation. The overview of our proposed system is shown in Fig. 1.

There are five threads running in parallel in our system. The framework is illustrated in Fig. 2. The tracking thread first extracts ORB features from the RGB image, then performs the pose estimation based on the combination of feature matching and the global ground constraint. When the current frame is considered to be a new keyframe, it enters the semantic segmentation thread where FCN8s [28] is adopted. After that, the semantic segmentation result is processed in both mapping thread and dense semantic map building thread, along with its corresponding depth image and the estimated pose. In the mapping thread, current ground plane is extracted and merged with the existed global ground plane. The updated global ground plane is subsequently used to refine the 3D points belonging to the semantic ground. The semantic map building thread generates the dense semantic map.

3.2 Semantic Ground Plane Building

Before describing the key idea of our method, we first introduce some mathematical denotations used in our paper. An image pixel and a 3D point can be denoted as $\mathbf{x} = [x\ y]^{\mathrm{T}} \in \mathbb{R}^2$ and $\mathbf{X} = [X\ Y\ Z]^{\mathrm{T}} \in \mathbb{R}^3$ whose homogeneous forms are $\dot{\mathbf{x}} = [x\ y\ 1]^{\mathrm{T}}$ and $\dot{\mathbf{X}} = [X\ Y\ Z\ 1]^{\mathrm{T}}$ respectively. We use subscripts w and

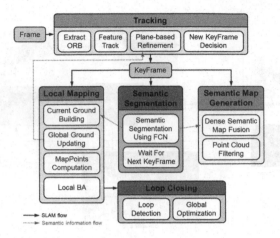

Fig. 2. The framework of our proposed system. Loop closing is the same as in ORB-SLAM. It searches for loops based on the image appearance and performs a graph optimization if a loop is detected.

c to represent the world (global) and camera (local) coordinates respectively. Then each pixel $\mathbf{x}$ and its local point $\mathbf{X}_c$ have the relation: $\mathbf{x} = \pi(\mathbf{K}\mathbf{X}_c)$. $\pi(\cdot)$ is the projection function set as $\pi(x, y, z) = (x/z, y/z)$. $\mathbf{K}$ is the camera intrinsic matrix.

The camera pose $\mathbf{T}_{cw} \in \mathbb{R}^{4 \times 4}$ can be decomposed into the rigid transformation of rotation $\mathbf{R}_{cw} \in \mathbb{R}^{3 \times 3}$ and translation $\mathbf{t}_{cw} \in \mathbb{R}^3$. Since the local point $\mathbf{X}_c$ can be transformed from the global point $\mathbf{X}_w$ through camera pose, we have $\dot{\mathbf{X}}_c = \mathbf{T}_{cw}\dot{\mathbf{X}}_w$ and $\dot{\mathbf{X}}_w = \mathbf{T}_{wc}\dot{\mathbf{X}}_c$. Moreover, we define the plane parameters as $\boldsymbol{\pi} = [\mathbf{n}^{\mathrm{T}}, d]^{\mathrm{T}} \in \mathbb{R}^4$, where $\mathbf{n} \in \mathbb{R}^3$ is the plane normal vector, and d is the distance from origin to the plane. For a 3D point $\mathbf{X}$ lying on the plane, we have $\mathbf{n}^{\mathrm{T}}\mathbf{X} + d = 0$. Similar to the 3D point, a global plane $\boldsymbol{\pi}_g$ and its corresponding local plane $\boldsymbol{\pi}_c$ have the relation: $\boldsymbol{\pi}_c = \mathbf{T}_{cw}^{-\mathrm{T}}\boldsymbol{\pi}_g$ and $\boldsymbol{\pi}_g = \mathbf{T}_{wc}^{-\mathrm{T}}\boldsymbol{\pi}_c$.

After carrying out semantic segmentation, the local ground plane is extracted from the keyframe in mapping thread. When semantically labeled ground regions are available, an 3D local ground plane is built based on the depth values located in the regions. We first randomly recover some 3D points in the regions. These points constitute the candidate points from which a local ground plane is extracted using RANSAC algorithm [29]. At each RANSAC iteration, three candidate points are randomly sampled and used to generate a plane hypothesis. We then compute the point-to-plane distances for all the candidate points and sort them from the smallest to the largest. Subsequently, the optimal plane is decided by the smallest value under median ranking. The inliers of the optimal plane are obtained from the candidate points whose point-to-plane distances are under a small threshold. All the inliers are then used to fit the local ground plane under the least square method.

For the global ground plane updating, the extracted local ground plane is further transformed to the world coordinate through the obtained camera pose from tracking. The new plane is fused with the existed global ground plane only if the angle between the plane normals is small and their distances to the origin are similar. The global ground plane is incrementally constructed through a weighting average strategy:

$$\pi_g = \frac{\pi_g + w\pi_g^{new}}{1 + w},\tag{1}$$

where w is defined as $e^{-\alpha D}$ and D measures the plane difference through the combination of normals and spatial distances.

3.3 Semantic Constraint Based Refinement

Although the ground plane can be incrementally maintained in the system, how to leverage this semantic constraint for the SLAM is still not solved. Intuitively, planes in addition to a point-based representation could provide extra constraints for the points that lie on the plane and in turn refine the camera pose. Different from the existed systems, we develop a straight-forward but efficient way to track each frame through matching towards both the keyframe and the semantic global ground plane. In addition, the semantic constraint is also used in 3D point computation to provide the geometric consistency. Our method efficiently refines the state estimation and effectively uses low-texture regions.

First, how to use the semantic constraint for consistent 3D point computation is given. The 3D map points are defined in the world coordinate. Each time after the global ground plane is updated, we use the current global ground plane to further refine 3D map points generated through SLAM since they may not be accurate merely through feature matching and triangulation. The points that need the refinement should satisfy the following conditions: (1) Their 2D image projections are semantically labeled as the ground; (2) Their spatial distances to the global ground plane are not too large. Therefore, we can derive $\lambda\dot{x} = K(R_{cw}X_w + t_{cw})$ and $n_g^T X_w + d_g = 0$ from the above conditions respectively and obtain:

$$X_w = \frac{-d_g}{n_g^T \left(R_{cw}^{-1}(K^{-1}\dot{x} - t_{cw})\right)} R_{cw}^{-1}(K^{-1}\dot{x} - t_{cw}).\tag{2}$$

We then mark the refined map point as ground feature point.

Then the use of semantic ground plane in the pose estimation is presented. In order to estimate the camera pose, the SLAM system establishes the correspondences between image pixels of the current frame and 3D map points of the keyframe through feature matching. The pose is then estimated by minimizing the geometric error which is computed as the sum of pixel distances between the projections of transformed 3D points and their corresponding 2D pixels. This process is known as frame-to-keyframe tracking which lacks long-term global performance and induces large error accumulation. Furthermore, feature-based tracking neither takes advantage of depth information nor effectively deals with low-texture environments. Incorporating the semantic ground plane into pose estimation could overcome these weakness. The semantic ground plane provides a high-level frame-to-model tracking process since each frame associates the local observations with the global model. Furthermore, the semantic ground constraint effectively utilizes the depth information and handles featureless regions to a certain extent. Here, an overall energy function is constructed through a weighted combination of different criteria which can be written as:

$$E = E_{fea} + \lambda E_{gnd},$$

$$E_{fea} = \left\| \pi \left(\mathbf{K} \left[\mathbf{R}_{cw} | \mathbf{t}_{cw} \right] \dot{\mathbf{X}}_w^f \right) - \mathbf{x} \right\|^2, E_{gnd} = \left(\mathbf{n}_g^{\mathrm{T}} \left[\mathbf{R}_{wc} | \mathbf{t}_{wc} \right] \dot{\mathbf{X}}_c^d + d_g \right)^2, \tag{3}$$

where E_{fea} measures the feature reprojective residual between a 3D map point $\mathbf{X}_w^f$ and a 2D pixel $\mathbf{x}$, while E_{gnd} is defined by the point-to-plane distance between a 3D point $\mathbf{X}_c^d$ and the global ground plane $[\mathbf{n}_g^{\mathrm{T}}, d_g]^{\mathrm{T}}$. The overall function presents a mixed nature of both frame-to-keyframe and frame to global model tracking. The frame to global model criteria is the key to improve the tracking accuracy since it could establish constraints between non-overlapping frames. The feature reprojective criteria is also necessary due to the limitation of ground plane observations in some scenes.

The 3D ground point $\mathbf{X}_c^d$ is obtained from the current depth image whose pixel corresponds to the projection of the already known ground point under the pose estimation. The type of ground points varies depending on whether the texture of ground regions is rich or not. The ground points come from the feature points if the ground regions have rich texture, otherwise the ground points are recovered from depth image of the reference keyframe.

Both point to plane and pixel projection distances are functions of the current camera pose $\mathbf{T}_{cw}$. We should note that $\mathbf{X}_c^d$ is also the function of camera pose but is neglected in our method. In order to handle outliers obtained due to the neglection and incorrect semantic segmentation, we exclude the point-to-plane distances which are larger than a threshold from the energy function minimization. We use the iterative gradient descent method to optimize the non-linear least-squares cost (3).

3.4   Dense Semantic 3D Map Building

In order to build the dense semantic 3D map, new keyframes from tracking thread and segmentation results from semantic segmentation thread are needed. The keyframe's depth image is first used to generate a local point cloud where each point is associated with a specific color representing the corresponding semantic label. The semantic local point cloud can then be converted to the global coordinate through the keyframe's camera pose. The filtering method, which is implemented by PCL library[1], is proceeded to fuse this point cloud with the global point cloud. The dense semantic point cloud is incrementally built over keyframes. The final semantic 3D map improves the perception level and could be employed for high-level tasks.

4   Experimental Results

In this section, experiments are carried out in order to demonstrate the improvement of our method. Both the TUM dataset and a real scene are used for evaluation. All the experiments are performed on a computer with Intel Xeon E5 CPU, Quadro M5000 GPU, and 32 GB memory where only semantic segmentation uses GPU.

Since our method is built upon ORB-SLAM, we make a comparison between them. The Absolute Trajectory Error (ATE) and Relative Pose Error (RPE) are used for

---

[1] http://www.pointclouds.org/.

quantitative evaluation. ATE represents the accuracy of the trajectory, while RPE measures the translational and rotational drift. Tracking performance is then assessed through the root mean square error (RMSE) and standard deviation (SD) since they can indicate the performance in accuracy and robustness.

4.1 The TUM Dataset

The public TUM RGB-D dataset [1] is first considered. It provides the sequences in indoor environments and obtains the ground truth by a motion capture system. The sequences in the dataset that show clear ground planes are chosen for evaluation. They are $fr1/360$, $fr1/floor$, $fr1/room$, $fr2/desk$, $fr3/office$, $fr3/notexture\_far$ and $fr3/notexture\_near$. The first four sequences have rich visual features on the ground, while the grounds in the last three sequences are featureless. Moreover, the last two sequences only show featureless grounds and walls which increase the level of difficulty.

Table 1. Tracking performance comparison between ORB-SLAM and our proposed method.

Sequences	ATE				Translation RPE				Rotation RPE			
	RMSE		SD		RMSE		SD		RMSE		SD	
	ORB	Ours	ORB	Ours	ORB	Ours	ORB	Ours	ORB	Ours	ORB	Ours
fr1/360	0.1637	**0.1141**	0.0664	**0.0594**	0.2396	**0.1692**	0.1138	**0.0912**	6.3896	**5.0249**	2.8404	**2.4700**
fr1/floor	0.0144	**0.0127**	0.0068	**0.0062**	0.0434	**0.0390**	0.0264	**0.0230**	2.9810	**2.7659**	1.3580	**1.2029**
fr1/room	0.0461	**0.0413**	0.0266	**0.0206**	0.0850	**0.0812**	0.0474	**0.0436**	3.3780	**3.2070**	1.6512	**1.4520**
fr2/desk	0.0100	**0.0091**	0.0040	**0.0034**	0.0377	**0.0361**	0.0210	**0.0206**	1.2358	**1.1953**	0.5414	**0.5342**
fr3/office	0.0114	**0.0102**	0.0052	**0.0044**	0.0287	**0.0285**	0.0145	**0.0142**	0.8565	0.8760	**0.3515**	0.3541
fr3/notexture_far	0.0354	**0.0235**	0.0129	**0.0085**	0.0538	**0.0356**	0.0281	**0.0164**	0.9750	**0.6874**	0.4121	**0.3061**
fr3/notexture_near	0.0327	**0.0189**	**0.0085**	0.0107	0.0582	**0.0359**	0.0323	**0.0195**	2.3896	**1.6661**	1.1429	**0.7668**

The quantitative comparison results are shown in Table 1 which presents the values of RMSE and SD of the absolute trajectory error (m), relative translational error (m) and relative rotational error (deg) between ORB-SLAM and our proposed method. Since sequences $fr3/notexture\_far$ and $fr3/notexture\_near$ have little texture which is difficult for SLAM, we decrease the required amount of feature points in the tracking thread to make ORB-SLAM work in the scenarios. The comparison results demonstrate that the tracking performance is improved with respect to accuracy and robustness through incorporating the semantic ground plane constraint. Specifically, the comparison results in sequences of both $fr3/notexture\_far$ and $fr3/notexture\_near$ show the significantly reduced tracking errors. Since neither of the two sequences contains a visual loop, we can get the conclusion that our method can properly refine the pose estimation for each frame. The textureless scenarios further indicate the effectiveness of our proposed method to handle low-texture regions.

In order to illustrate the performance more intuitively, Fig. 3 gives a visual comparison of the accumulated point clouds of the $fr1/360$ sequence. The left point cloud is generated through ORB-SLAM where the regions within the red circle show obvious errors induced by tracking inaccuracy. However, our method can alleviate the error. The results qualitatively illustrate that our method improves the trajectory accuracy and generates better geometric maps. What's more, our method only needs a little

Fig. 3. Visual comparison of the accumulated point clouds between ORB-SLAM and our proposed method.

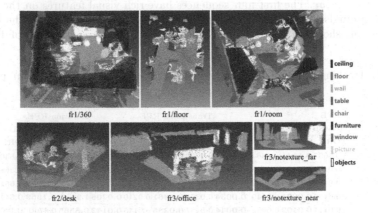

Fig. 4. Semantic maps from different sequences. Different colors represent different semantic labels.

more computation which is not very time consuming. Finally, the constructed semantic maps for different test sequences are shown in Fig. 4 which can be used for high-level tasks.

4.2 The Real Scene

The real scene is further considered for evaluation. We use the turtlebot, which is shown in Fig. 5, as the mobile platform to collect real data. Both Kinect RGB-D camera and Hokuyo 2D lidar are mounted on the turtlebot. The Kinect is used to capture RGB-D images to test our algorithm. While the 2D lidar data is the input to google cartographer algorithm[2] whose output trajectory is deemed as ground truth in the real scene. We record a trajectory with 2 loops in the real scene. In order to compare tracking accuracy, we record the RMSE and SD of ATE. They are 0.6907 and 0.2689 in ORB-SLAM, while 0.3602 and 0.1815 in our method. The trajectory comparison is also presented in Fig. 6. From the above results, our method does demonstrate an improved performance in the real scene.

---

[2] https://github.com/googlecartographer/cartographer.

Fig. 5. The mobile platform used in the real scene to collect data.

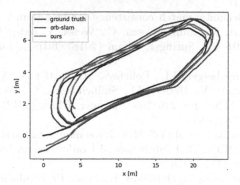

Fig. 6. Tracking trajectory comparison in the real scene.

5   Conclusions

In this paper, we develop a novel semantic SLAM in which the semantic information is used to not only produce the semantic map but also provide an extra constraint to improve the performance of SLAM. The proposed method utilizes a semantically meaningful ground plane to establish the semantic constraint. A novel formulation is also developed to incorporate the semantic constraint into the feature-based RGB-D SLAM. The proposed method can consistently refine the state estimation and effectively utilize low-texture regions. Both quantitative and qualitative results demonstrate the improvements of our proposed method. However, our current work only focuses on utilizing semantic information to boost SLAM where we use a trained network to implement the semantic segmentation. We will explore how to make visual SLAM and semantic segmentation be mutually beneficial in the future.

References

1. Sturm, J., Engelhard, N., Endres, F., et al.: A benchmark for the evaluation of RGB-D SLAM systems. In: IEEE/RSJ International Conference on Intelligent Robots and Systems, pp. 573–580 (2012)
2. McCormac, J., Handa, A., Davison, A., et al.: SemanticFusion: dense 3D semantic mapping with convolutional neural networks. In: IEEE International Conference on Robotics and Automation, pp. 4628–4635 (2017)
3. Nakajima, Y., Tateno, K., Tombari, F., et al.: Fast and accurate semantic mapping through geometric-based incremental segmentation. In: IEEE/RSJ International Conference on Intelligent Robots and Systems, pp. 385–392 (2018)
4. Sünderhauf, N., Pham, T.T., Latif, Y., et al.: Meaningful maps with object-oriented semantic mapping. In: IEEE/RSJ International Conference on Intelligent Robots and Systems, pp. 5079–5085 (2017)
5. Kobyshev, N., Riemenschneider, H., Gool, L.V.: Matching features correctly through semantic understanding. In: IEEE International Conference on 3D Vision, pp. 472–479 (2014)
6. Toft, C., et al.: Semantic match consistency for long-term visual localization. In: Ferrari, V., Hebert, M., Sminchisescu, C., Weiss, Y. (eds.) ECCV 2018. LNCS, vol. 11206, pp. 391–408. Springer, Cham (2018). https://doi.org/10.1007/978-3-030-01216-8_24
7. Lianos, K.-N., Schönberger, J.L., Pollefeys, M., Sattler, T.: VSO: visual semantic odometry. In: Ferrari, V., Hebert, M., Sminchisescu, C., Weiss, Y. (eds.) ECCV 2018. LNCS, vol. 11208, pp. 246–263. Springer, Cham (2018). https://doi.org/10.1007/978-3-030-01225-0_15
8. Yu, C., Liu, Z., Liu, X.J., et al.: DS-SLAM: a semantic visual slam towards dynamic environments. In: IEEE/RSJ International Conference on Intelligent Robots and Systems, pp. 1168–1174 (2018)
9. Bowman, S.L., Atanasov, N., Daniilidis, K., et al.: Probabilistic data association for semantic SLAM. In: IEEE International Conference on Robotics and Automation, pp. 1722–1729 (2017)
10. Doherty, K., Fourie, D., Leonard, J.J.: Multimodal semantic SLAM with probabilistic data association. In: IEEE International Conference on Robotics and Automation (2019)
11. Li, J., Meger, D., Dudek, G.: Context-coherent scenes of objects for camera pose estimation. In: IEEE/RSJ International Conference on Intelligent Robots and Systems, pp. 655–660 (2017)
12. Li, J., Xu, Z., Meger, D., et al.: Semantic scene models for visual localization under large viewpoint changes. In: Conference on Computer and Robot Vision (2018)
13. Li, P., Qin, T., Shen, S.: Stereo vision-based semantic 3D object and ego-motion tracking for autonomous driving. In: Ferrari, V., Hebert, M., Sminchisescu, C., Weiss, Y. (eds.) ECCV 2018. LNCS, vol. 11206, pp. 664–679. Springer, Cham (2018). https://doi.org/10.1007/978-3-030-01216-8_40
14. Yang, S., Scherer, S.: CubeSLAM: monocular 3D object SLAM. IEEE Trans. Rob. 35(4), 925–938 (2019)
15. Yang, S., Scherer, S.: Monocular object and plane SLAM in structured environments. arXiv:180903415 (2018)
16. Hosseinzadeh, M., Li, K., Latif, Y., et al.: Real-time monocular object-model aware sparse SLAM. CoRR: abs/1809.09149 (2018)

17. Salas-Moreno, R.F., Newcombe, R.A., Strasdat, H., et al.: SLAM++: simultaneous localisation and mapping at the level of objects. In: IEEE Conference on Computer Vision and Pattern Recognition, pp. 1352–1359 (2013)
18. Parkhiya, P., Khawad, R., Murthy, J.K., et al.: Constructing category-specific models for monocular object-SLAM. In: IEEE International Conference on Robotics and Automation, pp. 1–9 (2018)
19. Runz, M., Buffier, M., Agapito, L.: MaskFusion: real-time recognition, tracking and reconstruction of multiple moving objects. In: IEEE International Symposium on Mixed and Augmented Reality, pp. 10–20 (2018)
20. Dou, M., Guan, L., Frahm, J.-M., Fuchs, H.: Exploring high-level plane primitives for indoor 3D reconstruction with a hand-held RGB-D camera. In: Park, J.-I., Kim, J. (eds.) ACCV 2012. LNCS, vol. 7729, pp. 94–108. Springer, Heidelberg (2013). https://doi.org/10.1007/978-3-642-37484-5_9
21. Taguchi, Y., Jian, Y.D., Ramalingam, S., et al.: Point-plane SLAM for hand-held 3D sensors. In: IEEE International Conference on Robotics and Automation, pp. 5182–5189 (2013)
22. Huang, J., Dai, A., Guibas, L., et al.: 3DLite: towards commodity 3D scanning for content creation. ACM Trans. Graph. **36**(6), 203-1 (2017)
23. Zhang, Y., Xu, W., Tong, Y., et al.: Online structure analysis for real-time indoor scene reconstruction. ACM Trans. Graph. **34**(5), 159 (2015)
24. Yang, S., Song, Y., Kaess, M., et al.: Pop-up SLAM: semantic monocular plane SLAM for low-texture environments. In: IEEE/RSJ International Conference on Intelligent Robots and Systems, pp. 1222–1229 (2016)
25. Hosseinzadeh, M., Latif, Y., Reid, I.: Sparse point-plane SLAM. In: Australasian Conference on Robotics and Automation (2017)
26. Ma, L., Kerl, C., Stückler, J., et al.: CPA-SLAM: consistent plane-model alignment for direct RGB-D SLAM. In: IEEE International Conference on Robotics and Automation, pp. 1285–1291 (2016)
27. Mur-Artal, R., Tardós, J.D.: ORB-SLAM2: an open-source SLAM system for monocular, stereo and RGB-D cameras. arXiv:161006475 (2016)
28. Long, J., Shelhamer, E., Darrell, T.: Fully convolutional networks for semantic segmentation. In: IEEE Conference on Computer Vision and Pattern Recognition (2015)
29. Fischler, M.A., Bolles, R.C.: Random sample consensus: a paradigm for model fitting with applications to image analysis and automated cartography. Commun. ACM **24**(6), 381–395 (1981)

Blind Quality Assessment Method to Evaluate Cloud Removal Performance of Aerial Image

Zhe Wei, Yongfeng Liu, Mengjie Li, Congli Li$^{(\boxtimes)}$, and Song Xue

PLA Army Academy of Artillery and Air Defense, Hefei, Anhui, China
lcliqa@163.com

Abstract. People often use image-inpainting-based methods to remove cloud from aerial images, but it lacks a targeted quantitative evaluator to assess the removal result. In order to solve this issue to some extent, we propose an assessment method that combines ranking learning and regression learning. The main framework consists of several CNN down-sampling steps layer-by-layer. Firstly, to learn the distortion features of the inpainted image, an image classification task is conducted to train the network. Secondly, we use the proposed joint loss function to regress the features into the FR-IQA evaluator SSIM by retraining the whole network. Through end-to-end training, the proposed model learns a priori of the aerial image and realizes the approximation of SSIM. Experimental results demonstrate that our method has achieved better performance on SROCC, RMSE and PLCC in comparison with other blind image quality assessment methods.

Keywords: Image quality assessment · Cloud removal · Image inpainting · Deep learning · Joint learning

1 Introduction

Aerial imagery is an important source of geographic information, but the occlusion from thick clouds during imaging will lead to image quality degradation and incomplete target information. When the spectrum and temporal information are insufficient to restore, people often use inpainting methods to reconstruct the information under thick clouds [1–4]. However, the current assessment methods for the quality of cloud removal mainly rely on artificial subjective visual judgment and full-reference (FR) evaluators such as SSIM, PSNR, and MSE [3–8]. The quantitative assessment of cloud removal effect for aerial images is an ill-posed problem. Since there are no effective cloudless reference images in practical applications, a targeted blind assessment method needs to be studied.

An aerial image has a long imaging distance, less depth information, and a continuous distribution of ground features and textures, which is very different from general ground imaging. The cloud removal of aerial images is mainly

© Springer Nature Switzerland AG 2020
Y. Peng et al. (Eds.): PRCV 2020, LNCS 12305, pp. 280–290, 2020.
https://doi.org/10.1007/978-3-030-60633-6_23

to restore color, texture and other structural information, and the key to its assessment is to identify the inconsistency between the part and the whole. This inconsistent distortion is different from the type of distortion in the public IQA dataset. Combining ranking-based and regression-based ideas in the above two types, we propose a no-reference evaluation model for the quality of aerial image cloud removal result, which includes multiple convolutional layers and fully connected layers. We conduct a classification task as driver, and combine the ranking loss and regression loss to realize the prediction of the FR evaluator SSIM only with the inpainted image (Fig. 1 shows the assessment results of our method and other BIQA methods of aerial images). Our contributions are summarized as follows:

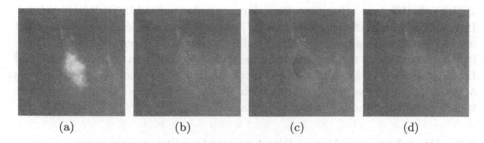

(a)   (b)   (c)   (d)

Fig. 1. The results of several inpainting methods of an aerial image. (a) is the original cloudy image, (b), (c) and (d) are the results of the inpainting method [5–7] respectively. The predicted scores of ILNIQE [12] are 59.1252, 62.6970, and 65.3415 (the lower the score, the higher the quality). Meanwhile, QAC are 0.7515, 0.7651 and 0.7984; RankIQA++ [22] are 8.1754, 8.2007, and 8.1949; MEON [19] are 8.3088, 8.3318, and 8.3302; DB-CNN [25] are 7.3386, 7.7626, and 6.6465. The proposed method are 8.2099, 8.6693 and 8.7995(the higher the score, the higher the quality), which is more in line with the order of human vision.

(1) We introduce BIQA into the assessment of cloud removal, and propose an effective quality predictor that does not target specific types of distortion. Through CNN end-to-end learning, the model can get blind assessment of the cloud removal effect of aerial images when reference images cannot be obtained.

(2) Combining ranking and regression training, a joint loss function is proposed. While obtaining a good approximation, we also take into account image quality differences and obtain good prediction performance.

(3) We adopt the idea of classification pretraining and regression-ranking combined training to predict the ground truth quality score, and prove that the model can effectively learn the a priori information of aerial images and perceive the inpainted area. Experimental results on the test samples demonstrate the model has good advantages in prediction accuracy and generalization ability over other BIQA methods.

2   Related Work and Our Contributions

2.1   Traditional Blind Image Quality Assessment

The traditional methods divide the assessment problem into two steps: feature extraction and quality score regression. An effective feature extraction method is the focus of research. Aiming at the degradation in image acquisition, transmission, and processing, corresponding detection and evaluation methods have been proposed for specific distortions such as JPEG compression [9], Gaussian blur [10], etc. But these methods can only deal with a single type of distortion. For complex mixed-distortion scenes, spatial and transform domain features have been proposed to describe the statistical laws of natural scene images, such as the normalized image block spatial features used in NIQE [11], ILNIQE [12], and BRISQUE [13], and other transform domain features used in BLIINDS [14,15]. In addition, CORNIA [16] and QAC [17] use unsupervised learning to construct a codebook from image samples, and build statistical features based on it. Quality score regression methods include multivariate Gaussian models, SVR, and Bayesian inference. Most traditional methods rely on manual designed features and statistical analysis, but gradually lose their advantages for more and more types of distortion and application scenarios.

2.2   Deep Learning-Based Blind Image Quality Assessment

Deep learning models can learn the features of the dataset through end-to-end supervised or unsupervised tasks, and gradually become a research hotspot of IQA. Kang [18] applies 5-layer CNN for the first time to predict the image quality score. Then people successively propose deeper quality assessment networks. But the manual annotation of image quality evaluation is usually expensive and time-consuming, so the number of annotations that can be provided is limited. Compared with the large number of parameters of the CNN model, training with only limited real value regression is easy to cause overfitting, so people usually expand the available data through various methods. Bosse [19] and Kim [20–22] cut the image into small blocks, and use the MOS of the entire image as the score of each block, and then average the score of each block during prediction, which increases the number of samples. Liu [23], Ma [24], etc. use a siamese network for image quality ranking training, which greatly expands the size of the data set. In addition, after pair-wise comparison training between samples, the network will also learn the features of image distortion, which will contribute to the next phase. Except for ranking learning, Ma [25], Zhang [26], etc. applies classification of distortion type and intensity to pre-train. In addition, Lin [27] attempts to recover the reference image information by using the GAN, so as to use the full reference assessment method to predict the quality score.

2.3   Image Inpainting

The purpose of image inpainting is to remove unwanted objects in the foreground and restore the consistency and naturalness to match it to human vision.

The traditional inpainting [8] is most commonly based on image patch-matching methods, which copy the most similar background patches of other areas to fill the hole. This kind of method sometimes has good subjective effects, but it needs to meet certain priori conditions, such as the consistency of lighting and structure. Therefore, it shows poor robustness and is easy to cause instabilityin aerial images. After the emergence of deep learning, modern inpainting methods often use GAN or auto-encoders for adversarial training [5–7], which helps to learn the distribution of image pixels from a certain dataset, and can synthesize high-level semantic information, which makes the inpainted results more realistic. However, how to assess the quality of the inpainted local area and to deal with the correlation of the overall image quality bring new challenges to the quality assessment method.

3   Proposed Method

Different from general ground-based images, the distortions in aerial images caused by inpainting usually appear near the inpainted region, so global statistical analysis and image block averaging are difficult to reflect the overall image degradation. We attempt to use CNN to learn the entire image to preserve the response to the inpainting distortion with the down-sampling and pooling structure layer by layer. Given the inpainted image I_c to be assessed, the FR-IQA algorithm can be expressed as the following equation:

$$S = f(I_s, I_c) \tag{1}$$

In the formula, $f(\bullet)$ represents a certain FR-IQA algorithm, I_s denotes a reference image, and S is the FR-IQA score. When I_s cannot be obtained, the formula (1) is degraded to a BIQA algorithm, which can be expressed by the following formula:

$$\bar{S} = g_\theta(I_c) \tag{2}$$

$g_\theta(\bullet)$ represents the BIQA model, θ denotes the model parameters, and $\bar{S}$ is the prediction result of the model. The next step is to train the model to let $\bar{S}$ approximates S. Due to the lack of I_s, the parameters of the prediction model $g_\theta(\bullet)$ must contain prior knowledge of the original image, which is usually difficult to model explicitly, but can be learned implicitly by training. To this end, we model $g_\theta(\bullet)$ as a CNN structure to predict the scores of inpainted images. In order to obtain the perception of the original image, a discriminative model $g'_\theta(\bullet)$ is trained to identify a large number of undistorted original images and inpainted images, which shares the same feature extraction part (convolutional layers) with $g_\theta(\bullet)$.After training, we consider that the feature extraction net has learned the main features from undistorted images as well as inpainted ones. Then the feature extraction part of $g'_\theta(\bullet)$ is retained, and the quality score is regressed with $g_\theta(\bullet)$. The flowchart of the proposed pipeline is depicted in Fig. 2. When predicting an unknown quality image, the trained $g_\theta(\bullet)$ takes a single image as input without any reference images.

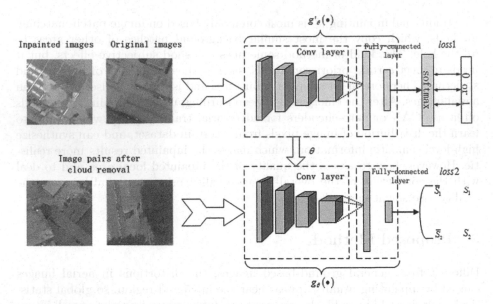

Fig. 2. Pipeline of blind quality assessment of cloud removal effect

3.1 Network Architecture

The network architecture consists of a 4 convolution-pooling-normalization layers and 2 fully-connected layers. The input is an image with shape $256 \times 256 \times 3$. After many experiments, we find that many normalized methods such as GDN, GN, and LN can effectively identify the inpainted and original images. Considering the wide application in other similar problems, we choose GDN as the normalization layer. The network structure is shown in Table 1, where conv(5,2,2,8) represents a convolution kernel with output channel of 8 and size of 5×5, performing a convolution operation with stride of 2, and a boundary padding of 2 (the same below); pool(2,2) represents the maximum pooling operation with size of 2×2 and stride of 2; fc(128) represents the fully connected layer of 128 output nodes; the output dimension C has different values due to different tasks, $C = 2$ in the classification task and $C = 1$ in the regression task.

Table 1. Model architecture and output dimensions

Layer	Operation	Output Dimensions
conv1	conv(5,2,2,8), gdn, pool(2,2)	$64 \times 64 \times 8$
conv2	conv(5,2,2,16), gdn, pool(2,2)	$16 \times 16 \times 16$
conv3	conv(5,2,2,32), gdn, pool(2,2)	$4 \times 4 \times 32$
conv4	conv(3,2,0,64), gdn, pool(2,2)	$1 \times 1 \times 64$
fc1	fc(128), gdn	128
fc2	fc(C)	C

3.2 Comparison Pairs

Similar to some methods [23,24], we also uses ranking learning to accord the output values from $g_\theta(\bullet)$ of any two images to their FR assessment score rankings. Because SSIM is more in line with the subjective perception of the human eye, and the output value is continuous between 0 and 1, which is convenient for regression. So we use it as the FR-IQA method $f(I_s, I_c)$ to be regressed.

Assuming that the amount of original image data is N, m_1 kinds of algorithm are used to inpaint with m_2 kinds of mask, we obtain Nm_1m_2 groups image data and corresponding SSIM values. Arranging these data in pairs, we can get $C^2_{Nm_1m_2}$ groups data, which meets the data scale of deep learning model. groups data, which meets the data scale of deep learning model.

3.3 Loss Function

The training of our method is divided into two phases. In the 1st phase, we conduct the classification task to drive model to learn the distortion features after image inpainting. Treating the undistorted (uninpainted) as positive examples and the distorted (inpainted) as negative, the classification network $g'_\theta(\bullet)$ maps the input image into a 2-dimensional vector:

$$g_\theta'(I) = \begin{bmatrix} \hat{y}_0 & \hat{y}_1 \end{bmatrix} \tag{3}$$

The labels of the training samples are denoted as $[y_0 \quad y_1]$, and both the positive and negative examples are encoded in one-hot format. We use softmax to map the output vector to confidence of each category:

$$p_0 = \frac{\exp(\hat{y}_0)}{\exp(\hat{y}_0) + \exp(\hat{y}_1)}, p_1 = \frac{\exp(\hat{y}_1)}{\exp(\hat{y}_0) + \exp(\hat{y}_1)} \tag{4}$$

We measure the classification performance with a cross-entropy loss function:

$$loss1 = \frac{1}{m} \sum_m -y_0 \log p_0 - y_1 \log p_1 \tag{5}$$

where m is training batch size. Since the number of negative examples is much larger than the positive, the positive examples are oversampled to balance. After the 1st phase of training, we use the comparison pairs generated in Sect. 3.2 to perform the training of ranking and regression. For any image comparison pair I_{c1} and I_{c2}, assuming $f(I_s, I_{c1}) > f(I_s, I_{c2})$, the network $g_\theta(\bullet)$ must not only make the corresponding output approximate the full reference score, but also make the two outputs meet the same score ranking:

$$\begin{cases} g_\theta(I_{c1}) = f(I_s, I_{c1}) \\ g_\theta(I_{c2}) = f(I_s, I_{c2}) \\ g_\theta(I_{c1}) > g_\theta(I_{c2}) \end{cases} \tag{6}$$

Therefore we design the following loss function for the 2nd phase:

$$loss2 = \lambda_1 \max[0, g_\theta(I_{c2}) - g_\theta(I_{c1}) + \varepsilon]$$
$$+ \lambda_2 \left[\frac{|g_\theta(I_{c1}) - f(I_s, I_{c1})| + |g_\theta(I_{c2}) - f(I_s, I_{c2})|}{2} \right] \qquad (7)$$
$$+ \lambda_3 \sum |\omega|$$

The loss function consists of 3 terms. The first one is the ranking loss. When $g_\theta(I_{c2}) > g_\theta(I_{c1})$, the term is positive. After gradient descent, the response to I_{c2} will gradually decrease and I_{c1} increase. A positive number ε makes the response to I_{c2} fall below the threshold, to avoid the output value to many samples from centering within a small interval. The second term is the regression loss. The third term is a regular term for all weights of the neural network, which also uses the L1 norm. To balance these 3 terms, hyperparameters λ_1, λ_2 and λ_3 are used as the weighting coefficients, and set to $\lambda_1 = \lambda_2 = 1$, $\lambda_3 = 10^{-5}$, $\varepsilon = 1.9 \times 10^{-4}$ after cross-validation.

4   Experimental Results

4.1   Benchmark Dataset

Our experimental data are aerial images of a region in Asia. We crop them into 16,542 non-overlapping with size of 256×256 patches, and then divide them into a training set and a test set at a ratio of 3:1. Three recent inpainting methods based on deep learning *DCGAN* [5], *context encoder* [6] and *generative inpainting* [7] are selected to learn and generate testing examples. 4 types of mask with different sizes and positions are constructed to simulate cloud regions. We obtain a total of 49,632 samples as the BIQA dataset. The SSIM of the images in the dataset are calculated according to the original image and stretched to the interval of 0-10 as the ground truth value to regress.

In the 1st phase of training, we train $g'_\theta(\bullet)$ with $loss1$. Batch size is set to 32, and optimizer is *Adam*. We train 100 times for 1 verification, and stop training till when the verification accuracy reaches higher than 97%, considering to have learned the distortion features of the inpainted image enough.

In the 2nd phase, we retain the convolutional layer parameters, and train the entire network $g_\theta(\bullet)$ with $loss2$. Batch size is set to 1 pair, and the number of epoch is 10.

In order to compare the performance, 1 non-learning assessment method ILNIQE, 1 unsupervised learning method QAC, and 3 deep learning-based methods, RankIQA, MEON, and DB-CNN, are selected and compared. For RankIQA, we find in the experiment that the 3 main network structures proposed by the author cannot effectively fit our data features, so we use the main frame of MEON instead, and mark it as RankIQA++ to distinguish.

4.2   Testset Results

The Spearman rank order correlation coefficient (SROCC), Pearson linear correlation coefficient (PLCC), and root mean square error (RMSE) are adopted to evaluate the performance of each method.

Table 2 shows the comparison of scores of each method under each criterion. It can be seen that the ILNIQE and DB-CNN methods perform poorly. These two methods are aimed at global consistent distortions. QAC is slightly better, but still has a certain gap with our method. For RankIQA++ and MEON, the main difference from ours lies in the training and loss functions. Our method takes all these three criteria into consideration in the loss function, so it achieves the best of all 6 methods on PLCC, RMSE and SROCC.

Table 2. Comparison of each method under 3 criteria

Method	SROCC	RMSE	PLCC
ILNIQE	0.0877	2.7378	0.3020
QAC	0.6224	2.0491	0.7013
RankIQA++	0.7565	0.8953	0.9502
MEON	0.7661	1.2004	0.9085
DB-CNN	0.4251	1.9445	0.7359
Ours	0.7721	0.6283	0.9758

4.3 Ablation Study

In order to further verify the effectiveness of our model and method, two optional parts of the method are replaced. Firstly, we do not conduct the 1st phase classification training, and directly perform the regression and ranking training, marked as ①; secondly, we just remove the ranking loss and keep the other unchanged, marked as ②. The scores of each method under each criterion are shown in Table 3.

Table 3. Comparison of each method under 3 criteria

Criterion	SROCC	RMSE	PLCC
①	0.7407	0.6512	0.9740
Degradation	0.0314	0.0229	0.0018
②	0.7408	0.6828	0.9713
Degradation	0.0313	0.0545	0.0045

Without classification training, ① cannot effectively perceive the difference between the inpainted region and other parts of the image, which makes it difficult for regression training. Without ranking learning, it is difficult for ② to perceive the comparison between image samples with different SSIM, which also causes a decrease in accuracy. ① and ② have different degrees of performance loss on the 3 criteria, proving that classification training and ranking training are essential.

4.4   Trend Analysis

We further extend these assessment methods to changing mask area. Masks with a rectangular hole at the center and an area of 5%, 10%, 15%, 20% and 25% is generated. The overall effect of the 3 inpainting methods is averaged, as shown in Table 4.

Table 4. Average assessment results of the 3 inpainting methods under 5 BIQA methods and SSIM

Method	DCGAN	Context encoder	Generative inpainting
RankIQA++	8.14234	8.20614	8.2125
MEON	8.30458	8.3633	8.36544
ILNIQE	52.58732	46.53446	50.92692
DB-CNN	5.77286	4.56722	4.24428
Ours	7.60582	8.76636	8.755
SSIM	0.93692	0.98806	0.9873

The SSIM ground truth value shows that *context encoder* is equivalent to the *generative inpainting*, and the *DCGAN* is inferior. Compared with several other assessment methods, our method reflects this trend more accurately.

5   Conclusions

In order to properly assess the cloud removal effect of an aerial image, we attempt to combine ranking learning and regression learning technique, and propose a blind assessment method of inpainted aerial image based on CNN. The training of the model is divided into two phases. Driven by the classification task, the proposed model learns the distortion features. Then we perform the end-to-end training from the image pairs to SSIM values with the joint loss function. We achieve better results on 3 evaluation criteria in the testset, and further verify the validity of the classification task and the joint loss function. Finally, we manifest that the method has a good approximation effect on the real SSIM value in the sense of more image averages.

The proposed CNN model treats the input image as a whole, which can easily discriminate local distortion caused by cloud removal. For global distortion in image processing such as SR, deblur and de-raining, a local model seems to be better. In future studies, we will try to unify the local and global models to adapt to more image processing tasks.

References

1. Siravenha, A.C., Sousa, D., Bispo, A., Pelaes, E.: Evaluating inpainting methods to the satellite images clouds and shadows removing. In: International Conference on Signal Processing, Image Processing, and Pattern Recognition, pp. 56–65 (2011)

2. Lorenzi, L., Melgani, F., Mercier, G.: Inpainting strategies for reconstruction of missing data in VHR images. IEEE Geosci. Remote Sens. Lett. **8**(5), 914–918 (2011)
3. Singh, P., Komodakis, N.: Cloud-Gan: cloud removal for sentinel-2 imagery using a cyclic consistent generative adversarial networks. In: IGARSS 2018–2018 IEEE International Geoscience and Remote Sensing Symposium, pp. 1772–1775 (2018)
4. Song, C., Xiao, C.: Single aerial photo cloud removal. J. Comput. Aided Des. Comput. Graph. **31**(1), 76 (2019)
5. Yeh, R.A., Chen, C., Yian Lim, T., Schwing, A.G., Hasegawa-Johnson, M., Do, M.N.: Semantic image inpainting with deep generative models. In: Proceedings of the IEEE Conference on Computer Vision and Pattern Recognition, pp. 5485–5493 (2017)
6. Pathak, D., Krahenbuhl, P., Donahue, J., Darrell, T., Efros, A.: Context encoders: feature learning by inpainting. In: Proceedings of the IEEE Conference on Computer Vision and Pattern Recognition, pp. 2536–2544 (2016)
7. Yu, J., Lin, Z., Yang, J., Shen, X., Lu, X., Huang, T.S.: Generative image inpainting with contextual attention. In: Proceedings of the IEEE Conference on Computer Vision and Pattern Recognition, pp. 5505–5514 (2018)
8. Darabi, S., Shechtman, E., Barnes, C., Goldman, D.B., Sen, P.: Image melding: combining inconsistent images using patch-based synthesis. ACM Trans. Graph. **31**(4), 1–10 (2012)
9. Wang, Z., Sheikh, H.R., Bovik, A.C.: No-reference perceptual quality assessment of JPEG compressed images. In: Proceedings of the International Conference on Image Processing, p. 1 (2002)
10. Wang, Z., Simoncelli, E.P.: Local phase coherence and the perception of blur. In: Advances in Neural Information Processing Systems, pp. 1435–1442 (2004)
11. Mittal, A., Soundararajan, R., Bovik, A.C.: Making a "completely blind" image quality analyzer. IEEE Signal Process. Lett. **20**(3), 209–212 (2012)
12. Zhang, L., Zhang, L., Bovik, A.C.: A feature-enriched completely blind image quality evaluator. IEEE Trans. Image Process. **24**(8), 2579–2591 (2015)
13. Mittal, A., Moorthy, A.K., Bovik, A.C.: No-reference image quality assessment in the spatial domain. IEEE Trans. Image Process. **21**(12), 4695–4708 (2012)
14. Saad, M.A., Bovik, A.C., Charrier, C.: Blind image quality assessment: a natural scene statistics approach in the DCT domain. IEEE Trans. Image Process. **21**(8), 3339–3352 (2012)
15. Hassen, R., Wang, Z., Salama, M.M.: Image sharpness assessment based on local phase coherence. IEEE Trans. Image Process. **22**(7), 2798–2810 (2013)
16. Ye, P., Kumar, J., Kang, L., Doermann, D.: Unsupervised feature learning framework for no-reference image quality assessment. In: 2012 IEEE Conference on Computer Vision and Pattern Recognition, pp. 1098–1105 (2012)
17. Xue, W., Zhang, L., Mou, X.D.: Learning without human scores for blind image quality assessment. In: 2013 IEEE Conference on Computer Vision and Pattern Recognition, pp. 995–1002 (2013)
18. Kang, L., Ye, P., Li, Y., Doermann, D.: Convolutional neural networks for no-reference image quality assessment. In: Proceedings of the IEEE Conference on Computer Vision and Pattern Recognition, pp. 1733–1740 (2014)
19. Bosse, S., Maniry, D., Müller, K.R., Wiegand, T., Samek, W.: Deep neural networks for no-reference and full-reference image quality assessment. IEEE Trans. Image Process. **27**(1), 206–219 (2017)

20. Kim, J., Lee, S.: Deep learning of human visual sensitivity in image quality assessment framework. In: Proceedings of the IEEE Conference on Computer Vision and Pattern Recognition, pp. 1676–1684 (2017)
21. Kim, J., Nguyen, A.D., Lee, S.: Deep CNN-based blind image quality predictor. IEEE Trans. Neural Netw. Learn. Syst. **30**(1), 11–24 (2018)
22. Kim, J., Lee, S.: Fully deep blind image quality predictor. IEEE J. Sel. Top. Sig. Process. **11**(1), 206–220 (2016)
23. Liu, X., van de Weijer, J., Bagdanov, A.D.: RankIQA: learning from rankings for no-reference image quality assessment. In: Proceedings of the IEEE International Conference on Computer Vision, pp. 1040–1049 (2017)
24. Ma, K., Liu, W., Liu, T., Wang, Z., Tao, D.: dipIQ: blind image quality assessment by learning-to-rank discriminable image pairs. IEEE Trans. Image Process. **26**(8), 3951–3964 (2017)
25. Ma, K., Liu, W., Zhang, K., Duanmu, Z., Wang, Z., Zuo, W.: End-to-end blind image quality assessment using deep neural networks. IEEE Trans. Image Process. **27**(3), 1202–1213 (2017)
26. Zhang, W., Ma, K., Yan, J., Deng, D., Wang, Z.: Blind image quality assessment using a deep bilinear convolutional neural network. IEEE Trans. Circuits Syst. Video Technol. **30**(1), 36–47 (2020)
27. Lin, K. Y., and Wang, G.: Hallucinated-IQA: no-reference image quality assessment via adversarial learning. In: Proceedings of the IEEE Conference on Computer Vision and Pattern Recognition, pp. 732–741 (2018)

Learning Multi-scale Retinex with Residual Network for Low-Light Image Enhancement

Long Ma[1], Jie Lin[1], Jingjie Shang[2], Wei Zhong[2], Xin Fan[2], Zhongxuan Luo[1], and Risheng Liu[2(✉)]

[1] School of Software, Dalian University of Technology, Dalian 116024, China
[2] DUT-RU International School of Information Science and Engineering, Dalian University of Technology, Dalian 116024, China
rsliu@dlut.edu.cn

Abstract. Existing network-based techniques addressing low-light image enhancement are developed by using the complex network architectures. However, the performance is not ideal, and they cannot give a reasonable interpretation of the effects of each layer in the network. To settle these issues, we design a residual network to learn multi-scale Retinex for handling low-light image enhancement. To be concrete, inspired by multi-scale Retinex, we define a new residual-type multi-scale Retinex model to gradually remove the illumination generated by the convolutional procedure. Thanks to the progressive mechanism, we can build an intuitive and explicit relationship between our residual-type multi-scale Retinex model and residual network. This enables us can directly utilize the residual network to learn a residual-type multi-scale Retinex by integrating the data distribution. Precisely because of our transparent modeling procedure, we can recognize the effects of each layer in our learnable architecture. It is valuable for more effectively exploit the network layers to handle this task. Extensive analytical experiments are performed to verify the effectiveness of our proposed method. A series of evaluative experiments are conducted to illustrate our superiority against other state-of-the-art methods.

Keywords: Low-light image enhancement · Multi-scale retinex · Residual nework

1 Introduction

Enhancing the low-light images to obtain more effective and valuable information is an important topic in computer vision areas. It has drawn much attention in

Supported by the National Natural Science Foundation of China (Nos. 61922019, 61733002, and 61672125), LiaoNing Revitalization Talents Program (XLYC1807088), and the Fundamental Research Funds for the Central Universities (DUT19TD19). The first author is a student.

Y. Peng et al. (Eds.): PRCV 2020, LNCS 12305, pp. 291–302, 2020.
https://doi.org/10.1007/978-3-030-60633-6_24

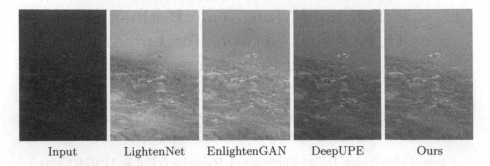

| Input | LightenNet | EnlightenGAN | DeepUPE | Ours |

Fig. 1. Visual comparison among state-of-the-art low-light image enhancement methods and our proposed LMSR. We can easily see that LightenNet [1] (*the numbers of parameters is* 27,648) generates the non-uniform illumination to degrade the image quality. EnlightenGAN [2] (*the numbers of parameters is* 8,646,656) brings about some unknown haze-like artifacts. The color distortion occurs in the result of DeepUPE [3] (*the numbers of parameters is* 2,997,248). By comparison, the result of our LMSR (*the numbers of parameters is* **2,048**) is more natural and satisfying.

the community. In the following, we will briefly introduce existing works related to low-light image enhancement. We also describe our main contributions.

1.1 Related Works

In the past few decades, many approaches are developed to handle low-light image enhancement. They can be roughly divided into two categories, model-based and data-driven methods.

In the early stage, utilizing the Gaussian filter to act on the low-light inputs is a commonly-used strategy. Single-Scale Retinex (SSR) [4] is a basic version that generated the illumination by performing the once Gaussian filtering on the low-light input in the log-domain, then obtained the enhanced results by subtraction operation. Multi-scale Retinex (MSR) [5] considered to multiply execute the Gaussian filtering to generate multiple the illumination in the log-domain, then derived the final outputs by presenting the weighted sum on the obtained the enhanced results based on SSR. MSRCR [6] further considered the color deviation problem. This method designed an effective weighting function to correct the color.

Some model-based methods view the design of the model and iterative algorithm as the basic appeal. Defining the prior regularizations is an important composition. A variety of regularization forms have been proposed for this task. For instance, ℓ_2-norm is a simple convex form to constrain the illumination in some works [7,8]. However, the edge-preserving property of ℓ_2-norm is weak so that the estimated illumination is not ideal. ℓ_1-norm and its variants [9–11] are then developed to perform the stronger edge-aware ability. However, limit to the assumption of these defined regularizations, these model-based methods always generate the unideal performance with insufficient details and inappropriate exposure. Additionally, their computational procedures are complex and time-consuming.

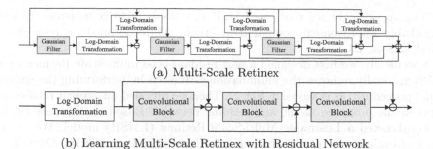

(a) Multi-Scale Retinex

(b) Learning Multi-Scale Retinex with Residual Network

Fig. 2. Comparing the computational flow between Multi-scale Retinex (MSR) [5] and our proposed Learnable Multi-Scale Retinex (LMSR). MSR removes different illuminations which are generated by using different Gaussian filters to act on the low-light input. And sum up these outputs to obtain the final result. Our LMSR progressively removes the illumination. We build an explicit relationship in these estimation modules. Obviously, our LMSR can be directly implemented by utilizing the well-known Residual Convolutional Neural Network.

Data-driven methods focus on designing the network architectures and learning the network parameters by the massive training data. RetinexNet [12] built a new dataset with noises, called the LOL dataset. They designed the network architecture based on the Retinex theory and trained it by the built dataset. KindNet [13] further ameliorated the issues existing in RetinexNet, they added some additional training loss and tuned up the network architecture. Lighten-Net [1] considered generating the training data using the Retinex theory by performing the uniform illuminations. They designed a lightweight network and trained it based on the generated training pairs. DeepUPE [3] defined an illumination estimation network. This work utilized the well-known MIT-Adobe 5 K dataset [14] to train the designed network, and it also provided a new dataset by using the same generation way with MIT-Adobe 5 K dataset [14]. EnlightenGAN [2] constructed a generator with the attention mechanism to generate enhanced results. They used the global-local discriminator [15] to constrain the outputs of the generator.

Indeed, these above-mentioned data-driven works are well-designed, but they cannot recognize the effects of each layer in the network. In other words, their designed network architectures possibly contain so many redundancies that are avoidable and unnecessary.

1.2 Our Contributions

In this work, inspired by the classical Multi-Scale Retinex (MSR), we proposed a novel and simple network architecture to handle low-light image enhancement. We called our method as LMSR (Learnable Multi-Scale Retinex). We demonstrated a group of comparison in terms of different state-of-the-art data-driven works in Fig. 1. Obviously, our result is prominent that perfectly ameliorating the exposure level without causing the color distortion and details loss. Additionally, we also reported the numbers of parameters with respect to each approach.

We can surprisingly see that our LMSR just needs 2048 parameters, this is a remarkably improvement compared to others. More detailed analyses can be found in Sect. 4.3.

Specifically, we first designed a new residual-type multi-scale Retinex model which gradually removes the illumination estimated by performing the convolutional procedure (as is shown in Fig. 2). Then we revealed an explicit relationship between the proposed model and residual network. Based on this relationship, we constructed a Learnable Multi-Scale Retinex (LMSR) model. We make a series of evaluations to indicates our effectiveness and superiority. Overall, our main contributions can be concluded as

- Inspired by multi-scale Retinex, we establish a new residual-type multi-scale Retinex model to gradually remove the illumination estimated by executing the convolutional procedure in the log-domain.
- Thanks to the conciseness of our designed model, we explicitly reveal the relationship between the proposed model and residual network. Based on this observation, we construct a Learnable Multi-Scale Retinex model to achieve the integration of data distribution.
- Extensive comprehensive and elaborated experiments are conducted to illustrate our effectiveness and superiority. Notably, we provide a groundbreaking analysis in terms of the intermediate layers to account for their effects. This analysis is considerably valuable for better exploiting the network layers to more effectively handle this task.

2   Residual-Type Multi-scale Retinex

In this section, we described the well-known multi-scale Retinex model. Inspired by MSR, we proposed our residual-type multi-scale Retinex model.

2.1   Multi-scale Retinex Model

In the community of low-light image enhancement, Retinex theory [16] is a basic physical principle that depicts a deterministic contact between low-light inputs and the ideal outputs. This Retinex theory can be formulated as $\mathbf{Z} = \mathbf{X} \odot \mathbf{Y}$, where $\mathbf{Z}, \mathbf{X}, \mathbf{Y}$ represent the low-light input, the estimated illumination, and the estimated reflectance (i.e., the output), respectively. $\odot$ is the operation of pixel-wise multiplication.

Multi-scale Retinex [5] first converted the image-domain to log-domain for reducing the computational burden. The convolutional procedure by the Gaussian kernel was performed to generate the illumination in the log-domain. The final reflectance was obtained by summing the derived reflectances based on these illuminations. The formulation of MSR can be described as

$$\mathbf{y}_c = \sum_{n=1}^{N} w_n \left(\mathbf{z}_c - \mathbf{x}_c(\mathcal{G}_n) \right), c \in (r, g, b) \tag{1}$$

where N is the numbers of scales being used. w_n is the weighting parameter, which is empirically set as $\frac{1}{n}$. $\mathbf{y}_c = \log(\mathbf{Y}_c)$, $\mathbf{z}_c = \log(\mathbf{Z}_c)$, $\mathbf{x}_c(\mathcal{G}_n) = \log(\mathbf{Z}_c \otimes \mathcal{G}_n)$, where $\mathcal{G}_n$ is the n-th Gaussian filter. $\otimes$ represents the convolutional operation.

2.2   Residual-Type Multi-scale Retinex Model

Actually, Eq. (1) exists two distinct issues:

1) The illumination estimations in terms of different scales are irrelevant.
2) The Gaussian filter limits the solution space of the estimated illumination.

To address these issues, we design a new model by progressively removing the illumination in the log-domain. Our model can be formulated as

$$\mathbf{y}_c^{t+1} = \mathbf{y}_c^t - \mathbf{x}_c^t, \ \mathbf{x}_c^t = \mathbf{y}_c^t \otimes \mathcal{K}, \tag{2}$$

where t is the numbers of stages. $\mathbf{y}^0 = \mathbf{z}$ is the initialization. $\mathcal{K}$ is a certain filter, maybe Gaussian filter, maybe others.

In Eq. (2), we construct a cascade mechanism to fully utilize the information in each stage, rather than individually estimating the illumination in the work of multi-scale Retinex. In this way, we solve the first issue which is pointed out in the above. Aimed at the second issue, we utilize a general filter acting on the log-domain, to replace the Gaussian filter in the image-domain. Doing this avoids the domain transformation in the enhanced procedure, reducing the computational burden.

3   Residual-Type Multi-Scale Retinex

In our designed Residual-type Multi-Scale Retinex model (RMSR), there still exists an undefined factor, i.e., the definition of the filter. In the following, we expound the explicit relationship between our model with the residual network. Based on this relationship, we obtain a Learnable Multi-Scale Retinex to integrate the data distribution.

3.1   The Relationship Between RMSR and Residual Network

Residual network [17] is a well-known network architecture. In which, the residual connection is its key component for addressing the vanishing/exploding gradients. This connection considers learning residual function rather than unreferenced functions to outcome excellent performance. Fortunately, our designed RMSR model also contains the residual mechanism. More importantly, our model needs to execute the convolutional process. We can exactly utilize the convolutional block to perform it. Additionally, our mechanism of multi-stage precisely corresponds to the multiple residual layers. Until now, we have clearly described the relationship between RMSR and residual network.

3.2   Learning Multi-scale Retinex with Residual Network

As is described in the above subsection, we know that the relationship between RMSR and residual network. Based on this relationship, we build a Learnable Multi-Scale Retinex (LMSR) to integrate the data distribution.

In Fig. 2, we plotted the pipeline comparison between the traditional MSR and our proposed LMSR. It is easy to see that our scheme is more straight and clearer than MSR. More concretely, we can recognize that our scheme takes on transitive persistence so that we can optimize our filters in a stage-wise manner. Note that we used the convolutional block rather than the specific numbers of filters. The cause lies in that we can further improve the flexibility of our framework. We make an elaborated analysis in terms of this mechanism in Sect. 4.2.

In the training phase, we utilized the following loss function.

$$\mathcal{L} = \mathcal{L}_{MSE} + \mathcal{L}_{Perceptual}, \tag{3}$$

where $\mathcal{L}_{MSE}$ is the Mean Square Error loss, described as

$$\mathcal{L}_{MSE} = \|\mathbf{Y} - \mathbf{Y}_{gt}\|^2, \tag{4}$$

where $\mathbf{Y} = \exp(\mathbf{y}^T)$, T is the maximum number of the residual layer. $\mathbf{Y}_{gt}$ represents the ground truth of the latent clear image.

$\mathcal{L}_{Perceptual}$ is the perceptual loss, formulated as

$$\mathcal{L}_{Perceptual} = \frac{1}{W_{i,j}H_{i,j}} \sum_{x=1}^{W_{i,j}} \sum_{y=1}^{H_{i,j}} \left(\phi_{i,j}\left(\mathbf{Y}\right)_{x,y} - \phi_{i,j}\left(\mathbf{Y}_{gt}\right)_{x,y} \right), \tag{5}$$

where we set $i = 4$, $j = 2$. $W_{i,j}$ and $H_{i,j}$ are the dimensions of the extracted feature maps, respectively. $\phi_{i,j}$ is the VGG-based function for extracting feature.

4   Experimental Results

4.1   Implementation Details

We trained our network using 500 image pairs which are randomly sampled from MIT-Adobe 5 K benchmark [14]. We also tested the performance on 100 image pairs which are also randomly sampled from the same dataset.

The commonly-used Adam [18] solver was adopted to optimize the network. The mini-batch size was set as 4. The learning rate is fixed to 1e-4 in the whole training phase. For the other hyper-parameters, we used the default settings in Adam. All the network models were trained with the Tensorflow [19] package.

4.2   Evaluation on Three Versions of Our LMSR

Actually, since the flexibility of our LMSR, we can design it in each residual sub-module by using different numbers of basic layers. Here, we only considered three styles using three kernels with the size of 5, 3, 1. We named them as LMSR-v1 (with three convolutional blocks and three residual connections), LMSR-v2

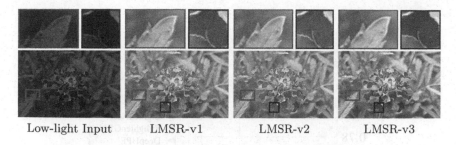

Low-light Input   LMSR-v1   LMSR-v2   LMSR-v3

Fig. 3. Visual comparison of three different versions of our LMSR.

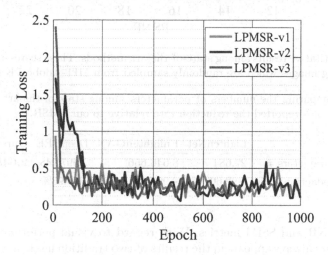

Fig. 4. Comparing the training procedures with regard to three versions of LMSR.

(with six convolutional blocks and three residual connections), LMSR-v3 (with six convolutional blocks and six residual connections).

In Fig. 4, we plotted the variations of their training loss. It is obvious that the LMSR-v3 has the fastest convergence speed (with about 780 steps). In terms of visual expression, as is shown in Fig. 3, we can observe that LMSR-v1 and v2 all generate some unknown artifacts to degrade the image quality, and the color saturation levels are not enough. Therefore we adopted LMSR-v3 as our method for evaluation and analysis in the following experiments.

4.3   Performance Evaluation

We compared our method with some existing state-of-the-art techniques including two representative model-based approaches (JIEP [10], RRM [11]), and three recently proposed data-driven methods (LightenNet [1], EnlightenGAN [2], DeepUPE [3]). Figures 5 and 6 reported the quantitative and qualitative comparison, respectively. It is noticeable that our LMSR obtains the best scores in

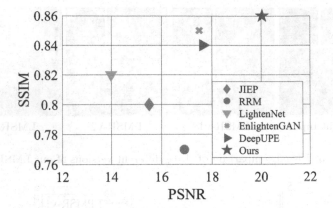

Fig. 5. Quantitative results among state-of-the-art methods. These scores are evaluated on 100 testing images which are randomly sampled from MIT-Adobe 5 K dataset.

Table 1. Comparing the numbers of parameters among state-of-the-art data-driven methods. We also reported the reduction rate relative to our LMSR.

	LightenNet	EnlightenGAN	DeepUPE	Ours
Parameters #	27,684	8,646,656	2,997,248	**2,048**
Reduction Rate	−92.60%	−99.98%	−99.93%	—

terms of PSNR and SSIM metrics. With regard to visual performance, insufficient exposure always appears in the results of two traditional schemes (i.e., JIEP and RRM). LightenNet brings about over-exposure and non-uniform illumination. The results of DeepUPE occurs the under-exposure and color distortion, especially the last row. In contrast, our LMSR achieves the most natural visual expression and competitive numeric scores.

As far as the execution efficiency, taking the difference in the operating platform of different methods into account, we compared the numbers of parameters, and the reduction rate relative to our LMSR. As is shown in Table 1, it is surprisingly that our LMSR just needs 2048 parameters. Our reduction rates compared to other works are conspicuous. This experiment fully indicates our practicability in real-world scenarios.

4.4　Effects of the Intermediate Layers

Because of our modeling procedure is transparent and deterministic, thus we can point out the effects of each layer in our designed LMSR. As is shown in Fig. 7, the training data indeed make the positive effects for this task. To be specific, we can easily see that intermediate illuminations present different colors and the reflectance gradually recovers the normal color. In terms of the exposure level, the reflectance of the first row is brighter and appears some details compared to the low-light input. The remaining intermediate reflectances nearly keep the

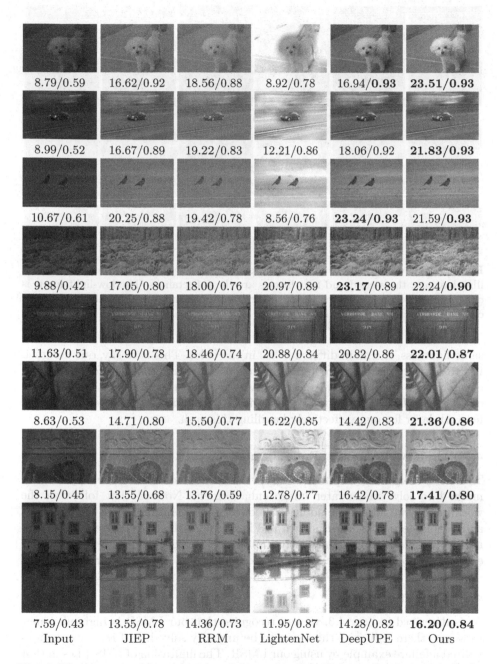

Fig. 6. Visual comparison with state-of-the-art low-light image enhancement methods. All these low-light images come from MIT-Adobe 5 K dataset [14]. PSNR/SSIM scores are reported below each image. The boldface represents the best scores. We can easily see that our LMSR achieves the competitive scores and the best visual performance.

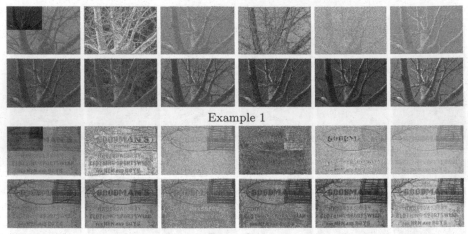

Example 1

Example 2

Fig. 7. Visual comparison of the immediate outputs. The first and third rows are the illuminations, the second and fourth rows are the reflectances. The low-light input is demonstrated in the top left corner of the first illumination layer. From the left to right column represent each stage in our LMSR.

same lightness, the main difference lies in the color. This is to say, our designed LMSR actually plays a more important role in correcting the color for the testing dataset. In a word, we have successfully recognized the effects of each layer in our designed LMSR. This observation will be helpful for us to better exploit the network layer for more effectively handling this task.

4.5   Underwater Image Enhancement

To further explore the potentiality of our proposed LMSR, we applied our method to solve underwater image enhancement. Noticed that following the work in [20], we added the additional converting operation acting on the input and output terminals. As is shown in Fig. 8, under the different real-world scenarios, our method all significantly improve the image quality. We will further exploit the potential application values of our LMSR in future works.

4.6   Limitations

As is described in Sect. 4.3, we can recognize that our LMSR is high-efficiency. However, there still exist the issues to be urgently solved. In Fig. 9, we demonstrated a failure example by using our LMSR. The limitation of LMSR lies in that over-exposure appearance always occurs in the sky (as is shown in red zoomed-in regions), although our LMSR can recover more details and improve the lightness in other places (as is shown in green and blue zoomed-in regions). The one possible reason is that our training loss is too simple so that the constraint force is not strong enough. We will ameliorate it in future work.

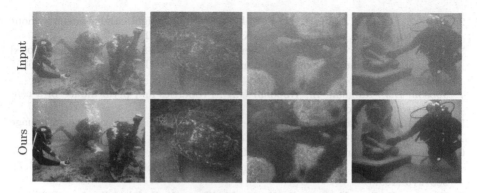

Fig. 8. Underwater image enhancement by using our proposed LMSR. Our method simultaneously corrects the color and recovers the details.

Input   Ours

Fig. 9. A failure example. Our method recovers the details and improves the exposure as is shown in the green and blue zoomed-in regions, but fails to keep the normal lightness in the sky as is shown in the red zoomed-in region. (Color figure online)

5 Concluding Remarks

This paper developed a novel and simple network for settling low-light image enhancement. We designed a residual-type multi-scale Retinex, then constructed a Learnable Multi-Scale Retinex to integrate the data distribution. Extensive analytical and comparative experiments verify our superiority.

In the future work, we will keep on exploring how to effectively exploit the network layers to handle low-light image enhancement in more complex and difficult real-world scenarios.

References

1. Li, C., Guo, J., Porikli, F., Pang, Y.: Lightennet: a convolutional neural network for weakly illuminated image enhancement. PRL **104**, 15–22 (2018)

2. Jiang, Y., Gong, X., Liu, D., et al.: Enlightengan: deep light enhancement without paired supervision. arXiv preprint arXiv:1906.06972 (2019)
3. Wang, R., Zhang, Q., Fu, C.-W., Shen, X., Zheng, W.-S., Jia, J.: Underexposed photo enhancement using deep illumination estimation. In: CVPR, pp. 6849–6857 (2019)
4. Jobson, D.J., Rahman, Z., Woodell, G.A.: Properties and performance of a center/surround retinex. IEEE TIP **6**(3), 451–462 (1997)
5. Rahman, Z., Jobson, D.J., Woodell, G.A.: Multi-scale retinex for color image enhancement. In: ICIP, pp.1003–1006 (1996)
6. Jobson, D.J., Rahman, Z., Woodell, G.A.: A multiscale retinex for bridging the gap between color images and the human observation of scenes. IEEE TIP **6**(7), 965–976 (1997)
7. Xueyang, F., Liao, Y., Zeng, D., Huang, Y., Zhang, X.-P., Ding, X.: A probabilistic method for image enhancement with simultaneous illumination and reflectance estimation. IEEE TIP **24**(12), 4965–4977 (2015)
8. Fu, X., Zeng, D., Huang, Y., Zhang, X.-P., Ding, X.: A weighted variational model for simultaneous reflectance and illumination estimation. In: CVPR, pp. 2782–2790 (2016)
9. Guo, X., Li, Y., Ling, H.: Lime: Low-light image enhancement via illumination map estimation. IEEE TIP **26**(2), 982–993 (2017)
10. Cai, B., Xu, X., Guo, K., Jia, K., Hu, B., Tao, D.: A joint intrinsic-extrinsic prior model for retinex. In: ICCV, pp. 4000–4009 (2017)
11. Li, M., Liu, J., Yang, W., Sun, X., Guo, Z.: Structure-revealing low-light image enhancement via robust retinex model. IEEE TIP **27**(6), 2828–2841 (2018)
12. Chen, W., Wang, W., Yang, W., Liu, J.: Deep retinex decomposition for low-light enhancement. In: BMVC, pp. 1–12 (2018)
13. Zhang, Y., Zhang, J., Guo, X.: Kindling the darkness: a practical low-light image enhancer. In: ACM MM, pp. 1632–1640 (2019)
14. Bychkovsky, V., Paris, S., Chan, E., Durand, F.: Learning photographic global tonal adjustment with a database of input/output image pairs. In: CVPR, pp. 97–104 (2011)
15. Mao, X., Li, Q., Xie, H., et al.: Least squares generative adversarial networks. In: ICCV, pp. 2794–2802 (2017)
16. McCann, J.: Retinex theory. Encyclopedia of Color Science and Technology, pp. 1118–1125 (2016)
17. He, K., Zhang, X., Ren, S., Sun, J.: Deep residual learning for image recognition. In: CVPR, pp. 770–778 (2016)
18. Kingma, D.P., Adam, J.B.: A method for stochastic optimization. In: ICLR, pp. 1–13 (2014)
19. Abadi, M., Barham, P., Chen, J., Chen, Z., et al.: Tensorflow: a system for large-scale machine learning. In: Symposium on Operating Systems Design and Implementation, pp. 265–283 (2016)
20. Galdran, A., Alvarez-Gila, A., Bria, A., Vazquez-Corral, J., Bertalmío, M.: On the duality between retinex and image dehazing. In: CVPR, pp. 8212–8221 (2018)

Enhanced Darknet53 Combine MLFPN Based Real-Time Defect Detection in Steel Surface

Xiao Yi, Yonghong Song[✉], and Yuanlin Zhang

Xi'an Jiaotong University, No.28, Xianning West Road, Xi'an, Shaanxi, China
1341451196@qq.com, {songyh,ylzhangxian}@mail.xjtu.edu.cn

Abstract. Real-time detection of wire surface defects is an important part of wire quality detection. Because traditional algorithms need lots of parameters and have weak universality, besides, time performance of detector based on candidate region is poor. To solve these problems, we study the effectiveness of single-stage detector in real-time detection, and propose a detection algorithm of wire surface defects combining enhanced darknet53 and feature pyramid (FPN). Firstly, we use CBAM_Darknet53 which introduces channel attention and spatial attention to extract more differentiated features. Secondly, considering the large change of defects, we use the multi-level feature pyramid (MLFPN) which adds the maximum pooling layer to fuse multi-level features to detect multi-scale defects. Then we reprocess the detector to improve the detection rate of defects and the accuracy of the detection box. Finally, network structure is optimized by modifying loss function. Experiments on defect datasets in real industrial environments show that recall and mAP of this method reach 94.49% and 88.46%, which is higher than state-of-the-art methods.

Keywords: Defect detection · Darknet53 · MLFPN · Attention

1 Introduction

The surface defect detection of wire is an important link in the quality detection. Traditional algorithms [1,2] need many parameters, they are not universal to defects and not sensitive to longitudinal defects with low contrast. Although the detector based on candidate region, represented by Faster RCNN [3] and mask RCNN [4] from r-cnn [5], has high detection performance, but time cannot meet the needs of actual production line. Single-stage detector such as YOLO [6–8] and SSD [9] is faster and more suitable for actual production line, and it has played a role in the industry in recent years. YOLO and SSD were applied to metal datasets with accuracy of 89.7% and 84.5% [10]. Attention-yolo [11] added attention to network, mAP reached 81.9% on Pascal VOC 2007. On the basis of SSD, Fu [12] used ResNet101 [13] and anti convolution to extract features and improve detection ability of small objects. Shen [14] proposed a stem block that can improve accuracy based on DenseNet [15], improved the situation that pre

© Springer Nature Switzerland AG 2020
Y. Peng et al. (Eds.): PRCV 2020, LNCS 12305, pp. 303–314, 2020.
https://doi.org/10.1007/978-3-030-60633-6_25

training weight needs to be loaded when training model. When the algorithm [16] based on SSD uses defect datas of the needle, accuracy of various types is good. The algorithm based on multi-scale SSD [17] introduces expanded convolution and attention residual module into SSD, it improves the effect of small objects.

Due to the variety of wire defects, we propose a method based on enhanced darknet53 and MLFPN. Production line is fast, so we select darknet53 as backbone network, spatial attention and channel attention are introduced to improve screening ability of key features without affecting time, then use improved MLFPN to fuse features to detect multi-scale defects. In the detection, design Min_iou to reduce overlapping boxes, and expand bounding box to ensure the integrity and accuracy of location. Finally, we use focal loss to distinguish planar defects those are difficult to distinguish. By testing the dataset under actual industrial production line, the detection performance of the method are proved.

The main contributions of this paper are summarized as follows. Firstly, we propose CBAM_Darknet53 with channel attention and spatial attention to screen more discriminative features without affecting time. Secondly, we use optimized MLFPN with maximum pooling layer to fuse multi-level features. Thirdly, we propose a detector post-processing strategy to ensure the accuracy of bounding box and filter overlapping boxes. Finally, we modify loss function strategy to optimize the network structure and ensure the accuracy of classification.

The rest of this paper is as follows. Sect. 2 review the research progress of attention and single-stage detector. Section 3 mainly introduce the detection algorithm of wire surface defects based on attention and MLFPN. The experimental evaluation is presented in Sect. 4 and this paper is concluded in Sect. 5.

2   Related Work

2.1   Attention Mechanism

Classical attention mechanism [18] includes position and item based attention, Squeeze and Excitation network (SENet) [19] and Convolutional Block Attention Module (CBAM) [20] are item based attention. SENet mainly includes squeeze and exception. Squeeze compresses the features into a real number sequence through global pooling. Exception is a dimension up and down operation based on a parameter, and a new sequence representing the importance of features is obtained. Finally, get a feature map which integrates the importance. CBAM proposes an attention module which combines spatial attention and channel attention. CBAM adds the maximum pooling layer to compute channel.

2.2   Single-Stage Detector

Single-stage detector classify and regress after extracting features, speed is fast. Due to the fast speed of production line, we analyze several classical detectors.

YOLO uses regression to realize detection, it has developed to YOLOv3. YOLOv3 includes darknet53 and YOLO layer, which are used to extract features and predict. In order to solve the gradient problem of deep network, darknet53 adds residual module composed of two convolution layers and one data link layer. YOLO layer adopts a structure similar to FPN for detecting more fine-grained features, and three feature maps of different scales are output for detection.

The network structure of SSD includes basic network and pyramid network. Only the first four layers of vgg16 are reserved in the basic network. After removing dropout layers, full connection layer is converted into the convolution layer, more feature maps are obtained by adding convolution layer for detection. Pyramid network is used to predict the classification and location from the feature map of different scales. SSD can get good detection effect in a small picture.

3 Method

3.1 Overall Architecture

Our network is composed of two parts: one is the backbone network, darknet53 with attention used to extract more differentiated features, the other is optimized MLFPN used to solve multi-scale problem. Its structure is shown in Fig. 1.

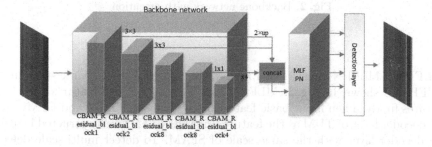

Fig. 1. The network structure.

In this paper, preprocessing is the first step. second, input image into backbone network to screen effective features, and use optimized MLFPN to fuse features, then detect defects. When bounding box is generated, expand boxes and use NMS to get results. The results are compared with the ground truth to get the accuracy of algorithm.

3.2 Detection Network Combining Attention and MLFPN

Attention Mechanism. Extracting excellent feature is important. We adopt darknet53 to extract features better and faster. Because shapes of defects are various, confidence of some indentation is low and edge is high, resulting in the omission and false recognition. We balance the confidence of feature map for filtering out effective feature map. Introduce attention in the feature extraction.

We test SENet and CBAM. Because the shortcut layer of darknet53 is used to aggregate multi-layer features. We add attention module to the shortcut layer to get good feature map. Experiments show CBAM is better, we add it to get CBAM_Darknet53. Backbone network with attention is shown in Fig. 2.

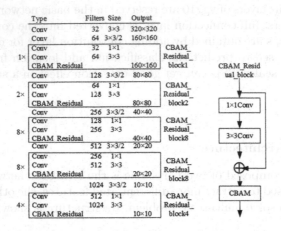

Fig. 2. Backbone network with attention.

MLFPN. MLFPN consists of FFM, TUM and SFAM. Network structure of MLFPN is shown in Fig. 3. MLFPN uses FFMv1 to fuse features into basic features firstly. Then input basic features into connected TUM and FFMv2, use the decoder layer of TUM as the feature. Finally, MLFPN is constructed by using the decoder layer with the same scale of SFAM. To detect multi-scale defects, when network extracts features, convolution layer5 is subsampling, convolution layer44 is up sampling and the layer10 is input into MLFPN to fuse features. And add the maximum pooling layer to SFAM.

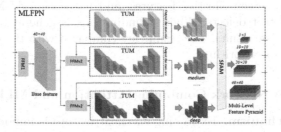

Fig. 3. The network structure of MLFPN.

LOSS Function. Loss function is the weighted sum of location loss (loss_l) and confidence loss (loss_c). Loss_l is smooth-l1 loss, loss_c is cross entropy loss. It is shown in formula (1). N is the number of boxes, α set as the weight of loss.

$$L(x, c, l, g) = 1/N(L_{conf}(x, c) + \alpha L_{loc}(x, l, g)) \tag{1}$$

Because loss_c is calculated for positive and negative samples, scab and indentation are similar, resulting in errors in positive and negative samples. In order to distinguish them, we add focal loss [21] to solve the problem of serious imbalance of positive and negative samples. Focal loss can alleviate the problem of sample imbalance. Due to the large difference between samples in training, it is necessary to control the weight of them. So focal loss expression of equilibrium is shown in formula (2).

$$FL(p_t) = -\alpha(1 - p_t)^y \log_(p_t) \tag{2}$$

Finally, loss function is shown in formula (3).

$$L(x, c, l, g) = 1/N(FL(x, c) + \alpha L_{loc}(x, l, g)) \tag{3}$$

3.3 Postprocessing of Detector

Min_iou. NMS is often used to filter the areas with the highest confidence in the post-processing. SoftNMS [22] only reduces confidence. We use softNMS to retain more boxes. However, because some indentation and scab is small, bounding box of some small defects and bounding box with the highest score have small IOU, resulting in the occurrence of overlapping boxes. In order to reduce overlapping boxes, we improved IOU by designing Min_iou, it is shown in formula (4).

$$Min\_iou = \frac{area(B_a \cap cB_b)}{\min area(B_a, B_b)} \tag{4}$$

Expand BoundingBox. Defect is various, planar defects are similar, as shown in Fig. 4. Red box is the ground truth, bounding box is green box, shape of bounding box in Fig. 4(a) and scab are similar, it is easy to cause a broad classification, low confidence and poor result. In Fig. 4(b), only part of the ground truth is detected, which may cause error detection and low accuracy. So the detection not only needs to detect the defects, but also needs box as accurate as possible.

According to the defect characteristics, we propose to predict a more complete box by expanding box. Because it is rectangular, we increase the width and height by a certain pixel to get a new box. Considering that each defect has its characteristics, we design different rules to get Algorithm 1.

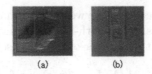

(a)   (b)

Fig. 4. Planar defects those are hard to distinguish. (Color figure online)

Algorithm 1: Algorithm description of expanding detection box

---

Input: 1024 * 1000 resolution wire gray image and detection boxes.
Output: expansion detection frame of each category.
for $c = 1$ to C do
  Choose the bbox with the highest score P_max;
  if $P\_max > 0.5 and\ Category = scratch$ **then**
  | extend 10 pixels vertically and 5 pixels horizontally, obtain P_max';
  end
  if $P\_max > 0.5 and\ Category = scab$ **then**
  | extend 5 pixels vertically and horizontally, obtain P_max';
  end
  if $P\_max > 0.5 and\ Category = indentation$ **then**
  | extend 10 pixels vertically and horizontally, obtain P_max';
  end
end

---

4   Experiment

Images are from actual industrial production line, we cut out wire areas to make dataset. The defects include scratch, scab and indentation, as shown in Fig. 5. Scratch is longitudinal, which runs through the whole surface of large-scale wire. Scab and indentation are planar defects. We use flip, translation and other methods to expand the data. See Table 1 for the division.

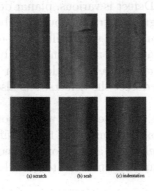

(a) scratch   (b) scab   (c) indentation

Fig. 5. Defect example.

Table 1. Division of datasets.

Dataset	Scratch	Scab	Indentation	Total
Training set	2097	3008	4709	9814
Test set	412	591	925	1928

Evaluation metrics including recall, precision and mean average precision (mAP) are employed to quantify the accuracy. The definitions are formula (5–6).

$$recall = tp/(tp + fn) \tag{5}$$

$$Precisiom = tp/(tp + fp) \tag{6}$$

Tp represents the number of image detected as defective correctly, fp represents the number of image detected as defective incorrectly, the object is not detected and fn is increased, the average precision for all classes can calculate the mAP.

The algorithm is implemented in Python under pytorch framework. The size of batch is 16, gamma is set to 0.1, and the initial value of learning rate is 0.004.

4.1 Module by Module Analysis

We explain the advantages of each module through six experiments. Table 2 lists the improvement of the model performance brought by the addition of modules.

Table 2. Module analysis.

Baseline	CBAM_Darknet53	MLFPN	Min_iou	Expandbox	Focalloss	Recall	Precision	mAP
√						78.23	58.74	60.56
√	√					79.89	69.26	71.28
√	√	√				80.06	71.29	73.60
√	√	√	√			85.47	85.29	82.22
√	√	√	√	√		90.11	86.29	85.33
√	√	√	√	√	√	94.49	85.25	88.46

In Table 2, baseline is the model of M2det [23] used in our dataset, it is a good one-stage detector based on SSD architecture. CBAM_Darknet53 increase mAP by 10.72%. The result shows attention can enhance important features, and confirm the advantage of full convolution network indirectly. When we modify MLFPN, mAP increases by 2.32%, which proves the superiority of MLFPN and the maximum pooling layer. Min_iou improve precision by 14%, which shows it can filter overlapping boxes. The expanded boxes can improve mAP by 3.11%, which shows that the method can further improve the accuracy. Finally, when loss_c is replaced by focal loss, recall is increased by 2.53%. It proves the superiority of focal loss in distinguishing difficult samples.

Impact of Backbone Network with Attention. Due to backbone network needs extract deep features well and time requirement of production line, full convolution network is suitable. we compare performance when resnet-50, resnet-101 and darknet53 are used in experiment. Through the experiment result we know that the performance of darknet53 are obviously better.

Due to the importance of feature map for subsequent processing. In Fig. 6, we introduce attention in the feature extraction, we compare performance and time of darknet53, with senet, with CBAM. From results, backbone network with attention can extract more effective features and improve the effect. Backbone network with CBAM is the best, and the maximum time of each image is about 60 ms, the time is allowed.

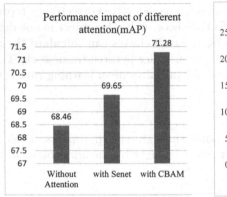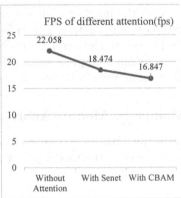

Fig. 6. Comprehensive comparison of different attention.

Influence of Optimizing Feature Pyramid. We take following experiments: Method1 inputs layer 5 and 10 features into MLFPN, method2 inputs layer 5, 10 and 44 features into MLFPN. Through experiments, method2 is better. On the basis of method2, Method3 changes average pooling layer of SFAM to the maximum pooling layer, method4 adds the maximum pooling layer in SFAM, and method5 adds spatial attention in SFAM. In Fig. 7, the results of each method for FPN optimization are compared.

From Fig. 7, we can see that mAP of method2 is increased by 0.68% compared with method1, indicating that fusion of features is conducive. The mAP of method3 is increased by 0.39%, indicating that the maximum pooling is effective. The mAP of method4 is further improved by 1.25%. It shows that maximum pooling and average pooling can improve the performance. The mAP of method5 increased by 0.59%, but FPS reduces to 11.186 fps, time consumption is large. So we add maximum pooling layer to SFAM, and then merge the results of average pooling layer and maximum pooling layer into the detection layer.

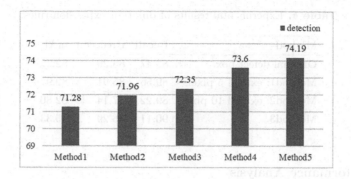

Fig. 7. Effect of FPN optimization methods on performance.

Impact of Min_iou. Softnms suppresses the confidence of the detection box according to IOU, and IOU is also one of the criteria to determine whether the detection box is correct or not, so the optimization of IOU is very important. We carry out the comparative experiments on IOU, Max_iou and Min_iou respectively. The experimental results are shown in Table 3. From the results, compared with IOU and Max_iou, precision and mAP of Min_iou are the best, which shows that Min_iou is effective and improves the detection performance.

Table 3. Impact of different IOU on detection.

Method of IOU	Recall	Precision	mAP
IOU	80.06	71.29	73.60
Max_iou	86.04	82.62	82.05
Max_iou	85.47	85.29	82.22

Impact of Expanding Box. We propose to predict more accurate defect box by expanding box. Method1 and method2 increase the width and height of box by 5 and 10 pixels respectively. Method3 designs different algorithm according to characteristics. We compare the results of different expansion rules in Table 4. We can see that compared with original method, mAP of method1 is increased by 0.68%. mAP of method2 is by 0.58%, which shows that the same extension method cannot be unified. The performance of method3 is the best, which shows that different defects need targeted optimization.

Impact of Loss Function Optimization. We compare the influence of smooth-l1 loss, confidence loss and focal loss. Experimental results show that increasing the weight of loss_c, recall and mAP are increased by 1.06% and 0.11%, increasing the weight of loss_c has certain effect. However, planar defects are difficult to distinguish, replace loss_c with focal loss. The weighted sum of smoothl1 loss and focal loss increases mAP to 88.46%.

Table 4. Experimental results of different expansion rules.

Method	Recall	Precision	mAP
Original defect box	85.47	85.29	82.22
Method1: extend 5 pixels	89.56	85.91	82.9
Method2: extend 10 pixels	89.22	87.14	82.80
Method3	90.11	86.29	85.33

4.2 Performance Analysis

In order to verify the superiority of our method, we replicate some state-of-the-art methods, and use the dataset for experiments. We analyze the performance between these methods [24,25] and our methods, results are shown in Table 5.

Table 5. Comparison results with state-of-the-art methods.

Method	Recall	Precision	mAP	FPS
YOLOv3	85.6	57.8	/	/
SSD	76.9	59.3	/	/
FSSD	81.36	65.45	67.79	22.727
RFB-SSD	73.48	61.33	61.68	20.610
Ours	94.49	85.25	88.46	16.178

Our method achieve better results compared with these methods, the slowest speed of single image meets the requirements of production line. In addition, performance of our method on different defects is listed in Table 6, which proves that our method has good detection effect on various types of defects.

Table 6. Comparison of performance on different categories of defects.

Classification	Recall	Precision	mAP
Scratch	96.33	97.87	90.75
Scab	92.62	84.23	87.52
Indentation	94.87	81.06	87.11
Dataset	94.49	85.25	88.46

In Fig. 8, we display the results of various defects under different algorithms, for scratches and special indentations, the result of our method is closer to the ground truth, and the detection box is better; for ordinary indentations and

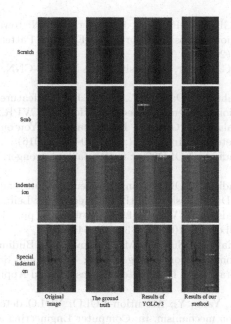

Fig. 8. Results of different algorithms on various defects

scabs, the result is good, but our method classification is more accurate. Compared with other methods, our method has better performance in the detection of wire surface defects.

5   Conclusion

Aiming at the surface defect detection of wire in industrial field, we design a single-stage detector combining enhanced darknet53 and MLFPN according to the characteristics of defects. Firstly, we add attention to extract effective features in backbone network, and use MLFPN with the maximum pooling layer to detect multi-scale defects, then we design Min_iou and expand bounding boxes to improve the accuracy of detection. Finally, we add focal loss to distinguish planar defects. In the experimental stage, we carried out modular experiments and compared state-of-the-art methods to prove the superiority of our method. Results show that our method has good performance.

References

1. Li, W., Lu, C., Zhang, J.C.: A lower envelope weber contrast detection algorithm for steel bar surface pit defects. Opt. Laser Technol. **45**, 654–659 (2013)
2. Li, C., Ji, D.: Research on surface defects detection of stainless steel spoon based on machine vision. In: Chinese Automation Congress, pp. 1096–1101 (2018)

3. Ren, S., He, K., Girshick, R., Sun, J.: Faster R-CNN: towards real-time object detection with region proposal networks. IEEE Trans. Pattern Anal. Mach. Intell. **39**(6), 1137–1149 (2017)

4. He, K., Gkioxari, G., Dollar, P., Girshick, R.: Mask R-CNN. In: ICCV, pp. 2980–2988 (2017)

5. Girshick, R., Donahue, J., Darrell, T., Malik, J.: Rich feature hierarchies for accurate object detection and semantic segmentation. In: CVPR, pp. 580–587 (2014)

6. Redmon, J., Divvala, S.K., Girshick, R., Farhadi, A.: You only look once: unified, real-time object detection. In: CVPR, pp. 779–788 (2016)

7. Redmon, J., Farhadi, A.: YOLO9000: better, faster, stronger. In: CVPR, pp. 6517–6525 (2017)

8. Redmon, J., Farhadi, A.: YOLOv3: an incremental improvement. In: CVPR (2018)

9. Liu, W., et al.: SSD: single shot multiBox detector. In: Leibe, B., Matas, J., Sebe, N., Welling, M. (eds.) ECCV 2016. LNCS, vol. 9905, pp. 21–37. Springer, Cham (2016). https://doi.org/10.1007/978-3-319-46448-0_2

10. Posilovic, L., Medak, D., Subasic, M., Petkovic, T., Budimir, M., Loncaric, S.: Flaw detection from ultrasonic images using YOLO and SSD. In: International Symposium on Parallel and Distributed Processing and Applications, pp. 163–168 (2019)

11. Xu, C., Wang, X., Yang, Y.: Attention-YOLO: YOLO detection algorithm that introduces attention mechanism. In: Computer Engineering and Applications, pp. 13–23 (2019)

12. Fu, C., Wei, L., Ananth, R., Ambrish, T., Berg, A.C.: DSSD: Deconvolutional single shot detector. In: CVPR (2017)

13. He, K., Zhang, X., Ren, S., Sun, J.: Deep residual learning for image recognition. In: CVPR, pp. 770–778 (2016)

14. Shen, Z., Liu, Z., Li, J., Jiang, Y., Chen, Y., Xue, X.: DSOD: learning deeply supervised object detectors from scratch. In: ICCV, pp. 1937–1945 (2017)

15. Huang, G., Liu, Z., Van Der Maaten, L., Weinberger, K.Q.: Densely connected convolutional networks. In: CVPR, pp. 2261–2269 (2017)

16. Yang, J., Li, S., Wang, Z., Yang, G.: Real-time tiny part defect detection system in manufacturing using deep learning. IEEE Access **7**, 89278–89291 (2019)

17. Yang, L., Wang, Z., Gao, S.: Pipeline magnetic flux leakage image detection algorithm based on multiscale ssd network. IEEE Trans. Ind. Inf. **16**(1), 501–509 (2020)

18. Borji, A., Itti, L.: State-of-the-art in visual attention modeling. IEEE Trans. Pattern Anal. Mach. Intell. **35**(1), 185–207 (2013)

19. Hu, J., Shen, L., Sun, G.: Squeeze-and-excitation networks.In: CVPR (2017)

20. Woo, S., Park, Lee, J., So Kweon, I.: CBAM: convolutional block attention module. In: ECCV, pp. 3–19 (2018)

21. Lin, T., Goyal, P., Girshick, R., He, K., Dollar, P.: Focal loss for dense object detection. In: ICCV, pp. 2999–3007 (2017)

22. Bodla, N., Singh, B., Chellappa, R., Davis, L.S.: Soft-NMS - improving object detection with one line of code. In: ICCV, pp. 5562–5570 (2017)

23. Zhao, Q., et al.: M2Det: a single-shot object detector based on multi-level feature pyramid network. AAAI **33**(01), 9259–9266 (2019)

24. Liu, S., Huang, D., Wang, Y.: Receptive field block net for accurate and fast object detection. In: CVPR (2017)

25. Li, Z., Zhou, F.: FSSD: feature fusion single shot multibox detector. In: CVPR (2017)

D-CrossLinkNet for Automatic Road Extraction from Aerial Imagery

Kang Huang⬤, Jun Shi⬤, Gaofeng Zhang⬤, Benzhu Xu⬤,
and Liping Zheng(✉)⬤

School of Computer Science and Information Engineering,
Hefei University of Technology, Hefei 230009, China
huangkang4321@sina.com, zhenglp@hfut.edu.cn
http://ci.hfut.edu.cn/

Abstract. Road extraction is of great importance to remote sensing image analysis. Compare to traditional road extraction methods which rely on handcrafted features, Convolutional Neural Network (CNN) uses deep hierarchical structure for automatically learning features and has superior performance of road extraction. In this paper, we propose a novel encoder-decoder architecture D-CrossLinkNet for extracting roads from aerial images. It consists of LinkNet, cross-resolution connections, and two dilated convolution blocks. As one of the representative semantic segmentation models, LinkNet uses the encoder-decoder architecture with skip connections for road extraction. However, the downsampling operation of LinkNet reduces the resolution of the feature maps, leading to the loss of spatial information. Therefore, we use the cross-resolution connections to supplement the spatial information to the decoder. Besides, the dilated-convolution blocks are introduced to increase receptive field of feature points while maintaining short-distance spatial feature information. Experimental results on two benchmark datasets (Beijing and Shanghai, DeepGlobe) demonstrate the proposed method has better performance than other CNN-based methods (e.g. DeepLab, LinkNet and D-LinkNet).

Keywords: Road extraction · Cross-resolution · Deep learning · Semantic segmentation

1 Introduction

Extracting roads from aerial images is a popular research topic in many fields such as geographic information update, automatic road navigation, disaster

This work was supported in part by the National Natural Science Foundation of China (No. 61972128) and the Youth Fund of National Natural Science Foundation of China (No. 61702155, 61906058) and Anhui Natural Science Foundation (No. 1808085MF176, 1908085MF210) and the Fundamental Research Funds for the Central Universities of China (Grant No. PA2019GDPK0071).

Y. Peng et al. (Eds.): PRCV 2020, LNCS 12305, pp. 315–327, 2020.
https://doi.org/10.1007/978-3-030-60633-6_26

response, digital mapping, etc. For example, it is time-consuming and laborious for manually making maps, for the reason that the accompanying map updates are very huge in volume since the city road changes continuously with the development of the society. If the road network can be automatically extracted from high-resolution aerial images, it will greatly reduce the workload and improve the efficiency of map generation. For another example, when a natural disaster occurs, the lack of available route information to navigate vehicles into the disaster area will hinder the rescue work, otherwise, with rapid generated route information, rescuers can ensure the efficiency and safety of the missions.

Owing to the development of Convolutional Neural Networks (CNN), it has achieved state-of-the-art performance on image segmentation tasks [1,2]. As one of the representative CNN model for semantic segmentation, LinkNet [3] is an efficient network which takes advantage of skip connections and encoder-decoder architecture. It has obtained excellent performance on CamVid [4] dataset, but it still has limitations when applied to road extraction tasks. One common drawback for the encoder-decoder network architecture is that the downsampling operation leads to the loss of spatial information which would affect segmentation accuracy. D-LinkNet[5] is a network model specifically designed for road extraction tasks. It uses a dilated convolution layer to increase the receptive field of feature points. Larger receptive field is conducive to learning long-distance spatial features. One shortage of D-LinkNet is that the dilated convolution layer contains a lot of dilated convolutions. Such multiple dilated convolutions will result in the loss of short-distance spatial feature details. To address the problems of LinkNet and D-LinkNet, we propose a novel approach, namely D-CrossLinkNet for road extraction. The main contributions of our work are as follows:

1. We introduce cross-resolution connections, inspired by the success of using fully-connected multi-branch convolutions for labeling pixels and regions tasks [6]. D-CrossLinkNet takes advatage of this idea by adding cross-resolution connections in the center. LinkNet sets skip connections between the encoder and the decoder, which feeds data to the decoder aim at recovering lost spatial information. D-CrossLinkNet reinforces this operation by utilizing crossresolution connections, instead of skip connections, to feed the multi-scale data into the decoder.

2. We optimize the dilated convolution layer of D-LinkNet by introducing dilated-convolution blocks. Considering the complexity, interlaced connectivity, and wide range of roads, D-LinkNet uses dilated convolution layer to deal with these problems. The layer contains multiple dilated convolutions, which increases the receptive field of feature points. However, multiple dilated convolutions also resulting in the loss of short-distance spatial feature information. Thus, we modify and extend dilated convolution layer by adding skip connections among dilated convolutions to supplement short-distance spatial feature information. Besides, we use a pair of dilated convolution blocks and make it into a parallel form for better accuracy against D-LinkNet.

The rest of this paper is organized as follows. Section 2 introduces the related works. Section 3 describes the network architecture of the proposed

D-CrossLinkNet. Section 4 demonstrates the experimental results and discussions. Section 5 presents our conclusion.

2   Related Works

The traditional methods of road extraction generally use image processing technique to extract the specific features of roads [7,8] (e.g. shape, texture and hybrid spectral–structural features) and then determine whether it is the road. One common drawback of these methods is their excessive reliance on feature selection.

Recent studies is trying to improve the accuracy of road extraction and similar semantic segmentation tasks. Deep learning based approaches, especially with pixel-level segmentation methods [9–11] have proven to be more effective than their traditional counterparts on a good deal of computer vision tasks.

DeepLab(v3+) [12] is a general-purpose segmentation architecture that has shown to be effective in wide variety of segmentation tasks (e.g., full scene segmentation in Cityscapes and foreground-background segmentation in PASCAL VOC 2012). It uses DeepLab (v3) [13] as the encoder and extends DeepLab (v3) by adding a decoder. DeepLab (v3+) applied Xception [14] model and depthwise separable convolution to both Atrous Spatial Pyramid Pooling and decoder. We apply DeepLab (v3+) to road extraction tasks and compare the segmentation performance with D-CrossLinkNet.

The fully-connected multi-branch convolutions for labeling pixels and regions are used for semantic segmentation tasks, and [5] proposed a network model with parallel low-resolution and high-resolution connection, it has multibranch convolutions in the network, the method refreshed three records in the COCO [15] dataset with state-of-the-art results.

A novel method [16] is developed to improve road extraction results. It utilizes vehicle GPS data from Beijing and Shanghai for data augmentation, which is to supplement the imagery data with GPS data. The method is tested on their Beijing and Shanghai dataset, and the results show that their method improves the segmentation performance of many models.

LinkNet is an efficient network model based on the encoder-decoder architecture, which is designed for semantic segmentation. It applies skip connections and residual blocks. LinkNet achieves state-of-the-art results on CamVid. Also, it gives comparable results on Cityscapes [17]. However, the drawback of LinkNet are also obvious. The downsampling operation will cause the spatial information of the feature map to be lost.

On the other hand, D-LinkNet uses LinkNet as its backbone. It applies dilated convolution layers to enlarge the receptive field of feature points and ensemble multi-scale features to keep the feature information for road extraction. In the DeepGlobe Road Extraction Challenge 2018 [18], D-LinkNet won the first place. But the dilated convolution layer of D-LinkNet has multiple dilated convolutions. One shortage for the D-LinkNet is that although the layer increases the receptive field of feature points, the pixels that are not involved in the computation will also cause the loss of short-distance spatial features as the number of dilated convolutions increases.

To deal with these issues and improve the performance of road extraction, we propose cross-resolution connections for D-CrossLinkNet by combining the ideas of skip connection and multi-branch convolutions. Cross-resolution connections provide different levels of spatial information to the decoder and assist the decoder in upsampling operations. Furthermore, we modify the dilated convolution layer of D-LinkNet and apply it to D-CrossLinkNet. We improved the dilated convolution layer by adding skip connections among dilated convolutions, which will supplement short-distance information between dilated convolutions. Also, in order to take advantage of different levels of feature information, We use dilated convolutions blocks in two levels to form a parallel structure.

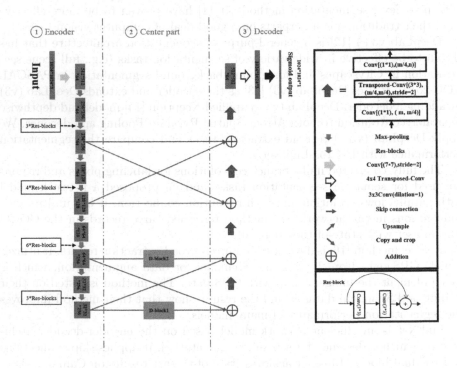

Fig. 1. D-CrossLinkNet model architecture, Part 1 is the encoder, Part 2 is the Center part, Part 3 is the decoder.

3   D-CrossLinkNet

3.1   General Architecture

Our proposed D-CrossLinket is divided into three parts, which are encoder, center part, and decoder, as shown in Fig. 1. Since the two datasets we use are aerial

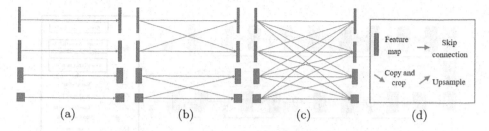

Fig. 2. (a) is commonly used in most models, we use (b) in D-CrossLinkNet, (c) is fully connected architecture, (d) is the description.

imageries with a resolution of 1024 * 1024, D-CrossLinkNet takes an image with resolution of 1024 * 1024 as input. The inputs goes through a series of encoders and pooling layers until the resolution of the feature map is 32 * 32. In the center part, we combine the cross-resolution connections and dilated convolution blocks(D-block1 and D-block2). Feature maps processed by Center part are sent to different decoders. In the decoder part, a series of upsampling restores the feature map resolution, and finally generates the binary image, namely the classification results of road extraction.

3.2   Encoder

In this paper, we use ResNet34 pretrained on ImageNet as encoder, same as D-LinkNet.

3.3   Center Part

In the center part, we add cross-resolution connections and our proposed dilated-convolution blocks.

Cross-Resolution Connections. Spatial information is important for semantic segmentation tasks. The encoder part of the network has reduced the feature map resolution through each pooling layer. Skip connection in LinkNet bypasses the input of each encoder layer to the corresponding decoder. The operation can bring in the features of lower convolution layers which contain rich low-level spatial information. However, for road extraction tasks, it is challenging for models with skip connections to obtain accurate road network segmentation results due to the complexity, the interlaced connectivity, and long span of the roads. Therefore, we introduce cross-resolution connections into our method. As shown in Fig. 2, there are three different types of connections. LinkNet, Unet [19] and FCN [20] use the form shown in (a). As for the form shown in (c), fully-connected connections will greatly increase the amount of computations and GPU memory requirements, hence we apply form (b) in our network model, unlike skip connection, we add connections between feature maps of different resolutions to make

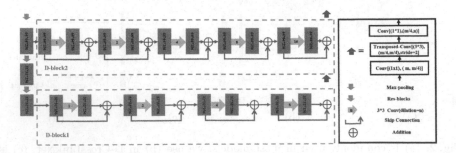

Fig. 3. Dilation blocks (D-block1 and D-block2) in D-CrossLinkNet.

information exchange between encoders and decoders at different levels. As a consequence, each decoder is provided with multi-scale of spatial information.

Dilated-Convolution Blocks. Many roads are usually connected to each other, so a wrong segmentation in one place may cause a chain of incorrect results. Thus, when training a network model, it is necessary to enlarge the receptive fields of feature points. Dilated convolution is widely used in semantic segmentation tasks, it is used to adjust receptive fields of feature points without reducing the resolution of feature maps.

The dilated-convolution blocks we proposed is shown in Fig. 3. The receptive field is increased after each dilated convolution which is useful for learning long-distance spatial features. However, multiple dilated convolutions unavoidably accompanies the loss of short-distance spatial information details. It is mainly reflected in the series of dilated convolutions will permanently lose pixels that are not involved in the computation due to the "holes" we filled in convolution kernel. Therefore, what we perform is to add the original data and the result of original data after dilated convolution, then send the result of the addition to the next dilated convolution. One such operation is used before each dilated convolution to supplement the short-distance spatial information. Same as D-LinkNet, the D-block1 use dilation rates of 1, 2, 4, 8. So the feature points on the last layer will see 31 * 31 points on the first feature map.

In order to enhance the effect of the dilated-convolution block and make better use of feature information at different levels, we do not make the dilated-convolution block into the same parallel structure as D-LinkNet. D-CrossLinkNet put a D-block2 between encoder 3 and decoder 3, the principle is the same as D-block1. But because the resolution of the feature map in this decoder is 64 * 64, we use dilation rates of 1, 2, 4, 8, 16. So the size of 63 * 63 points on the first feature map can be seen. The output of D-block1 is combined with the result of D-blcok2 through one upsamle operation. D-block1 and D-blcok2 combine form a parallel structure at different levels.

3.4 Decoder

The decoder layer of D-CrossLinkNet is kept the same design as the original LinkNet, The decoders restore the image from resolution of 32 * 32 to 1024 * 1024. Finally, a 3 * 3 convolution with a dilation of 1 is used to generate a 1024 * 1024 binary image as output.

4 Experiment

In this section, we demonstrates our experiments as follow. Part 1 presents implementation details about our experiments. Part 2 introduces the dataset used in the experiments. Part 3 introduces the performance metrics we used such as IoU (Intersection over Union) [21], Recall [22] and PA (Pixel Accuracy) [23] scores. At last, Part 4 presents the experimental results and discussions.

4.1 Implementation Details

D-CrossLinkNet is implemented by pytorch. The experiment is done using single GPU which is RTX 2080 TI with 11GB memory. The batch size of all models during training is selected as 1. The default learning rate is set to 2e-4. Same as D-LinkNet, we use BCE (binary cross entropy) and dice coefficient loss as loss function and use Adam [24] as optimizer. D-CrossLinkNet takes about 60 epochs to converge.

4.2 Datasets

In this paper, two benchmark datasets are used to evaluate the performance of D-CrossLinkNet, as shown in Fig. 4. The first one is Beijing and Shanghai dataset [16]. It is extracted from the Gaode map and manually labeled GroundTruth, which include 350 Beijing maps and 50 Shanghai maps. The samples of the Beijing and Shanghai dataset are shown in Fig. 4(a). The other dataset is Deep-Globe dataset from the DeepGlobe Road Extraction Challenge 2018 [18]. Since the organizer only provides the training set, we extract 500 images with masks from the training set for training, and 200 images with masks for testing. The samples of DeepGlobe dataset are shown in Fig. 4(b). In the experiment, we did not do data augmentation on these datasets.

4.3 Evaluation Metrics

IoU (Intersection over Union), Recall and PA (Pixel Accuracy) are commonly used performance metrics in semantic segmentation tasks. IoU is the most commonly used by researchers in road network extraction task [25,26]. In our experiments, TP (True Positive) is the number of road pixels that are correctly predicted. TN (True Negative) is the number of non-road pixels that are correctly predicted, FP (False Positive) is the number of road pixels that are incorrectly

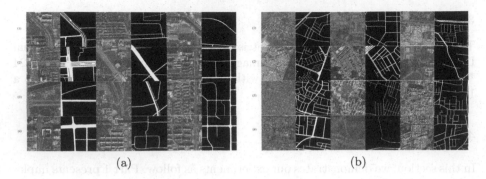

(a)   (b)

Fig. 4. (a) samples of Beijing and Shanghai dataset, (b) samples of DeepGlobe dataset.

predicted, and FN (False Negative) is the number of non-road pixels that are incorrectly predicted. IoU, Recall and Pixel Accuracy are then defined as:

$$IoU = \frac{TP}{TP + FN + FP} \tag{1}$$

$$Recall = \frac{TP}{TP + FN} \tag{2}$$

$$PA = \frac{TP + TN}{TP + TN + FP + FN} \tag{3}$$

4.4 Experiment Results

We have trained models on the Beijing and Shanghai dataset, DeepGlobe dataset, then calculated the IoU score in their testset, and the results are shown in Table 1. The IoU scores of DeepLab (v3+), LinkNet and D-LinkNet in Beijing and shanghai dataset are come from [14].

From Table 1 we can find that the IoU scores of D-CrossLinkNet in two datasets have reached 57.48 and 58.92 respectively, which are higher than other models, that is, the prediction results are closer to GroundTruth. In addition, it also can be seen from the results that all models scores in the Beijing and Shanghai datasets are lower than its scores in the DeepGlobe dataset, because there are more training samples in the DeepGlobe dataset.

Then, we calculate the Recall and PA scores in two datasets, and the results are shown in Table 1 as well. It can be seen that D-CrossLinkNet performs superiorly in Beijing and Shanghai dataset. Both Recall and PA scores are higher than other models. D-CrossLinkNet is more significant in Recall scores and slightly higher in PA scores.

Table 1. Segmentation evaluations

Models	Evaluation metrics	Beijing Shanghai dataset	DeepGlobe dataset
DeepLab(v3+)	IoU(%)	43.40	43.62
	Recall(%)	57.93	56.13
	PA(%)	92.51	92.15
LinkNet	IoU(%)	53.96	57.83
	Recall(%)	68.62	68.81
	PA(%)	93.53	94.60
D-LinkNet	IoU(%)	54.42	58.20
	Recall(%)	68.47	69.87
	PA(%)	93.47	94.58
D-CrossLinkNet	IoU(%)	**57.48**	**58.92**
	Recall(%)	**71.51**	**75.58**
	PA(%)	**93.67**	94.32

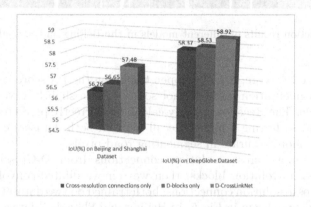

Fig. 5. Results of the Ablation Experiment.

In order to determine the contribution of cross-resolution connections and dilated-convolution blocks to the results, we have done a ablation experiment in both two datasets.

But when it comes to the DeepGlobe dataset, it can be seen that D-CrossLinkNet still performs well in the Recall scores, but the PA scores is 94.32, which is slightly lower than D-LinkNet's 94.58 and LinkNet's 94.60. We analyze the situation like this, the calculation method of the PA is to count the correctly predicted road pixels and correctly predicted non-road pixels into the result, but in the aerial image, the number of road pixels is much less than the nonroad pixels, which means the proportion of non-road pixels is much higher than road pixels, in this case, the accuracy of non-road segmentation will seri-

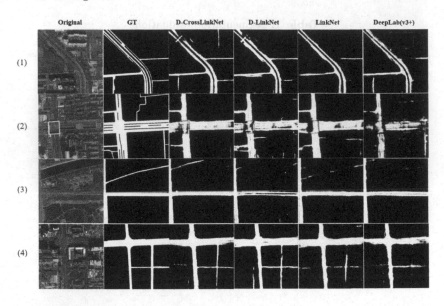

Fig. 6. Prediction results of different models in the Beijing and Shanghai dataset.

ously affect the PA score, it may happen that even the PA score is high but the actual road segmentation results is not so good, so a slightly lower PA score is understandable. The Recall score only calculates correctly predicted road pixels into the result, so having a higher Recall score in both datasets can show that our method is more accurate in segmenting road pixels.

We first remove cross-resolution connections from D-CrossLinkNet and reserve dilated-convolution blocks, then we remove dilated-convolution blocks and reserve crossresolution connections, both trained on two datasets. The experimental results are shown in Fig. 5. In Beijing and Shanghai dataset, The scores of the model with crossresolution connections and the model with D-blocks are 56.26 and 56.65 respectively, and the combined model score reaches 57.48. In DeepGlobe dataset, The scores of two models are 58.37 and 58.53 respectively, and the combined model score reached 58.92. It can be concluded that in our method, the contribution of D-blocks to the segmentation results is greater than cross-resolution connections.

We show the segmentation results of the Beijing and Shanghai dataset in Fig. 6, and the results of DeepGlobe dataset are shown in Fig. 7.

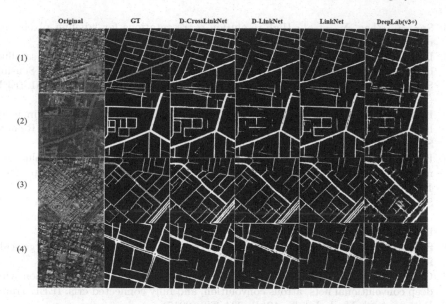

Fig. 7. Prediction results of different models in the DeepGlobe dataset.

5 Conclusion and Future Work

In this paper, we propose a method for road extraction based on the encoder-decoder structure, namely D-CrossLinkNet. By adding cross-resolution connections, the encoder part provides more image information to the decoder. By adding two dilated convolution blocks we propose to keep the short-distance feature information while increasing the receptive field of the network convolution kernel. We train and test our method on two datasets, and the results show the superiority of our method. In addition, since the distribution of the biological cell walls and the roads in the image has similarities, so we believe that our method is promising for such segmentation tasks.

References

1. Roth, H.R., Shen, C., et al.: Deep learning and its application to medical image segmentation. Med. Imaging Technol. **36**(2), 63–71 (2018)
2. Litjens, G., Kooi, T., et al.: A survey on deep learning in medical image analysis. Med. Image Anal. **42**, 60–88 (2017)
3. Chaurasia, A., Culurciello. E.: Linknet: exploiting encoder representations for efficient semantic segmentation. In: 2017 IEEE Visual Communications and Image Processing (VCIP), pp. 1–4. IEEE (2017)
4. Brostow, G.J., Fauqueur, J., et al.: Semantic object classes in video: a high-definition ground truth database. Pattern Recogn. Lett. **30**(2), 88–97 (2009)
5. Zhou, L., Zhang,C., et al.: D-linknet: linknet with pretrained encoder and dilated convolution for high resolution satellite imagery road extraction. In: CVPR Workshops, pp. 182–186 (2018)

6. Sun, K., Zhao, Y., et al.: High-resolution representations for labeling pixels and regions (2019). arXiv preprint arXiv:190404514
7. Li, X., Qiao, Y., et al.: The research of road extraction for high resolution satellite image. In: IGARSS 2003, 2003 IEEE International Geoscience and Remote Sensing Symposium. Proceedings (IEEE Cat. No. 03CH37477), vol. 6, pp. 3949–3951. IEEE (2003)
8. Huang, X., Zhang, L.: Road centreline extraction from high-resolution imagery based on multiscale structural features and support vector machines. Int. J. Remote Sens. **30**(8), 1977–1987 (2009)
9. Krizhevsky, A., Sutskever, I., et al.: Imagenet classification with deep convolutional neural networks. In: Advances in Neural Information Processing Systems, pp. 1097–1105 (2012)
10. Szegedy, C., Ioffe, S., et al.: Inception-v4, inception-resnet and the impact of residual connections on learning. In: Thirty-First AAAI Conference on Artificial Intelligence (2017)
11. Simonyan, K., Zisserman, A.: Very deep convolutional networks for large-scale image recognition (2014). arXiv preprint arXiv:14091556
12. Chen, L.-C., Papandreou, G., et al.: Deeplab: semantic image segmentation with deep convolutional nets, atrous convolution, and fully connected crfs. IEEE Trans. Pattern Anal. Mach. Intelli. **40**(4), 834–848 (2017)
13. Oh, H., Lee, M., et al.: Metadata extraction using deeplab v3 and probabilistic latent semantic analysis for intelligent visual surveillance systems. In: 2020 IEEE International Conference on Consumer Electronics (ICCE), pp. 1–2. IEEE (2020)
14. Chollet, F.: Xception: deep learning with depthwise separable convolutions. In: Proceedings of the IEEE Conference on Computer Vision and Pattern Recognition, pp. 1251–1258 (2017)
15. Lin, T.-Y., et al.: Microsoft COCO: common objects in context. In: Fleet, D., Pajdla, T., Schiele, B., Tuytelaars, T. (eds.) ECCV 2014. LNCS, vol. 8693, pp. 740–755. Springer, Cham (2014). https://doi.org/10.1007/978-3-319-10602-1_48
16. Sun,T., Di, Z., et al.: Leveraging crowdsourced GPS data for road extraction from aerial imagery. IN: Proceedings of the IEEE Conference on Computer Vision and Pattern Recognition, pp. 7509–7518 (2019)
17. Cordts, M., Omran, M., et al.: The cityscapes dataset for semantic urban scene understanding. In: Proceedings of the IEEE Conference on Computer Vision and Pattern Recognition, pp. 3213–3223 (2016)
18. Demir, I., Koperski,K., et al.: Deepglobe 2018: a challenge to parse the earth through satellite images. In: 2018 IEEE/CVF Conference on Computer Vision and Pattern Recognition Workshops (CVPRW), pp. 172–17209. IEEE (2018)
19. Ronneberger, O., Fischer, P., Brox, T.: U-Net: convolutional networks for biomedical image segmentation. In: Navab, N., Hornegger, J., Wells, W.M., Frangi, A.F. (eds.) MICCAI 2015. LNCS, vol. 9351, pp. 234–241. Springer, Cham (2015). https://doi.org/10.1007/978-3-319-24574-4_28
20. Long, J., Shelhamer, E., et al.: Fully convolutional networks for semantic segmentation. In: Proceedings of the IEEE Conference on Computer Vision and Pattern Recognition, pp. 3431–3440 (2015)
21. Berman, M., Rannen Triki, A., et al.: The lovász-softmax loss: a tractable surrogate for the optimization of the intersection-over-union measure in neural networks. In: Proceedings of the IEEE Conference on Computer Vision and Pattern Recognition, pp. 4413–4421 (2018)

22. Zhang, X., Feng, X., et al.: Segmentation quality evaluation using region-based precision and recall measures for remote sensing images. ISPRS J. Photogramm. Remote Sens. **102**, 73–84 (2015)
23. Wang, W., Yang, J., et al.: Modeling background and segmenting moving objects from compressed video. IEEE Trans. Circ. Syst. Video Technol. **18**(5), 670–681 (2008)
24. Chen, L.-C., Zhu, Y., et al.: Encoder-decoder with atrous separable convolution for semantic image segmentation. In: Proceedings of the European Conference on Computer Vision (ECCV), pp. 801–818 (2018)
25. Buslaev, A., Seferbekov, S.S., et al.: Fully convolutional network for automatic road extraction from satellite imagery. In: CVPR Workshops, vol. 207, p. 210 (2018)
26. Doshi, J.: Residual inception skip network for binary segmentation. In: CVPR Workshops, pp. 216–219 (2018)

PUI-Net: A Point Cloud Upsampling and Inpainting Network

Yifan Zhao[1,2], Jin Xie[1,2(✉)], Jianjun Qian[1,2], and Jian Yang[1,2]

[1] PCA Lab, Key Lab of Intelligent Perception and Systems for High-Dimensional Information of Ministry of Education, Nanjing University of Science and Technology, Nanjing, China
{yifanzhao,csjxie,csjqian,csjyang}@njust.edu.cn

[2] Jiangsu Key Lab of Image and Video Understanding for Social Security, School of Computer Science and Engineering, Nanjing University of Science and Technology, Nanjing, China

Abstract. Upsampling and inpainting sparse and incomplete point clouds is a challenging task. Existing methods mainly focus on upsampling low-resolution point clouds or inpainting incomplete point clouds. In this paper, we propose a unified framework for upsampling and inpainting point clouds simultaneously. Specifically, we develop a point cloud upsampling and inpainting network called PUI-Net, which consists of the attention convolution unit and non-local feature expansion unit. In the attention convolution unit, the channel attention mechanism is adopted to extract discriminative features of point clouds. With the extracted discriminative features, we employ the non-local feature expansion unit to generate dense feature maps of high-resolution point clouds. Furthermore, we formulate a novel inpainting loss to train the PUI-Net so that the holes in the incomplete clouds can be completed. Quantitative and qualitative comparisons demonstrate that our proposed PUI-Net can yield good performance in terms of the upsampling and inpainting tasks on the incomplete point clouds.

Keywords: 3D point cloud · Deep learning · Computer vision

1 Introduction

Point clouds, as one of 3D data representations, are widely used in different fields such as robotics, virtual reality and remote sensing. With the development of 3D imaging technology, in recent years, point clouds have been receiving more and more attention. However, due to the limitations of 3D scanning sensors, raw point clouds are often sparse and incomplete with noises and holes, which seriously affects the point cloud analysis and processing. The goal of the point cloud upsampling and inpainting task is to generate the high-quality dense point clouds from the low-resolution incomplete point clouds.

This work was supported by the National Science Fund of China (Grant No. 61876084).

Y. Peng et al. (Eds.): PRCV 2020, LNCS 12305, pp. 328–340, 2020.
https://doi.org/10.1007/978-3-030-60633-6_27

Recent research efforts have been devoted to point cloud upsampling. Particularly, deep learning based point cloud upsampling methods [1–4] were proposed. These methods mainly focus on generating high-resolution point clouds from the low-resolution complete point clouds. Actually, due to the incomplete scanning views of the sensor and occluded scenes, the raw point clouds often contain holes and incomplete parts. Nonetheless, the inpainting ability (i.e., the regression of coordinates of point clouds in the missing areas) of the existing point cloud upsampling methods is limited. It is desirable to generate the high-quality point clouds by upsampling and inpainting the low-resolution incomplete point clouds simultaneously.

Upsampling and inpainting point clouds is a very challenging task. The disordered point clouds with holes is sparse and unevenly distributed. Thus, compared to the image inpainting task, it is difficult to determine the reasonable areas of point clouds to fill. In addition, due to the topologic structure of the 3D object, there are the "real" holes in the point clouds. For the upsampling and inpainting task, it is difficult to distinguish the "real" holes and inpainting areas in point clouds.

In this paper, we propose an end-to-end network called PUI-Net that simultaneously performs the upsampling and completion task on point clouds. The network can upsample low-resolution point clouds and fill the incomplete parts in the formed high-resolution point clouds. In the proposed PUI-Net, in order to extract discriminative features of point clouds, by introducing the channel attention mechanism, we first develop a convolution attention unit. A non-local feature expansion unit is then proposed to generate the high-resolution point clouds and preserve the local geometric structures of the low-resolution point clouds. Finally, we formulate a complicated loss to train the network so that the incomplete parts in the formed high-resolution point clouds can be effectively filled. Experiments on our constructed point cloud dataset verify the effectiveness of the proposed PUI-Net, particularly in the cases of the incomplete point cloud upsampling.

In summary, the main contributions of this paper are three-fold:

1. To the best of our knowledge, this is the first work to upsample and inpaint the low-resolution point clouds simultaneously.
2. We propose a unified point cloud upsampling and inpainting framework, which consists of the convolution attention unit and non-local feature expansion uint.
3. Our proposed method can significantly improve performance of point cloud upsampling and inpainting.

2   Related Work

2.1   Deep Learning-Based Point Cloud Upsampling

In recent years, with the development of 3D deep learning, deep learning based point cloud networks such as PointNet [5], PointNet++ [6], DGCNN [7] have been proposed. Yu *et al.* first proposed a PointNet++ based point cloud upsampling network called PU-Net [2], which uses PointNet++ to extract features under different numbers of point clouds and expand them by duplication. Finally, the point clouds are reconstructed through several fully connected layers. Consequently, they presented EC-Net [1] by introducing the face and edge information of the 3d model, but EC-Net's labeling requirements are too high to be suitable for use. Wang *et al.* then proposed a patch-based progressive upsampling network [3] that captures detailed information at different levels through a cascaded multi-step structure. Recently, Li *et al.* proposed a PU-GAN [4] network by introducing the Gan structure, using adversarial training strategies to improve the quality of the upsampled point clouds. In general, the existing upsampling networks do not make full use of geometric information of the point clouds and cannot solve the point cloud inpainting problem well.

2.2   Optimization-Based Point Cloud Upsampling

Alexa *et al.* first proposed a method [8] of interpolation at the vertices of a Voronoi graph in a point's local tangent space. Subsequently, Lipman *et al.* proposed [9] a local optimization mapping operator (LOP) based on L_1 Norm to resample the point cloud. Also, Huang *et al.* proposed a progressive resampling method called Edge-aware point set resampling [10]. Among others, Wu *et al.* proposed a consolidation method [11] based on deep point representation. The key to this method is to represent the deep points by associating the surface points with an inner point on the meso-skeleton.

2.3   Optimization-Based Point Cloud Inpainting

So far, there are only a few inpainting methods for point clouds. Sahay *et al.* proposed an inpainting scheme [12] for online depth database. By projecting the point cloud into a depth map, they found the most similar in the depth database through dictionary learning, and then minimized the error with the known region. Dinesh *et al.* proposed a method [13] for inpainting by finding the most similar area on the point cloud itself. The order of inpainting holes is determined based on a given prior and the boundary of the holes. Only for dense indoor scanning point clouds, Quinsat *et al.* [14], Wang *et al.* [15] and Muraki *et al.* proposed [16] to inpainting the input point cloud by reconstructing it into a triangle mesh. Lozes *et al.* proposed [17] that partial different operators fill holes based on local geometric information near the holes. Fu *et al.* proposed [18] to use Non-local self-similarity to find the most similar region to the hole region on the point cloud, and to fill the hole by optimizing the error between the region and the hole region. The above methods relying heavily on the reconstruction quality or the quality of the normal estimation, and most of the source to inpainting is the area on the input point cloud itself, so the inpainting quality is limited.

3   Method

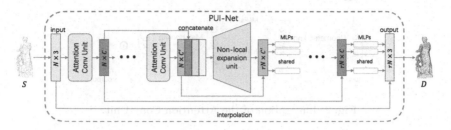

Fig. 1. Overview of the PUI-Net framework. The network inputs a sparse and incomplete point cloud S and outputs a dense point cloud D. Note that N is the number of points for the input patch, and r is the upsampling ratio.

For a sparse point cloud $S = \{s_i\}_{i=1}^{N}$ with holes of N points, where s_i is point coordinate, it is expected to obtain a dense and complete point cloud $D = \{s_i'\}_{i=1}^{rN}$ through network and the r is the upsampling rate. Figure 1 shows the details of the PUI-Net. In the PUI-Net, we propose two building blocks: *Attention Conv Unit* and *Non-local Expansion Unit*. In addition, we propose an inpainting loss function to give the network the ability to inpainting.

3.1   Network Architecture

The PUI-Net presents an encoder-decoder structure, extracting point cloud features through a series stacked *Attention Conv Units* and stitching features of different scales together before expansion. Based on the Non-local Neural Networks [19] we proposed a *Non-local expansion unit*, through which the features are expanded by a factor of r and then the features are restored to a point clouds through a series of multilayer perceptrons.

Attention Conv Unit. Figure 2 shows the details of the *Attention Conv Units*. In the beginning, we calculate the Knn relationship of points in euclidean space. In all subsequent units, we use this relationship to obtain the local geometric structure of the input feature map, which is similar to the pixel neighborhood in the image. Because this mechanism does not require to recalculate the K-neighbor relationship in the high-dimensional space, so it makes both training and testing faster. To better extract the features of the point clouds, we have introduced a channel attention mechanism [20].

As shown in Fig. 2, for the input feature map $X = [x_1, x_2, ..., x_C]$, there are N points for each c-dimensional channel. By summing the information of each channel on the feature map, we get a feature map $h \in R^c$ representing the information contained in each channel. Defined the element h_c of h as follows:

$$h_c = \frac{1}{N} \sum_{i=1}^{N} x_c(i). \tag{1}$$

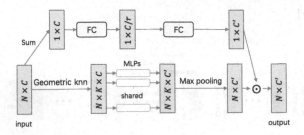

Fig. 2. Illustration of the Attention Conv Unit.

Here h_c is the mean value of the c-th channel and $x_c(i)$ is the c-th feature at i-th point. Such statistics for each channel can be regarded as collecting all the information about all points of each channel. Furthermore, we control the passage of features in each dimension through a gate mechanism. We formulate this mechanism as:

$$g = \gamma(W_d\sigma(W_ch)). \tag{2}$$

where W_c and W_d are parameters of the fully connected layer, γ and σ are Sigmoid and ReLU functions, respectively. Finally, the unit outputs o as:

$$o_j = g_j \cdot y_j. \tag{3}$$

where g_j and y_j are the gate value of the j-th channel and the feature of the j-th channel after pooling.

Non-local Expansion Unit. Both MPU [3] and PU-GAN [4] without considering the geometric relationship of the point clouds when expanding features. To make full use of the geometric relationship of the points in the patch, we propose a *Non-local feature expansion unit* as shown in the Fig. 3.   For the input

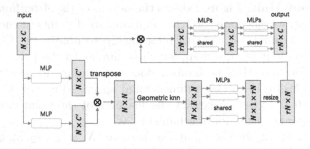

Fig. 3. Illustration of the non-local expansion unit.

sparse feature map $X = [x_1, x_2, ..., x_N]$, where N is the number of sparse point clouds. Our goal is to obtain a dense feature map $F = [f_1, f_2, ..., f_{rN}]$ through an expansion operation. First, we calculate the relationship between the i-th point and the k-th point in the input sparse feature map, defined as:

$$f(x_k, x_i) = \theta(x_k)^T \phi(x_i). \tag{4}$$

Here $\theta(x_k) = W_\theta x_k$ and $\phi(x_k) = W_\phi x_i$ are two MLPs. Then we collect the relationships in the K local neighborhood of each point and use these relationships to calculate a weight matrix P of size $rN \times N$. The elements in the matrix are defined as follows:

$$p_{ji} = \sigma(\sum_{k=1}^{K} \sum_{i=1}^{N} w_{ki} f(x_k, x_i) + b_j). \tag{5}$$

w_{ki} and b_j are learnable parameters. p_{ji} represents the weight of the i-th feature in the input sparse feature map to the j-th feature in the dense feature map. Then we can naturally define the *non-local expansion* operation as:

$$f_j = \frac{1}{N} \sum_{i=1}^{N} p_{ji} x_i. \tag{6}$$

Equation 6 means that the dense point cloud feature f_j is obtained by weighting each feature in the sparse feature map. Finally, we use two layers of MLPs to increase the non-linear capability of the unit to obtain a feature map of dense point cloud.

3.2 Joint Loss Functions

We propose a joint loss function, including the reconstruction loss, inpainting loss, hausdorff loss and uniform loss to train our PUI-Net.

Reconstruction Loss. We use Earth Mover's distance (EMD) to reduce the error between the output of the network and the ground truth as a whole. It should be noted that the EMD guarantees bijection mapping between two point clouds. Compared with the Chamfer distance, the EMD has better convergence for the point clouds with holes.

$$L_{rec} = \min_{\psi: D \to Q} \sum_{d_i \in D} \|d_i - \psi(d_i)\|^2. \tag{7}$$

Here D is the output of the upsampling network and Q is the dense ground truth.

Inpainting Loss. Upsampling and inpainting point clouds simultaneously is a very challenging task because it is difficult to define holes on the point cloud. We define the hole in the point cloud as the area enclosed by a closed loop of boundary edges after Delaunay Triangulation. Figure 4 shows a simple example. Since the Delaunay triangulation process is not differentiable, we try to find a

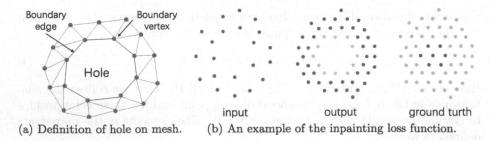

(a) Definition of hole on mesh.   (b) An example of the inpainting loss function.

Fig. 4. Examples of the hole and inpainting loss function. The red points are the repair target points selected by the inpainting loss function, and the orange points are the hole boundaries in the output point cloud and the nearest neighbors of the repair target points in the output point cloud. (Color figure online)

differentiable way to define the holes in the point cloud. Obviously, if there is a point near the center of the hole, its distance from any boundary vertex will be much greater than the length of almost any triangle edges. The reason is those edges obtained by triangulation have similar lengths because ground truth is uniformly sampled on the mesh surface and triangulation ensures that the nearest three points are used to form a triangle and the line segments do not intersect. We exploit this property and use the distance between the point and its nearest neighbor as a replacement for the edge to identify the hole. We think that those points in the ground truth, their distance from the nearest neighbor in the output point cloud is greater than the average of the distance between the point in the output and its nearest neighbor in the ground truth, are the inpainting target points. With the definition of the hole above the point cloud, we designed an inpainting loss function.

Figure 4 shows a simple example of how the inpainting loss function works. For an output of the network like (b) in Fig. 4, firstly, we calculate the distance between each point in the ground truth and the nearest neighbor in the output. Secondly we select the target point whose distance is greater than the threshold. Finally, we count and minimize the average distance between all patch target points and their nearest neighbors in the output. The Inpainting loss function are defined as:

$$L_{hole}(D, Q) = \frac{\sum_{q \in Q} \xi(\min_{d \in D} \|d - q\|^2)}{C}. \tag{8}$$

Here C represents the number of inpainting target points. When $C = 0$, the inpainting loss is zero. The threshold function and threshold value is defined as:

$$\xi(d) = \begin{cases} d, & d \geq \zeta \\ 0, & otherwise \end{cases}, \tag{9}$$

$$\zeta = \frac{\rho(\sum_{d \in D} \min_{q \in Q} \|q - d\|^2)}{N}. \tag{10}$$

Here N is the number of output point clouds. Hyperparameter ρ are generally set to 1.1. The threshold value is adaptively adjusted according to each batch of input data. When the threshold value is large, it mean the output quality is poor, and most points will be selected by the threshold function. In this case, the loss will restrict the whole point cloud. If the threshold value is small that mean quality of the output point cloud is high and the distribution approximates the ground truth, the threshold function will select unfilled target points to help the network inpainting.

Hausdorff Loss. For the inpainting task, the network needs to insert a point in the appropriate target space, it is easy to generate outliers for our network. Since the inpainting loss function only calculates the nearest error from the target inpainting point clouds, which cannot be constrained to the outliers too far away, so we introduce Hausdorff distance to constrain the outliers.

$$L_{hd} = \max_{d \in D} \|d - q\|^2 + \max_{q \in Q} \|d - q\|^2. \tag{11}$$

Uniform Loss. We use the loss function L_{uni} proposed in PU-GAN [4] to make the generated point cloud as uniform as possible.

Joint Loss. In summary, we use a joint loss function to train the upsampling and downsampling networks, defined as:

$$L = \lambda_{rec}L_{rec} + \lambda_{hold}L_{hole} + \lambda_{hd}L_{hd} + \lambda_{uni}L_{uni}. \tag{12}$$

where λ_{rec}, λ_{hole}, λ_{hd} and λ_{uni} are weights.

4 Experiments

In this section, we first introduce the data set we collected and the implementation details of the model. Secondly comparing our model with the existing point cloud upsampling methods and our model can achieve state-of-the-art results. Finally, we conduct the ablation study to demonstrate the effectiveness of our method.

4.1 Dataset and Implementation Details

Based on the datasets released by MPU [3], PU-Net [2], and PU-GAN [4], we add 3D human models in the FAUST [21] dataset to from the new dataset, including 3D models of 237 different categories. Then we randomly generate 5–30 inpainting areas of different sizes according to the number and complexity of the patches of the model. Finally, we randomly select 191 3D models as the training set and the rest as the test set. We show some models in the dataset in Fig. 5. In addition, we also test our model on real data.

All the experiments are implemented by TensorFlow. We use the same sampling strategy as MPU [3] to extract 195 patches for each 3D object with holes,

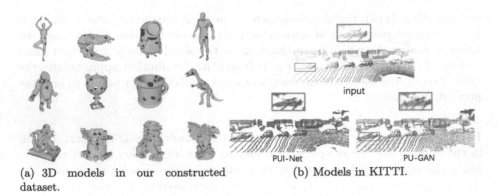

(a) 3D models in our constructed dataset.

(b) Models in KITTI.

Fig. 5. Visualization of 3D models in our constructed dataset and KITTI.

and a total of 37245 patches are obtained. Each patch contains 256 input points. Furthermore, the same strategy to extract the ground truth from 3D objects without holes and each patch contain 1024 points.

We use data augmentation in PU-GAN to prevent the network from over-fitting, which contains random rotation, scaling, and point noise. We use Adam as an optimizer and set the learning rate as 0.001 to train the PUI-Net for 100 epochs. Moreover, we set the batch size as 28 and the weights λ_{rec}, λ_{hole}, λ_{hd} and λ_{uni} is 100, 100, 20, and 20. Then the threshold hyperparameter ρ is 1.1. The entire framework is trained on a Nvidia 2080ti GPU.

4.2   Quantitative and Qualitative Results

Following the evaluation criterion in PU-Net we use the Chamfer distance (CD), Hausdorff distance (HD), and point-to-surface (P2F) distance for quantitative analysis. When testing, Poisson sampling is used to sample ground truth of 8192 points in a 3D model without holes and 2048 points of models with holes as input. Note that each model is still split into patches during testing. Table 1 shows the results of quantitative comparison. Our proposed PUI-Net can achieve state-of-the-art results.

Table 1. Quantitative comparison results of different methods.

Methods	CD(10^{-3})	HD(10^{-3})	P2F-Mean(10^{-3})	P2F-Std(10^{-3})
PU-Net	16.58	193.18	118.26	105.32
MPU	2.401	899.91	11.741	45.12
PU-GAN	0.412	12.67	6.906	8.823
PUI-Net	**0.327**	**7.06**	**6.741**	**8.429**

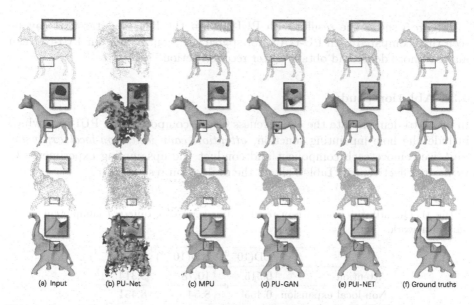

(a) Input   (b) PU-Net   (c) MPU   (d) PU-GAN   (e) PUI-NET   (f) Ground truths

Fig. 6. Comparison results of point cloud upsampling and inpainting with from 2048 points.

In addition to the quantitative comparison, we also visualize the upsampling and inpainting point clouds and surface reconstruction results in Fig. 6. The visualization results are generated from our PUI-Net (e), PU-Net(b), MPU(c) and PU-GAN(d). From the results, one can see that PU-Net cannot work well and generated points with MPU are concentrated near the input points and contain many noises like PU-GAN. In terms of surface reconstruction, the proposed PUI-Net can yield good inpainting performance. We show more visualization results in Fig. 7.

Table 2. Quantitative comparison results of different methods.

Methods	CD(10^{-3})	HD(10^{-3})	P2F(10^{-3})
PU-Net	0.342	3.89	74.576
MPU	**0.225**	1.74	**3.644**
PU-GAN	0.244	6.85	3.742
PUI-Net	0.242	**1.73**	4.282

In order to demonstrate the effectiveness of our model, Table 2 shows the quantitative results of PUI-Net in the case of upsampling point clouds without holes. It can be seen that our network has achieved competitive results.

Figure 5 shows the results with PUI-Net on the KITTI dataset. It can be seen that compared with PU-GAN, for sparse car scan points, our PUI-Net can make it more dense and obtain better reconstruction.

4.3 Ablation Study

In order to demonstrate the effectiveness of the components in PUI-Net, which include the hole inpainting function, *attention conv unit, non-local expansion unit*, we remove each component and conduct the upsampling experiments to verify its effectiveness. Table 3 shows the evaluation results.

Table 3. Quantitative comparison results after removing different components from our framework.

	$CD(10^{-3})$	$HD(10^{-3})$	$P2F(10^{-3})$
Attention	0.515	10.78	9.314
Non-local expansion	0.455	8.54	8.451
Inpainting loss	0.491	12.77	9.401
full pipeline	**0.327**	**7.06**	**6.741**

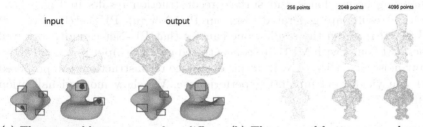

(a) The top and bottom rows show different inputs and the corresponding surface reconstruction results.

(b) The top and bottom rows show the inputs and upsampling results, respectively.

Fig. 7. The upsampling and inpainting results of PUI-Net.

It should be noted that the feature expansion unit (Non-local Expansion Unit) and the feature extraction unit (Attention Conv unit) are removed and replaced with the corresponding components in MPU and PU-GAN. It can be seen that after removing each component, the quantification results have become worse. Figure 7 shows the results of our upsampling network with different numbers of input points. It can be seen that our model is very stable for different inputs, even if the number of input point clouds is only 256 points.

5   Conclusion

In this paper, we proposed PUI-Net, an end-to-end point cloud upsampling and inpainting network. Specifically, we developed an *attention conv unit* and *non-local expansion unit* with an inpainting loss to train the network. Moreover we collected a wide range of datasets and used this dataset to train and test the network. Experiments demonstrate that our PUI-Net can achieve state-of-the-art results in the case of point clouds with holes, and is robust to upsampling multi-resolution input and noise point clouds. In addition, The network still achieved competitive results for point cloud data without holes.

References

1. Yu, L., Li, X., Fu, C.-W., Cohen-Or, D., Heng, P.-A.: EC-Net: an edge-aware point set consolidation network. In: ECCV, pp. 386–402 (2018)
2. Yu, L., Li, X., Fu, C.-W., Cohen-Or, D., Heng, P.-A.: Pu-Net: point cloud upsampling network. In: CVPR, pp. 2790–2799 (2018)
3. Yifan, W., Wu, S., Huang, H., Cohen-Or, D., Sorkine-Hornung, O.: Patch-based progressive 3D point set upsampling. In: CVPR, pp. 5958–5967 (2019)
4. Li, R., Li, X., Fu, C.-W., Cohen-Or, D., Heng, P.-A.: EC-Net: Pu-gan: a point cloud upsampling adversarial network. In: ICCV, pp. 7203–7212 (2019)
5. Qi, C.R., Su, H., Mo, K., Guibas, L.J.: PointNet: deep learning on point sets for 3D classification and segmentation. In: CVPR, pp. 652–660 (2017)
6. Qi, C.R., Yi, L., Su, H., Guibas, L.J.: Pointnet++: deep hierarchical feature learning on point sets in a metric space. In: Advances in Neural Information Processing Systems, pp. 5099–5108 (2017)
7. Wang, Y., Sun, Y., Liu, Z., Sarma, S.E., Bronstein, M.M., Solomon, J.M.: Dynamic graph CNN for learning on point clouds. Adv. Neural Inf, Process. Syst. **38**(5), 146 (2019)
8. Alexa, M., Behr, J., Cohen-Or, D., Fleishman, S., Levin, D., Silva, C.T.: Computing and rendering point set surfaces. IEEE Trans. Visual Comput. Graphics **9**(1), 3–15 (2003)
9. Lipman, Y., Cohen-Or, D., Levin, D., Tal-Ezer, H.: Parameterization-free projection for geometry reconstruction. ACM Trans. Graphics **26**(3), 22 (2007)
10. Huang, H., Wu, S., Gong, M., Cohen-Or, D., Ascher, U., Zhang, H.R.: Edge-aware point set resampling. ACM Trans. Graphics **32**(1), 9 (2013)
11. Wu, S., Huang, H., Gong, M., Zwicker, M., Cohen-Or, D.: Deep points consolidation. ACM Trans. Graphics **34**(6), 176 (2015)
12. Sahay, P., Rajagopalan, A.N.: Geometric in painting of 3D structures. In: CVPR Workshops, pp. 1–7 (2015)
13. Dinesh, C., Bajić, I.V., Cheung, G.: Exemplar-based framework for 3D point cloud hole filling. In: 2017 IEEE Visual Communications and Image Processing, pp. 1–4 (2017)
14. Quinsat, Y., et al.: Filling holes in digitized point cloud using a morphing-based approach to preserve volume characteristics. Int. J. Adv. Manuf. Technol. **81**(1–4), 411–421 (2015)
15. Wang, J., Oliveira, M.M.: Filling holes on locally smooth surfaces reconstructed from point clouds. Image Vis. Comput. **25**(1), 103–113 (2007)

16. Muraki, Y., Nishio, K., Kanaya, T., Kobori, K.,-I.: An automatic hole filling method of point cloud for 3D scanning. Václav Skala-UNION Agency (2017)
17. Lozes, F., Elmoataz, A., Lézoray, O.: PDE-based graph signal processing for 3-D color point clouds: opportunities for cultural heritage. IEEE Signal Process. Mag. **32**(4), 103–111 (2015)
18. Fu, Z., Hu, W., Guo, Z.: Point cloud in painting on graphs from non-local self-similarity. In: 2018 25th IEEE International Conference on Image Processing, pp. 2137–2141 (2018)
19. Wang, X., Girshick, R., Gupta, A., He, K.: Non-local neural networks. In: CVPR, pp. 7794–7803 (2018)
20. Zhang, Y., Li, K., Li, K., Wang, L., Zhong, B., Fu, Y.: Image super-resolution using very deep residual channel attention networks. In: ECCV, pp. 286–301 (2018)
21. Bogo, F., Romero, J., Loper, M., Black, M.J.: FAUST: dataset and evaluation for 3D mesh registration. In: CVPR, pp. 3794–3801 (2014)
22. Mao, X., Shen, C., Yang, Y.-B.: Image restoration using very deep convolutional encoder-decoder networks with symmetric skip connections. In: Advances in Neural Information Processing Systems, pp. 2802–2810 (2016)
23. Yu, J., Lin, Z., Yang, J., Shen, X., Lu, X., Huang, T.S.: Generative image in painting with contextual attention. In: CVPR, pp. 5505–5514 (2018)

An Automated Method with Feature Pyramid Encoder and Dual-Path Decoder for Nuclei Segmentation

Lijuan Duan[1,2,3], Xuan Feng[1,2,3], Jie Chen[4(✉)], and Fan Xu[1,2,3]

[1] Faculty of Information Technology, Beijing University of Technology, Beijing 100124, China
[2] Beijing Key Laboratory of Trusted Computing, Beijing 100124, China
[3] National Engineering Laboratory for Critical Technologies of Information Security Classified Protection, Beijing 100124, China
[4] Peng Cheng Laboratory, Shen Zhen 518055, China
chenj@pcl.ac.cn

Abstract. Nuclei instance segmentation is a critical part of digital pathology analysis for cancer diagnosis and treatments. Deep learning-based methods gradually replace threshold-based ones. However, automated techniques are still challenged by the morphological diversity of nuclei among organs. Meanwhile, the clustered state of nuclei affects the accuracy of instance segmentation in the form of over-segmentation or under-segmentation. To address these issues, we propose a novel network consists of a multi-scale encoder and a dual-path decoder. Features with different dimensions generated from the encoder are transferred to the decoder through skip connections. The decoder is separated into two subtasks to introduce boundary information. While an aggregation module of contour and nuclei is attached in each decoder for encouraging the model to learn the relationship between them. Furthermore, this avoids the splitting effect of independent training. Experiments on the 2018 MICCAI challenge of Multi-Organ Nuclei Segmentation dataset demonstrate that our proposed method achieves state-of-the-art performance.

Keywords: Pathology analysis · Nuclei segmentation · Deep learning

1 Introduction

Cancer is a severe disease that threatens human health. It caused about 9.6 million deaths in 2018 globally [1]. With developments of medicine and computer vision, digital pathology alleviates problems of associating cancer severity with morphological characteristics in images [2]. Nuclei instance segmentation, especially in a spanning range

Supported by the Key Project of Beijing Municipal Education Commission (No. KZ201910005008), Nature Science Foundation of China (No. 62081360152, No. 61972217), and Natural Science Foundation of Guangdong province in China (No. 2019B1515120049, 2020B1111340056).

© Springer Nature Switzerland AG 2020
Y. Peng et al. (Eds.): PRCV 2020, LNCS 12305, pp. 341–352, 2020.
https://doi.org/10.1007/978-3-030-60633-6_28

of patients and organs, has become one of the key modules in computational pipelines [3]. It contributes to tumor diagnosis and corresponding treatments with morphological features such as density, magnitude, and nucleus-cytoplasm ratio [4]. However, this task faces the following challenges. Firstly, uneven staining for images and coarse border for nuclei cause mislabeling, which leads to indistinguishable nuclei or even unreliable results. Meanwhile, it also hinders the progress of related works [5]. Secondly, popular threshold-based and morphology-based methods have little success in this work due to the inaccurate threshold in the complicated background [6, 7]. Moreover, this could cause under-segmentation or over-segmentation. Thirdly, nuclei generally appear in a state of clusters in images, which impacts the performance of techniques in instance-level. In addition, the diversity of nuclei among different organs raises another challenge to the generalization ability of solutions [8]. Fourthly, the limitation of traditional losses, like cross entropy (CE), that is committed to pixel-level classification results prevents the model from being trained with the complete position information of objects.

Most of the earlier algorithm, such as Otsu thresholding [9] and watershed segmentation [10], did not encounter the problem. Therefore, pathological image segmentation with the usage of convolutional neural networks (CNNs) has gradually become popular. Ronneberger et al. [11] employed a U-shaped network with a contracting path and an expansive path to ensure the combinations of low-level features, which lays the foundation for subsequent methods. However, this semantic-level approach cannot be applied to instance-level tasks, especially for adjacent objects. Chen et al. [12] proposed a deep contour-aware network (DCAN) to separate touching objects with the supplementary information of contour. Qu et al. [13] improved convolutional networks to enhance the learning of relationships among pixels with a variance constrained loss. Zhou et al. [8] utilized an information aggregation module with multi-scale features to solve difficulties in nuclei instance segmentation.

In this paper, to address these issues mentioned above, our main contributions are listed as follows:

1. We propose a novel U-shaped network with a dual-path structure in the decoder, which separates the task into two subtasks. Each decoder combines multi-scale features by corresponding skip connections. We use different loss functions in each path respectively.
2. We improve current aggregation methods of boundary information in the decoder to make sure that the model could learn the relationship between the subtasks.
3. We evaluate the effectiveness of our work on a dataset with various organs. In comparisons with others, our proposed method could achieve state-of-the-art performance.

2 Methodology

The framework of our proposed method based on the U-shaped structure is shown in Fig. 1. It consists of an encoder (left path) and a decoder (right path) with corresponding skip connections of different scales. The encoder follows the classical residual CNN with feature pyramid modules and the decoder is composed of dual-path expansive modules. The entire network is a fully convolutional network (FCN) [14] and could merge high-resolution features with contextual information and low-resolution features with semantic information together.

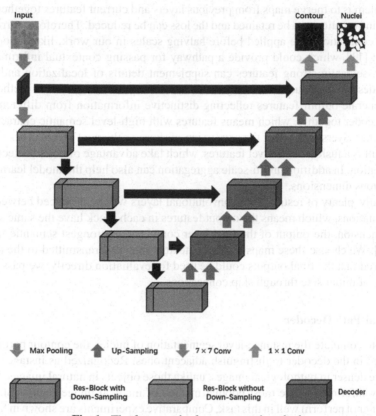

Fig. 1. The framework of our proposed method.

2.1 Feature Pyramid Encoder

In order to extract features in the encoder effectively, we introduce residual modules which have benefits of solving degradation problem when the network goes deeper [15]. For each block in the contracting path, the output y is defined as:

$$y = \mathcal{F}(x) + x \tag{1}$$

where x and $\mathcal{F}$ denote the input and the function to be learned in the network. We use a stack of three-layer-structure bottleneck architecture which contains $1 \times 1, 3 \times 3$ and 1×1 convolutions to ensure non-linearities and to reduce consumption. Shortcuts between inputs and outputs are also used in the building blocks. The specific structure of the encoder is the same as that in ResNet [15]. We adopt Conv-BN-ReLU [16, 17] structure for each convolutional layer. In particular, down-sampling is performed at the beginning of Res-Block 2, Res-Block 3, and Res-Block 4 with a stride of two.

Residual networks perform well in feature extraction. However, crucial details will be dropped because of the decrease of scales as the network deepens. The basic idea to solve this problem is to merge maps from previous layers and current features together so that detailed information can be retained and the loss can be reduced. Therefore, corresponding skip connections are applied before halving scales in our work, like the operation of U-Net [11], which could provide a pathway for passing contextual information. In this way, semantic-strong features can supplement details of localization and texture [8]. Besides, we introduce Feature Pyramid Networks (FPN) [18] into our method. FPN could generate output features reflecting distinctive information from different layers of the encoder together, which means features with high-level semantic characteristics from deeper layers will affect the expression of those in the next layers. Therefore, the final result is a fusion of all-level features, which take advantage of object detection and segmentation. In addition, multi-scale aggregation can also help the model learn objects with various dimensions.

Usually, plenty of results from convolutional layers will be produced between scale transformations, which means the output features in each block have the same size. For each dimension, the output of the final layer contains the strongest semantic information [18]. We choose these maps as our contextual features transmitted to the decoder. Considered that the final outputs could be used for evaluation directly, we pass $\{1, 1/2, 1/4, 1/8\}$ of initial size through skip connections.

2.2 Dual-Path Decoder

In order to complete the instance-level segmentation of nuclei, the contour part needs to be trained in the decoder to distinguish adjacent ones. Recognized as instance targets, nuclei are denser in pathological images, unlike those objects in natural images. Without boundary information, the most popular method of medical image segmentation, like U-Net, cannot perform well in this task. Comparative experiments are shown in Sect. 3.3.

Inspired by DCAN [12] and CIA-Net [8], we divide the task into two subtasks: contour segmentation and nuclei segmentation. The latter is for completing the segmentation task while the former is for fine-tuning the final results, which could provide crucial details of contour for segmenting touching nuclei. In order to avoid making contour information been just a corrective factor for the output, we introduce the contour path into the network instead of training it as an independent problem. This kind of multi-task training method is optimized together with an end-to-end way, which could increase the discriminative ability of intermediate features, thereby improving the robustness of performance [19]. With one forward propagation, the network can predict the results of nucleus objects and contours at the same time rather than generating contours through additional post-separating steps from the last layers [20, 21].

Both of targets in two subtasks are nuclei while they are separated into two branches artificially, which means they are strongly related in space and semantics. In this case, Zhou et al. [8] proposed Information Aggregation Module (IAM) to utilize such a correlation between two paths. In this paper, we make the following improvements to this module. One is altering the position of 3×3 convolutions, and the other is adding a parameter-free shortcut which is similar to the residual structure. Moreover, functions for merging different features are also variable. We have done ablation studies to prove that they are effective. Each decoder in this paper has the same structure shown in Fig. 2.

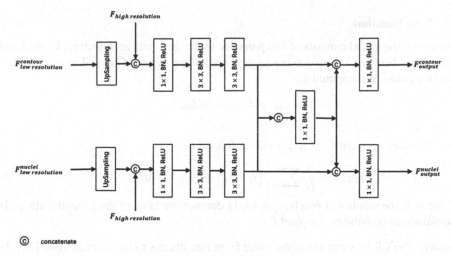

Fig. 2. The structure of a decoder.

2.3 Feature Pyramid and Dual-Path Network

The left path in Fig. 1 shows the process of data passing through backbone networks. Note that in this paper, we use pre-trained ResNet-101 [15] as our main backbone since our data volume is not enough to train a network of this size.

Encoder. The input to the network is a batch of pre-processed patches from images each epoch. Specifically, we attach a 7×7 convolution with a stride of one to make sure that the skip connection from this scale has the same dimension as initial patches. Therefore, the outputs of the decoder in the top could be used directly instead of being up-sampled, which can avoid the loss of information.

Decoder. To start the iteration, $F^{contour}_{lowresolution}$ and $F^{nuclei}_{lowresolution}$ (features from the previous decoder) are up-sampled to be merged with $F_{highresolution}$ (high resolution features from the encoder) by concatenation, followed by a 1×1 convolution to decrease channels for reducing the amount of calculation. Note that the channel of convolutional kernels for each layer in the decoder is the same as that in $F_{highresolution}$. Then, we append two 3×3 convolutions to ensure non-linearities, which could combine information from $F_{highresolution}$ and $F_{lowresolution}$ together. The structures of the two paths are symmetrical, while the parameters of convolutional layers are various. With concatenated features of nuclei and contour, we attach a 1×1 convolution to make sure that the dimension of the maps could be used by subsequent decoder directly. Similar to residual networks, a parameter-free shortcut exists in information aggregation to guarantee a positive mapping of this module. Both of element-wise addition and concatenation are methods of fusing the features of shortcut and those generated by the separate paths. In this paper, we compare the differences between them and details are shown in Sect. 3.3.

2.4 Loss Function

The loss of the model consists of two parts due to the dual-path architecture. Considered differences between the two subtasks, we define two types of loss. The overall loss function could be computed as:

$$L_{total} = L_{contour} + L_{nuclei} \tag{2}$$

Contour. CE is one of the most widely used losses in segmentation tasks [13]. In the case of binary segmentation, CE could be defined as:

$$L_{BCE} = -\frac{1}{N} \sum_{i=1}^{N} \left[y_i \log p_i + (1 - y_i) \log(1 - p_i) \right] \tag{3}$$

where N is the number of pixels, $y_i \in \{0, 1\}$ denotes the label of the ground truth, p_i is the estimated probability for pixel i.

Nuclei. The CE loss can cause the value from rare classes to be overwhelmed [22]. In the field of medical segmentation, the region of interest usually occupies only a small part of the scan, where the learning process can be trapped in local minima. As a result, the foreground is often missed [23]. In this case, we introduce soft dice loss [23] in this work, which could be written as:

$$L_{DICE} = 1 - DICE = 1 - \frac{2 \sum_{i=1}^{N} p_i y_i}{\sum_{i=1}^{N} p_i^2 + \sum_{i=1}^{N} y_i^2} \tag{4}$$

where N, p_i and y_i denote the same as those in CE loss.

Total. According to Eq. 4 and Eq. 5, Eq. 3 could be re-written as:

$$L_{total} = L_{BCE} + L_{DICE} \tag{5}$$

Note that we calculated the loss based on two corresponding maps separately.

3 Experiments

3.1 Dataset and Evaluation Metrics

Data. We evaluate our purposed method on the Multi-Organ Nuclei Segmentation (*MoNuSeg*) dataset [3] which consists of 30 hematoxylin and eosin (H&E) stained images from 2018 MICCAI challenge. Each image with a size of 1000 × 1000 is captured by The Cancer Genomic Atlas (TCGA) [24] from a whole slide image (WSI) that is unique to the patient for maximizing appearance variation. These images come from 18 hospitals and contain 7 organs, including benign and malignant cases. The dataset is divided into three parts: training (Train), testing for same organs (Test1) and testing for different organs (Test2). The compositions of *MoNuSeg* are shown in Table 1.

Table 1. The composition of *MoNuSeg*.

Subset	Organs							
	Total	Breast	Liver	Kidney	Prostate	Bladder	Colon	Stomach
Train	16	4	4	4	4	0	0	0
Test1	8	2	2	2	2	0	0	0
Test2	6	0	0	0	0	2	2	2

Evaluation Metrics. Current metrics for object detection, like F1-score, cannot measure shape matching for detected objects. In addition, it is incomparable between two segmentation techniques if one performs well in detection while the other has a better shape matching. In this case, to combine both pixel-level and object-level evaluation, we introduce Aggregated Jaccard Index (AJI) [3] as our evaluation metrics, which could be defined as:

$$AJI = \frac{\sum_{i=1}^{N} y_i \cap p_j}{\sum_{i=1}^{N} y_i \cup p_j + \sum_{k \in N} p_k} \tag{6}$$

where N, p_j and y_i denote the number of pixels, the prediction of the model and corresponding ground truth, and j could be written as:

$$j = arg\ \max_k \frac{y_i \cap p_k}{y_i \cup p_k} \tag{7}$$

3.2 Implementation Details

We complete our work with a pre-trained backbone from ImageNet dataset [25]. Before training, stain normalization [26] is utilized due to the large variation of colors in H&E images [27]. We randomly crop the initial images into patches with a size of 224×224 each. Moreover, data augmentations like random scale, flip and rotation are also performed. In order to obtain final results, the output from nuclei segmentation is subtracted from that in another path. In the end, we attach morphological operations, including dilation, small areas removal, and labeling of connected domains, which could help fill in the extracted areas that are used to separate neighboring nuclei. The mini-batch size and initial learning rate are set as 8 and 0.001. The strategy of stepped decay of learning rate and the strategy of early stopping are introduced when the validation error is not decreasing during the training process. The model is trained on NVIDIA TESLA V100 32 GB with Adam optimizer.

3.3   Results and Comparisons

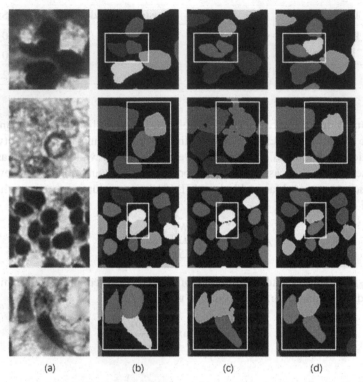

(a)   (b)   (c)   (d)

Fig. 3. Representative samples. (a) original image, (b) ground truth, (c) result of proposed without aggregation, (d) result of proposed. White rectangles highlight the differences. (Color figure online)

Effectiveness of Dual-Path and Aggregation in Decoder. We perform the effectiveness of the dual-path structure in the decoder, which is shown in Table 2. We can see that it has a significant improvement on Test2 compared with the performance of single-path. However, the little advantage of Test1 indicates the instability of the model without Aggregation. It is observed that the network with the aggregation module has a boost for AJI on both Test1 and Test2, which reflects the effectiveness of our proposed method. Representative samples are shown in Fig. 3.

Effectiveness of the Position of Convolutions and Concatenation in the Decoder. Firstly, we named a series of control groups shown in Table 3. Note that 3×3 Before means that convolutions exist before aggregation, while 3×3 After is on the contrary. Decoder E and F have 4 3×3 convolutions in each path of the decoder. Concatenation and Addition denote the methods of feature aggregation. In particular, Addition is followed by an extra 3×3 convolution to smooth generated maps.

Table 2. Results of structures in decoder with ResNet-101.

Structure	AJI	
	Test1	Test2
Single-path	0.5419	0.5622
Proposed w/o aggregation	0.5433	0.6046
Proposed	**0.6154**	**0.6455**

Table 3. Names of options in decoder.

Options	A	B	C	D	E	F
3 × 3 Before	√	√			√	√
3 × 3 After			√	√	√	√
Concatenation	√		√		√	
Addition		√		√		√

Then, with these combinations, we can get the results shown in Table 4. Compared with the results from A to F, we can see that our proposed method has the best results on both Test1 and Test2. According to C and D, concatenation has a negative impact when we adopt such architecture in the decoder. However, in this case, the element-wise addition introduces more parameters than concatenation by 25% due to extra convolutions. Meanwhile, the results of E and F explain that the performance of the model does not necessarily get better with the increase of convolutions, which could be illustrated as the accompanying redundancy of information, especially on small datasets. In general, our proposed method A has the best performance with relatively few parameters.

Table 4. Results of options in decoder.

Method	AJI	
	Test1	Test2
ResNet-101 + A (proposed)	**0.6154**	**0.6455**
ResNet-101 + B	0.6152	0.6384
ResNet-101 + C	0.5828	0.6375
ResNet-101 + D	0.6151	0.6431
ResNet-101 + E	0.5645	0.5953
ResNet-101 + F	0.6076	0.6368

Comparisons with Others. We conduct experiments to compare our work with the following techniques. Cell Profiler [28] is a python-based software for computational pathology with suggested pipelines. Fiji [29] is a java-based software for nuclei segmentation with watershed transform. CNN3 [3] is a three-class FCN with utilization of region growing as post-processing. U-Net [11] is one of the most popular methods for medical image analysis. DCAN [12] is a multi-task FCN of contours and objects for precise segmentation. FullNet [13] is a full resolution FCN with a variance constrained CE loss for the learning of spatial relationship. HoVer-Net [30] is a convolutional network for separating clustered nuclei with the vertical and horizontal distances of pixels to the centers of mass. CIA-Net [8], the reference structure and baseline for our work, is a contour-aware FCN with aggregation modules. The results are shown in Table 5.

Table 5. Results of comparisons.

Method	AJI	
	Test1	Test2
Cell profiler [28]	0.1548	0.0809
Fiji [29]	0.2507	0.3034
CNN3 [3]	0.5154	0.4989
U-Net [11]	0.5815	0.5481
DCAN [12]	0.6082	0.5449
FullNet [13]	0.5946	0.6164
Hover-Net [30]	0.6180[1]	
CIA-Net [8]	0.6129	0.6306
Ours	**0.6154**	**0.6455**

The author only shows the average of Test1 and Test2 in the paper.

The contrast between the first two lines and others illustrates that methods with deep learning and neural networks perform much better than those with traditional algorithm. According to comparisons among CNN-based techniques, we can see the improvement of introducing contour information with Dual-Path architecture. On the other hand, it confirms the significance of applying a deep network as an encoder. Moreover, comparisons of the last four lines perform the strength of our work. This could prove that our proposed method achieves state-of-the-art performance.

4   Conclusion

Nuclei segmentation is playing a crucial role in pathology analysis for cancer diagnosis and treatments. In this paper, we propose a method based on deep learning for nuclei segmentation in instance-level. It consists of feature pyramid encoders with residual

CNNs and dual-path decoders with aggregation modules. Experiments demonstrate that our work achieves state-of-the-art results in H&E stained histopathology images. We hope that our work could assist in this field.

References

1. Bray, F., Ferlay, J., Soerjomataram, I., Siegel, R.L., Torre, L.A., Jemal, A.: Global cancer statistics 2018: GLOBOCAN estimates of incidence and mortality worldwide for 36 cancers in 185 countries. CA: Cancer J. Clin. **68**(6), 394–424 (2018)
2. Gurcan, M.N., Boucheron, L.E., Can, A., Madabhushi, A., Rajpoot, N.M., Yener, B.: Histopathological image analysis: a review. IEEE Rev. Biomed. Eng. **2**(2), 147–171 (2009)
3. Kumar, N., Verma, R., Sharma, S., Bhargava, S., Vahadane, A., Sethi, A.: A dataset and a technique for generalized nuclear segmentation for computational pathology. IEEE Trans. Med. Imaging **36**(7), 1550–1560 (2017)
4. Veta, M.M., et al.: Prognostic value of automatically extracted nuclear morphometric features in whole slide images of male breast cancer. Mod. Pathol. **25**(12), 1559–1565 (2012)
5. Irshad, H., et al: Crowdsourcing image annotation for nucleus detection and segmentation in computational pathology: evaluating experts, automated methods, and the crowd. In: Pacific Symposium on Biocomputing, pp. 294–305. World Scientific (2014)
6. Cheng, J., Rajapakse, J.C.: Segmentation of clustered nuclei with shape markers and marking function. IEEE Trans. Biomed. Eng. **56**(3), 741–748 (2009)
7. Jung, C., Kim, C.: Segmenting clustered nuclei using h-minima transform-based marker extraction and contour parameterization. IEEE Trans. Biomed. Eng. **57**(10), 2600–2604 (2010)
8. Zhou, Y., Onder, O.F., Dou, Q., Tsougenis, E., Chen, H., Heng, P.-A.: CIA-Net: robust nuclei instance segmentation with contour-aware information aggregation. In: Chung, A.C.S., Gee, J.C., Yushkevich, P.A., Bao, S. (eds.) IPMI 2019. LNCS, vol. 11492, pp. 682–693. Springer, Cham (2019). https://doi.org/10.1007/978-3-030-20351-1_53
9. Xue, J.H., Titterington, D.M.: t-Tests, F-Tests and Otsu's methods for image thresholding. IEEE Trans. Image Process. **20**(8), 2392–2396 (2011)
10. Yang, X., Li, H., Zhou, X.: Nuclei segmentation using marker-controlled watershed, tracking using mean-shift, and Kalman filter in time-lapse microscopy. IEEE Trans. Circ. Syst. **53**(11), 2405–2414 (2006)
11. Ronneberger, O., Fischer, P., Brox, T.: U-Net: convolutional networks for biomedical image segmentation. In: Navab, N., Hornegger, J., Wells, W.M., Frangi, A.F. (eds.) MICCAI 2015. LNCS, vol. 9351, pp. 234–241. Springer, Cham (2015). https://doi.org/10.1007/978-3-319-24574-4_28
12. Chen, H., Qi, X., Yu, L., Dou, Q., Qin, J., Heng, P.A.: DCAN: deep contour-aware networks for object instance segmentation from histology images. Med. Image Anal. **36**(36), 135–146 (2017)
13. Qu, H., Yan, Z., Riedlinger, G.M., De, S., Metaxas, D.N.: Improving nuclei/gland instance segmentation in histopathology images by full resolution neural network and spatial constrained loss. In: Shen, D., Liu, T., Peters, T.M., Staib, L.H., Essert, C., Zhou, S., Yap, P.-T., Khan, A. (eds.) MICCAI 2019. LNCS, vol. 11764, pp. 378–386. Springer, Cham (2019). https://doi.org/10.1007/978-3-030-32239-7_42
14. Long, J., Shelhamer, E., Darrell, T.: Fully convolutional networks for semantic segmentation. In: CVPR, pp. 3431–3440 (2015)
15. He, K., Zhang, X., Ren, S., Sun, J.: Deep residual learning for image recognition. In: CVPR, pp. 770–778 (2016)

16. Ioffe, S., Szegedy, C.: Batch normalization: accelerating deep network training by reducing internal covariate shift. In: ICML, pp. 448–456 (2015)
17. Jia, Y., et al.: Caffe: convolutional architecture for fast feature embedding. In: Proceedings of the 22nd ACM international conference on Multimedia, pp. 675–678 (2014)
18. Lin, T.Y., Dollar, P., Girshick, R., He, K., Hariharan, B., Belongie, S.: Feature pyramid networks for object detection. In: CVPR, pp. 936–944 (2017)
19. Zhang, Z., Luo, P., Loy, C.C., Tang, X.: Facial landmark detection by deep multi-task learning. In: ECCV, pp. 94–108 (2014)
20. Gunduz-Demir, C., Kandemir, M., Tosun, A.B., Sokmensuer, C.: Automatic segmentation of colon glands using object-graphs. Med. Image Anal. **14**(1), 1–12 (2010)
21. Wu, H.S., Xu, R., Harpaz, N., Burstein, D., Gil, J.: Segmentation of intestinal gland images with iterative region growing. J. Microsc. **220**(3), 190–204 (2005)
22. Lin, T.Y., Goyal, P., Girshick, R., He, K., Dollar, P.: Focal loss for dense object detection. IEEE Trans. Pattern Anal. Mach. Intell. **42**(2), 318–327 (2020)
23. Milletari, F., Navab, N., Ahmadi, S.A.: V-net: fully convolutional neural networks for volumetric medical image segmentation. In: Fourth International Conference on 3D Vision, pp. 565–571 (2016)
24. The Cancer Genome Atlas. http://cancergenome.nih.gov/. Accessed 14 May 2016
25. Deng, J., Dong, W., Socher, R., Li, L.J., Li, K., Fei-Fei, L.: ImageNet: a largescale hierarchical image database. In: CVPR, pp. 248–255 (2009)
26. Macenko, M., et al.: A method for normalizing histology slides for quantitative analysis. In: 2009 IEEE International Symposium on Biomedical Imaging: From Nano to Macro, pp. 1107–1110 (2009)
27. Vahadane, A.: Structure-preserving color normalization and sparse stain separation for histological images. IEEE Trans. Med. Imaging **35**(8), 1962–1971 (2016)
28. Carpenter, A.E., Jones, T.R., Lamprecht, M.R., et al.: Cell profiler: image analysis software for identifying and quantifying cell phenotypes. Genome Biol. **7**(10), 1–11 (2006)
29. Schindelin, J., Arganda-Carreras, I., Frise, E., et al.: Fiji: an open-source platform for biological-image analysis. Nat. Methods **9**(7), 676–682 (2012)
30. Graham, S., et al.: Hover-Net: simultaneous segmentation and classification of nuclei in multi-tissue histology images. Med. Image Anal. **58**, 101563 (2019)

MD-ST: Monocular Depth Estimation Based on Spatio-Temporal Correlation Features

Xuyang Meng[✉], Chunxiao Fan, Yue Ming, Runqing Zhang, and Panzi Zhao

Beijing University of Posts and Telecommunications, Beijing 100876, China
mengxy@bupt.edu.cn

Abstract. Monocular depth estimation methods based on deep learning have shown very promising results recently, most of which exploit deep convolutional neural networks (CNNs) with scene geometric constraints. However, the depth maps estimated by most existing methods still have problems such as unclear object contours and unsmooth depth gradients. In this paper, we propose a novel encoder-decoder network, named Monocular Depth estimation with Spatio-Temporal features (MD-ST), based on recurrent convolutional neural networks for monocular video depth estimation with spatio-temporal correlation features. Specifically, we put forward a novel encoder with convolutional long short-term memory (Conv-LSTM) structure for monocular depth estimation, which not only captures the spatial features of the scene but also focuses on collecting the temporal features from video sequences. In decoder, we learn four scales depth maps for multi-scale estimation to fine-tune the outputs. Additionally, in order to enhance and maintain the spatio-temporal consistency, we constraint our network with a flow consistency loss to penalize the errors between the estimated and ground-truth maps by learning residual flow vectors. Experiments conducted on the KITTI dataset demonstrate that the proposed MD-ST can effectively estimate scene depth maps, especially in dynamic scenes, which is superior to existing monocular depth estimation methods.

Keywords: Monocular depth estimation · CNN-RNN · Convolutional long short-term memory · Spatio-temporal correlation features

1 Introduction

Monocular depth estimation is an important module of 3D reconstruction [20] and simultaneous localization and mapping (SLAM) [22] in computer vision, such as autonomous driving, augmented reality applications, and virtual travel [3].

The work presented in this paper is supported by Beijing Natural Science Foundation of China (Grant No. L182033), Fund for Beijing University of Posts and Telecommunications (2019PTB-001).

Y. Peng et al. (Eds.): PRCV 2020, LNCS 12305, pp. 353–364, 2020.
https://doi.org/10.1007/978-3-030-60633-6_29

With the rise and development of deep learning, the application of deep learning methods on monocular depth estimation is begun to be explored [11]. Most monocular depth estimation, deep learning-based methods utilize convolutional neural networks (CNNs) to predict scene depth maps in an end-to-end manner [15]. However, most existing CNN-based methods train their networks with spatial features yet ignore the enormous and important temporal features in monocular videos. What's more, most networks only regress the depth map from the last scale, which lose multi-scale information of feature maps.

In this paper, we propose a novel framework, named Monocular Depth estimation with Spatio-Temporal feasures (MD-ST), to capture spatio-temporal correlation features based on CNNs and recurrent neural networks (RNNs) for monocular depth estimation. First, the depth estimation network is improved from the existing deep single-view and multi-view depth estimation networks [14] by adding a convolutional long short-term memory (Conv-LSTM) unit [19] after each convolutional layer to encode the temporal features from multiple previous frames for each estimated depth map. The Conv-LSTM structure transfers the temporal information from the previous view to the depth estimation of the current frame. Second, we adopt the FlowNet [4] to learn the residual flow between adjacent estimated depth maps and ground-truth depth maps respectively, which makes sure that the temporal information is considered by the entire model during training. In order to maintain the temporal consistency, we design a flow consistency loss to penalize the errors between the flow f_d learned from estimated depth maps and f_{gt} learned from ground-truth maps. Last but not least, we apply multi-scale strategy and output four depth maps compared to the corresponding resized ground-truth depth maps to optimize the estimations. Therefore, our proposed framework can effectively learn spatial-temporal accurate depth maps from videos. The proposed pipeline is depicted in Fig. 1.

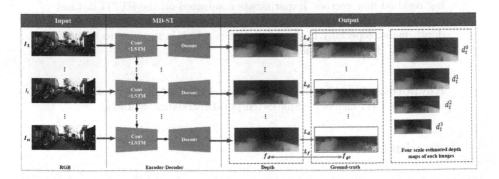

Fig. 1. The overview of MD-ST.

Our contributions are summarized as follows:

(1) **Depth Estimation**: We propose a novel CNN-RNN network, MD-ST, which learns spatio-temporal correlation features for monocular depth

estimation. Additionally, we learn multi-scale depth maps to maintain different scale depth and semantic information contributed for accurate depth estimation.

(2) **Spatio − Temporal Consistency**: The temporal information in video sequences is fully maintained by learning flow vectors between depth maps from a pre-trained flow network, which ensure that the temporal information in the RGB images is retained in the depth map. The spatial and temporal constraints can enforce the spatio-temporal consistency and improve the performance of monocular depth estimation in videos.

(3) **Experimental Feasibility**: Experiments on KITTI dataset [6] demonstrate the effectiveness of our proposed network in both outdoor and indoor scenes, and our method outperforms most existing methods in depth maps estimation, especially in dynamic scenes.

The rest of this paper is organized as follows. Section 2 reviews related works about depth estimation based on deep learning. Section 3 introduces our proposed method and spatial-temporal consistency loss functions. Section 4 presents our experiments and results on KITTI dataset about monocular depth estimation. Our conclusions are summarized in Sect. 5.

2   Related Work

Deep estimation based on deep learning was firstly proposed by Eigen et al. [5] in 2014, which used a multi-scale neural network with two components to generate a rough estimate globally and optimize the results locally. On this basis, more and more researchers began to recovery depth maps utilizing deep neural networks. In this section, we summarize some related works about depth estimation based on deep learning.

2.1   Depth Estimation Based on CNNs

Supervised Learning. Li et al. [10] added a jump connection to the network to accelerate the convergence of the network based on Eigen et al. [5], which fused the depth gradient with the estimated depth and improved the accuracy of the depth map effectively. Liu et al. [12] combined a convolutional neural network with continuous conditional random field (CRF) to propose a deep convolutional neural field model. Kim et al. [9] used a depth analogy to construct a depth change model for single-image depth estimation, and used global and local networks to capture relative information, respectively, to solve the problem of scale blur in static scene estimation. These methods all supervise the true value of the scene depth, and assume that the scene is static. When dynamic objects appear in the scene, the accuracy of the estimated depth map is poor, as shown in Fig. 2.

Unsupervised Learning . Godard et al. [7] used epipolar geometric constraints to generate parallax maps, and then used left and right parallax consistency to optimize performance and improve robustness. Zhan et al. [21] adopted stereo

Fig. 2. Some failed depth estimations in dynamic scenes [9].

vision to train the depth estimation and camera pose networks. Zhou et al. [23] used two independent networks to estimate the change of camera pose in a single-frame depth map and video sequence, respectively. Wang et al. [18] applied three parallel CNN networks to simultaneously predict stereo image depth, camera pose and scene optical flow, and calculated the geometric relationship among them during the training process. However, these methods all use CNNs for spatial depth prediction, ignoring the temporal information in the video sequence, which limits the performance of depth estimation.

2.2   Depth Estimation Based on RNNs

Most methods based on CNNs only consider the spatial information of the images, yet ignore the temporal information between frames in video. Recently, several methods have used RNNs to learn the temporal information for depth maps estimation. Mancini et al. [13] proposed an RNN-based network to predict scene depth, which was first trained on synthetic data and then generalized to real data. Their network adopted LSTM to exploit the input stream sequentiality, where the LSTM was behind the encoder layers. Kumar et al. [2] proposed a ConvLSTM-based network for monocular video sequence depth prediction, whose encoder network consisted of only multiple ConvLSTM [19] layers. Their structure learned the depth as an appearance function while implicitly learned the object pose and its smooth temporal variation. Atapour et al. [1] proposed a network based on multi-task learnings that could jointly perform deep estimation and semantic segmentation. They used the RNN-based feedback network to take the output of the previous time step as a cyclic input, and utilized the optical flow estimation model to fully consider the temporal information. Wang et al. [17] proposed an RNN-based network to simultaneously estimate visual odometry and depth maps, which adopted multi-view re-projection and forward-backward flow consistency losses for training. It intersected the convolutional and the LSTM layers, and could be trained in supervised and unsupervised ways. However, their optical flow was calculated based on the estimated depth and camera pose, which only affect the rigid objects. When non-rigid objects (e.g. pedestrians) appeared in the scene, the accuracy of depth map was decreased. Considering the optical flow estimation on the entire scene, our method can obtain a higher-precision depth maps, especially for dynamic scenes. In addition, these methods regressed depth maps only in the last scale, which ignore the information of multi-scale depth maps resulting in low quality depth maps. In this paper, we adopt multi-scale strategy to predict four scale depth maps to achieve an accurate estimation.

3 Our Method

In this section, we first elaborate an overview of the proposed framework, MD-ST, for monocular depth estimation; then we detail the architecture of our MD-ST; finally, we describe the spatial and temporal consistency constraints.

3.1 Overview of MD-ST

Our proposed encoder-decoder framework, MD-ST, as shown in Fig. 1, learns monocular depth maps in the supervised learning. The inputs of our pipeline are monocular sequence images, and the outputs are four scales estimated depth maps. Specially, the proposed depth estimation network is trained with ground-truth maps to learn the spatio-temporal correlation features for monocular depth estimation. And it predicts depth maps at different scales compared to the resized ground-truth maps for back-propagation. In addition, we adopt a depth-flow estimation network to learn residual flow vectors of estimated depth maps and ground-truth maps, constrained by a flow consistency loss to capture dynamic objects in the scene and maintain the temporal consistency of the video sequence. All parts are jointly trained in a supervised learning and interacting at the same time, making the understanding of 3D scenes easier.

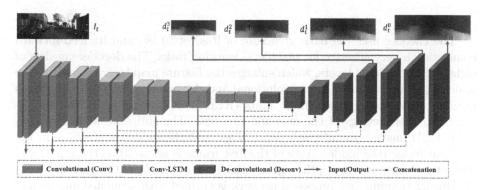

Fig. 3. The architecture of MD-ST. Parameters are similar to the DispNet [14].

Architecture of MD-ST. For the depth estimation module, we use a U-shaped encoder-decoder network to predict the scene depth maps, which is similar to DispNet [14], as shown in Fig. 3. The network takes color images as input and uses ground-truth depth maps as supervisory data to learn the depth map of each frame in a monocular video sequence. It is worth noting that each convolutional layer in the encoder of the MD-ST network is immediately followed by a convolutional LSTM (Conv-LSTM) unit [19]. This structure can save and transmit the captured 2D feature map to the depth estimation of the next frame. This convolutional neural network combined with the structure of a recurrent neural network can capture not only spatial features but also the temporal information in depth estimation. The spatio-temporal correlation features learned by the encoder are input to the decoder, and accurate depth map restoration can be performed by up-sampling.

Table 1. Parameters of MD-ST. "C+L" is convolutional layer and Conv-LSTM; "D+Co" is deconvolutional layer and concatenate.

Type	Filters	Size	Type	Filters	Size
Input		128 × 416 × 3			
C+L	16, 7×7×3, 2	64×208×16	D+Co	1024, 3×3×1024, 2	2×7×1024
C+L	32, 5×5×16, 2	32×104×32	D+Co	512, 3×3×1024, 2	4×13×512
C+L	64, 3×3×32, 2	16×52×64	D+Co	256, 3×3×512, 2	8×26×256
C+L	128, 3×3×64, 2	8×26×128	D+Co	128, 3×3×256, 2	16×52×128
C+L	256, 3×3×128, 2	4×13×256	D+Co	64, 3×3×128, 2	32×104×64
C+L	512, 3×3×256, 2	2×7×512	D+Co	32, 3×3×64, 2	64×208×32
C+L	1024, 3×3×512, 2	1×4×1024	D+Co	16, 3×3×32, 2	128×416×16
			Output	1, 3×3×16, 2	128×416×1

The encoder uses the basic structure of ResNet-50 [8], and its feed-forward connection is very effective for pixel-level learning tasks. The decoder consists of eight deconvolutional layers, which enlarges the feature map layer by layer to the scale of the input image. All convolutional and deconvolutional layers use batch normalization and rectified linear units (RELU) as excitation. Table 1 details the parameter settings of the network structure. The entire network structure takes RGB image, ground-truth depth map, and the hidden state h_{t-1} in the previous frame (the hidden state of the first frame is set to 0) as the input, and the network outputs the depth map and the hidden state h_t of the current frame.

During training, our proposed network is trained with sequence images and supervised with corresponding ground-truth depth maps. The encoder learns the spatio-temporal correlation features of image sequences, and the decoder estimates the depth maps at different scales to penalize multi-scale errors for back-propagation. The loss functions consist of spatial and temporal losses. In test stage, we only estimate the depth map at the last layer of the decoder.

3.2 Loss Functions

In order to estimate the accurate depth map and fully consider the spatio-temporal correlation features of the monocular video sequence, we use spatial consistency constraints and temporal consistency constraints to train the entire network, as shown in Fig. 1. This section introduces the loss functions applied in the MD-ST during training.

Spatial Consistency Constraints. The monocular depth estimation network in this paper is a supervised training network based on CNN-RNN, and its inputs are RGB images and ground-truth depth images. At t, its input is I_t and the predicted depth map is d_t^i ($i = 0,1,2,3$). In order to improve the training effect of the network, we apply a multi-scale loss function to penalize the errors and estimate the depth maps at the last four scales, and we down-sample the ground-truth depth maps to form four depth maps with different scales corresponding to the estimated depth maps scales. The loss function commonly used in regression problems is the L_1 norm. Thence, we use L_1 to minimize the errors between the predicted depth map d_t^i and the ground-truth depth map $\hat{d}_t^i$ at the same scale. The depth estimation loss function is:

$$L_d^i = \sum_{t=0}^{n-1} \|d_t^i - \hat{d}_t^i\|_1 \tag{1}$$

Therefore, the total depth estimation loss function is :

$$L_d = \sum_{i=0}^{3} L_d^i \tag{2}$$

The change of the image gray-level will cause the discontinuity of the image depth, so it is necessary to penalize the errors around edges. We use the gradient loss L_{grad} to express the edge-aware term constraint:

$$L_{grad} = \sum_{i=0}^{3} \sum_{t=0}^{n-1} \sum_h \sum_{u,v} \|g_h[d_t^i](u,v) - g_h[\hat{d}_t^i](u,v)\|_2 \tag{3}$$

where h is the spatial step used to calculate g in different scales. The vector g_h is the normalized discrete measurement of the proportion of the local change of f. The measurement is defined as

$$g_h[f](u,v) = (\frac{f(u+h,v) - f(u,v)}{|f(u+h,v)| + |f(u,v)|}, \frac{f(u,v+h) - f(u,v)}{|f(u,v+h)| + |f(u,v)|}) \tag{4}$$

Temporal Consistency Constraints. The monocular depth estimation network combined with Conv-LSTM can cyclically input the output of each frame into the depth estimation process of the next frame. Therefore, the network can implicitly learn the temporal continuity of the monocular video sequence.

In addition, in order to ensure that the estimated depth map also has temporal consistency when output, we use the trained flow estimation network, FlowNet [4], to enhance temporal consistency. It is worth noting that we only perform flow estimation on the highest resolution depth map.

We use a pre-trained flow estimation network to estimate the flow vector F $(d_t^0, d_{t+1}^0))$ of the adjacent predicted depth maps, and the flow F $(\hat{d}_t^0, \hat{d}_{t+1}^0)$ of the ground-truth depth maps corresponding to the predicted depth map. In order to capture the dynamic information in the scene and maintain temporal continuity, we predicted the errors of the predicted depth map flow and the ground truth depth map flow, that is, the flow consistency loss function is:

$$L_f = \sum_{t=0}^{n-1} \|F(d_t^0, d_{t+1}^0) - F(\hat{d}_t^0, \hat{d}_{t+1}^0)\|_2 \tag{5}$$

In this paper, the spatial consistency constraint and temporal consistency constraint are used to train the network at the same time. The interaction between the depth estimation and flow estimation is to obtain a highly accurate monocular depth maps. All loss functions are combined as:

$$L_{final} = \lambda_d L_d + \lambda_g L_{grad} + \lambda_f L_f \tag{6}$$

where $\lambda_d, \lambda_g, \lambda_f$ are parameters of each loss function.

4   Experiments and Results

We use KITTI [6] to train and test the proposed MD-ST for depth estimation, and evaluate its performance with both quantitative and qualitative aspects.

The KITTI dataset [6] is sampled and synchronized at a frequency of 10 Hz, including 61 video sequences in outdoor scenes such as urban areas and high speeds, 93,000 RGB-D training samples, and an image size of 1242×375. In order to reduce the running time and calculation cost, we adjust the image size to 416×128. For fair comparison, we apply KITTI$_{eigen}$ [5] dataset for training and testing, which contains 23,488 images of 33 scenes for training, and 697 images of 28 sequences for testing. According to Eigen et al. [5], the length of the sequence is set to 3 in our experiments.

4.1   Implementations and Metrics

The monocular depth estimation network, MD-ST, is built on the PyTorch [16] deep network framework. The hardware equipment is a 6-core i7-8750 H 2.2 GHz CPU and NVIDIA GTX 1060 GPU. We use the Adam optimizer to train the network, where $\beta_1 = 0.9$, $\beta_2 = 0.99$; the initial learning rate is set to 0.002, and manually reduced; the batch size is 16, and the minimum batch size is 8; the training process is gradually stabilized when carried out about 200 epochs. For the parameters in the joint loss function, we set $[\lambda_d, \lambda_g, \lambda_f] = [10,10,1]$ according to experimental debugging.

We use the evaluation metrics [5] to quantitatively evaluate the performance of our proposed network and compare it with existing methods. Evaluation metrics include error and accuracy metrics. The error metrics (smaller is better) include absolute relative error (Abs.rel), square relative error (Sq.rel), root mean square error (RMSE), and the logarithm root mean square error (log RMS); the accuracy rate metrics (the bigger the better) include $\delta < 1.25^t$, where t = 1,2,3.

4.2 Results of Depth Estimation

In order to verify the effectiveness of the proposed method in this paper and evaluate the performance of its monocular depth estimation, the KITTI$_{eigen}$ [5] dataset is used for training; during testing, the time for single image depth estimation is about 7.5 ms.

The comparison of the quantitative results of the depth estimation is shown in Table 2. It can be seen that our method can effectively estimate the scene depth maps of monocular video. Our results are superior to most existing methods, regardless of error or accuracy matrix, which proves the effectiveness and advancement of our proposed network. Our results are better than those based on CNNs, which shows that extracting the temporal information of video sequences can effectively improve the performance of depth estimation. It is worth noting that our results are worse than Wang et al. [17] on Sq.rel, RMSE, and $\delta < 1.25^3$, because their method utilizes the camera pose to improve the performance of depth estimation, yet we don't predict the camera motion.

Table 2. Comparisons of depth estimation on KITTI.

Methods	Super.	Abs. rel	Sq. rel	RMSE	Log RMS	$\delta < 1.25$	$\delta < 1.25^2$	$\delta < 1.25^3$
Eigen [5]	✓	0.203	1.548	6.307	0.282	0.702	0.890	0.958
Liu [12]	✓	0.202	1.614	6.523	0.275	0.678	0.895	0.965
Li [10]	✓	0.161	–	6.402	0.168	0.765	0.950	0.988
Zhou [23]		0.208	1.768	6.856	0.283	0.678	0.885	0.957
Godard [7]		0.148	1.344	5.927	0.247	0.803	0.922	0.964
Zhan [21]		0.144	1.391	5.869	0.241	0.803	0.928	0.969
Wang [18]		0.060	0.833	4.187	0.135	0.955	0.981	0.990
Mancini [13]	✓	0.312	1.072	5.654	0.366	0.512	0.786	0.911
Kumar [2]	✓	0.137	1.019	5.187	0.218	0.809	0.928	0.971
Atapour [1]	✓	0.193	1.438	5.887	0.234	0.836	0.930	0.958
Wang [17]	✓	0.077	**0.205**	**1.698**	**0.110**	0.941	0.990	**0.998**
Ours	✓	**0.058**	1.011	2.125	**0.110**	**0.956**	**0.990**	0.995

The visualization results of image depth estimation are shown in Fig. 4. We compare our estimated depth map with various methods, such as Kumar et al. [2]. It can be seen from the qualitative results that our proposed method can effectively predict scene depth maps. Compared to other methods, ours can retain boundaries for dynamic and thin objects (such as trees) and learn better depth map in smooth depth gradients (as shown in green rectangles).

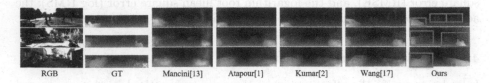

RGB   GT   Mancini[13]   Atapour[1]   Kumar[2]   Wang[17]   Ours

Fig. 4. Visualization results of monocular depth estimation. (Color figure online)

4.3   Ablation Studies

In order to prove the effectiveness of the network structure and temporal consistency constraints proposed in this paper, experiments are verified one by one. As shown in Table 3, Our Full represents the entire monocular depth estimation network, and Full\D represents the method doesn't apply multi-scale depth estimation and only calculate the depth errors on the highest resolution depth map. Full\D\F represents the method neither utilizes multi-scale depth estimation nor adopts the depth map flow consistency constraint. Full\D\F\LSTM represents the basic U-shaped monocular depth estimation network, which doesn't include Conv-LSTM layers and use the multi-scale depth and the flow consistency loss.

Table 3. Results of ablation studies.

Methods	Abs. rel	Sq. rel	RMSE	Log RMS	$\delta < 1.25$	$\delta < 1.25^2$	$\delta < 1.25^3$
Full\D\F\LSTM	0.201	1.584	6.471	0.273	0.705	0.899	0.968
Full\D\F	0.151	1.343	5.848	0.245	0.785	0.923	0.974
Full\D	0.109	1.211	4.986	0.175	0.848	0.950	0.980
Our Full	**0.058**	**1.011**	**2.125**	**0.110**	**0.956**	**0.990**	**0.995**

The results of Full\D\F is better than Full\D\F\LSTM's. It shows that the network combined with Conv-LSTM layers can not only capture spatial features but also learn and maintain temporal features, which keeps the temporal information between the forward and backward frames and is helpful for depth estimation. The results of Full\D outperforms Full\D\F's, which verifies the importance of depth map flow estimation. Flow consistency constraint makes sure that the temporal information in the input images is propagated to the estimated depth maps and enforces the temporal consistency. The results of our Full is superior to Full\D's, which demonstrate that multi-scale depth estimation can effectively retain the depth information of different scales and enable the network to remember the spatial relationship of feature maps with different resolutions, improving the accuracy of depth estimation.

In a word, the comparison of the experiment results shows the effectiveness of the proposed temporal consistency and the recurrent convolutional network, and the contribution of each term to the monocular depth estimation.

5   Conclusion

In this paper, we propose a monocular depth estimation framework based on CNN-RNN, namely MD-ST. The network is constructed using a CNN combined with the Conv-LSTM, which can fully transmit the spatio-temporal correlation features in the video sequence. Specially, in order to fuse depth features of different scales, we apply a multi-scale depth estimation loss to penalize the errors at last four estimated depth maps. In addition, we also propose a flow consistency loss function, which can maintain the temporal information in the output results. Experiments on the KITTI dataset prove that the proposed method can effectively predict the monocular depth maps, and its performance is better than most existing methods. In the future, it is planned to introduce a multi-scale feature fusion strategy to encode features at different scales to further improve the performance of depth estimation; in addition, pixel-level semantic segmentation can also be introduced to enrich the semantic information of the scene.

References

1. Atapour-Abarghouei, A., Breckon, T.P.: Veritatem dies aperit-temporally consistent depth prediction enabled by a multi-task geometric and semantic scene understanding approach. In: Proceedings of the IEEE Conference on Computer Vision and Pattern Recognition, pp. 3373–3384 (2019)
2. CS Kumar, A., Bhandarkar, S.M., Prasad, M.: Depthnet: a recurrent neural network architecture for monocular depth prediction. In: Proceedings of the IEEE Conference on Computer Vision and Pattern Recognition Workshops, pp. 283–291 (2018)
3. Doan, A.D., Latif, Y., Chin, T.J., Liu, Y., Do, T.T., Reid, I.: Scalable place recognition under appearance change for autonomous driving. In: Proceedings of the IEEE International Conference on Computer Vision, pp. 9319–9328 (2019)
4. Dosovitskiy, A., et al.: Flownet: learning optical flow with convolutional networks. In: Proceedings of the IEEE International Conference on Computer Vision, pp. 2758–2766 (2015)
5. Eigen, D., Puhrsch, C., Fergus, R.: Depth map prediction from a single image using a multi-scale deep network. In: Advances in Neural Information Processing Systems, pp. 2366–2374 (2014)
6. Geiger, A., Lenz, P., Urtasun, R.: Are we ready for autonomous driving? the kitti vision benchmark suite. In: 2012 IEEE Conference on Computer Vision and Pattern Recognition, pp. 3354–3361. IEEE (2012)
7. Godard, C., Mac Aodha, O., Brostow, G.J.: Unsupervised monocular depth estimation with left-right consistency. In: Proceedings of the IEEE Conference on Computer Vision and Pattern Recognition, pp. 270–279 (2017)

8. He, K., Zhang, X., Ren, S., Sun, J.: Deep residual learning for image recognition. In: Proceedings of the IEEE Conference on Computer Vision and Pattern Recognition, pp. 770–778 (2016)
9. Kim, Y., Jung, H., Min, D., Sohn, K.: Deep monocular depth estimation via integration of global and local predictions. IEEE Trans. Image Process. **27**(8), 4131–4144 (2018)
10. Li, J., Klein, R., Yao, A.: A two-streamed network for estimating fine-scaled depth maps from single rgb images. In: Proceedings of the IEEE International Conference on Computer Vision, pp. 3372–3380 (2017)
11. Li, Y., Zhu, J., Hoi, S.C., Song, W., Wang, Z., Liu, H.: Robust estimation of similarity transformation for visual object tracking. In: Proceedings of the AAAI Conference on Artificial Intelligence, vol. 33, pp. 8666–8673 (2019)
12. Liu, F., Shen, C., Lin, G., Reid, I.: Learning depth from single monocular images using deep convolutional neural fields. IEEE Trans. Pattern Anal. Mach. Intell. **38**(10), 2024–2039 (2015)
13. Mancini, M., Costante, G., Valigi, P., Ciarfuglia, T.A., Delmerico, J., Scaramuzza, D.: Toward domain independence for learning-based monocular depth estimation. IEEE Rob. Autom. Lett. **2**(3), 1778–1785 (2017)
14. Mayer, N., et al.: A large dataset to train convolutional networks for disparity, optical flow, and scene flow estimation. In: Proceedings of the IEEE Conference on Computer Vision and Pattern Recognition, pp. 4040–4048 (2016)
15. Meng, X., Fan, C., Ming, Y., Shen, Y., Yu, H.: Un-vdnet: unsupervised network for visual odometry and depth estimation. J. Electron. Imaging **28**(6), 063015 (2019)
16. Paszke, A., et al.: Automatic differentiation in pytorch (2017)
17. Wang, R., Pizer, S.M., Frahm, J.M.: Recurrent neural network for (un-) supervised learning of monocular video visual odometry and depth. In: Proceedings of the IEEE Conference on Computer Vision and Pattern Recognition, pp. 5555–5564 (2019)
18. Wang, Y., Wang, P., Yang, Z., Luo, C., Yang, Y., Xu, W.: Unos: unified unsupervised optical-flow and stereo-depth estimation by watching videos. In: Proceedings of the IEEE Conference on Computer Vision and Pattern Recognition, pp. 8071–8081 (2019)
19. Xingjian, S., Chen, Z., Wang, H., Yeung, D.Y., Wong, W.K., Woo, W.C.: Convolutional lstm network: a machine learning approach for precipitation nowcasting. In: Advances in Neural Information Processing Systems, pp. 802–810 (2015)
20. Yu, Z., Zheng, J., Lian, D., Zhou, Z., Gao, S.: Single-image piece-wise planar 3D reconstruction via associative embedding. In: Proceedings of the IEEE Conference on Computer Vision and Pattern Recognition, pp. 1029–1037 (2019)
21. Zhan, H., Garg, R., Saroj Weerasekera, C., Li, K., Agarwal, H., Reid, I.: Unsupervised learning of monocular depth estimation and visual odometry with deep feature reconstruction. In: Proceedings of the IEEE Conference on Computer Vision and Pattern Recognition, pp. 340–349 (2018)
22. Zhang, K., Chen, J., Li, Y., Zhang, X.: Visual tracking and depth estimation of mobile robots without desired velocity information. IEEE Trans. Cybern. **50**(1), 361–373 (2018)
23. Zhou, T., Brown, M., Snavely, N., Lowe, D.G.: Unsupervised learning of depth and ego-motion from video. In: Proceedings of the IEEE Conference on Computer Vision and Pattern Recognition, pp. 1851–1858 (2017)

A Remote Sensing Image Segmentation Model Based on CGAN Combining Multi-scale Contextual Information

Xili Wang[1]([✉]), Xiaojuan Zhang[1], Huimin Guo[2], Shuai Yu[1], and Shiru Zhang[3]

[1] School of Computer Science, Shaanxi Normal University, Xi'an, China
wangxili@snnu.edu.cn
[2] School of Electrical and Electronic Engineering, Nanyang Technological University, Singapore, Singapore
[3] College of Communication and Information Engineering, Xi'an University of Science and Technology, Xi'an, China

Abstract. In order to extract targets in different scale under complex scene, this paper proposes a remote sensing image segmentation model based on Conditional Generative Adversarial Network (CGAN) combing multi-scale contextual information. This end-to-end model consists of a generative network and a discriminant network. A SegNet model fusing multi-scale contextual information is proposed as the generative network. In order to extract multi-scale contextual information, the multi-scale features of the end pooling feature map in the encoder are extracted using different proportion of dilated convolution. The multi-scale features are further fused with the global feature. The discriminant network is a convolution neural network for two category classification, determines whether the input is a generated result or the ground truth. After alternate adversarial training, the experimental results on a remote sensing road dataset show that the road segmentation results of the proposed model are superior to those of the comparable models in terms of target integrity and details preserving.

Keywords: Road segmentation · Conditional Generative Adversarial Network · Multi-scale context · Feature fusion

1 Introduction

In recent years, deep convolution neural networks have made breakthroughs in pixel-level classification of natural images. Various image segmentation models based on fully convolutional networks have been proposed. Long et al. [1] proposed fully convolutional networks (FCNs) to complete pixel-level classification. They replaced the fully connected layer in CNN with a convolutional layer, and used the deconvolution layer to upsample the feature map to score each category. FCN creates a precedent for encoding-decoding structure based on pixel-level classification. Following this paradigm, many

This work is supported by the Second Tibetan Plateau Scientific Expedition and Research (2019QZKK0405).

© Springer Nature Switzerland AG 2020
Y. Peng et al. (Eds.): PRCV 2020, LNCS 12305, pp. 365–373, 2020.
https://doi.org/10.1007/978-3-030-60633-6_30

CNN architectures have been proposed and further improved the segmentation performance. U-Net [2] makes full use of low-level and high-level features by using jump connections to obtain more accurate segmentation results. SegNet [3] uses the pooling information of the encoding stage in the decoder to reduce the amount of network parameters. After that, in order to reduce the information loss caused by the pooling operation, DeepLab V1 [4] and V2 [5] were proposed with atrous convolution. Huang et al. [6] proposed to share atrous convolution, different receptive fields share the same convolution kernel, reduce the amount of network parameters, and improve the generalization ability of the network. The segmentation results of these kind of models are relatively coarse due to feature loss caused by repeated pooling and multi-strided convolutions. The segmentation results of these kind of models are relatively coarse due to feature loss caused by repeated pooling and multi-strided convolutions. Therefore, post-processing is proposed to optimize the segmentation results [5, 7]. But the post-processing is complicated, computational costly and is independent of deep convolution network models.

Generative Adversarial Networks (GAN) [8] is one of the research hotspot in computer vision. The structure of GAN includes two parts: generator and discriminator. Without assumption of the distribution of the training data, the data generated by the generator can be infinitely close to the actual distribution of training data after iterative adversarial training. GAN achieves the optimal state when discriminator cannot distinguish whether the input is the training data or the generated data. However, the training of the original GAN will become uncontrollable if the dimension of data is high and the data distribution is complex.

Conditional Generative Adversarial Network (CGAN) [9] is an extension of the original GAN. It introduces conditional variables in the generator and discriminator to constrain the model training. For image segmentation, the conditional variable is the class label at pixel level. In CGAN, generator accomplishes segmentation task, and discriminator distinguishes that the input is come from the generator or the ground truth. CGAN was first used by Luc et al. [10] in semantic segmentation task. Results showed that it can produce more smooth and accurate segmentation results compared with semantic segmentation models. For example, references [11, 12] proposed CGAN bases image segmentation methods. They leverage commonly used semantic segmentation models as the generative network and fully connected convolutional neural network as the discriminant network. These models performed well in medical image segmentation task. In such tasks, the differences between the lesion area and the adjacent background area is obvious, thus the extracted features are more distinguishable. However, it is more challenging to extract features from targets and backgrounds in remote sensing images due to the complex scenes - there are many targets and occlusion, shadows in image. In recent years, GAN has been used in target segmentation of remote sensing images. For example, Jin et al. [13] used FCN model as the generation network of GAN and three-layer convolutional neural network without pooling layer as the discriminant network to extract buildings from SAR images. Giorgio et al. [14] used GAN to extract shadows from high-resolution multispectral satellite images. The generative network was a two-layer convolution structure similar to the U-Net model. The discriminant network was a five-layer fully convolutional network without pooling layer. Zhu et al. [15] proposed a new adversarial segmentation model SegGAN. It consisted of semantic segmentation generative network and discriminant network, trained end-to-end on public datasets and achieved the best results. Though the above researches explored the application of GAN model in remote sensing image target segmentation, the existing semantic segmentation networks were used as the generative network. They cannot well deal with the problems such as multi-scale, occlusion or shadows of the targets in remote sensing images segmentation, hence result in coarse and inaccurate segmenting results. In order to improve the segmentation performance, more effective features are needed.

Based on this, this paper proposes a Conditional Generative Adversarial Network-Fusing multi-scale Context (CGAN-FMC) model for remote sensing image segmentation. The main contribution of this paper are as follows: aiming at the difficulties in object segmentation in remote sensing images, a new semantic segmentation model that based on SegNet model is proposed as the generator. In order to extract distinguishing features from complex scenes, different proportions of dilated convolution are used to extract the multi-scale features from the end-pooling layer feature map in the encoder. After fusion of the multi-scale features, the features are fused with global feature to obtain multi-scale contextual information. The proposed generative network is able to retain and fuse features of different resolution, thus deal with multi-scale image targets segmentation better in target integrity, details preserving and spatial continuity. Meanwhile, labels as conditions improve training controllability and the performance of the model.

2 Methodology

2.1 CGAN-FMC Model

CGAN-FMC consists of two sub-networks: generative network and discriminant network. The generative network is a semantic segmentation model that outputs the segmentation results. The discriminant network is a classification model based on convolutional neural network that judges whether the input comes from the ground truth images or the generated images. The model structure is shown in Fig. 1. The input of the generative network is three-channel color image and the corresponding pixel-level labeled images. The input of the discriminant network includes images generated by the generative network, the pixel-level labeled images and the original color images. The original images and the pixel-level labeled images form positive samples while the original images and the generated images form negative samples.

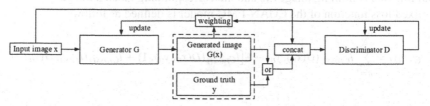

Fig. 1. The structure of CGAN-FMC model

The structure of the generative network is shown in Fig. 2. Its backbone network consists of an encoder and a decoder. The encoder consists of 13 convolution layers and 5 pooling layers. The decoder is mirror image of the encoder, it consists of 13 deconvolution layers and 5 unpooling layers. The perceptual fields of different proportions of dilated convolution are different and the local information contained in the extracted features is also different. To extract contextual information at different scales, the dilated convolution with proportions of 3, 6 and 10 are used in the feature map of the end pooling layer in encoder. In deep network, larger dilated proportion is help to extract more contextual information, but not help to capture smaller targets and is insensitive to the target edge details. Therefore, the feature maps obtained via different proportions of dilated convolution are fused via layer by layer and pixel by pixel. In this way the hierarchical dependencies of features at different scales are considered and the local consistency of feature is maintained. Finally, the fusion results are concatenated with the unprocessed global feature map (i.e. Pool5 feature map) to achieve the purpose of fusing global information. During the decoding process, feature maps containing multi-scale features and global features are first reduced in dimension and then fed into the decoder.

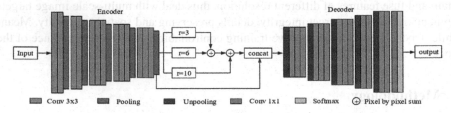

Fig. 2. Structure of the generative network

The discriminator consists of 10 convolution layers, 3 pooling layers, a fully connected layer and a softmax layer. Generated images or labeled images are concatenated with the original color images as the input of the discriminant network. The convolution layers are responsible for feature extraction, and the pooling layers are responsible for feature dimension reduction and mapping.

The fully connected layer converts the feature map extracted from the previous layer into a fixed length feature vector and sends it to the softmax layer for classification.

2.2 Objective Function and Model Optimization

Assuming that N training images x_n and the corresponding labeled images y_n are given, the mixed loss function of the CGAN-FMC model is defined as follows:

$$l(\theta_G, \theta_D) = \sum_{n=1}^{N} l_{CrossG}(G(x_n), y_n) - \lambda[l_{CrossG}(D(x_n, y_n), 1) + l_{CrossD}(D(x_n, G(x_n)), 0)]$$

(1)

Where θ_G, θ_D represents the parameters of the generative and the discriminant network respectively. $G(x_n)$ represents the image generated by the generator. l_{CrossG} stands for the loss function of the generator. l_{CrossD} stands for the loss function of the discriminator. The model is optimized by alternating optimization:

(1) Optimizing discriminant network: fix the parameters of the generative network, the loss function of the discriminant network is expressed as follows:

$$L_D = \sum_{n=1}^{N} l_{CrossD}(D(x_n, y_n), 1) + l_{CrossD}(D(x_n, G(x_n)), 0) \qquad (2)$$

There are two forms of input data for discriminant network: when input (x_n, y_n), the discriminant network is labeled as True, or '1'; when input $(x_n, G(x_n))$, the discriminant network is labeled as False, or '0'. Then the parameters of the discriminant network are updated using back propagation algorithm.

(2) Optimizing generative network: fix the parameters of the discriminant network, the loss function of the generative network is expressed as follows:

$$L_G = \sum_{n=1}^{N} l_{CrossG}(G(x_n), y_n) - \lambda l_{CrossD}(D(x_n, G(x_n)), 0) \qquad (3)$$

The second item in formula (3) lets the discriminant network playing a role in the training process of the generative network. In which λ represents the weight of the discriminator's loss function and determines the degree of information feedback from discriminator to generator. When $\lambda = 0$, it is equivalent to the traditional semantic segmentation networks training method. The parameters of the generative network are updated using back propagation algorithm.

For better convergence of the model, the second item is usually replaced by $+ \lambda l_{CrossD}(D(x_n, G(x_n)), 1)$. It maximizes the probability that $G(x_n)$ being predicted as the ground truth by the discriminant network, that is, makes the generated image closer to the ground truth image.

Both generator and discriminator can be regarded as two category classifier. Thus cross entropy function is used as the loss function of the two sub-networks. It is defined as follows:

$$Cross(l, p, \theta) = \frac{1}{N} \sum_{i=1}^{N} \sum_{k=1}^{K} -\sigma(l_i = k) \log p_{k,i} \qquad (4)$$

Where l_i represents the ground truth label of pixel i. $p_{k,i}$ represents the output probability of pixel i belonging to class k. K stands for the total number of classes. N stands for the total number of pixels in batch images. And

$$\sigma(l_i = k) = \begin{cases} 1 & l_i = k \\ 0 & otherwise \end{cases}.$$

3 Dataset and Evaluation Indicators

The Road Detection dataset [16] is used in the experiment. It is collected from Google Earth and the road segmentation reference map with its corresponding center line reference map have been manually labeled. The dataset contains 224 high-resolution images with spatial resolution of 1.2 m, each image has at least 600 x 600 pixels and the width of the road is around 12–15 pixels. The 224 images were randomly divided into 180 training, 14 validation and 30 test samples. The network was trained with 300 x 300 image blocks.

Precision (P), recall (R) and intersection over union (IOU) are used to evaluate the segmentation result. They are defined as follows:

$$P = \frac{TP}{TP + FP}$$

$$R = \frac{TP}{TP + FN}$$

$$IOU(P_m, P_{gt}) = \frac{|P_m \cap P_{gt}|}{|P_m \cup P_{gt}|}$$

Where: TP-positive class predicted as positive. FP-negative class predicted as positive. FN-positive class predicted as negative. TN-negative class predicted as negative. P_{gt}-the pixel set of ground truth images. P_m-the pixel set of the prediction images. $\cap$ and $\cup$-intersection and union operations. $|\cdot|$ means calculate the number of pixels.

4 Experimental Results and Analysis

The experiment was carried out on a workstation equipped with a NVIDIA TITAN Xp GPU and based on PyTorch deep learning framework. The results of CGAN-FMC on the dataset were compared with those of SegNet(with rate), SegNet(with GAN) and SegNet model. SegNet(with rate) means adding different proportions of dilated convolution to the SegNet model. SegNet(with GAN) means replacing generator in the CGAN-FMC model with SegNet network while other structures unchanged.

During training, the fixed learning rate was 0.0002, batch size was 3, gamma was 50, $\lambda = 10$, momentum was 0.5 and epoch was 200 with 5000 iterations per epoch.

Some of the results are shown in Fig. 3. It can be seen that the results of CGAN-FMC are closer to the ground truth images. There are fewer misclassified pixels and more details are retained, also, the road is smoother and integrated.

The average results of the test set of each model are shown in Table 1. We also compared the results to that of the reference 15, whose method is semantic segmentation by an improved Unet (i.e. Unet-S) network. Unet-S fuses multi-layer features but only has segmentation network. It can be seen that the results of CGAN-FMC are the best on all the three evaluation indicators. This is due to the characteristics of adversarial training and feature fusion in CGAN-FMC.

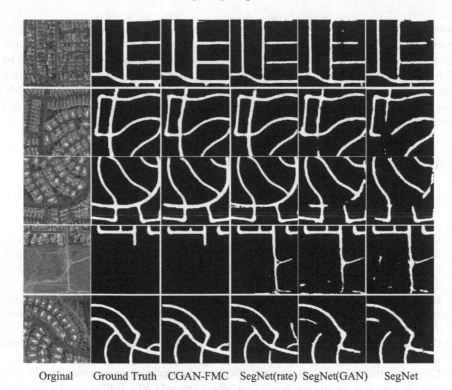

Orginal   Ground Truth   CGAN-FMC   SegNet(rate) SegNet(GAN)   SegNet

Fig. 3. Some of the segmentation results of CGAN-FMC and Other Models

Table 1. Results of different methods

Avg. of test set	P	R	IOU
CGAN-FMC	0.98	0.88	0.84
SegNet(with GAN)	0.95	0.81	0.76
SegNet(with rate)	0.96	0.84	0.80
SegNet	0.94	0.77	0.74
Unet-S [16]	0.97	0.83	0.82

5   Conclusion

Deep neural network segmentation models produce coarse segmentation results due to its structural characteristics. It is difficult to obtain ideal segmentation results for remote sensing images containing complex targets and backgrounds. This paper proposes a remote sensing image segmentation model CGAN-FMC. In the model, an encoding-decoding network that fuses multi-scale contextual information is proposed as the generator and a convolutional neural network with fully connected layer is used as the discriminator. CGAN-FMC provides more discriminated features by combining local feature information with global feature information, and implements better training by discriminator's feedback and the labeled information. After alternating adversarial training, segmented images which are close to the ground truth images can be obtained by CGAN-FMC. By performing training and testing on the Road dataset and comparing with other comparable semantic segmentation models, it can be seen that CGAN-FMC achieves higher segmentation accuracy, the results are more complete, smoother and contain more details.

References

1. Long, J., Shelhamer, E., Darrell, T.: Fully convolutional networks for semantic segmentation. In: Proceedings of the IEEE Conference on Computer Vision and Pattern Recognition, pp. 3431–3440 (2015)
2. Ronneberger, O., Fischer, P., Brox, T.: U-Net: convolutional networks for biomedical image segmentation. In: Navab, N., Hornegger, J., Wells, W.M., Frangi, A.F. (eds.) MICCAI 2015. LNCS, vol. 9351, pp. 234–241. Springer, Cham (2015). https://doi.org/10.1007/978-3-319-24574-4_28
3. Badrinarayanan, V., Kendall, A., Cipolla, R.: SegNet: a deep convolutional encoder-decoder architecture for image segmentation. IEEE Trans. Pattern Anal. Mach. Intell. **39**(12), 2481–2495 (2017)
4. Chen, L.C., Papandreou, G., Kokkinos, I.: Semantic image segmentation with deep convolutional nets and fully connected CRFs. arXiv preprint arXiv:1412.7062 (2014)
5. Chen, L.C., Papandreou, G., Kokkinos, I.: DeepLab: Semantic image segmentation with deep convolutional nets, atrous convolution, and fully connected CRFs. IEEE Trans. Pattern Anal. Mach. Intell. **40**(4), 834–848 (2018)
6. Huang, Y., Wang, Q., Jia, W.: See more than once – kernel-sharing atrous convolution for semantic segmentation. arXiv preprint arXiv:1908.09443 (2019)
7. Zhang, Y., Brady, M., Smith, S.: Segmentation of brain MR images through a hidden Markov random field model and the expectation-maximization algorithm. IEEE Trans. Med. Imaging **20**(1), 45–57 (2001)
8. Goodfellow, I., Pouget-Abadie, J., Mirza, M.: Generative adversarial nets. In: Advances in Neural Information Processing Systems, NIPS, pp. 2672–2680 (2014)
9. Mirza, M., Osindero, S.: Conditional generative adversarial nets. arXiv preprint arXiv:1411.1784 (2014)
10. Luc, P., Couprie, C., Chintala, S.: Semantic segmentation using adversarial networks. arXiv preprint arXiv:1611.08408 (2016)
11. Mahmood, F., Borders, D., Chen, R.: Deep adversarial training for multi-organ nuclei segmentation in histopathology images. arXiv preprint arXiv:1810.00236 (2018)
12. Lahiri, A., Ayush, K., Kumar Biswas, P.: Generative adversarial learning for reducing manual annotation in semantic segmentation on large scale microscopy images: automated vessel segmentation in retinal fundus image as test case. In: IEEE Conference on Computer Vision and Pattern Recognition Workshops, pp. 42–48 (2017)
13. Jin, F., Wang, F., Rui, J.: Residential area extraction based on conditional generative adversarial networks. In: 2017 SAR in Big Data Era: Models, Methods and Applications, pp. 1–5 (2017)
14. Morales, G., Arteaga, D., Huamán, S.G.: Shadow detection in high-resolution multispectral satellite imagery using generative adversarial networks. In: 2018 IEEE XXV International Conference on Electronics, Electrical Engineering and Computing, pp. 1–4 (2018)

15. Zhang, X., Zhu, X., Zhang, N.: SegGAN: semantic segmentation with generative adversarial network. In: 2018 IEEE Fourth International Conference on Multimedia Big Data, pp. 1–5 (2018)
16. Cheng, G., Wang, Y., Xu, S.: Automatic road detection and centerline extraction via cascaded end-to-end convolutional neural network. IEEE Trans. Geosci. Remote Sens. **55**(6), 3322–3337 (2017)

Feature Space Based Loss for Face Photo-Sketch Synthesis

Keyu Li[1], Nannan Wang[1](✉), and Xinbo Gao[2,3]

[1] State Key Laboratory of Integrated Services Networks, School of
Telecommunications Engineering, Xidian University, Xi'an 710071, China
likeyuchn@gmail.com, nnwang@xidian.edu.cn
[2] State Key Laboratory of Integrated Services Networks, School of Electronic
Engineering, Xidian University, Xi'an 710071, China
[3] Chongqing Key Laboratory of Image Cognition, Chongqing University of Posts and
Telecommunications, Chongqing 400065, China
gaoxb@cqupt.edu.cn

Abstract. Learning-based face photo-sketch synthesis has made great progress in the past few years because of the development of the generate adversarial networks (GANs) [1]. However, these existing GAN-based methods mostly yield poor texture and details on the synthesized sketch/photo which leads low perceptual similarity between the synthesized sketch/photo and real sketch/photo. In order to tackle this problem, we first introduce the perceptual loss into our objective loss function which can measure the difference of content and style between synthesized sketch/photo and real sketch/photo in feature level. Second, we propose a feature map based loss termed content feature loss, which are utilized to supervise the generator in our network to make the synthesized sketch/photo have more perceptual quality. To achieve this, we use the pre-trained VGG network to extract the feature maps of the sketch/photo as a feature extractor, and calculate the Euclidean difference between these feature maps and the feature maps from the hidden layers of the generator. Extensive experiments both synthesis quality and recognition ability assessment of the public face photo-sketch database are conducted to show that our method can obtain better results in comparison with existing state-of-the-art methods.

Keywords: Face photo-sketch synthesis · GAN · Neural style transfer.

Supported in part by the National Key Research and Development Program of China under Grant 2018AAA0103202, in part by the National Natural Science Foundation of China under Grant 61922066, Grant 61876142, Grant 61671339, Grant 61772402, Grant U1605252, Grant 61976166, and Grant 62036007, in part by the National High-Level Talents Special Support Program of China under Grant CS31117200001, in part by the Fundamental Research Funds for the Central Universities under Grant JB190117, in part by the Xidian University Intellifusion Joint Innovation Laboratory of Artificial Intelligence, and in part by the Innovation Fund of Xidian University.

Y. Peng et al. (Eds.): PRCV 2020, LNCS 12305, pp. 374–385, 2020.
https://doi.org/10.1007/978-3-030-60633-6_31

1 Introduction

Face photo-sketch synthesis process can be regarded as an image transformation problem which aims at synthesizing a face sketch from a given photo and reverse. It has a wide range of applications in law enforcement and digital entertainment.

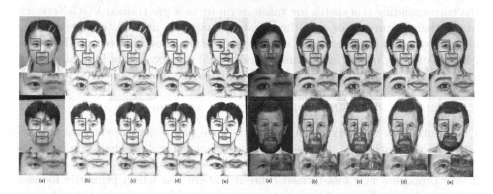

Fig. 1. Illustration results of exiting methods and the proposed methods. (a) Photo, (b) Pix2Pix, (c) Cycle-GAN, (d) Ours, (e) Sketch by artist. Our results show more drawing style and details.

Existing works of face photo-sketch synthesis can be broadly classified into two categories: Exemplar-based methods [2–6] and Learning-based methods [5,7,8]. The process of Exemplar-based methods can be subdivided into two steps: neighbor selecting and weight representation reconstructing, which cannot generate sketch/photo real time as learning-based methods. In learning-based methods, GAN-based methods have made a great success in the past years. However, these approaches remain a main drawback that their results often get blurred effects due to the pixel-wise loss function. As shown in the column (b) and column (c) of Fig. 1, the results of the Pix2Pix [9] and Cycle-GAN [10] have less texture detail and look like gray photo rather than a sketch in some regions. To our best knowledge, the neural style transfer(NST) [11,12] can provide a better solution for texture synthesis. However, NST requires a style image to provide the global statistic of the textures, it means that we need a matched sketch image(it doesn't exist in fact) if we want to use NST to generate the sketch for a given photo.

In this paper, We propose a new objective loss function inspired by neural style transfer(NST) which overcomes the aforementioned limitations of GAN-based methods. In our method, we first introduce the perceptual loss including the style reconstruction loss and content reconstruction loss from NST which utilized to measure the difference of the style and content between the synthesized sketch and the corresponding real sketch. Besides, we propose a feature map based loss termed content feature loss.

As shown in Fig. 2, our framework can seem as three parts, the former convolution part of generator is regarded as an extractor which used to extract the content feature maps of the input photo, and the latter deconvolution part is used to generate sketch from the feature maps. The rest part which composed of residual blocks aims at transferring the photo content feature maps to the sketch content feature maps. And in training stage, both the input photo and the corresponding real sketch are taken as input to a pre-trained VGG-Network to get their feature maps, and these feature maps are utilized to supervise the hidden layers of the generator. Experimental results show that our method outperform existing methods in terms of perceptual quality and additionally study show that the content feature loss can improve the quality of the synthesized sketch especially in structure and texture aspect.

In summary, the main contributions of this paper are three folds:

1. We introduce the perceptual loss to the face photo-sketch synthesis process which makes the generated sketch have higher perceptual quality in comparison with traditional L_1 norm loss or mean squared loss.
2. We propose a new loss termed content feature loss which primarily focuses on the hidden layers of the generator. It makes the synthesized sketch have more structure and texture details.
3. Extensive experiments are conducted to compare the proposed framework with state-of-the-art methods in terms of objective evaluations (objective image quality assessment and face recognition accuracy)

The rest of this paper is organized as follows. Section 2 introduces related works of photo-sketch synthesis and neural style transfer. Section 3 describes the details of the proposed framework. Experiment results of the quality evaluation are presented in Sect. 4 and Sect. 5 concludes this paper.

2   Related Work

2.1   Face Photo-Sketch Synthesis

Learning-based methods achieved a great success in the past years because of the development of the deep learning [10, 13–15]. Researchers have come up with a variety of methods, to name a few, Wang *et al..* [16] propose a novel synthesis framework called Photo-Sketch Synthesis using Multi-Adversarial Networks (PS²-MAN) which can generate high-resolution realistic images by using multi-adversarial networks to supervise the generator. Recently Zhang *et al..* [17] presented a method by multi-domain adversarial learning (termed MDAL) which introduce two adversarial process to reconstruct face photos and sketches, respectively. More recently, Yu *et al..* [18] propose a novel composition-aided generative adversarial network(CA-GAN), which employ facial composition information in the loop of learning the generator. Chen *et al..* [19] propose a semi-supervised learning framework(we call it 'semi' for short in this paper) for face sketch synthesis which only need a small reference set of photo-sketch pairs together with a large face photo dataset without ground truth sketches.

2.2   Image Style Transfer

Image style transfer is a process of transferring a style of one image(target style image) to another image(content image). Gatys *et al.* first presented the convolution neural network (CNN) based method [11,12] for image style transfer and proposed Gram matrix which can represent the style of image. In their method, they first extract the feature maps of the style image and the content image by using a pre-trained VGG-Network. Then, they compute the perceptual loss including content reconstruction loss and style reconstruction loss between the style image and the generated image and finally generate stylized image by updating the content image iteratively until minimizing the perceptual loss. However this approach needs 200 300 iterations could generate one image which cost too much time. To tackle this problem, Johnson *et al.* [20] further proposed a feed-forward CNN method which can achieve transferring image style in real-time by using a trained model. The feed-forward network is designed to fit the process of the style transfer which supervised by the content reconstruction loss and style reconstruction loss similar to [12].

3   Proposed Method

3.1   Preliminaries

In this subsection, the structure of our framework will be provided firstly. Our framework has two generators G_A and G_B which are used to generate sketch from photo and generate photo from sketch respectively. The basic structure of generator is consists of deep residual networks with skipping connection similar to Cycle-GAN. Besides, two discriminator D_A and D_B are used to discriminate an input photo/sketch is real or fake. Thus, the process of photo/sketch synthesis can be expressed as:

$$F_B = G_A(R_A), D_A(R_A/F_A) \rightarrow real/fake?$$

$$F_A = G_B(R_B), D_B(R_B/F_B) \rightarrow real/fake?$$

where R_A/R_B denotes the real photo and sketch which need to be transformed, and F_A/F_B denotes the generated photo and sketch, respectively.

Because these procedures of photo synthesis and sketch synthesis are symmetric, we only take face sketch synthesis as an example to introduce our method in the following subsection.

3.2   Perceptual Loss

The perceptual loss is first introduced in neural style transfer which includes two reconstruction loss used to measure the content difference and style difference respectively. Following the work of [11], we know that Gram matrices of VGG-16 feature maps can regard as a style representation of one style image(sketch), Similar to [12], a Gram matrix of the feature map in the lth layer is defined

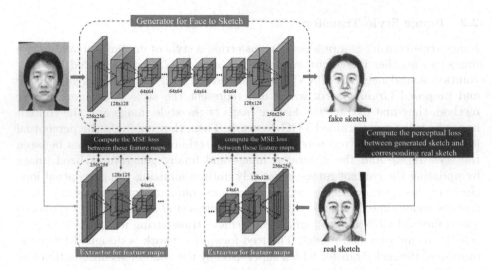

Fig. 2. Illustration of our proposed methods. The VGG-Network is utilized to extract the feature maps from the given photo and corresponding real sketch.

by an inner product between two channels of this feature maps in layer l, and it can capture the style density distribution of the sketch without any spatial information:

$$G_{ij}^l(X) = F_i^l(X) \cdot F_j^l(X)$$

where $F_i^l(X)$ denote the vectorized ith channel of the feature map in the lth layer of the sketch X.

Based on the aforementioned Gram matrices, we can define the content reconstruction loss and style reconstruction loss for face sketch synthesis.

Content Reconstruction Loss. The content reconstruction loss is a feature based loss which encourages the synthesized sketch and real sketch have similar feature representation rather than pixel-wise matching. Denote the feature map extracted from lth layer by $F^l(Y)$. And the content reconstruction loss is defined by the Euclidean distance between the feature map in *conv3_2* of the synthesized sketch and the corresponding real sketch:

$$\mathcal{L}_c = \left\| F^{conv3\_2}(F_B) - F^{conv3\_2}(R_B) \right\|_2^2$$

Style Reconstruction Loss. While the content reconstruction loss measures the difference of content between the synthesized sketch and real sketch, the style reconstruction loss is utilized to measure their difference in style: colors, textures, patterns, etc. Following [12], we define the style reconstruction loss between the synthesized sketch Y_s and the corresponding real sketch Y_r using Gram matrices. Here we use the feature maps which from *conv1_1*, *conv2_1*, *conv3_1*, *conv4_1* of VGG-16 to calculate the Gram matrix:

$$\mathcal{L}_s = \sum_{l \in L_s} \frac{1}{4M_l^2 N_l^2} \left\| G^l(F_B) - G^l(R_B) \right\|_2^2$$

where N_l denotes the channels of feature map at layer l, and M_l is the product of width and height of the feature map at layer l. Thus,

$$\mathcal{L}_p = \lambda_c \mathcal{L}_c + \lambda_s \mathcal{L}_s$$

3.3 Content Feature Loss

We know that in most GAN-based framework, the convolutional layers and the deconvolutional layers can generate different sizes of feature maps respectively. In proposed PS²-MAN method, the authors get these feature maps which from deconvolutional layers and forward them to a 3×3 convolutional layer to generate output images at different resolutions, then they resize the real sketch in the same resolutions as output images and calculate the difference between these images. In our method, we also supervise the hidden layers of the network based on the feature map. As shown in Fig. 2 we can get the feature maps whose size is 64×64, 128×128 and 256×256 from convolutional layers and the deconvolutional layers respectively. Denote the feature maps from convolutional layers and deconvolutional layers whose size is $N \times N$ by $G_C^N(X)$ and $G_D^N(X)$ respectively. Then, we can define the content feature loss as the Euclidean distance between these feature maps and other feature maps which extracted from the pre-trained VGG-16 having the same size. In our method, we get these extracted feature maps from *conv1_2*, *conv2_2* and *conv3_2*, respectively(the *convA_b* means the b_{th} ReLU of A_{th} stage of the VGG). Thus, it can be expressed as:

$$\mathcal{L}_{cf} = \sum_{N \in N_l} \left(\left\| G_C^N(R_A) - F^N(R_A) \right\|_2^2 + \left\| G_D^N(R_A) - F^N(R_B) \right\|_2^2 \right)$$

Where N_l includes the 64×64, 128×128 and 256×256 three kinds of size. $F^N(X)$ denote the feature map whose size is $N \times N$ which extracted from the pre-trained VGG-Network.

3.4 Generative and Adversarial Network Loss

For the mapping from photo to sketch, we have generator G_A and discriminator D_A. Similar to the normal GAN-based methods, the objective function is expressed as:

$$\mathcal{L}_{GAN_A} = \mathbb{E}_{A \sim p_{data}(A)}[(D_A(G_A(R_A))^2] + \mathbb{E}_{B \sim p_{data}(B)}[(D_A(R_B) - 1)^2]$$

Note that we replace logarithm by least loss for easier convergence when training the generator. We can also get the loss function for mapping from sketch to photo: $\mathcal{L}_{GAN_B}$ To make the synthesized sketches as close to real sketch as possible, we also minimize synthesis error $\mathcal{L}_{syn}$ which is defined as the L_1 difference between synthesized sketch and corresponding real sketch:

$$\mathcal{L}_{syn} = \|G_A(R_A) - R_B\|_1$$

In addition, a cycle forward-backward consistency loss $\mathcal{L}_{cyc}$ is defined to regularize the network for reducing the space of possible mapping functions. And to encourage spatial smoothness in the output image, we make use of *total variation regularization* $\mathcal{L}_{tv}$ Thus,

$$\mathcal{L}(G_A, D_A) = \mathcal{L}_{GAN_A} + \mathcal{L}_{syn} + \mathcal{L}_{cyc} + \lambda_{tv}\mathcal{L}_{tv}$$

To summarize, the final objective function is composed of three parts: the perceptual loss, the content feature loss and the GAN loss. The perceptual loss and the content feature loss are computed by difference based on feature maps, which improve the structure and texture of synthesized images. And the GAN loss can subdivided in adversarial loss, synthesize loss and cycle-consistency loss and enables the network to synthesize sketches which are closer to the real sketch in pixel level.

Finally, the total objective loss function of our method can expressed as:

$$\mathcal{L}_{total} = \lambda_p\mathcal{L}_p + \lambda_{cf}\mathcal{L}_{cf} + \mathcal{L}(G_A, D_A)$$

4   Experiment

4.1   Implementation Details

Dataset Setting. We conduct experiments on two public databases: CUFS database and CUFSF database. The CUFS database consists of 606 face photo-sketch pairs from three databases: the CUHK student database (188 persons), the AR database(123 persons), and the XM2VTS database (295 persons). The CUFSF database includes 1194 persons which are particularly challenging since these photos are taken under different lighting variation and sketches are drawn with shape exaggeration compared to the photos. Each person has a face photo and corresponding face sketch drawn by the artist in both the CUFS and CUFSF database. All these face photos and sketches are geometrically aligned relying on three points: two eye centers and the mouth center, and they are cropped to the size of 250 × 200.

Face Sketch Synthesis. For the CUHK student database, 88 pairs of face photo-sketch are taken for training and the rest for testing. For the AR database, we randomly choose 80 pairs for training and the rest 43 pairs for testing. For the XM2VTS database, we randomly choose 100 pairs for training and the rest 195 pairs for testing. Five state-of-the-art methods are compared: MRF [21], RSLCR [6], BP-GAN [8], Semi [19], MDAL [17] and PS²-MAN [16]. As shown in Fig. 3, results from our method have more detailed information and texture

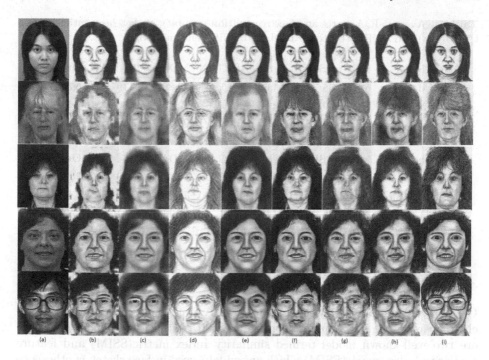

Fig. 3. Examples of synthesized face sketches on the CUFS(the first three rows) and CUFSF(the last two rows) database, (a) Photo, (b) MRF [21], (c) RSLCR [6], (d) Semi [19], (e) BP-GAN [8], (f) MDAL [17], (g) PS^2-MAN [16], (h) Ours, (i) Sketch by artist

than other methods. We further evaluate the objective image quality in the next section.

Face Photo Synthesis. We also utilize the proposed method to generate face photos by using the generator for mapping from sketch to photo. Because only a few numbers of methods released their generated photos, here we only compare the proposed method with two methods: Cycle-GAN [10] and PS^2-MAN [16]. Results are shown in Fig. 4. Obviously, there are serious blurred effects in the photos synthesized by Cycle-GAN and PS^2-MAN method while ours photos get more perceptual quality. Table 2 presents the average SSIM, FSIM and SCOOT scores of the synthesized photos in the CUFS and CUFSF database. We can see that our method obtains the best SCOOT values on these two databases. Thus, according to the Fig. 4 and Table 2, we can draw a conclusion that our method can generate face photos matching human perception closely.

Training Details. We trained our networks with a batch size of 1 for 200 epochs while the generator and discriminator are updated alternatively at every iteration. The trad-off weight λ_{tv} is set to 10^{-5} when we use CUFS database and 10^{-2} when use CUFSF database. We used Adam with a learning rate of 10^{-4}. We implemented our model using Pytorch, and trained it on a single GTX 2080Ti GPU.

Table 1. Average IQA scores and face recognition accuracy of sketches synthesized on CUFS(top) and CUFSF(bottom) database by different methods

Methods	MRF	RSLCR	Semi	BP-GAN	MDAL	PS^2-MAN	Ours
SSIM (%)	51.32	55.72	54.63	55.15	52.80	54.19	**57.25**
FSIM (%)	70.45	69.66	72.56	68.99	72.75	72.30	**74.44**
NLDA (%)	87.29	98.38	97.95	92.93	96.81	97.05	**98.54**
SCOOT (%)	51.50	44.99	48.78	46.80	–	48.75	**70.70**
Methods	MRF	RSLCR	Semi	BP-GAN	MDAL	PS^2-MAN	Ours
SSIM (%)	37.20	**44.91**	40.85	41.94	38.18	41.13	44.23
FSIM (%)	69.56	66.48	71.59	68.14	70.76	72.33	**74.10**
NLDA (%)	46.03	75.94	77.97	68.32	67.11	77.51	**78.99**
SCOOT (%)	61.67	45.32	50.39	49.36	–	53.86	**75.63**

4.2 Quantitative Comparison

Objective Image Quality Assessment. IQA aims to use computational models to measure the image quality consistently with subjective evaluations. There are two well-known model termed similarity index metric(SSIM) and feature similarity index metric(FSIM) which are widely used in face sketch synthesis to evaluate the quality of synthesized sketches. The SSIM is measure the difference of structure between two images while the FSIM concerns the difference of the low-level feature sets between two images. Besides, Fan *et al.* [22] proposed a new assessment metric named SCOOT [22] which is more closely related to the human subjective evaluation. In order to show the effect of our method comprehensively, we utilize the SSIM, FSIM and SCOOT metrics in this paper. In addition, for these methods which have no testing the metrics score in their paper, we test this result using the generated images released by their author. Results are listed in Table 1, we can see that our method achieves the best FSIM and SCOOT scores both on CUFS and CUFSF databases, and the SSIM scores are also comparable. This results show that our method can generate sketches which are more conform to the perception of human in comparison with other methods.

Face Sketch Recognition. Face sketch recognition is an important application of face sketch synthesis which always used to assist law enforcement and is frequently utilized to quantitatively evaluate the quality of the synthesized sketches. Following [6], we employ the null-space linear discriminant analysis (NLDA) [23] to conduct the face sketch recognition experiment. Recognition results are listed in Table 1, we can see that our method achieves the best result in comparison with other methods.

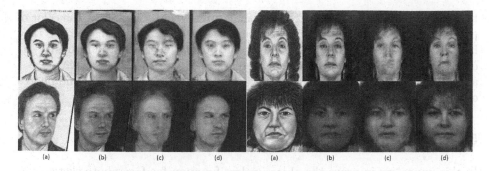

Fig. 4. Examples of synthesized face photos on the CUFS and CUFSF database, (a) Sketch by artist, (b) Cycle-GAN [10], (c) PS²-MAN [16], (d) Ours.

Table 2. Average IQA scores of the synthesized photos on CUFS and CUFSF database.

		Cycle-GAN	PS²-MAN	Ours
SSIM (%)	CUFS	55.24	65.23	**66.53**
	CUFSF	58.83	**63.12**	63.09
FSIM (%)	CUFS	74.50	78.19	**79.31**
	CUFSF	76.45	78.12	**78.95**
SCOOT (%)	CUFS	55.99	57.90	**64.57**
	CUFSF	56.75	39.38	**60.69**

4.3   Analysis of the Content Feature Loss

In this section, we train a new model without content feature loss on CUHK Student database to study the importance of this loss in our total objective loss function. Training configuration are same as above. We name it as **ours(w/o)**. Results are illustrated in Fig. 5. As we can see in Fig. 5, the sketches synthesized

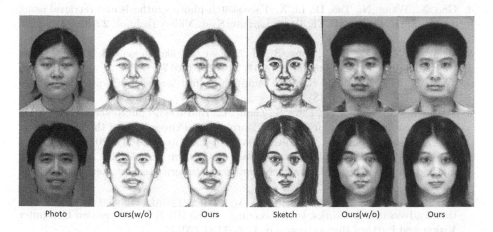

Photo   Ours(w/o)   Ours   Sketch   Ours(w/o)   Ours

Fig. 5. Examples of synthesized face sketches/photos on the CUHK Student database.

by **ours** have more texture and details than **ours(w/o)**. Obviously, the color of photos synthesized by **ours(w/o)** are turbid. For the IQA average scores, they have the similar SSIM and FSIM scores, but **ours** outperform than **ours(w/o)** on SCOOT values. The synthesized sketches and photos of **ours** have 62.73% and 61.94% SCOOT value respectively, and the synthesized sketches and photos of **ours(w/o)** have 60.78% and 57.50% SCOOT value respectively.

5   Conclusion

In this paper, we improve the objective loss function for face photo-sketch synthesis which major includes the perceptual loss and content feature loss. Instead of supervising the network only use the traditional L_1 norm loss, we calculate the difference between the synthesized sketch/photo and real sketch/photo based on feature space. This allows our generator can produce sketch/photo which have more texture details. Experiments show that the synthesized sketch/photo of our method have better perceptual quality compare with other state-of-the-art methods on two public benchmarks(in terms of SSIM, FSIM and SCOOT scores). And we will further investigate the influence of the feature maps of generator on synthesized sketch/photo in the future.

References

1. Goodfellow, I., et al.: Generative adversarial nets. In: Advances in Neural Information Processing Systems, pp. 2672–2680 (2014)
2. Tang, X., Wang, X.: Face photo recognition using sketch. In Proceedings. International Conference on Image Processing, vol. 1, pp. I. IEEE (2002)
3. Liu, Q., Tang, X., Jin, H., Lu, H., Ma, S.: A nonlinear approach for face sketch synthesis and recognition. In: 2005 IEEE Computer Society Conference on Computer Vision and Pattern Recognition (CVPR 2005), vol. 1, pp. 1005–1010. IEEE (2005)
4. Gao, X., Wang, N., Tao, D., Li, X.: Face sketch-photo synthesis and retrieval using sparse representation. IEEE Trans. Circuits Syst. Video Technol. **22**(8), 1213–1226 (2012)
5. Wang, N., Zhu, M., Li, J., Song, B., Li, Z., Fast face sketch synthesis: Data-driven vs. model-driven. Neurocomputing **257**, 214–221 (2017)
6. Wang, N., Gao, X., Li, J.: Random sampling for fast face sketch synthesis. Pattern Recogn. **76**, 215–227 (2018)
7. Zhang, L., Lin, L., Wu, X., Ding, S., Zhang, L.: End-to-end photo-sketch generation via fully convolutional representation learning. In: Proceedings of the 5th ACM on International Conference on Multimedia Retrieval, pp. 627–634 (2015)
8. Wang, N., Zha, W., Li, J., Gao, X.: Back projection: an effective postprocessing method for GAN-based face sketch synthesis. Pattern Recogn. Lett. **107**, 59–65 (2018)
9. Isola, P., Zhu, J.-Y., Zhou, T., Efros, A.A.: Image-to-image translation with conditional adversarial networks. In: Proceedings of the IEEE Conference on Computer Vision and Pattern Recognition, pp. 1125–1134 (2017)

10. Zhu, J.-Y., Park, T., Isola, P., Efros, A.A.: Unpaired image-to-image translation using cycle-consistent adversarial networks. In: Proceedings of the IEEE International Conference on Computer Vision, pp. 2223–2232 (2017)

11. Gatys, L., Ecker, A.S., Bethge, M.: Texture synthesis using convolutional neural networks. In: Advances in Neural Information Processing Systems, pp. 262–270 (2015)

12. Gatys, L.A., Ecker, A.S., Bethge, M.: Image style transfer using convolutional neural networks. In: Proceedings of the IEEE Conference on Computer Vision and Pattern Recognition, pp. 2414–2423 (2016)

13. Gong, C., Shi, H., Liu, T., Zhang, C., Yang, J., Tao, D.: Loss decomposition and centroid estimation for positive and unlabeled learning. IEEE Trans. Pattern Anal. Mach. Intell., 1 (2019)

14. Gong, C., Tao, D., Liu, W., Liu, L., Yang, J.: Label propagation via teaching-to-learn and learning-to-teach. IEEE Trans. Neural Netw. Learn. Syst. **28**(6), 1452–1465 (2016)

15. Gong, C., Tao, D., Maybank, S.J., Liu, W., Kang, G., Yang, J.: Multi-modal curriculum learning for semi-supervised image classification. IEEE Trans. Image Proces. **25**(7), 3249–3260 (2016)

16. Wang, L., Sindagi, V., Patel, V.: High-quality facial photo-sketch synthesis using multi-adversarial networks. In: 2018 13th IEEE International Conference on Automatic Face & Gesture Recognition (FG 2018), pp. 83–90. IEEE (2018)

17. Zhang, S., Ji, R., Jie, H., Xiaoqiang, L., Li, X.: Face sketch synthesis by multidomain adversarial learning. IEEE Trans. Neural Netw. Learn. Syst. **30**(5), 1419–1428 (2018)

18. Yu, J., Xu, X., Gao, F., Shi, S., Wang, M., Tao, D., Huang, Q.: Toward realistic face photo-sketch synthesis via composition-aided GANs. IEEE Trans. Cybern., 1–13 (2020)

19. Chen, C., Liu, W., Tan, X., Wong, K.-Y.K.: Semi-supervised learning for face sketch synthesis in the wild. In: Jawahar, C.V., Li, H., Mori, G., Schindler, K. (eds.) ACCV 2018. LNCS, vol. 11361, pp. 216–231. Springer, Cham (2019). https://doi.org/10.1007/978-3-030-20887-5_14

20. Johnson, J., Alahi, A., Fei-Fei, L.: Perceptual losses for real-time style transfer and super-resolution. In: Leibe, B., Matas, J., Sebe, N., Welling, M. (eds.) ECCV 2016. LNCS, vol. 9906, pp. 694–711. Springer, Cham (2016). https://doi.org/10.1007/978-3-319-46475-6_43

21. Wang, X., Tang, X.: Face photo-sketch synthesis and recognition. IEEE Trans. Pattern Anal. Mach. Intell. **31**(11), 1955–1967 (2008)

22. Fan, D.-P., et al.: Scoot: a perceptual metric for facial sketches. In: Proceedings of the IEEE International Conference on Computer Vision, pp. 5612–5622 (2019)

23. Chen, L.-F., Liao, H.-Y.M., Ko, M.-T., Lin, J.-C., Yu, G.-J.: A new LDA-based face recognition system which can solve the small sample size problem. Pattern Recogn. **33**(10), 1713–1726 (2000)

Cross-View Image Synthesis
with Deformable Convolution
and Attention Mechanism

Hao Ding[1], Songsong Wu[1(✉)], Hao Tang[2], Fei Wu[1], Guangwei Gao[1],
and Xiao-Yuan Jing[3]

[1] College of Automation and College of Atificial Intelligence,
Nanjing University of Posts and Telecommunications, Nanjing 210023, China
sswuai@126.com
[2] University of Trento, Trento, Italy
[3] Wuhan University, Wuhan, China

Abstract. Learning to generate natural scenes has always been a daunting task in computer vision. This is even more laborious when generating images with very different views. When the views are very different, the view fields have little overlap or objects are occluded, leading the task very challenging. In this paper, we propose to use Generative Adversarial Networks (GANs) based on a deformable convolution and attention mechanism to solve the problem of cross-view image synthesis (see Fig. 1). It is difficult to understand and transform scenes appearance and semantic information from another view, thus we use deformed convolution in the U-net network to improve the network's ability to extract features of objects at different scales. Moreover, to better learn the correspondence between images from different views, we apply an attention mechanism to refine the intermediate feature map thus generating more realistic images. A large number of experiments on different size images on the Dayton dataset [1] show that our model can produce better results than state-of-the-art methods.

Keywords: Cross-view image synthesis · GANs · Attention
mechanism · Deformable convolution

1 Introduction

Cross-view image synthesis aims to translate images between two distinct views, such as synthesizing ground images from aerial images, and vice versa. This problem has aroused great interest in the computer vision and virtual reality communities, and it has been widely studied in recent years [2–9]. Earlier work used encoder-decoder convolutional neural networks (CNNs) to study the viewpoint code included in the bottleneck representation for urban scene synthesis [10] and 3D object transformations [11]. Besides, when the view fields have little overlap or objects are occluded, and similar objects in one view may be

© Springer Nature Switzerland AG 2020
Y. Peng et al. (Eds.): PRCV 2020, LNCS 12305, pp. 386–397, 2020.
https://doi.org/10.1007/978-3-030-60633-6_32

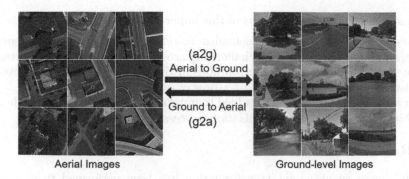

Fig. 1. Example images in overhead/aerial view (left) and street-view/ground-level (right). The images reflect the great diversity and richness of features in two views implying that the network needs to learn a lot for meaningful cross-view generation.

completely different from another view (i.e., view invariance issues), this task will be more challenging. For example, the aerial view of a building (i.e., the roof) tells very little about the color and design of the building seen from the street-view. The generation process is generally easier when the image contains a single object in a uniform background. In contrast, when the scene contains multiple objects, generating other view becomes much more challenging. This is due to the increase in underlying parameters that contribute to the variations (e.g., occlusions, shadows, etc). An example scenario, addressed here, is generating street-view (a.k.a ground level) image of a location from its aerial (a.k.a overhead) image. Figure 1 illustrates some corresponding images in the two different views.

To solve this challenging problem, Krishna and Ali [6] proposed a conditional GAN model that jointly learns the generation in both the image domain and the corresponding semantic domain, and the semantic predictions are further utilized to supervise the image generation. Although this method has been interestingly explored, there are still unsatisfactory aspects of the generated scene structure and details. Moreover, Tang et al. [12] recently proposed the multi-channel attention selection generation adversarial network (SelectionGAN), which can learn conditional images and target semantic maps together, and the automatically learned uncertainty map can be used to guide pixel loss to achieve better network optimization. However, we observe that there are still unsatisfactory aspects in the generated scene structure and details. For example, for the outline boundaries of some objects, there are obvious wrong marks and unclear.

To tackle this challenging problem, we add deformed convolution to the U-net network to improve the network's ability to extract features of objects at different scales. At the same time, we use the attention mechanism [13] to refine the feature map to obtain a more detailed feature map for generating more realistic images. A large number of experiments show that our model can produce better results than state-of-the-art models, i.e., Pix2Pix [2], X-Fork [6], X-Seq [6] and SelectionGAN [12].

In summary, our contributions of this paper are as follows:

- We employed the attention mechanism to refine the feature map to generate more realistic images for the challenging cross-view image translation tasks.
- We also embed deformable convolutions in the U-net network to improve the network's ability for extracting features of objects at different scales.
- An additional loss function is added to improve the network training, thereby achieving a more stable optimization process.

2   Related Work

Existing work on viewpoint transformation has been performed to synthesize novel views of the same object [14–16]. For example, Zhou et al. [16] proposed models learn to copy pixel information from the input view and uses them to retain the identity and structure of the object to generate a new view. Tatarchenko et al. [15] trained a network of codes to obtain 3D representation models of cars and chairs, which were subsequently used to generate different views of unseen images of cars or chairs. Dosovitskiy et al. [14] learned to generate models by training 3D renderings of cars, chairs, and tables, and synthesize intermediate views and objects by interpolating between views and models. Zhai et al. [17] explored the semantic layout of predicting ground images from their corresponding aerial images. They synthesized ground panoramas using the predicted layouts. Previous work on aerial and ground images has addressed issues such as cross-view co-localization [18,19], ground-to-aerial geo-localization [20] and geo-tagging the cross-view images [21].

Compared with existing methods such as Restricted Boltzmann Machines [22] and Deep Boltzmann Machines [23], generative adversarial networks (GANs) [24] have shown the ability to generate better quality images [25–28]. The vanilla GAN model [24] has two important components, i.e., the generator G and the discriminator D. The generator G aims to generate realistic from the noise vector, while D tries to distinguish between real image and image generated by G. Although it has been successfully used to generate high visual fidelity images [26,29–31], there are still some challenges such as how to control the image generation process under specific settings. To generate domain-specific images, the conditional GAN (CGAN) [28] has been proposed. CGAN usually combines vanilla GAN with some external information.

Krishna and Ali [6] proposed two structures (X-Fork and X-Seq) based on Conditional GANs to solve the task of image translation from aerial to street-view using additional semantic segmentation maps. Moreover, Tang et al. [12] proposed the multi-channel attention selection generation adversarial network (SelectionGAN), which consists of two generation stages. In the first stage, a cyclic semantically guided generation sub-net was proposed. This network receives images and conditional semantic maps in one view, while synthesizing images and semantic maps in another view. The second stage uses the rough predictions and learned deep semantic features of the first stage, and uses the suggested multi-channel attention selection the module performs fine-grained generation.

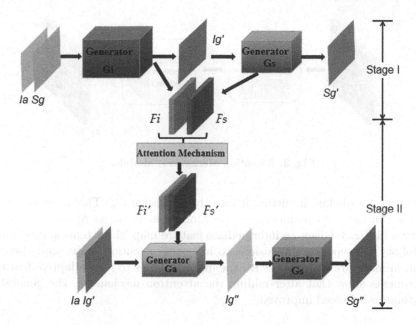

Fig. 2. Architecture of the proposed network.

3 Network Design

The network structure we proposed is based on the SelectionGAN model, which consists of three generators (i.e., G_i, G_a, G_s), two discriminators (i.e., D_1, D_2), and an attention mechanism module. The network structure can be divided into two stages, as shown in Fig. 2.

In the first stage, an image I_a of one perspective and a semantic map S_g of another perspective are input to the generator G_i to generate an image I'_g of another perspective and the feature map F_i of the last convolution layer. Then the generated image I'_g is input into the generator G_s to generate the corresponding semantic map S'_g.

In the second stage, the feature maps F_i and F_s generated in the first stage are refined through the attention mechanism module to obtain the refined feature maps F'_i and F'_s. Next, they are combined with the image I_a and the generated image I'_g and inputted to the generator G_a to generate a refined image I''_g as the final output. This refined image I''_g is then input to the generator G_s to generate the corresponding semantic map S''_g.

Note that we use only one generator G_s in both the first and second stages, since the purpose is to generate a corresponding semantic image from an image.

3.1 Attention Mechanism

Since the SelectionGAN model takes the coarse feature map as input of the second stage. So we consider that we can use the attention mechanism to refine

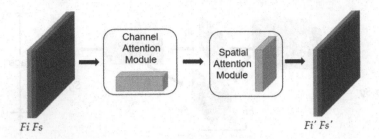

Fig. 3. Attention Mechanism Module.

the feature map before inputting it into the generator G_a. The attention mechanism is consisted of Channel Attention Module and Spatial Attention Module, as shown in Fig. 3. Given an intermediate feature map, the attention mechanism will follow two separate dimensions to infer the attention maps, and then the attention maps are multiplied with the input features to map adaptive features. Experiments show that after adding the attention mechanism, the generation performance is indeed improved.

3.2 Deformable Convolution

Deformable convolution [32] adds spatial sampling positions with additional offsets and learns offsets in the target task without additional supervision. The new module can easily replace the ordinary peers in existing CNNs and a large number of experiments have verified that this method learns dense spatial transformations in deep CNNs and is effective for complex visual tasks such as object detection and semantic segmentation.

Therefore, we embed deformable convolutions into U-net. The outermost layer of the network can better extract the features from the input maps. The network structure is shown in Fig. 4.

3.3 Overall Optimization Objective

Adversarial Loss. SelectionGAN [12] uses one discriminator D_1 for the generated images on two stages. D_1 takes the input and the generated fake image as input, however, the semantic map is not take into consideration. Therefore, we propose a new discriminator D_2, which also takes the semantic map as input. The proposed semantic-guided adversarial losses can be expressed as follows,

$$
\begin{aligned}
\mathcal{L}_{cGAN} &\left(I_a \oplus S_g, I_g' \oplus S_g'\right) \\
&= \mathbb{E}\left[\log D_2\left(I_a \oplus S_g, I_g \oplus S_g\right)\right] \\
&+ \mathbb{E}\left[\log\left(1 - D_2\left(I_a \oplus S_g, I_g' \oplus S_g'\right)\right)\right],
\end{aligned}
\tag{1}
$$

$$
\begin{aligned}
\mathcal{L}_{cGAN} &\left(I_a \oplus S_g, I_g'' \oplus S_g''\right) \\
&= \mathbb{E}\left[\log D_2\left(I_a \oplus S_g, I_g \oplus S_g\right)\right] \\
&+ \mathbb{E}\left[\log\left(1 - D_2\left(I_a \oplus S_g, I_g'' \oplus S_g''\right)\right)\right],
\end{aligned}
\tag{2}
$$

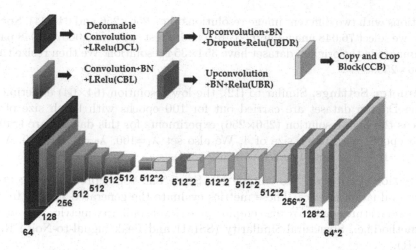

Fig. 4. Network structure of the proposed generator. BN means batch-normalization layer.

where the symbol $\oplus$ denotes the channel-wise concatenation operation. Thus, the total adversarial loss can be formulated as follows,

$$\begin{aligned}\mathcal{L}_{cGAN} =&\mathcal{L}_{cGAN}\left(I_a, I'_g\right) + \lambda\mathcal{L}_{cGAN}\left(I_a, I''_g\right) \\ &+ \mathcal{L}_{cGAN}\left(I_a \oplus S_g, I'_g \oplus S'_g\right) \\ &\lambda\mathcal{L}_{cGAN}\left(I_a \oplus S_g, I''_g \oplus S''_g\right),\end{aligned} \tag{3}$$

where $\mathcal{L}_{cGAN}\left(I_a, I'_g\right)$ and $\mathcal{L}_{cGAN}\left(I_a, I''_g\right)$ are the adversarial losses defined in SelectionGAN.

Overall Loss. The total optimization loss is a weighted sum of several losses. The generators G_i, G_s, G_a and discriminators D_1, D_2 are trained in an end-to-end fashion optimizing the following min-max function,

$$\min_{\{G_i, G_s, G_a\}} \max_{\{D_1, D_2\}} \mathcal{L} = \sum_{i=1}^{4} \lambda_i \mathcal{L}_p^i + \mathcal{L}_{cGAN} + \lambda_{tv}\mathcal{L}_{tv}, \tag{4}$$

where $\mathcal{L}_p^i$ uses the $L1$ reconstruction to separately calculate the pixel loss between the generated images I'_g, S'_g, I''_g and S''_g and the corresponding real ones. $\mathcal{L}_{tv}$ is the total variation regularization on the final synthesized image I''_g. λ_i and λ_{tv} are the trade-off parameters to control the relative importance of different objectives.

4   Experiments

Datasets. We follow [6,12,33] and perform extensive experiments on the challenging Dayton dataset in a2g (aerial-to-ground) and g2a (ground-to-aerial) directions with two different image resolutions (i.e., 256×256 and 64×64). Specifically, we select 76,048 images and create a train/test split of 55,000/21,048 pairs. The images in the original dataset have 354×354 resolution. We then resize them to 256×256.

Parameter Settings. Similar to [12], the low resolution (64×64) experiments on the Dayton dataset are carried out for 100 epochs with batch size of 16, whereas the high resolution (256×256) experiments for this dataset are trained for 35 epochs with batch size of 4. We also set λ_1=100, λ_2=1, λ_3=200, λ_4=2 and λ_{tv}=1$e-6$ in Eq. (4), and λ=4 in Eq. (3).

Evaluation Protocol. We employ KL Score and top-k prediction accuracy as the evaluation metrics. These metrics evaluate the generated images from a high-level feature space. We also employ pixel-level similarity metrics to evaluate our method,i.e., Structural-Similarity (SSIM) and Peak Signal-to-Noise Ratio (PSNR).

State-of-the-Art Comparisons. We compare the proposed model with exiting cross-view image translation methods, i.e., Pix2Pix [2], X-Fork [6], X-Seq [6] and SelectionGAN [12]. Quantitative results of different metrics are shown in Tables 1 and 2.

We compute top-1 and top-5 accuracies in Table 1. As we can see, for lower resolution images (64×64) our method outperforms the existing leading cross-view image translation methods. For higher resolution images (256×256), our method also achieves the best results on top-1 and top-5 accuracies. This shows the effectiveness of our method and the necessity of the proposed modules.

Moreover, we provide results of SSIM, PSNR, and KL scores Table 2. We observe that the proposed method is consistently superior to other leading methods, validating the effectiveness of the proposed method.

Qualitative Evaluation. Qualitative results compared with the most related work, i.e., SelectionGAN [12] are shown in Fig. 5, 6 and 7. We can see that our method generates sharper details than SelectionGAN on objects/scenes,

Table 1. Accuracies of different methods.

Dir ⇌	Method	Dayton (64×64)				Dayton (256×256)			
		Top-1 Accuracy(%)		Top-5 Accuracy(%)		Top-1 Accuracy(%)		Top-5 Accuracy(%)	
a2g	Pix2pix [2]	7.90	15.33	27.61	39.07	6.80	9.15	23.55	27.00
	X-Fork [6]	16.63	34.73	46.35	70.01	30.00	48.68	61.57	78.84
	X-Seq [6]	4.83	5.56	19.55	24.96	30.16	49.85	62.59	80.70
	SelectionGAN [12]	45.37	79.00	83.48	97.74	42.11	68.12	77.74	92.89
	Ours	**47.61**	**81.24**	**86.12**	**98.44**	**45.07**	**77.12**	**80.04**	**94.54**
g2a	Pix2pix [2]	1.65	2.24	7.49	12.68	10.23	16.02	30.90	40.49
	X-Fork [6]	4.00	16.41	15.42	35.82	10.54	15.29	30.76	37.32
	X-Seq [6]	1.55	2.99	6.27	8.96	12.30	19.62	35.95	45.94
	SelectionGAN [12]	14.12	51.81	39.45	74.70	20.66	33.70	51.01	63.03
	Ours	**14.26**	**52.17**	**52.55**	**78.72**	**20.81**	**38.41**	**55.51**	**65.84**

Table 2. SSIM, PSNR, and KL score of different methods.

Dir ⇌	Method	Dayton(64×64)			Dayton(256×256)		
		SSIM	PSNR	KL	SSIM	PSNR	KL
a2g	Pix2pix [2]	0.4808	19.4919	6.29±0.80	0.4180	17.6291	38.26±1.88
	X-Fork [6]	0.4921	19.6273	3.42±0.72	0.4963	19.8928	6.00±1.28
	X-Seq [6]	0.5171	20.1049	6.22±0.87	0.5031	20.2803	5.93±1.32
	SelectionGAN [12]	0.6865	24.6143	1.70±0.45	0.5938	23.8874	2.74 ±0.86
	Ours	**0.7100**	**24.9674**	**1.55±0.51**	**0.6524**	**24.4012**	**2.47±0.76**
g2a	Pix2pix [2]	0.3675	20.5135	6.39±0.90	0.2693	20.2177	7.88±1.24
	X-Fork [6]	0.3682	20.6933	4.55±0.84	0.2763	20.5978	6.92±1.15
	X-Seq [6]	0.3663	20.4239	7.20±0.92	0.2725	20.2925	7.07±1.19
	SelectionGAN [12]	0.5118	23.2657	2.25±0.56	0.3284	21.8066	3.55±0.87
	Ours	**0.6116**	**24.5445**	**2.13±0.48**	**0.3924**	**22.7143**	**3.17±0.82**

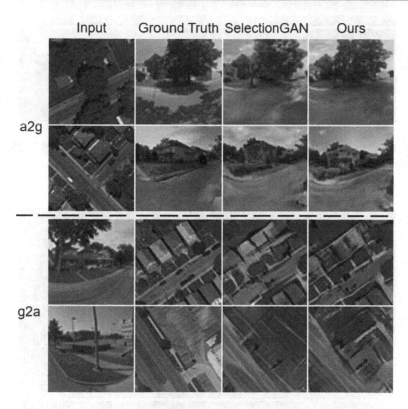

Fig. 5. Results generated by the proposed method and SelectionGAN [12] in 64×64 resolution in both a2g (top) and g2a (bottom) directions on the Dayton dataset.

e.g., houses, buildings, roads, clouds, and cars. For example, we can see that the houses generated by our method are more natural than those generated by SelectionGAN as shown in Fig. 6.

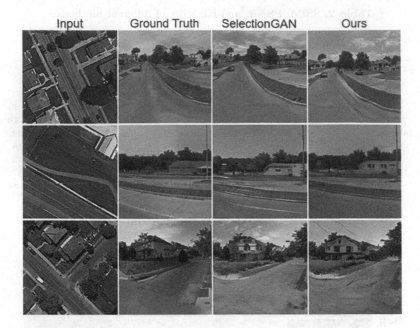

Fig. 6. Results generated by the proposed method and SelectionGAN [12] in 256×256 resolution in a2g direction on the Dayton dataset.

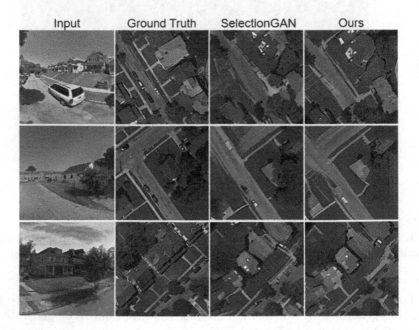

Fig. 7. Results generated by the proposed method and SelectionGAN [12] in 256×256 resolution in g2a direction on the Dayton dataset.

Ablation Study. We also conduct an ablation study in a2g (aerial-to-ground) direction on the Dayton dataset. To reduce the training time, we follow Selection-GAN and randomly select 1/3 samples from the whole 55,000/21,048 samples, i.e., around 18,334 samples for training and 7,017 samples for testing. The proposed model consists of 4 baselines (A, B, C, D) as shown in Table 3. Baseline A uses SelectionGAN (SGAN). Baseline B combines SGAN and the proposed attention mechanism (AM). Baseline C employs deformable convolution (DC) on baseline B. Baseline D adopts the proposed loss function (LS). It is obvious that as each module is added, we can obtain better results of both SSIM and PSNR metrics. This means by adding the proposed attention mechanism, deformable convolution, and the proposed loss function, the overall performance can be further boosted.

Table 3. Ablations study of the proposed method.

Baseline	Method	PSNR	SSIM
A	SGAN [12]	23.9310	0.6176
B	SGAN + AM	24.0539	0.6309
C	SGAN + AM + DC	24.3345	0.6507
D	SGAN + AM + DC + LS	**24.6421**	**0.6927**

5 Conclusion

In this paper, we propose a novel generative adversarial network based on deformable convolution and attention mechanisms for solving the challenging cross-view image generation task. We propose a novel attention mechanism to refine the feature maps, thus improving the ability of feature representation. We also embed deformed convolution in our generator to improve the network's ability for extracting object features at different scales. Moreover, a novel semantic-guide adversarial loss is proposed to improve the whole network training, thus achieving a more robust and stable optimization. Extensive experimental results show that the proposed method obtains better results than state-of-the-art methods.

References

1. Vo, N.N., Hays, J.: Localizing and orienting street views using overhead imagery. In: Leibe, B., Matas, J., Sebe, N., Welling, M. (eds.) ECCV 2016. LNCS, vol. 9905, pp. 494–509. Springer, Cham (2016). https://doi.org/10.1007/978-3-319-46448-0_30
2. Isola, P., Zhu, J.Y., Zhou, T., Efros, A.A.: Image-to-image translation with conditional adversarial networks. In: Proceedings of the IEEE Conference on Computer Vision and Pattern Recognition, pp. 1125–1134 (2017)

3. Krizhevsky, A., Sutskever, I., Hinton, G.E.: Imagenet classification with deep convolutional neural networks. In: Advances in Neural Information Processing Systems, pp. 1097–1105 (2012)
4. Kim, T., Cha, M., Kim, H., Lee, J.K., Kim, J.: Learning to discover cross-domain relations with generative adversarial networks. In: Proceedings of the 34th International Conference on Machine Learning-Volume 70, pp. 1857–1865 (2017)
5. Zhu, X., Yin, Z., Shi, J., Li, H., Lin, D.: Generative adversarial frontal view to bird view synthesis. In: International Conference on 3D Vision, pp. 454–463 (2018)
6. Regmi, K., Borji, A.: Cross-view image synthesis using conditional gans. In: Proceedings of the IEEE Conference on Computer Vision and Pattern Recognition, pp. 3501–3510 (2018)
7. Mathieu, M., Couprie, C., LeCun, Y.: Deep multi-scale video prediction beyond mean square error. arXiv preprint arXiv:1511.05440 (2015)
8. Huang, R., Zhang, S., Li, T., He, R.: Beyond face rotation: global and local perception gan for photorealistic and identity preserving frontal view synthesis. In: Proceedings of the IEEE International Conference on Computer Vision, pp. 2439–2448 (2017)
9. Park, E., Yang, J., Yumer, E., Ceylan, D., Berg, A.C.: Transformation-grounded image generation network for novel 3d view synthesis. In: Proceedings of the IEEE Conference on Computer Vision and Pattern Recognition, pp. 3500–3509 (2017)
10. Cordts, M., et al.: The cityscapes dataset for semantic urban scene understanding. In: Proceedings of the IEEE Conference on Computer Vision and Pattern Recognition, pp. 3213–3223 (2016)
11. Yang, J., Reed, S.E., Yang, M.H., Lee, H.: Weakly-supervised disentangling with recurrent transformations for 3d view synthesis. In: Advances in Neural Information Processing Systems, pp. 1099–1107 (2015)
12. Tang, H., Xu, D., Sebe, N., Wang, Y., Corso, J.J., Yan, Y.: Multi-channel attention selection gan with cascaded semantic guidance for cross-view image translation. In: Proceedings of the IEEE Conference on Computer Vision and Pattern Recognition, pp. 2417–2426 (2019)
13. Woo, S., Park, J., Lee, J.Y., So Kweon, I.: Cbam: convolutional block attention module. In: Proceedings of the European Conference on Computer Vision, pp. 3–19 (2018)
14. Dosovitskiy, A., Springenberg, J.T., Tatarchenko, M., Brox, T.: Learning to generate chairs, tables and cars with convolutional networks. IEEE Trans. Pattern Anal. Mach. Intell. **39**(4), 692–705 (2016)
15. Tatarchenko, M., Dosovitskiy, A., Brox, T.: Multi-view 3D models from single images with a convolutional network. In: Leibe, B., Matas, J., Sebe, N., Welling, M. (eds.) ECCV 2016. LNCS, vol. 9911, pp. 322–337. Springer, Cham (2016). https://doi.org/10.1007/978-3-319-46478-7_20
16. Zhou, T., Tulsiani, S., Sun, W., Malik, J., Efros, A.A.: View synthesis by appearance flow. In: Leibe, B., Matas, J., Sebe, N., Welling, M. (eds.) ECCV 2016. LNCS, vol. 9908, pp. 286–301. Springer, Cham (2016). https://doi.org/10.1007/978-3-319-46493-0_18
17. Zhai, M., Bessinger, Z., Workman, S., Jacobs, N.: Predicting ground-level scene layout from aerial imagery. In: Proceedings of the IEEE Conference on Computer Vision and Pattern Recognition, pp. 867–875 (2017)
18. Lin, T.Y., Belongie, S., Hays, J.: Cross-view image geolocalization. In: Proceedings of the IEEE Conference on Computer Vision and Pattern Recognition, pp. 891–898 (2013)

19. Nelson, B.K., Cai, X., Nebenführ, A.: A multicolored set of in vivo organelle markers for co-localization studies in arabidopsis and other plants. Plant J. **51**(6), 1126–1136 (2007)
20. Lin, T.Y., Cui, Y., Belongie, S., Hays, J.: Learning deep representations for ground-to-aerial geolocalization. In: Proceedings of the IEEE Conference on Computer Vision and Pattern Recognition, pp. 5007–5015 (2015)
21. Workman, S., Souvenir, R., Jacobs, N.: Wide-area image geolocalization with aerial reference imagery. In: Proceedings of the IEEE International Conference on Computer Vision, pp. 3961–3969 (2015)
22. Hinton, G.E., Osindero, S., Teh, Y.W.: A fast learning algorithm for deep belief nets. Neural Comput. **18**(7), 1527–1554 (2006)
23. Hinton, G.E., Salakhutdinov, R.R.: A better way to pretrain deep boltzmann machines. In: Advances in Neural Information Processing Systems, pp. 2447–2455 (2012)
24. Goodfellow, I., et al.: Generative adversarial nets. In: Advances in Neural Information Processing Systems, pp. 2672–2680 (2014)
25. Wang, X., Gupta, A.: Generative image modeling using style and structure adversarial networks. In: Leibe, B., Matas, J., Sebe, N., Welling, M. (eds.) ECCV 2016. LNCS, vol. 9908, pp. 318–335. Springer, Cham (2016). https://doi.org/10.1007/978-3-319-46493-0_20
26. Karras, T., Aila, T., Laine, S., Lehtinen, J.: Progressive growing of gans for improved quality, stability, and variation. arXiv preprint arXiv:1710.10196 (2017)
27. Gulrajani, I., Ahmed, F., Arjovsky, M., Dumoulin, V., Courville, A.C.: Improved training of wasserstein gans. In: Advances in Neural Information Processing Systems, pp. 5767–5777 (2017)
28. Tang, H., Wang, W., Xu, D., Yan, Y., Sebe, N.: Gesturegan for hand gesture-to-gesture translation in the wild. In: Proceedings of the 26th ACM International Conference on Multimedia, pp. 774–782 (2018)
29. Zhang, H., Goodfellow, I., Metaxas, D., Odena, A.: Self-attention generative adversarial networks. arXiv preprint arXiv:1805.08318 (2018)
30. Radford, A., Metz, L., Chintala, S.: Unsupervised representation learning with deep convolutional generative adversarial networks. arXiv preprint arXiv:1511.06434 (2015)
31. Liu, G., Tang, H., Latapie, H., Yan, Y.: Exocentric to egocentric image generation via parallel generative adversarial network. In: IEEE International Conference on Acoustics, Speech and Signal Processing, pp. 1843–1847 (2020)
32. Dai, J., et al.: Deformable convolutional networks. In: Proceedings of the IEEE International Conference on Computer Vision, pp. 764–773 (2017)
33. Tang, H., Xu, D., Yan, Y., Torr, P.H., Sebe, N.: Local class-specific and global image-level generative adversarial networks for semantic-guided scene generation. In: Proceedings of the IEEE Conference on Computer Vision and Pattern Recognition, pp. 7870–7879 (2020)

TriSpaSurf: A Triple-View Outline Guided 3D Surface Reconstruction of Vehicles from Sparse Point Cloud

Hanfeng Zheng[1], Huijun Di[1(✉)], Yaohang Han[1], and Jianwei Gong[2]

[1] Beijing Laboratory of Intelligent Information Technology, School of Computer Science and Technology, Beijing Institute of Technology, Beijing, China
{zhf_cs,ajon,hanyaohang}@bit.edu.cn
[2] Intelligent Vehicle Research Center, School of Mechanical Engineering, Beijing Institute of Technology, Beijing, China
gongjianwei@bit.edu.cn

Abstract. Vehicle is one of the important subjects studied in the domain of computer vision, autonomous driving and intelligent transportation system. 3D models of vehicles are used widely in the literature of vehicle categorization, pose estimation, detection, and tracking. However, previous work uses only a small set of 3D vehicle models either from CAD design or multi-view reconstruction, limiting their representation ability and performance. A feasible approach to acquire extensive 3D vehicle models is desired. In this paper, we are interested in 3D surface reconstruction of on-road vehicles from sparse point cloud captured by laser scanners equipped ubiquitously on autonomous driving platforms. We propose an innovative reconstruction pipeline and method, called TriSpaSurf, which could reconstruct unbroken and smooth surface robustly from just a single frame of noisy sparse point cloud. In the TriSpaSurf, triple-view 2D outlines are first fitted on the 2D points from the projection of 3D point cloud under each view, and then 2.5D surface reconstruction is carried out under the guidance from triple-view outlines. By projecting 3D point cloud onto 2D views, 2D outlines could be estimated robustly due to the reduced complexity and higher signal-to-noise ratio in 2D views, and could provide fairly stable and tight multi-view constraints for 3D surface reconstruction. The effectiveness of our method is verified on the KITTI and Sydney dataset.

Keywords: Surface reconstruction · Triple-View outline · 2.5D representation

1 Introduction

Vehicle is one of the important subjects studied in the domain of computer vision, autonomous driving and intelligent transportation system. It offers a rich test bed for a number of interesting topics, such as fine-grained vehicle categorization and verification [1–9], vehicle detection, pose estimation and tracking [10–18].

H. Zheng–This is a student paper.

© Springer Nature Switzerland AG 2020
Y. Peng et al. (Eds.): PRCV 2020, LNCS 12305, pp. 398–409, 2020.
https://doi.org/10.1007/978-3-030-60633-6_33

| (a) Image | (b) Point cloud input | (c) TriSpaSurf | (d) TriSpaSurf (textured) | (e) Ball Pivoting | (f) Screened Poisson |

Fig. 1. 3D surface reconstruction from sparse point cloud. Our TriSpaSurf result (c) is compared with Ball Pivoting (e) and Screened Poisson (f).

3D models of vehicles are used widely in the literature [1, 4, 7–13], due to the ability to handle the view-point variations. Detailed 3D representations of vehicles are learnt from a small set of 3D CAD models in [1, 4, 8–11], and from multi-view reconstruction in [7, 13]. A simple 3D wireframe model of vehicles is defined in [12]. However, vehicles possess a tremendous amount of designs and models. A simple model would be inadequate to capture sufficient geometry for fine-grained categorization or accurate estimation of vehicle's location and pose. While on the other hand, CAD models are not possible to be available for every vehicle, and it is tedious and time consuming to build a great deal of vehicle models by multi-view reconstruction. A feasible approach to acquire 3D models of vehicles is therefore desired.

In this paper, we are interested in 3D surface reconstruction of on-road vehicles which are scanned by a mobile platform equipped with sparse laser scanner, e.g., Velodyne LiDAR. Owing to the gold rush of autonomous driving, such platforms are ubiquitous and massive 3D data of vehicles can be obtained during the platforms running on the road. However, the point cloud captured by the sparse laser scanner is with low density, especially in vertical direction (e.g., only 16/32/64 lines, see Fig. 1b for an examples). On the other hand, the point cloud is also noisy. While under the sparsity, smoothing cannot be applied straightforwardly to resist the noise. As discussed in Sect. 2, it is challenging for previous work of 3D reconstruction to handle such sparse and noisy point cloud, the reconstructed surface is often fragmented (see Fig. 1e), with spicule (see Fig. 1e and 1f), or over smooth (see Fig. 1f).

To deal with above challenges, this paper proposes an innovative reconstruction pipeline and method, called **TriSpaSurf**, which could reconstruct unbroken and smooth surface robustly from just a single frame of noisy sparse point cloud (see Fig. 1). The TriSpaSurf contains two important ingredients to handle the above challenges:

1) Multi-view 2D outlines fitting (three orthographic 2D views are used currently). There are two advantages by projecting 3D point cloud onto 2D views. On the one hand the complexity of 2D curve fitting is lower than 3D surface fitting. It is easier to regularize the 2D fitting to handle the missing data and noise. On the other hand, the points in 2D views are denser and with higher signal-to-noise ratio, making the 2D fitting more stable. As a result, multi-view outlines could be estimated robustly, which provide fairly stable and tight multi-view constraints of vehicles' 3D surface.
2) Triple-view outline guided 2.5D surface reconstruction. Based on the triple-view 2D outlines, a reference 3D surface is reconstructed. We adopt a 2.5D surface representation which represents the object's surface as an evolution of the reference surface along the normal. And surface reconstruction is finally posed as an inference problem where optimal evolution is inferred from the surface under the constraint of piecewise smoothness and reference prior.

With the guidance from triple-view 2D outlines, the reference surface can be con-structed almost with high fidelity, which makes later 2.5D surface optimization much easier and robust, especially in handling the point sparsity and noise. By 2.5D surface representation, the object's surface is constrained as an evolution of the reference sur-face, which affords a great degree of regularization. Such regularization could help in resisting the distraction of the noised point observations, and could prevent object's sur-face from collapsing into the neighboring or inner points due to massive data missing. As a result, our TriSpaSurf is capable of reconstructing unbroken and smooth surface from just single noisy sparse point cloud.

The remainder of this paper is organized as follows. After discussing related work, our reconstruction pipeline in overviewed in Sect. 3. The framework of shape representation and optimization for 2D outline fitting and 3D surface reconstruction is discussed in Sect. 4. In the two succeeding sections, technical details of the TriSpaSurf are presented. We provide experimental results in Sect. 7 and conclude in Sect. 8.

2   Related Work

In the following, we first discuss how 3D models of vehicles are used and con-structed in the literature of vehicle categorization, detection and pose estimation. The frameworks and methods of surface reconstruction are subsequently reviewed.

2.1   3D Vehicle Modeling

Due to the ability to handle the viewpoint variations, 3D models of vehicles are used widely in the literature [1, 4, 7–13]. Literature [1] promotes the spatial pyramids (SPM) and bubble bank (BB) from 2D to 3D on the level of feature appearance and location for fine-grained categorization. A 3D deformable car model represented by a collection of 3D points and edges is built on the Active Shape Model (ASM) formulation in [4]. Sim-ilarly, the work of [7–9, 12] also use a 3D wireframe model to recognize and locate road vehicles. A denser mesh model than wireframe is used in [10, 11] for vehicle surveil-lance and detection. The work of [13] stores the features of different positions obtained by sampling patches from different viewpoints and scales on the car for detection and pose estimation. Due to the lack of extensive high-quality 3D models of vehicles, the above methods use only a limited set of detailed models or are based on coarse models, limiting their representation ability and performance. Our method can be used to con-struct abundant high-quality 3D models of vehicles to facilitate the researches in related domain.

2.2   Surface Reconstruction

Surface reconstruction technology can be generally divided into implicit and explicit methods. Poisson surface reconstruction in [20] is a typical implicit function method. The implicit equation represented by the surface information described by the point cloud model is obtained by solving the Poisson equation. The advantage is that the reconstructed model has closed characteristics of water tightness, and has good geometric surface characteristics and detail characteristics. The Screened Poisson surface reconstruction algorithm in [21] introduces constraints of the point set and gradient on the basis of the original Poisson surface reconstruction algorithm, and the input equation extracted by the isosurface is converted from the original Poisson equation to the Screened Poisson equation. But both Poisson and Screened Poisson reconstruction are not suitable for non-watertight surfaces reconstruction like the incomplete on-road vehicle point cloud.

Another method for 3D reconstruction is the explicit methods. The Delaunay triangulation in [22] is the fundamental and representative technique. Delaunay triangulation has the characteristics of maximizing the minimum angle and the triangle network closest to the regularization. A new Voronoi-based technique referred as Curst algorithm is proposed in [23], which is an improved version of Voronoi diagram and Delaunay triangulation. Ball-Pivoting algorithm in [24] has a different ideology to reconstruct 3D surface. The algorithm starts with a seed triangle, and then rotates the ball along one side of the seed triangle until the ball touches the next point. The main disadvantage of this algorithm is that when the point cloud data is uneven, the ball will not touch the point during the rolling process, so it will produce holes. Moreover, one limitation of explicit methods is that they are sensitive to the noise in point cloud, and may construct the surface with spicule.

3   TriSpaSurf Overview

Figure 2 provides an overview of the whole pipeline of the TriSpaSurf, which consists of the following three major components:

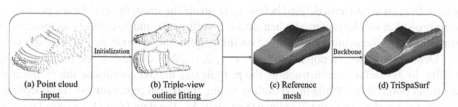

Fig. 2. Overall pipeline of the TriSpaSurf.

Triple-View Outline Fitting. The input 3D point cloud is first projected onto three orthographic 2D views. Object's outline is then estimated in each view to fit the projected points (Sect. 5).

Reference Mesh Construction. Based on the triple-view 2D outlines, a reference mesh is constructed (Sect. 6.1), and pruned according to the visibility. The reference mesh is also augmented with triple-way slice contours, which are used to estimate the mesh normal for 2.5D mesh and defines the message passing directions during 2.5D mesh optimization.

2.5D Mesh Optimization. Object's surface reconstruction is posed as a problem of MRF inference from the reference mesh. The MRF is defined on the reference mesh, and loopy belief propagation is adopted to infer the optimal evolution, under the constraint of piecewise smoothness and reference prior (Sect. 6.2).

4   Shape Representation and Optimization

In the TriSpaSurf, we adopt a unified framework of shape representation and optimization for 2D outline fitting and 3D surface reconstruction. More concretely, 1.5D/2.5D shape representation is used for the object's 2D outline/3D surface which is represented as an evolution of a reference contour/mesh along the normal.

Suppose the reference contour/mesh is represented by a graph $\mathcal{G}$ comprising a set of vertices $\mathcal{V} = \{v_j; j = 1, 2, ..., M\}$ along with a collection of edges $\mathcal{E} = \{(p, q); p$ and q is the index of $\mathcal{V}\}$. The v_j denotes the 2D/3D coordinate of the jth vertex on the reference contour/mesh. The edge set $\mathcal{E}$ characterizes the connections between vertices in the contour/mesh, and defines the neighborhood structure.

The object's 2D outline/3D surface is expressed by another graph $\mathcal{G}_u$ with the same neighborhood structure $\mathcal{E}$ but a new vertex set $\mathcal{V}_u = \{v_j + n_j * u_j; j = 1, 2, ..., M\}$, where n_j is the normal at the vertex v_j in the reference contour/mesh, and u_j is an evolution along the normal n_j. The goal of the 2D outline fitting and 3D surface reconstruction is to infer the optimal evolution $\tilde{u}_j$ from the observed point data.

The optimal evolution is inferred by minimizing a following objective function:

$$E(u) = \sum_{j=1}^{M} \left[\lambda_p |u_j| - D(u_j) \right] + \lambda_s \sum_{(p,q) \in \mathcal{E}} \left(u_p - u_q \right)^2 \tag{1}$$

where $D(u_j)$ is a data term defined below, λ_p controls the strength of reference prior (i.e., zero evolution will be preferred when data is insufficient), and λ_s controls the piecewise smoothness of the evolutions enforced on the neighborhood structure of the graph $\mathcal{G}$.

Denote the observed point data $\{p_i; i = 1, 2, ..., N\}$, where N is the number of the points. In 3D surface reconstruction, the p_i indicates the 3D coordinate the ith point in the input point cloud. While in 2D outline fitting, the p_i will represent the 2D coordinate the ith projected point in the 2D views (See Sect. 5). The data term $D(u_j)$ calculates the score of the evolution u_j over the observed point data, and is defined as

$$D(u_j) = \sum_{i=1}^{N} exp\left(-v_j + n_j * u_j - p_i^2 / \sigma^2 \right) \tag{2}$$

$D(u_j)$ can be understood as a weighted number of the point p_i which supports the evolution u_j. The weight is higher when p_i is closer to the evoluted vertex $v_j + n_j * u_j$, and σ controls the spread of the weighing. The term $D(u_j)$ can also be view as a fusion process that integrates point observations into the evolutions along the normal of the vertices.

We discretize the possible evolutions, and optimize the energy function in Eq. 1 via loopy belief propagation (LBP). Once the optimal evolutions $\tilde{u}_j$ got, the object's 2D outline/3D surface is finally represented by a graph $\mathcal{G}_{\tilde{u}}$ with a vertex set $\mathcal{V}_{\tilde{u}} = \{v_j + n_j * \tilde{u}_j; j = 1, 2, ..., M\}$ and the same edge set $\mathcal{E}$ as the one of the reference graph $\mathcal{G}$.

5 Triple-View Outline Fitting

Given the input 3D point cloud of the target object, the orientation for triple view projection is determined first, then outlines are fitted in each projected 2D views.

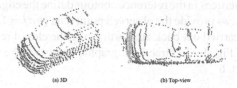

(a) 3D   (b) Top-view

Fig. 3. The main plane (red points) and another plane (green points). (Color figure online)

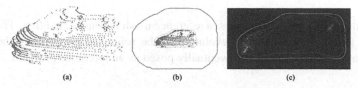

(a)   (b)   (c)

Fig. 4. (a) is the original point cloud. The blue line in (b) is an isocontour of the DT at higher level, and the green line is the reference contour. The red in (c) line is the optimal evolution of the reference contour. (Color figure online)

The orientation for triple view projection should be chosen to meet two requirements: 1) the projected points could be overlapped sufficiently, to make them dense on the 2D views, 2) and the bounding rectangle of the projected points should be as minimal as possible, to make the reference 3D surface reconstructed from the triple-view outlines tight.

As shown in Fig. 3, to meet the first requirement, a main plane is first detected from the input 3D point cloud via RANSAC. Afterwards, the main plan will be rectified to be parallel to the x-z plane by a rotation matrix R_1 defined by principle components of PCA. To meet the second requirement, two rotation matrixes R_2 and R_3 are subsequently determined. R_2 is defined as a rotation around x-axis, and is chosen such that the bounding rectangle of the projected points on x-z view is minimal. R_3 is defined as a rotation around y-axis, and is chosen such that the bounding rectangle of the projected points on y-z view is minimal. The overall rotation matrix R is finally calculated as $R = R_1 \cdot R_2 \cdot R_3$. The input 3D point cloud is then transformed by R. The transformed point cloud is ultimately projected onto three orthographic views, i.e., top view (x-y), front view (y-z), and side view (x-z). Outline fitting is carried out in each view.

Outline fitting is carried out by first generating a smooth reference contour, and then inferring an optimal evolution of the reference contour to fit the projected points in current 2D view. The reference contour is generated based on distance transform (DT) of the projected points in current 2D view. To ensure the smoothness of the reference contour, an isocontour of the DT at a higher level (say, 0.5 m) is extracted, and then shrink to enclose the projected points in current 2D view (see Fig. 4b). The reference contour as well as associated normal is ultimately generated from the isocontour via B-Spline curve fitting.

Once the reference contour is generated, the framework discussed in Sect. 4 is applied for 1.5D outline representation and optimization. We sample a sequence of M points on the reference contour to form the vertex set $\mathcal{V}$ in the reference graph $\mathcal{G}$, and the connections between vertices in the reference contour define the edge set $\mathcal{E}$. The projected points in current 2D views provide the observed point data $\{p_i; i = 1, 2, ..., N\}$ for outline optimization. LBP is carried out back and forth along the closed reference contour. The obtained outlines (see Fig. 4c) in three orthographic views will be used to construct the reference mesh in Sect. 6.

6 Surface Reconstruction and Optimization

Based on the triple-view 2D outlines, a reference mesh is constructed first. The object's surface is then represented as an evolution of the reference mesh along the normal. Object's surface reconstruction is eventually posed as an inference problem.

6.1 Reference Mesh from Triple-View Outlines

Constructing the reference mesh from triple-view 2D outlines is similar to the problem of shape from silhouette [19], but under orthographic projection. A visual hull is constructed as the intersection of the triple viewing cylinders, each formed by back projecting the silhouette image generated from the outline into the 3D space through the orientation of orthographic projection.

Once the visual hull is constructed, we slice the volume in three orthographic directions, and generate a set of triple-way slice contours. The reference mesh is then constructed according to the vertexes and connections in these slice contours. The normal of each vertex can be estimated as the cross product of the tangents of the slice contours which intersect at that vertex. Once the normal is estimated, the reference mesh is pruned according to the visibility. The pruned reference mesh will be served as a backbone for 2.5D mesh optimization.

6.2 2.5D Mesh Optimization

Once the reference mesh is generated, the framework discussed in Sect. 4 is applied for 2.5D mesh representation and optimization. The reference graph G is formed according to the pruned reference mesh, where the mesh's vertices define the vertex set V, and the connections between vertices in the mesh constitute the edge set E.

The input 3D point cloud provides the observed point data $\{p_i; i = 1, 2, ..., N\}$. In addition to constructing the reference mesh, the triple-way slice contours are also used to define the message passing directions of LBP during 2.5D mesh optimization, where LBP is carried out back and forth along each slice contour, and is iterated alternately in triple-way slice contours. The inference result of LBP gives the final surface reconstruction.

7 Experimental Results and Comparison

We conducted many experiments on the KITTI and Sydney dataset, and the experimental results confirmed the performance and effectiveness of our method. Note that the KITTI dataset provides a complete road environment point cloud, so we extract the car point cloud from the bounding box provided and remove the extra ground points. And the vehicles in the Sydney dataset are not collected on real roads, so their posture is very messy. In order to confirm the robustness of the experiment, we also test our method on different point cloud densities. Next, we will introduce the experimental results of the TriSpaSurf and then compare our results with other methods.

7.1 Detailed Results

Figure 5 is the surface reconstruction results on different point cloud densities (from the KITTI dataset). The point cloud densities gradually decrease from Fig. 5a to Fig. 5e. It can be seen from the figure that our TriSpaSurf is still stable even when there are few

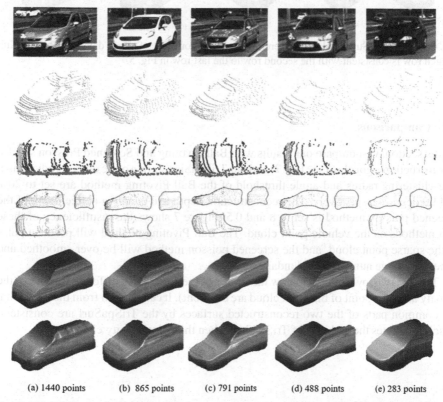

(a) 1440 points   (b) 865 points   (c) 791 points   (d) 488 points   (e) 283 points

Fig. 5. Our TriSpaSurf results on different point cloud densities (the KITTI dataset). First row is the car image. Second row is original point cloud. Third row is the top view, where the red dot is the main plane and the green dot is another plane. Red curve in the fourth row is the triple-view outline fitting results. Fifth row is the reference surface from triple-view outlines. The last row is the final TriSpaSurf result. (Color figure online)

points. Figure 6 is the results on different vehicle categories (from the Sydney dataset). As can be seen from the figure, our TriSpaSurf performs well on different type of vehicles.

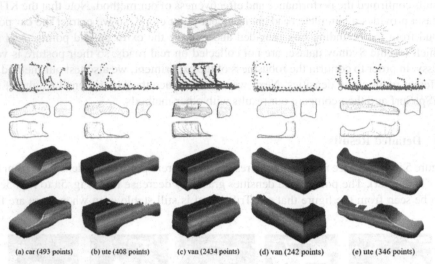

(a) car (493 points)   (b) ute (408 points)   (c) van (2434 points)   (d) van (242 points)   (e) ute (346 points)

Fig. 6. Our TriSpaSurf results on different vehicle categories (the Sydney dataset). The meaning of each row is consistent with the second row to the last row in Fig. 5.

7.2 Comparisons

In this section, we compare our results with Ball Pivoting and Screened Poisson surface reconstruction algorithm (see Fig. 7) and our TriSpaSurf show a quite obvious advantage. The clustering radius and angle threshold of the Ball Pivoting method are set to 20% and 90 degrees respectively. The reconstruction depth and adaptive octree depth of the screened poisson method is set to 8 and 0.5. Figure 7 shows the insufficiency of these two methods on the vehicle point cloud. The Ball Pivoting method will produce holes in the sparse point cloud, and the screened poisson method will be over smoothed and generate a large number of redundant surfaces.

The last two rows of Fig. 7 show the results of the same car at different distances (the density and viewpoint of the point cloud are different). It can be seen from the figure that the common parts of the two reconstructed surfaces by the TriSpaSurf are consistent, which illustrates the stability of TriSpaSurf when the point density change.

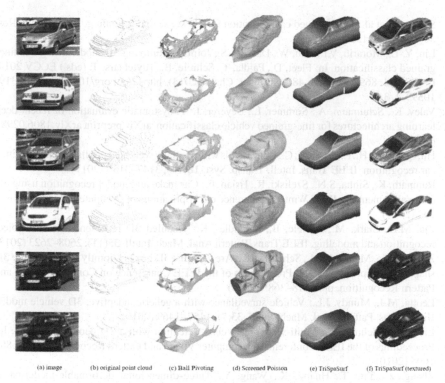

(a) image   (b) original point cloud   (c) Ball Pivoting   (d) Screened Poisson   (e) TriSpaSurf   (f) TriSpaSurf (textured)

Fig. 7. Comparison with Ball Pivoting and Screened Poisson.

8 Conclusion

This paper proposes a novel method of 3D surface reconstruction that uses multi-view outlines as guidance and is able to reconstruct surfaces from noisy sparse point clouds. By projecting 3D point cloud onto 2D views, 2D outlines could be estimated robustly due to the reduced complexity and higher signal-to-noise ratio in 2D views, and could provide fairly stable and tight multi-view constraints for 3D surface reconstruction. Compared with other traditional surface reconstruction algorithms on sparse point cloud, such as Ball Pivoting and Screened Poisson surface reconstruction, our TriSpaSurf shows the superior performance. In future work, we plan to extend our method by fusing multi-frame to get a more complete result of surface reconstruction.

References

1. Krause, J., Stark, M., Deng, J., Fei-Fei, L.: 3D object representations for fine-grained categorization. In: Proceedings of the IEEE International Conference on Computer Vision Workshops, pp. 554–561 (2013)
2. Yang, L., Luo, P., Change Loy, C., Tang, X.: A large-scale car dataset for fine-grained categorization and verification. In: Proceedings of the IEEE Conference on Computer Vision and Pattern Recognition, pp. 3973–3981 (2015)

3. Stark, M., et al.: Fine-grained categorization for 3D scene understanding. Int. J. Robot. Res. **30**(13), 1543–1552 (2011)

4. Lin, Y.-L., Morariu, V.I., Hsu, W., Davis, L.S.: Jointly optimizing 3D model fitting and fine-grained classification. In: Fleet, D., Pajdla, T., Schiele, B., Tuytelaars, T. (eds.) ECCV 2014. LNCS, vol. 8692, pp. 466–480. Springer, Cham (2014). https://doi.org/10.1007/978-3-319-10593-2_31

5. Valev, K., Schumann, A., Sommer, L., Beyerer, J.: A systematic evaluation of recent deep learning architectures for fine-grained vehicle classification. arXiv preprint arXiv:1806.02987 (2018)

6. Hu, Q., Wang, H., Li, T., Shen, C.: Deep CNNs with spatially weighted pooling for fine-grained car recognition. IEEE Trans. Intell. Transp. Syst. **18**(11), 3147–3156 (2017)

7. Ramnath, K., Sinha, S.N., Szeliski, R., Hsiao, E.: Car make and model recognition using 3D curve alignment. In: IEEE Winter Conference on Applications of Computer Vision, pp. 285–292 (2014)

8. Zia, M.Z., Stark, M., Schiele, B., Schindler, K.: Detailed 3D representations for object recognition and modeling. IEEE Trans. Pattern Anal. Mach. Intell. **35**(11), 2608–2623 (2013)

9. Zeeshan Zia, M., Stark, M., Schindler, K.: Are cars just 3D boxes?–jointly estimating the 3D shape of multiple objects. In: Proceedings of the IEEE Conference on Computer Vision and Pattern Recognition, pp. 3678–3685 (2014)

10. Leotta, M.J., Mundy, J.L.: Vehicle surveillance with a generic, adaptive, 3D vehicle model. IEEE Trans. Pattern Anal. Mach. Intell. **33**(7), 1457–1469 (2010)

11. Liebelt, J., Schmid, C.: Multi-view object class detection with a 3D geometric model. In: Proceedings of the IEEE Conference on Computer Vision and Pattern Recognition, pp. 1688–1695 (2010)

12. Zhang, Z., Tan, T., Huang, K., Wang, Y.: Three-dimensional deformable-model-based localization and recognition of road vehicles. IEEE Trans. Image Process. **21**(1), 1–13 (2011)

13. Glasner, D., Galun, M., Alpert, S., Basri, R., Shakhnarovich, G.: Viewpoint-aware object detection and pose estimation. In: Proceedings of the IEEE Conference on Computer Vision, pp. 1275–1282 (2011)

14. Geiger, A., Lenz, P., Urtasun, R.: Are we ready for autonomous driving. In: Proceedings of the IEEE Conference on Computer Vision and Pattern Recognition, pp. 3354–3361 (2012)

15. Ozuysal, M., Lepetit, V., Fua, P.: Pose estimation for category specific multi-view object localization. In: Proceedings of the IEEE Conference on Computer Vision and Pattern Recognition, pp. 778–785 (2009)

16. Li, Y., Gu, L., Kanade, T.: Robustly aligning a shape model and its application to car alignment of unknown pose. IEEE Trans. Pattern Anal. Mach. Intell. **33**(9), 1860–1876 (2011)

17. Sun, Z., Bebis, G., Miller, R.: On-road vehicle detection: a review. IEEE Trans. Pattern Anal. Mach. Intell. **28**(5), 694–711 (2006)

18. Sivaraman, S., Trivedi, M.M.: Looking at vehicles on the road: a survey of vision-based vehicle detection, tracking, and behavior analysis. IEEE Trans. Intell. Transp. Syst. **14**(4), 1773–1795 (2013)

19. Newcombe, R.A., et al.: KinectFusion: real-time dense surface mapping and tracking. In: Proceedings of IEEE International Symposium on Mixed and Augmented Reality, pp. 127–136 (2011)

20. Kazhdan, M., Bolitho, M., Hoppe, H.: Poisson surface reconstruction. In: Proceedings of the Fourth Eurographics Symposium on Geometry Processing, pp. 61–71 (2006)

21. Kazhdan, M., Hoppe, H.: Screened poisson surface reconstruction. ACM Trans. Graph. (ToG) **32**(3), 1–13 (2013)

22. Du, D.Z., Hwang, F.: Computing in Euclidean Geometry, pp. 193–233. World Scientific Publishing, Singapore (1995)

23. Amenta, N., Bern, M., Kamvysselis, M.: A new Voronoi-based surface reconstruction algorithm. In: Proceedings of the 25th Annual Conference on Computer Graphics and Interactive Techniques, pp. 415–421 (1998)

24. Bernardini, F., Mittleman, J., Rushmeier, H., Silva, C., Taubin, G.: The ball-pivoting algorithm for surface reconstruction. IEEE Trans. Visual Comput. Graphics 5(4), 349–359 (1999)

3D Human Body Shape and Pose Estimation from Depth Image

Lei Liu, Kangkan Wang$^{(\boxtimes)}$, and Jian Yang

PCA Lab, Key Lab of Intelligent Perception and Systems for High-Dimensional
Information of Ministry of Education, and Jiangsu Key Lab of Image and Video
Understanding for Social Security, School of Computer Science and Engineering,
Nanjing University of Science and Technology, Nanjing, China
{liuloijs,wangkangkan,csjyang}@njust.edu.cn

Abstract. This work addresses the problem of 3D human body shape
and pose estimation from a single depth image. Most 3D human pose
estimation methods based on deep learning utilize RGB images instead
of depth images. Traditional optimization-based methods using depth
images aim to establish point correspondences between the depth images
and the template model. In this paper, we propose a novel method to
estimate the 3D pose and shape of a human body from depth images.
Specifically, based on the joints features and original depth features,
we propose a spatial attention feature extractor to capture spatial local
features of depth images and 3D joints by learning dynamic weights of
the features. In addition, we generalize our method to real depth data
through a weakly-supervised method. We conduct extensive experiments
on SURREAL, Human3.6M, DFAUST, and real depth images of human
bodies. The experimental results demonstrate that our 3D human pose
estimation method can yield good performance.

Keywords: Human shape and pose estimation · Deep learning · Weak
supervision

1 Introduction

Human pose estimation has numerous applications in robotics, augmented reality
(AR), and virtual reality (VR). With the rapid development of computer vision,
3D human shape and pose estimation from depth images has gained popularity
in the 3D computer vision community. However, estimating 3D human models
directly from depth images is still a challenging problem since the human bodies
have large deformations and self-occlusions in motion.

Recently, a few methods use deep learning to directly predict 3D human models from depth images. Most 3D human body reconstruction methods based on
a sequence of depth images aim to establish the point correspondences between

This work was supported by the Natural Science Foundation of China under Grant
Nos. 61602444, U1713208, and Program for Changjiang Scholars.

Y. Peng et al. (Eds.): PRCV 2020, LNCS 12305, pp. 410–421, 2020.
https://doi.org/10.1007/978-3-030-60633-6_34

the consecutive frames, which results in the error accumulation. At the same time, most human pose estimation methods use RGB images as the inputs. However, the RGB images lack depth information naturally, which makes it a classical ill-posed inverse problem. In addition, directly extending these methods to depth images cannot obtain high-precision 3D human models. Another problem with 3D human pose estimation is the lack of a large number of labeled 3D human pose datasets, and labeling 3D human pose datasets is very difficult, which results in great difficulty for the training of deep neural networks.

In this paper, we propose a 3D human body shape and pose estimation method from a single depth image. First, we employ ResNet50 [1] to extract the latent features of depth images. Then, we simultaneously estimate the 3D joints and the 3D human model parameters. Besides, we introduce the attention mechanism to improve the accuracy of 3D human pose estimation. Based on the 3D joints features and the original depth features, we construct a spatial attention feature extractor to learn the dynamical attention weights, which can capture local geometric structure and assess the feature map actively. Similar to [2], we use a cyclic regression network to regress the human model parameters from features. Further, to improve the performance of the network on real depth data, we introduce a weakly-supervised mechanism to fine-tune the network. In addition to using traditional joints information, we also introduce a differentiable rendering layer to render the predicted human models to depth images and silhouettes. Intuitively, the rendered depth images and silhouettes should be consistent with the original inputs. The experimental results on SURREAL [3], Human3.6M [4], DFAUST [5], and the real data of human bodies demonstrate the effectiveness of our proposed method. In summary, the main contributions of our method are as follows:

1. We propose a novel 3D human pose and shape estimation framework based on depth images to predict the 3D human model parameters, which achieves the state-of-the-art performance on human pose and shape recovery.
2. We propose a spatial attention feature extractor for extracting more effective features, which effectively improves the accuracy of the human shape and pose estimation.
3. We fine-tune the network by using joints information and a differentiable render layer in weakly-supervised manner, which improves the performance of the network on real depth data.

2 Related Work

2.1 3D Human Model Estimation from RGB Images

With the simplicity and extensibility of the SMPL model [6], there has been substantial recent work in estimating the parameters of this model. Bogo et al. [7] propose SMPLify, which estimates the position of the corresponding 2D human joints. Then, they recover SMPL parameters by minimizing the 2D projection of model joints and the detected 2D joints. Huang et al. [8] expand the SMPLify and propose MuVS. MuVS uses multi-view images as inputs and a deep neural network is used to segment the human body, which can eliminate the effects of background pictures. In addition to using joints projection as a constraint, this method also matches the 3D human models with the segmentation results to improve the accuracy of the results. Kanazawa et al. [2] propose an end-to-end framework HMR for recovering a 3D human model from an image, which directly extracts features from the images and recovers the SMPL pose parameters through regression. In addition to directly regressing SMPL model parameters, some recent works also use other methods to predict 3D human models. For example, Kolotouros et al. [9] propose a Graph CNN method, which first attaches the encoded features from an input RGB image to 3D vertex coordinates of a template mesh and then predicts the mesh vertex coordinates using a convolutional mesh regression.

2.2   3D Human Model Estimation from Depth Images

In recent years, there have been some studies on human pose estimation based on depth images. Guo et al. [10] use a L_0 based motion regularizer with an iterative optimization solver to deform the pre-scanned template model to each input depth images. There are some template-less methods which can create the 3D human models without any prior knowledge about the human shape and fuse all depth maps to reconstruct 3D human models with slow motion. These methods aim to build point correspondences for each depth frame by searching the closest 3D point, which will be invalid when the input depth image is very different from the template model. Wei et al. [10] build the point correspondences by matching the learned feature descriptors for depth images of human bodies. Then, the 3D models are generated by fitting the template SMPL model to learned point correspondences using [10]. Kadkhodamohammadi et al. [11] propose a multi-view RGB-D approach for human pose estimation, which proves the advantages of using depth data. Pavlakos et al. [12] use ordinal depth constraints as a form of weak-supervision to train a network which can predict the 3D pose. Li et al. [13] propose a dynamic fusion module that enables training models with RGB-D data to address the ambiguity problem.

3   Approach

For a given depth image, our method can be used to estimate the 3D human models aligned with the depth images. The algorithm flow is shown in Fig. 1. First of all, we use ResNet50 [1] to extract features from depth images. Then, we use a joints estimation network to learn the human pose information. We propose a spatial attention feature extractor to learn the refined features. In this way, the network could suppress irrelevant regions and highlights useful features for the human pose and shape estimation. Therefore, the accuracy of 3D human pose and shape estimation can be effectively improved. In addition, we also introduce a differentiable rendering layer [14], which can render the predicted human models to silhouettes and depth images. We exploit joints projection, silhouettes, and depth images as constraints to keep our prediction models consistent with the input information. By this means, we could fine-tune the network through a weakly-supervised method and improve its performance on the real data. Therefore, our method does not require real labeled 3D human pose datasets.

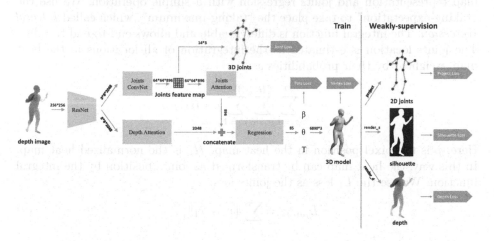

Fig. 1. Overview of the proposed framework. Our framework can predict the 3D human shape and pose from a single depth image. The joints prediction module is used to explicitly extract pose information from the features. The depth attention and joints attention module are used to extract more effective depth features and pose features. We exploit a weakly-supervised method to fine-tune our network on the real depth data. The numbers in the figure show the size of the input of each module.

3.1 3D Body Model

We follow the previous 3D human pose estimation methods, using the Skinned Multi-Person Linear (SMPL) model of Loper et al. [6] to represent the human body. The SMPL model is a statistical parametric differentiable model that can accurately represent various human shapes under natural conditions, and it is compatible with existing graphics pipelines. It uses shape parameter $\beta \in \mathbb{R}^{10}$ to control the appearance of the human body model. Pose parameter $\theta \in \mathbb{R}^{72}$ are used to represent different poses of the human body. For the given shape parameter β and pose parameter θ, the SMPL model provides a function $M(\beta, \theta)$, which can map the shape and pose parameters to 6890 vertexes V of the human body. At the same time, SMPL also provides a $6890 * 24$ joints regression coefficient R, which can map the 3D human model to the corresponding 24 joints through $X(V, R)$.

3.2   3D Human Joints Estimation

As a simplified representation of human pose, 3D joints contain a large amount of pose information. In this paper, the 3D joints estimation module is used to explicitly extract human pose information, which is beneficial to the convergence of the network. We use the same method as [15], which relates and unifies the heat map representation and joints regression with a simple operation. We use the "taking-expectation" to take place the "taking-maximum", which called *integral regression*. The integral function is differentiable and allows end-to-end training. The joints location is estimated as the integration of all locations in the heat map, weighted by their probabilities as:

$$J_k = \sum_{p_z=1}^{D} \sum_{p_y=1}^{H} \sum_{p_x=1}^{W} p * \widetilde{H}_k(p). \tag{1}$$

Here, p is the pixel position in the heat map, $\widetilde{H}_k$ is the normalized heat map. In this way, the heat map can be transformed as joints position by the integral function. We use the L_1 loss as the joints loss:

$$L_{joints} = \sum_i \|x_i - \hat{x}_i\|_1. \tag{2}$$

3.3   Spatial Attention Feature Extracting

We introduce the attention mechanism to learn dynamic weights for the 3D joints features and original depth features which improve the effectiveness and generalization of the features and highlights the useful features. As far as we know, this is the first time that the attention mechanism been used in 3D human pose and shape estimation task. The architecture of joints attention and depth attention is shown in Fig. 2. The network builds relationships between spatial regions that are far from each other by providing a dynamic weighted features. Moreover, the network with attention layer can autonomously assess the effectiveness of the feature on the human pose estimation and adjust the weights of different features dynamically. Specifically, the depth attention is used to capture the local structured information in depth features while the joints attention is used to capture the pose information in joints features. We define the operation as: $\psi = W * x$, where W is the weight matrix learned by a MLP. Then, we get the attention map ω_{ij} by computing the softmax of ψ:

$$\omega_{ij} = \frac{exp(\psi_{ij})}{\sum_{i=1}^{N} \psi_{ij}}. \tag{3}$$

To obtain the final attention feature map, we apply the matrix multiplication between the attention map ω_{ij} and the original feature x_j,

$$\hat{x}_i = \sum_{i=1}^{N} \omega_{ij} x_j. \tag{4}$$

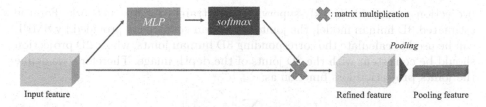

Fig. 2. An illustration of our attention architecture. The depth attention and joints have similar architecture. The input feature map is passed to the attention module which learns the dynamic weights ω. We apply the matrix multiplication between the dynamic weights and original features to obtain the refined features. Finally, the refined features are aggregated by max pooling and average pooling for joints attention and depth attention, respectively.

To obtain their respective global feature descriptors, the new joints features and depth features obtained from the attention layer are aggregated using max pooling and average pooling, respectively.

3.4 Regression the Parameters of Human Model

For the human model parameters regression module, we use a deep neural network, whose architecture has the same design with Kanazawa et al. [2]. We concatenate the joints feature descriptors, depth feature descriptors, and the mean model parameters together as the inputs. Then, we regress the parameters β, θ, and T in an iterative error feedback (IEF) loop, where progressive changes are made recurrently to the current estimation. After that, we map the parameters into 6890 vertexes of 3D human body with the function $M(\beta, \theta)$ provided by SMPL. In this setting, the parameter loss and vertex loss are provided:

$$L_{para} = \left\| [\beta, \theta] - \left[\hat{\beta}, \hat{\theta} \right] \right\|_2^2, \tag{5}$$

$$L_{vertex} = \| v_i - \hat{v}_i \|_2^2. \tag{6}$$

In summary, the loss function of the entire network is

$$L = \varphi L_{joints} + \gamma L_{para} + \sigma L_{vertex}. \tag{7}$$

Among them, φ, γ, and σ are the hyperparameters of the corresponding loss functions, which are used to balance the value of each loss function.

3.5 Fine-Tuning for Real Depth Data with Weakly-Supervision

Since the training data used in this paper comes from synthetic data and does not contain any real data, the performance of the network on real data is not satisfactory. Therefore, we fine-tune the network on real depth data in a weakly-supervised manner. Similar to other weakly-supervised methods, we use 2D joints projection to provide weakly-supervised constraints for the network. For the estimated 3D human model, the joints regression coefficients provided by SMPL can be used to calculate the corresponding 3D human joints, whose 2D projection should be consistent with the 2D joints of the depth image. Therefore, we define the joints projection loss function as:

$$L_{project} = \sum_i \left\| \vartheta_i - \hat{\vartheta}_i \right\|_1. \tag{8}$$

Here, ϑ_i is the 2D projection of predicted 3D human joints, $\hat{\vartheta}_i$ is the ground truth joints of the depth image. However, the joints projection can only constrain the pose of the human body without considering the shape of the 3D human body. Because of that, we introduce a differentiable rendering layer, which can render the 3D models to 2D images. Similar to the joints projection, the 2D rendered silhouettes of the human models should be consistent with the silhouettes of the input depth images. The silhouette loss function is defined as

$$L_{sil} = \left\| p_{sil}(M(\beta, \theta) + T) - \hat{s} \right\|_2^2. \tag{9}$$

Here, p_{sil} is the differentiable rendering silhouette function, $\hat{s}$ is the silhouette corresponding to the input depth image. Whether the joints projection loss or the silhouette loss are essentially 2D constraints on the 3D human models. As a result, there may still be multiple 3D human body models that match the 2D information. Therefore, in addition to the above two constraints, we introduce a depth loss to constrain the network with 3D information. We use the differentiable rendering layer mentioned above to render the 3D human models to depth images. Intuitively, the rendered depth images should be consistent with the input depth images, thus the depth loss of this paper is defined as

$$L_{depth} = \left\| p_d(M(\beta, \theta) + T) - \hat{d} \right\|_2^2. \tag{10}$$

Here, p_d is the differentiable rendering depth function, $\hat{d}$ is the input depth image. The depth loss function can be used to constrain the depth of the network output model, so that the model is aligned with the input depth image.

Above all, the weakly-supervised loss is

$$L = \lambda L_{project} + \eta L_{sil} + \mu L_{depth}. \tag{11}$$

Among them, λ, η, and μ are the hyperparameters of the corresponding loss functions, which are used to balance the value of each loss function.

4   Empirical Evaluation

4.1   Datasets

Here is a brief description of the training data and test data used in this paper. We conduct extensive experiments on SURREAL [3], Human3.6M [4],

DFAUST [5], and real human depth images. SURREAL is a large synthetic human dataset. The dataset contains more than 55,000 sets of 3D human models, each of which contains 100 frames of different actions. We uniformly sample 200,000 human models from this data for training. The DFAUST dataset contains scan data of more than 40,000 real human bodies and the corresponding SMPL model. We uniformly sample 10,000 human body models from this data. Human3.6M is a RGB indoor dataset, we obtain ground truth SMPL parameters for the training images using MoSh [16] from the raw 3D MoCap markers. We also sample 10,000 human models from it. For all the sampled human models, we render them to depth images and obtain the corresponding 3D joints.

4.2 Implementation Details and Error Metrics

Implementation Details. We use the $256 * 256$ depth images as inputs. The ResNet50 is used for feature extraction. The 3D joints estimation module consists of 3 deconvolution layers with filter $4 * 4$ (stride 2), which channel number is fixed to 256, followed by the $1 * 1$ deconvolution layer. The architectures of joints attention module and the depth attention layer are similar, the set of MLP is (32, 32, 896/2048), the hyperparameters used in the network are set to $\varphi = 1$, $\gamma = 60$, $\sigma = 60$, $\lambda = 100$, $\eta = 10$, and $\mu = 1 * 10^{-4}$. We use Adam [17] optimizer with batch size of 32 and the learning rate is $1 * 10^{-4}$. We train and test our model on a single NVIDIA GTX2080Ti GPU, and the total number of iterations is set to 200.

Error Metrics. We conduct a quantitative and qualitative evaluation of the network. We use the Mean Average Vertex Error (MAVE) [18] over all vertexes of the recovered 3D human models in millimeter (mm) to quantify the reconstruction error:

$$\varepsilon = \frac{1}{N} \sum_{i=1}^{N} \sqrt{\|v_i - \hat{v}_i\|_2^2}. \tag{12}$$

Here, N is the number of vertexes of the 3D model, v is the vertex on the predicted 3d human models, and $\hat{v}$ is the vertex of the corresponding ground truth value.

4.3 Experiment Results

Quantitative Results. In order to verify the performance of the method proposed in this paper, we compare it with other methods on the same test set. Bogo et al. [7] propose SMPLify that first detects 2D joints on the image, and then matches the SMPL model's joints projection to the detected 2D joint points. Wei et al. [10] obtain the correspondences by matching the feature descriptors of the depth image and the template 3D human model. After that, they deform the SMPL models according to the correspondences. We deform the estimated models of [10] and [7] to input depth images by searching the nearest points. The experimental results are shown in Table 1.

Table 1. Reconstruction errors (mm) with different methods on SURREAL, Human3.6M, and DFAUST datasets. Results on the top part are used for comparison, while the results of ablation study about weak-supervision are at the bottom of this table.

Methods	SURREAL	Human3.6M	DFAUST
Bogo et al. [7]	75.3	87.2	91.5
Wei et al. [10]	68.3	80.2	66.5
Kolotouros et al. [19]	59.5	65.1	63.7
Kanazawa et al. [2]	50.1	57.6	56.3
Kolotouros et al. [9]	48.1	52.4	51.8
Zhu et al. [20]	46.8	50.3	49.2
Our method (no joints attention)	27.5	29.2	28.9
Our method (no depth attention)	29.4	31.8	30.1
Our method (no depth or joints attention)	38.8	40.0	39.5
Our method	22.4	24.9	23.6

In addition, We also compare our method with the methods of Kanazawa et al. [2], Kolotouros et al. [9], Kolotouros et al. [19], and Zhu et al. [20]. Since these methods are based on RGB images, we re-train them use depth images to make the comparison more fair. The comparison results are shown in Table 1. Besides, we also provide the visualization results of all the above methods on SURREAL in Fig. 3, all results are presented in the form of heat map.

Qualitative Results. We also evaluate our method qualitatively on real depth data. Among them, "Kungfu" from [21], "crane" from [22] "BUFF" from [23], and "Taiji" is collected in the laboratory with Kinect v2. Because our training data does not contain any real depth data, so we fine-tune our network in a

error(in cm):0 ▬▬▬ 20

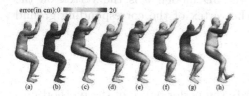
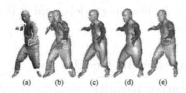

Fig. 3. The visualization of reconstruction accuracy using different methods on the SURREAL data. (a) The input scan. (b) our method. (c) Zhu et al. [20]. (d) Kolotouros et al. [9]. (e) Kanazawa et al. [2]. (f) Kolotouros et al. [19]. (g) Wei et al. [10]. (h) Bogo et al. [7].

Fig. 4. An example of weakly-supervised fine-tuning on "Kungfu" data [21]. (a) The input scan. (b) The results before fine-tuning. (c, d, e) The results after weakly-supervised fine-tuning with $L_{project}$, $L_{project} + L_{sil}$, and $L_{project} + L_{sil} + L_{depth}$, respectively.

weakly-supervised way to improve the performance of the network on real depth data. The sampled results of "Kungfu" before and after fine-tuning are shown in the Fig. 4. One can see that the network can output a 3D human model aligned with the input depth image after fine-tuning the network. More results using our method are shown in Fig. 5. From the experimental results, one can see that our method can deal well with the occlusion and random noise of real depth data, and generate corresponding 3D human models.

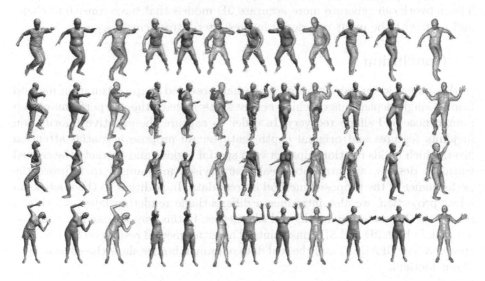

Fig. 5. Some recovered 3D models using our method on real data. For each result, we show the extracted raw depth scan, the reconstructed model, and the overlay with alignment between the reconstructed model and the raw depth scan. From top to bottom: "Kungfu" from [21], "crane" from [22], "Taiji" from [23], and "BUFF".

4.4 Ablation Study

Spatial Attention Layers. We first evaluate the effectiveness of our spatial attention feature layers. In Table 1, we provide the results for four different settings of our approach, one where the network is trained with all the spatial attention layers, a second where the network is trained without the module of depth attention, a third where the network is trained without the module of joints attention, a fourth where the network is trained without the both attention module. As we can see from the Table 1, the error of the network which uses all the spatial attention layers is better than the network without the module. It shows that the spatial attention feature extracting module can extract the spatial local features from the original depth features and 3D joints features, thus leading to higher recovery accuracy of 3D human pose and shape estimation.

Weakly-Supervision for the Real Depth Data. We also evaluate the effectiveness of our weakly-supervised method on real depth data, we compare the alignment results of the 3D human model and the original input depth image before and after weakly-supervised fine-tuning the network with different weakly-supervised loss functions. We provide the results of "Kungfu" in Fig. 4. As shown in Fig. 4, the alignment results of the predicted 3D human model are not satisfactory because the training data does not contain any real labeled data. The $L_{project}$ or $L_{project} + L_{sil}$ can only supervise the network with 2D information. The network can generate more accurate 3D models that have consistent shape and pose with the input real depth images after introducing L_{depth}.

5 Conclusion

In this paper, we propose a novel 3D human pose and shape estimation method from a single depth image, which achieves the state-of-the-art performance on human pose and shape recovery. In order to capture the effective information in joints features and original depth features, we propose a spatial attention layer, which builds relationships between spatial regions and extracts the refined features. Besides, we introduce weakly-supervised mechanism to improve the performance of the proposed method on real data. In addition to the traditional joints projection, we also introduce a differentiable rendering layer, which can render the 3D models to depths and silhouettes. In this way, we can fine-tune the network in both 2D and 3D constraints. The experimental results on SURREAL, Human3.6M, DFAUST, and the real data of human bodies show the effectiveness of our method.

References

1. He, K., Zhang, X., Ren, S., Sun, J.: Deep residual learning for image recognition. In: Proceedings of the IEEE Conference on Computer Vision and Pattern Recognition, pp. 770–778 (2016)
2. Kanazawa, A., Black, M.J., Jacobs, D.W., Malik, J.: End-to-end recovery of human shape and pose. In: The IEEE Conference on Computer Vision and Pattern Recognition (CVPR), pp. 7122–7131, June 2018
3. Varol, G., et al.: Learning from synthetic humans. In: The IEEE Conference on Computer Vision and Pattern Recognition (CVPR), pp. 109–117, July 2017
4. Ionescu, C., Papava, D., Olaru, V., Sminchisescu, C.: Human3.6M: large scale datasets and predictive methods for 3D human sensing in natural environments. IEEE Trans. Pattern Anal. Mach. Intell. **36**(7), 1325–1339 (2014)
5. Bogo, F., Romero, J., Pons-Moll, G., Black, M.J.: Dynamic faust: registering human bodies in motion. In: The IEEE Conference on Computer Vision and Pattern Recognition (CVPR), pp. 6233–6242, July 2017
6. Loper, M., Mahmood, N., Romero, J., Pons-Moll, G., Black, M.J.: SMPL: a skinned multi-person linear model. ACM Trans. Graph. **34**, 1–16 (2015)
7. Bogo, F., Kanazawa, A., Lassner, C., Gehler, P., Romero, J., Black, M.J.: Keep it SMPL: automatic estimation of 3D human pose and shape from a single image. In: Leibe, B., Matas, J., Sebe, N., Welling, M. (eds.) ECCV 2016. LNCS, vol. 9909, pp. 561–578. Springer, Cham (2016). https://doi.org/10.1007/978-3-319-46454-1_34

8. Huang, Y., et al.: Towards accurate marker-less human shape and pose estimation over time. In: 2017 International Conference on 3D Vision (3DV), pp. 421–430. IEEE (2017)

9. Kolotouros, N., Pavlakos, G., Daniilidis, K.: Convolutional mesh regression for single-image human shape reconstruction. In: Proceedings of the IEEE Conference on Computer Vision and Pattern Recognition, pp. 4501–4510 (2019)

10. Wei, L., Huang, Q., Ceylan, D., Vouga, E., Li, H.: Dense human body correspondences using convolutional networks. In: Proceedings of the IEEE Conference on Computer Vision and Pattern Recognition, pp. 1544–1553 (2016)

11. Kadkhodamohammadi, A., Gangi, A., de Mathelin, M., Padoy, N.: A multi-view RGB-D approach for human pose estimation in operating rooms. In: 2017 IEEE Winter Conference on Applications of Computer Vision (WACV), pp. 363–372. IEEE (2017)

12. Pavlakos, G., Zhou, X., Daniilidis, K.: Ordinal depth supervision for 3D human pose estimation. In: Proceedings of the IEEE Conference on Computer Vision and Pattern Recognition, pp. 7307–7316 (2018)

13. Li, R., Cai, C., Georgakis, G., Karanam, S., Chen, T., Wu, Z.: Towards robust RGB-D human mesh recovery. arXiv preprint arXiv:1911.07383 (2019)

14. Kato, H., Ushiku, Y., Harada, T.: Neural 3D mesh renderer. In: The IEEE Conference on Computer Vision and Pattern Recognition (CVPR), pp. 3907–3916 (2018)

15. Sun, X., Xiao, B., Wei, F., Liang, S., Wei, Y.: Integral human pose regression. In: Ferrari, V., Hebert, M., Sminchisescu, C., Weiss, Y. (eds.) ECCV 2018. LNCS, vol. 11210, pp. 536–553. Springer, Cham (2018). https://doi.org/10.1007/978-3-030-01231-1_33

16. Loper, M., Mahmood, N., Black, M.J.: MoSh: motion and shape capture from sparse markers. ACM Trans. Graph. (TOG) **33**(6), 1–13 (2014)

17. Kingma, D.P., Ba, J.: Adam: a method for stochastic optimization. arXiv preprint arXiv:1412.6980 (2014)

18. Yu, R., Saito, S., Li, H., Ceylan, D., Li, H.: Learning dense facial correspondences in unconstrained images. In: Proceedings of the IEEE International Conference on Computer Vision, pp. 4723–4732 (2017)

19. Kolotouros, N., Pavlakos, G., Black, M.J., Daniilidis, K.: Learning to reconstruct 3D human pose and shape via model-fitting in the loop. In: Proceedings of the IEEE International Conference on Computer Vision, pp. 2252–2261 (2019)

20. Zhu, H., Zuo, X., Wang, S., Cao, X., Yang, R.: Detailed human shape estimation from a single image by hierarchical mesh deformation. In: Proceedings of the IEEE Conference on Computer Vision and Pattern Recognition (CVPR), pp. 4491–4500 (2019)

21. Guo, K., Xu, F., Wang, Y., Liu, Y., Dai, Q.: Robust non-rigid motion tracking and surface reconstruction using L0 regularization. In: Proceedings of the IEEE International Conference on Computer Vision, pp. 3083–3091 (2015)

22. Vlasic, D., Baran, I., Matusik, W., Popović, J.: Articulated mesh animation from multi-view silhouettes. In: ACM SIGGRAPH 2008 papers, pp. 1–9 (2008)

23. Zhang, C., Pujades, S., Black, M.J., Pons-Moll, G.: Detailed, accurate, human shape estimation from clothed 3D scan sequences. In: The IEEE Conference on Computer Vision and Pattern Recognition (CVPR), pp. 4191–4200

Highlight Removal in Facial Images

Ting Zhu[1], Siyu Xia[1(✉)], Zhangxing Bian[2], and Changsheng Lu[3]

[1] School of Automation, Southeast University, Nanjing 210096, China
xsy@seu.edu.cn
[2] University of Michigan, Ann Arbor, MI 48109, USA
[3] Shanghai Jiao Tong University, Shanghai, China

Abstract. In this paper, we present a method for highlight removal in facial images. In contrast to previous works relying on physical models such as dichromatic reflection models, we adopt the structure of conditional generative adversarial network (CGAN) to generate highlight-free images. By taking the facial images with specular highlight as the condition, the network predicts the corresponding highlight-free images. Meanwhile, a novel mask loss is introduced through highlight detection, which aims to make the network focus on more on the highlight regions. With the help of multi-scale discriminators, our method generates highlight-free images with high-quality details and fewer artifacts. We also built a dataset containing both real and synthetic facial images, which is, to our best knowledge, the largest image dataset for facial highlight removal. By comparing with the state-of-the-arts, our method shows high effectiveness and strong robustness in different lighting environments.

Keywords: Highlight removal · Facial image · GAN

1 Introduction

Over-exposure and oily skin can often cause the specular highlight, which conceals the original skin color and texture. This artifact has a negative impact on the downstream computer vision tasks and medical examinations, which can give rise to false detections and recognition errors. In the dermatological examination, highlight on the face interferes the diagnosis of pigmentation diseases due to the difficulty of diagnosing the skin regions where the specular highlight exhibits. As such, a high-quality highlight removal method is necessary.

Nevertheless, the complexity and diversity of human skin make normal highlight removal methods ineffective. Some models [1,2] are not applicable to human skin. Also, the geometry of the human face make it more complicated. When it comes to restoring a face with freckles or moles, it is not only necessary to remove the highlight but to recover the original textures. As a surface-related feature, the strength and shape of highlight can be easily affected by the light position and viewing angles.

The first author is a student.

© Springer Nature Switzerland AG 2020
Y. Peng et al. (Eds.): PRCV 2020, LNCS 12305, pp. 422–433, 2020.
https://doi.org/10.1007/978-3-030-60633-6_35

A large body of work for highlight removal is based on physical models [3] whose results are limited to prior knowledge or assumptions and always fail to fully express the reflection in real scenarios. As for the neural network methods, there is no existing available dataset of facial highlight. It is almost impossible to capture the face images in the same illumination with and without highlight. The lack of ground truth on real data makes the network training a challenging task.

Fig. 1. Examples of highlight removal in facial images. The first row presents facial images with highlight while the second row shows the corresponding highlight-free images.

In this paper, we aim to remove the highlight parts in a facial image through an end-to-end and fully automatic framework, as shown in Fig.1. To address the difficulty of lacking in ground truth, we propose a dataset that consists of 150 pairs of real faces and 1170 synthetic faces, where the highlight-free real images are obtained by using cross-polarized imaging while the synthetic faces are rendered by using graphics models. We adopt two-step training strategy in the network, namely being pretrained in the synthetic dataset and then finetuned by the real dataset, which could effectively improve the output of network. To further enhance the quality of the prediction, the multi-scale discriminators [4] and highlight detection [5] are also employed.

2   Related Work

Highlight removal from a single input image has been studied for decades. In this section, we briefly review previous highlight removal methods based on various ideas.

Most existing highlight removal approaches are based on physical models. As highlight removal is an ill-posed problem, prior knowledge or assumptions of natural images should be applied to simplify the problem. Early highlight removal approaches relies on color space analysis [6–8] which has certain limits. This kind of approaches perform badly on light color objects and the restored highlight regions are easily covered by black pixels. Recently methods are based on dichromatic reflection models. Li et al. [3] utilize a skin color model alone with a dichromatic reflection model to guide the removal of specular highlights in facial images. Kim et al. [2] exploit the dark channel prior to derive a pseudo specular-free image. Suo et al. [9] remove the specular highlight by defining a new L2 chromaticity. Yang et al. [10] propose a fast real-time implementation by applying bilateral filtering. Liu et al. [11] make use of an additional compensation step to optimize the highlight-free images. Guo et al. [12] remove highlight from a single image with a sparse and low-rank reflection model. These methods based on physical models are only applicable to simple surfaces. When it comes to the complex surface such as skin, it will lead to artificial highlight removal. Meanwhile, physical models cannot completely model the real-world environments, especially for the high-frequency details such as highlight. Several other methods [11,13,14] use pixel clustering to restore the diffuse colors covered by highlight.

Besides the methods based on physical models, neural networks are also applied in specular highlight removal. A recent approach [15] train a deep neural network to extract the highlight component from a face image and utilize a low-rank feature among the diffuse chromaticity of multiple same faces to refine the network. Inpainting techniques are applied to remove specular highlight [5,16,17] by using the neighboring information to guiding the inpainting process. Besides, relighting techniques [18–20] is used to generate a new portrait image under a target lighting condition, which is able to restore the diffuse colors of skin and change the specular highlight. Notably, style transfer is similar to relighting. Photorealistic image stylization [21,22] enables to switch the texture between different faces. However, the switched skin texture is not preferred in the medical scenario. With regard to methods using deep learning, the main problem is that difficulty in data collection makes it hard to train models.

In contrast to previous methods, we proposes an end-to-end and fully automatic framework to remove the highlight in facial images. Synthetic faces and real faces captured by polarizers are used to train the network. Highlight detection and multi-scale discriminators are employed in our two-step training framework to improve the quality of the highlight-free results.

3   Dataset

In practice, it is difficult to capture the same faces with and without highlight in an identical illumination setting. Moreover, there is no existing available dataset of this sort. Thus, in order to address data problem, we built a dataset that consists of both real and synthetic images. To our best knowledge, it is the largest image dataset for facial highlight removal so far. We train the network in two steps on our dataset. Firstly, the network is pretrained on synthetic data. The synthetic ground-truth images help to minimize the distance between the predicted highlight-free image and the corresponding ground truth. Then we use real data to finetune the pretrained network, which could make the predictions more natural.

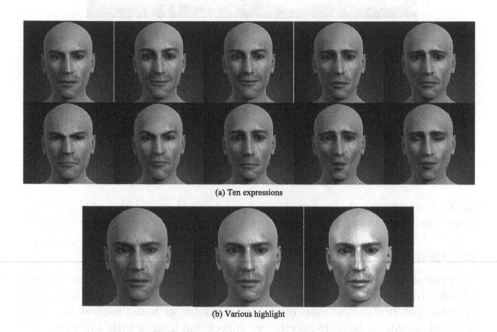

(a) Ten expressions

(b) Various highlight

Fig. 2. A set of example images of synthetic faces. (a) ten expressions within a same face; (b) three kinds of light to one face.

3.1   Synthetic Facial Images

For the pretraining of our network, we generate synthetic faces by using 3D software Character Creator [23] whereby we can change the material and roughness to add or remove highlight in the 3D model under the same light condition. With these settings, we could get exactly the same faces except for the highlight. Headshot, a plug-in of Character Creator, is employed to generate 3D faces automatically with facial images input. After generating the 3D faces, we set up the skin properties such as roughness. In the end, we add lights as needed. It should note that the light in different positions can improve the diversity of highlight.

Using 39 faces collected from the internet, we add three distinct kinds of light to each 3D face. And each 3D face has 10 different expressions, as shown in Fig. 2, including happy faces, sad faces, angry faces, fear faces, kissing, and disgusting. In total, we render 1170 facial images with and without highlight.

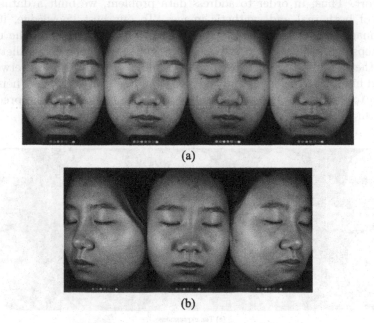

(a)

(b)

Fig. 3. A set of example images of real faces. (a) faces under three kinds of light intensity and without highlight. (b) pictures taken from three different camera angles.

3.2 Real Facial Images

As for the artificiality of synthetic data, we finetuned the network on the real data to bridge the discrepancy. We employ polarizers to filter out the highlight when capturing pictures so that pairs of real facial photos with and without highlight can be obtained. To be more specific, we collect real images with a device called Visia [24], which is used for skin analysis in dermatology hospitals. Visia captures high quality, standardized facial images that reserve great detailed skin textures. Images are taken under three different light intensities and cross-polarized light. The quick shutter speed minimizes movement influence that could possibly occur during the capturing.

The real dataset contains 150 pairs of facial images. It has a large variety in camera angles and age, gender, expression and skin condition. For data augmentation, we apply the image flip. A few samples are shown in Fig. 3.

4 Method

As summarizes in Fig. 4, the proposed network architecture consists of two parts: the generative network and the discriminative network. Taking a facial image with highlight as input, the network predicts the image free from highlight as real as possible. The purpose of the network is to find the highlight parts in a facial image and remove them to restore the original color of the skin in a natural way. In this section, we first describe the loss functions and then show the training details.

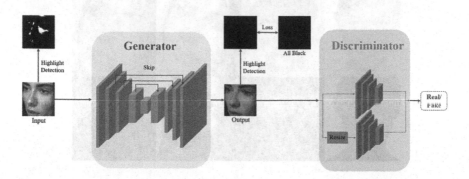

Fig. 4. The architecture of our proposed network.

4.1 Highlight Removal Using GAN

As all Conditional Generative Adversarial Networks, our proposed network is composed of two parts: a generator G and a discriminator D. Based on an image input x, the generator G attempt to produce a realistic image, trying to confuse the discriminator D for distinguishing the real image and the generated image. The generator G and the discriminator D promote each other in training as a result of adversarial competition.

The purpose of our network is to generate a natural facial image without highlight. The generator G and discriminator D are conditioned on the input RGB facial image x which has highlight. Given the input, the generator G is trained to generate the corresponding output $G(x)$ which contains no specular highlight. We use y to represent the ground truth of $G(x)$. The generator G aims to make $G(x)$ as close to y as possible. We use the U-Net [25] structure as the generator. The skip connections of the U-net are capable of integrating features of different scales which contribute to more fine features. As the architecture illustrated in Fig. 4, our generator has 16 convolution-BatchNorm-ReLu blocks and skip connections connect the intermediary layers. The skip connections help to output an edge preserved image. Our generative adversarial loss can be expressed as:

$$L_{GAN}(G, D) = \boldsymbol{E}_{x,y}[logD(x, y)] + \boldsymbol{E}_x[log(1 - D(x, G(x)))] \tag{1}$$

Apart from the adversarial loss, we introduce another two types of losses. The first loss is the reconstruction loss which compares the generated image $G(x)$ with the corresponding ground truth y. We use L1 distance as the reconstruction loss since it encourages less blurring [26]:

$$L_1(G) = \boldsymbol{E}_{x,y}\left[\|y - G(x)\|_1\right]. \tag{2}$$

Fig. 5. Examples of highlight detection. The first row shows source images while the second row shows the detection results. The highlight detection can approximately locate the highlight regions, and the detection results of the generated highlight-free images are all black (no highlight is detected), which verifies the effectiveness of our method.

The other loss is based on highlight detection. The detection loss is used to encourage no highlight to be detected in the generated image. We apply a simple highlight detection method [5] to extract the mask of highlight which makes the network pay more attention to the highlight regions, as presented in Fig. 5. L1 distance is used to compare the mask with an all-black mask which represents that there is no highlight. We have

$$L_{mask}(G) = \boldsymbol{E}_x\left[\|M(G(x)) - m_{black}\|_1\right], \tag{3}$$

where $M(G(x))$ denotes the mask extracted from the generated image while m_{black} denotes an all black mask.

Considering the receptive field of a single discriminator is limited, we are inspired by [4]. Multi-scale discriminators are conducive to extract information from different scales. We adopt the two discriminators which have the same structure and the only difference between them is the scale of input.

Overall, the total loss can be written as:

$$arg\min_G\max_D L_{GAN}(G, D) + \lambda_1 L_1(G) + \lambda_2 L_{mask}(G) \tag{4}$$

4.2 Network Architecture and Training Details

The architecture of our network is based on U-Net [25] while the structure of discriminators is adapted from PatchGAN [26]. We optimize our network with the Adam solver [27]. The network is initialized with $\lambda_1 = 100, \lambda_2 = 100$ and learning rate = 0.0005. Our code is based on pytorch.

5 Experiments

In order to examine that our proposed network can remove specular highlight in facial images, we compare our highlight-free results to those of prior works [10,13–16,28] in Fig. 6. Highlight-net [15] utilizes a low-rank feature among the diffuse chromaticity of multiple same faces to constrain the network. Other methods are based on physical models. Besides, we also compare some parts of our network architecture to verify the effectiveness of highlight detection and multi-scale discriminators.

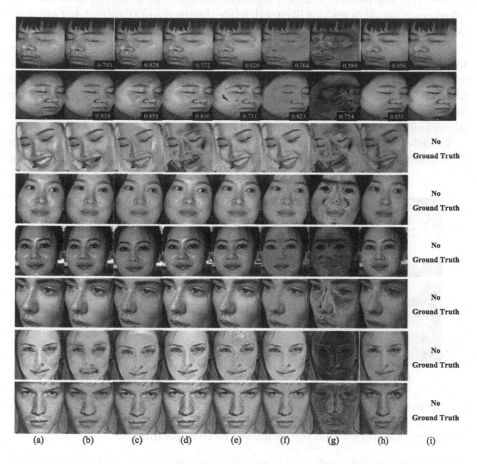

(a)   (b)   (c)   (d)   (e)   (f)   (g)   (h)   (i)

Fig. 6. Here we show comparisons between prior works and our method. (a) Source facial images. (b) Tan2003 [16]. (c) Highlight-net [15]. (d) Souza2018 [14]. (e) Shen2008 [13]. (f) Shen2009 [28]. (g) Yang2010 [10]. (h) Ours. (i) Ground truth captured by cross-polarization. We use SSIM [29] as similarity metrics. In the first two rows, SSIM values are given at the bottom-right corner.

Figure 6 displays several qualitative comparisons to other methods on facial images under different lighting environments. The first two rows show highlight removal results with ground truth captured by cross-polarization and the other rows are results on natural images. The 3–5 rows show clear-skinned faces and the 6–8 rows present faces with obvious texture such as wrinkles, freckles and moles. In these examples above, our method attains the purpose of removing highlights naturally without destroying the origin texture in the highlight regions. The highlight removal method which is capable of preserving the skin texture is superior to those based on inpainting, because most inpainting approaches fail to recover the skin features. As shown in Fig. 6, our approach is more effective in highlight removal and is visually much better than other approaches. As can be observed, our method removes highlight more comprehensively and have less influence on non-skin regions like teeth while the results of Highlight-net are more reddish than the normal skin color. Highlight-net utilizes landmarks to crop face region as the input, so it fails on a facial image which has closed eyes. We also notice that those techniques which are based on color analysis or dichromatic reflection model have unsatisfied performance on facial images, the highlight reflections still exist, especially around the nose. Not all highlight regions on the faces are restored. As for the restored parts, either highlight is remained or the original skin colors are wrongly inpainted by black pixels (Table 1).

In order to evaluate the effectiveness of highlight detection and multi-scale discriminators, we trained three distinct models and compared them on our

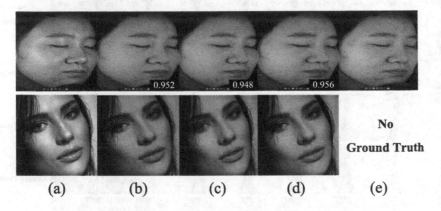

Fig. 7. Here we show comparisons of some parts of our network architecture. (a) Source facial images. (b) Without multi-scale discriminators. (c) Without highlight detection. (d) Our method. (e) Ground truth captured by cross-polarization.

Table 1. Quantitative evaluation results. We use SSIM [29] to show comparisons of structural similarities.

Model	SSIM
Only highlight detection	0.952
Only multi-scale discriminators	0.948
Our model	**0.956**

dataset. These three models are: a model using highlight detection only, a model using multi-scale discriminators only, and a model using both highlight detection and multi-scale discriminators. In Fig. 7, we show the highlight removal results of the three models. We use SSIM [29] as similarity metrics. As shown in the evaluation table, the model using both highlight detection and multi-scale discriminators performs better than the others. This is the evidence that both the highlight detection and multi-scale discriminators improve the quality of highlight removal.

Fig. 8. Examples of our model being used to remove highlight on faces with different angles and skin color. Rows (a) presents the source facial images with highlight while rows (b) shows the corresponding highlight-free results predicted by our method.

One of the advantages of our approach is that it has strong robustness for different light environments and different expressions, as displayed in Fig. 8. For the faces under uneven illumination, only the highlight on the faces is removed while the uneven light with shading is retained. The superiority of our highlight removal comes from the combination of highlight detection and CGAN which overcomes the difficulty in completely removing highlight regions especially the small highlight spots on the protrusions like noses and cheeks. Moreover, multi-scale discriminators improve the quality of the generated images. It is the multi-scale discriminators that pursue both global structure and local details. However, the performance of our work is limited to the diversity of real data which only has Asian faces. As a result, the highlight-free images have a yellow tint similar to the skin color of Asians.

6 Conclusion

In this work, we propose a GAN based network for highlight removal in facial images. A novel mask loss is introduced to GAN through highlight detection, which guides the network to focus on more on the highlight region. We also construct a facial highlight dataset, including both synthetic and real images, which makes data-driven algorithms possible for future study. We present qualitative comparisons to prior works and our method produces satisfactory results. In the future, we plan to investigate removing highlight using the properties of highlight and intrinsic decomposition. Also, the potential connection underlying between highlight and material is worth further exploring.

References

1. Fu, G., Zhang, Q., Song, C., Lin, Q., Xiao, C.: Specular highlight removal for real-world images. Comput. Graph. Forum **38**, 253–263 (2019)
2. Kim, H., Jin, H., Hadap, S., Kweon, I.: Specular reflection separation using dark channel prior. In: Proceedings of the IEEE Conference on Computer Vision and Pattern Recognition, pp. 1460–1467 (2013)
3. Li, C., Lin, S., Zhou, K., Ikeuchi, K.: Specular highlight removal in facial images. In: Proceedings of the IEEE Conference on Computer Vision and Pattern Recognition, pp. 3107–3116 (2017)
4. Wang, T.C., Liu, M.Y., Zhu, J.Y., Tao, A., Kautz, J., Catanzaro, B.: High-resolution image synthesis and semantic manipulation with conditional GANs. In: Proceedings of the IEEE Conference on Computer Vision and Pattern Recognition, pp. 8798–8807 (2018)
5. Meslouhi, O., Kardouchi, M., Allali, H., Gadi, T., Benkaddour, Y.: Automatic detection and inpainting of specular reflections for colposcopic images. Open Comput. Sci. **1**(3), 341–354 (2011)
6. Klinker, G.J., Shafer, S.A., Kanade, T.: The measurement of highlights in color images. Int. J. Comput. Vision **2**(1), 7–32 (1988)
7. Schluns, K., Teschner, M.: Analysis of 2D color spaces for highlight elimination in 3d shape reconstruction. Proc. ACCV. **2**, 801–805 (1995)
8. Bajcsy, R., Lee, S.W., Leonardis, A.: Detection of diffuse and specular interface reflections and inter-reflections by color image segmentation. Int. J. Comput. Vision **17**(3), 241–272 (1996)
9. Suo, J., An, D., Ji, X., Wang, H., Dai, Q.: Fast and high quality highlight removal from a single image. IEEE Trans. Image Process. **25**(11), 5441–5454 (2016)
10. Yang, Q., Wang, S., Ahuja, N.: Real-time specular highlight removal using bilateral filtering. In: Daniilidis, K., Maragos, P., Paragios, N. (eds.) ECCV 2010. LNCS, vol. 6314, pp. 87–100. Springer, Heidelberg (2010). https://doi.org/10.1007/978-3-642-15561-1_7
11. Liu, Y., Yuan, Z., Zheng, N., Wu, Y.: Saturation-preserving specular reflection separation. In: Proceedings of the IEEE Conference on Computer Vision and Pattern Recognition, pp. 3725–3733 (2015)

12. Guo, J., Zhou, Z., Wang, L.: Single image highlight removal with a sparse and low-rank reflection model. In: Ferrari, V., Hebert, M., Sminchisescu, C., Weiss, Y. (eds.) ECCV 2018. LNCS, vol. 11208, pp. 282–298. Springer, Cham (2018). https://doi.org/10.1007/978-3-030-01225-0_17

13. Shen, H.L., Zhang, H.G., Shao, S.J., Xin, J.H.: Chromaticity-based separation of reflection components in a single image. Pattern Recogn. **41**(8), 2461–2469 (2008)

14. Souza, C.S., Macedo, M.C.F., Nascimento, V.P., Oliveira, B.S.: Real-time high-quality specular highlight removal using efficient pixel clustering. In: 2018 31st SIBGRAPI Conference on Graphics, Patterns and Images, pp. 56–63. IEEE (2018)

15. Yi, R., Zhu, C., Tan, P., Lin, S.: Faces as lighting probes via unsupervised deep highlight extraction. In: Ferrari, V., Hebert, M., Sminchisescu, C., Weiss, Y. (eds.) ECCV 2018. LNCS, vol. 11213, pp. 321–338. Springer, Cham (2018). https://doi.org/10.1007/978-3-030-01240-3_20

16. Tan, P., Lin, S., Quan, L., Shum, H.Y.: Highlight removal by illumination constrained inpainting. In: Proceedings Ninth IEEE International Conference on Computer Vision, pp. 164–169. IEEE (2003)

17. Tan, P., Quan, L., Lin, S.: Separation of highlight reflections on textured surfaces. In: 2006 IEEE Computer Society Conference on Computer Vision and Pattern Recognition, pp. 1855–1860. IEEE (2006)

18. Sun, T., et al.: Single image portrait relighting. ACM Trans. Graph. **38**(4), 79:1–79:12 (2019)

19. Zhou, H., Hadap, S., Sunkavalli, K., Jacobs, D.W.: Deep Single-image portrait relighting. In: Proceedings of the IEEE International Conference on Computer Vision, pp. 7194–7202 (2019)

20. Meka, A., et al.: Deep reflectance fields: high-quality facial reflectance field inference from color gradient illumination. ACM Trans. Graph. **38**(4), 1–12 (2019)

21. Li, Y., Liu, M.-Y., Li, X., Yang, M.-H., Kautz, J.: A closed-form solution to photorealistic image stylization. In: Ferrari, V., Hebert, M., Sminchisescu, C., Weiss, Y. (eds.) ECCV 2018. LNCS, vol. 11207, pp. 468–483. Springer, Cham (2018). https://doi.org/10.1007/978-3-030-01219-9_28

22. Luan, F., Paris, S., Shechtman, E., Bala, K.: Deep photo style transfer. In: Proceedings of the IEEE Conference on Computer Vision and Pattern Recognition, pp. 4990–4998 (2017)

23. Character Creator. https://www.reallusion.com/character-creator/

24. VISIA. https://www.canfieldsci.com/imaging-systems/visia-cr/

25. Ronneberger, O., Fischer, P., Brox, T.: U-net: convolutional networks for biomedical image segmentation. In: Navab, N., Hornegger, J., Wells, W.M., Frangi, A.F. (eds.) MICCAI 2015. LNCS, vol. 9351, pp. 234–241. Springer, Cham (2015). https://doi.org/10.1007/978-3-319-24574-4_28

26. Isola, P., Zhu, J.Y., Zhou, T., Efros, A.A.: Image-to-image translation with conditional adversarial networks. In: Proceedings of the IEEE Conference on Computer Vision and Pattern Recognition, pp. 1125–1134 (2017)

27. Kingma, D.P., Adam, J.B.: A method for stochastic optimization. arXiv preprint arXiv:1412.6980, pp. 1–15 (2014)

28. Shen, H.L., Cai, Q.Y.: Simple and efficient method for specularity removal in an image. Appl. Opt. **48**(14), 2711–2719 (2009)

29. Wang, Z., Bovik, A.C., Sheikh, H.R., Simoncelli, E.P.: Image quality assessment: from error visibility to structural similarity. IEEE Trans. Image Process. **13**(4), 600–612 (2004)

Global Context Enhanced Multi-modal Fusion for Referring Image Segmentation

Jianhua Yang[1], Yan Huang[2], Linjiang Huang[2], Yunbo Wang[2], Zhanyu Ma[1], and Liang Wang[2,3,4(✉)]

[1] Pattern Recognition and Intelligent System Laboratory, Beijing University of Posts and Telecommunications, Beijing, China
{youngjianhua,mazhanyu}@bupt.edu.cn
[2] Center for Research on Intelligent Perception and Computing, NLPR, CASIA, Beijing, China
{yhuang,wangliang}@nlpr.ia.ac.cn,
{linjiang.huang,yunbo.wang}@cripac.ia.ac.cn
[3] Center for Excellence in Brain Science and Intelligence Technology, CAS, Beijing, China
[4] Artificial Intelligence Research (CAS-AIR), Chinese Academy of Sciences, Beijing, China

Abstract. The referring image segmentation is a challenging task which aims to segment the object of interest in an image according to a natural language expression. Most existing works directly concatenate the global language representation with local visual features, and follow by a convolutional operation to fuse two modalities. These works ignore that the global contextual information from vision is essential for vision-language fusing and inferring the referred objects. The global context can establish a perception of the full image, thus it's fusion with global language representation is beneficial to reduce mislabeled pixels of similar objects in an image. To address aforementioned issue, we propose a global fusion network (*GFNet*), which is composed of visual guided global fusion module and language guided global fusion module. By modeling the expression-region interactions, two modules can aggregate the expression-related visual contextual information and fuse it with global representation of language expression. Moreover, to alleviate the distribution differences between two modalities, we introduce a channel-wise self-gate on visual-language concatenated features. We validate the proposed network on four standard datasets, the experimental results show that our approach outperforms state-of-the-art methods.

Keywords: Semantic segmentation · Natural language expression · Attention mechanism.

This work is jointly supported by National Key Research and Development Program of China (2016YFB1001000), Key Research Program of Frontier Sciences, CAS (ZDBS-LY-JSC032), National Natural Science Foundation of China (61525306, 61633021, 61721004, 61806194, U1803261, 61976132), Shandong Provincial Key Research and Development Program (2019JZZY010119), HW2019SOW01, and CAS-AIR.
J. Yang—The first author is a student.

© Springer Nature Switzerland AG 2020
Y. Peng et al. (Eds.): PRCV 2020, LNCS 12305, pp. 434–446, 2020.
https://doi.org/10.1007/978-3-030-60633-6_36

1   Introduction

Beyond conventional image semantic segmentation [1–3], in which the semantic labels are restricted to a set of predefined object classes (*e.g.*, "person" and "car"), the referring image segmentation (*RIS*) [4,5] aims at segmenting the object referred by the natural language expression. Since the language expression is not limited to specific object classes but more detailed descriptions, such as visual attributes, actions, spatial relationships and interaction with context objects, it is accurate to refer to the object of interest in an image. This challenging task usually requires semantic understandings of both image and language contents, and appropriate association between the two modalities.

Expression: *boy in black shirt front right*

(a) Input image   (b) *LSTM-CNN*   (c) Our method

Fig. 1. Given an input image (a) and a language expression *"boy in black shirt front right"*, our method (c) can reject the ambiguity caused by similar local objects and generate more precise segmentation result than *LSTM-CNN* [4] model (b).

Many works [4,6–8] have been proposed to tackle this task, these works usually apply Long Short-Term Memory network (*LSTM*) [9] to extract the representation of natural language expression, and use the convolutional neural network (*CNN*) to extract visual features. The language representation is tiled to the same spatial sizes as visual feature map and fuse with visual features via a "concatenation-convolution" procedure [4]. Recently, some works [5,10–13] indicate that the performance could be improved via modeling fine-grained word-region interactions between language expression and image. These proposed fine-grained manners, *e.g.*, convolutional multimodal *LSTM* [5], region-wise word attention [11], cross-modal self-attention [12], can associate each word representation with each region representation of image and make the two modal representations fuse better. However, these works ignore that the semantic information from two modalities are asymmetric, as language feature vector is a global representation of entire language expression while the visual features at each position of map are local region representations of an image. Such semantically asymmetric fusion can not understand the entire image scene and may introduce the ambiguities caused by similar local objects which described by the same language expression. Considering the asymmetric fusion may lead to mislabeled pixels of similar objects and suppress the improvement of segmentation performance, it is important to introduce visual global contextual information for referring image segmentation. In Fig. 1, we illustrate an example, given an image in which the appearances of two boys on the right are similar, and a language expression *"boy in black shirt front right"*, the *LSTM-CNN* [4] model can not precisely localize and segment the referred boy by the language expression, because of lacking visual global contextual information to infer the position and relation of two boys during two modalities fusion process.

To alleviate aforementioned problem and further promote the performance of *RIS*, we propose a global fusion network (*GFNet*), which is composed of two modules, *i.e.*, visual guided global fusion module (*VGGF*) and language guided global fusion module (*LGGF*). By modeling the expression-region and region-region interactions, the *VGGF* enables a single feature from any position to perceive the features of all the other positions, thus obtaining the global context for each position of feature map. The *LGGF* can model the expression-region interaction and aggregate the visual features into a global visual representation. The visual representations from two modules contains global contextual information, and can symmetrically fuse with the global representation of expression. Both the *VGGF* and the *LGGF* are based on self-attention [12,14] but have significant differences. The first is that the *VGGF* and the *LGGF* are asymmetric self-attention models, which can promote the correlation between visual information and language expression for aggregating expression-related contextual information. The second is we introduce a cross channel attention (*CCA*) in the *VGGF* to make the key and query to focus on distinguished features and obtain a better attention map. Besides, to reduce the distribution differences between two modalities, we introduce a channel-wise self-gate (*CWSG*) to reweight the visual-language concatenated features along channel dimension. We perform experiments on four available *RIS* datasets and achieve state-of-the-art results. Our main contributions can be summarized as follows:

- We propose a global fusion network, which consists of visual guided global fusion module and language guided global fusion module, to aggregate the visual global contextual information and perform semantic symmetric fusion with global representation of language expression.
- We introduce a cross channel attention in visual guided global fusion module to obtain a better attention map.
- We introduce a channel-wise self-gate to reduce the distribution differences between visual and language concatenated features.
- Our proposed approach outperforms previous state-of-the-art methods on four *RIS* datasets.

2   Related Work

Hu et al. [4] first proposed the task of *RIS*, they concatenated the visual feature map produced by semantic segmentation networks with the language representation produced by an *LSTM*, then performed convolution operations to fuse two modalities along channel dimension and predict the segmentation mask.

After that, [5,7,8,10–12] follow such "concatenation-convolution" procedure to solve this task. Liu et al. [5] and Margffoy-Tuay et al. [10] applied the sequential natural of expression to gradually fuse each word representation with visual features. Shi et al. [11] introduced a region-wise word attention to model the keyword-aware contextual information. Ye et al. [12] used self-attention to encode expression and model long-range correlations between visual and language representations. Chen et al. [13] used visual features at each state and segmentation heatmap from previous stage to attend and update the language representation. In addition, Skip connections [10], convolutional *RNN* [7,13] and gate mechanism [12] are investigated to fuse the low level features and further improve the segmentation performance.

Different from aforementioned approaches, another perspective [15,16] to solve the *RIS* task is first localize referred object with bounding box and then perform segmentation within bounding box. However, these approaches rely on complicated data processing and additional object detector, which are not effective and efficient as box-free approaches. Our work follow the "concatenation-convolution" procedure to address this task, but focus on solving the semantic asymmetric problem from two modalities during fusion process, which has not been investigated in previous works.

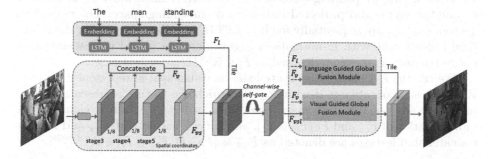

Fig. 2. An overview of our proposed global fusion network (*GFNet*) for *RIS* task, which consists of visual guided global fusion module and language guided global fusion module. Two modules can establish the correlation between visual information and language expression for aggregating language-related contextual information. (Color figure online)

3   Proposed Method

The overall architecture of our proposed global fusion network is illustrated in Fig. 2. Given an image and a language expression, we first use a *CNN* to extract visual representation from input image, and adopt an *LSTM* to encode input expression into a holistic language representation. We follow "concatenation-convolution" procedure to fuse two modalities and feed the fusion features into a classifier to predict the segmentation mask. In order to aggregate visual context and alleviate the semantic asymmetric fusion problem, we propose a global fusion model, which is composed of visual guided global fusion module and language guided global fusion module. Moreover, to reduce the distribution differences of visual and language features, we introduced a channel-wise self-gate to re-weight the concatenated futures. We will detail our framework in the following.

3.1 Visual and Language Representations

Following previous work [5], we adopt *DeepLab ResNet-101* [3] to generate visual feature maps. The *conv* layer of this model can be divided into five stages, in which the output feature maps are down-sampled by 4, 4, 8, 8, 8 with respect to input image. Considering the low level features can provide more precise location information while the high level features contains more semantic information, we fuse multi-scale features to enhance visual representation. Specifically, we concatenate the visual features from states 3, 4 and 5 which have identical spatial sizes, and follow by a *conv* layer with *l2*-normalization to obtain a multi-scale visual representation F_v. To allow the model to reason about the spatial relationship, 8-dimensional spatial coordinates also fused with F_v and obtain a visual-spatial representation $F_{vs} \in \mathbb{R}^{h \times w \times c}$.

For each language expression, we keep it has fixed length ($N = 20$) as previous work [4,5] by padding-zeros or cut the excess. Each word w_n is encoded as one-hot vector and projected into a word embedding space. The embedded vectors $\{e_n\}_{n=1}^{N}$ are sequentially fed into *LSTM* to update the hidden state. The final hidden state with *l2*-normalized is considered as global representation of entire language expression, denoted as $F_l \in \mathbb{R}^{1 \times 1 \times c}$.

We tile the F_l to have same spatial size as visual-spatial representation (*i.e.*, $\mathbb{R}^{h \times w \times c}$) and concatenate it with F_{vs}. To enrich the feature representation, we introduce additional feature map (blue block in Fig. 2), which is element-wise multiplication of F_{vs} and F_l ($F_{vs} \otimes F_l$), and concatenate it with F_{vs} and F_l. The concatenated features are denoted as $F_{vsl} = [F_{vs}, F_l, F_{vs} \otimes F_l] \in \mathbb{R}^{h \times w \times 3c}$.

3.2 Channel-wise Self-gate

In existing works [4–7,10], the concatenated features are directly fused with an 1×1 *conv* layer. This operation may introduce distracting noise in fusion process as the features from two modalities have different distribution in channel dimensions. To alleviate distribution gap and focus on more important information, we introduce a channel-wise self-gate. This gate allows us to adjust the weights of different channels in concatenated features and learn discriminant information. The channel-wise self-gate is formulated as:

$$\alpha = sigmoid\{C_{1 \times 1}(F_{vsl})\} \quad (\alpha \in [0, 1]^{h \times w \times 3c}) \qquad (1)$$

where $sigmoid\{\cdot\}$ and $C_{1 \times 1}\{\cdot\}$ denote sigmoid function and 1×1 *conv* layer, respectively. The gate α determines which channel is more important at different positions of feature map. Finally, α performs element-wise multiplication with F_{vsl} to re-weight the concatenated features, *i.e.*, $F_{vsl} = \alpha \otimes F_{vsl}$.

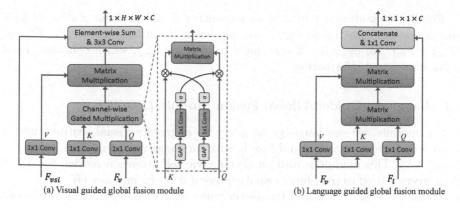

Fig. 3. (a) the visual guided global fusion module and (b) the language guided global fusion module.

3.3 Visual Guided Global Fusion Module (*VGGF*)

Using an 1×1 *conv* operation to concatenated features will lead to semantic asymmetric fusion, as language representation is global representation of input language expression while the visual representation at each position is local region representation of image. To aggregate global contextual information of visual features and encourage it to symmetrically fuse with language representation, we design a visual guided global fusion module, as shown in Fig. 3(a). Different from general self-attention model [14], this module is an asymmetric self-attention model, which allows us to model the expression-region and region-region interactions. These interactions can associate the visual information with language representation and aggregate the expression-related visual context. This module can be described by the Q-K-V term. The key ($K \in \mathbb{R}^{h \times w \times c_1}$) and value ($V \in \mathbb{R}^{h \times w \times c_0}$) are transformed from F_{vsl} via two 1×1 *conv* layers, respectively. The query ($Q \in \mathbb{R}^{h \times w \times c_1}$) is transformed from F_v by another 1×1 *conv* layer. The general self-attention will generate attention map from matrix multiplication of Q and K. Here, to obtain more accurate attention map, we introduce cross channel attention between Q and K to enable two different features focus on more discriminative features. The cross channel attention can be defined as:

$$\alpha_g^i = sigmoid\{C_{1 \times 1}(GAP(i))\} \quad (\alpha_g^i \in \mathbb{R}^{1 \times 1 \times c_1}; i \in \{Q, K\}) \tag{2}$$

$$Q' = \alpha_g^K \otimes Q; \quad K' = \alpha_g^Q \otimes K \tag{3}$$

where GAP denotes the global average pooling. Then Q' and K' are used to formulate a self-attention as follows:

$$F_{out} = softmax(Q'K'^T)V \quad (F_{out} \in \mathbb{R}^{h \times w \times c_0}) \tag{4}$$

Finally, we multiply it by a scale parameter β and perform a element-wise sum operation with the features F_{vsl} to obtain the final output, *i.e.*, $F_{out}^v = \beta F_{out} + F_{vsl}$. We use a 3×3 *conv* layer to fuse the F_{out}^v, which contains visual global contextual information.

3.4   Language Guided Global Fusion Module *(LGGF)*

In this module, we use language as query to aggregate visual features into a global visual representation and fuse it with language representation, as shown in Fig. 3(b). This module is also an asymmetric self-attention model, the 1×1 *conv* layers is used to transform visual representation F_v into key ($K \in \mathbb{R}^{h \times w \times c}$) and value ($V \in \mathbb{R}^{h \times w \times c}$), and transform language representation F_l into query ($Q \in \mathbb{R}^{1 \times 1 \times c}$). The global visual representation are computed as follows:

$$F_v' = softmax(QK^T)V \quad (F_v' \in \mathbb{R}^{1 \times 1 \times c}) \tag{5}$$

The aggregated visual representation F_v' then concatenates with language representation F_l and followed by an 1×1 *conv* layer to fuse them. We tile the fused features in space and denotes it as $F_{out}^l \in \mathbb{R}^{h \times w \times c}$. Finally, we concatenate the features from two modules and use an $1 \times 1 conv$ layer to fuse them. A 3×3 *conv* layer is attached to produce the probability map of foreground f as

$$f = sigmoid\{C_{3 \times 3}\{C_{1 \times 1}\{[F_{out}^v, F_{out}^l]\}\}\} \quad (f \in \mathbb{R}^{h \times w}) \tag{6}$$

During training, we up-sample the f to the same size as input image with bi-linear interpolation, and minimize the binary cross-entropy loss function of

$$L = -\frac{1}{WH} \sum_{i=1}^{W} \sum_{j=1}^{H} (g_{ij} log(f_{ij}) + (1 - g_{ij} log(1 - f_{ij}))) \tag{7}$$

where g is the binary ground-truth label, the W and H are width and height of the input image, respectively.

4   Experiments

4.1   Experimental Settings

Datasets. We evaluate our model on four available *RIS* datasets: 1) *UNC* [17]: this dataset consists of 19,994 images with 142,209 referring expressions for 50,000 objects, and two or more objects of the same object category appear in each image. 2) *UNC+* [17]: this dataset contains 141,564 expressions for 49,856 objects in 19,992 images, the location words are not allowed in the expressions. 3) *G-Ref* [18]: this dataset contains 104,560 expressions for 54,822 objects in 26,711 images, the average length of expressions is 8.4 words which is longer than other datasets (less than 4 words). 4) *ReferIt* [19]: this dataset composed of 130,525 expressions referring to 96,654 distinct regions in 19,894 natural images, the segmentation targets in this dataset consists of object and stuff (*e.g.,* "*water*" *and* "*ground*").

Evaluation Metrics. Following previous works [4,5,7], we adopt intersection-over-union (*IoU*) and precision (*P@X, X* ∈ [0.5, 0.6, 0.7, 0.8, 0.9]) as evaluation metrics. The *IoU* calculates total intersection region over total union regions of all test images. The *P@X* measures the percentage of test images with an *IoU* score high than the threshold *X* (*X* ∈ {0.5, 0.6, 0.7, 0.8, 0.9}).

Implementation Details. We implement our method based on TensorFlow. The *DeepLab ResNet-101* [3] is pre-trained on *Pascal VOC* [20]. All input images are resized and zero-padded to $W = H = 320$ in our experiments. We employ a poly learning rate policy where the initial learning rate is multiplied by $(1 - \frac{iter}{total\_iter})^{0.9}$ after each iteration, the initial learning rate is set to $2.5e^{-4}$. The network is trained using Adam optimizer [19] with a weight decay of $5e^{-4}$.

Table 1. Ablation study on the *UNC* val set. The segmentation results are not post-processed with *DenseCRF*.

Models	S5	S54	S543	$F_{vs} \otimes F_1$	CWSG	VGGF	LGGF	IoU
1	✓							45.95
2		✓						49.11
3			✓					49.64
4			✓	✓				51.06
5			✓	✓	✓			51.81
6			✓	✓	✓	✓		60.33
7			✓	✓	✓		✓	53.00
8	✓			✓	✓	✓	✓	56.45
9		✓		✓	✓	✓	✓	59.74
10			✓	✓	✓	✓	✓	60.47

Table 2. Comparison based on metric of *P@X* on the *UNC* val set.

Methods	P@0.5	P@0.6	P@0.7	P@0.8	P@0.9	IoU
RMI [5]	42.99	33.24	22.75	12.11	2.23	45.18
RRN [7]	61.66	52.50	42.40	28.13	8.51	55.33
CMSA [12]	66.44	59.70	50.77	35.52	10.96	58.32
STEP [13]	70.15	63.37	53.15	36.53	10.45	59.13
Ours (w/o *CCA*)	71.97	65.31	55.98	38.56	11.34	59.64
Ours (w/ *CCA*)	**71.99**	**65.98**	**56.71**	**40.83**	**13.73**	**60.84**

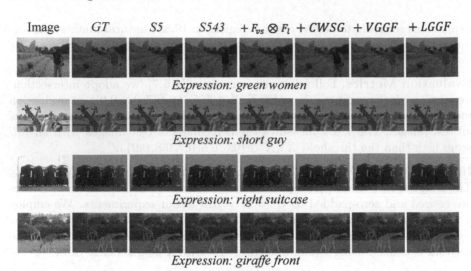

Fig. 4. Visualization of ablation study on *UNC* val set. From the first column to the fourth column, we show the input images, ground truth masks, baseline results with visual features from stage 5 and combination of stages 3, 4 and 5. From the fifth column to the last column, the results are based on the fourth column and gradually add the components of $F_{vs} \otimes F_l$, *CWSG*, *VGGF*, and *LGGF*.

4.2 Ablation Study

To validate the effectiveness of each component in our proposed model, we conduct ablation experiments on *UNC* val set and summarize the main results in Table 1 and Table 2.

In Table 1, we first explore the impact of fusing different scale visual features. For simplicity, the *S5* indicates the visual features of stage 5, the *S54* indicates the fused visual features of stages 4 and 5, the *S543* indicates the fused visual features of stages 3, 4 and 5. Lines 1–3 show that combination with low level features can improve the segmentation results. It brings 3.69% *IoU* improvement from *S5* to *S543*. In Fig. 4, the third and fourth columns show the segmentation results from *S5* and *S543*, respectively. It can be observed that the fused multi-scale features are helpful to localize and recognize the referred objects by the language expressions. In the following experiments, we choose *S543* as visual representation to investigate the relative contributions of other components. In Line 4, the element-wise multiplication features ($F_{vs} \otimes F_l$) can enrich the feature representation and generate 1.42% *IoU* improvement. The corresponding qualitative performance of this component is shown in the fifth column in Fig. 4. The impact of channel-wise self-gate is shown in Line 5, the result demonstrates that it can reduce the differences of two modalities and brings 0.75% *IoU* improvement. The corresponding qualitative performance of this component is shown in the sixth column in Fig. 4.

The impact of visual guided global fusion model (*VGGF*) and language guided global fusion model (*LGGF*) are shown in Lines 6 and 7, respectively. The values of *IoU* are improved by 8.52% and 1.19% corresponding to *VGGF* and *LGGF*. The best performance is obtained from the combination of *VGGF* and *LGGF*, which brings 8.66% *IoU* improvement, as shown in Line 10. The qualitative performance when add components of *VGGF* and *LGGF* are shown in the last two columns, respectively. It can be observed that when gradually add *VGGF* and *LGGF*, the model can well localize and segment the referred objects and reduce the impact of other objects. Finally, we also provide the results of *S4* and *S54* to compare with *S543* in full model, as shown in Lines 8 and 9, respectively. The comparison of Lines 8, 9 and 10 further demonstrates that fusing different scale features is helpful to improve the performance in our model.

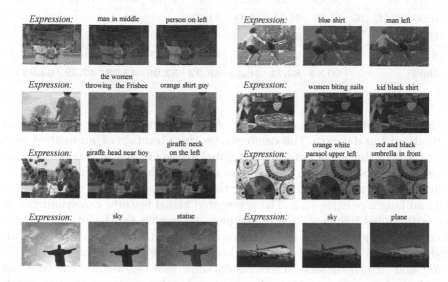

Fig. 5. Some qualitative segmentation results of *RIS* with our method. From top to down, we show the samples from *UNC*, *UNC+*, *G-Ref* and *ReferIt* datasets.

In Table 2, we compares our model with/without cross channel attention (*CCA*) in *VGGF* and the results on evaluation of *P@X* against other state-of-the-art methods. For fair comparison, the final results are post-processed with *DenseCRF* [21]. This comparison demonstrates that the *CCA* can focus on distinguished features during generating attention map and further improve the segmentation results. In addition, this comparison also show that our method outperforms other methods in terms of *P@X* and *IoU* on *UNC* val set.

4.3 Comparison with State-of-the-Art Methods

We conduct experiments on four available *RIS* datasets and compare the segmentation results refined by *DenseCRF* with other state-of-the-art methods,

Table 3. Comparison of segmentation performance (IoU) with state-of-the-art methods on four evaluation datasets.

Methods	UNC			UNC+			G-Ref	ReferIt
	val	testA	testB	val	testA	testB	val	test
LSTM-CNN [4]	–	–	–	–	–	–	28.14	48.03
LSTM-CNN [6]	–	–	–	–	–	–	34.06	49.91
RMI [5]	45.18	45.69	45.57	29.86	30.48	29.50	34.52	58.73
DMN [10]	49.78	54.83	45.13	38.88	44.22	32.29	36.76	52.81
KWAN [11]	–	–	–	–	–	–	36.92	59.19
RRN [7]	55.33	57.26	53.95	39.75	42.15	36.11	36.45	63.63
MAttNet [15]	56.51	62.37	51.70	46.67	52.39	40.41	-	-
ASGN [8]	50.46	51.20	49.27	38.41	39.79	35.97	41.36	60.31
REOS-CAC [16]	58.90	61.77	53.81	44.32	–	–	–	–
CMSA [12]	58.32	60.61	55.09	43.76	47.60	37.89	39.98	63.80
STEP [13]	60.04	63.46	57.97	48.19	52.33	40.41	46.40	64.13
Ours	**60.84**	**63.71**	**58.43**	**48.32**	**52.96**	**42.23**	**47.22**	**64.93**

as shown in Table 3. The performance of our method is outperforms previous methods on four datasets. This demonstrates the advantage of our method, especially the contribution of visual guided global fusion module and language guided global fusion module. These two modules aggregate the expression-related visual contextual information by establishing the correlation between visual information and language expression. The aggregated visual features can symmetrically fuse with global representation of language expression and facilitate to predict the referred objects. The Fig. 5 presents some qualitative segmentation results produced by our method. From top to down, the samples are from *UNC*, *UNC+*, *G-Ref* and *ReferIt* datasets. These results show that our method can take arbitrary language expression (*e.g.*, short or long, with or without location words) as input and precisely segment the referred objects (*e.g.*, "man", "plane" and "giraffe") or background (*e.g.*, "sky") in the image.

5 Conclusions

We have proposed a global fusion network to solve the *RIS* task in this work. To make the visual and language features fusion better, the semantic information from two modalities should be symmetric and aligned. Thus we proposed a visual guided global fusion module and a language guided global fusion module to aggregate visual global contextual information and fuse it with global language representation. Moreover, a channel-wise self-gate is introduced to reduce the distribution differences between two modalities. We have performed extensive experiments to validate the effectiveness of our proposed model. The experimental results show that our model achieves superior performance on four available *RIS* datatsets.

References

1. Long, J., Shelhamer, E., Darrell, T.: Fully convolutional networks for semantic segmentation. In: CVPR, pp. 3431–3440 (2015)
2. Chen, L.-C., Yang, Y., Wang, J., Xu, W., Yuille, A.L.: Attention to scale: scale-aware semantic image segmentation. In: CVPR, pp. 3640–3649 (2016)
3. Chen, L.-C., Papandreou, G., Kokkinos, I., Murphy, K., Yuille, A.L.: DeepLab: semantic image segmentation with deep convolutional nets, atrous convolution, and fully connected CRF's. TPAMI **40**(4), 834–848 (2017)
4. Hu, R., Rohrbach, M., Darrell, T.: Segmentation from natural language expressions. In: Leibe, B., Matas, J., Sebe, N., Welling, M. (eds.) ECCV 2016. LNCS, vol. 9905, pp. 108–124. Springer, Cham (2016). https://doi.org/10.1007/978-3-319-46448-0_7
5. Liu, C., Lin, Z., Shen, X., Yang, J., Lu, X., Yuille, A.: Recurrent multimodal interaction for referring image segmentation. In: ICCV, pp. 1271–1280 (2017)
6. Hu, R., Rohrbach, M., Venugopalan, S., Darrell, T.: Utilizing large scale vision and text datasets for image segmentation from referring expressions. arXiv preprint arXiv:1608.08305 (2016)
7. Li, R., et al.: Referring image segmentation via recurrent refinement networks. In: CVPR, pp. 5745–5753 (2018)
8. Qiu, S., Zhao, Y., Jiao, J., Wei, Y., Wei, S.: Referring image segmentation by generative adversarial learning. TMM **22**(5), 1333–1344 (2020)
9. Hochreiter, S., Schmidhuber, J.: Long short-term memory. Neural Comput. **9**(8), 1735–1780 (1997)
10. Margffoy-Tuay, E., Pérez, J.C., Botero, E., Arbeláez, P.: Dynamic multimodal instance segmentation guided by natural language queries. In: ECCV, pp. 630–645 (2018)
11. Shi, H., Li, H., Meng, F., Wu, Q.: Key-word-aware network for referring expression image segmentation. In: ECCV, pp. 38–54 (2018)
12. Ye, L., Rochan, M., Liu, Z., Wang, Y.: Cross-modal self-attention network for referring image segmentation. In: CVPR, pp. 10502–10511 (2019)
13. Chen, D.-J., Jia, S., Lo, Y.-C., Chen, H.-T., Liu, T.-L.: See-through-text grouping for referring image segmentation. In: CVPR, pp. 7454–7463 (2019)
14. Vaswani, A., et al.: Attention is all you need. In: Advances in Neural Information Processing Systems, pp. 5998–6008 (2017)
15. Yu, L., et al.: MattNet: modular attention network for referring expression comprehension. In: CVPR, pp. 1307–1315 (2018)
16. Chen, Y.-W., Tsai, Y.-H., Wang, T., Lin, Y.-Y., Yang, M.-H.: Referring expression object segmentation with caption-aware consistency. arXiv preprint arXiv:1910.04748 (2019)
17. Yu, L., Poirson, P., Yang, S., Berg, A.C., Berg, T.L.: Modeling context in referring expressions. In: Leibe, B., Matas, J., Sebe, N., Welling, M. (eds.) ECCV 2016. LNCS, vol. 9906, pp. 69–85. Springer, Cham (2016). https://doi.org/10.1007/978-3-319-46475-6_5

18. Mao, J., Huang, J., Toshev, A., Camburu, O., Yuille, A.L., Murphy, K.: Generation and comprehension of unambiguous object descriptions. In: CVPR, pp. 11–20 (2016)
19. Kazemzadeh, S., Ordonez, V., Matten, M., Berg, T.: Referitgame: referring to objects in photographs of natural scenes. In: EMNLP (2014)
20. Everingham, M., Gool, L.V., Williams, C.K.I., Winn, J., Zisserman, A.: The pascal visual object classes (VOC) challenge. IJCV 88(2), 303–338 (2010)
21. Krähenbühl, P., Koltun, V.: Efficient inference in fully connected CRFs with Gaussian edge potentials. In: Advances in Neural Information Processing Systems, pp. 109–117 (2011)

Underwater Enhancement Model via Reverse Dark Channel Prior

Yue Shen, Haoran Zhao, Xin Sun$^{(\boxtimes)}$ (iD), Yu Zhang, and Junyu Dong

Ocean University of China, Qingdao, Shandong Province, China
sunxin1984@ieee.org

Abstract. The interference of suspended particles causes the problems of color distortion, haze effect and visibility reduction in complex underwater environment. However, existing methods for enhancement often result in overexposure of the low contrast area or distortion of the seriously turbid area. The main factor is the diversity of the underwater images. In this paper, we propose a novel underwater imaging model which can recover different kinds of underwater images, especially the image from the murky water. Specifically, we develop a reverse dark channel prior algorithm which can separate the image with large degradation by setting a maximum filter dark channel threshold. Using the algorithm, we can calculate the adaptive transmissivity and restore the blurred images. The restored image will be sent into an end-to-end generative model for coloring transformation and enhancement. Furthermore, we build an edge deepening module for edge recovery. Comprehensive experimental evaluations show that our method performs promising on image enhancement for different kinds of turbid water. Compared with other previous methods, our method is more universal and more suitable for practical applications.

Keywords: Underwater imaging model · Image restoration · Image enhancement · Dark channel prior · Generative Adversarial Network

1 Introduction

Recent years, machine vision has played an important role in the field of underwater robots and ocean research. However, the complex underwater environment cause serious degradation of underwater images, such as color deviation and contrast reduction. One of the main reasons for this degradation of image quality is the absorption of light by the substances in the water. The absorption consumes the energy of the imaging light, which leads to the color deviation of underwater image. In addition, the scattering effect of light in water will also degrade the

This work was supported by the National Natural Science Foundation of China (No. 61971388, U1706218, 41576011, L1824025), Key Research and Development Program of Shandong Province (No. GG201703140154), and Major Program of Natural Science Foundation of Shandong Province (No. ZR2018ZB0852).

Y. Peng et al. (Eds.): PRCV 2020, LNCS 12305, pp. 447–459, 2020.
https://doi.org/10.1007/978-3-030-60633-6_37

qualification of underwater images. When the light hits the suspended particles in the water, the backward scattering occurs. And the backward scattering light reduces the contrast of the image. Apart from these factors, another challenge in underwater image enhancement is the complexity of water surroundings. Turbid water causes serious scattering and blurs the underwater images. It is difficult to find a general method to recover the different kinds of turbid underwater images. It is not appropriate to deal with the underwater images in serious turbidity and the shallow water images with less degradation in a same way.

Currently, previous methods have been proposed to solve the problems of light attenuation and light scattering. However, most methods still use the traditional hazy-removal algorithms to predict underwater imaging model parameters. Due to the differences of properties of imaging and illumination, the traditional hazy-removal models perform poorly in serious turbidity underwater scenes. Previous works cannot effectively handle the challenge of image degradation. In addition, it is challenging to apply these algorithms to different underwater scenes flexibility. To address these problems, we perform a novel method on a synthetic underwater dataset [1]. There are 8 types of underwater images in the synthetic dataset. With our underwater recovery algorithm, we separate the underwater images in serious turbidity, and recover the images specifically. We further propose a novel underwater enhancement model which performs color and structure restoration for multiple types of underwater images. Our method achieves superior performance on the underwater dataset compared to the previous methods.

Our main contributions are summarized as follows:

- We propose a reverse dark channel prior algorithm that can separate the turbid underwater images by threshold. The separation of the underwater images in serious turbidity water and in shallow water enables us to handle different types of data separately.
- We propose a underwater imaging model to recover the separated turbid images and overcome the overexposure and severe degradation problems. It performs promising on dealing with the serious blur caused by scattering.
- We propose an adversarial model with an edge recovery module. For the recovered images and the separated weak turbid images, the model builds the mapping between the recovered image and ground truth. It can nicely recover the color and structure of the images.

The rest of the paper is organized as follows. In Sect. 2, we provide the related works. In Sect. 3, we provide the derivation of algorithm and the structures of underwater enhancement model. In Sect. 4, we evaluate the proposed method through qualitative and quantitative analysis. Finally, conclusions are given in Sect. 5.

2   Related Works

2.1   Underwater Imaging Model

A large number of underwater image restoration algorithms based on underwater imaging models have been proposed to solve the problems of light attenuation and scattering. The key to restore the underwater image is to establish a corresponding restoration model according to the image degradation process. Reverse the solving process and estimate the model parameters. These models are able to restore the underwater image reasonably. In this line of research, many works are primarily inspired by the dark channel (DCP) prior [2]. DCP [2] is a hazy-removal algorithm based on the dark channel prior which is discovered by statistical analysis. Due to the differences of the imaging and lighting properties, the DCP is not suitable for underwater scenes. Serikawa et al. [3] improve the DCP [2] and introduces a joint triangular filter to control the medium transmission. With the proposed method, the exposure of dark areas can be improved to some extent. Although this method achieves greater improvement than the original DCP [2] for underwater image restoration, it does not solved the problems of large estimated transmittance caused by the low value of the red channel in the underwater images. Yang et al. [4] set a threshold to determine the use of red channel information for dark prior and consider the effect of artificial light source which can bring more natural colors. Drews et al. [5] propose an underwater dark channel prior method (UDCP) according to the properties of underwater imaging. However, it will failed when there are white objects in the underwater scene. Peng et al. [6] propose a generalized dark channel prior which estimates the medium transmission by calculating the difference between the intensity and ambient light, and finally restores the degraded image according to the imaging model. Carlevaris et al. [7] propose a prior based on the differences between the maximum value of red channel and other channels. However, it has some deficiencies in estimating medium transmission in shallow waters. The backward scattering in shallow water is not serious, so it is much necessary to recover the missing color caused by light absorption. Serious scattering and turbidity will occur in the deep-water environment. After enhancement, the image is easy to maintain low contrast in the area of insufficient enhancement and high contrast in the area of excessive enhancement. The main disadvantage of the existing methods is that they cannot restore and enhance the images well in turbid waters and cannot deal with the diversity underwater images.

2.2   Dark Channel Haze Removal Algorithm

The atmospheric scattering model [8] widely used in the field of image dehazing is shown as:

$$I(x) = J(x)t(x) + A[1 - t(x)] \qquad (1)$$

where $I(x)$ is the light intensity of each pixel which means the brightness obtained from a image, $J(x)$ is the scene radiance, which is the image after dehazing and restoration, $t(x)$ is the transmission from scene to camera. A is the atmospheric ambient light. The images can be restored from the input haze images by estimating $t(x)$ and A. The dark channel prior [2] is a statistical prior. In most of the in-air images with non-haze, at least one color channel has some pixels with very low intensity. The minimum intensity value is almost close to zero:

$$J_{dark}(x) = \min_{c=r,g,b} (\min_{y=\Omega(x)} J^c(y)) \rightarrow 0 \qquad (2)$$

where $\Omega(x)$ is a local patch centered at x, $\min_{y=\Omega(x)}$ is a minimum filter, J^c is a color channel of J, J_{dark} is the dark channel of J. To estimate the atmospheric A, He et al. [2] pick the top 0.1% brightest pixels in the dark channel. Among these pixels, the pixel with highest intensity in the input image is selected as the atmospheric light. Putting Eq. 2 into Eq. 1, we can estimate the transmission:

$$t(x) = 1 - \min_{c=r,g,b} (\min_{y=\Omega(x)} \frac{I^c(y)}{A}). \qquad (3)$$

We can get the fog-free images by implementing A and $I(x)$ into Eq. 1. The dark channel prior algorithm provides excellent results in scene dehazing.

2.3   Generative Adversarial Networks

Deep learning networks have developed rapidly in recent decade. Generative Adversarial Networks (GAN) have achieved excellent performances in fields like style translation [9–12] and super-resolution [13,14]. GAN consists of two models, i.e., the generation model G and the discriminative model D. The role of the generator is to learn the mapping from the generated samples to the real samples and then generate images as real as possible to make the discriminant model indistinguishable. The two models are trained at the same time, against witH each other to achieve a dynamic balance. The original GAN formulation can be expressed as:

$$\min_G \max_D V(D,G) = E_{x \sim p_{data}}[logD(x)] + E_{z\ p_z}[log(1 - D(G(z)))]. \qquad (4)$$

Recent years, deep learning methods based on GAN and other convolutional networks have been widely used for underwater image restoration. These methods are very effective at reconstructing images, and make images better consistent with the distribution of real images through adversarial training. Water GAN [15] takes in-air images and depth maps as input, generates the synthetic images through the attenuation generator, the scattering generator and the vignette generator. The color correction network trained by these generated images can correct the real underwater image well. Sun et al. [16] propose a knowledge transfer framework for underwater target recognition, which solved the problem of extracting identification features from low-contrast underwater images effectively. Fabbri et al. [17] use U-Net structure [18] and Patch GAN [19] discriminator for underwater image restoration. The loss functions based on Wasserstein distance [20] and gradient difference restore details and edge better. Wang et al. [21] propose an unsupervised GAN based on U-Net to generate underwater images from clear aerial images. And they train the U-Net with different loss functions on a synthetic underwater dataset [1] for underwater image enhancement. Sun et al. [22] design an encoder-decoder framework, which enhances underwater images by setting a pixel-by-pixel refinement network.

3   Method

In this section, we introduce our proposed framework. The overview of the proposed framework is shown in Fig. 1. The specific framework is composed by two parts. The first part is an underwater imaging model with a novel reverse dark channel prior algorithm. The model separates the images with severe degradation, and feeds them to a prior network for image restoration. In the second part, the images are fed into a network which is based on Generative Adversarial Networks for coloring transforms and enhancement. In addition, an edge recovery module is implemented for edge recovery.

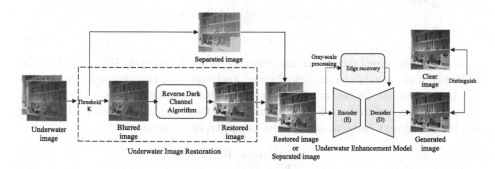

Fig. 1. Underwater imaging model.

3.1   Underwater Image Restoration via Reverse Dark Channel Algorithm

Underwater haze phenomena mainly results from the backward scattering in the water. The absorption of light associated with backward scattering causes the degradation of light at different channels. Most underwater images are dominated by a bluish or greenish hue. Thus, it is unsuitable to use the original dark channel algorithm directly in the underwater scene. In addition, underwater images will be seriously blurred in the turbid water. To address this problem, we reverse the dark channel prior algorithm through calculation and statistics of different types of underwater images. We perform reverse dark channel algorithm on more than four thousand underwater images and normal in-air images. The proposed dark channel is obtained by maximum value filtering:

$$I'(x) = \max_{c=g,b}(\max_{y=\Omega(x)} I^c(y)). \tag{5}$$

In contrast to the dark channel prior, our prior is proposed for the turbid patches. Through analysis and statistics, we find that the average values of the dark channel prior filtered by maximum filter have different intervals between severe turbidity images and shallow water images. In addition, the average value of the dark channel prior from the images in the air is close to a constant β:

$$J'(x) = \max_{c=g,b}(\max_{y=\Omega(x)} J^c(y)) \to \beta. \tag{6}$$

The turbidity of the images depend on the difference between the dark channel value and the constant β. We introduce a threshold parameter K. If β is greater than the threshold K, it indicates that the image comes from turbid water. In addition, we modify the Eq. 1 according to the different attenuation rates of three channels in the underwater scene. The improved imaging model can be given as:

$$I'(x) = J'(x)t(x) + A'[1 - t(x)] \tag{7}$$

$$A' = \alpha A \tag{8}$$

where α corresponds to the attenuation rate of each channel. We select the top 0.1% brightest pixels in the dark channel and estimate the atmospheric A with the highest pixel. To better obtain the adaptive transmission, we set different values of α in different channels when calculating the transmission. Then, we use the estimated ambient light A' to calculate the transmittance:

$$t(x) = \frac{\max_{c=g,b}(\max_{y=\Omega(x)} I^c(y)) - \alpha A}{\beta - \alpha A}. \tag{9}$$

The restored images are obtained according to Eq. 7 for the subsequent steps.

3.2   Underwater Enhancement Model Based on U-Net

To generate the realistic underwater images, we propose an underwater enhancement network model which consists of a main network and a sub-network. The network is used for color restoration and the restoration of the details. The edge recovery module use the adversarial strategy to make the network confident favor the real label class, so as to generate images with clearer edges. The main network is used for image enhancement. The sub-network is used for edge information enhancement. Each network contains a generator and a discriminator. GAN are very effective at reconstructing images, and make images better consistent with the distribution of real images through adversarial training. To retain some low-level information (i.e., the main network needs to retain color information and the sub-network needs to retain contour edge information) during the enhancement process, the overall architecture of the generator uses a U-net with skip connections, which has an encoder-decoder unit. The discriminator uses a multi-layer CNN classifier with cross-entropy loss. The detailed generation framework is shown in Fig. 2.

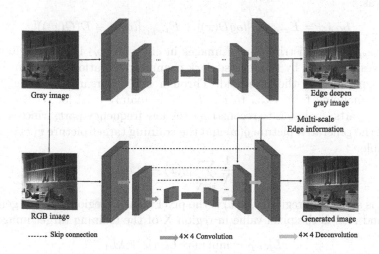

Fig. 2. The generation architecture for underwater image enhancement.

The network inputs consist of the images restored by the reverse dark channel restoration algorithm and the images separated by the threshold. Firstly, the size of RGB images is resized to 256×256. In the encoder, the image is down-sampled through a series of convolutional operations. In each down-sampling stage, 4×4 convolution with a stride of 2 is performed with a rectified linear unit (ReLU) activation function. In the decoder, as the down-sampling causes the loss of information in each layer, the skip connection is done after each up-sampling. We concatenate the corresponding down-sampling output tensor of each layer in the encoder. Our discriminant model of the main network is composed of a convolution layer, a pooling layer, and a fully connected layer. The discriminator distinguishes the generated images and ground truth images.

In the sub network, the encoder takes the gray-scale maps of underwater images as input. The layers of decoder and encoder in sub-network are as same as the layers in main network. To get the edge information, we connect the shallow layers of the decoder to the corresponding layers of the main network decoder. The discriminator of sub-network discriminates the generated gray-scale maps, gray-scale maps of ground truth, and gray-scale maps with Gaussian blur. The Gaussian blurred gray-scale maps are considered as negative samples. The purpose of taking edge Gaussian blurred maps as negative samples for discrimination is to make the generated edge clear.

3.3   Loss Functions

In order to improve the performance of model, we formulate the loss functions for the main network and the sub network. The main network computes a loss L_{NET}, which is the weighted addition of adversarial loss L_{GAN} and L_1 loss. Adversarial loss helps the main network generate realistic in-air images, which is given as:

$$L_{GAN} = E_{x \sim p_{data}}[logD(x)] + E_{x \sim p_G}[log(1 - D(G(x)))] \qquad (10)$$

where, p_{data} is the distribution of images in clear air, p_G is the distribution of the images recovered by the reverse dark channel restoration algorithm and the images separated by the threshold. Through the adversarial training, we can learn the mapping from p_{data} to p_G. L_{GAN} is mainly used for recovering high frequency parts. L_1 loss is to reconstruct the low frequency part, which is mainly to make the generated picture $g(x)$ and the training target picture $r(x)$ as similar as possible.

$$L_1 = \frac{1}{N} \sum_{x \in X} |g(x) - r(x)| \qquad (11)$$

where x is a pixels in region X, $g(x)$ is the pixel value in region X of the generated image and $r(x)$ is the pixel value in region X of the training target image.

$$L_{NET} = \min_G \max_D L_{GAN} + \lambda L_1 \qquad (12)$$

The sub-network loss function uses adversarial loss. Inspired by Cartoon GAN [23], $E_{x \sim p_G}[log(1 - D(x))]$ is added in the ordinary GAN adversarial loss.

$$L_{GAN2} = E_{x \sim p_{data}}[logD(x)] + E_{x \sim p_G}[log(1 - D(G(x)))] + E_{x \sim p_E}[log(1 - D(x))] \qquad (13)$$

$$L_{NET2} = \min_G \max_D L_{GAN2} \qquad (14)$$

where p_E is the distribution of underwater images processed by edge Gaussian blur and gray scale. We add the edge Gaussian blurred image as negative sample. The goal is to generate clear edges by the generator G of sub-network.

4   Experiments

4.1   Datasets

Many methods have been proposed to synthesize underwater images from clear images. We train our model on such a synthetic underwater image dataset. We use the NYU RGB dataset [24] to provide us with the in-air images. The synthetic underwater image dataset is built by the methods mentioned in [1] on the NYU RGB dataset. Following the experimental setup and classification method proposed by Uplavikar et al. [25], we combine the high-similarity image types 1 and 3, types II and III from the synthetic dataset, reduce the total number of original types to 6. In addition, the type 7 and type 9 in synthetic dataset are underwater images with serious turbidity. We also test our model on a real-world dataset. Figure 3 shows six types of a given image in the synthetic dataset.

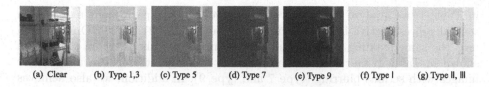

(a) Clear   (b) Type 1,3   (c) Type 5   (d) Type 7   (e) Type 9   (f) Type I   (g) Type II, III

Fig. 3. Types of the synthetic underwater image dataset.

4.2 Settings

All networks are trained on NVIDIA GeForce GTX TITAN. We implement the model with TensorFlow. Our model is trained for 60 epochs, and the batch size is 16 with output images set to 256×256. In addition, we use Adam optimizer with a learning rate of 0.0001.

Table 1. Comparison of our model to previous methods with SSIM and PSNR. Higher values mean better performances. Bold font indicates the best performance.

	Type	RAW	RED	UDCP	ODM	UIBLA	UWCNN	Ours
PSNR	1	15.535	15.596	15.757	16.085	15.079	21.790	**25.270**
	3	14.688	12.789	14.474	14.282	13.442	20.251	**25.270**
	5	12.142	11.123	10.862	14.123	12.611	17.517	**20.198**
	7	10.171	9.991	9.467	12.266	10.753	12.219	**19.985**
	9	9.502	11.620	9.317	9.302	10.090	13.232	**20.636**
	I	17.356	19.545	18.816	18.095	17.488	**25.927**	19.289
	II	20.595	20.791	17.204	17.610	18.064	**24.817**	19.510
	III	16.556	16.690	14.924	16.710	17.100	**22.633**	19.510
SSIM	1	0.7065	0.7406	0.7629	0.7240	0.6957	0.8558	**0.8898**
	3	0.5788	0.6639	0.6614	0.6765	0.5765	0.7951	**0.8898**
	5	0.4219	0.5934	0.4269	0.6441	0.4748	0.7266	**0.8314**
	7	0.2797	0.5089	0.2628	0.5632	0.3052	0.6070	**0.8290**
	9	0.1794	0.3192	0.1624	0.4178	0.2202	0.4920	**0.8540**
	I	0.8621	0.8816	0.8172	0.4178	0.7449	**0.9376**	0.7885
	I	0.8761	0.8837	0.8387	0.8251	0.8017	**0.9236**	0.7934
	III	0.5726	0.7911	0.7587	0.7546	0.7655	**0.8795**	0.7934

4.3 Results

To verify the effectiveness of our method, we employ quantitative assessment metrics, namely the structural similarity (SSIM) and peak signal to noise ratio (PSNR), for the generated images relative to their clear counterparts. As shown in Table 1, we analyze the model quantitatively by comparing our model with several representative image enhancement methods [5,21,26–28]. Our method uses the same training parameters as the compared methods. From the results shown in Table 1, Our method has achieved the highest scores in underwater images with serious blurring (type 7 and type 9). In addition, we also achieves the best results on other kinds of underwater images.

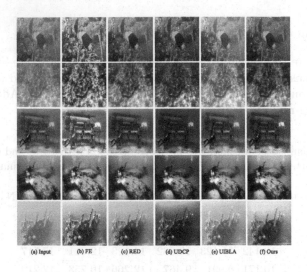

(a) Input   (b) FE   (c) RED   (d) UDCP   (e) UIBLA   (f) Ours

Fig. 4. Illustration of the results on the real-word images.

We also quantitatively compare our method with other existing methods on the real-world underwater dataset U45. Visual comparisons with competitive methods can be seen in the Fig. 4. The compared methods include fusion enhancement (FE) [29], RED [26], UDCP [5] and UIBLA [28]. Our proposed method has good performance of color recovery and natural effect on degraded underwater images. Both FE [29] and UIBLA [28] over-enhance the blue effect, while RED [26] and UDCP [5] have unnatural contrast. In addition, FE [29] have regional overexposure problems in the underwater environment with bright light and white objects. UDCP [5] make the image darker. Our proposed method can recover the images with appropriate contrast on the basis of removing haze effectively. To show the effectiveness of the dark channel haze removal algorithm, the ablation experiment is presented on the degraded images(type 9). The results are shown in Table 2.

Table 2. Ablation experiment.

	U-net Model Only	All
PSNR	14.820	20.636
SSIM	0.7033	0.8540

5   Conclusions

According to the complexity of water environment, we analyze the imaging principle of underwater images. We propose a reverse dark channel prior algorithm and set a threshold to separate the turbid underwater images. Through the separation of different types of images, our model is able to process the data in a targeted way. We achieve the goal of recovery of turbid images by improving the atmospheric scattering model. Then, we use a architecture based on generative adversarial network for coloring transformation and enhancement. Also, we perform qualitative and quantitative evaluations. Experimental results show that our algorithm perform better compared to the previous methods. And it is able to get high-quality images by recovering images in turbid water. We believe our method is a valuable complement to the state-of-the-arts.

References

1. Li, C., Anwar, S.: Underwater scene prior inspired deep underwater image and video enhancement. Pattern Recogn. **98**, 107038 (2019)
2. He, K., Sun, J., Tang, X.: Single image haze removal using dark channel prior. IEEE Trans. Pattern Anal. Mach. Intell. **33**(12), 2341–2353 (2011)
3. Serikawa, S., Lu, H.: Underwater image dehazing using joint trilateral filter. Comput. Electr. Eng. **40**(1), 41–50 (2014)
4. Aiping, Y., Chang, Q., Jian, W., Liyun, Z.: Underwater image visibility restoration based on underwater imaging model. J. Electron. Inf. Technol. **40**, 298–305 (2018)
5. Drews, P.L.J., Nascimento, E.R., Botelho, S.S.C., Campos, M.F.M.: Underwater depth estimation and image restoration based on single images. IEEE Comput. Graph. Appl. **36**(2), 24–35 (2016)
6. Peng, Y.T., Cao, K., Cosman, P.C.: Generalization of the dark channel prior for single image restoration. IEEE Trans. Image Process. **27**(6), 2856–2868 (2018)
7. Carlevaris-Bianco, N., Mohan, A., Eustice, R.M.: Initial results in underwater single image dehazing. In: OCEANS 2010 MTS/IEEE SEATTLE, pp. 1–8 (2010)
8. McCartney, E.: Scattering phenomena (book reviews: optics of the atmosphere scattering by molecules and particles). Science **196**, 1084–1085 (1977)
9. Zhu, J.Y., Park, T., Isola, P., Efros, A.A.: Unpaired image-to-image translation using cycle-consistent adversarial networks. In: IEEE International Conference on Computer Vision (ICCV), pp. 2242–2251 (2017)
10. Yi, Z., Zhang, H., Tan, P., Gong, M.: DualGAN: unsupervised dual learning for image-to-image translation. In: 2017 IEEE International Conference on Computer Vision (ICCV), pp. 2868–2876 (2017)

11. Kim, T.: Learning to discover cross-domain relations with generative adversarial networks. In: Proceedings of the 34th International Conference on Machine Learning, PMLR, vol. 70, pp. 1857–1865 (2017)

12. Liu, T.Y.: Conditional image-to-image translation. In: 2018 IEEE/CVF Conference on Computer Vision and Pattern Recognition (CVPR), pp. 5524–5532 (2018)

13. Ledig, C., et al.: Photo-realistic single image super-resolution using a generative adversarial network. In: Computer Vision and Pattern Recognition (CVPR), pp. 4681–4690 (2016)

14. Wang, X., et al.: ESRGAN: enhanced super-resolution generative adversarial networks. In: Leal-Taixé, L., Roth, S. (eds.) ECCV 2018. LNCS, vol. 11133, pp. 63–79. Springer, Cham (2019). https://doi.org/10.1007/978-3-030-11021-5_5

15. Li, J., Skinner, K.A., Eustice, R.M., Johnson-Roberson, M.: WaterGAN: unsupervised generative network to enable real-time color correction of monocular underwater images. IEEE Robot. Autom. Lett. 3(1), 387–394 (2017)

16. Sun, X., Shi, J., Liu, L., Dong, J., Plant, C., Wang, X., Zhou, H.: Transferring deep knowledge for object recognition in low-quality underwater videos. Neurocomputing 275, 897–908 (2017)

17. Fabbri, C., Islam, M.J., Sattar, J.: Enhancing underwater imagery using generative adversarial networks. In: 2018 IEEE International Conference on Robotics and Automation (ICRA), Brisbane, QLD, 2018, pp. 7159–7165 (2018)

18. Ronneberger, O., Fischer, P., Brox, T.: U-Net: convolutional networks for biomedical image segmentation. In: International Conference on Medical Image Computing and Computer-Assisted Intervention, vol. 9351, pp. 234–241 (2015)

19. Isola, P., Zhu, J.Y., Zhou, T., Efros, A.: Image-to-image translation with conditional adversarial networks. In: IEEE Conference on Computer Vision and Pattern Recognition (CVPR), pp. 5967–5976, July 2017

20. Arjovsky, M., Chintala, S., Bottou, L.: Wasserstein generative adversarial networks. In: Proceedings of the 34th International Conference on Machine Learning (PMLR), vol. 70, pp. 214–223, January 2017

21. Wang, N., Zhou, Y., Han, F., Zhu, H., Zheng, Y.: UWGAN: underwater GAN for real-world underwater color restoration and dehazing. In: ICLR 2020 Conference Blind Submission (2020). https://openreview.net/forum?id=HkgMxkHtPH

22. Sun, X., Liu, L., Li, Q., Dong, J., Lima, E., Yin, R.: Deep pixel-to-pixel network for underwater image enhancement and restoration. IET Image Process. 13(3), 469–474 (2019)

23. Chen, Y., Lai, Y.K., Liu, Y.J.: CartoonGAN: generative adversarial networks for photo cartoonization. In: IEEE/CVF Conference on Computer Vision Pattern Recognition, pp. 9465–9474 (2018)

24. Silberman, N., Hoiem, D., Kohli, P., Fergus, R.: Indoor segmentation and support inference from RGBD images. In: Proceedings of the 12th European conference on Computer Vision - Volume Part V, pp. 746–760 (2012)

25. Uplavikar, P., Wu, Z., Wang, Z.: All-in-one underwater image enhancement using domain-adversarial learning. In: Proceedings of the IEEE/CVF Conference on Computer Vision and Pattern Recognition (CVPR) Workshops, pp. 1–8 (2019)

26. Galdran, A., Pardo, D., Picon, A., Alvarez-Gila, A.: Automatic red-channel underwater image restoration. J. Vis. Commun. Image Representation 26, 132–145 (2015)

27. Li, C.Y., Guo, J.C., Cong, R.M., Pang, Y.W., Wang, B.: Underwater image enhancement by dehazing with minimum information loss and histogram distribution prior. IEEE Trans. Image Process. 25(12), 5664–5677 (2016)

28. Peng, Y.T., Cosman, P.C.: Underwater image restoration based on image blurriness and light absorption. IEEE Trans. Image Process. **26**(4), 1579–1594 (2017)
29. Ancuti, C., Codruta, A., Haber, T., Bekaert, P.: Enhancing underwater images and videos by fusion. In: 2012 IEEE Conference on Computer Vision and Pattern Recognition, pp. 81–88, June 2012

Underwater Image Processing by an Adversarial Network with Feedback Control

Kangming Yan and Yuan Zhou[✉]

Tianjin University, Tianjin 300072, China
zhouyuan@tju.edu.cn

Abstract. Owing to absorption and scattering of light in water, raw underwater images suffer from low contrast, blurred details, and color casts. These characteristics significantly interfere the visibility of underwater images. In this paper, we propose a new robust adversarial learning framework via physics model based feedback control and domain adaptive mechanism for underwater image enhancement. A new method for simulating underwater-like training dataset by underwater image formation model is proposed. Upon this, a novel framework, which introduces a domain adaptive mechanism and a physics model constraint feedback control, is trained to enhance the underwater scenes. Extensive experiments demonstrate the significant superiority of the proposed framework compared with the state-of-the-art methods.

Keywords: Underwater image enhancement · Domain adaptation · Degradation model

1 Introduction

Different from common images taken on the ground, underwater images suffer from poor visibility such as blurred details and color casts, resulting from the attenuation of propagated light, mainly due to wavelength-dependent light absorption and scattering [1]. The poor visibility also brings great difficulty to the subsequent image processing.

To solve this difficult task, Various processing methods for images degraded by the underwater environment have been developed. For example, traditional methods including underwater image restoration methods and underwater image enhancement methods. The restoration methods [2–6] take the degradation model of underwater imaging into consideration while restoring the input images, but these methods are difficult to recover the complex underwater imaging process by estimating only a few parameters. While underwater image enhancement methods [1,7–10] focus on adjusting image pixel values to acquire satisfactory results, but these methods show poor generalization performance for different underwater images.

© Springer Nature Switzerland AG 2020
Y. Peng et al. (Eds.): PRCV 2020, LNCS 12305, pp. 460–468, 2020.
https://doi.org/10.1007/978-3-030-60633-6_38

Alternatively, deep neural networks [11] have been shown excellent performance in vision tasks [12,13]. Some researchers try to perform underwater image enhancement tasks through similar methods [14–19]. However, due to the complexity of the problem (e.g., lack of ground truth), a simple network with a random initialization is unable to estimate the solution well, these algorithms are still less effective.

In this paper, we design a novel end-to-end enhancing framework to solve the complex problem of improving underwater image quality, as illustrated in Fig. 1, which shows excellent performance. The key contributions of this paper lies in:

1. We propose a physics model constrained learning algorithm so that it can guide the estimation of the GAN framework, which acts as the feedback controller to provides explicit constraints for the enhancement of image and also improves the interpretability of the model.
2. This is the first attempt to introduce a domain adaptive mechanism to eliminate the domain gap between synthetic training images and real-world testing images, which helps the network effective enough for enhancing real scenes.
3. We synthesize a novel image training set based on the physical imaging model in underwater scenario, which is more in line with the visual effects of multiple real-world underwater images and more suitable for the training of network.

Through a large number of experiments on both synthetic scenes and real underwater images, we have proved the excellent performance of the proposed algorithm.

2   Method

2.1   Synthesizing Underwater Images

According to Jaffe-McGlamery underwater imaging model [20], the image formulation procedure in underwater scenario can be formulated as:

$$I(x) = J(x) \cdot t_\lambda(x) + B_\lambda(1 - t_\lambda(x)),$$
$$\lambda \in \{red, green, blue\}, \tag{1}$$

where $I(x)$ is the captured image; $J(x)$ is the clear underwater scene; λ is the wavelength of the light for the red, green and blue channels, B_λ is the global background light, the medium energy ratio $t_\lambda(x)$ represents the percentage of the scene radiance reaching at the camera after reflecting: $t_\lambda(x) = Nrer(\lambda)^{d(x)}$, where $d(x)$ is the distance from the camera lens to the object surface, $Nrer(\lambda)$ is the ratio of residual energy to the initial energy per unit of distance and is dependent on the wavelength of light. This imaging model has been widely used in the field of imaging through scattering media.

Basing on this model, we adopt a novel underwater image synthesis algorithm. NYU-V2 [21] is a versatile dataset and includes the depth information for each image, which is important for us to convert images taken in the air into

Table 1. Parameter setting and randomization methods for $Nrer(\lambda)$ and B_λ for different water types.

Type	Param	Red	Green	Blue
(1)	$Nrer(\lambda)$	0.79 ± 0.06	0.92 ± 0.06	0.94 ± 0.05
	B_λ	0.05 ± 0.15	0.60 ± 0.30	0.70 ± 0.29
(2)	$Nrer(\lambda)$	0.71 ± 0.04	0.82 ± 0.06	0.80 ± 0.07
	B_λ	0.05 ± 0.15	0.60 ± 0.30	0.70 ± 0.29
(3)	$Nrer(\lambda)$	0.67 ± 0.03	0.73 ± 0.03	0.67 ± 0.03
	B_λ	0.15 ± 0.05	0.80 ± 0.05	0.70 ± 0.05

underwater styles. We apply Eq. (1) to synthesize our underwater image training set. For the parameter setting of $Nrer(\lambda)$ and B_λ in the model, we made lots of attempts based on [16,22], and compared the synthesis results with variety of real underwater images. Finally, we selected three sets of parameter as presented in Table 1. They cover different turbidity conditions from clear waters in the offshore to more turbid waters in the deep ocean. All the degradation images are combined together to form a synthesize training set.

2.2 Proposed Enhancement Framework

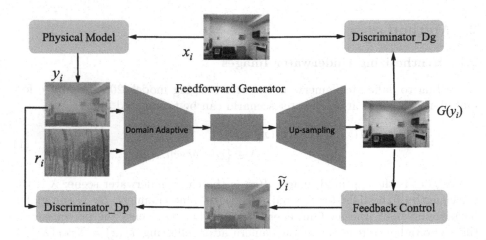

Fig. 1. Schematic of the proposed framework.

Feedforward Generator G. As shown in Fig. 1, the generator is an feedforward network whose purpose is to convert the underwater scenes into clear ones. The impressive performance of the image-to-image translation method such as

[23] encourages us to explore the similar structure. By means of a nine-residual-block stack, the downsample-upsample model learns the essence of input, and a synthesized version will emerge at the original resolution after the image reconstruction operations.

We train our network by the synthetic data mentioned above. However, though the synthetic images seem similar to real underwater images, there still have domain gap between them for they are from different underlying distribution. If the domain gap between training data and testing data were ignored, the generalization ability of the model will be seriously affected. To dill with this problem, a differentiable loss function [24] is incorporated into the downsampling module of our feedforward generator:

$$\mathcal{L}_{\text{domain}} = \frac{1}{4d^2} \| C_S - C_T \|_F^2 \tag{2}$$

where $\| \cdot \|_F^2$ denotes the squared matrix Frobenius norm, d is the number of channels for the features, and C_S (C_T) are the feature covariance matrices of source and target domains:

$$C_S = \frac{1}{n_S - 1} (F_y^\mathsf{T} F_y - \frac{1}{n_S} (1^\mathsf{T} F_y)^\mathsf{T} (1^\mathsf{T} F_y)), \tag{3}$$

$$C_T = \frac{1}{n_T - 1} (F_r^\mathsf{T} F_r - \frac{1}{n_T} (1^\mathsf{T} F_r)^\mathsf{T} (1^\mathsf{T} F_r)), \tag{4}$$

where n_S and n_T are the number of pixels in feature maps of source and target images. F_y and F_r are the feature matrices of synthetic image and real image generated by the down-sampling module.

Feedback Controller P. In order to guarantee our solutions satisfy the physics model (1) and thus generate clear and real enough images, we develop a new method to improve the estimation result of GAN under the guidance of the physics model. The feedforward generator mentioned above learns the mapping function G and generates the intermediate enhanced image $G(y_i)$ from the input y_i. Then, we re-apply the physics model (1) to $G(y_i)$:

$$\widetilde{y}_i = G(y_i) \cdot t_{i_\lambda} + B_{i_\lambda}(1 - t_{i_\lambda}),$$
$$\lambda \in \{red, green, blue\} \tag{5}$$

This model based feedback controller provides explicit constraint for our enhancement framework, ensures that the final results should be consistent with the observed images.

In our framework, there are two discriminators. D_g takes the x_i and $G(y_i)$ as the inputs and it is used to classify whether $G(y_i)$ is clear. While D_p takes the y_i and $\widetilde{y}_i$ as the inputs, which is used to classify whether the generated results satisfy the physics model. So, the adversary loss of our framework is:

$$\mathcal{L}_a = \mathbb{E} \left\{ [\log D_g(x_i)] + [\log(1 - D_g(G(y_i)))] \right\}$$
$$+ \mathbb{E} \left\{ [\log D_p(y_i)] + [\log(1 - D_p(\widetilde{y}_i)]\right\} \tag{6}$$

We also use a L_1 norm regularized pixel-wise loss function:

$$\begin{aligned}\mathcal{L}_{\text{pixel}} &= \mathcal{L}_{\text{G}} + \mathcal{L}_{\text{P}} \\ &= \mathbb{E}\left\{\|G(y_i) - x_i\|_1\right\} + \mathbb{E}\left\{\|\widetilde{y}_i - y_i\|_1\right\}.\end{aligned} \tag{7}$$

Finally, the total optimization objective we proposed is the linear combination of the above-mentioned losses:

$$G^* = arg\min_{G}\max_{D_g, D_p}(\mathcal{L}_{\text{a}} + \lambda_1\mathcal{L}_{\text{pixel}} + \lambda_2\mathcal{L}_{\text{domain}}), \tag{8}$$

3   Experiments

In this part, we perform qualitative and quantitative comparisons with the state-of-the-art underwater image enhancement methods on synthetic scenes and real images. These compared methods include UDCP [2], IBLA [5], Fusion [1], Cyclegan [23] and DUIENet [17]. They are representative in tradition underwater image restoration and enhancement, and deep learning based underwater image enhancement, respectively.

We train our network using the Adam optimizer with an initial learning rate 0.0002, and the batch size is set to be 1. And we implemented our network with the PyTorch framework and trained it using 64 GB memory and 11 GeForce Nvidia GTX 1080 Ti GPU.

3.1   Evaluation on Synthetic Underwater Image

We first evaluate the result of our method using the synthesized underwater images. Some of the subjective results of different methods are shown in Fig. 2. As you can see, UDCP [2] nearly fail to correct the underwater color meanwhile Fusion [1], IBLA [5], CycleGan [23] and DUIENet [17] introduce artificial color obviously, the results of these methods still retain the haze blur effect. While our method not only enhances the visibility of the images but also restores an aesthetically pleasing texture and colors, which nearly the same as the ground-truth.

Quantitative results in terms of three different metrics: MSE, PSNR and SSIM are shown in Table 2. Here, we evaluate the performance on 80 synthetic underwater images, and the average scores are presented. As seen, our method achieves the best scores in all metrics.

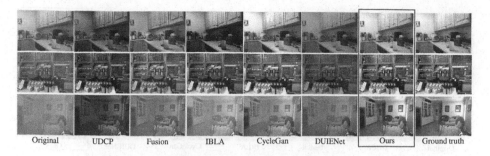

Fig. 2. Subjective result comparison for samples of synthetic underwater images.

Table 2. Quantitative evaluations on the test set. As seen, our method achieves the best scores in all metrics.

Method	Original	UDCP	Fusion	IBLA	CycleGan	DUIENet	Ours
MSE	3864.3577	5788.5313	2528.0611	2291.2073	506.4429	2700.4965	**90.2791**
PSNR	12.6176	10.9260	14.5415	15.3116	21.7324	14.1571	**28.8487**
SSIM	0.7524	0.6553	0.8211	0.7941	0.8707	0.8254	**0.9532**

3.2 Evaluation on Real-World Underwater Image

We also evaluate the our framework on real-world underwater images. We collected a test set of 80 real underwater images acquired from [1, 4], which have different characteristics of degradation. Some of the subjective results are presented in Fig. 3. As it shows, although the haze in raw images are removed by UDCP [2], IBLA [5] and Fusion [1], the color and details are not good enough. CycleGAN [23] and DUIENet [17] works well in changing the overall styles and removes the haze effects, but the contents and the structures are slightly distorted, affects the subjective visual effect. In contrast, our method shows promising results on all of the images. The underwater tone is totally removed and greatly matches the nature scenes. And it is obvious that our algorithm can enhance the detailed information as clean as possible.

We use UIQM [25] to evaluate the underwater image processing methods, which can provide a comprehensive assessment of the effectiveness of various methods. Table 3 lists the average values on our test set. As we can see, our method is significantly better than the compared methods with absolute superiority.

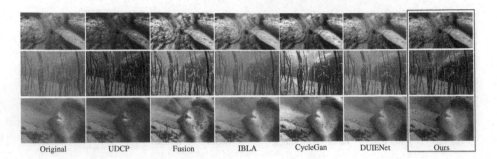

Original   UDCP   Fusion   IBLA   CycleGan   DUIENet   Ours

Fig. 3. Qualitatively comparison on real-world underwater images.

Table 3. Underwater image quality evaluation of different processing methods on real-world underwater images.

Method	Original	UDCP	Fusion	IBLA	CycleGan	DUIENet	Ours
UISM	3.7823	3.7733	5.1587	3.9638	6.1765	4.5795	**6.2318**
UICM	0.8239	2.0890	**3.7882**	2.9803	2.1636	3.0063	1.6609
UIConM	0.5607	0.7712	0.8005	0.5809	0.7036	0.6844	**0.8264**
UIQM	3.1449	3.9303	4.4923	3.3314	4.4006	3.8842	**4.8418**

4 Conclusion

This paper proposes an underwater image enhancement framework inspired by underwater scene prior and the knowledge of domain adaption. Base on the synthetic underwater images, an adversarial learning architecture with domain adaptative mechanism and physics model constraint feedback control introduced in was trained to enhance the real underwater scenes. Consequently experiences on synthetic and real underwater images demonstrate the significant superiority of the proposed method.

References

1. Ancuti, C., Ancuti, C.O., Haber, T., Bekaert, P.: Enhancing underwater images and videos by fusion. In: IEEE Conference on Computer Vision and Pattern Recognition, pp. 81–88 (2012)
2. Drews, P.L., Nascimento, E.R., Botelho, S.S., Campos, M.F.M.: Underwater depth estimation and image restoration based on single images. IEEE Comput. Graph. Appl. **36**(2), 24–35 (2016)
3. Galdran, A., Pardo, D., Picón, A., Alvarez-Gila, A.: Automatic red-channel underwater image restoration. J. Vis. Commun. Image Represent. **26**, 132–145 (2015)
4. Chiang, J.Y., Chen, Y.C.: Underwater image enhancement by wavelength compensation and dehazing. IEEE Trans. Image Process. **21**(4), 1756–1769 (2011)
5. Peng, Y.T., Cosman, P.C.: Underwater image restoration based on image blurriness and light absorption. IEEE Trans. Image Process. **26**(4), 1579–1594 (2017)

6. Zhou, Y., Wu, Q., Yan, K., Feng, L., Xiang, W.: Underwater image restoration using color-line model. IEEE Trans. Circuits Syst. Video Technol. **29**(3), 907–911 (2019)

7. Iqbal, K., Odetayo, M., James, A., Salam, R.A., Talib, A.Z.H.: Enhancing the low quality images using unsupervised colour correction method. In: IEEE International Conference on Systems, Man and Cybernetics, pp. 1703–1709 (2010)

8. Fu, X., Zhuang, P., Huang, Y., Liao, Y., Zhang, X.P., Ding, X.: A retinex-based enhancing approach for single underwater image. In: IEEE International Conference on Image Processing (ICIP), pp. 4572–4576 (2014)

9. Ancuti, C.O., Ancuti, C., De Vleeschouwer, C., Bekaert, P.: Color balance and fusion for underwater image enhancement. IEEE Trans. Image Process. **27**(1), 379–393 (2017)

10. Li, C.Y., Guo, J.C., Cong, R.M., Pang, Y.W., Wang, B.: Underwater image enhancement by dehazing with minimum information loss and histogram distribution prior. IEEE Trans. Image Process. **25**(12), 5664–5677 (2016)

11. LeCun, Y., Bengio, Y., Hinton, G.: Deep learning. Nature **521**(7553), 436–444 (2015)

12. Pan, J., et al.: Physics-based generative adversarial models for image restoration and beyond. IEEE Trans. Pattern Anal. Mach. Intell. **1**(1), 1–14 (2020)

13. Zhang, H., Sindagi, V., Patel, V.M.: Image de-raining using a conditional generative adversarial network. IEEE Trans. Circuits Syst. Video Technol. **30**, 1–14 (2019)

14. Li, J., Skinner, K.A., Eustice, R.M., Johnson-Roberson, M.: WaterGAN: unsupervised generative network to enable real-time color correction of monocular underwater images. IEEE Robot. Autom. Lett. **3**(1), 387–394 (2017)

15. Fabbri, C., Islam, M.J., Sattar, J.: Enhancing underwater imagery using generative adversarial networks. In: 2018 IEEE International Conference on Robotics and Automation (ICRA), pp. 7159–7165. IEEE (2018)

16. Li, C., Anwar, S., Porikli, F.: Underwater scene prior inspired deep underwater image and video enhancement. Pattern Recogn. **98**, 1–11 (2020)

17. Li, C., et al.: An underwater image enhancement benchmark dataset and beyond. IEEE Trans. Image Process. **29**, 4376–4389 (2020)

18. Yu, X., Qu, Y., Hong, M.: Underwater-GAN: underwater image restoration via conditional generative adversarial network. In: Zhang, Z., Suter, D., Tian, Y., Branzan Albu, A., Sidère, N., Jair Escalante, H. (eds.) ICPR 2018. LNCS, vol. 11188, pp. 66–75. Springer, Cham (2019). https://doi.org/10.1007/978-3-030-05792-3_7

19. Guo, Y., Li, H., Zhuang, P.: Underwater image enhancement using a multiscale dense generative adversarial network. IEEE J. Oceanic Eng. **45**, 862–870 (2019)

20. Jaffe, J.S.: Computer modeling and the design of optimal underwater imaging systems. IEEE J. Oceanic Eng. **15**(2), 101–111 (1990)

21. Silberman, N., Hoiem, D., Kohli, P., Fergus, R.: Indoor segmentation and support inference from RGBD images. In: Fitzgibbon, A., Lazebnik, S., Perona, P., Sato, Y., Schmid, C. (eds.) ECCV 2012. LNCS, vol. 7576, pp. 746–760. Springer, Heidelberg (2012). https://doi.org/10.1007/978-3-642-33715-4_54

22. Berman, D., Treibitz, T., Avidan, S.: Diving into haze-lines: color restoration of underwater images. In: Proceedings of the British Machine Vision Conference (BMVC), vol. 1, pp. 1–12 (2017)

23. Zhu, J.Y., Park, T., Isola, P., Efros, A.A.: Unpaired image-to-image translation using cycle-consistent adversarial networks. In: Proceedings of the IEEE International Conference on Computer Vision, pp. 2223–2232 (2017)

24. Sun, B., Saenko, K.: Deep CORAL: correlation alignment for deep domain adaptation. In: Hua, G., Jégou, H. (eds.) ECCV 2016. LNCS, vol. 9915, pp. 443–450. Springer, Cham (2016). https://doi.org/10.1007/978-3-319-49409-8_35
25. Panetta, K., Gao, C., Agaian, S.: Human-visual-system-inspired underwater image quality measures. IEEE J. Oceanic Eng. **41**(3), 541–551 (2015)

Inception Parallel Attention Network for Small Object Detection in Remote Sensing Images

Shuojin Yang[1,3,4], Liang Tian[1,3,4], Bingyin Zhou[2,3], Dong Chen[1,3,4],
Dan Zhang[1,3], Zhuangnan Xu[1,3], Wei Guo[2,3], and Jing Liu[1,3,4,5(✉)]

[1] College of Computer and Cyber Security, Hebei Normal University, Shijiazhuang
050024, Hebei, China
liujing01@ict.ac.cn
[2] School of Mathematical Sciences, Hebei Normal University, Shijiazhuang 050024,
Hebei, China
guowei@chmiot.net
[3] The Key Laboratory of Augmented Reality, School of Mathematical Sciences,
Hebei Normal University, Shijiazhuang 050024, Hebei, China
[4] Hebei Provincial Engineering Research Center for Supply Chain Big Data Analytics
and Data Security, Hebei Normal University, Shijiazhuang 050024, Hebei, China
[5] The Beijing Key Laboratory of Mobile Computing and Pervasive Device, Institute
of Computing Technology, Chinese Academy of Sciences, Beijing 100190, China

Abstract. Small object detection in remote sensing images is a major
challenge in the field of computer vision. Most previous methods detect
small objects using a multiscale feature fusion approach with the same
weights. However, experiments shows that the inter feature maps and
the feature map in different scales have different contribution to the network. To further strengthen the effective weights, we proposed an inception parallel attention network (IPAN) that contains three main parallel
modules i.e. a multiscale attention module, a contextual attention module, and a channel attention module to perform small object detection
in remote sensing images. In addition, The network can extract not only
rich multiscale, contextual features and the interdependencies of global
features in different channels but also the long-range dependencies of the
object to another based on the attention mechanism, which contributes
to precise results of small object detections. Experimental results shows
that the proposed algorithm significantly improves the detection accuracy especially in complex scenes and/or in the presence of occlusion.

Keywords: Parallel · Small objects · Attention mechanism · Complex
scene · Occlusion

The research was jointly supported by the National Natural Science Foundation of
China (Grant No.: 61802109, 61902109), the Science and Technology Foundation of
Hebei Province Higher Education (Grant No.: QN2019166), the Natural Science Foundation of Hebei province (Grant No.: F2017205066, F2019205070), the Science Foundation of Hebei Normal University (Grant No.: L2017B06, L2018K02, L2019K01).

© Springer Nature Switzerland AG 2020
Y. Peng et al. (Eds.): PRCV 2020, LNCS 12305, pp. 469–480, 2020.
https://doi.org/10.1007/978-3-030-60633-6_39

1 Introduction

With the development of earth observation technology and the improvement in the performance of computation processing of large-scale data, object detection in remote sensing images has attracted increasing attention [2,7,18,19,21]. However, for remote sensing images with complex backgrounds, small size and in the presence of occlusion, it is still a significant challenge. Deep learning algorithms have demonstrated good performance in object detection in recent years, including region based methods and single shot methods. e.g., faster region-based CNN (faster-RCNN), spatial pyramid pooling networks (SPP-Net) [5], you only look once (YOLO) [15] and the single-shot multibox detector (SSD) [10]. However, these innovations usually fail to detect very small objects, as small object features are always lost in the downsampling process of deep CNNs (DCNNs). The above DCNN methods cannot effectively extract the features of small objects. especially from remote sensing images in which objects usually small and blurry, which has created considerable challenges.

To increase the accuracy of the networks in detecting small objects in remote sensing images, one method is to employ features of the middle layers of the CNN and then exploit multiscale features with multilevel information. Ding et al. [2] directly concatenated multiscale features of a CNN to obtain fine-grained details for the detection of small objects. Lin et al. adopted pixelwise summation to incorporate the score maps generated by multilevel contextual features of different residual blocks for the segmentation of small objects such as a vehicle. Mou et al. [13] proposed a top-down pathway and lateral connection to build a feature pyramid network with strong semantic feature maps at all scales. The method assigned the feature maps of different layers to be responsible for objects at different scales. Yang et al. [21] introduced the dense feature pyramid network (DFPN) for automatic ship detection; each feature map was densely connected and merged by concatenation. Furthermore, Li et al. [7] proposed a hierarchical selective filtering layer that mapped features of multiple scales to the same scale space for ship detection at various scales. Gao et al. [4] designed a tailored pooling pyramid module (TPPM) to take advantage of the contextual information of different subregions at different scales. Qiu et al. proposed A2RMNet [14] to improve the problem of wrong positioning caused by excessive aspect ratio difference of objects in remote sensing images. Xie et al. proposed a fully connect network model NEOONet [20], which solved the problem of category imbalance in remote sensing images.

Although the above methods have achieved promising detection results by aggregating multiscale features, there are still some problems. When the objects are in the complex scene or partly exposed the previous methods can not extract discriminative features for detection.

To address these issues, First, we proposed a multiscale attention module to guide the network to learn more useful information in low-level feature maps with long-range dependencies, which is important for the detection of small objects. Then we presented a contextual attention module to extract interdependencies of foreground and background in feature maps, So the module can learn more of the correlated background and foreground information with long-range dependencies which can efficiently detect objects in complex scenes and in the presence of occlusion. In addition, we designed a channel attention module to model the importance of each feature map, which suppresses the background or unrelated category features with long-range dependencies, and improves the accuracy of the model.

2   Proposed Work

Previous methods cannot effectively extract the features of small objects because of the contradiction between resolution and high-level semantic information; i.e., low-level feature maps have high resolution but little semantic information, and high-level feature maps have low resolution but rich semantic information, which are all useful for accurately detecting objects. In addition, when an object is in a complex scene and in the presence of occlusion, the accuracy of the previous algorithms will decrease. Therefore, motivated by DANet [3], the proposed method uses a parallel self-attention module to enhance the accuracy of small object detection in remote sensing images.

As illustrated in Fig. 1, three types of attention modules are designed, and they contain a series of matrix operations to synthetically consider multiscale features, local contextual features and global features in the residual network after every residual block (resblock). In addition, in order to extract the feature of direction sensitive objects we use Deformable convolution and upsampling bottleneck operations to add deep feature maps to the shallow feature maps and we use four-scale feature maps to predict which are advantageous for detecting small objects. Before prediction, 3 3 convolutional layers are used to prevent aliasing effects.

2.1   Multiscale Attention Module

Feature maps of different scales have different semantic and spatial information; in high-level feature maps, semantic information are rich, and in the low-level feature maps, spatial information are rich, however, both types of information are useful for detecting small objects in remote sensing images. To strengthen the small object feature representation, a multiscale attention module is proposed to combine deep and shallow features. The structure of the multiscale attention module is illustrated in Fig. 2.

The feature maps of A and B are used obtained by resblock1 and resblock2 to calculate the attention map. H, W, and C represent the height, width and the number of channels of the feature maps respectively. $\mathbf{B_1} \in \mathbb{R}^{C \times H \times W}$, $\mathbf{B_2} \in \mathbb{R}^{2C \times H/2 \times W/2}$, 1×1 convolution is used to set B_2 to $\mathbf{B} \in \mathbb{R}^{C \times H/2 \times W/2}$. We reshape A to $\mathbb{R}^{C \times N}$ and B to $\mathbb{R}^{C \times N/4}$, and $N = H \times W$ is the number of pixels. Then, a matrix multiplication of the transpose of A and B is performed and a softmax layer is applied to normalize the weights and calculate the multiscale

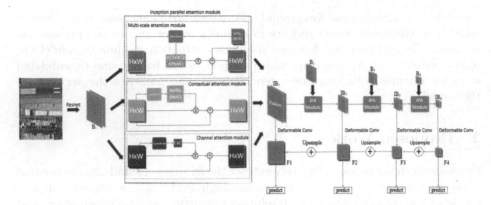

Fig. 1. The architecture of the IPAN. The proposed method use three parallel modules to obtain rich multiscale, contextual features and global features in different channels. Next, an element-wise summation operation is performed on the feature maps of the above three modules to obtain the final representations reflecting the long-range contexts. Finally, four scales with a multiscale feature fusion operation are used for prediction.

attention map $\mathbf{M} \in \mathbb{R}^{N \times H/4}$:

$$M_{ji} = \frac{\exp\left(A_i \cdot B_j\right)}{\sum_{i=1}^{N} \exp\left(A_i \cdot B_j\right)} \tag{1}$$

where M_{ji} measures the i th position's impact on the j th position at different scales. Then, we perform a matrix multiplication of B and the transpose of M and reshape the result to $R^{C \times H \times W}$. Finally, the result is multiplied by a scale parameter α and perform an elementwise summation is performed with A to obtain the final output $\mathbf{E} \in \mathbb{R}^{C \times H \times W}$ as follows:

$$E_j = \alpha \sum_{i=1}^{N} (M_{ji} B_i) + B_{1j} \tag{2}$$

The feature map B is deeper than A in resnet, which has more semantic information. Therefore, the attention map can guide B to learn more low-level information. As a result, E combines the low-level position information and high-level semantic information, which is effective for small object detection. From the heat map in E, we can see that more small object features are activated.

2.2   Contextual Attention Module

A discrimination of foreground and background features is essential for object detection and can be achieved by capturing contextual information. Some studies [6,17] have used contextual information to improve the detection result. To

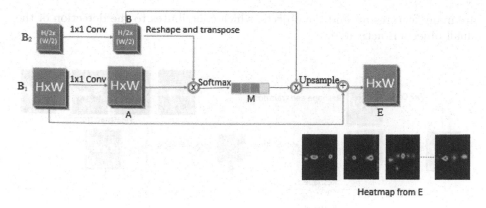

Heatmap from E

Fig. 2. The architecture of multiscale attention module. "×" represents matrix multiplication.

model rich contextual relationships over local features, we introduce a contextual attention module. The contextual attention module encodes a wide range of contextual information into the local features, thus enhancing the representation capability. Next, we elaborate the process to adaptively aggregate spatial contexts.

As illustrated in Fig. 3, given a local feature $\mathbf{B_1} \in \mathbb{R}^{C \times H \times W}$, the local feature is first fed into convolution layers to generate two new feature maps C and D the size of the filter is 7×7 (for example, B_1). With the deepening of the network, the filter sizes are 7×7, 5×5, and 3×3 at different scales, and large-scale filters can extract more contextual features, where $\mathbf{C}, \mathbf{D} \in \mathbb{R}^{C \times H \times W}$. Then, reshape them $\mathbb{R}^{C \times N}$, where $N = H \times W$ is the number of pixels. Then, a matrix multiplication of the transpose of C and B is performed, and a softmax layer is applied to calculate the contextual attention map $\mathbf{P} \in \mathbb{R}^{N \times N}$:

$$P_{ji} = \frac{\exp\left(C_i \cdot D_j\right)}{\sum_{i=1}^{N} \exp\left(C_i \cdot D_j\right)} \tag{3}$$

$$F_j = \beta \sum_{i=1}^{N} \left(P_{ji} B_{1i}\right) + B_{1j} \tag{4}$$

where M_{ji} measures the ith position's impact on the jth position with contextual information. The more similar feature representations of the two positions contribute to a greater correlation with contextual information between them. Then, we perform a matrix multiplication of B_1 and the transpose of M and reshape the result to $\mathbb{R}^{C \times H \times W}$. Finally, we multiply the result by a scale parameter β and perform an elementwise summation operation with the features B_1 to obtain the final output $\mathbf{F} \in \mathbb{R}^{C \times H \times W}$.

The attention map from B and C contains contextual information with long-range dependencies. From the heat map shown in F, it can be observed that there are many features around the objects, which contributes to the detection of the small objects is activated.

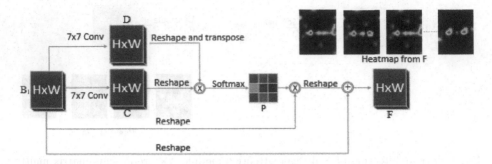

Fig. 3. The architecture of contextual attention module. "×" represents matrix multiplication.

2.3   Channel Attention Module

Each channel map has a different class and spatial position response, some contribute to object detection but others do not. To strengthen the positive response, weaken the negative response and exploit the interdependencies between channel maps, it can be emphasized that interdependent feature maps can improve the feature representation of the class and specific position. Therefore, we build a channel attention module to explicitly model the interdependencies between the channels and learn the long-range dependencies in the feature maps.

The structure of the channel attention module is illustrated in Fig. 4. Different from the contextual attention module, we directly calculate the channel attention map $\mathbf{M} \in \mathbb{R}^{C \times C}$ from the original features $\mathbf{B_1} \in \mathbb{R}^{C \times H \times W}$. Specifically, B_1 is reshaped to $\mathbb{R}^{C \times N}$, and then a matrix multiplication of original B_1 and the transpose of B_1 is performed. Finally, a softmax layer is applied to obtain the channel attention map $\mathbf{Q} \in \mathbb{R}^{C \times C}$:

$$Q_{ji} = \frac{\exp\left(B_{1i} \cdot B_{1j}\right)}{\sum_{i=1}^{C} \exp\left(B_{1i} \cdot B_{1j}\right)} \tag{5}$$

$$G_j = \gamma \sum_{i=1}^{N} \left(Q_{ji} B_{1i}\right) + B_{1j} \tag{6}$$

where M_{ji} measures the i th channel's impact on the j th channel. In addition, we perform a matrix multiplication of the transpose of M and B_1 and reshape the result to $\mathbb{R}^{C \times H \times W}$. Then, we multiply the result by a scale parameter γ and perform an elementwise summation operation with B_1 to obtain the final output $\mathbf{G} \in \mathbb{R}^{C \times H \times W}$. From the heat map shown in G from Fig. 4, we can see that more features strongly associated with the objects are activated.

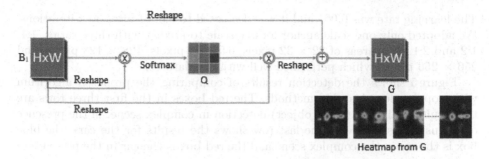

Fig. 4. The architecture of channel attention module. "×" represents matrix multiplication.

2.4 Embedding with Networks

As shown in Fig. 1, we set our parallel attention module after every resblock. After every IPA module, the feature maps have multiscale contextual and global information with long-range dependencies. Positive feature response achieves mutual gains, thus increasing differences in the objects and backgrounds. Then, every scale feature map is fed to RPN [16] for a further prediction. In addition, the model can exploit spatial information at all corresponding positions to model the channel correlations. Notably, the proposed attention modules are simple and can be directly inserted into the existing object detection pipeline. The modules do not require too many parameters yet effectively strengthen the feature representations.

3 Experiment

We performed a series of experiments with RSOD [11] including objects that are small in a complex scene and in the presence of occlusion. The proposed method achieved state-of-the-art performance: 95.9% for aircrafts and 95.5% for cars. We first conducted evaluations of the remote sensing object detection dataset to compare the performance to that of other state-of-the-art object detection methods, and evaluated the robustness of the proposed method using the COCO [9] dataset.

3.1 Comparison with State-of-the-Arts Methods in Remote Sensing Object Detection Dataset

We chose the ResNet-101 model as the backbone network. The model was initialized by the ImageNet classification model and fine-tuned with the RSOD [11]. We randomly split the samples into 80% for training and 20% for testing. In all experiments, we trained and tested the proposed method based on the TensorFlow deep learning framework. First, we resized the images to 800 × 800 pixels and applied stochastic gradient descent for 50k iterations to train our model.

The learning rate was 0.001 and it was decreased to 0.0001 after 30 k iterations. We adopted only one scale anchor for one scale to predict with three ratios, 1:1, 1:2 and 2:1, with areas of 32 × 32 pixels, 64 × 64 pixels, 128 × 128 pixels, and 256 × 256 pixels, which performed well with the dataset.

Figure 5 shows the detection results of comparing the proposed algorithm with popular deep learning methods. The red boxes in the first three rows are the results of aircraft (small object) detection in complex scene, in the presence of occlusion respectively. The last row shows the results for the cars, the blue box is the car in the complex scene and the red box is the car in the presence of occlusion.

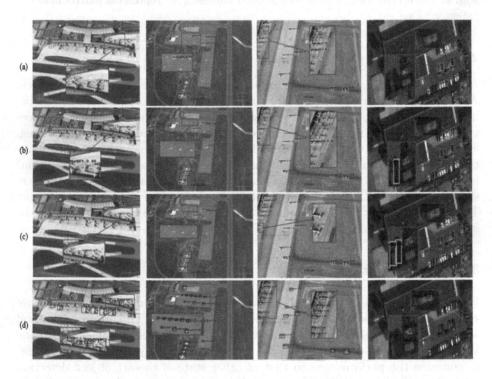

Fig. 5. Visualization of the small object detection results for a complex scene in the presence of occlusion for the proposed algorithm with popular deep learning methods on the aircraft and car datasets: (a) is the results of FPN [8], (b) is the results of Modified faster-RCNN [17], (c) is the results of R-FCN [1] and (d) the proposed method. (Color figure online)

Firstly, we compared the proposed approach with seven common state-of-the-art methods for object detection and five state-of-the-art methods for remote sensing object detection. As shown in Table 1, the deep learning method is obviously better in terms of accuracy than traditional methods such as the HOG [19] and SIFT [18]. Because of the superiority of the proposed method in small object detection and the strong ability to handle complex scene with occlusion, the proposed method has higher accuracy and robustness when detecting aircrafts and cars. From the AP and recall (R) shown in Table 1, we can observe that in terms of AP over all the aircraft and the car categories, the proposed approach improve about 7% on average to other deep learning methods, and outperforms the previously most accurate USB-BBR [11] method and AARMN [14] method by 1% and 1%, respectively. The proposed method can obtain the highest AP on the premise of relatively high recall.

Table 1. Compares the proposed method to other methods in the remote sensing object detection datasets. The symbol "–" indicates that the method does not provide relevant results.

Methods	Aircraft AP (%)	Aircraft R (%)	Car AP (%)	Car R (%)
YOLO [15]	84.92	82.81	71.73	70.93
faster-RCNN [16]	80.15	78.91	82.52	81.21
FPN [8]	91.22	90.18	85.12	86.33
R-FCN [1]	84.3	95.26	89.3	88.2
USB-BBR [11]	94.69	93.09	–	–
NMMDPN [12]	–	–	91.36	90.56
NEOON [20]	94.49	–	93.22	–
MFast-RCNN [17]	84.5	85.1	80.1	80.5
AARMN [14]	94.27	94.18	94.65	92.16

3.2 Ablation Study

An ablation study is performed to validate the contribution of each module of the proposed method. In each experiment, one part of our method is omitted, and the remaining parts are retained. The APs are listed in Table 2. It can be seen that the AP increased by almost 1% by including the contextual attention module and the channel attention module, because the contextual attention module contains contextual information about the object. Second, we used the proposed channel attention module and the multiscale attention module after the resblock, and the result improved by 2.1%. Which indicates that global information and multiscale information are important for detection. Finally, we interposed the contextual attention module and the multiscale attention module into the network, and the AP improved by 2.3%. Thus, the contextual information and multiscale information are effective in helping the network detect small objects. It is obvious that when all modules are applied, the proposed method can improve the AP by 4.7%.

Figure 6 shows the visualized results of the ablation study for situations 1–5 in Table 3, corresponding to (b)–(f). It can be easily seen that due to the use of the multiscale attention module, the proposed method obtains more spatial structural information and semantic information about small objects, so it can enhance the feature representation of small objects. The contextual attention module and the channel attention module enable the network to learn the correspondence between the background and foreground and the global interdependence in different channels and further strengthen the feature representation, so the network can efficiently detect the objects in a complex scenes and in the presence of occlusion.

Table 2. The result of the ablation study for the proposed framework in the aircraft detection task. N: No, Y: Yes.

IPAN	1	2	3	4	5
Multiscale attention	N	N	Y	Y	Y
Contextual attention	N	Y	N	Y	Y
Channel attention	N	Y	Y	N	Y
Aircraft AP(%)	91.2	92.1	93.3	93.5	95.9

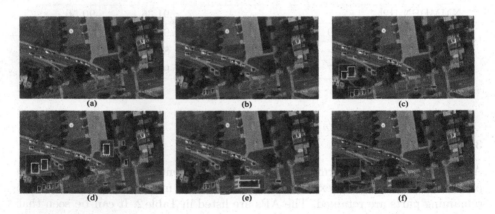

Fig. 6. The visualized results of the ablation study. (a) is the original image and (b)–(f) are situations from "1" to "5" in Table 2.

3.3 Robustness Experiments

In order to verify the robustness of the proposed method, we evaluated the performance with the COCO [9] dataset. As shown in Table 3, the dataset has more classes and considerably smaller objects so the proposed algorithm has better performance than other methods.

Table 3. Compares the proposed method in the COCO dataset to the state-of-the-art methods.

Methods	COCO mAP(%)
R-FCN [1]	29.9Y
FPN [8]	36.2
SSD [10]	31.2
YOLO [15]	33
Proposed method	38.3

4 Conclusion

In this paper, an IPAN is presented for small object detection in remote sensing images, which enhances the feature representation and improves the detection accuracy for small objects, especially in complex scenes and in the presence of occlusion. Specifically, we introduced three parallel modules: a multiscale attention module guiding the model extract more information of small objects, a contextual attention module to capture contextual correlation, and a channel attention module to learn global interdependencies in different channels, In addition, the IPAN can capture long-range dependencies which helps to detect objects. The experiments and ablation study show that the network is effective and achieve precise detection results. The proposed IPAN consistently achieves outstanding performance with remote sensing object detection datasets. The future work will focus on further enhancing the robustness of our model.

References

1. Dai, J., Li, Y., He, K., Sun, J.: R-FCN: object detection via region-based fully convolutional networks. In: Advances in Neural Information Processing Systems, pp. 379–387 (2016)
2. Ding, J., Chen, B., Liu, H., Huang, M.: Convolutional neural network with data augmentation for SAR target recognition. IEEE Geosci. Remote Sens. Lett. **13**(3), 364–368 (2016)
3. Fu, J., et al.: Dual attention network for scene segmentation. In: Proceedings of the IEEE Conference on Computer Vision and Pattern Recognition, pp. 3146–3154 (2019)
4. Gao, X., et al.: An end-to-end neural network for road extraction from remote sensing imagery by multiple feature pyramid network. IEEE Access **6**, 39401–39414 (2018)
5. Kaiming, H., Xiangyu, Z., Shaoqing, R., Jian, S.: Spatial pyramid pooling in deep convolutional networks for visual recognition. IEEE Trans. Pattern Anal. Mach. Intell. **37**(9), 1904–1916 (2015)
6. Li, K., Cheng, G., Bu, S., You, X.: Rotation-insensitive and context-augmented object detection in remote sensing images. IEEE Trans. Geosci. Remote Sens. **56**(4), 2337–2348 (2017)

7. Li, Q., Mou, L., Liu, Q., Wang, Y., Zhu, X.X.: HSF-NET: multiscale deep feature embedding for ship detection in optical remote sensing imagery. IEEE Trans. Geosci. Remote Sens. **56**(12), 7147–7161 (2018)
8. Lin, T.Y., Dollar, P., Girshick, R., He, K., Hariharan, B., Belongie, S.: Feature pyramid networks for object detection. In: Proceedings of the IEEE Conference on Computer Vision and Pattern Recognition, pp. 2117–2125 (2017)
9. Lin, T.-Y., et al.: Microsoft COCO: common objects in context. In: Fleet, D., Pajdla, T., Schiele, B., Tuytelaars, T. (eds.) ECCV 2014. LNCS, vol. 8693, pp. 740–755. Springer, Cham (2014). https://doi.org/10.1007/978-3-319-10602-1_48
10. Liu, W., et al.: SSD: single shot MultiBox detector. In: Leibe, B., Matas, J., Sebe, N., Welling, M. (eds.) ECCV 2016. LNCS, vol. 9905, pp. 21–37. Springer, Cham (2016). https://doi.org/10.1007/978-3-319-46448-0_2
11. Long, Y., Gong, Y., Xiao, Z., Liu, Q.: Accurate object localization in remote sensing images based on convolutional neural networks. IEEE Trans. Geosci. Remote Sens. **55**(5), 2486–2498 (2017)
12. Ma, W., Guo, Q., Wu, Y., Zhao, W., Zhang, X., Jiao, L.: A novel multi-model decision fusion network for object detection in remote sensing images. Remote Sensing **11**(7), 737 (2019)
13. Mou, L., Zhu, X.X.: Vehicle instance segmentation from aerial image and video using a multitask learning residual fully convolutional network. IEEE Trans. Geosci. Remote Sens. **56**(11), 6699–6711 (2018)
14. Qiu, H., Li, H., Wu, Q., Meng, F., Ngan, K.N., Shi, H.: A2RMNet: adaptively aspect ratio multi-scale network for object detection in remote sensing images. Remote Sens. **11**(13), 1594 (2019)
15. Redmon, J., Divvala, S., Girshick, R., Farhadi, A.: You only look once: unified, real-time object detection. In: Proceedings of the IEEE Conference on Computer Vision and Pattern Recognition, pp. 779–788 (2016)
16. Ren, S., He, K., Girshick, R., Sun, J.: Faster R-CNN: towards real-time object detection with region proposal networks. In: Advances in Neural Information Processing Systems, pp. 91–99 (2015)
17. Ren, Y., Zhu, C., Xiao, S.: Small object detection in optical remote sensing images via modified faster R-CNN. Appl. Sci. **8**(5), 813 (2018)
18. Singh, B., Davis, L.S.: An analysis of scale invariance in object detection snip. In: Proceedings of the IEEE Conference on Computer Vision and Pattern Recognition, pp. 3578–3587 (2018)
19. Xiao, Z., Liu, Q., Tang, G., Zhai, X.: Elliptic Fourier transformation-based histograms of oriented gradients for rotationally invariant object detection in remote sensing images. Int. J. Remote Sens. **36**(2), 618–644 (2015)
20. Xie, W., Qin, H., Li, Y., Wang, Z., Lei, J.: A novel effectively optimized one-stage network for object detection in remote sensing imagery. Remote Sens. **11**(11), 1376 (2019)
21. Yang, X., et al.: Automatic ship detection in remote sensing images from google earth of complex scenes based on multiscale rotation dense feature pyramid networks. Remote Sens. **10**(1), 132 (2018)

Multi-human Parsing with Pose and Boundary Guidance

Shuncheng Du, Yigang Wang(✉), and Zizhao Wu

School of Media and Design, Hangzhou Dianzi University, Hangzhou, China
`yigang.wang@hdu.edu.cn`

Abstract. In this work, we present a novel end-to-end semantic segmentation framework for multi-human parsing, which integrates both the high- and low-level features. Our framework includes three modules: segmentation module, pose estimation module, and boundary detection module. The pose estimation model will guide high-level feature extraction to identify the part location; meanwhile, the boundary detection module will concentrate on the low-level feature extraction so as to distinguish the boundary. Both modules are united in the backbone segmentation module to generate the desired accurate high-resolution prediction for multi-human parsing. Experiment results on the PASCAL-Person-Part dataset demonstrate that our method achieves superior results over state-of-the-art methods. Code has been made available at https://github.com/scmales/MHP-with-Pose-and-Boundary-Guidance.

Keywords: Human parsing · Pose estimation · Multi-task learning

1 Introduction

Human parsing refers to assigning image pixels from the human body to a semantic category, e.g. hair, dress, leg. This task can also be considered as fine-grained semantic segmentation for human body. At present, human parsing is critical to the understanding of humans and supports lots of industrial applications, such as virtual try-on [1], dressing style recognition [2], human behavior analysis [3], and so on.

Recently, with the advance of deep convolutional neural networks, multi-human parsing has attracted a huge amount of interest [4,5] and achieved great progress. Most of the previous works [6,7] focus on single-human parsing, which simplifies the scene and ignores the human interaction, occlusion, and rich human body posture. However, in the real complex scenarios, there exist the diverse background composition, the physical contact between people, and the occlusion between human body parts, these kinds of approaches may fail to deal with these issues. As shown in Fig. 1, the inaccurate detection of edge and large areas of missing body parts are common problems.

The first author is a student.

© Springer Nature Switzerland AG 2020
Y. Peng et al. (Eds.): PRCV 2020, LNCS 12305, pp. 481–492, 2020.
https://doi.org/10.1007/978-3-030-60633-6_40

To address these obstacles, researchers [8–11] have further studied on the multi-human parsing problem, and significant progress have been made. For example, researchers [8,9] have tried to pre-process the scene by detecting each person, and then performing human parsing individually, thus reducing the complexity of the problem. Following these approaches, researchers have also found that human parsing and human pose estimation are two popular applications of human body analysis [12–14], despite the differences in detail, these two tasks are highly correlated and complementary. On one hand, pose estimation focuses on the joint detection of human body, which can be regarded as the high-level abstraction of human body, while neglecting the other details. On the other hand, human parsing considers the categories of every pixel, while flaws in modeling high-level abstraction. Inspired from these points, they suggested to jointly analyze both tasks [8,10,15].

In this paper, rather than mutual learning human parsing and pose estimation, we propose an end-to-end unified framework for simultaneous human parsing, pose estimation, and boundary detection. Specifically, our framework includes three modules, they are semantic segmentation module, pose estimation module, and edge detection module. These modules are unified with our framework, i.e., in the forward propagation, we obtain the multi-resolution features of the pose estimation module and the edge detection module. Then, we integrate these features as the residual input to the semantic segmentation module based on the encoder-decoder structure, to achieve the accurate high-resolution multi-human parsing results. We evaluate our method on the PASCAL-Person-Part dataset, as well as the ablation studies, which demonstrate that our method is competitive among existing multi-human parsing models.

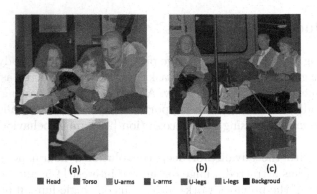

(a)   (b)   (c)

■ Head   ■ Torso   ■ U-arms   ■ L-arms   ■ U-legs   ■ L-legs   ■ Backgroud

Fig. 1. The problem of multi-human parsing. (a) and (b) show the inaccurate edges for segmentation. (c) shows the lack of human body parts.

2   Related Work

2.1   Human Parsing

Human parsing aims to segment a human image into multiple semantic parts, which can also be considered as fine-grained semantic segmentation for human body. Benefiting from the success of deep learning and convolutional neural networks, significant progress have been achieved. For example, Fully Convolutional Network (FCN) [16] had made the first attempt to realize the end-to-end training of semantic segmentation. More recent works [17,18] have been proposed based on a variety of encoder-decoder networks. Among the existing models, the representation learning of the network played a key role to the results. Ideally, the learned feature representation should capture both the global and local information of the image so as to make an accurate prediction. In order to achieve this goal, some specific modules have been devised to aggregate the context features on different regions, such as Atrous Spatial Pyramid Pooling (ASPP) module [19], Pyramid Scene Parsing (PSP) module [20]. More recently, researchers focused their attention on the concrete human parsing problem by exploiting human parsing benchmark dataset [21,22] and explored the potential connections between human body. For example, Liu et al. suggested a multi-path enhance network [5], which extracts multi-dimensional feature based on the different scales of input pictures to enhance the final results, but the complexity of the network is high, which leads the network hard to train.

2.2   Pose Estimation and Edge Detection

There are abundant of works [23,24] have been investigated to the pose estimation problem. In the early stage of this task, CNN has been used to get the joint coordinates directly [25], but the results are not very satisfying. Later, Wei et al. [26] suggest a novel framework which combined the ideas of multi-stage and post-process refinement, to solve the problem of partial occlusion and invisibility. Following their work, Cao et al. [27] proposed a real-time multi-human pose estimation network based on the Part Affinity Fields (PAFs), which abstracts the connections between the joints. This network carried out human joint detection, the regression training, and then conducted the Hungarian algorithm to separate the different joints. Different from this method, researchers [28,29] suggested to firstly perform the human detection to separate the single human, and then performed the pose estimation individually, thus cast the multi-human pose estimation into the traditional single human pose estimation problem.

Edge detection has also been regarded as an important research field of image processing. The current mainstream work is based on CNN due to its high performance. For example, Xie et al. [30] proposed a holistically-nested edge detection method, which implements the end-to-end training of edge detection through a full convolution neural network and depth supervision, significant performance has been achieved of this method.

2.3   Multi-task Learning

Multi-task learning aims to learn multiple different tasks simultaneously while maximizing performance on one or all of the tasks. In the human parsing field, researchers [31–33] noticed that human pose estimation and human parsing are related tasks since both problems are strongly dependent on the human body representation and analysis, which inspired them to propose many multi-task learning approaches in the field. Particularly, Xia et al. [8] unified the pose estimation and semantic segmentation in their framework, Conditional Random Field (CRF) was further employed to fuse both tasks, and obtained significant performance. Gong et al. [10] explored a new self-supervising structure-sensitive learning method that does not require additional monitoring information but the joint coordinates, to improve the analytical results. Liang et al. [11] presented a novel joint human parsing and pose estimation network which incorporates the multi-scale feature connections and iterative location refinement in an end-to-end framework, to investigate efficient context modeling to facilitate the task of multi-task learning.

There exist some methods [34,35] aim to proceed with edge detection and semantic segmentation simultaneously, while maximizing performance on one or all of the tasks. Among them, Ruan et al. [36] suggested a Context Embedding with Edge Perceiving (CE2P) framework which unifies several useful properties, including feature resolution, global context information, edge details, and leverage them to benefit the human parsing task.

In this paper, comparing with the traditional method [21,22,36], we fuse the multi-level features by using the multi-task learning structure. What's more, comparing with multi-task methods [8,11], our advantage is that we construct a synchronous end-to-end network which is easy to train and fuse multiple related tasks for multi-human parsing.

3   Algorithm

Our architecture consists of three modules: backbone semantic segmentation modules, pose estimation module, edge detection module, and semantic segmentation module. Figure 2 illustrates the details. In our architecture, we employ the Residual Network (ResNet-101 [37]) as the backbone architecture. The output of the backbone network is fed into the followed three modules to learn features globally and locally.

Given an input picture with a fixed size of $H \times W \times 3$, the final output of the entire model is the result of semantic segmentation defined as P, which contains c channels of confidence maps, and is consistent with the number of human semantic segmentation labels. Our network is multi-module based, which will generate additional outputs include m channels of the Part Confidence Maps (PCMs) K, n channels of the Part Affinity Fields (PAFS) F from the pose estimation module, and two channels of edge confidence map E from the edge detection module, where m denotes the number of labels for joints and background, and n

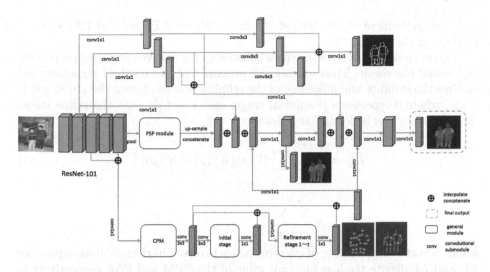

Fig. 2. The framework of our approach for multi-task learning. The architecture has three modules: semantic segmentation module, pose estimation module, and edge detection module.

denotes the double number of part vectors between the joints. For more details about PAFs, we refer the reader to OpenPose [27].

In our case, we set $c = 7, m = 15, n = 28$ to the PASCAL-Person-Part dataset, the values are defined as is shown in Fig. 2, given the output K and F of the pose estimation module, we stitch it with the output features of the pose estimation layer, and further concatenate them with the intermediate features of the edge detection network, finally, they are fed into the semantic segmentation network to guide the generation of human parsing results.

In the following, we describe the details of the modules in our network.

3.1 Pose Estimation Module

We use the bottom-up strategy to design the structure of the module. To reduce the cost of training, the depth-wise separable convolution layers have been introduced to construct a light-weight pose estimation network [38].

Figure 2 illustrates the structure of our network, we simplify the description of refinement stages and only retain the last feature representation of PCMs and PAFs. Specifically, the module takes the features extracted from the third and last layer of ResNet-101 as the input, then we proceed the input through the conventional pose machine (CPM) layer, the initial stage and a total number of t refinement stages, these three stages are composed in the convolutional layers and the exponential linear units layers. Furthermore, the input of each refinement stage is the concatenated features between the output of the previous stage and the output of the CPM layer. PCMs and PAFs will be output at the initial stage and each refinement stage. We get the final output of PCMs and PAFs of this module on the final layer.

The purpose of this structure is to construct an iterative optimization process and refine the result. Thus exploit the potential of the residual structure and ensures the stability and efficiency of the refining process. Given the $s(0 \leq s \leq t)$ stage, where 0 represents the initial stage, and $1 \sim t$ denotes the refine stage, the regression loss is defined as follows:

$$L_K^s = \frac{1}{N} \sum_{i=0}^{m} \sum_{p} W(p) \|K_i^s(p) - K_i'(p)\|^2 \tag{1}$$

$$L_F^s = \frac{1}{N} \sum_{i=0}^{n} \sum_{p} W(p) \|F_i^s(p) - F_i'(p)\|^2 \tag{2}$$

where N is the total number of effective pixels involved in the calculation process; L_K^s and L_F^s denote the loss function value of the PPM and PAF respectively in phase s; i represents the $i-th$ channel; p denotes the location of an image; W is a binary mask with $W(p) = 0$ when the pixel is added during data enhancement at location p; $K_i^s(p)$ represents the joint confidence maps prediction value in the i-th channel at the location p; $K_i'(p)$ denotes the label value as location p. The variables related to PAFs can also be described similarly.

In summary, the loss function of our pose estimation module is defined as:

$$L_{pose} = \sum_{s=0}^{t} (L_K^s + L_F^s) \tag{3}$$

3.2 Edge Detection Module

In the edge detection module, we aim to extract features in different spatial dimensions and enhance the performance of the edge detection, thus we emphasize more on the low-level features. Our module takes the output of the three modules in ResNet-101 as the input. As shown in Fig. 2, three modules of input are processed separately; then they have been fused in the last convolutional layer; finally, we obtain the output of the edge detection. In summary, the edge detection module consists of convolutional submodules, where each submodule contains the convolutional layer, the batch normalization layer, and the rectified linear unit layer. Overall, the edge detection can be regarded as the binary classification problem. We construct the loss function of edge L_{edge} based on cross-entropy loss of category balance.

3.3 Semantic Segmentation Module

The Semantic Segmentation module is based on the encoder-decoder framework. As shown in Fig. 2, in the encoder stage, PSP module [20] is employed as the method to obtain context features of the image through four different scale pooling operations, to effectively increase the network receptive field, and then samples and splices the multi-scale features, and finally sends them to the decoder stage.

The content of the decoder stage is composed of several convolutional submodules, which contain the convolution layer, the normalization layer, and the rectified linear unit layer. Specifically, we use edge features and pose features for feature enhancement, and exploit the residual structure to ensure the learning performance. As shown in Fig. 2, we observe that the edge features are the concatenated features from all convolution outputs of the first layer of the edge detection module; the pose features are the concatenated features from the output of CPM layer, PCMs, and PAFs. The characteristics of feature enhancement are reflected as follows: firstly, the input to the first convolutional module is based on the output of the PSP module, the second layer of ResNet-101, and the residual features; secondly, the input to the intermediate convolutional submodule is based on the output of the last layer, the residual of edge module, and the residual of pose module. We argue that the features of pose estimation and semantic segmentation correspond to high-level semantics. To learn the transformation correlation between the high-level semantics better, we use the residual features twice.

Figure 3 illustrates PCMs, PAFs, the intermediate features of edge detection, and their analyzable visualization after feeding input to the model. We can observe that the PCMs and PAFs are able to identify the abstract location of the individual body parts, and thus provide prior information for human part segmentation. Besides, the intermediate features of edge detection contain rich details of human body parts and can effectively supplement the details deficiency in semantic segmentation.

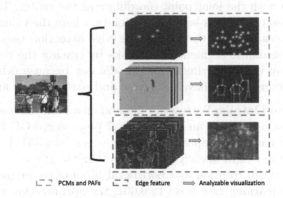

⊏⊐ PCMs and PAFs   ⊏⊐ Edge feature   ⟹ Analyzable visualization

Fig. 3. A figure illustrating PCMs, PAFs and edge features.

The output of the module is the maps of human body with the model parameters. So the total loss of semantic segmentation module L_{seg} is a sum of two components, which is similarly defined as multi-class cross-entropy loss.

Finally, the loss defined on the whole model is defined as:

$$L = \alpha L_{pose} + \beta L_{edge} + \gamma L_{seg} \tag{4}$$

where α, β, γ are the weights defined on different losses. L_{edge} and L_{seg} belong to classification loss, and it has been proved in the paper [36] that adding the two in the same proportion is effective. Besides, inspired by the paper [39], we unify the regression loss L_{pose} and classification loss L_{seg} to the same order of magnitude in that avoiding the loss with a small gradient being carried away by the loss with a large gradient. We train our end-to-end model by minimizing the above loss function.

4   Experiments

Dataset. We evaluate our results on the PASCAL-Person-Part dataset [40], which is developed for human parsing evaluation and comes from PASCAL VOC dataset. The training set of PASCAL-Person-Part contains 1716 images, with a validate set contains 1817 pictures. Each image corresponds to a labeled image which is regarded as the benchmark segmentation results. Besides the segmentation results, the dataset also contains the pose results [8]. But we found that the open-source dataset lacked 40 pose labels for training and verification, so we removed them according to the method [8]. Overall, there are 7 semantic segmentation categories in the dataset, including background, head, trunk, upper arm, lower arm, upper leg, and lower leg; there also exist 14 types of joint point coordinates, including head, neck, left shoulder point, left elbow joint, left wrist, left hip, left knee, left ankle, right shoulder point, right elbow joint, right wrist, right hip point, right knee, and right ankle. The label of the PCMs is from the Gaussian distribution with the joint point coordinate as the center. The labels of the PAFs are composed of vectors between the joints, where the value can be defined as the L_2 distances between the joints. The edge detection tags mentioned are generated from semantic segmentation masks by tracing the 8-pixel neighborhood changes. We only experiment on this dataset because other datasets for multi-human parsing do not have both pose and part segment annotations.

Implementation Details. We test our method on a PC, which is equipped with an Ubuntu operating system, an Intel 4 GHz i7 processor, 8 GB RAM, and RTX 2080 Ti graphics card. The input size of images is set to 384. For our joint loss function, we set the weight of each term as $\alpha = 0.5; \beta = 1; \gamma = 1$. To augment the data, we also perform rotation and horizontal flipping to augment the original data. The initial learning rate is set to 0.001, the optimization strategy used in this work is the stochastic gradient descent (SGD) with a linear learning rate decay. We also employed the two norm regularization, the weight decay value is set to 0.0005.

Ablation Studies. We further evaluate the effectiveness of coarse-to-fine schemes of our model, including three situations as follows: only semantic segmentation module, edge detection module with semantic segmentation module, the complete architecture with three modules. Figure 4 illustrates the comparison between the prediction with the ground-truth. For instance, as we can see from the second row, the result of Fig. 4(b) shows that the foot is missing or even detected by mistake, the result of Fig. 4(c) is better than that of Fig. 4(b), while the result of Fig. 4(d) correctly predicts the foot. Then from the third row of the figure, the accuracy of the three experimental results is also increasing compared with the part of the upper arm of the human body covered by the dog. In the same way, the results of the third image also show that the integrity of the prediction results of the network with pose estimation and edge detection is better than that of the other strategies after ablation study.

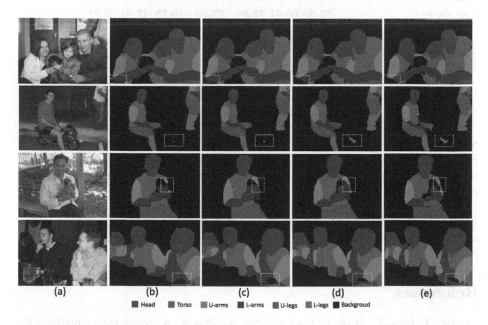

Fig. 4. Visual comparison of ablation study. (a) input images. (b) predictions based solely on the semantic segmentation module. (c) predictions based on semantic segmentation with edge detection module. (d) predictions based on complete architecture. (e) ground truth.

Results. The evaluation criterion used in this paper is mainly based on the Mean Intersection Over Union (MIoU) of each class. The higher the values are, the more accurate the results are. We conduct the prediction and the evaluation of the validation set of the PASCAL-Person-Part dataset. The experiment results are shown in Table 1. We note that our model achieves superior performance, when comparing with the existing mainstream algorithms [4,7,8,10,15,19], the MIoU of our model is 66.11%. We also note that our method gets the highest score on the cross-ratio of all body parts. The result demonstrates the efficiency of our method for multi-human parsing.

Table 1. Mean pixel IOU (mIOU) of human parsing on PASCAL-Person-Part.

Method	Head	Torso	U-arms	L-arms	U-legs	L-legs	Background	Ave
Attention [4]	81.47	59.06	44.15	42.50	38.28	35.62	93.65	56.39
HANZ [7]	80.76	60.50	45.65	43.11	41.21	37.74	93.78	57.54
LIP-SSL [10]	83.26	62.40	47.80	45.58	42.32	39.48	94.68	59.36
DeepLabv3* [19]	84.06	66.96	54.26	52.80	48.08	43.59	94.79	63.50
Joint [8]	85.50	67.87	54.72	54.30	48.25	44.76	95.32	64.39
MuLA [15]	–	–	–	–	–	–	–	65.10
Our model (w/o pose/edge)	84.98	66.85	54.52	53.72	47.77	44.22	95.13	63.88
Our model (w/o pose)	85.45	68.28	56.78	56.63	49.31	46.56	**95.34**	65.48
Our model	**85.99**	**68.61**	**58.08**	**57.50**	**49.78**	**47.61**	95.23	**66.11**

5   Conclusion

In this work, we have proposed a multi-human parsing algorithm that integrates the tasks of pose estimation and edge detection, and realized synchronous end-to-end learning. Specifically, our method has suggested multi-module architecture, which composes of the pose estimation module and edge detection module, and provides the auxiliary features for the multi-human parsing tasks. The experiment results in the PASCAL-Person-Part dataset demonstrated the effectiveness of our method.

Acknowledgements. This work was supported in part by the National Natural Science Foundation of China (61602139), and in part by the Zhejiang Province Science and Technology Planning Project (2018C01030).

References

1. Lin, J., Guo, X., Shao, J., Jiang, C., Zhu, Y., Zhu, S.: A virtual reality platform for dynamic human-scene interaction. In: SIGGRAPH ASIA, Virtual Reality meets Physical Reality: Modelling and Simulating Virtual Humans and Environments, pp. 11:1–11:4. ACM (2016)
2. Wang, Y., Tran, D., Liao, Z., Forsyth, D.A.: Discriminative hierarchical part-based models for human parsing and action recognition. J. Mach. Learn. Res. **13**, 3075–3102 (2012)
3. Liang, X., Wei, Y., Shen, X., Yang, J., Lin, L., Yan, S.: Proposal-free network for instance-level object segmentation. CoRR abs/1509.02636 (2015)
4. Chen, L., Yang, Y., Wang, J., Xu, W., Yuille, A.L.: Attention to scale: scale-aware semantic image segmentation. CoRR abs/1511.03339 (2015)
5. Lin, G., Milan, A., Shen, C., Reid, I.D.: RefineNet: multi-path refinement networks for high-resolution semantic segmentation. CoRR abs/1611.06612 (2016)
6. Liang, X., et al.: Human parsing with contextualized convolutional neural network. IEEE Trans. Pattern Anal. Mach. Intell. **39**(1), 115–127 (2017)
7. Xia, F., Wang, P., Chen, L., Yuille, A.L.: Zoom better to see clearer: human part segmentation with auto zoom net. CoRR abs/1511.06881 (2015)

8. Xia, F., Wang, P., Chen, X., Yuille, A.L.: Joint multi-person pose estimation and semantic part segmentation. CoRR abs/1708.03383 (2017)
9. Zhao, J., et al.: Understanding humans in crowded scenes: deep nested adversarial learning and a new benchmark for multi-human parsing. CoRR abs/1804.03287 (2018)
10. Gong, K., Liang, X., Shen, X., Lin, L.: Look into person: self-supervised structure-sensitive learning and a new benchmark for human parsing. CoRR abs/1703.05446 (2017)
11. Liang, X., Gong, K., Shen, X., Lin, L.: Look into person: joint body parsing & pose estimation network and a new benchmark. IEEE Trans. Pattern Anal. Mach. Intell. **41**(4), 871–885 (2019)
12. Bourdev, L.D., Malik, J.: Poselets: body part detectors trained using 3D human pose annotations. In: IEEE 12th International Conference on Computer Vision, pp. 1365–1372. IEEE Computer Society (2009)
13. Dong, J., Chen, Q., Xia, W., Huang, Z., Yan, S.: A deformable mixture parsing model with parselets. In: IEEE International Conference on Computer Vision, pp. 3408–3415. IEEE Computer Society (2013)
14. Yang, Y., Ramanan, D.: Articulated pose estimation with flexible mixtures-of-parts. In: The 24th IEEE Conference on Computer Vision and Pattern Recognition, pp. 1385–1392. IEEE Computer Society (2011)
15. Nie, X., Feng, J., Yan, S.: Mutual learning to adapt for joint human parsing and pose estimation. In: Ferrari, V., Hebert, M., Sminchisescu, C., Weiss, Y. (eds.) ECCV 2018. LNCS, vol. 11209, pp. 519–534. Springer, Cham (2018). https://doi.org/10.1007/978-3-030-01228-1_31
16. Long, J., Shelhamer, E., Darrell, T.: Fully convolutional networks for semantic segmentation. CoRR abs/1411.4038 (2014)
17. Ronneberger, O., Fischer, P., Brox, T.: U-Net: convolutional networks for biomedical image segmentation. CoRR abs/1505.04597 (2015)
18. Badrinarayanan, V., Kendall, A., Cipolla, R.: SegNet: a deep convolutional encoder-decoder architecture for image segmentation. IEEE Trans. Pattern Anal. Mach. Intell. **39**(12), 2481–2495 (2017)
19. Chen, L., Papandreou, G., Schroff, F., Adam, H.: Rethinking atrous convolution for semantic image segmentation. CoRR abs/1706.05587 (2017)
20. Zhao, H., Shi, J., Qi, X., Wang, X., Jia, J.: Pyramid scene parsing network. CoRR abs/1612.01105 (2016)
21. Liang, X., et al.: Deep human parsing with active template regression. IEEE Trans. Pattern Anal. Mach. Intell. **37**(12), 2402–2414 (2015)
22. Liu, S., et al.: Matching-CNN meets KNN: quasi-parametric human parsing. CoRR abs/1504.01220 (2015)
23. Güler, R.A., Neverova, N., Kokkinos, I.: DensePose: dense human pose estimation in the wild. CoRR abs/1802.00434 (2018)
24. Li, J., Wang, C., Zhu, H., Mao, Y., Fang, H., Lu, C.: CrowdPose: efficient crowded scenes pose estimation and a new benchmark. CoRR abs/1812.00324 (2018)
25. Toshev, A., Szegedy, C.: DeepPose: human pose estimation via deep neural networks. CoRR abs/1312.4659 (2013)
26. Wei, S., Ramakrishna, V., Kanade, T., Sheikh, Y.: Convolutional pose machines. CoRR abs/1602.00134 (2016)
27. Cao, Z., Hidalgo, G., Simon, T., Wei, S., Sheikh, Y.: OpenPose: realtime multi-person 2D pose estimation using part affinity fields. CoRR abs/1812.08008 (2018)
28. Fang, H., Xie, S., Lu, C.: RMPE: regional multi-person pose estimation. CoRR abs/1612.00137 (2016)

29. He, K., Gkioxari, G., Dollár, P., Girshick, R.B.: Mask R-CNN. IEEE Trans. Pattern Anal. Mach. Intell. **42**(2), 386–397 (2020)
30. Xie, S., Tu, Z.: Holistically-nested edge detection. Int. J. Comput. Vis. **125**(1–3), 3–18 (2017)
31. Liu, S., Liang, X., Liu, L., Lu, K., Lin, L., Cao, X., Yan, S.: Fashion parsing with video context. IEEE Trans. Multimedia **17**(8), 1347–1358 (2015)
32. Yamaguchi, K., Kiapour, M.H., Ortiz, L.E., Berg, T.L.: Parsing clothing in fashion photographs. In: 2012 IEEE Conference on Computer Vision and Pattern Recognition, pp. 3570–3577. IEEE Computer Society (2012)
33. Yang, W., Luo, P., Lin, L.: Clothing co-parsing by joint image segmentation and labeling. CoRR abs/1502.00739 (2015)
34. Bertasius, G., Shi, J., Torresani, L.: Semantic segmentation with boundary neural fields. CoRR abs/1511.02674 (2015)
35. Chen, L., Barron, J.T., Papandreou, G., Murphy, K., Yuille, A.L.: Semantic image segmentation with task-specific edge detection using CNNs and a discriminatively trained domain transform. CoRR abs/1511.03328 (2015)
36. Liu, T., et al.: Devil in the details: towards accurate single and multiple human parsing. CoRR abs/1809.05996 (2018)
37. He, K., Zhang, X., Ren, S., Sun, J.: Deep residual learning for image recognition. CoRR abs/1512.03385 (2015)
38. Osokin, D.: Real-time 2D multi-person pose estimation on CPU: lightweight Open-Pose. CoRR abs/1811.12004 (2018)
39. Kendall, A., Gal, Y., Cipolla, R.: Multi-task learning using uncertainty to weigh losses for scene geometry and semantics. CoRR abs/1705.07115 (2017)
40. Chen, X., Mottaghi, R., Liu, X., Fidler, S., Urtasun, R., Yuille, A.L.: Detect what you can: detecting and representing objects using holistic models and body parts. CoRR abs/1406.2031 (2014)

M2E-Net: Multiscale Morphological Enhancement Network for Retinal Vessel Segmentation

Le Geng[1], Panming Li[1], Weifang Zhu[1], and Xinjian Chen[1,2](✉)

[1] School of Electronic and Information Engineering, Soochow University,
Suzhou 215006, China
xjchen@suda.edu.cn

[2] State Key Laboratory of Radiation Medicine and Protection, Soochow University,
Suzhou 215123, China

Abstract. Accurate retinal vessel segmentation plays a critical role in the early diagnosis of diabetic retinopathy. The uneven background distribution and complex morphological structure are still the challenges for segmenting the vessels, especially the capillaries. In this paper, we propose a novel deep network called Multiscale Morphological Enhancement Network (M2E-Net). First, we develop a multiscale morphological enhancement block, specifically the Top-hat/Bottom-hat module. Then, the block is embedded in the shallow layers of the U-shape fully convolutional network, which aims to reduce the influence of the uneven background distribution and enhance the texture contrast of fundus images. It is the first time introducing explainable morphological operation into CNN. Furthermore, the Structural Similarity (SSIM) loss guides the network to pay more attention to predicting the fine structures and clear boundaries during training. The proposed M2E-Net is evaluated on two digital retinal databases, DRIVE and CHASEDB1, and achieves the state-of-the-art performance with AUC of 98.40% and 98.94% respectively. The experimental results demonstrate the effectiveness of our proposed method.

Keywords: Retinal vessel segmentation · Multiscale morphological enhancement · Structural similarity loss

1 Introduction

Retinal vessel segmentation in fundus images is of great value in the diagnosis, screening, treatment, and evaluation of various cardiovascular and ophthalmic diseases, such as diabetes, hypertension, arteriosclerosis, and choroidal neovascularization [18]. However, due to the uneven background distribution and complex morphological structure, accurate vessel segmentation is a remaining challenge.

© Springer Nature Switzerland AG 2020
Y. Peng et al. (Eds.): PRCV 2020, LNCS 12305, pp. 493–502, 2020.
https://doi.org/10.1007/978-3-030-60633-6_41

Recent retinal vessel segmentation algorithms published in the literature can be divided into two categories. The first category of methods is generally following traditional image processing approaches, including filter-based and model-based techniques [15]. For example, different types of filters are jointly employed to extract 41-dimensional visual features to represent retinal vessels for subsequent classification [7]. To suppress background structure and image noise, various filtering methods have been proposed, such as Hessian matrix-based filters [4] and symmetry filter [17]. Another category of methods is based on deep learning (DL), which has shown its advantages in many computer vision tasks, including retinal vessel segmentation. Fu et al. [6] proposed a fully convolutional neural network combined with a fully-connected Conditional Random Fields (CRFs). Wu et al. [13] designed an efficient inception-residual convolutional block for improved feature representation. Lei et al. [8] combined spatial attention and channel attention to further integrate local features with their global dependencies adaptively.

Recently, attention mechanisms are becoming prevailing in deep learning. In medical image segmentation, structural information is the most valuable and of great significance, such as edges, shapes, etc. In ET-Net, the authors embedded edge-attention representations to guide the segmentation network to focus on edge information [16]. Yan et al. [15] make the network increase the attention to the edge by polishing the loss function.

Standing on the shoulders of these studies, in this paper, we propose the Multiscale Morphological Enhancement network (M2E-Net) which aims to improve the performance of retinal vessel segmentation. To reduce the effect of unevenly distributed background and improve the contrast of capillaries in fundus images, we designed a multiscale morphological enhancement block. It follows the principle of top-hat operation, which is widely used in traditional algorithms. To the best of our knowledge, it is the first time that the theory of morphological operation is introduced into the convolutional neural network. We embed three multiscale morphological enhancement (M2E) blocks into the shallow layers of the baseline modified from LinkNet [2]. To predict fine structures and clear boundaries, we introduce the multiscale SSIM loss into the hybrid loss function to capture the structural characteristics of vessels. Experimental results on two digital retinal databases, DRIVE and CHASEDB1, show significant performance improvements with AUC of 98.40% and 98.94% respectively.

2   Method

The overview of the proposed retinal vessel segmentation framework M2E-Net is shown in Fig. 1, which is an improved U-shape encoder-decoder architecture. Specifically, the residual network (ResNet) block is employed as the backbone for each encoder block. The multiscale morphological enhancement blocks are inserted in the shallow layers of the encoder to reduce the influence of the uneven background distribution and enhance the texture contrast. Then the features from the encoder are added to the decoder path via skip connections to achieve

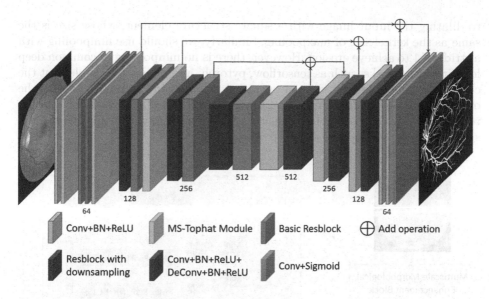

Fig. 1. The architecture of the multiscale morphological enhancement network: M2E-Net.

rich representation for better segmentation. Finally, the hybrid loss consisting of binary cross-entropy, dice loss, and multiscale structural similarity loss is employed to guarantee better performance.

2.1 Multiscale Morphological Enhancement Block

Motivation. When performing retinal vessel segmentation based on traditional algorithms, for the problems of inconspicuous contrast (such as capillaries) and uneven illumination and background, top-hat or bottom-hat transformation is a fast and efficient enhancement method. Inspired by this, we hope to embed such a trainable module in a network to achieve a similar effect during the training. In addition, with the research on the interpretability of deep learning, we can easily treat shallow convolution operations as some elementary filtering operations, such as low-pass filtering, high-pass filtering, and edge detection. However, the mathematical morphology operations which are common and efficient in image processing have not been applied or found in trained models. We consider this to be one of the gaps between traditional image processing algorithms and data-driven deep learning algorithms.

Implementation. We design a module based on top-hat transformation, which can extract the bright detail features in the image. The top-hat transformation mainly includes dilation, erosion, and subtraction. According to the definition of morphological operations based on the maximum and minimum values, we use maxpooling with a stride of 1 to implement dilation. This operation is equivalent to dilating the input image with a square structure element, whose size is the same as the kernel size of maxpooling. Similarly, we should use minpooling with a stride of 1 to achieve erosion. However, there is no minpooling in common deep learning frameworks, such as tensorflow, pytorch, and mxnet. We implement the equivalent minpooling for erosion operation by following steps: (1) reverse the original image by multiplying -1; (2) perform dilation; (3) reverse the dilation result. Then, we can easily implement a top-hat module with a specified kernel size of N using the module designed above.

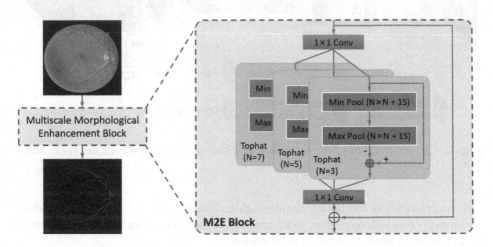

Fig. 2. The illustrations of proposed Multiscale Morphological Enhancement Block.

The top-hat module can only be used to extract bright details, but dark details are also significant features. We add the control factor before the top-hat module, which is implemented by a 1×1 convolution. It can automatically adjust and extract valuable light or dark details and suppress unnecessary details. To highlight the details of different scales, the outputs of different kernel size top-hat modules are fused via a 1×1 convolution, which also adjusts the fused result to the same size as the input. Finally, the input and the result after adaptive fusion are added as the overall output of the enhancement module. The whole structure of the M2E block is shown in Fig. 2. In order to ensure compatibility, the M2E block is entirely implemented by existing operators in the deep learning framework.

Contribution. Figure 2 shows that after the image or feature map is processed by the M2E block, the contrast between capillaries and the background is enhanced. It can also learn adaptive features which is more advanced than traditional morphological operations. The detailed features on the image or feature map mainly exist in the shallow layers of the neural network, so we embed it in the shallow layers of the convolutional network, specifically after the first three blocks of Resnet in Fig. 1. We also tried to embed the block deeper, and the experimental results show that the performance of the network will be degraded. To the best of our knowledge, this is the first time introducing an explainable morphological operation into the CNN.

2.2 Structural Similarity Loss

When designing a medical image segmentation algorithm based on deep learning, we pay more attention to pixel-wise segmentation, which can guarantee the convergence of the model. There are still some limitations, for example, models trained with BCE loss usually have low confidence in classifying boundary pixels, leading to blurry boundaries. In [3,16], edge loss is utilized to increase attention to edges, but it is still calculated at pixel-level, and the structural relevance is not fully considered. In the area of image quality assessment, structure similarity is a common evaluation index, which can reflect the structural similarity between two images [12].

Inspired by the design of dice loss, we convert the structural similarity, originally proposed for image quality assessment, into a loss function. We integrate it into our hybrid loss to learn the structural information of the vessel. Let x and y be two corresponding patches, with the same size $N \times N$, cropped from the predicted probability map S and the binary ground truth mask G respectively. The SSIM loss is defined as:

$$L_{ssim} = 1 - \frac{(2\mu_x\mu_y + C_1)(2\sigma_{xy} + C_2)}{(\mu_x^2 + \mu_y^2 + C_1)(\sigma_x^2 + \sigma_y^2 + C_2)} \tag{1}$$

where μ_x, μ_y and σ_x, σ_y are the mean and standard deviations of x and y respectively, σ_{xy} is their covariance, C_1 and C_2 are used to avoid dividing by zero. The patch size N is a hyper-parameter for SSIM and it determines the receptive field. We apply multiscale SSIM loss here which is a combination of three SSIM losses with the N fixed to 5, 7, 11. BCE loss, multiscale SSIM loss, and dice loss are fused together equally as the final employed hybrid loss, which calculates loss from pixel-level, patch-level, and image-level respectively. Hence, the trained model could obtain high-quality regional segmentation and clear boundaries.

$$L_{hybrid} = L_{bce} + L_{ssim} + L_{dice} \tag{2}$$

In addition, SSIM loss has a friendly side effect that it introduces edge attention, which is especially suitable for the segmentation of slender targets such as vessels. It considers a local neighborhood of each pixel and assigns higher weights to the boundary, i.e., the loss is higher around the boundary, even when the predicted probabilities on the boundary and the rest of the foreground are the same.

As shown in Fig. 3, we enumerate three different situations in the training process to compare the changes in the loss of BCE and SSIM at different pixel positions. The first row is the heatmap of BCE loss, and the second row is that of SSIM loss. These three columns represent three different situations. The

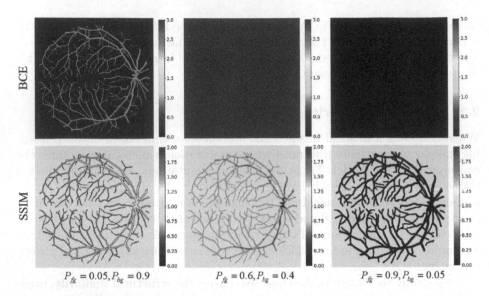

Fig. 3. The illustrations of the impact of losses. P_{fg} and P_{bg} denote the predicted probability of the foreground and background, respectively

heatmaps indicate that the BCE loss treats pixels at any position equally. For SSIM, even when the predicted probability on the boundary is the same as the probability of other pixel positions in the foreground, the loss of the boundary pixel position is higher. This is equivalent to automatically adding the attention mechanism to the boundary, which helps to optimize the boundary. Compared with a simple convex target, the edge pixels in the slender vessels occupy a larger proportion of the target pixels. Therefore, the characteristics of SSIM make it very suitable as a loss function for complex structures.

3   Experiments

3.1   Database

Two public databases DRIVE [10] and CHASEDB1 [5] are used to evaluate the proposed M2E-Net. The DRIVE database includes 40 color fundus images divided into two parts: 20 training images and 20 testing images. All fundus images have the same resolution of 565 × 584. Each image in the testing set has two manual labels, while each image in the training set has only one manual label. We use the labels provided by the first observer as the ground truth for performance evaluation.

The CHASEDB1 database has 28 color images of the retina and the resolution of each image is 999 × 960, which are taken from both eyes of 14 school children. Usually, the first 20 images are used for training, and the rest are used for testing, as described in [15]. The binary field of view (FOV) mask and segmentation ground truth is obtained by manual methods.

3.2   Implementation Details

Our proposed M2E-Net was implemented based on the PyTorch library (version 1.1.0). The network was trained on one NVIDIA K400 GPU which has a memory about 12 GB. Adam optimizer was applied to optimize the whole network and the learning rate was fixed to 2e−4. To improve the generalization capabilities of the model, we applied random data augmentation before training, including adjusting brightness, color, contrast, sharpness, and rotating the images. The enhancement factors were all following the log-normal distribution. After data enhancement, z-score processing was performed for normalizing each channel of images. In the testing stage, the data augmentation was removed while the z-score processing was still retained. During the training period, we used a hybrid loss function that is a weighted sum of three items to optimize the model together.

4   Results

Figure 4 shows two retinal fundus images and their ground truth, together with the segmentation results obtained by UNet and the proposed M2E-Net. The example from the DRIVE database is shown in the first row, and the one from the CHASEDB1 is shown in the second row. Due to the proposed enhancement block and SSIM loss, it can be observed that M2E-Net produces more accurate and detailed results than UNet. Vessels at low contrast can also be segmented, which is helpful for medical analysis and diagnosis.

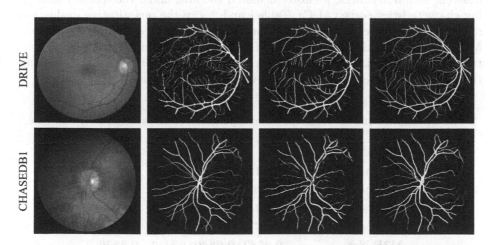

Fig. 4. Two test images from the DRIVE (1st row) and CHASEDB1 (2nd row) databases, ground truth (2nd column), the U-Net (3rd column), and the proposed M2E-Net (4th column).

Table 1. Results of M2E-Net and other methods on DRIVE database.

Methods	Acc	AUC	Sens	Spec
U-Net (2015) [9]	0.9531	0.9601	0.7537	0.9639
R2U-Net (2018) [1]	0.9556	0.9784	0.7792	0.9813
DE U-Net (2019) [11]	0.9567	0.9772	0.7940	0.9816
ACE-Net (2019) [18]	0.9569	0.9742	0.7725	0.9842
Vessel-Net (2019) [13]	0.9578	0.9821	0.8038	0.9802
Proposed method	**0.9682**	**0.9840**	**0.8149**	**0.9846**

Table 2. Results of M2E-Net and other methods on CHASEDB1 database.

Methods	Acc	AUC	Sens	Spec
R2U-Net (2018) [1]	0.9634	0.9815	0.7756	0.9712
MS-NFN (2018) [14]	0.9637	0.9825	0.7538	**0.9847**
LadderNet (2018) [19]	0.9656	0.9839	0.7978	0.9818
DE U-Net (2019) [11]	0.9661	0.9812	0.8074	0.9821
Vessel-Net (2019) [13]	0.9661	0.9860	0.8132	0.9814
Proposed method	**0.9751**	**0.9894**	**0.8510**	0.9835

As for the evaluation metrics, accuracy, sensitivity, and specificity are used to evaluate the performance of M2E-Net. To further evaluate the performance of different neural networks, we also calculated the area under the receiver operating characteristics curve (AUC). Table 1 and 2 give the quantitative evaluation results of several recently proposed state-of-the-art methods, and the proposed M2E-Net on the DRIVE and CHASEDB1 database respectively. The experimental result shows that the proposed model outperforms other models by achieving the highest AUC, accuracy, and sensitivity for each database.

Table 3. Ablation studies using the same experiment settings on DRIVE database.

Methods	Acc	AUC	Sens	Spec
Baseline	0.9518	0.9684	0.8133	0.9651
Baseline + M2E (S)	0.9665	0.9830	**0.8568**	0.9813
Baseline + M2E (E)	0.9615	0.9819	0.8136	0.9732
Baseline + MS-SSIM	0.9677	0.9834	0.8290	0.9832
M2E-Net	**0.9682**	**0.9840**	0.8149	**0.9846**

In order to further understand the performance gain of the proposed block and the effect of the SSIM loss, we conducted the ablation experiments on the

DRIVE database. Table 3 shows, from top to bottom, the performance of baseline, baseline with M2E blocks embedded in shallow layers (M2E-Net without MS-SSIM loss), baseline with M2E blocks embedded entirely, baseline trained with multiscale SSIM loss and M2E-Net trained with multiscale SSIM loss (proposed method). It reveals that: (1) the multiscale morphological enhancement block leads to improvement; (2) the block is more suitable for shallow layers; (3) multiscale SSIM loss helps boost the performance. When the M2E block is embedded entirely from shallow to deep layers, the performance of the model declines instead. This is because the M2E module is originally designed to extract detailed features from images or feature map, which only exist in the shallow layer of the model. The application of morphological operation in deep layers will destroy the semantic information and lead to poor performance.

5   Discussion and Conclusions

In this paper, we propose a Multiscale Morphological Enhancement Network (M2E-Net) for retinal vessel segmentation. The newly designed M2E block is effective in extracting useful detailed information and reducing the influence of the uneven background distribution. The multiscale SSIM loss, increasing edge attention during training, also boosts the segmentation performance. Experimental results on DRIVE and CHASEDB1 indicate that the proposed method has significantly outperformed the state-of-the-art for retinal vessel segmentation.

References

1. Alom, M.Z., Hasan, M., Yakopcic, C., Taha, T.M., Asari, V.K.: Recurrent residual convolutional neural network based on U-Net (R2U-Net) for medical image segmentation. arXiv preprint arXiv:1802.06955 (2018)
2. Chaurasia, A., Culurciello, E.: LinkNet: exploiting encoder representations for efficient semantic segmentation. In: 2017 IEEE Visual Communications and Image Processing (VCIP), pp. 1–4. IEEE (2017)
3. Fang, Y., Chen, C., Yuan, Y., Tong, K.: Selective feature aggregation network with area-boundary constraints for polyp segmentation. In: Shen, D., et al. (eds.) MICCAI 2019. LNCS, vol. 11764, pp. 302–310. Springer, Cham (2019). https://doi.org/10.1007/978-3-030-32239-7_34
4. Frangi, A.F., Niessen, W.J., Vincken, K.L., Viergever, M.A.: Multiscale vessel enhancement filtering. In: Wells, W.M., Colchester, A., Delp, S. (eds.) MICCAI 1998. LNCS, vol. 1496, pp. 130–137. Springer, Heidelberg (1998). https://doi.org/10.1007/BFb0056195
5. Fraz, M.M., et al.: An ensemble classification-based approach applied to retinal blood vessel segmentation. IEEE Trans. Biomed. Eng. 59(9), 2538–2548 (2012)
6. Fu, H., Xu, Y., Lin, S., Kee Wong, D.W., Liu, J.: DeepVessel: retinal vessel segmentation via deep learning and conditional random field. In: Ourselin, S., Joskowicz, L., Sabuncu, M.R., Unal, G., Wells, W. (eds.) MICCAI 2016. LNCS, vol. 9901, pp. 132–139. Springer, Cham (2016). https://doi.org/10.1007/978-3-319-46723-8_16
7. Lupascu, C.A., Tegolo, D., Trucco, E.: FABC: retinal vessel segmentation using AdaBoost. IEEE Trans. Inf Technol. Biomed. 14(5), 1267–1274 (2010)

8. Mou, L., et al.: CS-Net: channel and spatial attention network for curvilinear structure segmentation. In: Shen, D., et al. (eds.) MICCAI 2019. LNCS, vol. 11764, pp. 721–730. Springer, Cham (2019). https://doi.org/10.1007/978-3-030-32239-7_80

9. Ronneberger, O., Fischer, P., Brox, T.: U-Net: convolutional networks for biomedical image segmentation. In: Navab, N., Hornegger, J., Wells, W.M., Frangi, A.F. (eds.) MICCAI 2015. LNCS, vol. 9351, pp. 234–241. Springer, Cham (2015). https://doi.org/10.1007/978-3-319-24574-4_28

10. Staal, J., Abràmoff, M.D., Niemeijer, M., Viergever, M.A., Van Ginneken, B.: Ridge-based vessel segmentation in color images of the retina. IEEE Trans. Med. Imaging 23(4), 501–509 (2004)

11. Wang, B., Qiu, S., He, H.: Dual encoding U-Net for retinal vessel segmentation. In: Shen, D., et al. (eds.) MICCAI 2019. LNCS, vol. 11764, pp. 84–92. Springer, Cham (2019). https://doi.org/10.1007/978-3-030-32239-7_10

12. Wang, Z., Bovik, A.C., Sheikh, H.R., Simoncelli, E.P.: Image quality assessment: from error visibility to structural similarity. IEEE Trans. Image Process. 13(4), 600–612 (2004)

13. Wu, Y., et al.: Vessel-Net: retinal vessel segmentation under multi-path supervision. In: Shen, D., et al. (eds.) MICCAI 2019. LNCS, vol. 11764, pp. 264–272. Springer, Cham (2019). https://doi.org/10.1007/978-3-030-32239-7_30

14. Wu, Y., Xia, Y., Song, Y., Zhang, Y., Cai, W.: Multiscale network followed network model for retinal vessel segmentation. In: Frangi, A.F., Schnabel, J.A., Davatzikos, C., Alberola-López, C., Fichtinger, G. (eds.) MICCAI 2018. LNCS, vol. 11071, pp. 119–126. Springer, Cham (2018). https://doi.org/10.1007/978-3-030-00934-2_14

15. Yan, Z., Yang, X., Cheng, K.T.: Joint segment-level and pixel-wise losses for deep learning based retinal vessel segmentation. IEEE Trans. Biomed. Eng. 65(9), 1912–1923 (2018)

16. Zhang, Z., Fu, H., Dai, H., Shen, J., Pang, Y., Shao, L.: ET-Net: a generic edge-attention guidance network for medical image segmentation. In: Shen, D., et al. (eds.) MICCAI 2019. LNCS, vol. 11764, pp. 442–450. Springer, Cham (2019). https://doi.org/10.1007/978-3-030-32239-7_49

17. Zhao, Y., et al.: Automatic 2-D/3-D vessel enhancement in multiple modality images using a weighted symmetry filter. IEEE Trans. Med. Imaging 37(2), 438–450 (2017)

18. Zhu, Y., Chen, Z., Zhao, S., Xie, H., Guo, W., Zhang, Y.: ACE-Net: biomedical image segmentation with augmented contracting and expansive paths. In: Shen, D., et al. (eds.) MICCAI 2019. LNCS, vol. 11764, pp. 712–720. Springer, Cham (2019). https://doi.org/10.1007/978-3-030-32239-7_79

19. Zhuang, J.: LadderNet: multi-path networks based on U-Net for medical image segmentation. arXiv preprint arXiv:1810.07810 (2018)

DUDA: Deep Unsupervised Domain Adaptation Learning for Multi-sequence Cardiac MR Image Segmentation

Yueguo Liu[2] and Xiuquan Du[1,2(✉)]

[1] Key Laboratory of Intelligent Computing and Signal Processing,
Ministry of Education, Anhui University, Hefei, China
dxqllp@163.com
[2] School of Computer Science and Technology, Anhui University, Hefei, China

Abstract. Automatic segmentation of ventricles and myocardium in late gadolinium-enhanced (LGE) cardiac images is an important assessment for the clinical diagnosis of myocardial infarction. The most segmentation methods rely on the annotation of ventricles and myocardium. However, pathological myocardium in LGE images makes its annotation difficult to be obtained. In this paper, we propose a new deep unsupervised domain adaptation (DUDA) framework, which can effectively solve the segmentation problem of LGE images without any annotations through cross domain based adversarial learning. A CycleGAN network is used to construct the LGE image training set, meanwhile, a encoder and classier is used as a segmentor to generate the prediction mask. To keep our domain adaption model segmentation results consistent, a new loss function named cross-domain consistency loss is introduced. We tested our method on the dataset of the MICCAI2019 Multi-Sequence cardiac MR image Segmentation Challenge. The average dice of the myocardium, left ventricle, and right ventricle segmentation is 0.762, 0.857, and 0.841, respectively. Compared with other domain adaptation methods, our method can well segment myocardium and right ventricle structures in LGE images. Experimental results prove the superiority of our proposed framework and it can effectively help doctors diagnose myocardial infarction faster without any manual intervention in the clinic.

Keywords: Domain adaptation · Adversarial learning · Cross-domain consistency · Cardiac MR image

1 Introduction

Cardiac image segmentation is of great significance in medical applications and accurate diagnosis of cardiovascular diseases, which are the leading cause of death in many countries. Magnetic resonance (MR) technology provides an important means for imaging anatomical and functional information of the heart,

© Springer Nature Switzerland AG 2020
Y. Peng et al. (Eds.): PRCV 2020, LNCS 12305, pp. 503–515, 2020.
https://doi.org/10.1007/978-3-030-60633-6_42

in particular LGE cardiac magnetic resonance (CMR) sequence which visualizes myocardial infarction (MI), the T2-weighted (T2) CMR which images the acute injury and ischemic regions, and the balanced-Steady State Free Precession (bSSFP) cine sequence which captures cardiac motions and presents clear boundaries [12,29]. Compared with healthy tissue, LGE CMR enhances the infarcted myocardium with unique brightness, as shown in Fig. 1(a), make it difficult to be annotated. In contrast, bSSFP/T2 images show clear boundaries and can be easily annotated. Therefore, it would be helpful if we could learn the LGE segmentation model from bSSFP/T2 data.

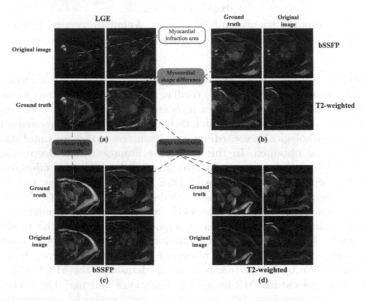

Fig. 1. Distribution differences between multiple sequential cardiac MR images. The red, green, and blue lines represent the left ventricle, myocardium, and right ventricle, respectively. (Color figure online)

Deep convolutional neural networks (DCNN) have achieved remarkable success in medical image segmentation [19], especially in cardiac image segmentation. Unfortunately, common CNN-based segmentation methods rely on manual annotation, and the training/test data sets belong to the same distribution. When the distribution of test data is different from the training data, domain shifts will occur [24], which will cause significant performance degradation. In multi-sequence cardiac MR images, the domain shift is more serious. In the face of this challenge, one feasible way is domain adaptation. Essentially, domain adaptation methods learn the marginal distribution [28] to transform the source domain into the target domain. However, the appearance and anatomical structure of multi-sequence cardiac MR images are different. As shown in Fig. 1(a), (b), (c), (d), there are differences in the shape of the right ventricle and myocardial structure in different sequence images. In addition, there is no right ventricular structure in some sequence images. The myocardium structure with scars in the LGE image, as shown in Fig. 1(a), is different from the normal myocardial structure, which also makes its segmentation more challenging. Therefore, the domain adaptation method only learns the poor marginal distribution in the LGE images.

Cross-domain adaptation has been used in cardiac image analysis, such as image synthesis and segmentation with unpaired images [10,27]. However, cross-domain adaptation is very difficult with significant domain shifts. Common cross-domain adaptation methods use CNN to learn the shared feature space between different domains [10]. Although it solves unlabeled target domain segmentation, it may ignore pixel-level information. Another effective method is to use CycleGAN [28] to realize the cross-domain transfer of unpaired images, which be used for cross-domain adaptation in segmentation tasks [20]. Compared with the domain adaptation method, it does not necessarily maintain semantic feature level information. Considering the possible shortcomings in the above methods, some researchers have proposed to further improve the segmentation performance of the target domain by combining image transfer and shared feature learning [2,6]. Although joint image transfer and feature learning can achieve pixel-level and feature-level information complementarity. Unfortunately, joint learning does not take into account that the anatomical structure of the cardiac image should keep consistent among different domains.

Motivated by [6], in this paper, we design a new deep unsupervised domain adaptation (DUDA) framework for LGE image segmentation, which could well narrow the domain shift between the LGE images and the bSSFP/T2 images to solve unlabeled LGE images segmentation. The proposed deep unsupervised domain adaptive segmentation framework includes several parts, namely the synthesis module (SM), synthesis and reconstruction module (SRM), LGE images segmentation module (Seg_T), bSSFP/T2 image segmentation module (Seg_S). (1) The SM is composed of a generator G_1 and a discriminator D_1, which synthesizes LGE images on the condition of bSSFP/T2 images with labels. (2) The SRM consists of a shared encoder SE and decoder D, which synthesizes bSSFP/T2 images based on real LGE images, and reconstructs the synthesized LGE images into bSSFP/T2 images. (3) The Seg_T Module consists of a shared encoder SE and a classifier C, which is used for LGE image segmentation. (4) The Seg_S consists of an encoder E_s and a classifier C_s, which is used for bSSFP/T2 image segmentation. We apply cross-domain consistency loss to the synthetic bSSFP/T2 image prediction results and the real LGE image prediction results and provide other supervision for the Seg_T module. The deep features obtained by the encoder in [6] are directly used in the classifier, resulting in the loss of shallow information. Feature fusion is used in the segmentation module to effectively reduce information loss. In addition, we add Maximum Mean Discrepancy (MMD) constraints to the probability maps to enhance the transferability of the deep features used for segmentation. Experimental results show that this method can effectively solve the problem of LGE image segmentation without labels, and has certain advantages compared with some other unsupervised domain adaptation methods.

2   Related Work

Unsupervised Domain Adaptation. Unsupervised domain adaptation methods can be divided into two categories: 1) pixel-level adaptation and 2) feature-level adaptation. The pixel-level domain adaptation method solves the domain gap problem by using image-to-image transfer to perform data augmentation in the target domain. According to the idea of pre-training [5,26], the target domain is transformed into the source domain to fine-tune the pre-trained source domain segmentation model. On the contrary, some studies have tried to transform the source image to the target image [1,13]. Then the transformed image is used to train the segmentation model of the target images. However, in the study of pixel-level domain adaptation, these methods presume to target images that have paired cross-domain data. In fact, in the real world, it is difficult to obtain many such data pairs. Therefore, Zhu have proposed CycleGAN [28], which is innovative in that it can transfer image content from the source domain to the target domain without paired training data. Many subsequent studies are based on the modifications made by CycleGAN, which also have very important applications in medical image segmentation [5,14,26].

Enhancing domain-invariant feature learning through adversarial learning is another method for unsupervised domain adaptation, which can be understood as a feature-level domain adaptation method. The key is to extract domain invariant features while ignoring the appearance differences between the source and target domains. Following such kind of principle, several previous works have proposed to learn transferable features with deep convolutional neural networks (DCNNs) [17,22] by minimizing a distance metric of domain discrepancy, such as Maximum Mean Discrepancy (MMD) [18]. In recent years, with the successful application of adversarial learning. Based on the above-mentioned adversarial learning idea, [11] and [21] are representative methods, which use a discriminator to distinguish feature spaces across domains and promote feature adaptation. This method has also been successfully applied in medical image processing [9,16].

Both pixel-level and feature-level domain adaptation solve the unsupervised domain adaptation problem from different angles. In recent years, the combination of pixel-level and feature-level domain adaptation strategies has been increasing. In [13], the authors proposed to combining CycleGAN with feature domain adaptation, which adapts representations at both the pixel and feature levels, enhances cycle-consistency while exploiting specific task losses. In [6], the author proposes to combine the synthetic data with the real data to learn the domain invariant features.

Cycle Consistency. Cycle consistency constraints have been successfully applied to various problems. In image-to-image translation, enforcing cycle consistency allows the network to learn the mappings without paired data [1,28].

According to the semantic prediction space of the heart image, the anatomical structure of the human heart should be consistent between different imaging methods.

3 Method

The overview of the unsupervised domain adaptation method is proposed to solve the LGE image segmentation without ground truth is shown in Fig. 2. We introduce MMD constraints and cross-domain consistency loss based on collaborative pixel and feature adaptation methods to effectively narrow the performance gap due to domain shift.

3.1 Network Architecture

Synthesis Module (SM). Given a set of labeled samples $\{x_i^S\}_{i=1}^N$ and $\{y_i^S\}_{i=1}^N$ from bSSFP/T2 images X^S, as well as unlabeled samples $\{x_i^l\}_{i=1}^M$ from LGE images X^l. We denote the data distribution as $x^S \sim P_{data}(X^S)$ and $x^l \sim P_{data}(X^l)$. In the SM, we aim to transform the appearance of X^S to X^l, and the fake LGE images $X^{l\text{-}fake}$ are used to train the segmentation model Seg_T. In practice, we apply a generative adversarial network (GAN) consists of a generator G_1 and a discriminator D_1. Since we do not have any paired bSSFP/T2 images with the LGE images, this real LGE image is randomly selected from the training dataset. The generator $G_1(X^S) = X^{l\text{-}fake}$ mainly synthesizes $X^{l\text{-}fake}$ under the condition of X^S. The discriminator competes with the generator to correctly distinguish the fake LGE image and the real LGE image. A successful generation G_1 will confuse D_1 to make the wrong prediction.

Synthesis and Reconstruction Module (SRM). To ensure the original content of the fake LGE image $X^{l\text{-}fake}$, we apply the idea of cycle consistency loss in cycleGAN [28], a reverse generator is adopted to reconstruct bSSFP/T2 images. As shown in Fig. 2, the shared encoder (SE) and decoder (D) form LGE-to-bSSFP/T2 reverse generator $G_2 = SE \circ D$, which is used to reconstruct the $X^{l\text{-}fake}$ back to the bSSFP/T2 images. Also, the generator G_2 synthesizes $X^{S\text{-}fake}$ images on the condition of real LGE images to train the segmentation model Seg_S. The above-mentioned cycle consistency is realized by the cycle-consistency loss function L_{cycle}, which encourages $G_2(G_1(X^S)) \approx X^S$ and $G_1(G_2(X^l)) \approx X^l$ to recover the original image:

$$
\begin{aligned}
L_{cycle} = {} & E_{x^S \sim P_{data}(X^S)} \left[\left\| G_2(G_1(x^S)) - x^S \right\|_1 \right] \\
& + E_{x^l \sim P_{data}(X^l)} \left[\left\| G_1(G_2(x^l)) - x^l \right\|_1 \right]
\end{aligned}
\tag{1}
$$

In an ideal state, the appearance of bSSFP/T2 images can be transferred to LGE images by the loss of cycle consistency and adversarial learning ideas. Therefore, these fake LGE images obtained by the transformation can be used to train the segmentation network Seg_T.

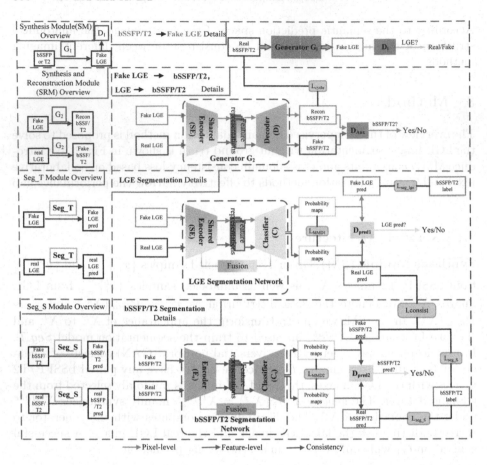

Fig. 2. Overview of our unsupervised domain adaptation framework. The SM serve the bSSFP/T2-to-LGE image transformation. The SRM form the reverse generator. The Seg_T and Seg_S serves the LGE image and bSSFP/T2 images segmentation. Blue and red arrows indicate data streams for pixel adaptation and feature adaptation, respectively, and dark green arrows indicate prediction result consistency. (Color figure online)

Seg_T Module. Considering that the LGE images without annotations, after extracting features from the fake LGE image $X^{l\text{-}fake}$ by a shared encoder (SE), the feature representation $SE(X^{l\text{-}fake})$ is forwarded to the classifier (C) to predict the segmentation mask. In other words, the composition $SE \circ C$ serves as a segmentation model for LGE images. The deep features $SE(X^{l\text{-}fake})$ contain abstract information of cardiac structure, and the segmentation results of LGE images obtained directly by the classifier may be incomplete, ignoring some detailed information, such as anatomical information of myocardium and right ventricle structures. With that in mind, when performing the segmentation task, we fuse shallow and deep information to capture more detailed information during training. The segmentation loss is defined as:

$$L_{Seg\_T}\left(SE, C\right) = CE\left(\hat{y}^{l\text{-}fake}, y^S\right) + Dice\left(\hat{y}^{l\text{-}fake}, y^S\right) \tag{2}$$

where $CE\left(\cdot, \cdot\right)$ is the cross-entropy loss function.

As shown in Fig. 2, to enhance the domain invariance of the obtained feature representation, we apply an adversarial learning strategy to the segmentation results obtained from the fake LGE image and the real LGE image. We apply the discriminator D_{pred1} to classify the predicted output corresponding to the fake LGE image and the real LGE image. We also use the discriminator D_{Aux} to distinguish whether the generated images are the fake bSSFP/T2 images or reconstructed bSSFP/T2 images.

To boost the transferability of feature representations and narrow the difference between feature representations of the fake LGE images and the real LGE images, MMD constraints are added to the segmentation task layer, which will increase the possibility of feature representation match between different domains. The constraint formula is expressed as follows:

$$L_{MMD1}\left(X^{l\text{-}fake}, X^l\right) = \left\| \underset{p}{E}\varphi\left(X^{l\text{-}fake}\right) - \underset{q}{E}\varphi\left(X^l\right) \right\|^2 \tag{3}$$

where φ is the feature mapping function, p and q represent the sample distributions of bSSFP/T2 images and LGE images, respectively.

Seg_S Module. In order to enable our model to produce more accurate predictions for unlabeled LGE images, we construct the bSSFP/T2 image segmentation model Seg_S. The network structure of the segmentation model $Seg\_S = E_s \circ C_s$ is the same as the segmentation model of the LGE image, as shown in Fig. 2, we define its segmentation loss function as:

$$\begin{aligned} L_{Seg\_S} = &CE\left(\hat{y}^S, y^S\right) + Dice\left(\hat{y}^S, y^S\right) \\ &+ \mu_S \cdot \left(CE\left(\hat{y}^{S\text{-}fake}, y^S\right) + Dice\left(\hat{y}^{S\text{-}fake}, y^S\right)\right) \end{aligned} \tag{4}$$

The third term is the cross-entropy loss and dice loss between the fake bSSFP/T2 image prediction result and the real bSSFP/T2 image label.

Cross-Domain Consistency Loss. Although the real LGE image is different from the fake LGE image in appearance or style, it should be consistent in anatomy. Based on this insight, we apply a cross-domain consistency loss between the real LGE image segmentation results and the fake bSSFP/T2 image segmentation results. This loss improves the consistency between the X^l segmentation result and the $X^{S\text{-}fake}$ segmentation result, thereby further improving the segmentation performance of the LGE images. We define the cross-domain consistency loss function as:

$$\begin{aligned} L_{consist} = &- E_{x^l \sim X^l} \sum f_S\left(x_i^{S\text{-}fake}\right) \log\left(f_l\left(x_i^l\right)\right) \\ &- E_{x^l \sim X^l} \sum f_l\left(x_i^l\right) \log\left(f_S\left(x_i^{S\text{-}fake}\right)\right) \end{aligned} \tag{5}$$

where f_S and f_l are the task predictions for bSSFP/T2 images and LGE images respectively.

3.2   Implement Details

Training and Parameter Settings. According to the method proposed in [6], the key points in the learning graph proposed in this paper are the shared encoder between pixels and feature adaptation and the introduction of cross-domain consistency loss. With the shared encoder enabling seamless integration of the image and feature adaptations, we can train the unified framework in an end-to-end manner. At each training iteration, all the modules are sequentially updated in the following order: $G_1 \rightarrow D_1 \rightarrow SE \rightarrow C \rightarrow D \rightarrow E_s \rightarrow C_s \rightarrow D_{Aux} \rightarrow D_{pred1} \rightarrow D_{pred2}$. The overall objective of the Seg_T and Seg_S modules in our framework is as follows:

$$
Ł_{total}^T = L_{Seg\_T} + \alpha_l \cdot L_{consist} + \beta_l \cdot (G_1, D_1) \\
+ \chi_l \cdot (SE, C, D_{pred1}) + \delta_l \cdot (SE, D_{Aux}) + \varepsilon_l \cdot L_{MMD1}
\tag{6}
$$

$$
Ł_{total}^S = L_{Seg\_S} + \alpha_S \cdot L_{consist} + \chi_S \cdot (E_s, C_s, D_{pred2}) + \varepsilon_S \cdot L_{MMD2}
\tag{7}
$$

For training practice, when updating with the adversarial learning losses, we used the Adam optimizer with a learning rate of 2×10^{-4}. For the segmentation task, the Adam optimizer was parameterized with an initial learning rate 1×10^{-3} and a stepped decay rate of 0.9 every 2 epochs.

Network Configurations of the Modules. The layer configuration of the G_1 follows the practice of CycleGAN [28]. For the D, we follow the configuration of SIFA [6]. For all the three discriminators $D_{Aux}, D_{pred1}, D_{pred2}$, we follow the configuration of PatchGAN [15]. The encoder SE and E_s uses residual connections and dilated convolutions (dilation rate = 2) to enlarge the size of the receptive field while preserving the spatial resolution for dense predictions [25]. The classifier C and C_s are both 1×1 convolutional layer, followed by an upsampling layer fusing shallow information to recover resolution of segmentation predictions to original image size.

4   Experiments

4.1   Dataset and Evaluation Metrics

Dataset. The Multi-sequence Cardiac MRI data by the challenge organizers [29] (includes bSSFP, LGE, and T2 images) used in the paper were collected from 45 patients, where ground truth (GT) of the myocardium (Myo), left ventricle (LV) and right ventricle (RV) in 35 patients were provided for bSSFP and T2 images, while for 5 patients ground truth (GT) of LGE images were provided for validation. LGE images of the rest 40 patients are used for the train. For each patient, the bSSFP images consist of 8–12 slices, with an in-plane resolution of 1.251.25 mm and a slice thickness of 8 to 13 mm. The T2 images have 3–7 slices. The LGE images have 10–20 slices. In this paper, we used bSSFP and T2-weights images as the source domain and LGE images as the target domain. Since the

Challenge no longer provides testing services, and no labels are given to the remaining 40 patients, in the following experiments, our evaluation is based on the validation set (only 5 patients). To train our model, we used a coronal view image slice that was resized to 256 × 256. To reduce overfitting, LGE images are augmented using rotation, scaling, and affine transformations.

Evaluation Metrics. In all our experiments, the Dice [20] was employed to compare the different models.

4.2   Results

Ablation Experiments 1. We designed some ablation experiments to confirm the effectiveness of our proposed method. (1) baseline network (joint pixel-level and feature-level) P/F; (2) Adding MMD constraints based on P/F (P/F+MMD). (3) adding cross-domain consistency loss based on P/F (P/F+consistency); (4) DUDA. Table 1 summarizes the comparison results. When the joint learning strategy is used, the dice average of the left, right ventricle and myocardial structure can be restored to 0.751. However, due to the differences in the distribution between bSSFP/T2 images and LGE images, especially the myocardial structure with the scar, the average dice of its segmentation recovered only to 0.644, and the segmentation effect was far from satisfactory. After the MMD constraint is added, the transferability of high-level features between different domains is enhanced. Compared with joint learning, the average dice of the segmentation results in the LGE image can be restored to 0.774. The dice of myocardium structure recovers to 0.667. When considering cross-domain consistency loss, it ensures the consistency of its anatomical information after image transfer, further improving the segmentation performance of the LGE image, and the average dice value of the three heart structures in the LGE image can be restored to 0.766. The average dice value of myocardial structure can be restored to 0.678. The DUDA considers both MMD constraints and cross-domain consistency loss to restore the average dice value to 0.778. The myocardial structure can be restored to 0.689. Two different constraints can effectively alleviate the differences between the two domains, indicating the necessity of each component in our proposed method.

Ablation Experiments 2. The DUDA fuses shallow and deep information to reduce information loss, especially for myocardium and right ventricular structures. Figure 3 shows the visualization of whether or not to integrate shallow information. As can be drawn, when the shallow information is not fused, the method of joint pixel-level and feature-level cannot capture the detailed information of the myocardium and right ventricle. The fusion of shallow and deep information can capture the detailed information of myocardium and right ventricle well.

Compared with the Other Methods. To further confirm the effectiveness of our proposed method, we compare DUDA with some commonly used unsupervised domain adaptation methods, including cycleGAN [28], CyCADA [13],

Table 1. Ablation study of our method on validation dataset of 5 patients LGE images. The average and std represent the average and standard deviation of the Dice segmentation score, respectively, and mean represents the average of the Dice segmentation scores of three different structures.

Methods	Dice						
	Myo		LV		RV		
	Average	Std	Average	Std	Average	Std	Mean
P/F	0.644	0.018	0.814	0.084	0.795	0.028	0.751
P/F + MMD	0.667	0.027	0.846	0.058	0.810	0.050	0.774
P/F + consistency	0.678	0.055	0.821	0.070	0.800	0.042	0.766
DUDA	**0.689**	**0.033**	**0.823**	**0.070**	**0.821**	**0.025**	**0.778**

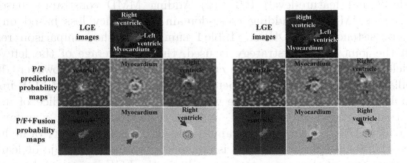

Fig. 3. Two examples of the LGE image segmentation probability map. The red, green, and blue arrows represent the left ventricle, myocardium, and right ventricle, respectively. (Color figure online)

CrDoCo [8]. The results are presented in Table 2, which indicates that our method significantly surpasses the previous method in terms of dice. It can be seen from the comparison that the DUDA can greatly improve the segmentation performance of unlabeled LGE images, especially for the segmentation of scarred myocardium and right ventricle with diversified shapes. Besides, we also compare DUDA with some studies published on the official website of the MICCAI 2019 multi-sequence cardiac MR image segmentation challenge. Table 3 counts the dice value of each method. In [4], the author proposes to combine the idea of multimodal style transfer with the cascade U-Net network, and use the information of bSSFP/T2 images to assist the segmentation of LGE images. In addition, the author also applied the idea of Gaussian distribution in the training process to achieve the best segmentation performance. From the average dice value of the three heart structures in the LGE image, we know that our method has exceeded most methods. The experimental results prove once again the effectiveness of our proposed method. Compared with [4], our proposed method needs to be improved. It is particularly worth noting that the label of the LGE image test set is not available. Our experimental results are statistically obtained on the validation set of 5 patients.

Table 2. Performance comparison between DUDA and other state-of-the-art methods for LGE-MR image segmentation on patients 1 to 5.

Methods	Adaptation		Dice			
	Pixel	Feature	Myo	LV	RV	Average
SIFA [6]	√	√	0.644	0.814	0.795	0.751
CycleGAN [28]	√		0.603	0.779	0.780	0.721
CyCADA [13]	√	√	0.557	0.779	0.759	0.700
CrDoCo [8]	√	√	0.604	0.797	0.742	0.714
DUDA	√	√	**0.762**	**0.857**	**0.841**	**0.820**

Table 3. Performance comparison between DUDA and other segmentation methods for LGE-MR image segmentation on patients 1 to 5.

Methods	Dice			
	Myo	LV	RV	Average
Chen J et al. [7]	0.596	0.760	0.717	0.691
Sulaiman et al. [23]	0.671	0.862	0.766	0.766
Vctor M et al. [3]	0.688	0.809	0.820	0.772
Tao X et al. [20]	0.607	0.764	0.738	0.703
Chen C et al. [4]	0.816	0.897	0.895	0.869
DUDA	**0.762**	**0.857**	**0.841**	**0.820**

5 Conclusion

Similar to most previous domain adaptation methods, the adaptation process of our method narrows the distribution difference between the bSSFP/T2 images and the LGE images by image appearance transformation and learning domain invariant features. Besides, we apply MMD constraints to narrow distribution differences. To improve the LGE image segmentation performance, the bSSFP/T2 image segmentation model is constructed to introduce cross-domain consistency loss. To reduce the loss of detailed information of the corresponding domain, the shallow information and the deep information are integrated into the segmentation module. The experimental results prove the feasibility of our proposed method.

Acknowledgment. This work was supported in part by the National Science Foundation of China (61702003), and in part by the Anhui Provincial Natural Science Foundation (1708085QF143, 1808085MF175). The authors also thank all the anonymous reviewers for their valuable comments and suggestions, which were helpful for improving the quality of the paper.

References

1. Bousmalis, K., Silberman, N., Dohan, D., Erhan, D., Krishnan, D.: Unsupervised pixel-level domain adaptation with generative adversarial networks. In: IEEE Conference on Computer Vision and Pattern Recognition (CVPR), pp. 95–104 (2017)
2. Cai, J., Zhang, Z., Cui, L., Zheng, Y., Yang, L.: Towards cross-modal organ translation and segmentation: a cycle- and shape-consistent generative adversarial network. Med. Image Anal. **52**, 174–184 (2019)
3. Campello, V.M., Martín-Isla, C., Izquierdo, C., Petersen, S.E., Ballester, M.A.G., Lekadir, K.: Combining multi-sequence and synthetic images for improved segmentation of late gadolinium enhancement cardiac MRI. In: Pop, M., et al. (eds.) STACOM 2019. LNCS, vol. 12009, pp. 290–299. Springer, Cham (2020). https://doi.org/10.1007/978-3-030-39074-7_31
4. Chen, C., et al.: Unsupervised multi-modal style transfer for cardiac MR segmentation. In: Pop, M., et al. (eds.) STACOM 2019. LNCS, vol. 12009, pp. 209–219. Springer, Cham (2020). https://doi.org/10.1007/978-3-030-39074-7_22
5. Chen, C., Dou, Q., Chen, H., Heng, P.-A.: Semantic-aware generative adversarial nets for unsupervised domain adaptation in chest x-ray segmentation. In: Shi, Y., Suk, H.-I., Liu, M. (eds.) MLMI 2018. LNCS, vol. 11046, pp. 143–151. Springer, Cham (2018). https://doi.org/10.1007/978-3-030-00919-9_17
6. Chen, C., Dou, Q., Chen, H., Qin, J., Heng, P.: Synergistic image and feature adaptation: towards cross-modality domain adaptation for medical image segmentation. In: Conference on Artificial Intelligence (AAAI), pp. 865–872 (2019)
7. Chen, J., Li, H., Zhang, J., Menze, B.: Adversarial convolutional networks with weak domain-transfer for multi-sequence cardiac MR images segmentation. In: Pop, M., et al. (eds.) STACOM 2019. LNCS, vol. 12009, pp. 317–325. Springer, Cham (2020). https://doi.org/10.1007/978-3-030-39074-7_34
8. Chen, Y., Lin, Y., Yang, M., Huang, J.: CrDoCo: pixel-level domain transfer with cross-domain consistency. In: IEEE Conference on Computer Vision and Pattern Recognition (CVPR), pp. 1791–1800 (2019)
9. Dou, Q., et al.: PnP-AdaNet: plug-and-play adversarial domain adaptation network with a benchmark at cross-modality cardiac segmentation. arXiv:1812.07907 (2018)
10. Dou, Q., Ouyang, C., Chen, C., Chen, H., Heng, P.: Unsupervised cross-modality domain adaptation of convnets for biomedical image segmentations with adversarial loss. In: International Joint Conference on Artificial Intelligence, pp. 691–697 (2018)
11. Ganin, Y., et al.: Domain-adversarial training of neural networks. J. Mach. Learn. Res. **17**(1), 189–209 (2016)
12. Han, W.K., Farzaneh-Far, A., Kim, R.J.: Cardiovascular magnetic resonance in patients with myocardial infarction: current and emerging applications. J. Am. Coll. Cardiol. **55**(1), 1–16 (2009)
13. Hoffman, J., et al.: CyCADA: cycle-consistent adversarial domain adaptation. In: International Conference on Machine Learning (ICML), pp. 1994–2003 (2018)

14. Huo, Y., Xu, Z., Bao, S., Assad, A., Abramson, R.G., Landman, B.A.: Adversarial synthesis learning enables segmentation without target modality ground truth. In: IEEE Conference on International Symposium on Biomedical Imaging (ISBI), pp. 1217–1220 (2018)
15. Isola, P., Zhu, J., Zhou, T., Efros, A.A.: Image-to-image translation with conditional adversarial networks. In: IEEE Conference on Computer Vision and Pattern Recognition (CVPR), pp. 5967–5976 (2017)
16. Joyce, T., Chartsias, A., Tsaftaris, S.A.: Deep multi-class segmentation without ground-truth labels. In: International conference on Medical Imaging with Deep Learning (MIDL) (2018)
17. Long, M., Cao, Y., Wang, J., Jordan, M.I.: Learning transferable features with deep adaptation networks. In: International Conference on Machine Learning (ICML), pp. 97–105 (2015)
18. Pan, S.J., Tsang, I.W., Kwok, J.T., Yang, Q.: Domain adaptation via transfer component analysis. IEEE Trans. Neural Networks 22(2), 199–210 (2011)
19. Ronneberger, O., Fischer, P., Brox, T.: U-Net: convolutional networks for biomedical image segmentation. In: Navab, N., Hornegger, J., Wells, W.M., Frangi, A.F. (eds.) MICCAI 2015. LNCS, vol. 9351, pp. 234–241. Springer, Cham (2015). https://doi.org/10.1007/978-3-319-24574-4_28
20. Tao, X., Wei, H., Xue, W., Ni, D.: Segmentation of multimodal myocardial images using shape-transfer GAN. In: Pop, M., et al. (eds.) STACOM 2019. LNCS, vol. 12009, pp. 271–279. Springer, Cham (2020). https://doi.org/10.1007/978-3-030-39074-7_29
21. Tzeng, E., Hoffman, J., Saenko, K., Darrell, T.: Adversarial discriminative domain adaptation. In: IEEE Conference on Computer Vision and Pattern Recognition (CVPR), pp. 2962–2971 (2017)
22. Tzeng, E., Hoffman, J., Zhang, N., Saenko, K., Darrell, T.: Deep domain confusion: maximizing for domain invariance. arXiv:1412.3474 (2014)
23. Vesal, S., Ravikumar, N., Maier, A.: Automated multi-sequence cardiac MRI segmentation using supervised domain adaptation. arXiv:1908.07726 (2019)
24. Yang, J., Dvornek, N.C., Zhang, F., Chapiro, J., Lin, M.D., Duncan, J.S.: Unsupervised domain adaptation via disentangled representations: application to cross-modality liver segmentation. In: Shen, D., et al. (eds.) MICCAI 2019. LNCS, vol. 11765, pp. 255–263. Springer, Cham (2019). https://doi.org/10.1007/978-3-030-32245-8_29
25. Yu, F., Koltun, V., Funkhouser, T.: Dilated residual networks. In: IEEE Conference on Computer Vision and Pattern Recognition (CVPR), pp. 636–644 (2017)
26. Zhang, Y., Miao, S., Mansi, T., Liao, R.: Task driven generative modeling for unsupervised domain adaptation: application to x-ray image segmentation. arXiv:1806.07201 (2018)
27. Zhang, Z., Yang, L., Zheng, Y.: Translating and segmenting multimodal medical volumes with cycle- and shape-consistency generative adversarial network. In: IEEE Conference on Computer Vision and Pattern Recognition (CVPR), pp. 9242–9251 (2018)
28. Zhu, J., Park, T., Isola, P., Efros, A.A.: Unpaired image-to-image translation using cycle-consistent adversarial networks. In: International Conference on Computer Vision (ICCV), pp. 2242–2251 (2017)
29. Zhuang, X.: Multivariate mixture model for myocardium segmentation combining multi-source images. IEEE Trans. Pattern Anal. Mach. Intell. 44(12), 2933–2946 (2019)

Learning from Rankings with Multi-level Features for No-Reference Image Quality Assessment

Yiqun Li[(✉)], Lan Ma, and Dahai Yu

TCL Hong Kong Research Center, Shenzhen, Hong Kong SAR
{yiqunli,rubyma,dahai.yu}@tcl.com

Abstract. Deep neural networks for image quality assessment have been suffering from a lack of training data for a long time, as it is expensive and laborious to collect sufficient subjective mean opinion scores (MOS). The Siamese network (learning from rankings) makes it possible to use images with only rough labels, such as the relative quality. On the other hand, features from intermediate layers, which possess local information highly sensitive to the quality degradation, are overlooked in the direct use of DNNs. In light of these findings, in this paper, we propose a framework for NR-IQA based on transferring learning from the Siamese network to the traditional CNNs by exploiting features from multiple layers. Specifically, we first train a Siamese network on a large artificially generated dataset. Then, we fine-tune the network with a small number of MOS-labeled images to match the perceptual scaling of human beings. In both steps, features from several layers are combined before being fed into the regression layer for the final score. Experimental results are presented, validating the effectiveness of this transfer-learning framework when considering multi-level information. Furthermore, performance comparable to state-of-the-art NR-IQA approaches on standard IQA datasets is achieved.

Keywords: No-reference image quality assessment · Deep neural network · Learning from rankings · Multi-level feature extraction

1 Introduction

The extreme lack of training data has been a long-lasting problem for training a deep neural network, which largely hinders the performance and generalization of DNNs. In particular, turning to the task of image quality assessment, no-reference IQA (NR-IQA) approaches aim to predict the perceptual quality of an image without any information from the original undistorted images. Currently, there are several image databases equipped with subjective mean opinion scores (MOS) of the perceptual quality. However, they all have limitations. For instance, both LIVE IQA [13] and TID2013 [12] are designed for synthetically distorted images only; the content of distorted images are highly correlated; each image

© Springer Nature Switzerland AG 2020
Y. Peng et al. (Eds.): PRCV 2020, LNCS 12305, pp. 516–526, 2020.
https://doi.org/10.1007/978-3-030-60633-6_43

possesses one type of distortions at most; and their MOSs are collected under the laboratory environment, which may not be representative enough. Database such as LIVE Challenge [3] is for authentic distortions, but only 1,162 images are contained with MOSs by crowdsourcing.

One of the common ways to deal with the problem of small training sets is transfer learning. Deep convolution neural networks have achieved great success in object recognition on the large-scale image dataset ImageNet [2]. Relatively recently, based on deep CNN architectures of MobileNet [5], VGG16 [14] and InceptionV2, [16] and their pre-trained weights on ImageNet, NIMA [17] proposed an approach for both technical and aesthetic quality assessment of images by replacing the last layer of the baseline CNNs with a fully connected one with expected number of neurons. After fine-tuning on the target databases, NIMA [17] gets good performance in terms of the correlation of rankings between the predicted quality scores and MOSs. MFIQA [7] adopts ResNet-50 [4] as the feature generator, and achieves even better results on LIVE IQA and LIVE Challenge databases.

Instead of only focusing on modifying the architecture and training strategy of deep CNNs, the Siamese network allows us to explore data with only ranking information, which greatly increases the number of images available to train a network for discriminating ability in regards to distortion levels. Aiming to learn from rankings, based on the Waterloo Exploration Database [10], RankIQA [9] first trains a Siamese network to rank images in accordance with the quality of degradation by using synthetically generated distortions for which relative quality of the images is known. Then, only one branch of the Siamese network is extracted for fine-tuning on target datasets. RankIQA gets excellent esti-mating results on LIVE IQA and TID2013. MT-RankIQA [8] further extends algorithms in RankIQA to a more general method, exploiting unlabeled data by self-supervised learning to rank, the framework of which is applied to IQA and crowd counting tasks. Both achieve state-of-the-art performance. What's more, Zhang et al. in 2019 [21] were the first to train blind IQA models based on Siamese network on multiple databases simultaneously without additional subjective testing for scale realignment. They aim to build a NR-IQA model which works for both synthetic and authentic distortions within one single set of parameters, and which outperforms most existing NR-IQA models.

As discussed in [20], in a convolution neural network, low-level layers basically learn local features of the input image, while features captured by high-level layers are more abstract and related to the semantic information, which suggests that a deep CNN can be regarded as a feature generator. GoogleNet [15] added an auxiliary classifier to encourage discrimination in the lower stages in the classifier. MFIQA [7] investigates the learning ability of features from different-level layers based on the ResNet-50.

In this paper, we propose a universal framework which works for both syn-thetic and authentic distortions for the NR-IQA problem, via learning-from-rankings with multi-level features. Specifically, we first train a Siamese network on a large-scale artificially generated dataset. Then, we fine-tune the network with a small number of MOS-labeled images to match the perceptual scaling of human beings. What differs from RankIQA [9] is that we adopt several combinations of features from multiple intermediate layers so that our framework can work well not only for synthetic distortions, but also for authentic ones. We further did a series of experiments validating that the Siamese network trained on a massive amount of automatically generated ranked image pairs does indeed capture the ability to discriminate between distortion levels, and multi-level features are effective in understanding distortions, which greatly enhances the performance of the network.

The paper is organized as follows: We shall introduce our proposed framework - RMFIQA in Sect. 2. Then, details of experiments and analysis of experimental results will be given in Sect. 3. Finally we will conclude the paper in Sect. 4 and discuss our the future work.

2   Framework of RMFIQA

As shown in Fig. 1, RMFIQA consists of a backbone network (such as the VGG16 in this paper) as feature generator, a number of encoding layers to extract and encode those features generated from multiple intermediate layers of the base network, and a regression head for the final score. A crop of size 224×224 will be randomly selected for an image before being fed into the network. Please note that the VGG16 can be replaced by another base network to generate features. Here, we take VGG16 as an example.

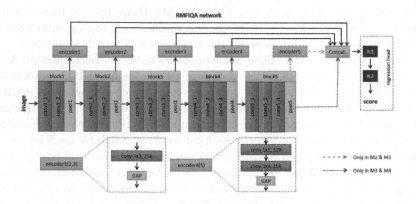

Fig. 1. Structure of RMFIQA

2.1   Encoder Block

The encoder blocks here are used to unify the size of the output from different convolution blocks in the backbone network, so that concatenation can be made before the regression layer. These encoders will reduce the dimension of channels, while simultaneously enhancing the expressibility of the extracted features. As presented in Fig. 1, besides a 3×3 convolution layer and a global average pooling (GAP) layer, for high-level encoders such as encoder 4 and 5, one more 1×1 convolution layer is added before the 3×3 one to cut down on the parameters needed to be learned.

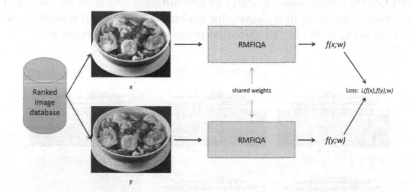

Fig. 2. Pipe line for training the Siamese network.

2.2 Learning from Ranking

To explore the unlabeled data, we adopt the scheme in [9] to train a Siamese network first and then transfer to the conventional CNNs for NR-IQA tasks for fine-tuning. Specifically, we first generate a large dataset with synthetic distortions at different levels added to all images. Note that relative quality of degradation is known for this dataset. We then train RMFIQA in the "Siamese" way (see Fig. 2 for details), which means that by letting the RMFIQA network serve as the two branches in the Siamese network (whose weights are shared), the network will be trained from ranked images pairs. Each time before calculating the loss for back propagation, a pair of images with relative quality known will pass through the whole network and obtain two estimated scores. Let x and y denote two image inputs. Without loss of generality, we assume that the quality of x is better than y. Then, the pairwise ranking hinge loss is defined as follows:

$$L(x, y; \omega) = max(0, f(y; \omega) - f(x; \omega) + \epsilon), \tag{1}$$

where ω denotes the parameters needed to be learned in the network. $f(\cdot)$ represents the predicted score of the input image, and ϵ is the margin.

After finishing training the Siamese network to rank distorted images, we then fine-tune the single RMFIQA network in a standard manner for DNNs. The mean squared error between predicted scores and MOSs is adopted as the loss function for tine-tuning.

2.3   Database

We here give more details on the three datasets we use in this paper. The Waterloo Exploration database [10], which consists of 4,744 natural images of high quality, are used as reference to generate ranked distorted image pairs for training the Siamese network. To be specific, we artificially added four types of distortions: Gaussian Blur (GB), White Noise (WN), JPEG Compression (JPEG) and JPEG2000 Compression (JP2K) at five different levels to all the 4,744 reference images, resulting in a dataset for ranking with 113,856 distorted images. Figure 3 gives examples of GN distorted images at five distortion levels.

Fig. 3. An example of GN distorted images from the artificially generated ranked dataset. The original image is from the Waterloo Database.

We use LIVE IQA [13] and LIVE Challenge Database [3] as the two datasets for fine-tuning and testing. LIVE IQA contains 779 distorted images (GB, WN, JEPG, JP2K and fast-fading) generated from 29 original images. Each image in LIVE IQA possesses one type of synthetic distortions at most. LIVE Challenge Database has 1,162 authentically mixed distorted images which are selected from the Flickr website, wherein photos are captured by a vast range of devices under different environmental conditions.

3   Experiments and Analysis

In this section, we shall analyze three points through a series of experiments: the capability of the Siamese network in taking advantage of rough labeled database to train for the discriminating ability, and the effectiveness of both the transfer learning framework and the fusion of multi-layer features.

3.1   Technical Details

To train and test the proposed framework, the target IQA database is randomly divided into training (80%) and test (20%) sets. For the LIVE database, we first divide it according to the reference images to make sure that the content of distorted images in two subsets are independent. To evaluate the models, two standard metrics are considered: Spearman's Rank Order Correlation Coefficient (SRCC) and Person's Linear Correlation Coefficient (PLCC), to measure the consistency between the MOSs and the predicted (relative) rankings and (absolute) scores, respectively. When testing an image, 30 patches of size 224×224 will be randomly cropped and go through the network for 30 scores individually, whose average will be assigned as the final score of this image.

To validate the efficacy of multi-level intermediate-layer features under the proposed training framework, we test four different methods of feature fusion for the model. We list detailed components of four tested models in Table 1, where the second column represents whether the model possesses encoders or not in addition to the number of encoders, and the third column is for the output of the *pool5* layer. For instance, model RMFIQA-M2 will concatenate features from encoders 1 to 5 together but leave the output of *pool5* layer alone as the input for the fully connected layer in regression head.

Table 1. Summary of models experimented in this paper.

Model	Encoder	pool5
M1	✗	✓
RMFIQA-M2	*1,2,3,4,5*	✗
RMFIQA-M3	*1,2,3,4,5*	✓
RMFIQA-M4	*1,2,3,4*	✓

Table 2. SRCC and PLCC comparison of the Siamese network according to the model type on the LIVE IQA database. Note that the Siamese network was trained with only four types of distortions in LIVE IQA except for the fast fading, but tested on all five types of distortions. Furthermore, the model M2, M3 and M4 are only trained for corresponding encoders and the regression head.

Model	SRCC	PLCC
M1	0.903	0.721
RMFIQA-M2	0.886	0.754
RMFIQA-M3	0.911	0.723
RMFIQA-M4	0.907	0.754

3.2 Performance of the Siamese Network

When training the Siamese network, except for the M1 model which possesses no encoders at all, we have all the convolution blocks in the backbone network (VGG16 in this case) frozen for other models, which means that only encoders and regression head were trained from scratch. The backbone network, as the feature generator, adopted weights pre-trained on ImageNet for the classification task. For the performance of the pre-trained Siamese network, as shown in Table 2, even though we do not fine-tune the backbone network for the IQA task particularly, models demonstrate reasonable SRCC and PLCC on the LIVE IQA database. As a result, we can use this trained Siamese network to further fine-tune on specific databases. For all models, the SRCC is much higher than the PLCC, which is natural at this stage as the Siamese network here learned from image pairs for which only relative quality is known. It can be seen later that after fine-tuning with clear MOSs, the PLCC is improved by a large margin.

Table 3. SRCC comparison of the fine-tuned model according to the model type. Note that the Siamese network was trained with four types of distortions as in LIVE IQA except for the fast fading, but tested on all five types of distortions. LIVE CL stands for the LIVE Challenge database.

Model	Backbone update	LIVE	LIVE CL
M1	✓	0.945	0.786
RMFIQA-M2*	✗	0.958	–
RMFIQA-M2	✓	**0.963**	**0.803**
RMFIQA-M3	✓	*0.963*	*0.806*
RMFIQA-M4	✓	0.954	0.793

Table 4. PLCC comparison of the fine-tuned model according to the model type. Note that the Siamese network was trained with four types of distortions as in LIVE IQA except for the fast fading, but tested on all five types of distortions. LIVE CL stands for the LIVE Challenge database.

Model	Backbone update	LIVE	LIVE CL
M1	✓	0.924	0.819
RMFIQA-M2*	✗	0.949	–
RMFIQA-M2	✓	**0.954**	**0.838**
RMFIQA-M3	✓	*0.953*	*0.836*
RMFIQA-M4	✓	0.938	0.823

3.3   Performance of RMFIQA Framework

Similar to the traditional CNN approaches for computer vision tasks, which predict or estimate a single image at a time, we fine-tuned the pre-trained network in the same manner, which means that the knowledge of the Siamese network is transferred through a number of single images, not image pairs.

The SRCC and PLCC results of the fine-tuned models on LIVE IQA and LIVE Challenge databases are presented in Table 3 and Table 4 respectively. The second column - "Backbone update" indicates whether the backbone network was also fine-tuned or not. We mark the fine-tuned RMFIQA-M2 model without update on the backbone by an asterisk to facilitate the presentation. It can first be found that transfer learning indeed takes the model to the next level. For example, the SRCC (PLCC) of RMFIQA-M2 for the LIVE IQA database increases from 0.886 (0.754) to 0.963 (0.954). Moreover, even without fine-tuning the backbone network, RMFIQA-M2* is able to pull up the two measure to 0.945 and 0.949 respectively.

Secondly, on the LIVE IQA database, the RMFIQA-M4 achieves better performance than M1 - 0.954 v.s. 0.945 in SRCC, which implies that features from inter-mediate layers do make a difference in understanding the quality degradation. Furthermore, the RMFIQA-M3 performs even better than RMFIQA-M4 by an almost equal margin ($0.009 = 0.963 - 0.954$ in SRCC, $0.015 = 0.953 - 0.938$ in PLCC) to the RMFIQA-M4 compared to M1 ($0.009 = 0.954 - 0.945$ in SRCC, $0.014 = 0.938 - 0.924$ in PLCC), indicating that the single encoder5 enhances the capability of the deep network more strongly than the four low-level ones, consistent with findings in [19] that the expressibility of the auto-encoder increases exponentially with the network depth.

On the LIVE Challenge database, from Tables 3 and 4, RMFIQA-M3 demonstrates the best performance, followed by the RMFIQA-M2, while the RMFIQA-M4 only gets a little bit better results than M1, which implies that images with authentic distortions can be more robust with lower-level features than images with synthetic distortions. On the other hand, for applications such as image compression and transmission which produce more synthetic-type distortions like "JPEG" and "JP2K", one may focus more on the extraction of lower-level features than high-level ones in quality assessment.

From Table 5, it can be found that, compared to benchmarks in [1, 3, 6–8, 11, 17, 18], our RMFIQA model achieves comparable performance in terms of SRCC and PLCC on both LIVE IQA and LIVE Challenge databases. We note that our results can also be further improved by selecting other deeper networks as feature generator (the backbone network). Finally, due to the limitation of distortion types generated for training the Siamese network (only four types at current stage), the generalizing ability for authentically distorted images of models that we have trained for this paper is limited. Thus, our model does not perform so well as the MFIQA on the LIVE Challenge database where all images are exposed to authentic distortions.

Table 5. SRCC and PLCC comparison of the RMFIQA model with benchmarks. LIVE CL stands for the LIVE Challenge database.

Model	LIVE		LIVE CL	
	SRCC PLCC		SRCC PLCC	
BRISQUE	0.940 -		–	
FRIQUEE	0.93 0.95		0.71 0.70	
HOSA	0.950 0.953		–	
CNN	0.956 0.953		–	
DIQaM	0.960 **0.972**		0.606 0.601	
WaDIQaM	0.954 0.963		0.671 0.680	
NIMA	–		0.637 0.698	
MT-RankIQA	*0.976 0.973*		–	
MFIQA	**0.964 0.967**		*0.835 0.867*	
RMFIQA	**0.963** 0.953		**0.806 0.836**	

4   Conclusion

In this paper, we propose a framework for NR-IQA by transfer learning, exploring unlabeled data through the Siamese network, and exploiting the expressibility of features from multiple intermediate layers. From a series of experiments, we validate that the Siamese network trained on the automatically generated ranked dataset captures the ability to discriminate distortion levels of images; multi-level features help the network understand the distortion better, and in particular, high-level features are more sensitive to authentic distortions because more changes in global statistics, such as color contrast and saturation, appear frequently in authentically distorted images. It is also noted that we believe that the experimental results presented here are not the best performance that our RMFIQA can achieve. One could replace the VGG16 feature generator with any other deeper CNNs and expect better predicting results, as more useful information will be gained by encoding features from deeper layers, especially for those images captured in a natural environment.

Last but not least, we want to point out that the idea of exploring unlabeled data is not only important in scientific research, but also essential in industrial applications. There is infinite image data around the world, and hundreds of thousands of new digital images are produced every day, which makes us impossible to label "enough" data. What's more, exploiting the usage of intermediate features can help networks understand the contextual content better so as to enhance the capability of DNNs in vision tasks. We shall continue work towards this direction in the future.

References

1. Bosse, S., Maniry, D., Müller, K.R., Wiegand, T., Samek, W.: Deep neural networks for no-reference and full-reference image quality assessment. IEEE Trans. Image Process. **27**(1), 206–219 (2017)
2. Deng, J., Dong, W., Socher, R., Li, L.J., Li, K., Fei-Fei, L.: ImageNet: a large-scale hierarchical image database. In: 2009 IEEE Conference on Computer Vision and Pattern Recognition, pp. 248–255. IEEE (2009)
3. Ghadiyaram, D., Bovik, A.C.: Massive online crowdsourced study of subjective and objective picture quality. IEEE Trans. Image Process. **25**(1), 372–387 (2015)
4. He, K., Zhang, X., Ren, S., Sun, J.: Deep residual learning for image recognition. In: Proceedings of the IEEE Conference on Computer Vision and Pattern Recognition, pp. 770–778 (2016)
5. Howard, A.G., et al.: MobileNets: efficient convolutional neural networks for mobile vision applications. arXiv preprint arXiv:1704.04861 (2017)
6. Kang, L., Ye, P., Li, Y., Doermann, D.: Convolutional neural networks for no-reference image quality assessment. In: Proceedings of the IEEE Conference on Computer Vision and Pattern Recognition, pp. 1733–1740 (2014)
7. Kim, J., Nguyen, A.D., Ahn, S., Luo, C., Lee, S.: Multiple level feature-based universal blind image quality assessment model. In: 2018 25th IEEE International Conference on Image Processing (ICIP), pp. 291–295. IEEE (2018)
8. Liu, X., Van De Weijer, J., Bagdanov, A.D.: Exploiting unlabeled data in CNNs by self-supervised learning to rank. IEEE Trans. Pattern Anal. Mach. Intell. **41**, 1862–1878 (2019)
9. Liu, X., van de Weijer, J., Bagdanov, A.D.: RankIQA: learning from rankings for no-reference image quality assessment. In: Proceedings of the IEEE International Conference on Computer Vision, pp. 1040–1049 (2017)
10. Ma, K., et al.: Waterloo exploration database: new challenges for image quality assessment models. IEEE Trans. Image Process. **26**(2), 1004–1016 (2017)
11. Mittal, A., Moorthy, A.K., Bovik, A.C.: No-reference image quality assessment in the spatial domain. IEEE Trans. Image Process. **21**(12), 4695–4708 (2012)
12. Ponomarenko, N., et al.: Image database TID2013: peculiarities, results and perspectives. Signal Process. Image Commun. **30**, 57–77 (2015)
13. Sheikh, H.R., Sabir, M.F., Bovik, A.C.: A statistical evaluation of recent full reference image quality assessment algorithms. IEEE Trans. Image Process. **15**(11), 3440–3451 (2006)
14. Simonyan, K., Zisserman, A.: Very deep convolutional networks for large-scale image recognition. arXiv preprint arXiv:1409.1556 (2014)
15. Szegedy, C., et al.: Going deeper with convolutions. In: Proceedings of the IEEE Conference on Computer Vision and Pattern Recognition, pp. 1–9 (2015)
16. Szegedy, C., Vanhoucke, V., Ioffe, S., Shlens, J., Wojna, Z.: Rethinking the inception architecture for computer vision. In: Proceedings of the IEEE Conference on Computer Vision and Pattern Recognition, pp. 2818–2826 (2016)
17. Talebi, H., Milanfar, P.: NIMA: neural image assessment. IEEE Trans. Image Process. **27**(8), 3998–4011 (2018)
18. Xu, J., Ye, P., Li, Q., Du, H., Liu, Y., Doermann, D.: Blind image quality assessment based on high order statistics aggregation. IEEE Trans. Image Process. **25**(9), 4444–4457 (2016)
19. Ye, J.C., Sung, W.K.: Understanding geometry of encoder-decoder CNNs. arXiv preprint arXiv:1901.07647 (2019)

20. Zeiler, M.D., Fergus, R.: Visualizing and understanding convolutional networks. In: Fleet, D., Pajdla, T., Schiele, B., Tuytelaars, T. (eds.) ECCV 2014. LNCS, vol. 8689, pp. 818–833. Springer, Cham (2014). https://doi.org/10.1007/978-3-319-10590-1_53
21. Zhang, W., Ma, K., Yang, X.: Learning to blindly assess image quality in the laboratory and wild. arXiv preprint arXiv:1907.00516 (2019)

Reversible Data Hiding Based on Prediction-Error-Ordering

Jianqiang Qin and Fangjun Huang[✉]

Guangdong Province Key Laboratory of Information Security Technology,
School of Data and Computer Science, Sun Yat-sen University, Guangzhou 510006, GD, China
huangfj@mail.sysu.edu.cn

Abstract. In this paper, we propose a new reversible data hiding scheme called prediction-error-ordering (PEO). Instead of sorting the pixels directly as that in pixel-value-ordering (PVO) series, the obtained prediction errors in each block are sorted in ascending (or descending) order. After sorting, the maximum/minimum value in each block is predicted with the second largest/smallest value. Since the prediction errors are around zero in general, a steeper histogram with a peak value of 0 can be easily obtained, which means that more elements can be expanded for data hiding in the next embedding process. Combined with a new adaptive embedding strategy, the marked images with better visual quality can be obtained compared with those popular PVO series.

Keywords: Prediction-error-ordering · Reversible data hiding · Local texture complexity

1 Introduction

Reversible data hiding (RDH) aims to embed secret information into a cover image by slightly modifying its pixels, and more importantly, the original image can be completely restored while extracting the embedded secret information from the marked image [1, 2]. This technique can be applied in many sensitive areas, such as law forensics, medical image processing, and military image processing.

Various RDH schemes have been proposed on the basis of three fundamental strategies, *i.e.*, lossless compression [3–5], difference expansion (DE) [6–9], and histogram shifting (HS) [10–14]. Recent studies demonstrate that DE and HS strategies can be integrated together to form a series of new RDH strategies, such as prediction-error expansion (PEE) [15–22], pixel-value-ordering (PVO) [23–26], pairwise prediction-error expansion (Pairwise-PEE) [27–29], and multiple histograms modification (MHM) [22, 26, 29]. In all these aforementioned strategies, PVO has received much attention because of its simple principle yet quite good performance, which was first proposed by Li *et al.* [23]. In PVO based method, the image is divided into blocks and pixels in each block are sorted in ascending (or descending) order. Then, the second biggest value and smallest value are used to predict the maximum value and minimum value, respectively. Finally, the secret information is embedded through modifying the maximum value and

© Springer Nature Switzerland AG 2020
Y. Peng et al. (Eds.): PRCV 2020, LNCS 12305, pp. 527–537, 2020.
https://doi.org/10.1007/978-3-030-60633-6_44

minimum value. After data embedding the order of pixel values in each block remains unchanged, thus ensuring the reversibility. In [24], Peng *et al.* proposed an Improved-PVO (IPVO) with considering the locations of the maximum/minimum and the second largest/smallest pixels. In [25, 26], Ou *et al.* further applied IPVO to two-dimensional (2D) histogram modification and multiple histograms modification (MHM). Since more redundancy can be exploited in these methods [24–26], the marked image with better visual quality can be obtained.

In this paper, motived by the philosophy behind PVO, we propose a new RDH scheme called prediction-error-ordering (PEO). In our new scheme, instead of sorting the pixels directly, the obtained prediction errors in each block are sorted in ascending/descending order. The maximum/minimum value in each block is predicted with the second largest/smallest value. Since the prediction errors are near zero in general, a steeper histogram with a peak value of 0 can be easily obtained. That is, more elements can be expanded for data hiding in the next embedding process. Moreover, a new adaptive embedding strategy with utilizing the characteristics of checkerboard is proposed to further improve the performance of our proposed method.

The rest of this paper is organized as follows. In Sect. 2, we briefly review the related work. In Sect. 3, the proposed PEO scheme is introduced in detail. Experimental results are provided in Sect. 4 and the conclusion is given Sect. 5.

2 Related Work

In this section we will introduce the IPVO briefly. As mentioned above, IPVO is an improved version of PVO and can be applied to 2D histogram modification and multiple histograms modification. In IPVO method, the image is divided into pixel blocks with the same size. Suppose that a pixel block B has n pixels $x_1, x_2, \ldots, x_{n-1}, x_n$. Firstly, these pixels are sorted in ascending order to obtain the new pixel sequence $\{x_{\sigma_1}, x_{\sigma_2}, \ldots, x_{\sigma_{n-1}}, x_{\sigma_n}\}$. Note that $x_{\sigma_1} \leq x_{\sigma_2} \leq \ldots \leq x_{\sigma_{n-1}} \leq x_{\sigma_n}$. Then predict x_{σ_1} with x_{σ_2}, and predict x_{σ_n} with $x_{\sigma_{n-1}}$. The prediction errors of x_{σ_1} and x_{σ_n} are denoted by p_m and p_M, respectively, where p_m is calculated as follows.

$$s = \min(\sigma_1, \sigma_2) \tag{1}$$

$$t = \max(\sigma_1, \sigma_2) \tag{2}$$

$$p_m = x_s - x_t \tag{3}$$

After computing the prediction error, the message bit can be embedded according to Eq. (4).

$$p'_m = \begin{cases} p_m + b & \text{if } p_m = 1 \\ p_m - b & \text{if } p_m = 0 \\ p_m + 1 & \text{if } p_m > 1 \\ p_m - 1 & \text{if } p_m < 0 \end{cases} \quad \Leftrightarrow$$

$$x'_{\sigma_1} = \begin{cases} x_{\sigma_1} - b \text{ if } p_m = 1, \sigma_1 > \sigma_2 \\ x_{\sigma_1} - b \text{ if } p_m = 0, \sigma_1 < \sigma_2 \\ x_{\sigma_1} - 1 \text{ if } p_m > 1, \sigma_1 > \sigma_2 \\ x_{\sigma_1} - 1 \text{ if } p_m < 0, \sigma_1 < \sigma_2 \end{cases} \tag{4}$$

where b represents the secret message bit to be embedded, p'_m represents the prediction error after the message embedding, and x'_{σ_1} represents the pixel value in the marked image. In a similar way, p_M is calculated according to Eqs. (5)–(7).

$$u = \min(\sigma_{n-1}, \sigma_n) \tag{5}$$

$$v = \max(\sigma_{n-1}, \sigma_n) \tag{6}$$

$$p_M = x_u - x_v \tag{7}$$

As before, the message bit is embedded as follows.

$$p'_M = \begin{cases} p_M + b & \text{if } p_M = 1 \\ p_M - b & \text{if } p_M = 0 \\ p_M + 1 & \text{if } p_M > 1 \\ p_M - 1 & \text{if } p_M < 0 \end{cases} \Leftrightarrow$$

$$x'_{\sigma_n} = \begin{cases} x_{\sigma_n} + b \text{ if } p_M = 1, \sigma_{n-1} > \sigma_n \\ x_{\sigma_n} + b \text{ if } p_M = 0, \sigma_{n-1} < \sigma_n \\ x_{\sigma_n} + 1 \text{ if } p_M > 1, \sigma_{n-1} > \sigma_n \\ x_{\sigma_n} + 1 \text{ if } p_M < 0, \sigma_{n-1} < \sigma_n \end{cases} \tag{8}$$

where b represents the secret message bit to be embedded, p'_M represents the prediction error after the message embedding, and x'_{σ_n} represents the pixel value in the marked image. In order to further improve the visual quality of the marked image, IPVO adopts an adaptive block selection strategy.

3 The Proposed Method

As described in Sect. 2, in PVO series the divided image blocks may have different sizes. In our new scheme, the prediction errors may also be divided into blocks with different sizes. For simplicity, we represent it as k-PEO, where k represents the number of prediction errors in each block. In this section, the 4-PEO method will be introduced in detail firstly, and then the general approach will be given.

3.1 4-PEO

As that in [17], the pixels in a cover image are categorized into two sets, denoted as shadow pixels and blank pixels, which are shown in Fig. 1(a). For ease of explanation, suppose that the image is visited in the order from left to right and then top to bottom.

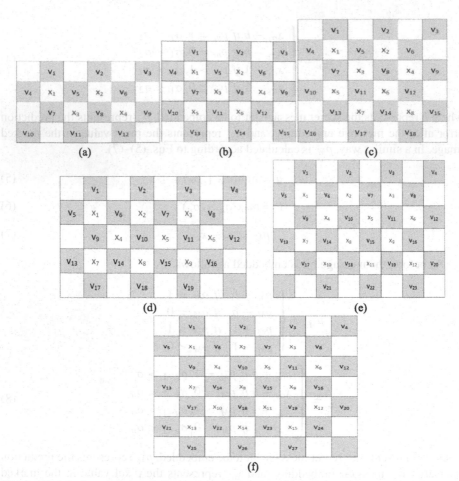

Fig. 1. The pixels and their neighbors of 4-PEO, 6-PEO, 8-PEO, 9-PEO, 12-PEO, and 15-PEO patterns.

The message bits are embedded into these two types of pixels separately, *i.e.*, in the first layer, the blank pixels are used for embedding data and the shadow pixels for computing prediction errors; in the second layer, the shadow pixels are used for embedding data and blank pixels for computing prediction errors.

The prediction values of x_1, x_2, x_3, x_4, denoted by p_1, p_2, p_3, p_4, are calculated as follows.

$$p_1 = \left\lfloor \frac{v_1 + v_4 + v_5 + v_7}{4} \right\rfloor \tag{9}$$

$$p_2 = \left\lfloor \frac{v_2 + v_5 + v_6 + v_8}{4} \right\rfloor \tag{10}$$

$$p_3 = \left\lfloor \frac{v_5 + v_7 + v_8 + v_{11}}{4} \right\rfloor \tag{11}$$

$$p_4 = \left\lfloor \frac{v_6 + v_8 + v_9 + v_{12}}{4} \right\rfloor \tag{12}$$

where, $\lfloor * \rfloor$ is the floor function. Then, the prediction errors corresponding to the pixels x_1, x_2, x_3, x_4, denoted by e_1, e_2, e_3, e_4, are computed according to Eqs. (13)–(16).

$$e_1 = x_1 - p_1 \tag{13}$$

$$e_2 = x_2 - p_2 \tag{14}$$

$$e_3 = x_3 - p_3 \tag{15}$$

$$e_4 = x_4 - p_4 \tag{16}$$

Different from that in PVO series the original pixels are sorted directly, in our algorithm the prediction errors e_1, e_2, e_3, e_4 are sorted in ascending order to obtain the sorted prediction-error sequence $\{e_{\sigma_1}, e_{\sigma_2}, e_{\sigma_3}, e_{\sigma_4}\}$, and $e_{\sigma_1} \le e_{\sigma_2} \le e_{\sigma_3} \le e_{\sigma_4}$. Then we use e_{σ_2} to predict e_{σ_1}, and e_{σ_3} to predict e_{σ_4}. The difference between e_{σ_1} and e_{σ_2} is represented with d_m, and the difference between e_{σ_3} and e_{σ_4} is represented with d_M. The difference d_m is calculated according to Eqs. (17)–(19).

$$s = \min(\sigma_1, \sigma_2) \tag{17}$$

$$t = \max(\sigma_1, \sigma_2) \tag{18}$$

$$d_m = e_s - e_t \tag{19}$$

In the embedding process, the values of difference d_m, prediction error e_{σ_1}, and pixel x_{σ_1} are changed as follows.

$$d'_m = \begin{cases} d_m + b & \text{if } d_m = 1 \\ d_m - b & \text{if } d_m = 0 \\ d_m + 1 & \text{if } d_m > 1 \\ d_m - 1 & \text{if } d_m < 0 \end{cases} \Leftrightarrow$$

$$e'_{\sigma_1} = \begin{cases} e_{\sigma_1} - b & \text{if } d_m = 1, \sigma_1 > \sigma_2 \\ e_{\sigma_1} - b & \text{if } d_m = 0, \sigma_1 < \sigma_2 \\ e_{\sigma_1} - 1 & \text{if } d_m > 1, \sigma_1 > \sigma_2 \\ e_{\sigma_1} - 1 & \text{if } d_m < 0, \sigma_1 < \sigma_2 \end{cases} \Leftrightarrow \tag{20}$$

$$x'_{\sigma_1} = \begin{cases} x_{\sigma_1} - b & \text{if } d_m = 1, \sigma_1 > \sigma_2 \\ x_{\sigma_1} - b & \text{if } d_m = 0, \sigma_1 < \sigma_2 \\ x_{\sigma_1} - 1 & \text{if } d_m > 1, \sigma_1 > \sigma_2 \\ x_{\sigma_1} - 1 & \text{if } d_m < 0, \sigma_1 < \sigma_2 \end{cases}$$

where b represents the secret message bit to be embedded, and d'_m, e'_{σ_1}, and x'_{σ_1} correspond to d_m, e_{σ_1}, and x_{σ_1} in the marked image.

Similarly, d_M can be calculated according to Eqs. (21)–(23).

$$u = \min(\sigma_3, \sigma_4) \tag{21}$$

$$v = \max(\sigma_3, \sigma_4) \tag{22}$$

$$d_M = e_u - e_v \tag{23}$$

The message bits are embedded as follows.

$$d'_M = \begin{cases} d_M + b & \text{if } d_M = 1 \\ d_M - b & \text{if } d_M = 0 \\ d_M + 1 & \text{if } d_M > 1 \\ d_M - 1 & \text{if } d_M < 0 \end{cases} \Leftrightarrow$$

$$e'_{\sigma_4} = \begin{cases} e_{\sigma_4} + b & \text{if } d_M = 1, \sigma_3 > \sigma_4 \\ e_{\sigma_4} + b & \text{if } d_M = 0, \sigma_3 < \sigma_4 \\ e_{\sigma_4} + 1 & \text{if } d_M > 1, \sigma_3 > \sigma_4 \\ e_{\sigma_4} + 1 & \text{if } d_M < 0, \sigma_3 < \sigma_4 \end{cases} \Leftrightarrow \tag{24}$$

$$x'_{\sigma_4} = \begin{cases} x_{\sigma_4} + b & \text{if } d_M = 1, \sigma_3 > \sigma_4 \\ x_{\sigma_4} + b & \text{if } d_M = 0, \sigma_3 < \sigma_4 \\ x_{\sigma_4} + 1 & \text{if } d_M > 1, \sigma_3 > \sigma_4 \\ x_{\sigma_4} + 1 & \text{if } d_M < 0, \sigma_3 < \sigma_4 \end{cases}$$

where b represents the secret information to be embedded, and d'_M, e'_{σ_4}, and x'_{σ_4} correspond to d_M, e_{σ_4}, and x_{σ_4} in the marked image.

In order to further improve the visual quality of the marked image, a new adaptive embedding strategy is proposed. Different from that in PVO series only the second largest and the second smallest pixels in each block can be used to estimate the local texture complexity of the corresponding image block, in our proposed method more pixels can be utilized to estimate the local texture complexity. Considering the characteristics of the checkerboard, the local texture complexity can be evaluated via the shadow pixels while modifying the blank pixels. When those shadow pixels are modified for data hiding, the local texture complexity is evaluated via the blank pixels. As shown in Fig. 1(a), the four pixels x_1, x_2, x_3, x_4 are divided into the same block, denoted it by B_{pro}. When these pixels are used to embed the secret message, the twelve shadow pixels $v_1, v_2, \ldots, v_{12}$ are used for evaluating the local complexity. The local texture complexity of the pixel block B_{pro}, represented with $LC(B_{pro})$, is defined as the sum of both vertical and horizontal absolute differences of every two consecutive pixels in the neighbors of x_1, x_2, x_3, x_4. The specific calculation method is as follows.

$$\begin{aligned} LC(B_{pro}) = & |v_1 - v_2| + |v_2 - v_3| + |v_4 - v_5| + |v_5 - v_6| + |v_7 - v_8| + |v_8 - v_9| \\ & + |v_{10} - v_{11}| + |v_{11} - v_{12}| + |v_4 - v_{10}| + |v_1 - v_7| + |v_5 - v_{11}| \\ & + |v_2 - v_8| + |v_6 - v_{12}| + |v_3 - v_9| \end{aligned} \tag{25}$$

As that in PVO series, a threshold T can be determined according to the length of the message bits to be embedded, and the pixel blocks with $LC(B_{pro}) > T$ will be selected for data hiding.

3.2 The General Approach of PEO

In fact, our proposed method is not limited to 4-PEO pattern. It can be easily extended to a series of k-PEO based methods. The 4-PEO, 6-PEO, 8-PEO, 9-PEO, 12-PEO, and 15-PEO patterns are exemplified in Figs. 1(a)–(f), respectively. For those pixels $x_1, x_2, \ldots, x_{n-1}, x_n$, the prediction errors $e_1, e_2, \ldots, e_{n-1}, e_n$ can be computed via the rhombus prediction [17]. Then the prediction errors are sorted in ascending order to obtain the sorted prediction-error sequence $\{e_{\sigma_1}, e_{\sigma_2}, \ldots, e_{\sigma_{n-1}}, e_{\sigma_n}\}$. As that in 4-PEO based method, the maximum value is predicted by the second largest value, and the minimum value is predicted by the second smallest value. Thus d_m and d_M can be obtained. Note that in all the k-PEO based method, d_m is computed as that in Eqs. (26)–(28), and d_M is computed as follows.

For each k-PEO pattern, the local texture complexity corresponding to the pixel block is defined as the sum of both vertical and horizontal absolute differences of every two consecutive pixels in the neighbors of $x_1, x_2, \ldots, x_{n-1}, x_n$, which can be easily extended from Eq. (25).

As that in [30], all the auxiliary information (i.e., the location map L, the length of location map L, the length of secret message bits to be embedded and the threshold T) bits can be embedded into the least significant bits (LSBs) of some predetermined pixels via LSB replacement, and the replaced LSBs of the pixels are embedded into the cover image as a part of the payload. Note that the auxiliary information should be extracted first at data extraction phase.

$$u = \min(\sigma_{n-1}, \sigma_n) \tag{26}$$

$$v = \max(\sigma_{n-1}, \sigma_n) \tag{27}$$

$$d_M = e_u - e_v \tag{28}$$

4 Experiment

The experiment is enforced on 6 standard 512×512 sized gray-scale images: Lena, Baboon, Airplane, Barbara, Lake, and Boat. All the images are downloaded from the USC-SIPI database [31].

Firstly, the histograms generated by PEO and IPVO are compared, which are shown in Fig. 2. The red curve represents the histogram generated by 4-PEO, and the black curve represents the histogram generated by 9-PEO. The blue curve represents the histogram generated by IPVO with 2×2 block, and the green curve represents the histogram generated by IPVO with 3×3 block. It is observed from Fig. 2 that generally a steeper histogram can be generated by our proposed PEO except that Lake is selected as the cover image.

Next, the capacity-distortion performance of PEO is evaluated compared with PVO and IPVO. In the experiment, 'PEO original complexity' represents that the local texture complexity is calculated as that in PVO and IPVO methods, and 'PEO new complexity' represents that the local texture complexity is computed according to our new adaptive

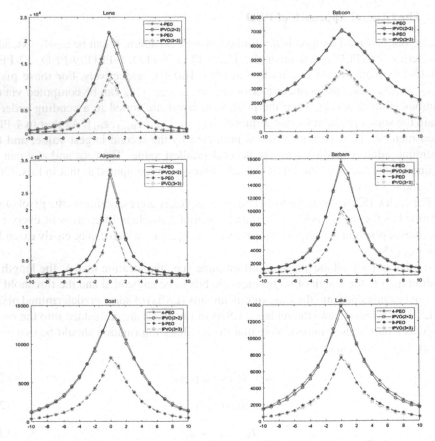

Fig. 2. Histogram generated by IPVO and PEO. (Color figure online)

embedding strategy (*i.e.*, according to Eq. (25)). The peak signal-to-noise ratio (PSNR), which is calculated between the original image and the marked image, is used as a measure to evaluate the capacity-distortion performance.

The different k-PEO (k = 4, 6, 8, 9, 12, 15) patterns are tested in our algorithm to find the one which may result in the highest PSNR value. The testing results corresponding to the optimal pattern are shown in Fig. 3, where the horizontal axis represents the embedding capacity (EC) and the vertical axis represents the PSNR value. Note that for PVO and IPVO series, the optimal block size is also exhaustively searched for a fair comparison. It is observed from Fig. 3 that with embedding the same number of information bits, 1) the PSNR values obtained by 'PEO original complexity' are larger than those obtained by IPVO and PVO in general except when image Lake is selected as the cover image, which demonstrates the efficiency of our new PEO strategy; 2) the PSNR values obtained by 'PEO new complexity' are larger than those obtained by 'PEO original complexity' in all cases, which demonstrates the efficiency of our new adaptive embedding strategy.

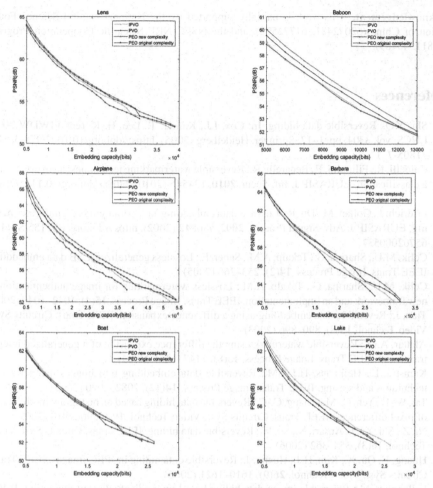

Fig. 3. Comparisons with PVO and IPVO.

5 Conclusion

In this paper, we propose a new RDH method called PEO. Instead of sorting the pixels according to their values directly, the prediction errors of the pixels are selected to be sorted in ascending/descending order, and the maximum/minimum value in each block is predicted with the second largest/smallest value. Since the prediction errors are around zero in general, a steeper histogram with a peak value of 0 can be obtained. Thus, more elements can be expanded for data hiding in the next embedding process. Combined with a new adaptive embedding strategy based on the characteristics of chessboard, the performance of our proposed method can be further improved. The experimental results demonstrate that with embedding the same number of information bits, our method can achieve better visual quality compared with the popular PVO series.

Acknowledgement. This work is partially supported by the National Natural Science Foundation of China (62072481, 61772572), and the NSFC-NRF Scientific Cooperation Program (61811540409).

References

1. Shi, Y.Q.: Reversible data hiding. In: Cox, I.J., Kalker, T., Lee, H.-K. (eds.) IWDW 2004. LNCS, vol. 3304, pp. 1–12. Springer, Heidelberg (2005). https://doi.org/10.1007/978-3-540-31805-7_1
2. Caldelli, R., Filippini, F., Becarelli, R.: Reversible watermarking techniques: an overview and a classification. EURASIP J. Inf. Secur. **2010**, 134546 (2010). https://doi.org/10.1155/2010/134546
3. Fridrich, J., Goljan, M., Du, R.: Lossless data embedding-new paradigm in digital watermarking. EURASIP J. Adv. Signal Process. **2002**, 986842 (2002). https://doi.org/10.1155/S1110865702000537
4. Celik, M.U., Sharma, G., Tekalp, A.M., Saber, E.: Lossless generalized-LSB data embedding. IEEE Trans. Image Process. **14**(2), 253–266 (2005)
5. Celik, M.U., Sharma, G., Tekalp, A.M.: Lossless watermarking for image authentication: a new framework and an implementation. IEEE Trans. Image Process. **15**(4), 1042–1049 (2006)
6. Tian, J.: Reversible data embedding using a difference expansion. IEEE Trans. Circuits Syst. Video Technol. **13**(8), 890–896 (2003)
7. Alattar, A.M.: Reversible watermark using the difference expansion of a generalized integer transform. IEEE Trans. Image Process. **13**(8), 1147–1156 (2004)
8. Kamstra, L., Heijmans, H.J.A.M.: Reversible data embedding into images using wavelet techniques and sorting. IEEE Trans. Image Process. **14**(12), 2082–2090 (2005)
9. Tai, W.-L., Yeh, C.-M., Chang, C.-C.: Reversible data hiding based on histogram modification of pixel differences. IEEE Trans. Circuits Syst. Video Technol. **19**(6), 906–910 (2009)
10. Ni, Z., Shi, Y.-Q., Ansari, N., Su, W.: Reversible data hiding. IEEE Trans. Circuits Syst. Video Technol. **16**(3), 354–362 (2006)
11. Huang, F., Qu, X., Kim, H.J., Huang, J.: Reversible data hiding in JPEG images. IEEE Trans. Circuits Syst. Video Technol. **26**(9), 1610–1621 (2006)
12. Fallahpour, M.: Reversible image data hiding based on gradient adjusted prediction. IEICE Electron. Exp. **5**(20), 870–876 (2008)
13. Hong, W., Chen, T.-S., Shiu, C.-W.: Reversible data hiding for high quality images using modification of prediction errors. J. Syst. Softw. **82**(11), 1833–1842 (2009)
14. Li, X., Li, B., Yang, B., Zeng, T.: General framework to histogram-shifting-based reversible data hiding. IEEE Trans. Image Process. **22**(6), 2181–2191 (2013)
15. Thodi, D.M., Rodriguez, J.J.: Expansion embedding techniques for reversible watermarking. IEEE Trans. Image Process. **16**(3), 721–730 (2007)
16. Hu, Y., Lee, H.-K., Li, J.: DE-based reversible data hiding with improved overflow location map. IEEE Trans. Circuits Syst. Video Technol. **19**(2), 250–260 (2009)
17. Sachnev, V., Kim, H.J., Nam, J., Suresh, S., Shi, Y.Q.: Reversible watermarking algorithm using sorting and prediction. IEEE Trans. Circuits Syst. Video Technol. **19**(7), 989–999 (2009)
18. Yi, S., Zhou, Y., Hua, Z.: Reversible data hiding in encrypted images using adaptive block-level prediction-error expansion. Signal Process. Image Commun. **64**, 78–88 (2018)
19. Gao, X., An, L., Yuan, Y., Tao, D., Li, X.: Lossless data embedding using generalized statistical quantity histogram. IEEE Trans. Circuits Syst. Video Technol. **21**(8), 1061–1070 (2011)
20. Li, X., Yang, B., Zeng, T.: Efficient reversible watermarking based on adaptive prediction-error expansion and pixel selection. IEEE Trans. Image Process. **20**(12), 3524–3533 (2011)

21. Wu, H.-T., Huang, J.: Reversible image watermarking on prediction errors by efficient histogram modification. Signal Process. **92**(12), 3000–3009 (2012)
22. Li, X., Zhang, W., Gui, X., Yang, B.: Efficient reversible data hiding based on multiple histograms modification. IEEE Trans. Inf. Forensics Secur. **10**(9), 2016–2027 (2015)
23. Li, X., Li, J., Li, B., Yang, B.: High-fidelity reversible data hiding scheme based on pixel-value-ordering and prediction-error expansion. Signal Process. **93**(1), 198–205 (2013)
24. Peng, F., Li, X., Yang, B.: Improved PVO-based reversible data hiding. Digit. Signal Process. **25**, 255–265 (2014)
25. Ou, B., Li, X., Wang, J.: High-fidelity reversible data hiding based on pixel-value-ordering and pairwise prediction-error expansion. J. Vis. Commun. Image Represent. **39**, 12–23 (2016)
26. Ou, B., Li, X., Wang, J.: Improved PVO-based reversible data hiding: a new implementation based on multiple histograms modification. J. Vis. Commun. Image Represent. **38**, 328–339 (2016)
27. Ou, B., Li, X., Zhao, Y., Ni, R., Shi, Y.-Q.: Pairwise prediction-error expansion for efficient reversible data hiding. IEEE Trans. Image Process. **22**(12), 5010–5021 (2013)
28. Li, X., Zhang, W., Gui, X., Yang, B.: A novel reversible data hiding scheme based on two-dimensional difference-histogram modification. IEEE Trans. Inf. Forensics Secur. **8**(7), 1091–1100 (2013)
29. Qin, J., Huang, F.: Reversible data hiding based on multiple two-dimensional histograms modification. IEEE Signal Process. Lett. **26**(6), 843–847 (2019)
30. Huang, F., Huang, J., Shi, Y.-Q.: New framework for reversible data hiding in encrypted domain. IEEE Trans. Inf. Forensics Secur. **11**(12), 2777–2789 (2016)
31. http://sipi.usc.edu/database/database.php?volume=mis

Aggregating Spatio-temporal Context for Video Object Segmentation

Yu Tao[1], Jian-Fang Hu[1,2(✉)], and Wei-Shi Zheng[1]

[1] Sun Yat-sen University, Guangzhou, China
[2] GuangDong Province Key Laboratory of Information Security Technology,
Guangzhou, China
hujf5@mail.sysu.edu.cn

Abstract. In this paper, we focus on aggregating spatio-temporal contextual information for video object segmentation. Our approach exploits
the spatio-temporal relationship among image regions by modelling the
dependencies among the corresponding visual features with a spatio-
temporal RNN. Our spatio-temporal RNN is placed on top of a pre-
trained CNN network to simultaneously embed spatial and temporal
information into the feature maps. Following the spatio-temporal RNN,
we further construct an online adaption module to adapt the learned
model for segmenting specific objects in given video. We show that our
adaption module can be optimized efficiently with closed-form solutions.
Our experiments on two public datasets illustrate that the proposed
method performs favorably against state-of-the-art methods in terms of
efficiency and accuracy.

Keywords: Video object segmentation · Spatio-temporal RNN ·
Online adaption · Feature extraction · Deep learning

1 Introduction and Related Work

Video object segmentation, which needs to segment objects in a pixel-level over
a sequence of video frames, has drawn more and more attention in recent years.
Typically, mask annotations indicating the objects to be segmented are assumed
to be given in the first frame. From this perspective, the video object segmentation problem can be viewed as a pixel-level tracking problem. The main challenge
for this problem is to handle the large appearance change caused by motions,
occlusion, deformation and interaction with other objects. Recent researches
[3,21] show that exploring useful contextual information is an effective way to
confront this challenge.

This work is partially supported by the National Key Research and Development Pro-
gram of China (2018YFB1004903), NSFC (61702567, 61628212), SF-China (61772570),
Pearl River S&T Nova Program of Guangzhou (201806010056), Guangdong Natural
Science Funds for Distinguished Young Scholar (2018B030306025), and FY19-Research-
Sponsorship-185.

Y. Peng et al. (Eds.): PRCV 2020, LNCS 12305, pp. 538–550, 2020.
https://doi.org/10.1007/978-3-030-60633-6_45

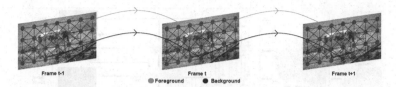

Fig. 1. Example illustrating the spatio-temporal dependencies in video. As shown in this figure, the contextual information depicted in the spatial and temporal neighboring image regions could be highly related with each other. This means that the spatio-temporal dependencies between corresponding features are helpful for distinguishing the foreground objects from background. With this in mind, we propose a spatio-temporal RNN to model the spatio-temporal dependencies for robust segmentation of foreground objects.

Many methods intend to capture some contextual information for improving video object segmentation. For example, the mask propagation based methods [6,8,11,21] model temporal dependencies by taking the mask predicted for the previous frame as an additional input to the model. Some methods [6,8,11] use optical flow to capture temporal information. However, the spatial dependencies among image regions are not specifically exploited in these methods. As illustrated in Fig. 1, the local contextual information depicted in neighboring image regions could be highly related with each other in both spatial and temporal directions. With this in mind, we argue that explicitly exploring spatial and temporal contexts jointly can benefit video object segmentation, which motivates our studies in this work. It is worth noticing that PML [3], which is the most related work to our approach, also utilizes spatial and temporal contextual information to improve video object segmentation. However, they simply take spatio-temporal coordinates as additional inputs and thus can not capture enough spatio-temporal information.

In this work, we propose a novel spatio-temporal Recurrent Neural Network (RNN) which exploits useful spatio-temporal contexts for video object segmentation. In our spatio-temporal RNN, the local context is explicitly propagated to the neighboring image regions along both spatial and temporal directions. We develop a spatio-temporal RNN layer that injects spatio-temporal context into the learned feature maps, and it can be architecturally integrated with existing Convolutional Neural Network (CNN) to construct an end-to-end network. Besides, we introduce dilated recurrent skip connections in spatio-temporal RNN to mitigate the information vanishing issue.

Recent studies show that tuning the model for each test video can obtain better segmentation results [1,6,8,11,18,19]. Following this pipeline, we develop an online adaption module for fast video object segmentation. In contrast to tuning all the parameters of the deep network as done in [1,11], in this module, we select to only update the parameters of the classification layer, which refers to 1×1 convolutions. More specifically, our adaption module subsumes two components: (i) tuning the model so that it is adapted to segment the objects

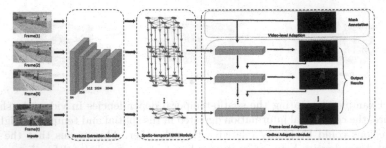

Fig. 2. An overview of the proposed framework. Our framework consists of three modules: a feature extraction module with several convolution layers for extracting feature maps from each frame; a spatio-temporal RNN module for embedding spatio-temporal contextual information into the learned feature maps; and finally, an online adaption module to adapt a 1 × 1 convolution layer online for segmenting specific objects in given video.

indicated by ground truth masks in the first frame; (ii) online updating the model to capture larger appearance variation based on the segmentation results of previous frames. We formulate each component in our online adaption module as an unconstrained quadratic problem, so that it can be optimized efficiently with closed-form solution.

Overall, our main contributions are: (i) a novel spatio-temporal RNN for aggregating spatial and temporal context simultaneously; (ii) an online adaption module for adapting the learned model to new video observations efficiently; and (iii) based on the Spatio-temporal RNN and online adaption, a novel video object segmentation system for segmenting objects at a fast speed (about 8.33 FPS).

2   Method

In this work, we treat the video object segmentation problem as a per-pixel classification task like most semantic segmentation models [2, 6, 21], and we present a novel framework which explicitly explores the spatial and temporal contextual information. Our framework mainly consists of three modules: (i) a feature extraction module for extracting basic visual feature from each individual frame; (ii) a spatio-temporal RNN module for aggregating dense local context in both spatial and temporal directions; and (iii) an online adaption module for updating the parameters of the classification layer, referring to a 1 × 1 convolution layer following the spatio-temporal RNN. The overall architecture of our framework is summarized in Fig. 2. In the following, we describe each module in detail.

2.1   Feature Extraction Module

Our feature extraction module is employed to extract basic visual features from each video frame. Here, we follow PML [3] to use the basic part of

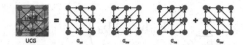

Fig. 3. The UCG is decomposed into 4 DAGs with different directions [16].

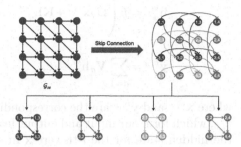

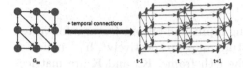

Fig. 4. Comparison of the DAG [16] (left) and our spatio-temporal RNN (right).

Fig. 5. DAG Graph with skip connections. In this example, the skip length is set as $S_x = 2, S_y = 2$.

Deeplab-v2[1] [2] as our feature extraction module. Deeplab-v2 is one of the most widely used deep model for image segmentation. Recent studies show that it performs well in capturing useful static contexts and generalizes well to other related segmentation tasks [22].

2.2 Spatio-temporal RNN Module

Following the feature extraction module, we develop a spatio-temporal RNN module to exploit useful spatio-temporal contexts in video segmentation.

Our spatio-temporal RNN is inspired by the DAG-RNN [16], which models the spatial contextual dependencies among image regions for scene segmentation. It is assumed that the neighboring elements interact with each other, which forms an undirected cyclic graph (UCG). As illustrated in Fig. 3, the UCG $\mathcal{G}^{\mathcal{U}}$ is then topologically approximated by 4 directed acyclic graphs (DAGs) as $\mathcal{G}^{\mathcal{U}} = \{\mathcal{G}_{se}, \mathcal{G}_{sw}, \mathcal{G}_{ne}, \mathcal{G}_{nw}\}$.

However, the UCG does not consider the temporal relationship of the image units, which makes it less effective for segmenting objects in videos. Here, we develop a spatio-temporal RNN by additionally adding connections along temporal direction to each DAG in $\mathcal{G}^{\mathcal{U}} = \{\mathcal{G}_{se}, \mathcal{G}_{sw}, \mathcal{G}_{ne}, \mathcal{G}_{nw}\}$, please refer to Fig. 4 for an example. In this way, our spatio-temporal RNN is able to aggregate both the spatial and temporal contexts among image regions. Specifically, we formulated our spatial-temporal RNN as:

---

[1] The Deeplab-v2 is pre-trained on COCO [10] with a ResNet-101 backbone.

$$\mathbf{h}_d^{v_{i,t}} = f\left(\mathbf{U}_d\mathbf{x}^{v_{i,t}} + \mathbf{K}\mathbf{h}_d^{v_{i,t-1}} + \sum_{v_{j,t}\in\mathcal{P}_{\mathcal{G}_d}(v_{i,t})} \mathbf{K}_d\mathbf{h}_d^{v_{j,t}} + \mathbf{b}_d\right),$$

$$\mathbf{y}^{v_{i,t}} = \sum_{d=1}^{|\mathcal{G}^{\mathcal{U}}|} \mathbf{V}_d\mathbf{h}_d^{v_{i,t}} + \mathbf{c}, \tag{1}$$

where $\mathbf{x}^{v_{i,t}}$ and $\mathbf{y}^{v_{i,t}}$ are the corresponding input and output features for vertex $v_{i,t}$, which form our input and output feature maps, respectively[2]. $\mathbf{h}_d^{v_{i,t}}$ indicates the hidden state for the i-th vertex in the t-th frame. $\mathbf{K}_d$ and $\mathbf{K}$ are matrices encoding the information propagation along the spatial and temporal directions, respectively. $\mathbf{U}_d$ and $\mathbf{V}_d$ are parameters to transform the input features and hidden states. $\mathcal{P}_{\mathcal{G}_d}(v_{i,t})$ denotes the direct predecessor set of vertex $v_{i,t}$ in DAG graph $\mathcal{G}_d$. f is an activation function. $\mathbf{c}$ and $\mathbf{b}_d$ are the bias terms.

Due to the context vanishing problem [16], long-term contextual dependencies could be hardly captured by the spatio-temporal RNN in Eq. (1). To mitigate this issue, we propose to use skip connections in our spatio-temporal RNN architecture. Taking the DAG graph $\mathcal{G}_{se}$ presented in Fig. 5 for example, the direct predecessor set of a vertex (i,j) is $\mathcal{P}_{\mathcal{G}_{se}}(i,j) = \{(i-1,j-1),(i-1,j),(i,j-1)\}$. When skip connections are applied, it can be formulated as $\mathcal{P}_{\mathcal{G}_{se}}(i,j) = \{(i-S_x,j-S_y),(i-S_x,j),(i,j-S_y)\}$, where S_x and S_y indicate the skip length along x-axis and y-axis, respectively. Then, the original DAG graph are splitted into $S_x \cdot S_y$ non-overlapping sub-graphs. The proposed skip connections enable our spatio-temporal RNN to capture long-term contextual dependencies and it has the advantage of reducing the computation overhead as well.

2.3   Online Adaption Module

Since the spatio-temporal RNN and feature extraction modules can only learn some object-independent spatio-temporal information from the training videos, they can not be directly used to segment specific objects in given videos. Recent researches [1,11] show that tuning the learned model for each specific video can obtain a better segmentation performance. However, their methods need to update all the parameters of the deep network, which often takes a lot of time. This makes their approaches less applicable in real-world applications.

Here, we tackle the video object segmentation as a pixel-wise classification problem and place a classification layer on top of the spatio-temporal RNN module. The classification layer is defined as a 1×1 convolution layer. Here, we only tune the classification layer and fix the parameters of the feature extraction and spatio-temporal RNN modules. We refer this procedure as online adaption module. Our adaption module contains two different components, i.e., video-level adaption and frame-level adaption. The video-level adaption is used to tune the model so that it is suitable for segmenting objects of interest, which are indicated by the masks in the first frame. While the frame-level adaption is to update the

---

[2] Each vertex corresponds to a pixel on the feature map.

model for capturing more appearance variation as more frames are provided and segmented. In the following, we will describe each component in detail. We also show that both of our adaption components have closed-form solutions and thus the adaption module can be computed very efficiently.

Video-Level Adaption. This component is developed to tune the parameters in the 1×1 convolution layer (denoted by $\mathbf{W}_1$). The video-level adaption is achieved by solving the following optimization problem:

$$\mathbf{W}_1^* = \arg \min_{\mathbf{W}_1} [L(\mathbf{W}_1) + \lambda_1 \|\mathbf{W}_1\|_F^2],$$
$$L(\mathbf{W}_1) = L_+(\mathbf{W}_1) + \lambda_0 L_-(\mathbf{W}_1), \tag{2}$$

where $L_+(\mathbf{W}_1) = \|\mathbf{W}_1^T \mathbf{X}^+ - \mathbf{Y}^+\|_F^2$ and $L_-(\mathbf{W}_1) = \|\mathbf{W}_1^T \mathbf{X}^- - \mathbf{Y}^-\|_F^2$ are two regression losses. They are employed to regress the positive and negative samples in first frame. $\mathbf{X}^+$ and $\mathbf{X}^-$ indicate the features extracted from the positive (objects of interest) and negative (background) samples in the first frame, respectively. $\mathbf{Y}^+$ and $\mathbf{Y}^-$ indicate the corresponding label information in one-hot format. λ_0 is used to control the contribution of the regression losses. $\lambda_1 \|\mathbf{W}_1\|_F^2$ is a regularization term.

This is an unconstrained convex minimization problem, whose solution is given by: $\mathbf{W}_1^* = (\mathbf{X}^+ \mathbf{X}^{+T} + \lambda_0 \mathbf{X}^- \mathbf{X}^{-T} + \lambda_1 \mathbf{I})^{-1} (\mathbf{X}^+ \mathbf{Y}^{+T} + \mathbf{X}^- \mathbf{Y}^{-T})$.

Frame-Level Adaption. The parameters $\mathbf{W}_1$ adapted from the mask annotation of the first frame could not be very robust for segmenting objects whose appearance vary drastically in the video. Here, we further update the parameters of the classification layer according to the new observed video frames and segmentation results in an online learning manner. Our frame-level adaption is achieved by solving the following problem:

$$\Delta \mathbf{W}^* = \arg \min_{\Delta \mathbf{W}} [L(\mathbf{W}_t) + \alpha_1 \|\mathbf{W}_t - \mathbf{W}_1\|_F^2 + \alpha_2 \|\Delta \mathbf{W}\|_F^2],$$
$$L(\mathbf{W}_t) = L_+(\mathbf{W}_t) + \lambda_0 L_-(\mathbf{W}_t),$$
$$\mathbf{W}_t = \mathbf{W}_{t-1} + \Delta \mathbf{W}, \tag{3}$$

where $\mathbf{W}_t$ is the parameter of the classification layer, adapted from the segmentation results for the t-th frame and it would be used for segmenting the next frame. Similar with formula (2), $L_+(\mathbf{W}_t)$ and $L_-(\mathbf{W}_t)$ are two regression losses for the positive and negative samples with high segmentation score, which are selected from the segmentation results of the t-th frame. Term $\alpha_1 \|\mathbf{W}_t - \mathbf{W}_1\|_F^2$ is used to control the deviation between the parameters with/without online learning. Term $\alpha_2 \|\Delta \mathbf{W}\|_F^2$ is a regularization term. We can also obtain a solution for this problem and thus the frame-level adaption can be accomplished very fast.

2.4 Model Learning

Here, we describe how to determine the model parameters using the given train-ing set and test videos. In the training stage, our model intends to learn some object-independent spatio-temporal cues that is useful for video object segmen-tation. While in the test stage, the model needs to learn some object-specific information for segmenting certain objects.

Model Training. Here, we mainly train the feature extraction module and Spatio-temporal RNN from the provided training videos. Since training the backbone CNN and spatio-temporal RNN jointly consumes a large amount of computing resources, we have to split the training stage into the following two steps:

- Step1: We discard the temporal connections in the spatio-temporal RNN and train the CNN and spatial RNN parts together.
- Step2: We fix the CNN part and tuning the parameters of spatio-temporal RNN module.

In training stage, we expected that the extracted features corresponding to the same object are closed to each other across different frames. Similar to [3], we randomly select several frames as anchor frames and pool frames. The foreground pixels in anchor frames make up the anchor set $\mathcal{A}$. The foreground pixels and background pixels in pool frames make up the positive pool $\mathcal{P}$ and the negative pool $\mathcal{N}$, respectively. It is expected that the extracted feature of the samples in $\mathcal{A}$ are pushed closer to the samples in $\mathcal{P}$, and stayed away from the samples in $\mathcal{N}$. Let $F(\cdot)$ denotes the feature extracted with the combination of the corresponding CNN and RNN networks described in Step1 and Step2. Then the loss function can be formulated as:

$$\sum_{a \in \mathcal{A}} \frac{\min_{p \in \mathcal{P}} \|F(a) - F(p)\|_2^2}{\min_{n \in \mathcal{N}} \|F(a) - F(n)\|_2^2 + \alpha}, \tag{4}$$

where α is a small constant used to avoid zero denominator.

Model Inference. Given a test video, our inference is quite straightforward. Specifically, we first use the feature extraction module to extract a feature map from each video frame. Then we feed the feature maps into the spatio-temporal RNN module to embed more spatio-temporal contextual information. Finally, we employ the proposed online adaption module to segment the objects indicated by the masks annotated in the initial frame.

3 Experiments

We evaluated our method on two datasets for video object segmentation. In the following, we will first describe some implementation details and then report the detailed results.

Table 1. Comparison results on the DAVIS-2016 validation set with other state-of-the-art approaches.

Method	$\mathcal{J}$	$\mathcal{F}$	FPS
MSK[11] (2017'CVPR)	79.7	75.4	0.08
OSVOS [1] (2017'CVPR)	79.8	80.6	0.11
OnAVOS [18] (2017'BMVC)	86.1	84.9	0.08
FAVOS [4] (2018'CVPR)	82.4	79.5	0.56
PML [3] (2018'CVPR)	75.5	79.3	3.63
OSMN [23] (2018'CVPR)	74.0	69.0	7.14
FEELVOS [17] (2019'CVPR)	80.3	83.1	2.22
FCVOS [5] (2019'TCSVT)	74.5	72	2.44
RSVOS [15] (2019'ICSPCC)	76.7	–	3.70
VAG [9] (2019'ICIEA)	66.2	61.4	10.00
Ours	79.0	77.6	8.33
Ours+FBS [14]	80.5	78.6	4.17

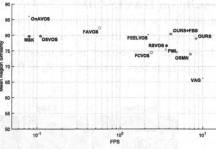

Fig. 6. A comparison of the quality and the speed with previous methods. We visualize the mean region similarity with respect to the FPS. Note that the FPS axis is in the log scale. Our method achieves a good trade-off between the segmentation accuracy and speed.

3.1 Implementation Details

We trained our model based on the training data provided in each set without using any auxiliary data. Our model is implemented using Tensorflow. We implement our feature extraction module based on the implementation in [20]. The dimensions of the hidden state and output in the spatio-temporal RNN are set as 256 and 128, respectively. We employed the widely used Adam algorithm to train our model. For the optimization of Step1, we randomly selected 3 frames from each training video and used one of them as anchor frame and the others as pool frames. We trained it 2 K iterations with a learning rate of 1e−5 and 1e−4 for the CNN and the spatial RNN, respectively. In Step2, in order to capture more temporal information, we randomly selected 12 ordering frames from each video and used 4 of them as anchor frames and the others as pool frames. We trained the model for 1k iterations with a learning rate of 1e−5.

It is worth noting that the computations of the basic feature extraction module for different video frames are independent with each other. Hence, we can compute the features of multiple frames in a parallel manner, which can further improve the inference speed.

3.2 Results for Video Object Segmentation

We evaluated our model on the DAVIS-2016 [12] and DAVIS-2017 [13] datasets and compared our results with the state-of-the-art approaches. We employed the metrics proposed in [12] (the region similarity $\mathcal{J}$ or IoU, contour accuracy $\mathcal{F}$) to measure the segmentation accuracy. We also compared the segmentation speed with other methods.

Results on DAVIS-2016. The DAVIS-2016 [12] is one of the most widely used dataset in the study of video object segmentation. This set contains a total of 50 videos, where 30 videos are used for training and the rest for validating. In this dataset, only one object is required to be segmented in each video. The model's hyper parameters were set as $\lambda_0 = 0.4, \lambda_1 = 0.1, \alpha_1 = 10^5$, and $\alpha_2 = 0.1$.

Fig. 7. Some visualization results on DAVIS-2016 set. As shown, for the samples that the objects are heavily occluded (Bmx-Trees and Libby), our method can achieve decent segmentation results. Our results for the sequences of *Drift-Straight* and *Paragliding-Launch* show that our method is also robust to the appearance change caused by motion and deformation. This figure is best viewed in color.

As summarized in Table 1, we compare our methods with the state-of-the-art approaches that do not use auxiliary data and other additional training datasets (such as YouTube-VOS). We observe that our network obtains decent segmentation performance in contrast to state-of-the-art methods, and its inference runtime is much faster. In comparison with PML [4], which models spatio-temporal dependencies via taking spatio-temporal coordinates as additional inputs, our network not only outperforms it by a significant margin of 3.5% in the term of Mean $\mathcal{J}$, but also more than twice faster than PLM. Note that PML use FBS [14] to refine the boundary which boost the performance. When we do the same, our performance is significantly better and the speed is still faster. This promising result validates that our proposed spatio-temporal RNN presents an effective prototype in capturing spatio-temporal contexts in video segmentation. Although VAG [9] performs a little bit faster than our method, it is far more behind other approaches in terms of Mean $\mathcal{J}$. Overall, our method achieves a good trade-off between the segmentation accuracy and speed. From Fig. 6, we can intuitively see the comprehensive advantage of our method.

By examining the visualization results in Fig. 7, we observe that our method achieves decent segmentation results when the objects of interest are heavily

Table 2. Comparison results on the DAVIS-2017 validation set with other state-of-the-art approaches.

Method	$\mathcal{J}$	$\mathcal{F}$	FPS
OSVOS [1] (2017'CVPR)	52.1	–	0.11
OnAVOS [18] (2017'BMVC)	61.0	66.1	0.08
FAVOS [4] (2018'CVPR)	54.6	61.8	0.56
OSMN [23] (2018'CVPR)	52.5	57.1	7.14
VideoMatch [7] (2018'ECCV)	56.5	–	2.86
FEELVOS [17] (2019'CVPR)	51.2	57.5	1.85
VAG [9] (2019'ICIEA)	44.2	48.4	–
Ours	54.2	57.9	8.33
Ours+FBS [14]	56.1	60.7	4.17

Table 3. The influence of spatio-temporal RNN and frame-level adaption on the system performance (%).

	Ours	no-Temp	no-STRNN	no-FLA
$\mathcal{J}$	79.0	77.6	72.5	77.4
Δ	–	-1.4	-6.5	-1.6

Table 4. The inference time (sec) of each module for segmenting 100 frames.

	Feature extraction	Spatio-temporal RNN	Online Adaption
Time	8.6	2.5	0.9

occluded in the video. For example, the riding person and running dogs in the first and second columns. Our results for the sequences of drift-straight (the third column) and paragliding-launch (the last column) show that our method is robust to the appearance variation caused by motion and deformation, which is logical as our method can capture some spatial and temporal dependencies among image regions for segmentation.

Results on DAVIS-2017. The DAVIS-2017 [13] is extended from DAVIS-2016 with more videos and objects considered. Overall, this set consists of 150 videos, a total of 10459 annotated frames and 376 objects to be segmented. Compared with DAVIS-2016, this set is more challenging in the following aspects: (i) the videos contains much smaller objects and fine structures, more occlusions and fast motions; (ii) some videos in this set have more than one objects and the objects could interact with each other in the scene. The model parameters were set as $\lambda_0 = 0.4, \lambda_1 = 0.1, \alpha_1 = 10^6$, and $\alpha_2 = 0.1$.

The result is tabulated in Table 2. As shown, our method achieves a good trade-off between accuracy and efficiency. In contrast to the online-tuning based methods [1,18], our method can obtain a comparable segmentation accuracy but the speed is about 100 times faster. Compared with other state of the art methods , our method can also achieve much faster speed while obtaining similar [7] or better [4,17,23] segmentation accuracy.

3.3   Ablation Study

In this section, we perform careful ablation experiments to justify the essential components in our system on the DAVIS-2016 set.

Spatio-temporal RNN. Here, we study the effect of the spatio-temporal RNN on the system performance. First, in order to understand how temporal dependency modeling contributes to the segmentation performance, we discard the temporal connections in the spatio-temporal RNN. We denoted this network architecture to be *no-Temp*. Furthermore, we replace the spatio-temporal RNN with a 1×1 convolution layer, which gives us insight to the effects of spatio-temporal contextual modelling. We use *no-STRNN* to indicate this network. As shown in Table 3, *no-STRNN* under-performs our network by a significant margin of 6.5% in terms of Mean $\mathcal{J}$ (72.5% vs 79.0%). This result demonstrates that our spatio-temporal RNN can capture rich contextual information both spatially and temporally to enhance the video segmentation performance. We observe that *no-Temp* also performs worse than our method (77.6% vs 79.0%), which demonstrates that the temporal dependencies captured by our spatio-temporal RNN is beneficial for video object segmentation.

Frame-Level Adaption. We evaluate the effect of the proposed frame-level adaption for capturing object appearance variation as more frames are segmented. Specifically, we re-implement our method without frame-level adaption and denote it as *no-FLA*. As shown in Table 3, the Mean $\mathcal{J}$ decreases by 1.6%, which demonstrates that the frame-level adaption contributes notably to the performance improvement.

Speed Analysis. Here, we investigate the detailed times used in each module for segmenting a video with 100 frames. The results in Table 4 show that the proposed spatio-temporal RNN and online adaption modules take up very little time (about 3.4 s in total) to segment the video. In contrast, the feature extraction module spends most of the time, which means that the speed of our system can be largely improved if a more light backbone model (e.g., MobileNet) is used to replace the Deeplab-v2 in our feature extraction module.

4   Conclusion

In this paper, we present a novel framework for fast video object segmentation. In the framework, a spatio-temporal RNN module is proposed to embed the spatio-temporal contextual information into the feature map computed by any CNN model. Meanwhile, we develop an online adaption module which can efficiently tune the model parameters for segmenting specific objects in given video. Our experiments show that the proposed method achieves a good trade-off between the segmentation accuracy and speed, and it performs favorably against state-of-the-art systems in terms of efficiency and accuracy.

References

1. Caelles, S., Maninis, K.K., Pont-Tuset, J., Leal-Taixé, L., Cremers, D., Van Gool, L.: One-shot video object segmentation. In: Proceedings of the IEEE Conference on Computer Vision and Pattern Recognition, pp. 221–230 (2017)

2. Chen, L.C., Papandreou, G., Kokkinos, I., Murphy, K., Yuille, A.L.: DeepLab: semantic image segmentation with deep convolutional nets, atrous convolution, and fully connected CRFs. IEEE Trans. Pattern Anal. Mach. Intell. **40**(4), 834–848 (2017)

3. Chen, Y., Pont-Tuset, J., Montes, A., Van Gool, L.: Blazingly fast video object segmentation with pixel-wise metric learning. In: Proceedings of the IEEE Conference on Computer Vision and Pattern Recognition, pp. 1189–1198 (2018)

4. Cheng, J., Tsai, Y.H., Hung, W.C., Wang, S., Yang, M.H.: Fast and accurate online video object segmentation via tracking parts. In: Proceedings of the IEEE Conference on Computer Vision and Pattern Recognition, pp. 7415–7424 (2018)

5. Gui, Y., Tian, Y., Zeng, D.J., Xie, Z.F., Cai, Y.Y.: Reliable and dynamic appearance modeling and label consistency enforcing for fast and coherent video object segmentation with the bilateral grid. IEEE Trans. Circ. Syst. Video Technol. (2019)

6. Hu, Y.T., Huang, J.B., Schwing, A.: MaskRNN: instance level video object segmentation. In: Advances in Neural Information Processing Systems, pp. 325–334 (2017)

7. Hu, Y.T., Huang, J.B., Schwing, A.G.: VideoMatch: matching based video object segmentation. In: Proceedings of the ECCV, pp. 54–70 (2018)

8. Khoreva, A., Benenson, R., Ilg, E., Brox, T., Schiele, B.: Lucid data dreaming for object tracking. In: The DAVIS Challenge on Video Object Segmentation (2017)

9. Liang, H., Tan, Y.: Visual attention guided video object segmentation. In: 2019 14th ICIEA, pp. 345–349. IEEE (2019)

10. Lin, T.-Y., et al.: Microsoft COCO: common objects in context. In: Fleet, D., Pajdla, T., Schiele, B., Tuytelaars, T. (eds.) ECCV 2014. LNCS, vol. 8693, pp. 740–755. Springer, Cham (2014). https://doi.org/10.1007/978-3-319-10602-1_48

11. Perazzi, F., Khoreva, A., Benenson, R., Schiele, B., Sorkine-Hornung, A.: Learning video object segmentation from static images. In: Proceedings of the IEEE Conference on Computer Vision and Pattern Recognition, pp. 2663–2672 (2017)

12. Perazzi, F., Pont-Tuset, J., McWilliams, B., Van Gool, L., Gross, M., Sorkine-Hornung, A.: A benchmark dataset and evaluation methodology for video object segmentation. In: Proceedings of the IEEE Conference on Computer Vision and Pattern Recognition, pp. 724–732 (2016)

13. Pont-Tuset, J., Perazzi, F., Caelles, S., Arbeláez, P., Sorkine-Hornung, A., Van Gool, L.: The 2017 davis challenge on video object segmentation (2017). arXiv preprint arXiv:1704.00675

14. Poole, B., Barron, J.T.: The fast bilateral solver. In: Proceedings of 14th European Conference on Computer Vision (ECCV), pp. 617–632 (2016)

15. Ren, X., Pan, H., Jing, Z., Gao, L.: Semi-supervised video object segmentation with recurrent neural network. In: 2019 IEEE International Conference on Signal Processing, Communications and Computing (ICSPCC), pp. 1–6. IEEE (2019)

16. Shuai, B., Zuo, Z., Wang, B., Wang, G.: Scene segmentation with dag-recurrent neural networks. IEEE Trans. Pattern Anal. Mach. Intell. **40**(6), 1480–1493 (2017)

17. Voigtlaender, P., Chai, Y., Schroff, F., Adam, H., Leibe, B., Chen, L.C.: FEELVOS: fast end-to-end embedding learning for video object segmentation. In: Proceedings of the IEEE Conference on Computer Vision and Pattern Recognition (2019)

18. Voigtlaender, P., Leibe, B.: Online adaptation of convolutional neural networks for video object segmentation. In: The British Machine Vision Conference (2017)

19. Wang, X., Hu, J.F., Lai, J.H., Zhang, J., Zheng, W.S.: Progressive teacher-student learning for early action prediction. In: Proceedings of the IEEE/CVF Conference on Computer Vision and Pattern Recognition (CVPR) (2019)

20. Wang, Z., Ji, S.: Smoothed dilated convolutions for improved dense prediction. In: Proceedings of the 24th ACM SIGKDD International Conference on Knowledge Discovery & Data Mining, pp. 2486–2495 (2018)
21. Wug Oh, S., Lee, J.Y., Sunkavalli, K., Joo Kim, S.: Fast video object segmentation by reference-guided mask propagation. In: Proceedings of the IEEE Conference on Computer Vision and Pattern Recognition, pp. 7376–7385 (2018)
22. Xia, F., Wang, P., Chen, X., Yuille, A.L.: Joint multi-person pose estimation and semantic part segmentation. In: CVPR (2017)
23. Yang, L., Wang, Y., Xiong, X., Yang, J., Katsaggelos, A.K.: Efficient video object segmentation via network modulation. In: Proceedings of the IEEE Conference on Computer Vision and Pattern Recognition, pp. 6499–6507 (2018)

Position and Orientation Detection of Insulators in Arbitrary Direction Based on YOLOv3

Run Ye, Qiang Wang$^{(\boxtimes)}$, Bin Yan, and Benhui Li

University of Electronic Science and Technology of China, Chengdu, China
{rye,uestcyan}@uestc.edu.cn, 604887956@qq.com, 1365295831@qq.com

Abstract. Although great progress has been made in using the method based on deep learning to detect insulators, the detection effect is still unsatisfactory due to the large aspect ratio and inclination of insulators. Therefore, we propose a method capable of detecting large aspect ratio and inclination of insulators, which uses YOLOv3 framework. Firstly, we use manual generation of the anchors to generate angular anchors and fit the label box as much as possible; then we modify the backbone to increase the attention of the detected objects with SE-block; finally, we add angular parameters in rectangular box regression to achieve inclined insulator detection. Compared with the traditional methods, our proposed method significantly improves the performance, and compared with the existing deep learning-based methods, the proposed can achieve real-time detection without degrading performance, and can achieve 82.9% accuracy on our dataset.

Keywords: Deep learning · Object detection · Inclined insulator

1 Introduction

In many methods of traditional insulator detection, the sample preparation of the insulator is manually labeled with a rectangular box as a label box, so the result of the labeled is in the form of a horizontal rectangular box, and does not have a rotating feature. However, considering the large aspect ratio of the insulator, when the width is wider or the length is longer, the angle is produced. But when labeling, we only choose the upper left and lower right corners, so a horizontal rectangular box is generated. The method will result in a large amount of background information while labeling the insulator, as shown in Fig. 1. Inspired by [22], we propose a similar method. Since the detected insulators have different angular properties, we also use rectangular box with angular properties to better label the insulators when labeling, so that the background information of the insulator will be greatly reduced. At the same time, the angled anchor is also used in the process of network training, which can predict the detection box with angular properties.

© Springer Nature Switzerland AG 2020
Y. Peng et al. (Eds.): PRCV 2020, LNCS 12305, pp. 551–563, 2020.
https://doi.org/10.1007/978-3-030-60633-6_46

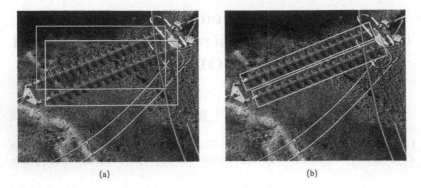

Fig. 1. Different marking methods for the same insulator. (a) is a traditional labeling method, which uses a horizontal labeling method; and (b) uses a rotated rectangle labeling method.

An important problem arises when detecting the insulators with a horizontal rectangular dataset. In the final NMS [12] phase, since the prediction box is horizontal, it will result in a large overlap of two inclined side-by-side insulators. When the threshold of NMS is set too large or too small, only one of the two insulators will be obtained, resulting in poor detection performance. However, if a detection box with an angular property is predicted, there will be no such an large overlap condition, and there will be no case where only one of the two insulators is obtained.

In the results of the above-mentioned insulator labeling, this is another problem because the labeling box is a large aspect ratio. When the width is very long and the height is very short or the width is very short and the height is very long, and because of the scale-invariant nature of the convolution network [16], the information of the feature map will become very small when the width or height is short. In order to solve this problem, we use a method of attention mechanism [6,8,18] to pay more attention to the rare feature information. This method will be described in detail later.

Our main work is as follows:

1. Modify the original Darknet53 feature extraction network, add an SE block after each residual block;
2. In the anchor selection, instead of using the traditional horizontal anchor, choose an angled anchor, or an anchor that can be rotated;
3. In the calculation method of the original regression rectangle, a new angle parameter is added, which can predict the angle information and merge with the horizontal rectangular frame to achieve the network prediction.

Our proposed method has been tested on the self-built dataset, and its performance has been improved by about 20% compared with the prevIoUs detection method, and compared with the existing detector that can detect the angle information, it can achieve real-time detection with state-of-the-art performance.

2   Related Work

Among the many methods of detecting insulators, traditional methods and deep learning have their own advantages. Before the popularization of the deep learning method, the traditional image methods [1,19,20,24] are used to achieve the detection of the insulator, which mainly use a certain feature of the insulator to detect, such as [24] color features. To achieve detection, [19] uses difference metrics of image to detect bad insulators, [1] achieves insulator detection using contour information of insulator, [20] achieves the detection of insulators with traditional Bayesian segmentation algorithms and mathematical morphology. The conventional methods for the detection of insulators have certain limitations, taking into account only one or several characteristics of the insulator. After the popularity of deep learning, the convolutional network is able to extract more features of the insulator and improve the detection performance of the insulator. [10] divides the detection into two modules, fast target detection and U-NET pixel classification, to get high accuracy and real-time performance, causing computational complexity due to using pixel-level classification; [23] uses a six-layer convolution network to achieve the detection of insulators, whose speed is fast, but the detection effect is general because of the simple network; [17] analyzes the difficulties of insulator detection, then proposes a data enhancement method, and designs a cascade network to enhance the network, which can significantly improve the detection accuracy of insulators, but reducing the speed a lot.

The detection method of rotating rectangular box has been applied in other scenes, such as text detection [9,11,21], remote sensing image detection. [11,22] uses the rotate anchor to more fully fit the detection objects, and RPN [14] is used to classify and regress these anchors, fine-tuning of the proposed anchors was made at last. The overall structure is similar to Faster RCNN, but it is also a pioneering article. Then [9] improved this method, proposed IF-Net, MDA-Net and rotate-branch to improve the original detection network, so that the detection results were greatly improved. [21] modify the ROI pooling and the NMS, make it adapt to the rotating targets. [22] focuses on detecting the remote sensing images, whose method can mark the direction of the ship's bow when detecting the ship, and improve the ROI align[4]. The adaptive ROI align and multi-task loss are proposed. Although the above methods were state-of-the-art algorithm in accuracy, the detection speed is far from real-time, and also textcolorrecrequire a lot of computing resources.

The backbone network based on deep learning is the most advanced network in classification tasks, such as ResNet [5], DenseNet [7], VGG [15]. The main model structure is realized by stacking residual networks. On the basis of these main model structures, some improvements are proposed, such as SE module [6], res2net[2].

3   Ours Approach

We will focus on the algorithms we have improved in this section. First, as shown in Fig. 2, the overall structure of our network, which is similar to the structure of

YOLOv3 [13], the difference lies in the backbone and the predictive output. In the backbone, the SE module is added to strengthen the model's sensitivity and attention to the location, it will be introduced in Sect. 3.1. In the anchor selection method, we no longer use clustering to get a fixed anchor, but use artificial inclined box as anchors, the calculation and training methods of the subsequent IoU will change due to the different strategies of the anchor generation. The detailed description is introduced in Sect. 3.2. In the network prediction, the angular output of the rectangular box is increased, so that the network can detect rectangular box in any direction, which is described in Sect. 3.3. During the training, we need to define the loss function, which will be detailed in Sect. 3.4.

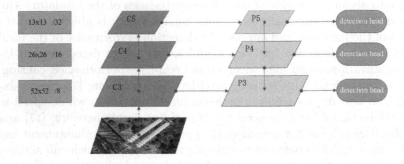

Fig. 2. The whole structure of our model. The model is mainly divided into backbone network, feature fusion network and prediction network. There are three scales in feature fusion network and prediction network. For large target, P5 is used to predict, for medium target, P4 is used to predict, and for small target, P3 is used to predict.

3.1   Our Network Structure

Our network consists of three main components, the improved backbone, the FPN and the detection network. Firstly, the improved backbone is used to extract the overall features of the image, and the three-layer feature map obtained by connecting the backbone with FPN is used for the fusion operation. Finally, the detection operation is performed on the merged three feature maps to obtain the predicted output.

In the backbone, it is considered that each channel on the feature obtained by the original backbone network after extracting the feature is average, but in reality, each channel of the feature map has different weights for each target which is confirmed in [6]. Therefore, we propose an improved method. By introducing the SE module, the channel used to detect the target is given a greater weight, and the channel that interferes with the detection is given a lower weight. As shown in Fig. 3, the main structure of the SE module, it can be added behind each residual network, and the flexible is very high.

Through the comparison of the left and right graphs, we can find that the main improvement is to add a network with different weights to different channels

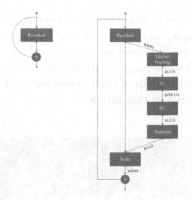

Fig. 3. Residual block and SE block. The left figure shows the residual block in ResNet network, while the right figure shows the SE block added to the residual block.

after the residual block. First, the input is globally pooled to obtain the vector of each channel, then goes through two fully connected networks, and finally the weight of each channel is obtained by activating sigmoid function, and the weight value of each channel is between $[0, 1]$. The final scale operation is to multiply the weight and the input by channel, and each channel can be given different weights.

3.2 Anchors with Rotating

In our algorithm, since the detection target is directional, we need to consider the directional anchor. The anchor of YOLOv3 obtains 9 kinds of anchors by clustering, and is assigned to three different size detection networks in order according to the size order. In our experiment, this method was adopted first, but the results are not very ideal. We think that the reason is that anchor has angular rotation. However, the horizontal direction obtained by the original clustering method is not universal for anchor in any direction.

On this basis, we adopt the method of [3] to set the rotation anchor manually. We analyze the size of the insulator of the dataset we use. For anchor, we select three scales, four aspect ratios and six angles at each scale. In OpenCV, the angular range of the rotating rectangle is $[-90, 0]$, and the definition of width is to form an angular edge according to the first edge of the horizontal line and rectangular frame. Therefore, even if the rectangular rotation is 360 degrees, the rotation can be reduced to $[-90, 0]$ degrees in this way, which greatly reduces the amount of calculation.

We divide the range of 90 into 18 points on average and take it every 5 degrees. However, these multi-angle values will not be applied to all three feature maps, but will be divided into three groups, six in each group. Finally, we can get three sets of angle values, one of which is $[-10, -25, -40, -55, -70, -85]$. According to the size of our insulators, we set three scales, 70, 60 and 40, respectively, corresponding to three detection networks. Then, according to the size of the insulator's aspect ratio, we set different aspect ratio for anchor of different sizes. For example, when the scale is 70, because of the larger insulator, our aspect ratio will be smaller, and we get [0.1, 0.125, 8, 10]. When the scale is small, the aspect ratio will be larger. Ultimately, we generate $1 \times 4 \times 6$ anchors at each point on the feature map of each detection network. As shown in Fig. 4 is the anchor type designed in this article.

Fig. 4. Anchor design. In the picture, red is the actual label box, and light blue is the 24 anchors generated at the center of the label box. They are 12 angles and 2 aspect ratios, respectively. (Color figure online)

After getting the anchor, a new problem will arise here, that is, the IoU calculation of the anchor and the label box. In the article [11], a more complicated IoU calculation method is adopted. Our improved IoU is defined as follows: If the angles of the two rectangular boxes differ by more than 15 degrees, then the IoU is 0; if the IoU of the two rectangular boxes is less than 15 degrees, then the two rectangular boxes IoU are calculated. The IoU method is the same as [11]. This calculation method can minimize the calculation of IoU, but only adds a lot of judgment process.

3.3   Bounding Box Regression

In the detection network, we will generate $24 \times (x, y, w, h, r, conf, class)$ vectors at each point on the feature map by regression calculation, where 24 represents each point on the feature map will generate 24 anchors, (x, y, w, h, r) represents the coordinates of the rotation rectangle, where (x, y) represents the offset of the center point, (w, h) represents the parameter amount of width and height, and finally r represents the parameter amount of the angle, $conf$ represents the confidence of each anchor at that point, class is always 1. Because there is only one class.

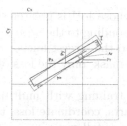

Fig. 5. Prediction regression diagram. On the way, the green rectangle represents the anchor, while the red rectangle represents the prediction box. The green box needs to get the red box according to the regression formula. (Color figure online)

Since the regression is predicted by the network as a parameter quantity, we need to convert it into the predicted rectangle on the feature map. The regression diagram is shown in Fig. 5. The parameter value of the anchors is (A_w, A_h, A_r), the network prediction value is $(B_x, B_y, B_w, B_h, B_r)$, and the coordinate value of the rectangle is $(P_x, P_y, P_w, P_h, P_r)$, and (C_x, C_y) represent the offset of the predicted center point from the upper left corner of the feature map. The specific formula is calculated as follows:

$$\begin{cases} P_x = \sigma(B_x) + C_x \\ P_y = \sigma(B_y) + C_y \\ P_w = A_w e^{B_w} \\ P_h = A_h e^{B_h} \\ P_r = A_r e^{B_r} \end{cases} \tag{1}$$

Since the offset of the predicted center point ranges from $[0, 1]$, so it is normalized to $[0, 1]$ using the sigmoid function, and then the c_x and c_y of the prediction grid relative to the upper left corner are added, we can get the rotation center point predicted by the grid relative to the feature map; the predicted width and height are converted using a nonlinear function, which is mainly based on the anchor value and the predicted width and height index values. The width and height of the feature map, that is, the formulas $P_w = A_w e^{B_w}$ and $P_h = A_h e^{B_h}$; finally, the prediction of the angle, the conversion method is the same as the width and height conversion method, the formula is $P_r = A_r e^{B_r}$, it also uses anchor values and network prediction value index to get the angle relative to the feature map.

3.4 Loss Calculation

In the training process, we need to get the target *mask* and the *no_mask* without the target. The *mask* and *no_mask* are one-to-one corresponding to the anchor on the feature map. The *mask* is initialized to 0, and the *no_mask* is initialized to 1. For the target to exist in A certain grid, then the anchor in this grid is used to calculate the IoU with the target. The IoU calculation method is introduced in Sect. 3.2. For an anchor whose IoU is greater than the threshold of 0.5, the position of the anchor corresponding to the *no_mask* is set to 0, and the position of the *mask* corresponding to the largest anchor of the IoU in the plurality of anchors is set to 1. With the mask, only the loss of the target can be calculated in the loss calculation.

We make an end-to-end training with multitasking loss. The total loss is mainly composed of three parts, coordinate loss, target score loss and category loss. The total loss formula is as follows:

$$Loss = L_{coord} + L_{cls} + L_{conf}$$

$$= \lambda_{coord} \sum_{i=0}^{s^2} \sum_{j=0}^{n} I_{ij}^{obj} [(\sigma(t_x) - \hat{t}_x)^2 + (\sigma(t_y) - \hat{t}_y)^2$$

$$+ (t_w - \hat{t}_w)^2 + (t_h - \hat{t}_h)^2 + (t_r - \hat{t}_r)^2]$$

$$+ \lambda_{cls} \sum_{i=0}^{s^2} \sum_{j=0}^{n} I_{ij}^{obj} BCE(t_c, (\hat{t}_c)) \tag{2}$$

$$+ \lambda_{obj} \sum_{i=0}^{s^2} \sum_{j=0}^{n} I_{ij}^{obj} (\sigma(t_o) - \hat{t}_o)^2$$

$$+ \lambda_{noobj} \sum_{i=0}^{s^2} \sum_{j=0}^{n} I_{ij}^{noobj} ((1 - \sigma(t_o)) - \hat{t}_{noobj})^2$$

Where L_{coord} represents the coordinate loss, including the central point loss, the width and the high loss and the angular loss, namely the second line formula in the above formula, and I_{ij}^{obj} indicates whether the *ith* anchor exists at the *jth* point of the feature map. The entire loss only calculates the loss of the target in the anchor. For the anchor without the target, it does not participate in the calculation, and the subsequent classification loss is the same. The coordinate prediction value is $(t_x, t_y, t_w, t_h, t_r)$, and the coordinate offset data obtained by the anchor and the label box is $(\hat{t}_x, \hat{t}_y, \hat{t}_w, \hat{t}_h, \hat{t}_r)$. The loss is calculated by using the squared error loss. L_{cls} indicates the classification loss. Since there are only one type of target categories, we do not need this loss. However, considering the other categories to be detected later, the category is added and the loss is calculated by means of binary cross entropy. L_{conf} represents the target score loss, which describes the confidence of the presence of the target on the first anchor at the *jth* point on the *ith* feature map. The loss is calculated to have a target on the *jth* anchor at the *ith* point, then the prediction confidence at that point is 1, and the confidence at other points is 0, and the loss is also calculated by the squared error loss. In the formula, $\lambda_{coord}, \lambda_{cls}, \lambda_{obj}, \lambda_{noobj}$ is the weight of each loss. We set it to 1, 1, 5, and 1. For the loss calculation of the existing target, we set the weight to be larger.

4   Experiments

4.1   Experiment Preparation

Our experimental data are taken by camera on the high voltage transmission line. Different cameras on different lines are shooting at a fixed time every day. More than 1000 images can be collected each month, and a total of 50000 images are collected for a total of 50000 sheets. The resolution of each image is 1920 * 1080. Since there are many images that are too similar in all images, the images are filtered to obtain 5000 images with different differences as our dataset, of which 800 are used as training sets, 100 sheets as a validation set and 100 sheets as a test set. In all the images, including the images of the three seasons of autumn, winter and spring, it has strong generalization ability.

A deep learning workstation is used to experiment, whose main configuration is as follows: CPU uses 2 Xeon 4114 series, GPU is Titan xp series, each graphics card 12g, a total of three.

Training Details: In our network we use the darknet group with the added SE module as the backbone network. Since our dataset is not very large, our backbone also uses the pre-trained weights on ImageNet as our initial weights, then uses the weights that YOLOv3 trains on the coco dataset as the weight of the new entire network, and finally uses our datasets to train. Adam optimizer is used, the learning rate is initialized to 0.01, and the learning rate is reduced by reducing the learning rate after the two epochs where the loss is no longer falling, and the rate of decline is 0.1. Training will be stopped when our learning rate no longer drops by 5 epoch, or when we reach the epoch we set. The input image of our network adopts multi-scale training. The input size is randomly transformed every 10 steps. The input size is randomly selected from 352 to 488, centered on 416.

Inference Details: Our input size is fixed at 416 when we are inference. The post-processing of the network is similar to that of YOLOv3. Adding an angle to the coordinate prediction part yields a 5-dimensional vector (x, y, w, h, r) and finally passes the threshold. And NMS to get the final test results.

4.2   Comparison with Horizontal Detection

On our self-built dataset, The first thing we do is to compare experiments with horizontal detection methods. We made horizontal and oblique annotations on the dataset, then used the original YOLOv3, SSD and the Faster RCNN method to train the horizontally labeled datasets and obtain the detection results. Our proposed method is used to train the tilted datasets and obtain the detection effect, the effect comparison is shown in Fig. 6. The data comparison is shown in Table 1.

In the comparison of the above figure, we can easily find that we do not use the rotated labeling method in the left picture. For the side-by-side insulators, there are large overlapping areas and background areas in the marked

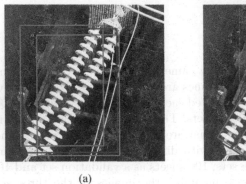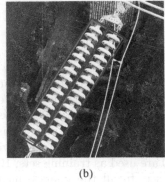

(a)   (b)

Fig. 6. Comparison of the horizontal detection frame and the rotation detection frame. (a) shows the result of horizontal mark level detection, and (b) shows the result of tilt mark tilt detection. The light blue is the actual label box and the red color is the prediction box. (Color figure online)

Table 1. Comparison of data with rotation detection and without rotation detection.

Method	Backbone	AP_{50}	AP_{75}	FPS
YOLOv3	Darknet53	0.553	0.324	31.1
SSD512	Resnet50	0.452	0.343	24.3
SSD512	Resnet101	0.526	0.396	15.4
Faster RCNN	Resnet50+FPN	0.638	0.428	5.2
Faster RCNN	Reanet101+FPN	0.701	0.471	2.5
Ours	Darknet53+SE	0.815	0.706	30.0

area. Although the detection results are not bad, but also just using a detection frame to contain the two insulators together, the performance is not satisfactory. In the right picture, the two insulators are marked by the method of rotating the label box, and it is obvious that there is no overlapping area between the two insulators. The area of the background area is also smaller, and the noise and interference are also significantly reduced. The two insulators are perfectly detected, and the case where the two insulators are framed together in the left figure does not occur. It is also one of the advantages of rotation detection.

From the above table, in addition to the methods proposed in this paper, we can find that the detected AP is not very high, whether it is a single-stage method or a two-stage method. The main reason is that during the final NMS, when the redundant frame is filtered out, larger detection frames in adjacent overlapping areas are also filtered out, resulting in a decrease in the recall rate. However, with our method, since there is only a small amount of overlap or no overlapping area after the detection, the two detection frames will be retained at the same time, so the detection effect is better. Compared with the Faster

RCNN that does not use rotation detection, it has increased by 11.4% points. Compared with our algorithm based on YOLOv3, it has increased by 25% points. While the speed has dropped a little, it can still reach real-time, which is a huge improvement over performance. A slight drop in speed does not affect the excellence of our algorithm.

4.3   Comparison with Advanced Rotation Detection Method

Our second experiment used our method to compare with the most advanced rotation detection method. While doing this experiment, we also compare the method of adding the SE module with the method of not adding the SE module. We also compared with state-of-the-art algorithms in the same dataset. The results are as follows: For Table 2, only the best two-stage algorithm R-DFPN exceeds our algorithm, the other two-stage algorithm is not as high as our algorithm, even if it exceeds our algorithm on AP, textcolorredonly more 0.006% points, but it is far from real time. Our algorithm is inherently superior in speed because it is single-stage. We can achieve real-time performance, meanwhile, the performance of the insulator detection algorithm can reach advanced standards. It is enough to illustrate the advantages of our algorithm.

Table 2. Comparison with other most advanced rotation detection algorithms.

Method	backbone	AP_{50}	AP_{75}	FPS
RRPN	VGG16	0.753	0.594	3.2
R2CNN	VGG16	0.772	0.614	1.9
R2CNN++	Resnet50	0.824	0.694	1.1
R-DFPN	Resnet50	0.835	0.724	1.8
Ours	Darknet53	0.813	0.693	30.9
Ours	Darknet53+SE	0.829	0.717	30.0

As shown in Fig. 7, we can clearly find that our algorithm has a better detection effect, containing well the insulators that need to be detected, and there is no missing detection for the side-by-side insulators, which is a good solution to the problem of horizontal detection method, and intuitively, more clear and accurate.

Fig. 7. The renderings on the dataset are obtained by using our proposed method

5 Conclusion

The proposed is a novel way to detect insulators, which can make the positioning of insulators more accurate and intuitive without decreasing the detection speed. Compared with the original method, ours can improve detection performance and detection accuracy, and compared to the most advanced rotation detection algorithms available today, our method can achieve real-time speed with the most advanced accuracy. Furthermore, our method belongs to the single-stage detection method in target detection. This method is also the first method for rotating target detection, which is a groundbreaking work. Our work is currently only used for insulators, our future work will be applied to other target inspection tasks.

Acknowledgments. We would like to thank the youth program of National Natural Science Foundation of China (Grant No. 61703060), science and technology program of Sichuan Province (Grant No. 2019yj0165) and general program of National Natural Science Foundation of China (Grant No. 61973055).

References

1. Di, W.: Research on insulator detection and location methods in aerial images. Master's thesis, North China Electric Power University (2017)
2. Gao, S.H., Cheng, M.M., Zhao, K., Zhang, X.Y., Yang, M.H., Torr, P.: Res2Net: a new multi-scale backbone architecture. arXiv preprint arXiv:1904.01169 (2019)
3. Girshick, R.: Fast R-CNN. In: Proceedings of the IEEE International Conference on Computer Vision, pp. 1440–1448 (2015)
4. He, K., Gkioxari, G., Dollár, P., Girshick, R.: Mask R-CNN. In: Proceedings of the IEEE International Conference on Computer Vision, pp. 2961–2969 (2017)

5. He, K., Zhang, X., Ren, S., Sun, J.: Deep residual learning for image recognition. In: Proceedings of the IEEE Conference on Computer Vision and Pattern Recognition, pp. 770–778 (2016)
6. Hu, J., Shen, L., Sun, G.: Squeeze-and-excitation networks. In: Proceedings of the IEEE Conference on Computer Vision and Pattern Recognition, pp. 7132–7141 (2018)
7. Huang, G., Liu, Z., Van Der Maaten, L., Weinberger, K.Q.: Densely connected convolutional networks. In: Proceedings of the IEEE Conference on Computer Vision and Pattern Recognition, pp. 4700–4708 (2017)
8. Jaderberg, M., Simonyan, K., Zisserman, A., et al.: Spatial transformer networks. In: Advances in Neural Information Processing Systems, pp. 2017–2025 (2015)
9. Jiang, Y., et al.: R2CNN: rotational region CNN for orientation robust scene text detection. arXiv preprint arXiv:1706.09579 (2017)
10. Ling, Z., Qiu, R.C., Jin, Z., Zhang, Y., Lei, C.: An accurate and real-time self-blast glass insulator location method based on faster R-CNN and U-net with aerial images. arXiv preprint arXiv:1801.05143 (2018)
11. Ma, J., et al.: Arbitrary-oriented scene text detection via rotation proposals. IEEE Trans. Multimedia **20**(11), 3111–3122 (2018)
12. Neubeck, A., Gool, L.V.: Efficient non-maximum suppression. In: International Conference on Pattern Recognition (2006)
13. Redmon, J., Farhadi, A.: YOLOv3: an incremental improvement. arXiv preprint arXiv:1804.02767 (2018)
14. Ren, S., He, K., Girshick, R., Sun, J.: Faster R-CNN: towards real-time object detection with region proposal networks. In: Advances in Neural Information Processing Systems, pp. 91–99 (2015)
15. Simonyan, K., Zisserman, A.: Very deep convolutional networks for large-scale image recognition. arXiv preprint arXiv:1409.1556 (2014)
16. Singh, B., Davis, L.S.: An analysis of scale invariance in object detection snip. In: Proceedings of the IEEE Conference on Computer Vision and Pattern Recognition, pp. 3578–3587 (2018)
17. Tao, X., Zhang, D., Wang, Z., Liu, X., Xu, D.: Detection of power line insulator defects using aerial images analyzed with convolutional neural networks. IEEE Trans. Syst. Man Cybern.: Syst., 1–13 (2018)
18. Wang, F., et al.: Residual attention network for image classification. In: Proceedings of the IEEE Conference on Computer Vision and Pattern Recognition, pp. 3156–3164 (2017)
19. Wu, Y.: Study on insulator identification and state detection method based on aerial image. Ph.D. thesis, North China Electric Power University (2016)
20. Xia, F., Tie, H.F., Shi, E.W., Zhang, Q., Wang, L.: Insulator recognition based on mathematical morphology and Bayesian segmentation. In: International Congress on Image & Signal Processing (2018)
21. Yang, X., et al.: R2CNN++: multi-dimensional attention based rotation invariant detector with robust anchor strategy. arXiv preprint arXiv:1811.07126 (2018)
22. Yang, X., Sun, H., Sun, X., Yan, M., Guo, Z., Fu, K.: Position detection and direction prediction for arbitrary-oriented ships via multitask rotation region convolutional neural network. IEEE Access **6**, 50839–50849 (2018)
23. Yue, L., Yong, J., Liang, L., Zhao, J., Li, Z.: The method of insulator recognition based on deep learning. In: International Conference on Applied Robotics for the Power Industry (2016)
24. Yuwei, H.: Extraction and recognition of aerial video insulator parts. Ph.D. thesis, Dalian Maritime University (2013)

R-PFN: Towards Precise Object Detection by Recurrent Pyramidal Feature Fusion

Qifei Jia[1], Shikui Wei[1(✉)], and Yufeng Zhao[2]

[1] Beijing Jiaotong University, Beijing, China
{18120303,shkwei}@bjtu.edu.cn
[2] China Academy of Chinese Medical Sciences, Beijing, China
snowmanzhao@163.com

Abstract. Object detection has been widely studied in the last few decades. However, handling objects with different scales is still marked as a challenging requirement. To solve this problem, we explore how to better utilize the features of multiple scales generated by a convolutional neural network. Specifically, we fuse a set of pyramidal features in a circular manner and propose a cascaded module, whose consideration is to enhance a single-scaled feature with information from another scale-different feature. Then, we make it recurrent to further facilitate the fusion of information among multi-scaled features. The proposed module can be integrated into any pyramid architecture. In this paper, we combine it with FPN-based Faster R-CNN, result in a framework named Recurrent Pyramidal Fusion Network (R-PFN). Experiments prove the effectiveness of R-PFN. We achieve new state-of-the-art performances, i.e., 82.0%, 43.3% on the PASCAL VOC 2007 benchmark and MS COCO benchmark in terms of mean AP, respectively.

Keywords: Object detection · Multi-scale instances · Feature pyramid · Recurrent forwarding

1 Introduction

Object detection aims at locating and recognizing objects of interest in given images of various scenes, which supports several applications like autonomous driving, intelligent transportation and augmented reality. For a long period of time, a quantity of methods were proposed to solve the challenges met in object detection [1–4]. Although remarkable progresses has been achieved, it still exists large space to be improved considering the hardly-solved problems, e.g., scale inconsistency, illumination change to name a few. In this paper, we focus on one of the widely concerned challenges, i.e., dig out all the instances with various scales in a single scene.

Student Paper.

© Springer Nature Switzerland AG 2020
Y. Peng et al. (Eds.): PRCV 2020, LNCS 12305, pp. 564–576, 2020.
https://doi.org/10.1007/978-3-030-60633-6_47

Recently, CNN-based approaches show impressive performances for finding objects in an image. However, a plain CNN is hard to extract acceptable features both for small-scale objects and large-scale objects. In other words, the researchers need to carefully tune the size of convolutional kernel to get a reasonable receptive field. A large receptive field may suppress the effective representation of an object with small scale, and a small one cannot catch the full information of a large-size object. Most of the existing methods concentrate on three aspects to try solving such challenge. The first starts from the perspective of the input of a CNN [2,5,6]. For example, MS-CNN [6] using pyramid structure to zoom in or out on it to different scales, which is equivalent to a kind of data enhancement. But this method is gradually abandoned because it greatly increasing the memory and computational complexity. The second explores multi-scale information at the feature level [1,4,7,8]. FPN [4] processes in a global manner, which provides low-level features with large-view information in high-level features, and prefers using multi-level features to perform subsequent detection. Contrast to FPN, SPPNet [9] directly crops out the feature of proposals on the latent output of a deep network. To avoid align size of each proposal, SPPNet applies adaptive pooling with different kernels on cropped features and generates features with fixed size; The last type makes explicit constraints during the training of a deep model [10,11]. For example, RetinaNet [10] use focal loss to explicit constraints to optimize the network.

In this paper, we first use detailed experiments to explore the advantages and disadvantages of FPN framework, i.e., the advantage is that the FPN uses multi-level fusion feature maps for training and testing, while the disadvantage is that the fusion mode of FPN loses a lot of information. In order to retain its advantage and weaken its disadvantage, we consider utilizing more scale-orientated information at feature level. Inspired by FCN [4], we design a simple yet effective module named Fusion-Adaptation-Extraction (FAE), which takes a feature pyramid as input and outputs a refined one that keeps exactly the same configuration. FAE is devoted to perform a better way to fuse features with different scale in the feature pyramid. Concretely, for two adjacent levels, we first upsample the smaller one to align the size of the larger one, then perform an element-wise summation, followed by several convolutional layers to better fusing the information from the two levels. Notice that we also regard the bottom level and the top level as adjacent. As the FAE does not change the configuration of the input, we further make FAE recurrent to better gather scale-varied information into a feature of single scale. Finally, we combine it with the FPN-based Faster R-CNN to form a complete pipeline, Recurrent Pyramidal Fusion Network(R-PFN), for general object detection. We conduct extensive experiments on the widely used benchmarks, i.e., PASCAL VOC and MS COCO to illustrate the effectiveness of the proposed R-PFN. Our contributions are summarized as follows:

1). We use experiments to explore the advantage and disadvantage of FPN framework, which makes people better understand the FPN and provides direction for the improvement of FPN.

2). We propose a simple yet effective module named FAE to perform better fusion of features with different scales. The proposed FAE can be made recurrent to further boost the performance, and can be inserted into any deep frameworks that produce multi-scale features.

3). We combine our FAE with FPN-based Faster R-CNN as a complete pipeline for object detection, and achieve the state-of-the-art performances on two popular benchmarks. We get 82.0% in terms of mean AP on PASCAL VOC, 41.7% in terms of mean AP under the single-scale test and 43.3% in terms of mean AP under the multi-scale test on MS COCO, respectively.

2   Method

2.1   Preliminaries

When we add FPN framework to Faster R-CNN [2] and compare it with the original Faster R-CNN, we observe an interesting phenomenon that using the FPN framework with Faster R-CNN on the MS COCO dataset can effectively increase the detection accuracy. However, on the PASCAL VOC dataset, using the FPN framework will not improve the detection effect.

Table 1. Detection accuracy comparisons in terms of mAP percentage on COCO-M, COCO-L and COCO-S datasets.

Method		Dataset			mAP
Faster R-CNN	Faster R-CNN + FPN	COCO-L	COCO-M	COCO-S	
✓		✓			44.5
✓			✓		36.9
✓				✓	15.7
	✓	✓			42.4
	✓		✓		35.1
	✓			✓	14.3

Starting from the phenomenon, we want to explore why the FPN framework fails on PASCAL VOC dataset. First, we should know the advantage of FPN. Since the main idea of FPN framework is to use multi-level fusion feature maps for training and testing, the advantages or disadvantages should be related to it. We assume that the advantage of FPN framework is to use multi-level layers to better deal with multi-scale objects. To test our hypothesis, we use the datasets whose objects have the similar scale to eliminate the advantage of FPN framework in dealing with multi-scale objects. Specifically, we divide the objects on MS COCO dataset into three parts according to the criteria of this dataset (large objects: area $> 96 \times 96$, medium objects: $32 \times 32 <$ area $< 96 \times 96$, small objects: area $< 32 \times 32$) and select 20,000 objects in each three part respectively to make train sets COCO-L, COCO-M and COCO-S. Since COCO test-dev includes separate test sets for different-scale objects, we directly use it as test set. The performance of Faster R-CNN (to eliminate the influence of anchor size, we adjust the size of the generated anchor to match different size of object) and Faster R-CNN + FPN on these three datasets is shown in Table 1, we can observe that mAP reduces when we add FPN framework to Faster R-CNN on both of the three datasets, i.e., reduces 2.1% on large objects, 1.8% on medium objects, 1.4% on small objects. The failure of FPN framework after losing the advantage of dealing with multi-scale objects confirms our hypothesis. In addition, since the mAP reduces significantly when we add FPN framework to Faster R-CNN, we think that FPN framework also has a disadvantage which drags the detection result down. A proof is the phenomenon that FPN framework can not improve the detection effect on PASCAL VOC dataset, i.e., PASCAL VOC dataset also include different-scale objects, the FPN should be able to deal with these objects better and increase detection effect, however, it's prevented by its shortcomings.

Table 2. The comparison of detection accuracy which using different feature layers in FPN on COCO-M, COCO-L and COCO-S datasets.

P2	P3	P4	P5	COCO-L	COCO-M	COCO-S	mAP
✓	✓	✓	✓	✓			42.4
	✓	✓	✓	✓			43.5
		✓	✓	✓			43.9
			✓	✓			44.3
✓	✓	✓	✓		✓		35.1
✓	✓	✓			✓		35.9
	✓	✓			✓		36.2
		✓			✓		36.4
✓	✓	✓	✓			✓	14.3
✓	✓	✓				✓	12.5
	✓	✓				✓	12.0
		✓				✓	11.7

Similarly, although the detection effect is improved on MS COCO, this problem also exists, but due to the more obvious scale-variation on MS COCO, the advantage of FPN framework in dealing with multi-scale objects plays an important role, this shortcoming is subtly concealed. Like the advantage of FPN, the disadvantage is also related to the main idea of FPN which uses multi-level layers to train and test. i.e., some feature maps we use are not effective, which drags down the overall detection accuracy. So can we find the bad layers and get rid of them? We carry out an experiment which use different feature layers (the feature layers from P2 to P5 are the bottom-up fusion of the feature maps from conv2 to conv5) of FPN + Faster R-CNN to train detectors and test on COCO-L, COCO-M and COCO-S datasets. It is worth mentioning that only using P5 does not mean the network is the same as Faster R-CNN, the size of the generated anchor is different. As shown in Table 2, with the number of feature layers we use increases, the mAP decreases on COCO-L and COCO-M datasets, while increases on COCO-S dataset. It indicates that the feature layers with poor detection effect are different when the scale of objects are different. For large-scale objects, P5 is the most useful feature layer for detection. While for small-scale objects, P2 is the most useful feature layer for detection. We can't simply remove some feature layers, because it may be beneficial to the detection of some scale objects.

Tracing back to the source, why are some multi-scale feature maps not good for detecting objects? We think it is related to insufficient information, the lack of information is mainly reflected in the following two points: (1) features in FPN are not fully integrated, some information is lost in the fusion process. (2) different feature layers lack different information. i.e., low-level feature map, like P2, lack semantic information which more helpful to detect large-scale objects, while high-level feature map, like P5, lack location information which more helpful to detect small-scale objects. So it is a good choice to adjust the fusion mode of FPN while keeping the main idea of using mutil-scale feature maps.

Our structure is shown in Fig. 1. This structure solves the two problems mentioned above in FPN framework. It achieves the goal of fully integrating information and supplementing all the feature layers with their required information, truly makes the information flow through our framework. Next, we will disassemble the structure and analyze the role of each component one by one. All components will be introduced in the following sections.

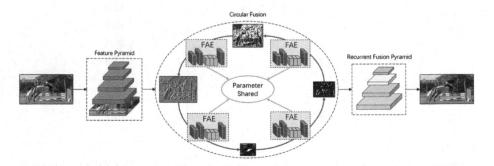

Fig. 1. The whole framework of R-PFN. The feature maps in the feature pyramid are fully fused through the circular fusion structure. This structure is built with multiple FAE modules and we make it recurrent to further facilitate the fusion of information among multi-scaled features. In addition, we share the convolution kernels weight of dilated convolution in each FAE modules.

2.2   Circular Fusion

Extracting features by using CNN in object detection can be seen as the process of transforming location information into semantic information. The low-level feature maps are not fully transformed, results in more location information but less semantic information, while the high-level feature maps are the opposite. The Feature Pyramid Network only adds the semantic information of the higher-level features to the lower-level features, the problem of lacking location information on higher layers is not completely solved. With this in mind, we fuse a group of pyramidal features in a circular manner and propose a structure which can be integrated into any pyramid architecture, as shown in Fig. 1. This structure not only brings the fused underlying information back to the top layer, supplements the required location information, but also brings information to each layer of features through the information flow formed by this circular structure.

2.3   FAE Module

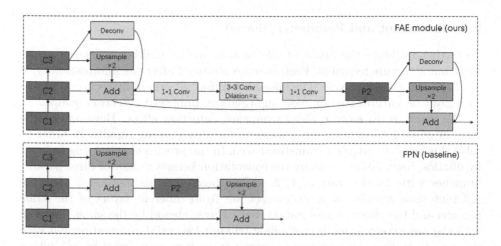

Fig. 2. FAE module (ours) and FPN (baseline) fusion mode comparison.

Fusion - Adaptation - Extraction (FAE) module is the key modules of R-PFN. As shown in Fig. 2, the whole structure can be regarded as a combination of three inseparable modules, feature fusion module, feature adaptation module and feature extraction module. The feature fusion module implemented by the basic operation of upsample and add is inherent in FPN [4]. However, the direct fusion causes serious information loss. Our feature adaptation and extraction module solve this problem.

For feature adaptation module, we not only add convolution to the fusion process, but also add deconvolution to the upsample process. Through the adjustment of convolution and deconvolution on feature adaptation and extraction module, the adaptability between different layers become stronger. So that the fusion process is smoother than direct fusion, which effectively reduces some information loss caused by direct fusion.

For feature extraction module, we use the convolution with different dilation rates. As mentioned in TridentNet [13], as the receptive field becomes larger, the accuracy of detecting small objects gradually decrease, while the accuracy of detecting large objects gradually increase. Based on this phenomenon, we naturally have the idea of constructing multiple parallel feature extractors to detect different scale objects. According to the characteristics of pyramid architecture—using the lower-level features to detect small objects and higher-level features to detect large objects. We use the dilation convolution extracting the lower-level features with lower dilation rate to control the receptive field and make it better for detecting small objects. Conversely, the higher-level features should be extracted with higher dilation rate. In addition, we also use residual framework, so that gradient will not disappear during the recurrent process. Generally, The FAE modules not only smooth the fusion process, but also extract features to complement the information loss.

2.4   Recurrent and Parameter Shared

To further facilitate the fusion of information among multi-scaled features, we make our structure recurrent. Feature maps obtained after one information cycle need to enter the next cycle. The information flow generated by recurrent makes each layer of features contain more information, improved detection quality.

We want to do more cycles to improve detection effect. However, as the number of cycles increases, the parameters is also growing rapidly. In order to solve this problem, we use parameter shared. In the process of extracting features by dilation convolution, we share the convolution kernels weight of three parallel branches with dilation rate of 1, 2, 3. The basis for this method is that the different scale feature maps we extract are from different layers of the same picture, and the effective information they express should be the same. Training with the same convolution kernel weight will not lose valid information and keep the detection effect. Meanwhile, our extra parameters are reduced by 2/3, which allows us to perform more cycles with limited parameter limits.

3   Experiments

In this section, we first conduct ablation experiments to prove the effectiveness of our framework, including the impact of recurrent number on detection effect, the advantages of our framework compared with FPN and the effectiveness of each component in our framework. Then, we compared with other good performance methods to prove that our structure achieved state-of-the-art performance.

3.1 Experimental Details

For experiments to explore the impact of recurrent number and validate the effectiveness of each component in the framework, we train and test our models on PASCAL VOC 2007 sets. To further illustrate the advantages of our R-PFN compared with the FPN, we conduct ablation study on COCO-L, COCO-M, COCO-S sets. Next, we train our R-PFN on VOC 07/12 trainval sets, test on VOC 07 test set and also train our model on COCO trainval35k set, test on COCO test-dev. Compared with other state-of-the-art methods to show the good performance of our detector.

For the experiments on PASCAL VOC, we use 2 Tesla K40 to train our models. The batch size is set to 4. We initialize the learning rate as 1×10^{-4}, and then decrease it to 1×10^{-5} at 5 epochs, eventually stop at 7 epochs. For the experiments on COCO, we use 4 NVIDIA GeForce GTX 1080Ti to train our models. The batch size is set to 8. We initialize the learning rate as 1×10^{-4}, and then decrease it to 1×10^{-5} at 5 epochs, eventually stop at 6 epochs.

3.2 Ablation Study

Recurrent Numbers: In order to verify the information flow formed by multiple recurrent can improve the detection results and explore the optimal solution of the recurrent number, we carried out this experiment.

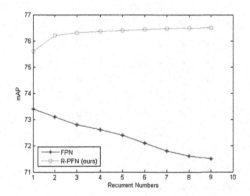

Fig. 3. Recurrent Number-mAP polyline for FPN and R-PFN (ours). Blue represents the FPN and red represents the R-PFN. (Color figure online)

The result is shown in Fig. 3, i.e., in the R-PFN, we set the number of recurrent to 1 to 9, and parameter sharing is used in the process of 4 to 9 recurrent to drop additional parameters. For fairly comparison, we also make the FPN recurrent and record the mAP in different recurrent numbers. The results show that in the R-PFN, add 1 recurrent can increase 0.6%, add 2 recurrent can increase 0.7%, we can observe that as the number of recurrent increases, mAP growth becomes slow, add 8 recurrent can only increase 0.9% and add more recurrent can not increase the mAP. Unlike R-PFN, the mAP drops significantly in FPN when we make it recurrent. This proves the effectiveness of our FAE module from another point, i.e., with the retention of the information by the FAE module that the loss of information during the recurrent is avoided.

Compared with the Baseline: In order to present the advantages of our R-PFN, we construct this experiment and verify the effectiveness of the R-PFN in dealing with single-scale objects by comparing it with FPN on the COCO-L, COCO-M, COCO-S datasets.

Table 3. Ablation study on MS COCO.

Methods		Layers				Dataset			mAP
FPN	R-PFN	P2	P3	P4	P5	COCO-L	COCO-M	COCO-S	
✓		✓	✓	✓	✓	✓			42.4
	✓	✓	✓	✓	✓	✓			46.7
✓			✓	✓	✓	✓			43.5
	✓		✓	✓	✓	✓			45.9
✓				✓	✓	✓			43.9
	✓			✓	✓	✓			45.3
✓					✓	✓			44.3
	✓				✓	✓			44.9
✓		✓	✓	✓	✓		✓		35.1
	✓	✓	✓	✓	✓		✓		38.5
✓			✓	✓	✓		✓		35.9
	✓		✓	✓	✓		✓		38.0
✓				✓	✓		✓		36.2
	✓			✓	✓		✓		37.4
✓					✓		✓		36.4
	✓				✓		✓		37.3
✓		✓	✓	✓	✓			✓	14.3
	✓	✓	✓	✓	✓			✓	17.2
✓			✓	✓	✓			✓	12.5
	✓		✓	✓	✓			✓	15.2
✓				✓	✓			✓	12.0
	✓			✓	✓			✓	14.5
✓					✓			✓	11.7
	✓				✓			✓	13.4

As shown in Table 3, the detection effect of R-PFN is better than the FPN when we use the same feature layers. We can observe that the more feature layers

Table 4. The performance of each component in the structure. The detection results are trained on VOC 07 trainval set and tested on VOC 07 test set.

FPN	FAE module (no dilation convolution)	Dilation convolution	Information cycle	Recurrent	Parameter Shared	mAP
✓						72.2
✓	✓					75.0
✓	✓	✓				75.6
✓	✓	✓	✓			76.1
✓	✓	✓	✓	✓		76.3
✓	✓	✓	✓	✓	✓	76.5

are used, the more obvious advantage of R-PFN is. This is mainly because the R-PFN solves the problem of FPN, i.e., R-PFN fully integrate information and supplement all the feature layers with their required information and guarantee the effectiveness of each feature.

The Effectiveness of Each Component: We also certificate the effectiveness of each component in our framework.

As Table 4 shows, we add our components on the basis of Faster R-CNN+FPN in turn, and finally form our complete structure. In this process, we fully demonstrate the effectiveness of each component. Our final detection result is 4.3% higher than the baseline. The most obvious gain is the FAE module, get 2.8% higher than the baseline. This is mainly due to the retention of information by this residual structure.

3.3   Experiment on PASCAL VOC

In order to verify the detection effect of our R-PFN, we conduct experiments on PASCAL VOC dataset and compared it with other state of the art methods. We selected two methods, R-PFN and R-PFN-shared. R-PFN is the method which recurrent 3 times, R-PFN-shared is the method which recurrent 9 times and use parameter shared to compress additional parameters.

Table 5. Detection results trained on PASCAL VOC 07/12 trainval set and tested on the PASCAL VOC 2007 test set.

Method	Backbone	Input size	mAP	aero	bike	bird	boat	bottle	bus	car	cat	chair	cow	table	dog	horse	mbike	person	plant	sheep	sofa	train	tv
Two Stage																						
Faster R-CNN [14]	ResNet-101	1000 × 600	76.4	79.8	80.7	76.2	68.3	55.9	85.1	85.3	89.8	56.7	87.8	69.4	88.3	88.9	80.9	78.4	41.7	78.6	79.8	85.3	72.0
MR-CNN [6]	VGG16	1000 × 600	78.2	80.3	84.1	78.5	70.8	68.5	88.0	85.9	87.8	60.3	85.2	73.7	87.2	86.5	85.0	76.4	48.5	76.3	75.5	85.0	81.0
R-FCN [3]	ResNet-101	1000 × 600	80.5	79.9	87.2	81.5	72.0	69.8	86.8	88.5	89.8	67.0	88.1	74.5	89.8	90.6	79.9	81.2	53.7	81.8	81.5	85.9	79.9
Anchor Free																						
CenterNet-DLA [15]	DLA-34	512 × 512	80.7	-	-	-	-	-	-	-	-	-	-	-	-	-	-	-	-	-	-	-	-
One Stage																						
SSD512 [1]	VGG16	512 × 512	79.5	84.8	85.1	81.5	73.0	57.8	87.8	88.3	87.4	63.5	85.4	73.2	86.2	86.7	83.9	82.5	55.6	81.7	79.0	86.6	80.0
DSSD513 [16]	ResNet-101	513 × 513	81.5	86.6	86.2	82.6	74.9	62.5	89.0	88.7	88.8	65.2	87.0	78.7	88.2	89.0	87.5	83.7	51.1	86.3	81.6	85.7	83.7
RefineDet320 [17]	VGG16	320 × 320	80.0	-	-	-	-	-	-	-	-	-	-	-	-	-	-	-	-	-	-	-	-
ScratchDet300 [18]	Root-ResNet-34	300 × 300	80.4	86.0	87.7	77.8	73.9	58.8	87.4	88.4	88.2	66.4	84.3	78.4	84.0	87.5	88.3	83.6	57.3	80.3	79.9	87.9	81.2
R-PFN(ours)	ResNet-101	512 × 512	81.6	89.2	86.7	82.6	78.3	67.1	89.8	89.5	91.9	58.0	86.8	73.1	89.8	89.2	87.3	86.6	56.9	86.5	76.6	82.0	83.9
R-PFN-shared(ours)	ResNet-101	512 × 512	82.0	89.7	87.0	82.5	78.3	67.7	89.9	89.4	92.1	59.4	86.9	73.5	89.8	89.4	88.3	87.2	57.9	86.8	77.1	83.3	84.1

As shown in Table 5, our method achieves the best mAP among state-of-the-art detectors when the input size is small (512×512). Compared to our baseline—Faster R-CNN, the two methods we selected improve the mAP by 5.2% and 5.6% respectively.

3.4   Experiment on MS COCO

We also conduct experiments on MS COCO and compared with some state-of-the-art methods. In order to fair comparison, we respectively conduct experiments with small-scale (512 × 512) and large-scale (1000 × 600) images as inputs. For MS-TEST, we use 1× and 2× scales to test our model.

Table 6. Detection results trained on MS COCO trainval35k set and tested on MS COCO test-dev set.

Method	Backbone	Input size	MS-TEST	FPS	AP	AP_{50}	AP_{75}	AP_S	AP_M	AP_L
Two Stage										
R-FCN [3]	ResNet-101	1000 × 600	False	9.0	29.9	51.9	-	10.8	32.8	45.0
Faster R-CNN + FPN [4]	ResNet-101	1000 × 600	False	4.5	36.2	59.1	39.0	18.2	39.0	48.2
Deformable R-FCN [19]	Inc-Res-v2	1000 × 600	False	-	37.5	58.0	40.8	19.4	40.1	52.5
Fitness-NMS [20]	ResNet-101	1024 × 1024	True	5.0	41.8	60.9	44.9	21.5	45.0	57.5
PANet [7]	ResNeXt-101	1000 × 600	True	-	42.0	63.5	46.0	21.8	44.4	58.1
R-DAD [21]	ResNet-101	1000 × 600	True	-	43.1	63.5	47.4	24.1	45.9	54.7
Libra R-CNN [11]	ResNet-101-FPN	1000 × 600	True	-	41.1	62.1	44.7	23.4	43.7	52.5
Anchor Free										
CornerNet [22]	Hourglass-104	1000 × 600	False	4.1	40.5	56.5	43.1	19.4	42.7	53.9
CenterNet-DLA [15]	DLA-34	512 × 512	False	28	39.2	57.1	42.8	19.9	43.0	51.4
FSAF [23]	ResNeXt-101	1000 × 600	False	2.7	42.9	63.8	46.3	26.6	46.2	52.7
One Stage										
YOLOv3 [24]	DarkNet-53	608 × 608	False	19.8	33.0	57.9	34.4	18.3	35.4	41.9
RetinaNet500 [10]	ResNet-101	832 × 500	False	11.1	34.4	53.1	36.8	14.7	38.5	49.1
ScratchDet300+ [18]	Root-ResNet-34	300 × 300	True	-	39.1	59.2	42.6	23.1	43.5	51.0
M2Det [8]	ResNet-101	512 × 512	False	15.8	38.8	59.4	41.7	20.5	43.9	53.4
R-PFN-shared(ours)	ResNet-101	512 × 512	False	6.5	39.2	61.2	41.3	18.3	42.6	52.3
R-PFN-shared(ours)	ResNet-101	512 × 512	True	-	40.6	62.4	43.0	21.4	44.7	52.9
R-PFN-shared(ours)	ResNet-101	1000 × 600	False	4.5	41.7	63.5	43.2	20.8	45.9	53.4
R-PFN-shared(ours)	ResNet-101	1000 × 600	True	-	43.3	64.3	45.4	22.2	46.7	56.4

As shown in Table 6, when the input size is 512×512, our R-PFN achieves AP of 39.2 on single-scale test, 40.6 on multi-scale test which higher than most object detectors. When the input size is 1000×600, our method achieves 41.7% AP on single-scale test, which is 5.5% higher than our baseline FPN on the same FPS and 43.3% AP on multi-scale test, higher than most object detectors. It is worth mentioning that under the same settings, compared with the same type of methods, R-PFN is 1.3% higher than PANet, 2.2% higher than Libra R-CNN and 2.3% higher than M2Det, respectively.

4   Conclusions

In this paper, a novel framework is proposed to facilitate the fusion of information among multi-scaled features to solve the problem of FPN. The proposed framework can be integrated into any pyramid architecture. We use several components and operations to build it, such as circular fusion, FAE module, recurrent and parameter shared, both of them play a crucial role. We also combined our framework with Faster R-CNN + FPN and named it R-PFN. Experiments show that our method achieves state-of-the-art performance.

Acknowledgements. This work is in part supported by National Key Research and Development of China (2017YFC1703503) and National Natural Science Foundation of China (61972022, 61532005).

References

1. Liu, W., et al.: SSD: single shot MultiBox detector. In: Leibe, B., Matas, J., Sebe, N., Welling, M. (eds.) ECCV 2016. LNCS, vol. 9905, pp. 21–37. Springer, Cham (2016). https://doi.org/10.1007/978-3-319-46448-0_2
2. Ren, S., He, K., Girshick, R., Sun, J.: Faster R-CNN: towards real-time object detection with region proposal networks. In: NIPS (2015)
3. Dai, J., Li, Y., He, K., Sun, J.: R-FCN: object detection via region-based fully convolutional networks. In: NIPS (2016)
4. Lin, T., Dollár, P., Girshick, R., He, K., Hariharan, B., Belongie, S.: Feature pyramid networks for object detection. In: CVPR (2017)
5. Girshick, R.: Fast R-CNN. In: ICCV (2015)
6. Gidaris, S., Komodakis, N.: Object detection via a multi-region and semantic segmentation-aware CNN model. In: ICCV (2015)
7. Liu, S., Qi, L., Qin, H., Shi, J., Jia, J.: Path aggregation network for instance segmentation. In: CVPR (2018)
8. Zhao, Q., et al.: M2det: a single-shot object detector based on multi-level feature pyramid network. In: AAAI (2019)
9. He, K., Zhang, X., Ren, S., Sun, J.: Spatial pyramid pooling in deep convolutional networks for visual recognition. IEEE Trans. Pattern Anal. Mach. Intell. **37**, 1904–1916 (2015)
10. Lin, T., Goyal, P., Girshick, R., He, K., Dollár, P.: Focal loss for dense object detection. In: ICCV (2017)
11. Pang, J., Chen, K., Shi, J., Feng, H., Ouyang, W., Lin, D.: Libra R-CNN: towards balanced learning for object detection. In: CVPR (2019)
12. Ghiasi, G. and Lin, T. and Le, Q. V.: NAS-FPN: learning scalable feature pyramid architecture for object detection. In: CVPR (2019)
13. Li, Y., Chen, Y., Wang, N., Zhang, Z.: Scale-aware trident networks for object detection (2019). arXiv:1901.01892
14. He, K., Zhang, X., Ren, S., Sun, J.: Deep residual learning for image recognition. In: CVPR (2016)
15. Zhou, X., Wang, D., Krähenbühl, P.: Objects as Points (2019). arXiv:1904.07850
16. Fu, C., Liu, W., Ranga, A., Tyagi, A., Berg, A.C.: DSSD: deconvolutional single shot detector (2017). arXiv:1701.06659

17. Zhang, S., Wen, L., Bian, X., Lei, Z., Li, S.Z.: Single-shot refinement neural network for object detection. In: CVPR (2018)
18. Zhu, R., et al.: ScratchDet: exploring to train single-shot object detectors from scratch (2018). arXiv:1810.08425
19. Dai, J., et al.: Deformable convolutional networks. In: ICCV (2017)
20. Tychsen-Smith, L., Petersson, L.: Improving object localization with fitness NMS and bounded IOU loss. In: CVPR (2018)
21. Bae, S.: Object Detection based on Region Decomposition and Assembly (2019). arXiv:1901.08225
22. Law, H., Deng, J.: CornerNet: detecting objects as paired keypoints. In: Ferrari, V., Hebert, M., Sminchisescu, C., Weiss, Y. (eds.) Computer Vision – ECCV 2018. LNCS, vol. 11218, pp. 765–781. Springer, Cham (2018). https://doi.org/10.1007/978-3-030-01264-9_45
23. Zhu, C., He, Y., Savvides, M.: Feature selective anchor-free module for single-shot object detection (2019). arXiv:1903.00621
24. Redmon, J., Farhadi, A.: YOLOV3: an incremental improvement (2018). arXiv:1804.02767

Image Super-Resolution Based on Non-local Convolutional Neural Network

Liling Zhao[1,2(✉)], Taohui Lu[1], and Quansen Sun[2]

[1] Nanjing University of Information Science and Technology, Nanjing 210044, Jiangsu, China
zhaoliling@nuist.edu.cn
[2] Nanjing University of Science and Technology, Nanjing 210094, Jiangsu, China

Abstract. Super-resolution (SR) is a widely used image processing technology. Recently, deep learning theory has been introduced into image SR task and achieved a significant improvement. However, feature extraction by convolutional mask is often operated in image local blocks, few attentions are given to non-local. Our aim is to insert a non-local network module into deep learning framework to further improve the performance of SR network. We choose ResNet as backbone and SRGAN as benchmark. The non-local network module is added into ResNet and a new generator network based on SRGAN is designed. To our knowledge, it is the first time to introduce non-local module into SR convolutional neural network. With a full discussion about the new generator network structure, our experiment shows that the introduction of non-local network for high resolution image generation has a superior performance to local or global convolutional neural network in image SR tasks.

Keywords: Non-local · SRGAN · Super-resolution · Deep learning

1 Introduction

Deep learning [1] has become a hot topic in computer vision research field. Many algorithms with good performance have been published in well-known international journals and meetings [2–5], such as *Pattern Recognition, Artificial Intelligence*, ICCV, ECCV, and CVPR. In 2014, Dong proposed SRCNN [6], which is considered to be the first successful attempt using convolutional layers for image super-resolution. During the past five years, network models for image super-resolution based on deep learning has developed fast. Many network frameworks with different type of modules and strategies, such as VDSR [7], LapSRN [8], SRGAN [9], EDSR [10], ZSSR [11] and RCAN [12], has been proposed one after another.

In these effective networks, Generative Adversarial Network for image SR (SRGAN) is viewed as one of the milestones achievements. It has become a popular model in image SR tasks. Different from other deep learning framework, SRGAN is comprised with two networks, one is the generator network (G net) and the other is discriminator network (D net). Because of the special network structure, SRGAN and its variate models, such as EDSR [10], EhanceNet [13], SRFeat [14] and ESRGAN [15], have improved image

© Springer Nature Switzerland AG 2020
Y. Peng et al. (Eds.): PRCV 2020, LNCS 12305, pp. 577–588, 2020.
https://doi.org/10.1007/978-3-030-60633-6_48

SR performance to a new level. However, with a careful analysis, we find that feature extraction is operated only in the local blocks in SRGAN and its derived network models, while the similar features in non-local blocks useful for SR tasks are never considered about.

We can easily find out that similar features in natural images may appear in local and non-local patches. It is defined as non-local similarity theory. Before deep learning was introduced into image SR tasks non-local filters based on this similarity theory was widely used to regularize the ill-posed problem with good effects [16]. Inspired by this traditional knowledge and by current research on non-local network blocks [17], it can be expected that there will be a better improvement by introducing non-local network blocks into neural network.

In this paper, different from the already improvements of SRGAN, we change the residual learning to non-local residual learning by insert a non-local network module in G net. For the similar features in non-local patches are extracted by our new convolution strategies, the receptive fields of network are effectively expanded. Figure 1 shows the result of our new algorithm. Obviously, based on the improved model, better results are achieved.

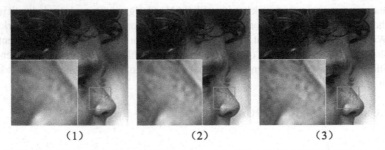

(1)   (2)   (3)

Fig. 1. Results of image SR (2×) (1) LR image (2) SRGAN (3) NL-SRGAN (proposed)

The major contributions of this paper are as follows:

1. We designed a new convolution neural network module named "non-local ResNet" based on non-local similar theory.
2. For image SR tasks, we proposed a generator network based on our new non-local ResNet module and the benchmark SRGAN.

2 Related Work

2.1 Non-local Features

As we all know, in digital image, one pixel and its local or non-local neighbour pixels are generally with similar features, such as line similarity, corner similarity and texture similarity. Thus, the local and non-local similarity are very useful for SR tasks. However, in SR algorithms based on deep learning, convolution is a processing operated just around the target pixel in local patches. This operation will lose many useful similar pixels in non-local patches. In summary, the convolutional operation based on local patches can't completely represent image features. In Fig. 2, we can easily find some non-local similarity patches. The non-local similarity is not a type of redundancy feature, on the contrary it is useful to solve ill-posed problems from the traditional super resolution methods to deep learning neural network. This non-local similarity idea was developed into [18], an improved projection onto convex sets (POCS) algorithm have the effect to restore and approach the high frequency of original image scene. In order to further improve the anti-noise performance and the expressive power of the SR method based on dictionary learning, paper [19] made full use of the uniqueness of the face image, and proposed a face SR method based on non-local similarity dictionary learning. In paper [20], a novel super-resolution reconstruction method based on non-local simultaneous sparse approximation was presented, which combines simultaneous sparse approximation method and non-local self-similarity. The sparse association between high-and low-resolution patches pairs of cross-scale self-similar sets via simultaneous sparse coding is defined, and the association as a priori knowledge was used for super-resolution reconstruction. Paper [27] presented a GA-GP approach via non-local similarity and deep convolutional neural network gradient prior. This method captured the structures of the underlying high-resolution image via the image non-local similarity and learned the gradient prior from external images by using deep CNN. The algorithm achieved better performance than other state-of-the-art SR methods. Therefore, we can predict that if the non-local similarity is introduced into a convolution neural network, features extraction and the image SR performance will be improved.

Fig. 2. Image similarity in non-local patches, the red rectangles are non-local similarity patches, the green rectangles are another non-local similarity patch. (Color figure online)

2.2 GAN for Image Super Resolution

The development of generative adversarial network (GAN) is one of the most important topics in machine learning since its first appearance in [21]. GAN employs a game-theoretic approach where two components of the model, namely a generator and discriminator, try to fool the later. The competition between generator network and discriminator network force the two sides to constantly optimize the network parameters until a balanced state is reached. Due to its flexibility when dealing with high dimensional data, GAN has obtained remarkable progresses on realistic image generation. In image SR task, the generator network creates SR image and the discriminator network cannot distinguish which is real HR image or an artificially super-resolved output. The super-resolution methods based on the GAN framework are explained by Eq. (1) as follows:

$$\min_G \max_D E_{I^{HR} \sim P_{train}(I^{HR})}[\log D(I^{HR})] + E_{I^{LR} \sim P_G(I^{LR})}[\log(1 - D(G(I^{LR})))] \quad (1)$$

ResNet is the main architecture in SRGAN, which is proposed originally for image classification to learn the relationship between low-resolution input and high-resolution output. In the generator network of SRGAN, image features are extracted by convolution module based on ResNet. However, the convolutional operation is only in local patches. Non-local information which is more important for image super-resolution is not considered. As analysed in Sect. 2.1, non-local pixels contributing to image SR should be paid more attention and a further improvement for SRGAN is still worth discussing. Considering that the non-local features are highly correlated with improving image visual quality, a natural idea is to add a non-local network module into ResNet to design a new generator network based on SRGAN. In the following section, we will focus on our proposed methods in details.

3 Proposed Method

In this paper, we proposed a new architecture named "ResNet + Non-local" replacing Residual Units in SRGAN. In this section, we first explain the non-local convolution processing, and then describe our "ResNet + Non-local" architecture in details. At last, we give a full discussion on the combination of our non-local network blocks and SRGAN.

3.1 Non-local Convolution

To introduce the non-local similarity patches into convolution operation, we give a strategy described in Fig. 3. In the traditional convolution processing, weights in convolution mask are only operated in the local patches of target pixel. While in our strategy, the non-local convolution processing is operated in all the similarity patches including local and non-local. In Fig. 3(1), the output pixel '1' is calculated by local similar patches with convolution kernel. In Fig. 3(2), the output pixel '0' is calculated by three non-local similar patches with convolution kernel. It can be clearly seen that the non-local convolution processing calculates comprehensively all the patches with similar features. With the participating of similar non-local patches in convolution processing, it will output more representative features and help the network learn better mapping relationship between HR and LR images.

Our non-local convolution strategy can be realized by a non-local neural network proposed by Wang XL et al. [17]. The non-local convolution operation is formulated by Eq. (2):

$$\mathbf{y}_i = \frac{1}{C(\mathbf{x})} \sum_{\forall j} f(\mathbf{x}_i, \mathbf{x}_j) g(\mathbf{x}_j) \quad (2)$$

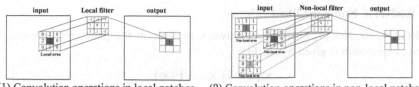

(1) Convolution operations in local patches   (2) Convolution operations in non-local patches

Fig. 3. Convolution processing operated in local and non-local patches.

where x_i is the input image patch (or features) and y_i is the output image patch (or features) with the same size as x_i, i is the index of image patches, j is the index of all possible patches. Function $f(x_i, x_j)$ computes the scalar (representing relationship such as affinity) between x_i and x_j. Function $g(x_j)$ is the representation of the input image patches at the position j and $C(x)$ is the normalized factor.

For simplicity, we only consider g in the form of a linear embedding: $g(x_j)=W_g x_j$, where W_g is a weight matrix to be learned. This is implemented as 1×1 convolution. Then f can be Gaussian, Embedded Gaussian and dot product function. The specific form of these function can be refered to the original paper [17].

Next, we wrap Eq. (2) the non-local operation into a non-local module that can be incorporated into the existing ResNet architectures. We define the non-local module as:

$$z_i = W_z y_i + x_i \tag{3}$$

Where W_z is a weight matrix, y_i is given in Eq. (2), "+ x_i" denotes a residual connection. The residual connection allows us to insert a new non-local module into any pre-trained model, without breaking its initial behavior. In this paper, we chose Embedded Gaussian as the pairwise function. We consider $f(x_i, x_j) = e^{\theta(x_i)^T \Phi(x_j)}$. Here $\theta(x_i) = W_\theta x_i$ and $\Phi(x_j) = W_\Phi x_j$ are two embeddings. We set the number of channels represented by W_θ, W_ϕ and W_g to be half of the number of channels in x. It follows the bottleneck design and reduces the computation of a block by about a half. The non-local module in our work is illustrated in Fig. 4.

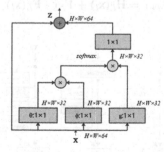

Fig. 4. The non-local network module. $H \times W$ is the product of high and wide. "$\otimes$" denotes matrix multiplication, and "$\oplus$" denotes element-wise sum. The softmax operation is performed on each row. The green boxes denote 1×1 convolutions. Here we show the embedded Gaussian version, with a bottleneck of 32 channels.

3.2   ResNet + Non-local Blocks Design

In SRGAN, ResNet [9] is the main network module in the generator network. The abstract expression is given by Eq. (4):

$$\mathbf{x}_{l+1} = \mathbf{H}(\mathbf{x}_l) + \mathbf{F}(\mathbf{x}_l) \tag{4}$$

where $\mathbf{x}_l$ and $\mathbf{x}_{l+1}$ denote the input and output of the l-th Residual Unit, $\mathbf{H}$ is an identity mapping and $\mathbf{F}$ is a non-linear residual function. As we know that, convolutional processing in ResNet is operated in local patch of image, Eq. (4) can be rewritten as Eq. (5):

$$\mathbf{x}_{l+1} = \mathbf{H}_L(\mathbf{x}_l) + \mathbf{F}_l(\mathbf{x}_l) \tag{5}$$

where the subscript L represents the filter operation in local patches, in the next section we use NL represents the filter operation in non-local patches.

To introduce the non-local information in convolution operations, appropriate combination between non-local neural network block and deep learning framework are very important. For a better improvement of image feature extraction and expression, two kinds of typical combination shown in Fig. 5. are discussed in this section. Considering that whether the non-local network is in ResNet or not, we design two structures in our modification. In the first structure, the non-local block lies in the trunk of ResNet. While, in the second structure the non-local network lies in the branch.

The details about the architectures ResNet + Non-local-A and ResNet + Non-local-B are provided respectively as following. The first one, called 'ResNet + Non-local-A', is combined with one non-local network and one ResNet. In this structure, non-local features are extracted first and the outputs of non-local network are used as the inputs of ResNet, which could be generally given by:

$$\mathbf{x}_{l+1} = \mathbf{F}_{NL}(\mathbf{H}_L(\mathbf{x}_l) + \mathbf{F}_L(\mathbf{x}_l)) = \mathbf{F}_{NL} \cdot \mathbf{H}_L(\mathbf{x}_l) + \mathbf{F}_{NL} \cdot \mathbf{F}_L(\mathbf{x}_l) \tag{6}$$

The second combination is called 'ResNet + Non-local-B'. The non-local network functions is acted as one part of ResNet. It is a type of close combination in which non-local networks is added into the branch of ResNet, which could be expressed as:

$$\mathbf{x}_{l+1} = \mathbf{H}_L(\mathbf{x}_l) + \mathbf{F}_{NL} \cdot \mathbf{F}_L(\mathbf{x}_l) \tag{7}$$

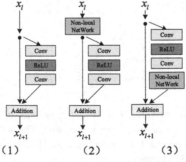

Fig. 5. The structure of non-local network (1) ResNet (2) ResNet+Non-local-A (3) ResNet + Non-local-B.

3.3 Build Non-local SRGAN

In this section, we will demonstrate our new SR network, called NL-SRGAN, with ResNet + Non-local blocks. We followed the architectural guidelines summarized by Feifei Li [9]. The convolution layer details, the structures and parameters of NL-SRGAN are described in Fig. 6. The overall network structure mainly includes two strategies: a ResNet + Non-local generator network G and a discriminator network D.

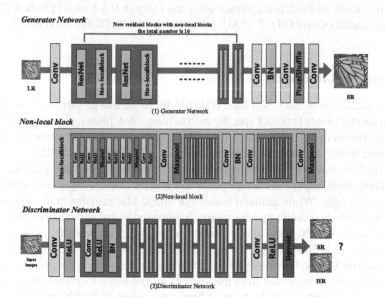

Fig. 6. The structural of NL-SRGAN (1) Our Generator Network (2) Our Non-local Convolutional Network (3) Discriminator Network.

In order to verify the advantage of these two ResNet + Non-local networks for image SR task, we develop these architectures in G net by replacing all the Residual Units with ResNet + Non-local. The comparisons of performance and time efficiency between the classical ResNet and the two new ResNet + Non-local network in SRGAN are given by our experiment.

4 Experiment

4.1 Experimental Setup

Dataset. We train our proposed network on DIV2K database [22] and select the public datasets include Set5 [23], Set14 [24] and BSD100 [25] for testing. We evaluate the performance of upscaling factors 2 and 4. For each upscaling factor, LR image is obtained by down sampling the HR image using bicubic kernel. The training dataset comprise HR images and the corresponding LR pairs. This can be expressed as $I = \{I_L, I_H\}$. Where $I_L = \{x_l^i\}_{i=1}^K$ and $I_H = \{x_h^i\}_{i=1}^K$ represent the LR and HR image datasets, x_h is HR images, x_l is LR images, i is the image index, K is the number of HR and LR image pairs.

Baselines and Evaluation Metric. Since our proposed approach can extract non-local features in network learning stage, we compare our model with the methods of local and global network. For local methods, we choose VDSR [7]. For global methods, we choose DSRN [26]. In the objective evaluation, we report the PSNR, SSIM and running time.

Implementation Details. In our experiment, we use Adam with adaptive learning rate. For optimization, all models are trained using the Pytorch 0.4.1 on a TITAN X NIVIDA GPU and Intel(R) Core(TM) i7-4790 3.6 GHz with 16 GB DDR3 1600 RAM.

4.2 Results and Discussion

In the classical dataset, Set5, Set14 and BSD100, there are a large variety of images ranging from natural image to object-specific such as baby, bird, plants, food. The experiment results are shown in Fig. 7 and Table 1, which demonstrates that our proposed approach can achieve better results and better performance on the PSNR and SSIM.

In Fig. 7, in column 3 and column 4, it can be observed that the generate images from VDSR considering local image patch and DSRN considering global image patch are with sharp edge. While artificial features or image blur remains in the reconstructed image. In column 5, it shows that the approach proposed in this paper can recover sharper edges and overcome the issue of over-smoothed regions. For example, in the third line, the first one is a LR Baby image, the second one is an HR image, the third one is an SR image based on VDSR. Obviously, the disadvantage in recovering details based on local SR algorithm can be observed from the textures on the Baby's eyelash. Its textures are still blurry. The fourth image in the third line is the result of DSRN, more sharp edges in the Baby's eyelash are reconstructed, but the blurs on the Baby's cap is still exist. For our result of ResNet + Non-local-B, the fourth image, details and textures are enhanced effectively, especially regarding the texture of the eyelash. Thus, our approach with non-local network can reconstruct image with sharper boundaries and richer textures. Our proposed single image super-resolution algorithm with non-local network and GAN has better performance.

Table 1 summarizes the PSNR, SSIM of VDSR, DSRN and our proposed method. When comparing the last 4 columns, it can be found that the performance of our network with non-local blocks is better than those without non-local blocks. Benefitting from the extending receptive field of non-local convolution, our trainable networks can achieve a higher evaluating indicator. By using non-local modules in SRGAN, the results firmly demonstrate the effectiveness of non-local block neural network and indicate the attentions to non-local features can really improves the performance. While for further research, other effective types of structure between non-local network and GAN should be discussed. Other factors influencing non-local network in SR task are also need a further investigation. These problems will be discussed in our next research.

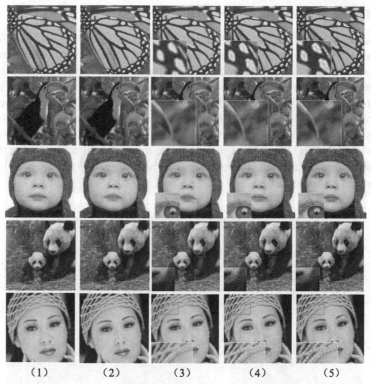

Fig. 7. Results of image 1 (2 ×) (1) LR image (2) HR image (3) VDER (4) DSRN (5) ResNet + Non-local-B.

Table 1. Comparison of PSNR and SSIM.

Dataset		Scale	VDSR	DSRN	ResNet +Non-local-A	ResNet +Non-local-B
Set5	PSNR	2×	37.53	37.63	37.92	**38.01**
		4×	31.35	31.53	31.89	**31.99**
	SSIM	2×	**0.9590**	0.9588	0.9490	0.9347
		4×	0.8830	0.8854	0.9042	**0.9561**
Set10	PSNR	2×	33.05	33.04	**33.89**	33.27
		4×	28.02	28.02	27.90	**29.18**
	SSIM	2×	0.9130	0.9118	**0.9213**	0.8579
		4×	0.7680	0.7670	0.7909	**0.7923**
BSD100	PSNR	2×	31.90	31.85	**32.09**	31.99
		4×	27.29	27.23	26.89	**28.43**
	SSIM	2×	0.8960	0.8942	0.8845	**0.8989**
		4×	0.7251	0.7233	0.7407	**0.7657**

4.3 Performance on the Real-World Data

Finally, we have tested our proposed technique on some meteorological satellite cloud images collected by the Meteorological Research Department. In this experiment, we use typhoon eye images from day and night satellite. Because of many non-local similar patches in satellite cloud images, our proposed algorithm is a perfect fit for this task, which shows the potential in real-world images super-resolution. The visual comparison results are shown in Fig. 8 for a scaling factor of 4. It can be seen that SRGAN approach is suffered from undesired artifacts especially around the edge of typhoon eyes. Our proposed NL-SRGAN method can overcome this difficulty and improve the visual quality. More fine details and sharper edges can be observed in Fig. 8. and the results are better on a computer screen.

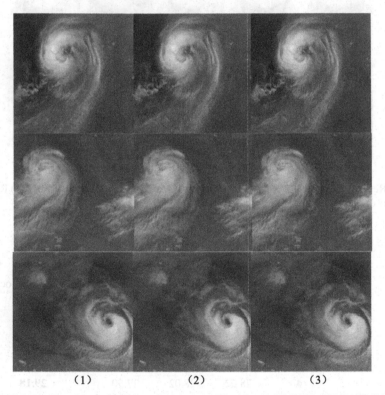

(1)   (2)   (3)

Fig. 8. Results of typhoon image SR (4×) (1) Bicubic (2) SRGAN (3) NL-SRGAN

5 Conclusions and Future Work

In this paper, a non-local convolutional module and the NL-SRGAN algorithm learning the end-to-end mapping between the LR and HR are proposed. The introduction of non-local convolution operation into the generator net of SRGAN make a considerable improvement in image feature representation. Our experimental results demonstrate that this new algorithm can produce better SR visual results than the other comparing methods. We hope the "ResNet + Non-local" will become an essential component of future SR network architectures. Next, we will discuss how to use the Non-local CNN to take the features from different images in the dataset as non-local information into account.

Acknowledgment. This work is supported by the National Natural Science Foundation of China (Grant No. 61802199).

References

1. Rumelhart, D.E., Hinton, G.E., Williams, R.J.: Learning internal representations by error propagation. Nature **323**(99), 533–536 (1986)
2. Yu, D., Deng, L.: Deep learning and its applications to signal and information processing. IEEE Sign. Process. Mag. **28**(1), 145–154 (2011)
3. Yu, D., Deng, L., Seide, F.: The deep tensor neural network with applications to large vocabulary speech recognition. IEEE Trans. Audio Speech Lang. Process. **21**(2), 388–396 (2012)
4. Hinton, G.E., Salakhutdinov, R.R.: Reducing the dimensionality of data with neural networks. Science **313**(5786), 504–507 (2006)
5. Lecun, Y., Bengio, Y., Hinton, G.: Deep learning. Nature **521**(7553), 436–444 (2015)
6. Dong, C., Loy, C.C., He, K., Tang, X.: Learning a deep convolutional network for image super-resolution. In: Fleet, D., Pajdla, T., Schiele, B., Tuytelaars, T. (eds.) ECCV 2014. LNCS, vol. 8692, pp. 184–199. Springer, Cham (2014). https://doi.org/10.1007/978-3-319-10593-2_13
7. Kim, J., Kwon Lee, J., Mu Lee, K.: Accurate image super resolution using very deep convolutional networks. In: IEEE Conference on Computer Vision and Pattern Recognition. IEEE Computer Society, Las Vegas, USA (2016)
8. Lai, W.S., Huang, J.B., Ahuja, N., et al.: Deep laplacian pyramid networks for fast and accurate super-resolution. In: IEEE Conference on Computer Vision and Pattern Recognition, pp. 5835–5843. IEEE Computer Society, Honolulu, Hawaii, USA (2017)
9. Ledig, C., Theis, L., Huszar, F., et al.: Photo-realistic single image super-resolution using a generative adversarial network. In: IEEE Conference on Computer Vision and Pattern Recognition, pp. 4681–4690. IEEE Computer Society, Honolulu, Hawaii, USA (2017)
10. Bee, L., Sanghyun, S., Heewon, K., et al.: Enhanced deep residual networks for single image super-resolution. In: IEEE Conference on Computer Vision and Pattern Recognition, pp. 136–144. IEEE Computer Society, Honolulu, Hawaii, USA (2017)
11. Shocher, A., Cohen, N., Irani, M.: Zero-shot super-resolution using deep internal learning. In: IEEE Conference on Computer Vision and Pattern Recognition, pp. 3118–3126. IEEE Computer Society, Honolulu, Hawaii, USA (2017)
12. Zhang, Y., Li, K., Li, K., Wang, L., Zhong, B., Fu, Y.: Image super resolution using very deep residual channel attention networks. In: European Conference on Computer Vision, pp. 294–310. Springer, Munich, Germany (2018)
13. Sajjadi, M.S., Schölkopf, B., Hirsch, M.: EnhanceNet: single image super-resolution through automated texture synthesis. In: IEEE International Conference on Computer Vision, pp. 4501–4510. IEEE Computer Society, Vines, Italy (2017)
14. Park, S.J., Son, H., Cho, S., Hong, K.S., Lee, S.: SRFeat: single image super-resolution with feature discrimination. In: 15th European Conference on Computer Vision, pp. 455–471. Springer, Munich, Germany (2018)

15. Wang, X., Yu, K., Wu, S., Gu, J., Liu, Y., Dong, C., Qiao, Y., Loy, C.C.: ESRGAN: enhanced super-resolution generative adversarial networks. In: Leal-Taixé, L., Roth, S. (eds.) ECCV 2018. LNCS, vol. 11133, pp. 63–79. Springer, Cham (2019). https://doi.org/10.1007/978-3-030-11021-5_5

16. Liu, D., Wen, B., Fan, Y., et al.: Non-local recurrent network for image restoration. In: Conference on Neural Information Processing Systems, pp. 1673–1682. MIT Press, Montréal, Canada (2018)

17. Wang, X., Girshick, R., Gupta, A., et al.: Non-local neural networks. In: IEEE Conference on Computer Vision and Pattern Recognition, pp. 7794–7803. IEEE Computer Society, Salt Lake City, Utah, USA (2018)

18. Yaqiong, F.: Research on image super-resolution reconst ruction based on non-local similarity. In: Nanjing University of Posts and Telecommunications, Nanjing, Jiangsu, China (2012)

19. Haibin, L., Youbin, C., Qinghu, C.: Non-local similarity dictionary learning based super-resolution for improved face recognition. Geomatics Inform. Sci. Wuhan Univ. **41**(10), 1414–1420 (2016)

20. Li, M., Li, S., Li, X., et al.: Super-resolution reconstruction algorithm based on non-local simultaneous sparse approximation. J. Electron. Inform. Technol. **33**(6), 1407–1412 (2011)

21. Goodfellow, I., et al.: Generative adversarial nets. In: Conference on Neural Information Processing Systems, pp. 2672–2680. MIT Press, Montréal, Canada (2014)

22. Agustsson, E., Timofte, R.: NTIRE 2017 challenge on single image super-resolution: dataset and study. In: IEEE Conference on Computer Vision and Pattern Recognition Workshops, pp. 1122–1131. IEEE Computer Society, Honolulu, Hawaii, USA (2017)

23. Bevilacqua, M., Roumy, A., Guillemot, C., Alberi-Morel, M.L.: Low-complexity single-image super resolution based on nonnegative neighbour embedding. In: Proceedings British Machine Vision Conference, pp. 135.1–135. 10. BMVA Press, Surrey, Britain (2012)

24. Zeyde, R., Elad, M., Protter, M.: On single image scale-up using sparse-representations. In: Boissonnat, J.D., Chenin, P., Cohen, A., Gout, C., Lyche, T., Mazure, M.L., Schumaker, L. (eds.) Curves and Surfaces 2010. LNCS, vol. 6920, pp. 711–730. Springer, Heidelberg (2012). https://doi.org/10.1007/978-3-642-27413-8_47

25. Martin, D., Fowlkes, C., Tal, D., Malik, J.: A database of human segmented natural images and its application to evaluating segmentation algorithms and measuring ecological statistics. In: IEEE International Conference on Computer Vision. IEEE Computer Society, Vancouver, Canada (2001)

26. Han, W., Chang, S., Liu, D., et al.: Image super-resolution via dual-state recurrent networks. In: IEEE Conference on Computer Vision and Pattern Recognition, pp. 1654–1663. IEEE Computer Society, Salt Lake City, Utah, USA (2018)

27. Ren, C., He, X., Pu, Y.: Non-local similarity modeling and deep CNN gradient prior for super resolution. IEEE Sign. Process. Lett. **25**(7), 916–920 (2018)

VH3D-LSFM: Video-Based Human 3D Pose Estimation with Long-Term and Short-Term Pose Fusion Mechanism

Wenjin Deng[1], Yinglin Zheng[1], Hui Li[1], Xianwei Wang[1], Zizhao Wu[2], and Ming Zeng[1(✉)]

[1] School of Informatics, Xiamen University, Xiamen, China
zengming@xmu.edu.cn
[2] Hangzhou Dianzi University, Hangzhou, China

Abstract. Following the success of 2D human pose estimation from a single image, a lot of work focus on video-based 3D human pose estimation by exploiting temporal information. In this scenario, several recent works have achieved significant advances via Temporal Convolution Network (TCN). However, the current TCN fashion suffers from lacking local coherence caused by excessive dependence on local frames and limited local dynamic range, failing to estimate poses correctly in real scenes, especially that with high-speed motions. To tackle this problem, we design a Long-term Bank to select and collect candidate key poses, and further provide a **LSFM** (**L**ong-term and **S**hort-term pose **F**usion Mechanism) to integrate long-term pose information into the short-term convolution window, thus to enhance the temporal coherence of local neighbor frames. Experimental results and ablation studies demonstrate that the proposed approach significantly promotes the accuracy and robustness of the state-of-the-art method.

Keywords: 3D human pose estimation · Video analysis

1 Introduction

Human pose estimation is a classic task in computer vision. In recent years, 2D pose estimation [1,2,4,6,15,19,24,26–29] has made significant progress and is gradually matured. Nowadays, the technique plays a critical role in many fields, including behavior detection in security monitoring, posture correction in

W. Deng—The first author is a student.
This work was supported by National Natural Science Foundation of China (No.61402387), Guiding Project of Fujian Province, China (No. 2018H0037), and Fundamental Research Funds for the Central Universities, China (No. 20720190003).

Electronic supplementary material The online version of this chapter (https:// doi.org/10.1007/978-3-030-60633-6_49) contains supplementary material, which is available to authorized users.

Y. Peng et al. (Eds.): PRCV 2020, LNCS 12305, pp. 589–601, 2020.
https://doi.org/10.1007/978-3-030-60633-6_49

medical health, to list a few. By extending the 2D applications to 3D scenarios, 3D pose estimation is attracting increasing attention in areas like 3D motion capture, body reconstruction, animation synthesis, etc.

Several attempts have been made to learn 3D pose directly from a single image, yet it remains challenging due to the inherent ambiguity of 2D information. In contrast, videos provide additional explicit and implicit temporal constraints for 3D skeleton estimation. To mine more temporal information, a large family of the existing work [5,12,16,22,23] employs time sequence models of Deep neural network. However, all these methods only use short-term local motions, which are usually discontinuous on high-speed conditions. This challenge is illustrated in Fig. 1. In this case, local poses($L_{t-2}, L_{t-1}, L_t, L_{t+1}, L_{t+2}$) are incoherent as reflected by the their similarities with the pose L_t of the center frame. This dramatics dynamics hinders the performance of convolution in the subsequent infer phase.

Our method is inspired by the idea of Wu et al. [30], where short-term and long-term features are utilized to capture local and global information for video understanding. With this motivation, we designed a long-term and short-term information fusion mechanism for video-based 3D pose estimation. Firstly, we build a Long-term Bank to store keyframes by a greedy strategy and update it dynamically. Secondly, the local poses are reconstructed with keyframes via a similarity-based rearrangement method before local convolution operation.

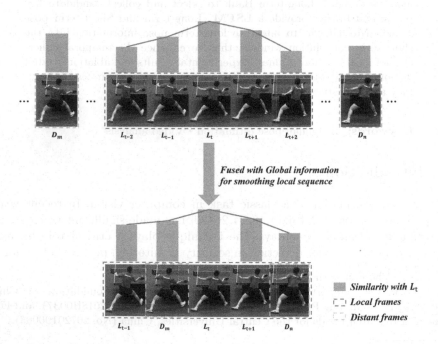

Fig. 1. Illustration of the Fast Motion Challenge and the solution idea. Origin local sequence ($L_{t-2}, L_{t-1}, L_t, L_{t+1}, L_{t+2}$) is fused with distant frames(D_m, D_n) for smoothing itself.

The contributions of this paper can be summarized:

- We develop a 3D human pose estimation algorithm that combines long-term and short-term information, which can effectively exploit local and global features to increase local continuity.
- We propose an effective example-based long-term pose extraction mechanism, and long-term and short-term pose fusion mechanism.
- We conduct comprehensive experiments to evidence the effectiveness of the proposed methods.

2   Related Work

2.1   2D Pose Estimation from a Single Image

In essence, 2D pose estimation is a regression task that learns coordinates of joints position from image pixels. The single-person 2D pose estimation output probabilistic heatmaps of human joints by taking a monocular RGB image as input, and then obtain the final coordinates from the heatmaps. For instance, CPM [29], Stacked Hourglass [19], etc. With the foundation of single-person estimation, the task can be extended to multi-person situations. Multi-person 2D pose estimation can be divided into top-down and bottom-up methods. The top-down methods first detect every single person's bounding box, and then performing single-person pose estimation for each person respectfully, e.g., Alpha-Pose [6], CPN [4], HRNet [28]. In contrast, the bottom-up method first estimates the joint points of various parts, and clusters joints of different people together to produce different individuals finally, such as work [3,8,18]. Those methods achieve excellent performance in estimating 2D poses from a single image. Despite the maturity of 2D pose estimation, the requirement of practical applications remains far from satisfied with only 2D poses. Thus quite a lot of work has emerged in the 3D pose estimation.

2.2   3D Pose Estimation from a Single Image

There are two branches of algorithms for 3D pose estimation from a single image. One of them is one-stage that directly predicts 3D pose from a monocular image, e.g., work [7,14,21,31]. Another branch requires two steps. More specifically, it first estimates 2D pose, and then predict 3D pose from the 2D pose. Powered by existing robust 2D pose detection models, the 3D pose estimation task can be optimized independently with low computational complexity, typical work are [13,17]. Due to the ambiguity that several 3D poses can be projected to the same 2D pose, several work [10,20] utilize the NRSFM (Non-rigid Structure from Motion) to recover 3D from 2D.

2.3   3D Pose Estimation on Video

Compared to a single monocular image, there are more temporal and spatial semantics in videos. The constraints of multi-view, spatial geometric, and temporal continuity can be used to mitigate the impact of self-occlusion and ambiguity. Sun et al. [25] proposed to estimate 3D poses from images in an end-to-end manner. However, methods learning from 2D pose sequences achieve state-of-the-art performance.

Different from the work using LSTM [12,23], Dario et al. [22] employ TCN [11] with stacked dilated convolution layers to predict 3D pose of the center frame by weighting adjacent frames. However, its accuracy largely depends on the size of local receptive field, the bigger the size, the better the performance. In their paper, the best size is 243 frames, which even larger than some videos.

A larger local receptive field means extra computational cost while a smaller local receptive field means information loss. To reduce the computation while taking advantage of the large receptive field, we design a Long-term Bank to store key poses extracted from the input sequence based on our selection strategy, and further propose a novel long-term and short-term information fusion algorithm to promote the local coherence.

3   Overview of Our Method

The pipeline of our method is shown in Fig. 2. Given a sequence of images, the pipeline first adopts a pre-trained 2D pose estimation network to infer the corresponding 2D positions of human joints frame by frame. Based on the estimated per-frame 2D pose, our method sequentially upgrades the temporal 2D poses to 3D poses using the Temporal Convocation Network (TCN). During the estimation, our method dynamically maintains a long-term pose bank to store diverse candidate 2D key poses, and adaptively select and integrate coherent long-term poses into the local convolution window (Short-term Bank).

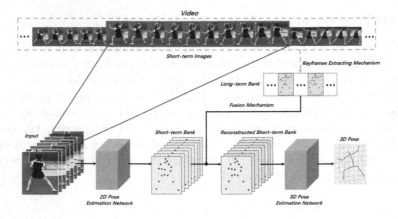

Fig. 2. The pipeline of our method

3.1   3D Pose Estimation Backbone Network

We employ the fashion of Temporal Convolution Network [22] as the backbone network of 3D pose estimator, as is illustrated in Fig. 3. In this network, the input is 2D poses $X = \{x_1^0, x_2^0, ..., x_T^0\}$, the model transforms them into 3D poses of the center frame through a series of residual blocks. Each residual block contains a dilation convolution layer with a kernel size of 3 and a linear layer with a kernel size of 1, formulated as

$$x_t^{i+1} = x_t^i + wl^i(w_0^i x_t^i + w_1^i x_{t-d}^i + w_2^i x_{t+d}^i) \tag{1}$$

where d is dilation rate, w^i is convolution weight, and wl^i is linear weight. The TCN model is trained in a semi-supervised manner, where supervised loss requires the estimated 3D poses to match the ground truth, and unsupervised loss ensures the projected estimated 3D poses are consistent with the input 2D poses [22].

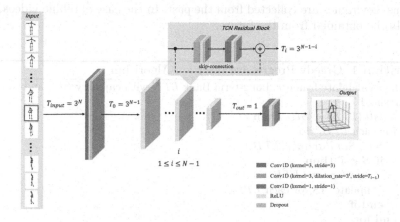

Fig. 3. Structure of temporal convolution network

3.2   Long-Term and Short-Term Poses Fusion Mechanism

Although Temporal Convolution Network captures smoothly continuous dynamics, it fails to model discontinuous motions, which commonly happens on fast-moving humans. To tackle this shortcoming, we collect a set of the Long-term key poses and integrate them into the short-term poses of the sliding temporal window of TCN, to reconstruct the local continuity of the human motion.

Construction of Long-Term Pose Bank. The Long-term Pose Bank (LTB) is designed to represent motion patterns of the whole sequence by storing key poses. To this end, the poses selected into LTB should be as diverse as possible. Therefore, we devise a similarity-based method to incrementally select diverse poses into the LTB from scratch as the Temporal convolution network proceeds in time sequence. The similarity between two 2D poses are measured based on their positions of joints:

$$S = 1 - \frac{\sum_{i=0}^{J-1} \|A_{transi} - B_i\|_2}{J \times avg\_bone\_length(B)} \qquad (2)$$

where A_{trans} is a translated pose from pose A to pose B by subtracting the offset of their centers, and J is the number of joint.

Based on the measurement of similarity between two poses, we adopt a greedy strategy to insert key poses into LTB progressively. In detail, when LTB is not full, the candidate frame x is inserted if the similarity S with all stored frames is less than the preset threshold T. While when reaching the max capacity $Size$, we dynamically maintain an accumulated similarity value AS for each keyframe pose and update poses by keeping the sum of AS of the entire bank as little as possible. The process is illustrated in Algorithm 1. In practice, for online video streams, keyframes are collected from the past. In the case of offline videos, they can also be obtained from the future.

Algorithm 1. Greedy Progressive Insertion Algorithm

Require: candidate frame x, Long-term Bank LTB with capacity $Size$ and similarity
threshold T
1: **if** $length(LTB) < Size$ **then**
2:    **for** each LTB_i **do**
3:        $S := Similarity(x, LTB_i)$
4:        **if** $S < T$ **then**
5:            insert x into LTB
6:            update AS for each LTB_i
7:        **end if**
8:    **end for**
9: **else**
10:    **for** each LTB_i **do**
11:        $S := Similarity(x, LTB_i)$
12:        **if** $S < T$ **then**
13:            delete item where AS is max
14:            insert x into LTB
15:            update AS for each LTB_i
16:        **end if**
17:    **end for**
18: **end if**

Fusion of Long-Term Bank and Short-Term Bank. The fusion mechanism aims to ensure that the Short-term Bank after fusion is more continuous than before. We design an Alternative Symmetric Pose Fusion Algorithm to

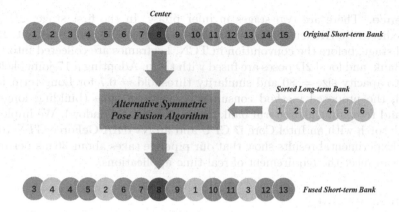

Fig. 4. Illustration of the procedure of Alternative Symmetric Pose Fusion Algorithm.

dynamically fuse key poses of LTB into STB, targeting at enhancing the motion coherency of the local spatial convolution window. Given the center frame of the convolution window of TCN, we first calculate the pose similarities between the center frame and 2D key poses that are stored in LTB. Then, these similarity values S_c are sorted in the descending order. After that, the 2D poses in LTB that are similar to center frame enough (exceed a threshold) are inserted into STB. To keep the balance between the two half sides of the STB window, the insertion is performed in an alternative symmetric manner. An example of the algorithm is shown in Fig. 4.

3.3 Multiple Rate Frame Sampling for Data Augmentation

Unlike the smooth motions collected in indoor environments with high-speed cameras, the poses in real scenes usually have multiple dynamic ranges. To encourage the network to learn this pattern, we perform augmentation on the training data by uniformly sampling original continuous sequence with varieties of sampling intervals to synthesize different motion dynamics. In our experiments, considering the possible dynamic ranges of human motion, the sampling intervals are set to 3, 5.

3.4 Implementation Details

Training. Regarding the efficiency of training, we directly train TCN on Human3.6M with ground-truth 2D keypoints. The loss function is defined as the mean per-joint position error between predictions and 3D annotations. Adam with momentum $= 0.1$ is adopted as the optimizer, and we train TCN with a receptive field of 27 frames for 80 epochs. The learning rate and batch size are set to $1e-3$ and 1024 respectively.

Inference. There are two stages in infer phase. In the first stage, 2D poses are predicted by HRNet which is pre-trained on the MS-COCO dataset. In the second stage, before the convolution in TCN, keyframes are collected into Long-term Bank, and local 2D poses are fused with them. Adopting a 17-joint skeleton, We set capacity size = 80 and similarity threshold = 0.7 for Long-term Bank. Indeed, the proposed method consumes some extra time (building long-term bank and rebuilding short-term bank for convolutional window). We implement it by Pytorch with an Intel Core i7 CPU and an NVIDIA GeForce GTX 1080Ti GPU. Experimental results show that our pipeline takes about 36 ms per frame, which can meet the requirement of real-time applications.

4   Experiments

4.1   Datasets and Evaluation Protocol

Following previous work [17,22,25], we conduct experiments on the Human3.6M dataset [9] as it contains abundant human poses with 2D and 3D annotations and includes 15 common actions. In addition to Human3.6M, we also collect a set of videos with fast-moving human, and we qualitatively compare our method with the state-of-the-art method on these challenging videos.

In our experiments, we utilize three metrics: MPJPE, P-MPJPE, MPJVE. **MPJPE** is the mean per-joint position error (also used for loss function). **P-MPJPE** is the error after alignment with the ground truth in translation, rotation, and scale. **MPJVE** is the mean per-joint velocity error. Note that all the following experiments are carried out in comparison with the recent temporal model of the work [22], abbreviated as *VideoPose*.

4.2   Comparison with the State-of-the-art Method

Effectiveness of Long-Term Bank. Since there is no large-scale human fast-moving dataset, we build a synthetic dataset to evaluate our methods by extracting frames from the Human3.6M origin sequences with a certain interval. With intervals of 3, 5 on testing data, the results in Table 1 and Table 2 show that the model with Long-term Bank (LTB) outperforms over the base model–*VideoPose*, which is without LTB. Note that since the original data is too smooth to insert key poses, the results with LTB are almost identical to the baseline results. In these experiments, the base model is pre-trained on Human3.6M with a receptive filed of 27. As for the Long-term Bank, the capacity is 80 and the similarity threshold is 0.7.

Effectiveness of Data Augmentation. To validate the effectiveness of data augmentation, we conduct ablation studies on the Human3.6M dataset regarding different sets of augmentation data. It is clear that incrementally adding augmented videos with different sampling rate indeed substantially improves the performance of the baseline model (Table 3).

Table 1. Effectiveness of Long-term Bank. MPJPE (mm) results on testing data of original (O) videos and synthesized videos with intervals of 3, 5 (S3, S5).

Testing data	LTB	Dir.	Disc.	Eat	Greet	Phone	Photo	Pose	Purch.	Avg
		Sit	SitD	Smoke	Wait	WalkD	Walk	WalkT		
Original (O)	w/o	38.91	43.03	34.71	38.66	**37.83**	41.31	**43.71**	38.17	39.02
		42.97	45.23	38.23	40.83	39.02	30.43	32.32		
	w/	**38.90**	**42.92**	**34.69**	38.66	37.86	**41.23**	43.78	**38.00**	**38.97**
		42.96	**45.16**	**38.19**	**40.69**	**38.90**	**30.37**	**32.27**		
Sample3 (S3)	w/o	51.94	58.51	49.00	58.35	50.45	54.59	55.63	61.09	55.24
		52.40	58.61	50.87	55.72	60.55	55.61	55.29		
	w/	**50.25**	**55.73**	**47.31**	**55.61**	**48.84**	**52.57**	**54.22**	**56.71**	**52.95**
		51.05	**56.38**	**48.90**	**53.48**	**56.71**	**53.52**	**52.90**		
Sample5 (S5)	w/o	60.05	69.65	55.80	71.25	56.49	64.46	64.22	74.95	64.56
		60.14	70.81	59.34	65.90	71.98	62.25	61.13		
	w/	**56.71**	**63.82**	**53.27**	**65.57**	**54.02**	**60.26**	**61.16**	**66.96**	**60.26**
		57.04	**66.15**	**55.79**	**61.39**	**65.04**	**58.91**	**57.87**		

Table 2. Effectiveness of Long-term Bank. P-MPJPE (mm) results on testing data of original (O) videos and synthesized videos with intervals of 3, 5 (S3, S5).

Testing data	LTB	Dir.	Disc.	Eat	Greet	Phone	Photo	Pose	Purch.	Avg
		Sit	SitD	Smoke	Wait	WalkD	Walk	WalkT		
Original (O)	w/o	**26.38**	31.00	**26.54**	**28.42**	**27.62**	31.13	**30.07**	27.01	28.75
		32.36	35.59	28.82	28.80	29.38	**23.11**	25.00		
	w/	26.43	31.00	26.56	28.43	27.67	31.13	30.15	**26.96**	28.75
		32.35	**35.56**	**28.81**	**28.76**	**29.31**	23.13	25.00		
Sample3 (S3)	w/o	37.64	40.56	37.78	43.05	36.48	38.53	39.09	40.32	40.47
		37.41	42.13	36.60	39.84	44.11	47.69	45.79		
	w/	**36.53**	**39.07**	**36.74**	**41.35**	**35.50**	**37.64**	**38.37**	**38.03**	**39.15**
		36.68	41.13	35.66	38.35	42.22	46.07	43.95		
Sample5 (S5)	w/o	44.82	47.24	42.79	52.52	40.31	43.62	45.37	48.45	46.60
		41.33	48.31	41.77	47.35	51.40	53.28	50.39		
	w/	**42.29**	**44.00**	**41.28**	**48.93**	**39.02**	**41.71**	**43.54**	**44.13**	**44.15**
		39.93	46.17	40.16	44.23	47.71	50.93	48.28		

4.3 Results of Challenging Videos with Fast-Moving Human

We collect 15 videos of four representative sports categories from the Internet with the keywords: "Badminton", "Tennis", "Table tennis", and "Skating". The total length of all the videos is approximately 6 h. Figure 5 and Fig. 6 visualize the comparison results between our method and *VideoPose*. The results show

Table 3. Effectiveness of data augmentation. Results on Human3.6M. The abbreviations in the first two columns corresponds to testing set and training set, respectively.

Testing data	Training data	MPJPE	P-MPJPE	MPJVE
Original (O)	O	39.0	28.7	1.89
	O+S3	39.2	28.6	1.8
	O+S3+S5	**38.7**	**28.2**	**1.74**
Sample3 (S3)	O	55.2	40.5	13.54
	O+S3	39.5	28.9	7.65
	O+S3+S5	**38.4**	**28.0**	**6.82**
Sample5 (S5)	O	64.6	46.4	21.80
	O+S3	41.6	30.7	12.84
	O+S3+S5	**39.0**	**28.5**	**10.36**

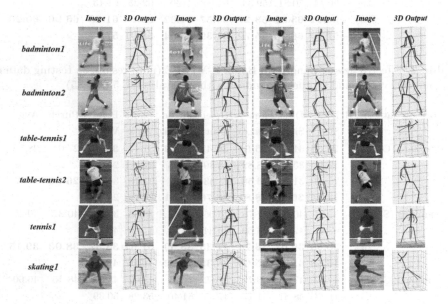

Fig. 5. Qualitative justification of long-term bank. The red skeleton is VideoPose, and the black is the model with Long-term Bank. (Color figure online)

that our method outperforms the benchmark model significantly. Even in cases with strong motion blur, our method can locate the fast-moving joints accurately, benefiting both from the support of long-term pose bank and multi-rate data augmentation.

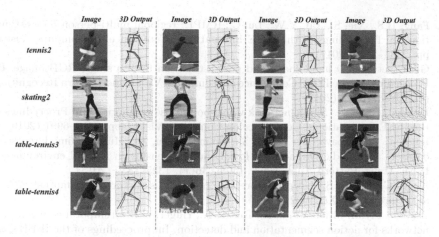

Fig. 6. Qualitative justification of multiple sampling rate augmentation. The red skeleton is VideoPose, while the black is trained on augmented data. (Color figure online)

5 Conclusion

In this paper, we propose a novel temporal information fusion mechanism that integrates short-term poses with long-term poses to improve the coherence of local pose sequence. We design an online method for constructing long-term pose bank and a greedy-based fusion mechanism for local key poses integration. Furthermore, to improve the ability of generalization of the basic TCN model, we also provide a data augmentation method that synthesizes sequences with different motion speeds by frame sampling. The qualitative and quantitative experimental results on the Human3.6M dataset and the fast-moving videos we collected demonstrate the superiority of the proposed method.

References

1. Alp Güler, R., Neverova, N., Kokkinos, I.: DensePose: dense human pose estimation in the wild. In: Proceedings of the IEEE Conference on Computer Vision and Pattern Recognition, pp. 7297–7306 (2018)
2. Bissacco, A., Yang, M.H., Soatto, S.: Fast human pose estimation using appearance and motion via multi-dimensional boosting regression. In: 2007 IEEE Conference on Computer Vision and Pattern Recognition, pp. 1–8. IEEE (2007)
3. Cao, Z., Simon, T., Wei, S.E., Sheikh, Y.: Realtime multi-person 2D pose estimation using part affinity fields. In: Proceedings of the IEEE Conference on Computer Vision and Pattern Recognition, pp. 7291–7299 (2017)
4. Chen, Y., Wang, Z., Peng, Y., Zhang, Z., Yu, G., Sun, J.: Cascaded pyramid network for multi-person pose estimation. In: Proceedings of the IEEE Conference on Computer Vision and Pattern Recognition, pp. 7103–7112 (2018)
5. Dabral, R., Mundhada, A., Kusupati, U., Afaque, S., Sharma, A., Jain, A.: Learning 3D human pose from structure and motion. In: Proceedings of the European Conference on Computer Vision (ECCV), pp. 668–683 (2018)

6. Fang, H.S., Xie, S., Tai, Y.W., Lu, C.: RMPE: regional multi-person pose estimation. In: Proceedings of the IEEE International Conference on Computer Vision, pp. 2334–2343 (2017)
7. Ge, L., et al.: 3D hand shape and pose estimation from a single RGB image. In: Proceedings of the IEEE Conference on Computer Vision and Pattern Recognition, pp. 10833–10842 (2019)
8. Hidalgo, G., et al.: Single-network whole-body pose estimation. In: Proceedings of the IEEE International Conference on Computer Vision, pp. 6982–6991 (2019)
9. Ionescu, C., Papava, D., Olaru, V., Sminchisescu, C.: Human3.6m: large scale datasets and predictive methods for 3D human sensing in natural environments. IEEE Trans. Pattern Anal. Mach. Intell. **36**(7), 1325–1339 (2013)
10. Kong, C., Lucey, S.: Deep interpretable non-rigid structure from motion (2019). arXiv preprint arXiv:1902.10840
11. Lea, C., Flynn, M.D., Vidal, R., Reiter, A., Hager, G.D.: Temporal convolutional networks for action segmentation and detection. In: proceedings of the IEEE Conference on Computer Vision and Pattern Recognition, pp. 156–165 (2017)
12. Lee, K., Lee, I., Lee, S.: Propagating LSTM: 3D pose estimation based on joint interdependency. In: Proceedings of the European Conference on Computer Vision (ECCV), pp. 119–135 (2018)
13. Li, C., Lee, G.H.: Generating multiple hypotheses for 3D human pose estimation with mixture density network. In: Proceedings of the IEEE Conference on Computer Vision and Pattern Recognition, pp. 9887–9895 (2019)
14. Li, S., Chan, A.B.: 3D human pose estimation from monocular images with deep convolutional neural network. In: Cremers, D., Reid, I., Saito, H., Yang, M.-H. (eds.) ACCV 2014. LNCS, vol. 9004, pp. 332–347. Springer, Cham (2015). https://doi.org/10.1007/978-3-319-16808-1_23
15. Li, W., et al.: Rethinking on multi-stage networks for human pose estimation (2019). arXiv preprint arXiv:1901.00148
16. Luo, Y., et al.: LSTM pose machines. In: Proceedings of the IEEE Conference on Computer Vision and Pattern Recognition, pp. 5207–5215 (2018)
17. Martinez, J., Hossain, R., Romero, J., Little, J.J.: A simple yet effective baseline for 3D human pose estimation. In: Proceedings of the IEEE International Conference on Computer Vision, pp. 2640–2649 (2017)
18. Newell, A., Huang, Z., Deng, J.: Associative embedding: end-to-end learning for joint detection and grouping. In: Advances in Neural Information Processing Systems, pp. 2277–2287 (2017)
19. Newell, A., Yang, K., Deng, J.: Stacked hourglass networks for human pose estimation. In: Leibe, B., Matas, J., Sebe, N., Welling, M. (eds.) ECCV 2016. LNCS, vol. 9912, pp. 483–499. Springer, Cham (2016). https://doi.org/10.1007/978-3-319-46484-8_29
20. Novotny, D., Ravi, N., Graham, B., Neverova, N., Vedaldi, A.: C3DPO: canonical 3D pose networks for non-rigid structure from motion. In: Proceedings of the IEEE International Conference on Computer Vision, pp. 7688–7697 (2019)
21. Pavlakos, G., Zhou, X., Derpanis, K.G., Daniilidis, K.: Coarse-to-fine volumetric prediction for single-image 3D human pose. In: Proceedings of the IEEE Conference on Computer Vision and Pattern Recognition, pp. 7025–7034 (2017)
22. Pavllo, D., Feichtenhofer, C., Grangier, D., Auli, M.: 3D human pose estimation in video with temporal convolutions and semi-supervised training. In: Proceedings of the IEEE Conference on Computer Vision and Pattern Recognition, pp. 7753–7762 (2019)

23. Rayat Imtiaz Hossain, M., Little, J.J.: Exploiting temporal information for 3D human pose estimation. In: Proceedings of the European Conference on Computer Vision (ECCV), pp. 68–84 (2018)
24. Sun, X., Shang, J., Liang, S., Wei, Y.: Compositional human pose regression. In: Proceedings of the IEEE International Conference on Computer Vision, pp. 2602–2611 (2017)
25. Sun, Y., Ye, Y., Liu, W., Gao, W., Fu, Y., Mei, T.: Human mesh recovery from monocular images via a skeleton-disentangled representation. In: Proceedings of the IEEE International Conference on Computer Vision, pp. 5349–5358 (2019)
26. Tang, W., Yu, P., Wu, Y.: Deeply learned compositional models for human pose estimation. In: Proceedings of the European Conference on Computer Vision (ECCV), pp. 190–206 (2018)
27. Toshev, A., Szegedy, C.: DeepPose: human pose estimation via deep neural networks. In: Proceedings of the IEEE Conference on Computer Vision and Pattern Recognition, pp. 1653–1660 (2014)
28. Wang, J., et al.: Deep high-resolution representation learning for visual recognition (2019). arXiv preprint arXiv:1908.07919
29. Wei, S.E., Ramakrishna, V., Kanade, T., Sheikh, Y.: Convolutional pose machines. In: Proceedings of the IEEE Conference on Computer Vision and Pattern Recognition, pp. 4724–4732 (2016)
30. Wu, C.Y., Feichtenhofer, C., Fan, H., He, K., Krahenbuhl, P., Girshick, R.: Long-term feature banks for detailed video understanding. In: Proceedings of the IEEE Conference on Computer Vision and Pattern Recognition, pp. 284–293 (2019)
31. Zhou, X., Huang, Q., Sun, X., Xue, X., Wei, Y.: Towards 3D human pose estimation in the wild: a weakly-supervised approach. In: Proceedings of the IEEE International Conference on Computer Vision, pp. 398–407 (2017)

Unregistered Hyperspectral and Multispectral Image Fusion with Synchronous Nonnegative Matrix Factorization

Wenjing Chen[1,2] and Xiaoqiang Lu[1(✉)]

[1] Key Laboratory of Spectral Imaging Technology CAS, Xi'an Institute of Optics and Precision Mechanics, Chinese Academy of Sciences, Xi'an 710119, China
chenwenjing2017@opt.cn, luxq666666@gmail.com
[2] University of Chinese Academy of Sciences, Beijing 100049, China

Abstract. Recently, many methods have been proposed to generate a high spatial resolution (HR) hyperspectral image (HSI) by fusing HSI and multispectral image (MSI). Most methods need a precondition that HSI and MSI are well registered. However, in practice, it is hard to acquire registered HSI and MSI. In this paper, a synchronous nonnegative matrix factorization (SNMF) is proposed to directly fuse unregistered HSI and MSI. The proposed SNMF does not require the registration operation by modeling the abundances of unregistered HSI and MSI independently. Moreover, to exploit both HSI and MSI in the endmember optimization of the desired HR HSI, the unregistered HSI and MSI fusion is formulated as a bound-constrained optimization problem. A synchronous projected gradient method is proposed to solve this bound-constrained optimization problem. Experiments on both simulated and real data demonstrate that the proposed SNMF outperforms the state-of-the-art methods.

Keywords: Image fusion · Nonnegative matrix factorization · Hyperspectral image · Multispectral image

1 Introduction

Hyperspectral images (HSIs) record the observed scenes over massive contiguous spectral bands [1,2]. As they contain abundant spectral information, HSIs play important roles in many applications, *e.g.* land-cover recognition [3] and change detection [4]. However, due to the limitations of optical systems, the spatial resolution of captured HSIs is generally low, which limits the use of HSIs [5,6]. Different from HSIs, multispectral images (MSIs) have a high spatial resolution [7]. Naturally, fusing HSI and MSI from the same observed scene is a feasible way to generate a high spatial resolution (HR) HSI [8].

The first author is a PhD student.

© Springer Nature Switzerland AG 2020
Y. Peng et al. (Eds.): PRCV 2020, LNCS 12305, pp. 602–614, 2020.
https://doi.org/10.1007/978-3-030-60633-6_50

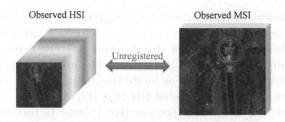

Fig. 1. Observed HSI and MSI are usually unregistered.

Recently, spectral unmixing-based methods [6,8,9] have drawn broad attention for fusing HSI and MSI, due to their physical interpretability [2]. Spectral unmixing-based methods rely on the linear mixture model [10], where a HSI can be represented as a linear combination of endmembers weighted by corresponding abundances. Spectral unmixing-based methods fuse the HSI and MSI by estimating the endmembers and abundances of the desired HR HSI [2]. Especially, the spectral response function (SRF) and the spatial degradation between HSI and MSI are usually adopted as the priors for simplifying the HSI and MSI fusion [8]. Recent spectral unmixing-based methods can be simply divided into joint optimization-based methods and separate optimization-based methods. Joint optimization-based methods couple the abundances of HSI and MSI via the spatial degradation [8,11]. The endmembers and abundances of the desired HR HSI are optimized jointly. However, the real spatial degradation is often unavailable in practice and hard to be accurately estimated [12]. Separate optimization-based methods independently estimate the endmembers of the desired HR HSI from HSI and the abundances of the desired HR HSI from MSI without requiring the spatial degradation [13,14]. The endmembers extracted from HSI are transformed into the endmembers of MSI through the SRF for estimating the abundances of MSI [14]. When the spatial degradation is difficult to obtain, separate optimization-based methods are more effective than joint optimization-based methods [15].

Although plentiful methods have been proposed for HSI and MSI fusion, most methods still rely on a precondition that HSI and MSI are well registered [8,9,11]. In practice, as shown in Fig. 1, HSI and MSI captured from different imaging sensors are usually not registered [15]. To fuse unregistered HSI and MSI, several methods [2,16] take the registration as a preprocessing operation. Owing to the great differences in the spatial-spectral resolution between HSI and MSI, registering HSI and MSI still faces challenges [17]. Consequently, designing an unregistered HSI and MSI fusion method that does not need registration operations is meaningful in practical applications. Since the spatial degradation between unregistered HSI and MSI is difficult to be estimated, joint optimization-based methods perform worse in unregistered HSI and MSI fusion [15,17]. In separate optimization-based methods, the abundances of HSI and MSI are estimated independently. The abundances of MSI will not be influenced by the abundance of HSI. Separate optimization-based methods have the potential for unregistered

HSI and MSI fusion without introducing registration operations. However, in recent separate optimization-based methods, only HSI is exploited for the optimization of endmembers [13]. The transformed HSI endmembers may not be the optimal endmembers for MSI [6,17]. And the inferior endmembers of MSI can make it difficult to accurately estimate the abundances of MSI, resulting in decreased quality of the reconstructed HR HSI [17,18].

In this paper, a synchronous nonnegative matrix factorization (SNMF) is proposed to directly fuse unregistered HSI and MSI. The proposed SNMF independently estimates the abundances of HSI and MSI for avoiding introducing registration operations. This paper assumes that the non-overlapping scene and the overlapping scene have similar underlying materials. Different from previous separate optimization-based methods, the proposed SNMF utilizes both HSI and MSI in the endmember optimization of the desired HR HSI. To exploit both HSI and MSI, the proposed SNMF formulates the unregistered HSI and MSI fusion as a bound-constrained optimization problem. The endmembers and abundances of the desired HR HSI are synchronously optimized in SNMF. Moreover, a synchronous projected gradient method is proposed to solve this bound-constrained optimization problem. Both simulated and real data are utilized to evaluate the proposed SNMF. Experiments demonstrate the effectiveness of SNMF in both registered and unregistered HSI and MSI fusion.

The main contributions of this paper are as below:

- To fuse unregistered HSI and MSI without introducing registration operations, a synchronous nonnegative matrix factorization is proposed to independently estimate the abundances of HSI and MSI. The abundances of MSI will not be influenced by the abundance of HSI.
- To exploit both HSI and MSI in the endmember optimization, the proposed SNMF formulates the unregistered HSI and MSI fusion as a bound-constrained optimization problem. SNMF can extract the optimal endmembers for the desired HR HSI.
- To keep the endmembers and abundances nonnegative, a synchronous projected gradient method is proposed to solve this bound-constrained optimization problem.

Subsequent sections are organized as below. Section 2 introduces the proposed SNMF. Section 3 reports experimental results. Section 4 summarizes this paper.

2 Proposed Method

The framework of the proposed SNMF is shown in Fig. 2. In this paper, all hyperspectral and multispectral images are expressed in the two-dimensional form [12]. Similar to recent methods [2,17], HSI and MSI are assumed to be captured under the same illumination and atmospheric conditions.

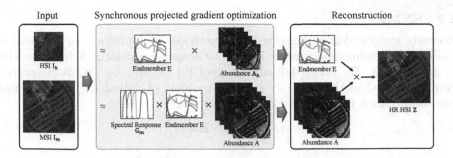

Fig. 2. Framework of the proposed SNMF. × represents the matrix multiplication.

2.1 Problem Formulation

The aim of HSI and MSI fusion is to reconstruct a HR HSI $\mathbf{Z} \in \mathbb{R}^{B \times L}$ based on the given observed HSI $\mathbf{I}_h \in \mathbb{R}^{B \times l}$ and MSI $\mathbf{I}_m \in \mathbb{R}^{b \times L}$. B and b denote the band number ($b < B$). l and L denote the pixel number ($l < L$). In recent methods [8,9], $\mathbf{I}_h$ is modeled as the degraded version of $\mathbf{Z}$ in the spatial domain. And $\mathbf{I}_m$ is modeled as the degraded version of $\mathbf{Z}$ in the spectral domain. $\mathbf{I}_h$ and $\mathbf{I}_m$ are formulated as

$$\mathbf{I}_h = \mathbf{Z}\mathbf{G}_h + \mathbf{R}_h, \tag{1}$$

$$\mathbf{I}_m = \mathbf{G}_m\mathbf{Z} + \mathbf{R}_m, \tag{2}$$

where $\mathbf{G}_h \in \mathbb{R}^{L \times l}$ is the spatial degradation that represents blurring and down-sampling. $\mathbf{G}_m \in \mathbb{R}^{b \times B}$ is the SRF. $\mathbf{R}_h \in \mathbb{R}^{B \times l}$ and $\mathbf{R}_m \in \mathbb{R}^{b \times L}$ are residuals.

In accordance with the linear mixture model [10], HSIs can be taken as the linear mixture of endmembers representing different underlying materials. The desired HR HSI $\mathbf{Z}$ can be formulated as

$$\mathbf{Z} = \mathbf{E}\mathbf{A} + \mathbf{R}_z, \tag{3}$$

where $\mathbf{E} \in \mathbb{R}^{B \times N}$ is the endmember matrix of $\mathbf{Z}$, $\mathbf{A} \in \mathbb{R}^{N \times L}$ is the abundance matrix of $\mathbf{Z}$, $\mathbf{R}_z \in \mathbb{R}^{B \times L}$ is the residual, and N is the number of endmembers. Each column of $\mathbf{E}$ represents the spectrum of each endmember. Each column of $\mathbf{A}$ represents the abundance fractions of the entire endmembers at each pixel. Besides, all elements of $\mathbf{E}$ and $\mathbf{A}$ are nonnegative [9].

By submitting Eq. (3) into Eq. (1) and Eq. (2), the HSI $\mathbf{I}_h$ and MSI $\mathbf{I}_m$ can be approximated as

$$\mathbf{I}_h \approx \mathbf{E}\mathbf{A}\mathbf{G}_h, \tag{4}$$

$$\mathbf{I}_m \approx \mathbf{G}_m\mathbf{E}\mathbf{A}. \tag{5}$$

Equation (4) and Eq. (5) show that both $\mathbf{I}_h$ and $\mathbf{I}_m$ involve the endmember matrix $\mathbf{E}$ and the abundance matrix $\mathbf{A}$ of the desired HR HSI $\mathbf{Z}$. Therefore, to precisely estimate $\mathbf{E}$ and $\mathbf{A}$ from $\mathbf{I}_h$ and $\mathbf{I}_m$ is the key to reconstructing $\mathbf{Z}$, where $\mathbf{Z} \approx \mathbf{E}\mathbf{A}$.

2.2 SNMF

Recently, many spectral unmixing-based methods [11,19] have been proposed to estimate the endmember matrix $\mathbf{E}$ and the abundance matrix $\mathbf{A}$ from the HSI $\mathbf{I}_h$ and MSI $\mathbf{I}_m$. By taking full advantage of Eq. (4) and Eq. (5), most spectral unmixing-based methods simply split the problem of optimizing $\mathbf{E}$ and $\mathbf{A}$ into two subproblems Eq. (6) and Eq. (7) [11,19].

$$\underset{\mathbf{E}\geq 0, \mathbf{A}\geq 0}{\arg\min} \left\| \mathbf{I}_h - \mathbf{E}\mathbf{A}_h \right\|_F^2, \tag{6}$$

$$\underset{\mathbf{E}\geq 0, \mathbf{A}\geq 0}{\arg\min} \left\| \mathbf{I}_m - \mathbf{E}_m \mathbf{A} \right\|_F^2, \tag{7}$$

where $\mathbf{A}_h \in \mathbb{R}^{N \times l}$ is regarded as the abundance matrix of $\mathbf{I}_h$ and $\mathbf{E}_m = \mathbf{G}_m \mathbf{E} \in \mathbb{R}^{b \times N}$ is regarded as the endmember matrix of $\mathbf{I}_m$. $\mathbf{E}_m$ and $\mathbf{E}$ are coupled through the SRF $\mathbf{G}_m$ [9].

In joint optimization-based methods, $\mathbf{A}_h$ is usually coupled with $\mathbf{A}$ through the spatial degradation $\mathbf{G}_h$, where $\mathbf{A}_h = \mathbf{A}\mathbf{G}_h$ [8,11]. Then, Eq. (6) and Eq. (7) are jointly optimized for obtaining $\mathbf{E}$ and $\mathbf{A}$ [14]. However, in practice, the real $\mathbf{G}_h$ is often unavailable and hard to be accurately estimated [12,13]. In separate optimization-based methods, the endmember matrix $\mathbf{E}$ of the desired HR HSI is extracted with Eq. (6). Then, $\mathbf{E}$ is transformed into $\mathbf{E}_m$ through the SRF $\mathbf{G}_m$ for subsequently estimating $\mathbf{A}$ from $\mathbf{I}_m$. Equation (6) and Eq. (7) are optimized separately [14]. The MSI $\mathbf{I}_m$ is not exploited in the optimization of $\mathbf{E}$, resulting in that $\mathbf{E}_m$ may not be the optimal endmember matrix of $\mathbf{I}_m$ [6,17]. In addition, using the inferior $\mathbf{E}_m$ can increase the difficulty of accurately estimating the abundance matrix $\mathbf{A}$, which will reduce the reconstruction quality of the desired HR HSI $\mathbf{Z}$ [17,18].

This paper proposes a novel SNMF to directly fuse the unregistered HSI and MSI. Different from separate optimization-based methods [8,12] that only HSI $\mathbf{I}_h$ is utilized in the optimization of $\mathbf{E}$, the proposed SNMF attempts to exploit both HSI $\mathbf{I}_h$ and MSI $\mathbf{I}_m$ in the optimization of $\mathbf{E}$. The optimization problem of the proposed SNMF can be expressed without $\mathbf{G}_h$ as

$$\underset{\mathbf{E}\geq 0, \mathbf{A}_h\geq 0, \mathbf{A}\geq 0}{\arg\min} \mathbf{F}(\mathbf{E}, \mathbf{A}_h, \mathbf{A}) = \frac{\lambda}{2} \left\| \mathbf{I}_h - \mathbf{E}\mathbf{A}_h \right\|_F^2 + \frac{1}{2} \left\| \mathbf{I}_m - \mathbf{G}_m \mathbf{E}\mathbf{A} \right\|_F^2, \tag{8}$$

where $\|\cdot\|_F$ denotes the Frobenius norm, and λ is the balancing coefficient. By exploiting both HSI $\mathbf{I}_h$ and MSI $\mathbf{I}_m$ in the optimization of the endmember matrix $\mathbf{E}$, the transformed $\mathbf{G}_m \mathbf{E}$ is optimal for $\mathbf{I}_m$. Meanwhile, the optimal abundance matrix $\mathbf{A}$ for $\mathbf{I}_m$ is estimated. Therefore, the proposed SNMF can reconstruct the high-quality $\mathbf{Z}$ by utilizing the optimal $\mathbf{E}$ and $\mathbf{A}$.

Owing to that SNMF attempts to directly estimate $\mathbf{E}, \mathbf{A}_h$ and $\mathbf{A}$ from unregistered HSI and MSI without relying on the spatial degradation, the classic multiplicative update method [11] is not fit for optimizing Eq. (8). In this paper, Eq. (8) is considered as a bound-constrained optimization problem [20]. A synchronous projected gradient method is proposed to simultaneously optimize $\mathbf{E}$, $\mathbf{A}_h$ and $\mathbf{A}$ in Eq. (8). The details of the proposed synchronous projected gradient are described in the following subsection.

2.3 Synchronous Projected Gradient Optimization

In this paper, Eq. (8) is seen as a bound-constrained optimization problem and a synchronous projected gradient method is proposed to optimize Eq. (8). In the proposed synchronous projected gradient method, $\mathbf{E}$, $\mathbf{A}_h$ and $\mathbf{A}$ are optimized synchronously to maintain the correlation between $\mathbf{E}$ and $\mathbf{A}$.

In the proposed synchronous projected gradient method, the solution of the k-th iteration is defined as $(\mathbf{E}^{(k)}, \mathbf{A}_h^{(k)}, \mathbf{A}^{(k)})$. In the next iteration, the solution is updated to $(\mathbf{E}^{(k+1)}, \mathbf{A}_h^{(k+1)}, \mathbf{A}^{(k+1)})$ synchronously as follows.

$$
\begin{aligned}
& \left(\mathbf{E}^{(k+1)}, \mathbf{A}_h^{(k+1)}, \mathbf{A}^{(k+1)}\right) \\
& = \left(\mathbf{E}^{(k)}(\alpha^{(k)}), \mathbf{A}_h^{(k)}(\alpha^{(k)}), \mathbf{A}^{(k)}(\alpha^{(k)})\right) \\
& = P\left[\left(\mathbf{E}^{(k)}, \mathbf{A}_h^{(k)}, \mathbf{A}^{(k)}\right) - \alpha^{(k)}\left(\frac{\partial \mathbf{F}}{\partial \mathbf{E}}, \frac{\partial \mathbf{F}}{\partial \mathbf{A}_h}, \frac{\partial \mathbf{F}}{\partial \mathbf{A}}\right)\right],
\end{aligned}
\tag{9}
$$

where

$$
P\left[(x_1, x_2, x_3)\right] = \left(max\left\{0, x_1\right\}, max\left\{0, x_2\right\}, max\left\{0, x_3\right\}\right), \tag{10}
$$

$$
\frac{\partial \mathbf{F}}{\partial \mathbf{E}} = \lambda\left(\mathbf{E}\mathbf{A}_h - \mathbf{I}_h\right)\mathbf{A}_h^T - \mathbf{G}_m^T \mathbf{I}_m \mathbf{A}^T + \mathbf{G}_m^T \mathbf{G}_m \mathbf{E}\mathbf{A}\mathbf{A}^T, \tag{11}
$$

$$
\frac{\partial \mathbf{F}}{\partial \mathbf{A}_h} = \lambda \mathbf{E}^T(\mathbf{E}\mathbf{A}_h - \mathbf{I}_h), \tag{12}
$$

$$
\frac{\partial \mathbf{F}}{\partial \mathbf{A}} = (\mathbf{G}_m \mathbf{E})^T (\mathbf{G}_m \mathbf{E}\mathbf{A} - \mathbf{I}_m), \tag{13}
$$

where $(\cdot)^T$ denotes the matrix transpose, and $\alpha^{(k)}$ is the step size calculated by "Armijo rule" [20,21]. According to "Armijo rule", $\alpha^{(k)}$ is set to β^{n_k}, where n_k is the smallest nonnegative integer n that satisfies Eq. (14).

$$
\begin{aligned}
& \mathbf{F}\left(\mathbf{E}^{(k)}(\beta^n), \mathbf{A}_h^{(k)}(\beta^n), \mathbf{A}^{(k)}(\beta^n)\right) - \mathbf{F}\left(\mathbf{E}^{(k)}, \mathbf{A}_h^{(k)}, \mathbf{A}^{(k)}\right) \leqslant \\
& \sigma\left(\frac{\partial \mathbf{F}}{\partial \mathbf{E}}, \frac{\partial \mathbf{F}}{\partial \mathbf{A}_h}, \frac{\partial \mathbf{F}}{\partial \mathbf{A}}\right)^T \left(\left(\mathbf{E}^{(k)}(\beta^n), \mathbf{A}_h^{(k)}(\beta^n), \mathbf{A}^{(k)}(\beta^n)\right) - \right. \\
& \left. \left(\mathbf{E}^{(k)}, \mathbf{A}_h^{(k)}, \mathbf{A}^{(k)}\right)\right),
\end{aligned}
\tag{14}
$$

where σ and β are fixed scalars. According to the literature [20], σ is set to 0.01, and β is set to 0.1 for SNMF. By iterating Eq. (9), the optimal $\mathbf{E}$, $\mathbf{A}_h$ and $\mathbf{A}$ can be estimated to reconstruct the desired HR HSI $\mathbf{Z}$. For bound-constrained optimization problems, the convergence of the projected gradient method with a positive step size has been proved in the literature [21]. In the proposed synchronous projected gradient method, the step size $\alpha^{(k)}$ is greater than 0. Thus, the limit point of $\{(\mathbf{E}^{(k)}, \mathbf{A}_h^{(k)}, \mathbf{A}^{(k)})\}_{k=1}^{\infty}$ exists and each limit point of $\{(\mathbf{E}^{(k)}, \mathbf{A}_h^{(k)}, \mathbf{A}^{(k)})\}_{k=1}^{\infty}$ is a stationary point of Eq. (8). During the optimization process, $\mathbf{E}$ is initialized with the vertex component analysis [22]. All elements of $\mathbf{A}_h$ and $\mathbf{A}$ are initialized with $1/N$. Then, Eq. (9) is repeated until convergence.

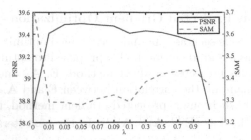

Fig. 3. PSNR and SAM of the SNMF with different λ for the registered HSI and MSI fusion on the Pavia University.

3 Experiments

3.1 Databases and Evaluation Metrics

In this paper, the proposed SNMF is evaluated on simulated and real data, respectively. First, the Pavia University and Washington DC Mall databases are utilized to simulate the observed HSI and MSI for experiments. Original HSIs of these two databases serve as the ground truth [9]. The observed HSI is simulated by blurring the ground truth and then downsampling with the scaling factor of 4 [5]. The observed MSI is simulated by spectrally downsampling the ground truth with uniform spectral responses corresponding to Landsat TM bands [9, 11]. Similar to [9], a 200×200 subscene of the Pavia University and a 240×240 subscene of the Washington DC Mall are selected to conduct experiments.

Second, the Paris database [12], which consists of both real HSI and MSI, is employed to further evaluate the proposed SNMF. The HSI is captured by the Hyperion sensor and the MSI is captured by the ALI sensor. Both HSI and MSI have the same spatial resolution of 30 m. Similar to [12], the original HSI of the Paris serves as the ground truth and is spatially degraded with the scaling factor of 4 to simulate the observed HSI. The original MSI is directly regarded as the observed MSI.

The performance of all methods is evaluated visually and qualitatively. Five evaluation metrics including peak signal-to-noise ratio (PSNR), universal image quality index (UIQI), root mean squared error (RMSE), spectral angle mapper (SAM) and erreur relative globale adimensionnelle de synthèse (ERGAS) [2] are adopted to evaluate the proposed SNMF. The reconstructed HR HSI is converted to 8-bit data for calculating performance.

3.2 Parameter Setting

The proposed SNMF involves the balancing coefficient λ that controls the trade-off of the two terms in Eq. (8). To explore the impact of λ on the performance of the SNMF, the Pavia University is employed to conduct experiments. The registered HSI and MSI fusion is utilized to find an appropriate value of λ. The

Table 1. Registered HSI and MSI Fusion on the Pavia University.

Methods	PSNR	UIQI	RMSE	ERGAS	SAM	Time
CNMF	33.072	0.963	5.828	3.654	3.710	**8.9**
BSR	**40.343**	0.982	3.339	2.109	3.711	72.3
HySure	38.710	0.983	3.226	2.037	3.453	44.8
CO-CNMF	36.787	0.976	4.194	2.556	3.523	29.9
FuVar	30.939	0.934	7.453	4.756	5.164	1171.8
SNMF	40.230	**0.984**	**2.907**	**2.031**	**3.058**	64.4

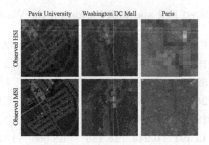

Fig. 4. Observed HSI and MSI on the Pavia University (R: 60, G: 30, and B: 10), Washington DC Mall (R: 50, G: 30, and B: 20) and Paris (R: 24, G: 14, and B: 3).

setting range of λ is from 0.01 to 1 in experiments. The PSNR and SAM of the reconstructed HR HSI are shown in Fig. 3. When λ is set to 0.05, the proposed SNMF shows superior performance. For convenience, the balancing coefficient λ of the SNMF is set to 0.05 in all experiments.

3.3   Comparisons with the State-of-the-Art Methods

To fully evaluate the proposed SNMF, both the registered and unregistered image fusion experiments are conducted. Several state-of-the-art methods are employed as the compared methods, including CNMF [11], BSR [7], HySure [12], CO-CNMF [9] and FuVar [8]. For a fair comparison, all methods use the known SRF as a prior. According to [2,8], the number of endmembers is set to 30 for all methods. The proposed SNMF focuses on unsupervised HSI and MSI fusion. Thus, deep learning-based methods are not employed as compared methods, since these methods usually require a sea of HR HSIs for supervised training.

Registered HSI and MSI Fusion. This paper mainly focuses on the unregistered HSI and MSI fusion. For simplicity, only the Pavia University is utilized to evaluate SNMF for the registered HSI and MSI fusion. Experimental results are reported in Table 1. BSR introduces a non-parametric Bayesian framework [7]. HySure proposes a subspace-based regularization to reconstruct the desired HR HSI [12]. BSR and HySure show superior performance in the registered HSI and

Table 2. Unregistered HSI and MSI Fusion on the Pavia University.

Methods	PSNR	UIQI	RMSE	ERGAS	SAM	Time
CNMF	20.083	0.458	25.819	15.108	14.581	**4.6**
BSR	**39.628**	0.984	3.518	2.113	3.917	60.3
HySure	32.126	0.956	6.594	3.749	6.339	46.4
CO-CNMF	22.886	0.641	20.453	11.563	13.884	24.2
FuVar	18.865	0.274	30.191	17.032	16.420	1250.4
SNMF	39.160	**0.986**	**3.216**	**2.089**	**3.217**	48.4

Table 3. Unregistered HSI and MSI Fusion on the Washington DC Mall.

Methods	PSNR	UIQI	RMSE	ERGAS	SAM	Time
CNMF	16.388	0.381	15.434	119.290	12.462	**9.4**
BSR	32.834	0.945	1.420	57.920	1.798	98.7
HySure	24.857	0.867	4.906	**56.080**	5.068	54.5
CO-CNMF	18.372	0.497	13.855	74.522	12.237	36.1
FuVar	15.391	0.244	18.186	68.566	13.346	2613.8
SNMF	**33.838**	**0.955**	**1.232**	64.407	**1.463**	76.6

Table 4. Unregistered HSI and MSI Fusion on the Paris.

Methods	PSNR	UIQI	RMSE	ERGAS	SAM	Time
CNMF	22.546	0.235	14.359	6.724	5.962	**1.3**
BSR	28.256	0.822	7.255	3.519	3.673	10.8
HySure	28.174	0.821	7.373	3.560	3.501	5.1
CO-CNMF	25.416	0.636	10.935	5.060	5.358	13.2
FuVar	23.063	0.409	13.104	6.002	4.994	367.1
SNMF	**28.444**	**0.835**	**7.079**	**3.405**	**3.244**	6.5

MSI fusion. Different from CNMF, CO-CNMF and FuVar, the proposed SNMF models the abundances of HSI and MSI independently. Both HSI and MSI are exploited to estimate the optimal endmembers and abundances to reconstruct the HR HSI. Table 1 shows that SNMF can compete with the compared methods in registered HSI and MSI fusion.

Unregistered HSI and MSI Fusion. Three databases are employed to evaluate SNMF for the unregistered HSI and MSI fusion. The observed HSI is rotated at $10°$ using the nearest neighbor interpolation to construct the unregistered HSI and MSI pair. The rotated HSI is further cropped to remove the areas that are not filled by pixels, and the observed MSI is also cropped to ensure that the

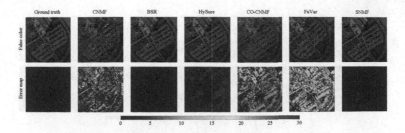

Fig. 5. Visual results and error maps on the Pavia University.

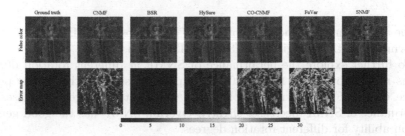

Fig. 6. Visual results and error maps on the Washington DC Mall.

Table 5. HSI with Different Rotation Degrees on the Pavia University.

Degrees	PSNR	UIQI	RMSE	ERGAS	SAM
0	40.230	0.984	2.907	2.031	3.058
5	39.638	0.986	3.119	2.080	3.093
10	39.160	0.986	3.216	2.089	3.217
15	39.179	0.986	3.342	2.131	3.305
20	38.338	0.986	3.519	2.187	3.338

scaling factor between HSI and MSI is 4. Unregistered HSI and MSI on the Pavia University, Washington DC Mall and Paris are shown in Fig. 4. Qualitative results are reported in Tables 2, 3 and 4. Visual results of the reconstructing HR HSIs are shown in Figs. 5, 6 and 7. CNMF, CO-CNMF and FuVar perform worse in the unregistered HSI and MSI fusion, due to that these methods rely on the spatial degradation to couple the abundances of HSI and MSI. However, the relationship of the abundances between unregistered HSI and MSI is complex and can not be simply coupled through the spatial degradation [15,17]. HySure estimates the spatial degradation between unregistered HSI and MSI. And the errors of estimated spatial degradation decrease the performance of HySure. BSR performs worse than the proposed SNMF, since that BSR only exploits HSI to estimate the endmembers of HSI. In the BSR, endmembers of HSI are transformed into the endmembers of MSI through the SRF for subsequently estimating the abundances of MSI. However, the transformed endmembers may

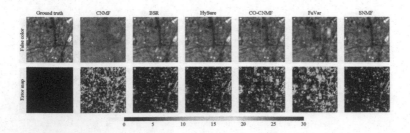

Fig. 7. Visual results and error maps on the Paris.

not be the optimal endmembers of MSI. The proposed SNMF shows superior performance in the unregistered HSI and MSI fusion.

To further explore the performance of SNMF at different rotation degrees, HSI is rotated by 5, 10, 15 and 20° for experiments, respectively. The performance of SNMF is shown in Table 5. As the rotation degrees increases, the performance of SNMF decreases slightly. Overall, SNMF has the good generalization ability for different rotation degrees.

In this paper, all methods are implemented on MATLAB R2014b with the Intel Core i7 5930k CPU and 64 GB RAM. The time cost (seconds) of different methods is reported in Tables 1, 2, 3 and 4. The proposed SNMF not only achieves excellent reconstruction results, but also takes less time than BSR.

4   Conclusions

This paper proposes a novel SNMF to directly fuse unregistered HSI and MSI. The proposed SNMF does not require the registration as a preprocessing operation. To utilize both HSI and MSI in the endmember optimization of the desired HR HSI, the unregistered HSI and MSI fusion is formulated as a bound-constrained optimization problem in SNMF. A synchronous projected gradient method is proposed to synchronously optimize the endmembers and abundances of the desired HR HSI. Experiments in unregistered HSI and MSI fusion verify the effectiveness of SNMF. However, SNMF still relies on the assumption that non-overlapping scenes and overlapping scenes have similar underlying materials, which limits the generalization performance of SNMF. In the future, we will study unsupervised convolutional networks that do not rely on this assumption.

Acknowledgments. This work was supported in part by the National Key R&D Program of China under Grant 2017YFB0502900, in part by the National Natural Science Foundation of China under Grant 61772510, in part by the Innovation Capability Support Program of Shaanxi under Grant 2020TD-015.

References

1. Sun, H., Zheng, X., Lu, X., Wu, S.: Spectral-spatial attention network for hyperspectral image classification. IEEE Trans. Geosci. Remote Sens. **58**(5), 3232–3245 (2020)
2. Yokoya, N., Grohnfeldt, C., Chanussot, J.: Hyperspectral and multispectral data fusion: a comparative review of the recent literature. IEEE Geosci. Remote Sens. Mag. **5**(2), 29–56 (2017)
3. Sun, H., Li, S., Zheng, X., Lu, X.: Remote sensing scene classification by gated bidirectional network. IEEE Trans. Geosci. Remote Sens. **58**(1), 82–96 (2020)
4. Lu, X., Yuan, Y., Zheng, X.: Joint dictionary learning for multispectral change detection. IEEE Trans. Cybern. **47**(4), 884–897 (2017)
5. Zhang, K., Wang, M., Yang, S., Jiao, L.: Spatial-spectral-graph-regularized low-rank tensor decomposition for multispectral and hyperspectral image fusion. IEEE J. Sel. Top. Appl. Earth Obs. Remote Sens. **11**(4), 1030–1040 (2018)
6. Gao, D., Zhentao, H., Ye, R.: Self-dictionary regression for hyperspectral image super-resolution. Remote Sens. **10**(10), 1574 (2018)
7. Akhtar, N., Shafait, F., Mian, A.: Bayesian sparse representation for hyperspectral image super resolution. In: Proceedings of the IEEE Conference on Computer Vision and Pattern Recognition, pp. 3631–3640, June 2015
8. Borsoi, R.A., Imbiriba, T., Bermudez, J.C.M.: Super-resolution for hyperspectral and multispectral image fusion accounting for seasonal spectral variability. IEEE Trans. Image Process. **29**, 116–127 (2020)
9. Lin, C.-H., Ma, F., Chi, C.-Y., Hsieh, C.-H.: A convex optimization-based coupled nonnegative matrix factorization algorithm for hyperspectral and multispectral data fusion. IEEE Trans. Geosci. Remote Sens. **56**(3), 1652–1667 (2018)
10. Zhou, Y., Feng, L., Hou, C., Kung, S.: Hyperspectral and multispectral image fusion based on local low rank and coupled spectral unmixing. IEEE Trans. Geosci. Remote Sens. **55**(10), 5997–6009 (2017)
11. Yokoya, N., Yairi, T., Iwasaki, A.: Coupled nonnegative matrix factorization unmixing for hyperspectral and multispectral data fusion. IEEE Trans. Geosci. Remote Sens. **50**(2), 528–537 (2012)
12. Simões, M., Bioucas-Dias, J., Almeida, L.B., Chanussot, J.: A convex formulation for hyperspectral image superresolution via subspace-based regularization. IEEE Trans. Geosci. Remote Sens. **53**(6), 3373–3388 (2015)
13. Lin, B., Tao, X., Mai, X., Dong, L., Jianhua, L.: Bayesian hyperspectral and multispectral image fusions via double matrix factorization. IEEE Trans. Geosci. Remote Sens. **55**(10), 5666–5678 (2017)
14. Akhtar, N., Shafait, F., Mian, A.: Sparse spatio-spectral representation for hyperspectral image super-resolution. In: Fleet, D., Pajdla, T., Schiele, B., Tuytelaars, T. (eds.) ECCV 2014. LNCS, vol. 8695, pp. 63–78. Springer, Cham (2014). https://doi.org/10.1007/978-3-319-10584-0_5
15. Dian, R., Li, S., Fang, L., Lu, T., Bioucas-Dias, J.M.: Nonlocal sparse tensor factorization for semiblind hyperspectral and multispectral image fusion. IEEE Trans. Cybern. **50**, 4469–4480 (2019)
16. Zhou, Y., Rangarajan, A., Gader, P.D.: An integrated approach to registration and fusion of hyperspectral and multispectral images. IEEE Trans. Geosci. Remote Sens. **58**(5), 3020–3033 (2020)
17. Qu, Y., Qi, H., Kwan, C.: Unsupervised and unregistered hyperspectral image super-resolution with mutual Dirichlet-Net. arXiv preprint arXiv:1904.12175 (2019)

18. Karoui, M.S., Deville, Y., Benhalouche, F.Z., Boukerch, I.: Hypersharpening by joint-criterion nonnegative matrix factorization. IEEE Trans. Geosci. Remote Sens. **55**(3), 1660–1670 (2017)
19. Lanaras, C., Baltsavias, E., Schindler, K.: Hyperspectral super-resolution by coupled spectral unmixing. In Proceedings of the IEEE International Conference on Computer Vision, pp. 3586–3594, December 2015
20. Lin, C.-J.: Projected gradient methods for nonnegative matrix factorization. Neural Comput. **19**(10), 2756–2779 (2007)
21. Bertsekas, D.P.: On the Goldstein-Levitin-Polyak gradient projection method. IEEE Trans. Autom. Control **21**(2), 174–184 (1976)
22. Nascimento, J.M.P., Dias, J.M.B.: Vertex component analysis: a fast algorithm to unmix hyperspectral data. IEEE Trans. Geosci. Remote Sens. **43**(4), 898–910 (2005)

Cloud Detection Algorithm Using Advanced Fully Convolutional Neural Networks in FY3D-MERSI Imagery

Yutong Ding[1], Xiuqing Hu[2], Yuqing He[1(✉)], Mingqi Liu[1], and Saijie Wang[1]

[1] Key Laboratory of Photoelectronic Imaging Technology, Ministry of Education of China, School of Optoelectronics, Beijing Insitute of Technology, Beijing 100081, China
yuqinghe@bit.edu.cn
[2] National Satellite Meteorological Center, Beijing 100081, China

Abstract. Cloud detection plays a very important role in the development of satellite remote sensing products and influences the accuracy of satellite products that characterize the properties of clouds, aerosols, trace gases, and ground surface parameters. However, current existing cloud detection methods rely heavily on the data of visible bands. It makes FY3D MERSI, which lacks visible band data at night, difficult to use these methods with high accuracy. In this paper, we proposed a cloud detection method based on deep learning termed CM-CNN for FY-3D MERSI. In order to ensure the effect of the network, the data has been strictly selected and consequently preprocessed. The method can automatically extract identified target features and fuse multi-level feature information, and adjust the parameters in the network without setting a threshold. Besides, this method proves to be better and more robust while only using mid-infrared and long-infrared band data in different cases.

Keywords: Remote sensing · FY-3D/MERSI · Cloud detection · CM-CNN · Deep learning

1 Introduction

Remote sensing image have been widely used in many science research and social application fields. Among them, cloud detection is an important step in remote sensing image processing. For example, it can influences the accuracy of satellite products that characterize the properties of clouds, aerosols, trace gases, and ground surface parameters. The exploration of cloud detection technology was started since last century.

The traditional threshold cloud detection method extracts various spectral characteristics of each pixel through various types of radiation band tests. At the same time, the radiation data obtained by the sensor will be combined with the surface radiation and other environmental information (such as the solar zenith angle, etc.) to establish a radiation transmission model. Retrieving the radiation characteristics under clear sky through the established radiation transmission model. Finally, one or more radiation tests are performed on these two radiation characteristics to determine whether there is

© Springer Nature Switzerland AG 2020
Y. Peng et al. (Eds.): PRCV 2020, LNCS 12305, pp. 615–625, 2020.
https://doi.org/10.1007/978-3-030-60633-6_51

a cloud. ISCCP method [1] and APOLLO method [2] are the two classic ones in the traditional threshold cloud detection method. The Fmask [3, 4] algorithm for Landsat series satellite images proposed in 2012 is one of the most famous and reliable threshold-based methods for cloud layer detection. They use the idea of a decision tree to mark each pixel as a cloudy or a clear. The determination of these threshold functions covers multiple physical tests, taking into account almost all frequency band information, and is also carefully designed by means of projection analysis.

With the development of the deep learning, deep neural network is applied in the image recognition, image segmentation and speech recognition, etc. In this task, the cloud mask is a label matrix with the same size as observation data and each pixel indicates the possibility of cloud in this area. So the cloud mask method is similar with an image segmentation task. For image segmentation, a series of significant successes has been achieved. Jonathan [5] proposed a fully convolutional network(FCN) for semantic segmentation and it achieved an end-to-end segmentation procedure. On the basis of FCN, more CNN models are designed to improve the segmentation accuracy such as DeepLab [6], PSPNet [7], RefineNet [8], etc.

In recent years, a large number of deep learning cloud detection methods have been proposed. Fengying Xie [10] proposed a multi-level cloud detection method based on deep learning and an improved simple linear iterative clustering method. Based on ResNet, and inspired by PSPNet, Zhiyuan Yan [11] proposed a full convolutional network (MFFSNet) with multi-level feature fusion segmentation. Z. Shao [12] designed a multi-scale feature convolutional neural network (MF-CNN) to extract multi-scale features and combine them with low-level features. They finally divide pixels into three types: thick cloud, thin cloud, and clear. Alistair Francis [13] used the Inception module to extract features on the U-Net structure to design CloudFCN, but its accuracy is slightly inferior to MF-CNN in the test. Michal Segal-Rozenhaimer [14] proposed a cloud detection method based on a domain adaptive neural network, which can better adapt to the needs between two satellite platforms in the prediction without training for each platform separately. Unfortunately, it is difficult to use existing cloud detection methods with high accuracy and precision for night cloud detection of FY3D MERSI because of the missing visible band data at night. Apart from this, the existing threshold methods and deep learning cloud detection methods are very dependent on the visible band data, which undoubtedly limits the universality of the methods.

In response to the deficiencies of the cloud detection methods discussed above, in this paper, we designed the CM-CNN which use ResNet [9] Block as a feature extraction module to obtain multi-level features and make full use of low-level features for feature fusion to generate the cloud mask for FY-3D MERSI imagery in different cases. We also selected data strictly and consequently preprocessed it to ensure the effect of the network. The remainder of this paper is organized as follows. Dataset description are arranged in Sects. 2. In Sect. 3, the proposed method for cloud detection are described. Section 4 presents the cloud detection experiment and corresponding results. Finally, Sect. 5 gives a conclusion and outlook for our work.

2   Data Selection and Preprocessing

2.1   FY-3D/MERSI Data

FY-3D satellite is the fourth of the Chinese second-generation polar-orbiting meteorological satellite and plays an important role in weather forecast, climate prediction and environmental monitoring. The satellite is equipped with a variety of remote sensing detection instruments, of which the medium-resolution spectral imager (MERSI) is the core optical load of FY-3D. The FY-3D MERSI can provide remote sensing data in 25 bands, including 19 visible and near-infrared bands, and 6 mid-infrared and long-infrared bands. Unfortunately, the visible and near infrared data of FY3D MERSI at night are missing. In order to obtain a network model that can be applied to both day and night, we selected 6 mid-long-wave infrared data for model training and verification. The correction level of the experimental data is L1. Table 1 shows the information of the bands we used.

Table 1. Information of FY-3D MERSI bands for cloud detection

Band number	Band width (μm)	Spatial resolution (m)	Main purpose
20	3.80	1000	Land, water cloud temperature
21	4.05	1000	
22	7.20	1000	Atmosphere, water vapor
23	8.55	1000	
24	10.80	250	Land, water cloud temperature
25	12.00	250	

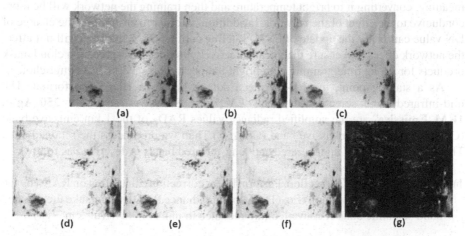

Fig. 1. MERSI infrared images used for cloud detection. (a) 3.80 μm band. (b) 4.05 μm band. (c) 7.20 μm band. (d) 8.55 μm band. (e) 10.80 μm band. (f) 12.00 μm band. (g) Corresponding visible light RGB image

The image size of the every experimental data is 2000 × 2048 as shown in Fig. 1. It includes the brightness temperature image of the six bands we selected and the corresponding visible light RGB image.

At the same time, we selected cloud mask product as the ground truth.The cloud mask has the same size with the experimental data as shown in Fig. 2. Cloud masks are obtained from meteorological products and are available during the day and night.The pixels in cloud mask indicate the possibility of the cloud and we set the label 0–3 as the representation of cloudy, probably cloudy, probably clear, clear. The above data are provided by the National Satellite Meteorological Center.

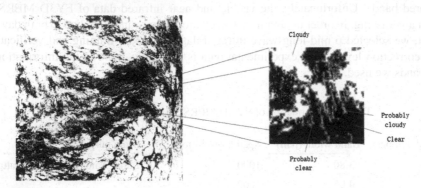

Fig. 2. Cloud mask image display.

2.2 Data Preprocessing

Compared with the digital number (DN) in the L1 level file that has no clear physical meaning, converting it to bright temperature and then training the network will be more conducive to the effect of the network. In addition, in order to ensure that the change of DN value caused by the update of FY3D satellite calibration technology will not affect the network effect, and ensure that it can effectively produce high-precision cloud mask products for a long time, training with bright temperature is obviously a better choice.

As a starting point, the FY-3D mersi data is saved in HDF5 file format. The mid-infrared band scientific datasets "**EV_1KM_Emissive**" and "**EV_250_Aggr. 1KM_Emissive**" are the amplified radiance values RAD_0 of the 1 km emissive band (CH_{20-23}) and 250 m emissive band (CH_{24-25}).Then we preprocess the retained data. The data processing of mid-infrared and long-infrared bands is described as follows:

1) **Linear Radiance Correction**. Perform linear correction calculation on RAD_0 by the linear correction equation (Eq. (1)) to obtain radiance RAD. It can make the infrared band scientific dataset converted into radiance in units of "mW/(m^2.cm^{-1}.sr)".

$$RAD = RAD_0 \times Slope + Intercept \tag{1}$$

where the correction coefficient "Slope"and "Intercept" are the **Slope** and **Intercept** attribute parameter of the respective dataset. One value for each band, and the slopes of CH_{20-25} are 0.0002, 0.0002, 0.01, 0.01, 0.01, 0.01.Intercepts all are 0.

2) **Equivalent Blackbody Brightness Temperature**. Based on the equivalent center wave number **MERSI_EquivMid_wn** and the band radiance, the equivalent blackbody brightness temperature T^*_{BB} is calculated by the Plank inverse transformation formula (Eq. (2)):

$$T^*_{BB} = \frac{c_2 v_c}{\ln[1 + (\frac{c_1 v_c^3}{RAD})]} \tag{2}$$

where $c_1 = 1.1910427 \times 10^{-5}$, $c_2 = 1.4387752$, v_c donate the infrared band Equivalent Mid Wave number (**EMW**) obtained by ground calibration.

3) **Blackbody Brightness Temperature Correction**. Reuse the band brightness temperature correction coefficient (**TCC** as shown in the Table 2) to convert to the blackbody brightness temperature, using Eq. (3) to convert:

$$T_{BB} = A \times T^*_{BB} + B \tag{3}$$

where A and B are attribute variables **TCC_A** and **TCC_B**. Respectively, the corresponding values of each band are shown in Table 2.

Table 2. Data calibration process parameter table.

Band	EMW (cm^{-1})	TCC	
		A	B
20	2634.359	1.00103	−0.4759
21	2471.654	1.00085	−0.3139
22	1382.621	1.00125	−0.2662
23	1168.182	1.00030	−0.0513
24	933.364	1.00133	−0.0734
25	836.941	1.00065	0.0875

3 Cloud Mask Detection Network

Compared with the semantic segmentation task of daily scenes, one of the difficulties of cloud detection is the accuracy of segmentation of edge details. Looking at multiple semantic segmentation networks, the multi-level feature extraction and fusion effect brought by FCN's skip connection is very potential, and it can keep more low-level feature information, which is more complicated for objects such as clouds than objects in daily scenes. Adding low-level features with more detailed information undoubtedly improves the accuracy of cloud detection. For this reason, it is extremely helpful to extract features by directly performing convolution operations on the initial image that has not been convoluted.

Cloudmask Convolutional Neural Network (CM-CNN) is designed based on FCN and ResNet. ResBlock used in the network for feature extraction is shown in Fig. 3. The architecture of the proposed CM-CNN is shown in Fig. 4.

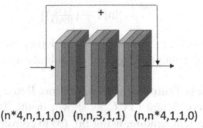

$$(n*4,n,1,1,0)\quad(n,n,3,1,1)\quad(n,n*4,1,1,0)$$

Fig. 3. ResBlock. The parameters in brackets are the convolutional layer parameters, from left to right are the number of layers of the input image, the number of layers of the output image, the size of the convolution kernel, the stride and the padding. The specific value of n is marked in the brackets below the yellow ResBlock in Fig. 4. (Color figure online)

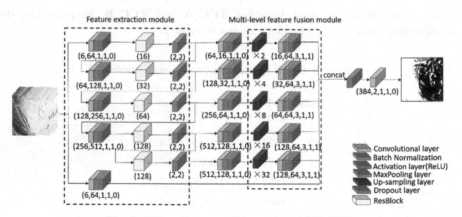

Fig. 4. The architecture of the proposed CM-CNN.

In the feature extraction module, ResBlock can effectively improve the depth of the neural network, at the same time make the optimization of the entire model structure easier, and reduce the problem of convergence deterioration of the very deep neural network due to the gradient dissipation phenomenon. In the multi-level feature fusion module, the low-level features need to be used as the input of the next layer of neurons in the neural network backbone to extract high-level features. The features at different levels may vary greatly.

In order to improve the fusion effect, we add a "Conv-ReLU-BN" unit with 3×3 kernel to perform a feature transformation on the multi-level features before fusion, so as to ensure the effect of feature fusion. In CM-CNN, the convolutional layer with a convolution kernel size of 1×1 is used to change the output of the previous layer to the number of layers we need. The feature maps of multi-scale are upsampled to original resolution and then fused together with concat layer. Finally, the fully connected layer is replaced by the convolutional layer with the 1×1 kernel, which is the key insight of the FCN. The output number of the last layer is 2 because the model just detects the cloudy and clear area.

The parameters under the convolution layer are expressed from left to right as the number of layers of the input image, the number of layers of the output image, the size of the convolution kernel, the stride and the padding. The parameters under the maxpooling layer are the size of the convolution kernel and the stride. This method of representation is very useful for users of the pytorch framework to reproduce the network. The number below ResBlock is the specific size of n in Fig. 4. In addition, we used bilinear interpolation for upsampling and dropout ratio is 0.5.

4 Experimental Results and Discussion

4.1 Dataset

In order to allow the training dataset to include multiple seasons, climates and terrains, we chose images obtained in February 1–10, May 1–10, September 1–10, October 10–20 (2019). A total of 3643 scenes of valid data were used for the experiment. Among them, 3500 scene data are used for model training, and 143 scene data are used for model accuracy test. Because of the limited video memory and RAM, the FY-3D MERSI data and cloud mask is too huge to read into the convolutional neural network. We crop the preprocessed data and cloud mask into 256×256 randomly as training and testing dataset. Finally, the training dataset consists of 120000 samples and the testing dataset consists of 12000 samples.

The probably cloudy and probably clear areas are not involved in training because of its low reliability. In addition, the number of probably cloudy and probably clear areas is much less than cloudy and clear areas, accounting for less than 5% of the whole training dataset. The probably areas may interfere the training.

It should be noted that the existing deep learning cloud detection method can't be compared with the traditional threshold cloud detection method in the judgment of probably cloudy and probably clear cloud mask. Therefore, the results of the two classifications, cloudy and clear, are used to compare and display the results. Our method has satisfactory results in the classification of cloudy and clear.

4.2 Experimental Results

In addition to the FCN-8s [5], we also reproduced the MF-CNN [12], which is one of the best cloud detection neural networks in recent years, for experiment comparison. Due to the different channels used by the training data and the type of network output, a few parameters of the current MF-CNN are different from the original one.

We use three metrics to evaluate the method performance for cloud detection using three networks including accuracy, precision and recall. These indicators can show the correctness of the model objectively.

The definitions of precision and recall are given here. Suppose that for category i, A is the pixel set that divides the correct category i, B is the pixel set of all categories i obtained by the split, and C is the set of all the categories of category i in the true value, then the precision and recall rate are defined as:

$$precision_i = \frac{A}{B} \tag{4}$$

$$recall_i = \frac{A}{C} \tag{5}$$

It can be seen from the definition that the precision rate and the recall rate can well evaluate whether there is a phenomenon of over-detection or missed detection in a certain category in image segmentation. If the precision is high and the recall rate is low, the model has a missed detection of the category. If the recall rate is high and the precision is low, the model has an over-detection of this category. Precision and recall rate are mutually influential.

In training, CM-CNN is optimized by stochastic gradient descent method. The batch size is 10. We set base learning rate to 0.00001 and the "step" learning rate policy is used to decrease the learning rate automatically. The momentum is 0.9 and the weight decay is 0.0005. The model is converging after 20,000 iterations. Our CNN model is built and optimized in Pytorch.

Table 3 and Table 4 shows the statistical results. It can be seen that the indicators of CM-CNN are far superior to the basic FCN-8 s network and also better than MF-CNN. At the same time, CM-CNN performs very well in different cases, with a minimum accuracy rate of more than 91%, which is comparable to other existing deep learning methods that use more radiation bands data.

We also use FCN-8s, MF-CNN and our CM-CNN to generate the cloud mask for integral FY-3D MERSI data. Our method can applied to images with different image size. But the preprocessed data can't be read into the GPU because its size is too huge.

Table 3. The precision, recall and accuracy in testing dataset.

Network	Cloud Mask	Precision(%)	Recall(%)	Accuracy(%)
FCN-8s	Cloudy	83.96	95.51	77.03
	Clear	61.86	85.91	
MF-CNN	Cloudy	90.75	94.17	87.14
	Clear	70.80	85.13	
CM-CNN	Cloudy	96.68	98.39	**94.02**
	Clear	95.12	90.23	

Table 4. The precision, recall and accuracy of CM-CNN in testing dataset in different cases.

Time-Surface type	Cloud Mask	Precision(%)	Recall(%)	Accuracy(%)
Day-Land	Cloudy	95.84	93.72	94.31
	Clear	92.53	95.03	
Day-Sea	Cloudy	98.81	95.82	95.89
	Clear	87.36	96.16	
Night-Land	Cloudy	91.10	93.45	91.71
	Clear	92.42	89.75	
Night-Sea	Cloudy	95.43	96.332	93.59
	Clear	87.09	84.31	

So we use the model to generate the cloud mask in a block every time and then stitch them together to a complete cloud mask. Some results of the integral FY-3D MERSI data are shown in Fig. 5 and Fig. 6.

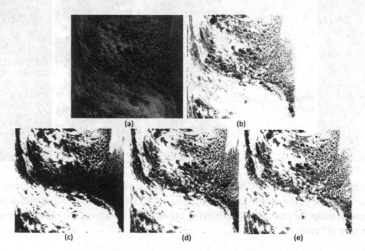

Fig. 5. Visual comparisons of different cloud detection methods in the partial scene of FY3D/MERSI(20190601_0225). (a) RGB image. (b) Existing cloud mask products. (c) Cloud detection result of FCN-8s. (d) Cloud detection result of MF-CNN. (e) Cloud detection result of CM-CNN.

As can be seen from Fig. 5, the cloud detection effect of CM-CNN and MF-CNN is better than that of FCN-8s, but the detection effect of CM-CNN on the thin cloud edge of the cloud is more outstanding, producing a more detailed cloud mask, which is also shown in the statistical result.

As for Fig. 6, in the example of the first line, compared with the more intuitive RGB image, the right corner of the existing cloud mask shows the misjudgment of the clouds above the land; in the example of the second line, the lower part of the existing cloud mask is affected by the sun glint and the obvious clear sky is divided into possible clouds; the third line example shows the streak phenomenon mentioned. There is also a clear misjudgment in the lower left corner of the existing cloud mask in the fourth line example. CM-CNN has effectively improved the above problems and produced higher quality cloud mask products. To a certain extent, the effect of the proposed method is superior to existing methods.

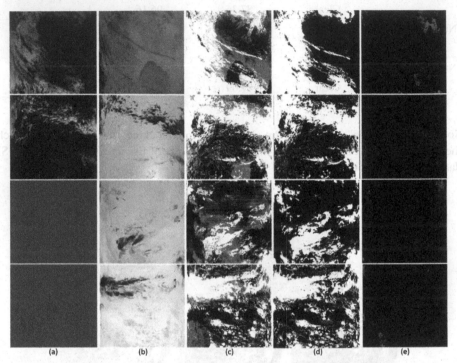

(a)   (b)   (c)   (d)   (e)

Fig. 6. Image examples and CM-CNN effect of different time and surface types. From top to bottom are Day-Land, Day-Sea, Night-Land, Night-Sea. (a) RGB image (b) 3.8 μm mid-wave infrared band brightness temperature image (c) Existing cloud mask products (d) network prediction (e) Pixel prediction difference.

5   Conclusion

In this paper, we proposed a cloud detection method based on the deep learning for FY-3D MERSI data. We designed a CNN model to learn the targets feature of FY-3D MERSI infrared bands data and generate the cloud mask products. The model can adjust its parameters automatically rather than set the thresholds artificially before. The experiment results show the robust and reasonable accuracy of our cloud detection method using deep learning technique.

In the future, we will design a more advanced CNN model based on CM-CNN to improve its detail expression capability and the probably areas will suitable for the model. Besides, we will apply the CM-CNN method to more other images taken by domestic satellites such as FY-4A, and investigate the possibility of the detection and identification of various subdivided forms of clouds using a single model.

Acknowledgments. This work is supported by the National Key Research and Development Plan of China (Grant No. 2018YFB0504901) and the National Natural Science foundation of China (Grant No. 41871249). The authors also acknowledge the National Satellite Meteorological Center for the imagery of FY-3D MERSI data.

References

1. Rossow, W.B.: ISCCP cloud data products. Bull. Am. Meteorol. Soc. 72(1), 2–20 (1991)
2. Gesell, G.: An method for snow and ice detection using AVHRR data an extension to the APOLLO software package. Int. J. Remote Sens. 10(4-5), 897–905 (1989)
3. Zhu, Z.: Object-based cloud and cloud shadow detection in landsat imagery. Remote Sens. Environ. 118, 83–94 (2012)
4. Qiu, S.: Fmask 4.0: improved cloud and cloud shadow detection in landsats 4–8 and sentinel-2 imagery. Remote Sens. Environ. 231, 1–20 (2019)
5. Long, J., Shelhamer, E.: Fully convolutional networks for semantic segmentation. TPAMI 39(4), 640–651 (2017)
6. Chen, L.C., Papandreou, G., Schroff, F., et al.: Rethinking atrous convolution for semantic image segmentation. ECCV 2018, 801–818 (2018)
7. Zhao, H., Shi, J., Qi, X.: Pyramid scene parsing network. In: CVPR 2017, pp. 6230–6239 (2017)
8. Lin, G., Milan, A.: Refinenet: multi-path refinement networks for high-resolution semantic segmentation. In: CVPR 2017, pp. 5168–5177 (2017)
9. He, K., Zhang, X.: Deep residual learning for image recognition. In: CVPR 2016, pp. 770–778 (2016)
10. Xie, F.: Multilevel cloud detection in remote sensing images based on deep learning. IEEE J. Sel. Top. Appl. Earth Observations Remote Sens. 10(8), 3631–3640 (2017)
11. Yan, Z.: Cloud and cloud shadow detection using multilevel feature fused segmentation network. IEEE Geosci. Remote Sens. Lett. 15(10), 1600–1604 (2018)
12. Shao, Z.: Cloud detection in remote sensing images based on multiscale features-convolutional neural network. IEEE Trans. Geosci. Remote Sens. 57(6), 4062–4076 (2019)
13. Francis, A.: CloudFCN: accurate and robust cloud detection for satellite imagery with deep learning. Remote Sens. 11(19), 2312–2333 (2019)
14. Segal-Rozenhaimer, M.: Cloud detection algorithm for multi-modal satellite imagery using convolutional neural-networks. Remote Sens. Environ. 237, 111446–111462 (2020)

Multi-layer Pointpillars: Multi-layer Feature Abstraction for Object Detection from Point Cloud

Shangfeng Huang, Qiming Xia, Yanhao Lin, Haiyan Lian, Zongyue Wang[✉],
Guorong Cai, and Jinhe Su

JiMei University, Xiamen 361021, China
wangzongyue@jmu.edu.cn

Abstract. In order to extract the spatial structure features of the original point cloud, multi layers pointpillars model, a fast and efficient one-stage network, is proposed for object detection from point cloud. Firstly, point cloud are divided into multi layers along z axis, by each layer to generate pillars in the vertical direction, and multi layers pseudo-image representing for multi layers are created by the method of pointpillars. Then, the multi layers and complete pseudo-image are fused as the input of RPN, and the feature maps with context information and multi-scale features are obtained. Finally, the detection boxes and classification score were obtained by SSD head according to the feature maps. We get a high quality prediction box and classification results. The experimental results show that multi-layer pointpillars can get higher precision than the original pointpillars.

Keywords: Object detection · Point clouds · Multi layers pointpillars

1 Introduction

In recent years, the research on unmanned driving technology has been very hot, and the object detection in front of the vehicle is an important part. Because 3D point cloud data is rich in spatial structure information, many researchers have turned their attention from image-based 2D object detection to 3D point cloud-based object detection [2,12,19,20]. The research on the two-dimensional detection technology started early, and the technical results it produced are relatively mature, but the 2D images and the 3D point cloud data are very different: 1) The 3D point cloud is naturally distributed and sparse; 2) 3D point cloud has a large amount of data; Therefore, the method in 2D object detection cannot be directly applied to 3D point cloud object detection.

The first author is a student. The work is supported by the National Natural Science Foundation of China (No. 41971424, No. 61701191), Xiamen Science and Technology Project (No. 3502Z20183032, No. 3502Z20191022, No. 3502Z20203057).

Y. Peng et al. (Eds.): PRCV 2020, LNCS 12305, pp. 626–637, 2020.
https://doi.org/10.1007/978-3-030-60633-6_52

Since 3d object detection is an analysis of 3d objects. Therefore, there are two popular methods in 3D object detection, voxelization operation and point-wise of raw point cloud. VoxelNet [12] first used Voxel-based data for object detection and achieved good accuracy at that time. But due to the large amount of point cloud data, it takes a lot of time to process it by using CNN networks and some data will be inevitably lost during voxelization.

Since PointNet [5] is able to learn the raw lidar point cloud features, some 3D object detection networks like Frustum-pointnets and PointRCNN [7] predict boxes after directly deal with raw point clouds. Since the point-wise feature extraction preserves the 3D information of the point cloud better, they yield good results in terms of detection results.

Pointpillars [9] proposed the concept of a pseudo-image that speeds up the feature extraction phase. Through using a simple structure, it cleverly transforms the point cloud into a two-dimensional structure and extracts the point cloud features using a two-dimensional CNN approach. Although there is a significant increase in speed, that inevitably lost a lot of features in the vertical direction.

Our method is based on PointPillars [9], and we hope to build on it. Point-Pillars generate the pseudo-image of the point features in the column through MAX. The disadvantage is that the method still lose a lot of information in the vertical direction. The improvement of our method use the multi-layered semantic information of point cloud, so that we can obtain multiple-layers feature information. Then fuse these feature maps to obtain contextual information, Multi-scale feature maps solve the problem of feature loss. Finally, the detection frame and classification scores are obtained through the SSD [15] head.

Contribution: 1) Multi-layer pseudo-image feature map extraction for point clouds is proposed to solve the problem of the loss of spatial structure information in pointpilllars. 2) Fusion of layered pseudo-image feature maps is to obtain feature maps with context information which improve object detection accuracy. 3) Experiments on the mission benchmark of the KITTI data set show that Multi-Layer pointpillars can reach advanced levels in the detection of cars, bicycles, and people.

2   Related Work

2.1   Point Cloud Data Processing

In the existing work on the processing of 3D point cloud data, some methods use the object detection method of the 2D image, such as projecting a sparse point cloud to a bird's eye view or a front view. [14] uses the projection function transforms lidar points into bev maps, and then classifies them by pixel level 2D semantic segmentation [21,25]. [2,24,26] use a mature 2D detection framework to fuse the two information to detect the object. In [2–4,18], the method of object detection and feature extraction is to project point cloud into perspective view, and use image-based feature extraction in perspective view. [5,6] proposed a series of end-to-end neural networks that can directly learn features from row point cloud, complete segmentation and classification tasks: pointnet series. [7,

28,29] also used that to directly segment the original point cloud data. [30] proposed a SANNet network to achieve precise segmentation on small targets. The emergence of Ponitnet [5] makes more and more people adopt the method of directly processing point clouds. No matter how these methods deal with point clouds, the overall framework is nothing more than one or two stages. The two-stage detection network includes [7,17] PointRcnn, STD, etc. In the first stage, a bottom-up method is used to generate a rough candidate frame. In the second stage, the local and global features are fused in its redefined coordinate system to optimize detection frame. Part A2 [8], which has been reconstructed on the basis of PointRCNN, utilizes the free position and label of internal objects provided by ground-truth boxes to construct a two-stage network. Both papers have good performance on the KIITI dataset [22] with high accuracy, but the two-stage network is slow in object detection tasks and difficult to achieve real-time detection. In contrast, the advantage of the one-stage network is that the detection speed is fast. The PointPillars proposed by [9] can achieve very fast speed while ensuring high precision, which fully reflects the advantages of the one-stage network.

Voxel method [2,3,10,11,18,23] divide the point cloud into a 3D voxel grid, and encodes each voxel with manual features, leading to information bottleneck problems. Voxelnet [12], which realizes the advancement from manual feature coding to machine learning features, effectively solves the problem of information bottlenecks, but there are a large number of 3D convolutional layers in Voxelnet [12], which makes the entire network process a huge point cloud at a very slow speed. Yan et al. proposed an optimized sparse 3D convolution algorithm SECOND [13], which replaced all common 3D convolutions in voxelnet with 3D sparse convolutions, which accelerated the speed of the entire network. Although SECOND [13] has the advantage of speeding up compared to Voxel-Net [12], it still can't get rid of the constraints of large volume and slow speed of 3D convolution. In order to solve this problem, Alex et al. proposed new network pointpillars, a kind of fast encoding based on point cloud data. The entire network uses 2D convolution, which is superior to the previous network in speed performance.

3   Multi-layer Pointpillars Network

Our method is a single-stage network, which absorbs the advantages of voxelization and multi-layer feature fusion. As shown in the Fig. 1, our network is mainly divided into three parts. (1) The conventional voxelization operation take a raw point cloud as input, and divides the space into pillars along the horizontal direction. In order to get more comprehensive local information, we further divide each pillar into two parts. And see the whole scene as three parts. (2) For making the feature more global, we still extract the feature of the non-layered pillars, and finally fuse the three parts of the feature to get a pseudo image. (3) A box prediction network is applied for fifinal prediction.

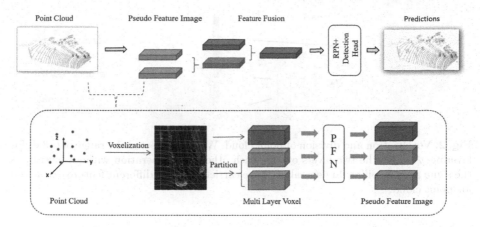

Fig. 1. The overall framework for 3D object detection. First divides the input raw point cloud into pillars horizontally and then for two parts vertically in space. We regard the whole point cloud scene as two parts. Next, We extract features from voxels of each part and transform them into feature-images. In addition, we also generate feature-image for the extracted features of the non-layered pillars. These feature-images will be further fused to get a new pseudo-image and ensure that the features correspond to the original position. Finally, a RPN generates the 3D detection.

3.1 Voxelization and Partition

Referring to the usual methods, pointpillar and second, we find that the combination of voxelization and sparse convolution can achieve significant results. We think that it is easy to lose the information that the object is perpendicular to the ground when the voxels are generated only in the horizontal direction, so we consider that the point cloud scene is divided into some parts in the Z axis. Through the experiment, we found that the effect of two layers is the best. In order to avoid the loss of context information in the scene, there are intersections among the two parts when we divide the scene (Fig. 2).

The Z axis of the point cloud ranges from - 3 to 1, and the number of points in each voxel is 100. Experiments show that the number of points in each voxel is far less than 100. In all voxels, about 97% of voxels have less than 50 points, and about 70% of voxels have 1 or 2 points (none of which is more than 10). If we divide the point cloud into three layers, we will get a large number of points with coordinates of $(0, 0, 0)$, which indirectly increases the feature of 0 point and weakens the feature of real point. Vehicle borne radar is placed on the roof, so there are very few point clouds in (0,1), and many empty grids will be obtained when voxels are divided. Therefore, our improved multiscale method only stratifies point clouds with z-axis range (- 3,0). The final results show that this method is better than the three-tier method.

The purpose of multi-scale method is to get better local feature information by dividing point clouds into smaller voxel blocks, but it often loses the global feature information. So we also extract the features of the original pointpillars.

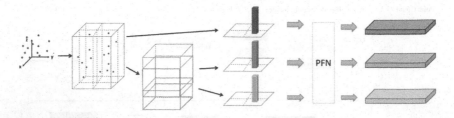

Fig. 2. Voxelization and division of point cloud. We only selected the range of (- 3,0) for layering. And let the two layers overlap each other. The operation without layering is the same as pointpillar. In general, we have carried out two different feature extraction for point cloud.

We use a similar PFN and the structure is shown in Fig. 3. We use a simplified PointNet for the feature tensor obtained in the previous step, through a series of transformations to generate a (C, P, N) sized tensor. After using the maximize function, the tensor is changed to (C, P). The difference is that we haven't coded yet, so the features obtained at this time are still scattered.

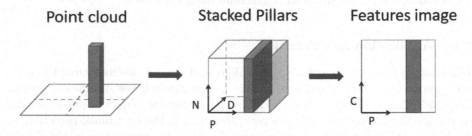

Fig. 3. SPFN layer. We get the layered two parts and the primitive voxels, then each part gets a feature-image through the SPFN layer. The features in the feature-image are all scattered, not corresponding to the original grid position.

3.2   Feature-Image Fusion and Detection Head

We get their features for each voxel that were selected of each layer and transform them to generate the feature map. We make each layer intersect with each other so that they can aggregate more contextual information. Our layered behavior gets more local information while losing the integrity of the whole. We propose a method to solve the problem of lack of integrity.

In this paper, we do tensor stitching for three feature-image, as shown in Fig. 4. After stitching the feature-images of the two layers, the features are aggregated by an MLP operation. We also extract feature information for the voxel box without layering. Then the feature information after hierarchical fusion is combined with the feature information obtained by extracting complete voxels. Finally, by encoding the feature information, the features are scattered back to the original location and get the final pseudo-image. Through the final Feature-image fusion, we can get the features including local and context information, which makes the features richer than the original features of pointpillar.

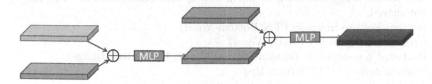

Fig. 4. The process of feature-image fusion. This is a further aggregation of information, making the context of information contact more compact. The features in the final feature-image will be mapped back to the original location, and then the next step of prediction will be carried out.

After obtaining the feature map, we continue to use the traditional object detection backbone network RPN and SSD detection head.

3.3 Network Pseudocode

We use pseudocode to explain our algorithm concisely. The input is the coordinates of the N points in the point cloud dataset, the range of the point cloud scene and the size of the divided voxels. The output is the result of our prediction. The first step is the preparation phase, where the data is layered. The second, third and fourth steps are to get the feature map of each layer. The fifth, sixth, seventh, and eighth steps are feature fusion. The last step is to use the detection head to get the prediction result.

3.4 LOSS

In this article, the loss use the classic loss of the 3D object detection task, which is composed of three parts. This classic loss has a stable and mature effect in the use of 3D object detection. The first part is localization loss, the second part is object classification loss, and the third part is discretized direction loss.

In order to complete the matching of anchor and ground truth, we define the residual amount between ground truth and anchor as the descriptor of localization offset loss:

$$\Delta x_e = \frac{x^g - x^a}{d_a}; \Delta y_e = \frac{y^g - y^a}{d_a}; \Delta z_e = \frac{z^g - z^a}{h_a}$$
$$\Delta w_e = log\frac{w^g}{w^a}; \Delta l_e = log\frac{l^g}{l^a}; \Delta h_e = log\frac{h^g}{h^a}$$
$$\Delta \theta_e = sin(\theta^g - \theta^a); d^a = \sqrt{(w^a)^2 + (l^a)^2}$$

where $(x, y, z, w, h, l, \theta)$ describes the center coordinate, length, width, height and the yaw rotation around the z-axis; the subscripts e, a, and g indicate the encoded value, the anchor, and the ground truth, respectively.

Algorithm 1 Multi-Layer pointpillars
***
Input: p(N,3);PointCloud-Range;voxel-size
Output: pre-box; pre-score;
 1: Total-Voxel, Bottom-Voxel, Up-Voxel = Voxelization-and-Partition(P, PointCloud-Range, voxel-size)
 2: Total-Feature-Image = PFN(Total-Voxel) // PFN function can make voxel features into pseudo image feature.It is worth noting that parameters of PFN function is not shared.
 3: Bottom-Feature-Image = PFN(Bottom-Voxel)
 4: Up-Feature-Image = PFN(Up-Voxel)
 5: Bottom-Up = connect(Bottom-Feature-Image, Up-Feature-Image)
 6: Scale-feature = MLP(Bottom-Up)
 7: Total-Scale = connect(Total-Scale, Scale-feature)
 8: feature-fusion = MLP(Total-Scale)
 9: pre-box, pre-score = RPN-Detection-Head(feature-fusion)
10: **return** pre-box, pre-score

We use the SmoothL1 function to deal with these residual amount to get the localization loss:

$$L_{loc} = \sum_{b \in (x,y,z,w,l,h,\theta)} SmoothL1(\Delta b) \tag{1}$$

Because focal loss function can effectively solve the problem of category imbalance, the object classification loss is

$$L_{cls} = -\alpha(1 - p^a)^\gamma log(p^a) \tag{2}$$

Where p^a is class probability of an anchor; $\alpha = 0.25$, $\gamma = 2$.

The total Loss is therefore:

$$L = \frac{1}{N_{pos}}(\beta_{loc}L_{loc} + \beta_{cls}L_{cls} + \beta_{dir}L_{dir}) \tag{3}$$

Where L_{dir} is the discretized direction loss; N_{pos} is the number of positive anchor; the hyperparameter $\beta_{loc}, \beta_{cls}, \beta_{dir}$, are 2, 1, 0.2.

4   Experiment

In this part of the experiment setup, we explain the detailed process of the experiment.

4.1   Dataset

Our algorithm verification is performed on the KITTI dataset. Although there are two types of data in the KITTI dataset, radar point cloud and image, the verification of the multi-layer pointpillars network only uses radar point cloud data. The network proposed in this paper adopts the same data set segmentation ratio as pointpillars. The 7481 samples of the original data set are used as the training set, and the remaining 7518 samples are used as the test set. Then 7481 training sets are divided again, of which 3712 samples are used as training sets, and the remaining 3769 samples are used as evaluation sets. The task of the KITTI dataset is to detect objects in front of the vehicle, including cars, bicycles, and pedestrians. The detection of three targets in our network is divided into two networks, one is the detection of cars, and the other is the detection of bicycles and pedestrians.

4.2   Experiment Details

We divide the KITTI dataset into two layers along the z axis. The point cloud spatial range of the KITTI dataset is $[(0, 70.4), (-40, 40), (-3, 1)]$, so we divide into two layers with the range of $[(-3, -1), (-2, 0)]$. We have adopted overlapping division methods. Such a layered method can make the context information more closely connected, and then through the later feature fusion, the effective intermediate region information is more abundant. In the experiment, we first voxelized the original point cloud space, and then used the pointpillars method to divide the columns based on the x, y plane grid, and finally layered the columns. After such operations, we can get the same tensor dimension for each layer of features, which is conducive to the experiment.

According to the standard of pointpillars, the resolution of grid division on the x and y planes is 0.16 m * 0.16 m. The maximum number of points inside each layer of pillars is set to 50, and the maximum number of pillars on each layer is 12000.

The feature map obtained by the network proposed in this paper will go through a typical RPN network, and then the detection results will be obtained through the detection heads of various categories. We use the same method of VOXELNET for the frame matching strategy. For anchor size setting, we refer to pointpillars. For cars, the length, width and height of anchor are: (1.6, 3.9, 1.5) m, and the center of z is -1 m; for bicycles, the size of anchor is: (0.6, 1.76, 1.73) m, z center is -0.6 m; for pedestrians, the size of the anchor is: (0.6, 0.8, 1.73) m, z center is -0.6 m. The matching positive and negative thresholds set by the car category and the non-car category are: (0.65, 0.45), (0.5, 0.35).

4.3   Data Augmentation

Before conducting the experiment, by randomly flipping, rotating, and scaling the original point cloud globally, the experimental results will have better performance. The setting of data enhancement refers to the standard of pointpillars, the probability of flipping is set to 0.5, the rotation angle is subject to a uniform distribution $U(\frac{\pi}{2}, \frac{\pi}{2})$. As with SECOND data enhancement, we randomly select ground truth samples, 15 for car, 8 for 8 for Pedestrian and 8 for Cyclist in each sample.

4.4 Experimental Results

Table 1. Results on the KITTI test BEV detection benchmark of car.

Method	Modality	Easy	Mod	Hard
MV3D	Lidar and Img	86.02	76.90	68.49
Cont-Fuse	Lidar and Img	88.81	85.83	77.33
Roarnet	Lidar and Img	88.20	79.41	70.02
AVOD-FPN	Lidar and Img	88.53	83.79	77.90
F-PointNet	Lidar and Img	88.70	84.00	75.33
HDNET	Lidar and Map	89.14	86.57	78.32
PIXOR++	Lidar	89.38	83.70	77.97
VoxelNet	Lidar	89.35	79.26	77.39
SECOND	Lidar	88.07	79.37	77.95
PointPillars	Lidar	88.35	**86.10**	79.83
Ours	Lidar	**89.18**	84.98	**80.67**

Quantitative Analysis. The KITTI dataset is stratified into easy, moderate and hard difficulties. All detection results are measured using the official KITTI evaluation detection metrics which are: bird's eye view (BEV), 2D, 3D, and average orientation similarity (AOS). The BEV detection is done in the point clouds without z axis, the 2D detection is done in the image plane and AOS assesses the average orientation (measured in BEV) similarity for 2D detection. We select the BEV and 3D to evaluate our method in this paper.

We compare with all detection results on the KITTI test set in car object. As show in Table 1, Compared to all method, ours have an advantage in accuracy. This is due to better feature extraction at multiple scales. The feature after the multiple scale can generate a pseudo-image with more information. With the help of these information, network can easily learn and generate the target boxes.

Ablation Studies. In this part, we compare two multi-layer pointpillars algorithms: two-layer method and three-layer method (Figs. 5 and 6).

From Table 2 and Table 3 it is clear that two-layer method performs better than two-layer method, because two-layer method aggregates global features, which make network more effective. Table 2 and Table 3 are the result of Pedestrian detection task. Whether it is a bev, 3d, or aos benchmark, the method of two-layer pointpillars is significantly better than the there-layer method.

Fig. 5. Qualitative analysis of KITTI results set with Kitti Viewer. We show the 3D bounding box on the bird's-eye view of the point cloud, as well as the 3D bounding box on the image. We use the green line to represent the real box and the red line to represent the predicted box. This is about the analysis of the car. (Color figure online)

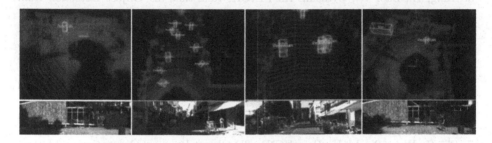

Fig. 6. Qualitative analysis of KITTI results. This is the result of pedestrian and cyclist.

Table 2. The accuracy of there benchmarks on the KITTI Pedestrian validation set of the two-layer method.

Benchmark	Easy	Middle	Hard
bev	72.78	66.22	60.35
3d	65.71	58.99	53.36
aos	34.96	37.01	35.72

Table 3. The accuracy of three benchmarks on the KITTI Pedestrian validation set of the there-layer method.

Benchmark	Easy	Middle	Hard
bev	70.01	63.59	58.92
3d	61.02	56.10	50.98
aos	36.08	34.93	33.30

5   Conclusion

In this paper, we introduce Multi-layer PointPillars, a novel deep network. Multi-layer PointPillars is an one stage network. It effectively solves the problem of loss of spatial structural features in PointPillars by fusing global features and layered features. Multi-layer PointPillars offer a high detection performance. In future work, we prepare to consider using the feature pyramid method to obtain multi-scale semantic features to improve the detection ability of the network.

References

1. Qi, C.R., et al.: Frustum PointNets for 3D Object Detection from RGB-D Data (2017)
2. Ku, J., et al.: Joint 3D Proposal Generation and Object Detection from View Aggregation (2017)
3. Yang, B., Luo, W., Urtasun, R.: PIXOR: Real-time 3D Object Detection from Point Clouds (2018)
4. Liang, M., Yang, B., Wang, S., Urtasun, R.: Deep continuous fusion for multi-sensor 3D object detection. In: Ferrari, V., Hebert, M., Sminchisescu, C., Weiss, Y. (eds.) ECCV 2018. LNCS, vol. 11220, pp. 663–678. Springer, Cham (2018). https://doi.org/10.1007/978-3-030-01270-0_39
5. Qi, C.R., et al.: PointNet: Deep Learning on Point Sets for 3D Classification and Segmentation (2016)
6. Qi, C.R., et al.: PointNet++: Deep Hierarchical Feature Learning on Point Sets in a Metric Space (2017)
7. Shi, S., Wang, X., Li, H.: PointRCNN: 3D Object Proposal Generation and Detection from Point Cloud (2018)
8. Shi, S., et al.: From Points to Parts: 3D Object Detection from Point Cloud with Part-aware and Part-aggregation Network (2019)
9. Lang, A.H., et al.: PointPillars: Fast Encoders for Object Detection from Point Clouds (2018)
10. Riegler, G., Ulusoy, A.O., Geiger, A.: OctNet: Learning Deep 3D Representations at High Resolutions (2016)
11. Song, S., et al.: Semantic scene completion from a single depth image. In: 2017 IEEE Conference on Computer Vision and Pattern Recognition (CVPR). IEEE (2017)
12. Zhou, Y., Tuzel, O.: VoxelNet: End-to-End Learning for Point Cloud Based 3D Object Detection (2017)
13. Yan, Y., Mao, Y., Li, B.: SECOND: sparsely embedded convolutional detection. Sensors 18(10), 3337 (2018)
14. Wu, B., et al.: SqueezeSeg: Convolutional Neural Nets with Recurrent CRF for Real-Time Road-Object Segmentation from 3D LiDAR Point Cloud (2017)
15. Liu, W., et al.: SSD: Single Shot MultiBox Detector (2016)
16. Chen, X., et al.: Multi-View 3D Object Detection Network for Autonomous Driving (2016)
17. Yang, Z., et al.: STD: Sparse-to-Dense 3D Object Detector for Point Cloud (2019)
18. Engelcke, M., et al.: Vote3Deep: Fast Object Detection in 3D Point Clouds Using Efficient Convolutional Neural Networks (2016)

19. Shin, K., Kwon, Y.P., Tomizuka, M.: RoarNet: A Robust 3D Object Detection based on RegiOn Approximation Refinement (2018)
20. Simon, M., et al.: Complex-YOLO: Real-time 3D Object Detection on Point Clouds (2018)
21. Chen, L.C., et al.: DeepLab: semantic image segmentation with deep convolutional nets, atrous convolution, and fully connected CRFs. IEEE Trans. Pattern Anal. Mach. Intell. **40**(4), 834 (2018)
22. Geiger, A., Lenz, P., Urtasun, R.: Are we ready for autonomous driving? The KITTI vision benchmark suite. In: Proceedings of the 2012 IEEE Conference on Computer Vision and Pattern Recognition (CVPR). IEEE (2012)
23. Song, S., Xiao, J.: Deep sliding shapes for amodal 3D object detection in RGB-D images. In: 2016 IEEE Conference on Computer Vision and Pattern Recognition (CVPR). IEEE (2016)
24. Xu, D., Anguelov, D., Jain, A.: PointFusion: deep sensor fusion for 3D bounding box estimation. In: 2018 IEEE/CVF Conference on Computer Vision and Pattern Recognition (CVPR). IEEE (2018)
25. Zhao, H., et al.: Pyramid scene parsing network. In: 2017 IEEE Conference on Computer Vision and Pattern Recognition (CVPR). IEEE (2017)
26. Gonzalez, A., et al.: Multiview random forest of local experts combining RGB and LIDAR data for pedestrian detection. In: 2015 IEEE Intelligent Vehicles Symposium (IV). IEEE (2015)
27. Jiang, M., et al.: PointSIFT: A SIFT-like Network Module for 3D Point Cloud Semantic Segmentation (2018)
28. Li, Y., et al.: PointCNN: Convolution on $\mathcal{X}$-Transformed Points (2018)
29. Li, J., Chen, B.M., Lee, G.H.: SO-Net: Self-Organizing Network for Point Cloud Analysis (2018)
30. Dai, A.: ScanNet: richly-annotated 3D reconstructions of indoor scenes. In: IEEE Conference on Computer Vision and Pattern Recognition. IEEE (2017)

Building Detection via Complementary Convolutional Features of Remote Sensing Images

Zeshan Lu[1], Kun Liu[1,2], Yongwei Zhang[4], Zhen Liu[1], Jiwen Dong[1,2], Qingjie Liu[3(✉)], and Tao Xu[1,2(✉)]

[1] School of Information Science and Engineering, University of Jinan, Jinan, China
xutao@ujn.edu.cn
[2] Shandong Provincial Key Laboratory of Network Based Intelligent Computing, Jinan, China
[3] Hangzhou Innovation Institute, Beihang University, Hangzhou, China
qingjie.liu@buaa.edu.cn
[4] Shandong Aerospace Electro-Technology Institute, Yantai, China

Abstract. Building detection in remote sensing image plays an important role in urban planning and construction, management and military fields. Most of the building detection methods utilized deep neural networks to achieve better detection results. However, these methods either train their models on panchromatic images or on the fused images of panchromatic and multispectral images, which do not take into full consideration the advantages of high resolution and multispectral of earth observation images. In this paper, a large-scale building dataset is presented, more than 10 million single buildings were annotated both on panchromatic images and the corresponding multispectral images. Moreover, a building detection method is proposed based on complementary convolutional feature via fusion on panchromatic and multispectral images. Experimental results on the proposed dataset and ResNet-101 demonstrate that, complementary convolutional feature based building detection methods outperform the methods that only use panchromatic convolution feature and multispectral convolution feature.

Keywords: Building dataset · Complementary convolutional feature · Remote sensing · Deep neural network

1 Introduction

With the development of high-resolution remote sensing satellites, the quality of remote sensing images has been constantly improved, facilitating great success in land and water segmentation [1], airport detection [2], built-up area extraction [3], etc. Applications of high-resolution remote sensing images in various fields, have greatly promoted the quality of production and living.

© Springer Nature Switzerland AG 2020
Y. Peng et al. (Eds.): PRCV 2020, LNCS 12305, pp. 638–647, 2020.
https://doi.org/10.1007/978-3-030-60633-6_53

In recent years, techniques have been successfully applied to the task of object detection. Deep neural networks are simulations of human brains thus have powerful representation ability, which make them good at object detection when analyzing complex scenes. For object detection from high-resolution remote sensing images, deep learning provides a good solution.

High-resolution remote sensing images have abundant information, traditional methods are unable to achieve satisfactory performance and are inefficiency. So it is very essential to use advanced techniques to solve the image interpretation problem from remote sensing images, automatically. Researchers have developed building detection methods based on deep neural networks to achieve better detection results in remote sensing images. However, these methods either train their models on panchromatic images or on the RGB bands of fused images, which do not take into full consideration the advantages of high resolution and multispectral of earth observation images. Panchromatic images have higher spatial resolution however lower spectral resolution than multispectral images. Most of the remote sensing image datasets and detection methods are based on Google Earth images, which only have three channels and will affect the accuracy of target detection. In order to facilitate development building detection of remote sensing images, this paper introduces a large-scale dataset and proposes a deep neural network structure with dual-branch input to make full use information of panchromatic and multispectral images, in which TFNet [4] was utilized for the design of network structure with an encoding-decode fusion process.

The main contributions of this paper are: (1) We created a building dataset for building detection in remote sensing images, in which buildings were labeled manually in both panchromatic images and multispectral images, and the number of building samples is up to 10 million. (2) A complementary convolutional feature based building detection method in remote sensing images is proposed.

The rest of the paper is organized as follows: we review and summarize related works in Sect. 2. The proposed dataset and complementary convolutional feature based method are introduced in Sect. 3. Section 4 presents the experiments with analysis, and conclusions are given in Sect. 5.

2   Related Work

Building detection in remote sensing images could provide targets of interest in urban management and military action. Researchers have proposed many states of the art methods of building detection and building dataset with detailed annotations [5].

In recent years, methods based on deep neural network have obtained good detection results in building detection. Sevo [6] proposed a convolutional neural network based two-stage network for aerial image classification and target detection, which achieved an accuracy of about 98.6% on UC Merced aerial image dataset. Zhang et al. [7] proposed a method based on deep convolutional neural network to automatically detect suburban buildings from high-resolution

Google Earth images. Their method could extract building areas and achieved 89% accuracy in the challenging Google Earth images. A convolutional neural network based multi-layer feature fusion method was proposed in Ref. [8], which designed a fusion module and contained two different levels of feature map fusion methods to solve fusion problems instead of only fusing the feature map of the last layer. Deng et al. propose a remote sensing image fusion method based on tensors for non-convex sparse modeling [9], commonly known as full sharpening. In this method, the multispectral images were sampled by interpolation method, and the weighted fidelity terms based on tensors were modeled for panchromatic and low-resolution multispectral images, in order to recover more spatial details. In addition, total change regularization is also used to describe the sparsity of potentially high-resolution multispectral images over gradient fields. FusionCNN [10] was proposed by Ye et al. The input is a pair of raw images, while the output is a fused image with end-to-end training.

Large-scale building dataset is required for training of deep neural network models, and research works have reported many building datasets [11]. The commonly used building datasets include UC Merced land-use dataset [12], WHU-RS19 dataset [13] and RSSCN7 dataset [14]. UC Merced land-use dataset contains 2100 images, each image is with 256×256 pixels. The images were from the USGS National Map Urban Area Imagery. The dataset consists of various urban areas around the country, and the pixel resolution of is 1 ft. The WHU-RS19 dataset contains 1005 images. The image sizes are 600×600 pixels. All images in the dataset are collected from Google Earth with spatial resolution up to 0.5 m. The RSSCN7 dataset contains 2800 images also collected from Google Earth with a size of 400×400.

Although building detection has attracted many researchers' attention, there are still many challenging problems: (1) the number and diversity of samples in database could not meet the needs of deep neural network based model training, and most of the samples in the dataset were only annotated in panchromatic image or fused image; (2) many building detection methods could not fully consider the high-resolution advantage of panchromatic image and the multispectral advantage of multispectral image. Aiming at the above problems, this paper proposed a large-scale building dataset and a complementary convolutional feature based method.

3   Complementary Convolutional Feature Based Method

3.1   Dataset

A few building datasets has been released. However, most of them are built on Google Earth images, and the number of samples in existing dataset is relatively small. These datasets does not take into consideration of multi-modalities nature of remote sensing data and thus can not take full advantage of spatial and spectral information of them.

A large-scale building dataset is presented in this paper. The remote sensing images were selected from the GaoFen-2 satellite, which have a spatial resolution of 0.8 m. More than 20000 images with the same size of 512×512 pixels were selected and all buildings in images were manually annotated. The number building instances is up to 10 million. During the experiment, we randomly selected 1090 images as the training set, including 16789 building instances, and 1090 images as the test set, including 15789 building instances. The dataset will be available for scientific research in the future. Some annotation samples of dataset on panchromatic images are shown in Fig. 1 and the corresponding annotation results of multispectral images are shown in Fig. 2.

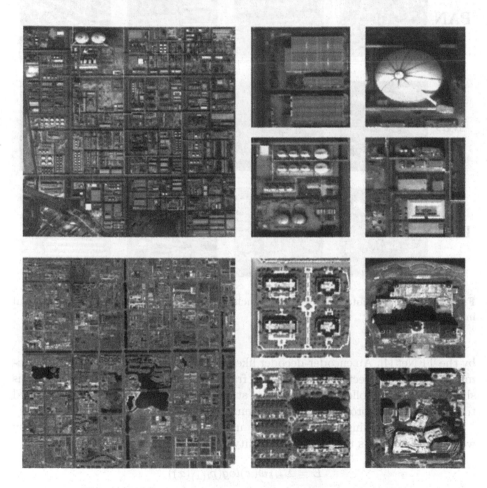

Fig. 1. Building samples annotated on panchromatic image.

3.2   Complementary Convolutional Feature Based Method

The complementary convolutional features here refer to both spatial information on panchromatic images and spectral information of multispectral images we used in our detection method, and the network structure of which is inspired

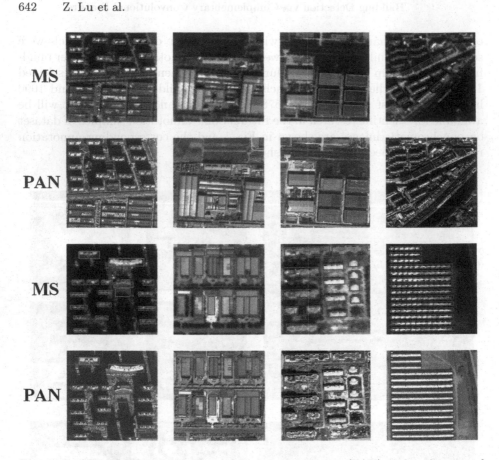

Fig. 2. Building annotation result on panchromatic image (PAN) and multispectral image (MS).

by the TFNet [4]. An encoder-decoder like structure is employed to fuse panchromatic and multispectral images. The framework of the proposed network is shown in Fig. 3. Following the fusion stage, a Faster RCNN head is applied to perform region proposal and detection.

Encoding-decoding network model uses the loss function with boundary weights, which gives the model the ability to separate boundaries.

$$E = \Sigma_{x \epsilon \delta} w(x) log(p_{l(x)}(x)) \tag{1}$$

$p_{l(x)}(x)$ is loss function of softmax, l is the label value of the pixel, and w is the weight of the pixel point in order to give higher weight to the pixel close to the boundary point in the image.

$$w(x) = w_c(x) + w_0 * exp(-(d_1(x) + d_2(x))^2/2\sigma^2) \tag{2}$$

Where w_c is the weight of the balance category proportion, d_1 is the distance from the pixel to the nearest target, and d_2 is the distance from the pixel to the second closest target. w_0 and σ are constant values.

Since the corresponding spectral bands of the two input images are different, the corresponding feature information is also different, so different networks are used to extract the features of the H_{MS} and H_{PAN}, and the two networks do not share weight during training and testing. As a result of the input image channel number is different, multispectral image feature extraction is a multi-channel network, panchromatic image feature extraction is a single channel network.

The main reason why different networks are used to extract features of panchromatic image H_{PAN} and multispectral image H_{MS} is that the corresponding spectral bands of the two input images are different, and the corresponding characteristic information is also different. If the same network or two networks sharing features are used to extract the feature information of the two images at the same time, despite the advantage in the number of network parameters, this method cannot distinguish the features of the two images. In the process of feature extraction, the feature information will be lost, which will affect the final fusion effect.

After feature extraction, the corresponding multispectral feature map h_{MS} and panchromatic feature map h_{PAN} are superimposed on the channel dimension to form the feature map, which is sent into the feature fusion network F. After encoding and decoding, the final output of the fused image is a fusion image H_F with color information and high resolution.

4   Experiment

4.1   Experiment Scheme

To evaluate the performance of the proposed method and the effectiveness of the fusion stage, we conduct experiments under three settings: building detection without fusion, building detection after an independent fusion, joint fusion and detection in an end-to-end learning manner.

In the first setting, there is no image fusion step. The original panchromatic and multispectral images are directly sent into the detection network as input. Before that, an indispensable part is concatenating the panchromatic and the up-sampled multispectral images to form new five-band images. We use this way to combine the spectral and spatial information of the two images.

In the second setting, a fusion algorithm is applied first to fuse the panchromatic and multispectral images to obtain the high resolution multispectral images. Then we feed them into the Faster RCNN network for detection.

In the third setting, the fusion and detection is achieved jointly. We firstly input panchromatic and multispectral images to the network which fuses features of them for detection. In the experiments, all detection networks use ResNet-101 network as backbone network.

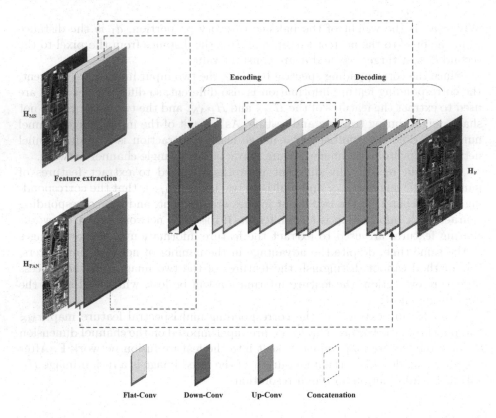

Fig. 3. Complementary convolutional feature based building detection methods. H_{MS} represent the input multispectral images, H_{PAN} represent the input panchromatic images, and H_F represent the final fusion images.

4.2   Experimental Results

Method1 is building object detection without fusion process, Method2 is building object detection after fusion, Method3 is end-to-end fusion and building detection.

The detection results are shown in Table 1. Experimental results show that the end-to-end building fusion and detection scheme has the best performance, and the complementary convolutional feature based building detection method in high-resolution remote sensing image plays a great role in promoting the accuracy of building target detection. The experimental scheme without image fusion for direct target detection lacks targeted and effective extraction of multispectral image information and panchromatic image information, resulting in the worst effect. Therefore, image fusion is an important part to improve experimental performance.

Figure 4 shows some examples of building target detection results of end-to-end fusion and detection method, which has the best performance of the three schemes.

Table 1. Comparison of detection results.

Method	Network	AP^{bb}	AP^{bb}_{50}	AP^{bb}_{75}	AR^{bb}
Method1	ResNet-101	22.8	48.1	20.1	30.4
Method1	ResNeXt-101	24.6	50.4	21.5	32.7
Method2	ResNet-101	24.5	50.7	22.3	33.1
Method2	ResNeXt-101	26.9	52.1	23.8	34.9
Method3	ResNet-101	25.6	51.2	23.0	34.6
Method3	ResNeXt-101	27.6	53.3	24.9	35.8

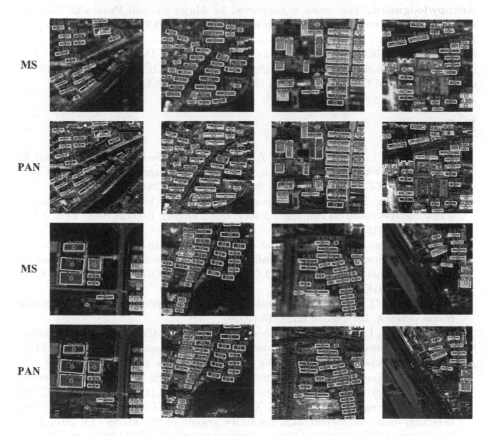

Fig. 4. Building detection results. Building detection results of Method3 on the multispectral image (MS) and panchromatic image (PAN).

The experimental results show that the remote sensing image fusion network is necessary and effective and can promote the accuracy of target detection. Although the end-to-end fusion detection method performs better, it is not much better than the fusion first and then detection method, which needs further improvement.

5 Conclusion

In this paper, two main contributions, i.e., a large-scale building dataset and a complementary convolutional feature based building detection method, was presented. There are 5 categories and 10 million building instances in the dataset. The proposed building detection method utilized complementary convolutional feature fused by convolutional feature of panchromatic images and multispectral images. Experimental results on ResNet-101 show the validity of the proposed method.

Acknowledgment. This work is supported by Major Special Provincial Industrialization Application Project of High-Resolution Earth Observation System (No. 79-Y40G03-9001-18/20), the Key Research and Development Program of Shandong Province (No. 2018JMRH0102), Science and Technology Development Program of Jinan, Doctoral Foundation of University of Jinan (No. XBS1653) and Natural Science Foundation of University of Jinan (Nos. XKY1803, XKY1804, XKY1928).

References

1. Liu, W., Ma, L., Chen, H., Han, Z., Soomro, N.Q.: Sea–land segmentation for panchromatic remote sensing imagery via integrating improved MNcut and Chan–Vese model. IEEE Geosci. Remote Sens. Lett. **14**(12), 2443–2447 (2017)
2. Zhu, D., Wang, B., Zhang, L.: Airport target detection in remote sensing images: a new method based on two-way saliency. IEEE Geosci. Remote Sens. Lett. **12**(5), 1096–1100 (2015)
3. Tan, Y., Xiong, S., Li, Y.: Automatic extraction of built-up areas from panchromatic and multispectral remote sensing images using double-stream deep convolutional neural networks. IEEE J. Sel. Top. Appl. Earth Obs. Remote Sens. **11**(11), 3988–4004 (2018)
4. Liu, X., Liu, Q., Wang, Y.: Remote sensing image fusion based on two-stream fusion network. Inf. Fusion **55**, 1–15 (2020)
5. Lu, Z., Liu, K., Liu, Z., Wang, C., Shen, M., Xu, T.: An efficient annotation method for big data sets of high-resolution earth observation images. In: Proceedings of the 2nd International Conference on Big Data Technologies, pp. 240–243 (2019)
6. Ševo, I., Avramović, A.: Convolutional neural network based automatic object detection on aerial images. IEEE Geosci. Remote Sens. Lett. **13**(5), 740–744 (2016)
7. Zhang, Q., Wang, Y., Liu, Q., Liu, X., Wang, W.: CNN based suburban building detection using monocular high resolution google earth images. In: 2016 IEEE International Geoscience and Remote Sensing Symposium (IGARSS), pp. 661–664. IEEE (2016)
8. Ma, C., Mu, X., Sha, D.: Multi-layers feature fusion of convolutional neural network for scene classification of remote sensing. IEEE Access **7**, 121685–121694 (2019)
9. Deng, L.J., Feng, M., Tai, X.C.: The fusion of panchromatic and multispectral remote sensing images via tensor-based sparse modeling and hyper-laplacian prior. Inf. Fusion **52**, 76–89 (2019)
10. Ye, F., Li, X., Zhang, X.: FusionCNN: a remote sensing image fusion algorithm based on deep convolutional neural networks. Multimed. Tools Appl. **78**(11), 14683–14703 (2018). https://doi.org/10.1007/s11042-018-6850-3

11. Cheng, G., Han, J., Lu, X.: Remote sensing image scene classification: benchmark and state of the art. Proc. IEEE **105**(10), 1865–1883 (2017)
12. Zhu, Q., Zhong, Y., Zhao, B., Xia, G.S., Zhang, L.: Bag-of-visual-words scene classifier with local and global features for high spatial resolution remote sensing imagery. IEEE Geosci. Remote Sens. Lett. **13**(6), 747–751 (2016)
13. Luus, F.P., Salmon, B.P., Van den Bergh, F., Maharaj, B.T.J.: Multiview deep learning for land-use classification. IEEE Geosci. Remote Sens. Lett. **12**(12), 2448–2452 (2015)
14. Zou, Q., Ni, L., Zhang, T., Wang, Q.: Deep learning based feature selection for remote sensing scene classification. IEEE Geosci. Remote Sens. Lett. **12**(11), 2321–2325 (2015)

Hyperspectral Image Super-Resolution via Self-projected Smooth Prior

Yuanyang Bu[1,2], Yongqiang Zhao[1,2(✉)], and Jonathan Cheung-Wai Chan[3]

[1] School of Automation, Northwestern Polytechnical University, Xi'an, China
buyuanyang@mail.nwpu.edu.cn, zhaoyq@nwpu.edu.cn
[2] Research and Development Institute of Northwestern Polytechnical University in Shenzhen, Shenzhen, China
[3] Department of Electronics and Informatics, Vrije, Universiteit Brussel, 1050 Brussels, Belgium
jcheungw@etrovub.be

Abstract. High spectral correlations and non-local self-similarities, as two intrinsic characteristics underlying hyperspectral image (HSI), have been widely used in HSI super-resolution. However, existing methods mostly utilize the two intrinsic characteristics separately, which still inadequately exploit spatial and spectral information. To address this issue, in this study, a novel self-projected smooth prior (SPSP) is proposed for the task of HSI super-resolution. SPSP describes that two full-band patches (FBPs) are close to each other and then the corresponding subspace coefficients are also close to each other, namely smooth dependences of clustered FBPs within each group of HSI. Suppose that each group of FBPs extracted from HSI lies in smooth subspace, all FBPs within each group can be regarded as the nodes on an undirected graph, then the underlying smooth subspace structures within each group of HSI are implicitly depicted by capturing the linearly pair-wise correlation between those nodes. Utilizing each group of clustered FBPs as projection basis matrix can adaptively and effectively learn the smooth subspace structures. Besides, different from existing methods exploiting non-local self-similarities with multispectral image, to our knowledge, this work represents the first effort to exploit the non-local self-similarities on its spectral intrinsic dimension of desired HSI. In this way, spectral correlations and non-local self-similarities of HSI are incorporated into a unified paradigm to exploit spectral and spatial information simultaneously. As thus, the well learned SPSP is incorporated into the objective function solved by the alternating direction method of multipliers (ADMM). Experimental results on synthetic and real hyperspectral data demonstrate the superiority of the proposed method.

Keywords: Hyperspectral image · Super-resolution · Non-local Self-similarity · Spectral correlation

Student paper.

1 Introduction

Hyperspectral imaging simultaneously acquires the inherent spectral characteristics across contiguous bands. The fine spectral and spatial details of HSI, which can be beneficial to greatly improve the performance of object classification [1–3] and detection [4, 5]. However, it is challenging to maintain acceptable signal-to-noise ratio of each band while obtain high spectral resolution due to the hardware limitations of sensors in optical imaging. There always exists a tradeoff between the spectral resolution and spatial resolution of HSI. Recently, combining a low spatial resolution hyperspectral image (LR-HSI) with a high spatial resolution multispectral image (HR-MSI) into a high spatial resolution hyperspectral image (HR-HSI) has become a popular scheme to enhance the spatial resolution of HSI [6–17].

Non-local self-similarities are actively investigated and proved to be effective to deal with HSI super-resolution [12–17]. Existing HSI super-resolution methods modeling non-local self-similarities can be roughly divided into two classes: sparsity-based [12–15] and low-rank based methods [16, 17]. Sparsity-based methods are to impose a proper sparsity prior for groups of clustered FBPs of HSI by assuming that each group of clustered FBPs lies in its sparse subspace. Specifically, each group of reconstructed clustered FBPs can be represented as a redundant projection basis matrix multiplied by its sparse coefficients. However, the reconstruction accuracy of HSI is susceptible to the selection of the redundant projection basis matrix, which mainly relies on the quality and quantity of training samples. The inappropriate selection of projection basis matrix always results in the deterioration of reconstructed results of HSI. Low-rank based methods exploit the repetitive structures in spatial domain and obtain the plausible reconstructed results in general. To be exact, low-rank based methods hold the hypothesis that each group of clustered FBPs lies in its low-rank subspace. Consequentially, the non-local low-rank property resorts to solving an optimization problem of rank minimization. Actually, the desired HSI are approximately low rank in general rather than exactly low rank. Discarding small singular values can barely preserve the spatial details. In addition, whether sparsity-based or low-rank based methods, all of them exploit non-local self-similarities with HR-MSI. However, the spectral dimension of HR-MSI is inconsistent with the intrinsic dimension of the desired HR-HSI in general and this may lead to inevitable reconstruction errors. Therefore, more appropriate and effective ways to model simultaneously the non-local self-similarities and spectral correlations of HSI are urgently demanded in HSI super-resolution.

To address above issues, in this study, two intrinsic characteristics underlying HSI are fully exploited, i.e., the high spectral correlations and non-local self-similarities. With regard to the high spectral correlations, it is well-known that the spectral dimension of HSI is far more than the intrinsic dimension of its spectral subspace. Besides, since the spectral information is mainly contained in LR-HSI, LR-HSI and HR-HSI should reside on a common spectral low-rank subspace. More specifically, the target HR-HSI is decomposed as an orthogonal projection basis matrix multiplied by its low-rank subspace coefficients obtained by singular value decomposition (SVD) to characterize the spectral correlations of HSI. Inspired by the subspace clustering, suppose that when the features are close to each other and then the corresponding subspace coefficients are also close to each other, namely smooth dependences of similar clustered FBPs within each group of HSI. In this way, a novel SPSP is proposed for tactfully modeling non-local self-similarities on its spectral subspace by assuming that each group of clustered FBPs lies in its spatial smooth subspace individually. Utilizing each group of clustered FBPs as projection basis matrix can adaptively and effectively learn the spatial subspace structures. As thus, the deficiencies of existing methods modeling non-local self-similarities are circumvented. In addition, different from existing methods exploiting non-local self-similarities with input HR-MSI, to our knowledge, this work represents the first effort to exploit the non-local self-similarities property on its spectral intrinsic dimension of desired HR-HSI. Then the learned SPSP is incorporated into the super-resolution framework in the ADMM scheme to obtain more accurate reconstructed results. Experimental results on synthetic and real hyperspectral data demonstrate the superiority of the proposed method.

The main contributions to this study are summarized as follows:

- This work represents the first effort to tactfully exploit the non-local self-similarities on its spectral intrinsic dimension of desired HSI. In this way, spectral correlations and non-local self-similarities of HSI are incorporated into a unified paradigm to exploit spectral and spatial information simultaneously.
- A novel self-projected smooth prior is proposed for the task of HSI super-resolution by assuming that each group of clustered FBPs of HSI lies in its spatial smooth subspace individually, namely smooth dependences of similar clustered FBPs within each group of HSI.
- We propose an efficient algorithm via ADMM framework to formulate the super-resolution problem and outperform state-of-the-art methods on several synthetic and real datasets.

2 HSI Super-Resolution via SPSP

In this section, we present the proposed SPSP model for HSI super-resolution.

2.1 Observation Model

In this study, we denote the desired HR-HSI as $\mathcal{H} \in \mathbb{R}^{X \times Y \times Z}$, which has Z bands and $X \times Y$ pixels. The observed LR-HSI is denoted as $\mathcal{L} \in \mathbb{R}^{x \times y \times Z}$, which has $x \times y$ pixels and Z bands. $\mathcal{M} \in \mathbb{R}^{X \times Y \times z}$ denotes the observed HR-MSI, which has $X \times Y$ pixels and z bands. $\mathcal{L}$ is regarded as the spatially down-sampled version of $\mathcal{H}$, while $\mathcal{M}$ can be generated by down-sampling $\mathcal{H}$ in spectral domain, i.e.,

$$L_{(3)} = H_{(3)}S, M_{(3)} = RH_{(3)} \tag{1}$$

where $L_{(3)} \in \mathbb{R}^{Z \times xy}$, $M_{(3)} \in \mathbb{R}^{z \times XY}$ and $H_{(3)} \in \mathbb{R}^{Z \times XY}$ denote the unfolding of $\mathcal{L}$, $\mathcal{M}$ and $\mathcal{H}$ along 3-mode, respectively. $S \in \mathbb{R}^{XY \times xy}$ and $R \in \mathbb{R}^{z \times Z}$ represent the spatial down-sampling matrix and spectral response matrix, respectively. Then the HSI super-resolution task can be formulated as follows according to the observation model in Eq. (1), i.e. (Fig. 1),

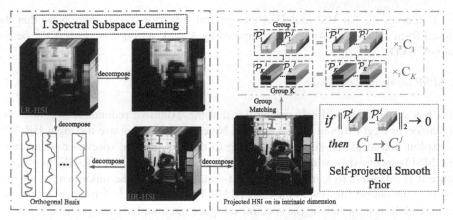

Fig. 1. Flowchart of self-projected smooth prior learning.

$$\min_{H_{(3)}} \left\| L_{(3)} - H_{(3)} S \right\|_F^2 + \left\| M_{(3)} - R H_{(3)} \right\|_F^2 + \lambda \phi(\mathcal{H}) \qquad (2)$$

where $\| \bullet \|_F$ represents the Frobenius norm, and $\phi(\mathcal{H})$ is the appropriate regularization to further constrain the solution domain, generally derived from the intrinsic structure of HR-HSI. λ is a parameter which balances the importance between data-fit terms and the regularization term. In the following, the novel priors on the HR-HSI will be introduced and then be incorporated into the HSI super-resolution framework in Eq. (2).

2.2 Self-projected Smooth Prior Learning

- Spectral subspace learning

HSI illustrates the high spectral correlations. As pointed out in [18], the spectral dimension of HSI is far more than the intrinsic dimension of its spectral subspace. Inspired by the study in [19], the HR-HSI can be assumed to reside on a global spectral low-rank subspace, i.e.,

$$\mathcal{H} = \mathcal{P} \times_3 A \qquad (3)$$

where $A \in \mathbb{R}^{Z \times z'} (z' \ll Z)$ denotes spectral low-rank subspace orthogonal basis, $\times_3$ denotes the n-mode product [20], $\mathcal{H} \in \mathbb{R}^{X \times Y \times Z}$ denotes the desired HR-HSI, $\mathcal{P} \in \mathbb{R}^{X \times Y \times z'}$ denotes the projected image on the spectral subspace and z' is the intrinsic dimension of $\mathcal{H}$.

Since the target $\mathcal{H}$ is unknown, the LR-HSI $\mathcal{L}$, which contains most of the spectral information of $\mathcal{H}$, is utilized to learn the spectral subspace orthogonal basis A. In this way, the orthogonal basis matrix A is estimated by solving the following problem

$$\min_A \frac{1}{2} \| \mathcal{L} - \mathcal{D} \times_3 A \|_F^2, s.t. A^T A = I \qquad (4)$$

$\mathcal{L}$ denotes the LR-HSI and $\mathcal{D}$ is the spectral subspace coefficients. Note that the closed form solution to (4) can be computed by $A = V$ letting $L_{(3)} = U \Sigma V^T$ be the rank-N SVD.

- SPSP on spectral subspace

HSI has strong non-local self-similarities that the extensive redundant structures are contained in local FBPs in spatial domain. Note that we do not use the input HR-MSI to exploit non-local self-similarities owing to the fact that the spectral dimension of HR-MSI is inconsistent with the intrinsic dimension of the desired HR-HSI in general. Since it is infeasible to learn non-local self-similarities from the unknown $\mathcal{H}$, we first spatially up-sample LR-HSI to the reduced HR-HSI $\mathcal{H}' \in \mathbb{R}^{X \times Y \times Z}$ by bicubic interpolation and $\mathcal{H}'$ is projected on the spectral subspace, then we have the projected reduced image $\mathcal{P}' \in \mathbb{R}^{X \times Y \times z'}$. In this paper, overlapped FBPs of size $\sqrt{d} \times \sqrt{d} \times Z$ from $\mathcal{H}'$ with a fixed interval r are extracted and divided into K groups. Suppose that there exists m similar FBPs in the i-th group of $\mathcal{H}'$, and they are stacked into tensor $\mathcal{H}'_i \in \mathbb{R}^{d \times m \times Z}$. So as to exploit the underlying structure among similar FBPs in $\mathcal{H}'_i$, inspired by the subspace clustering, a self-projected strategy is employed to depict the pair-wise similarities among FBPs, i.e.,

$$\mathcal{H}'_i = \mathcal{H}'_i \times_2 C_i \tag{5}$$

$C_i \in \mathbb{R}^{m \times m}$ are the subspace coefficient matrices, and each coefficient in C_i depicts the correlation between any two FBPs in the i-th group, namely, each FBP can be represented by a linear combination of all FBPs. $\mathcal{P}'_i \in \mathbb{R}^{d \times m \times z'}$ is the similar FBPs within the i-th group extracted from projected reduced image $\mathcal{P}'$ by group matching. As shown in Eq. (3), note that the whole groups of non-local FBPs should also reside on the common low-rank spectral subspace, namely $\mathcal{H}'_i = \mathcal{P}'_i \times_3 A$. And as shown in Eq. (5), we have $\mathcal{P}'_i = \mathcal{H}'_i \times_3 A^T = \mathcal{H}'_i \times_2 C_i \times_3 A^T = \mathcal{P}'_i \times_2 C_i$. Then for a given $\mathcal{H}'_i$ and corresponding $\mathcal{P}'_i$, they share the same coefficient matrix C_i. Based on the relationships, subspace coefficient matrix C_i is directly learned from the corresponding i-th group of FBPs of $\mathcal{P}'_i$ residing on the spectral intrinsic dimension of the desired HR-HSI. Assuming that each group of clustered FBPs lies in its spatial smooth subspace individually, we claim that if the FBPs are close to each other and then their subspace coefficients are close to each other as well. A novel self-projected smooth prior is adopted to fully exploit the correlations among similar FBPs in the specific group. We give the concrete definition of the SPSP as follows.

Definition (Self-projected Smooth Prior). Given a set of dZ-dimensional samples constituting a matrix $\widehat{H} = [h_1, h_2, \cdots, h_m] \in \mathbb{R}^{dZ \times m}$, $\widehat{C} = [c_1, c_2, \cdots, c_m] \in \mathbb{R}^{m \times m}$ is the self-projected smooth representation of $\widehat{H}$ if $\|h_i - h_j\|_2 \to 0 \Rightarrow \|c_i - c_j\|_2 \to 0, \forall i \neq j$.

The SPSP can be understood as the smooth dependences of similar FBPs extracted from $\mathcal{P}'_i$. Suppose that each group of FBPs extracted from HSI lies in smooth subspace individually, all FBPs within each group can be regarded as the nodes on an undirected graph. SPSP can implicitly depict the underlying subspace structures within each group of HSI by capturing the linearly pair-wise correlation between those nodes. Thus, we formulate the SPSP as the following optimization problem

$$\min_{C_i} \sum_{i=1}^{K} \left\| \mathcal{P}_i' - \mathcal{P}_i' \times_2 C_i \right\|_F^2 + \eta tr\left(C_i L_i C_i^T \right) \tag{6}$$

where L_i denote the Laplacian matrices, η is a parameter balancing fidelity term and regularization term, and tr calculates the trace of the matrix. For simplicity, we use k-nearest neighbor graph with 0-1 weights in our experiments. It is worth noting that Eq. (6) meets the condition of SPSP, the proof can be found in [21]. And Eq. (6) is a smooth convex program and has a unique solution by solving a standard Sylvester equation.

2.3 Proposed Super-Resolution Framework via SPSP

In general, the orthogonal basis matrix A is learnt in advance, and then the spatial non-local self-similarities of $\mathcal{P}'$ are fully exploited on this basis. As thus, we transform the non-local self-similarities in the image domain to the spectral intrinsic domain. Then a novel self-projected smooth prior is adopted to fully exploit the smooth dependences of similar FBPs in the specific group. Based on the learned SPSP, we incorporate it into the HSI super-resolution framework in Eq. (2) as follows

$$\min_{H_{(3)},\mathcal{G}_i} \left\| L_{(3)} - H_{(3)}S \right\|_F^2 + \left\| M_{(3)} - RH_{(3)} \right\|_F^2 + \lambda \sum_{i=1}^{K} \left\| \mathcal{Q}_i(H_{(3)}) - \mathcal{G}_i \right\|_F^2,$$
$$s.t.\ \mathcal{G}_i = \mathcal{G}_i \times_2 C_i, \forall i \tag{7}$$

where $\mathcal{Q}_i(H_{(3)}) = \mathcal{P}_i$ represents the group matching operation extracting similar FBPs within the i-th group in the spectral subspace. And the intermediate variable $\mathcal{G}_i$ complies with the self-projected smooth prior. To solve this problem, we utilize the standard ADMM scheme [22] to optimize Eq. (7) by introducing two splitting variables $W = H_{(3)}, Y = H_{(3)}$ and the following augmented Lagrangian function is obtained:

$$\mathcal{F}\left(H_{(3)}, \mathcal{G}_i, W, Y, V_1, V_2\right) = \left\| L_{(3)} - H_{(3)}S \right\|_F^2 + \left\| M_{(3)} - RY \right\|_F^2$$
$$+ \lambda \sum_{i=1}^{K} \left\| \mathcal{Q}_i(W) - \mathcal{G}_i \right\|_F^2 + \mu \left\| H_{(3)} - W + \frac{V_1}{2\mu} \right\|_F^2 + \mu \left\| H_{(3)} - Y + \frac{V_2}{2\mu} \right\|_F^2 \tag{8}$$

where V_1 and V_2 are Lagrangian multipliers. μ is a penalty parameter. The solution to optimization problem in (8) can be obtained via alternatively minimizing each manageable sub-problem as follows

1. Update $\mathcal{G}_i$:

$$\min_{\mathcal{G}_i} \left\| \mathcal{Q}_i(W) - \mathcal{G}_i \right\|_F^2, s.t. \mathcal{G}_i = \mathcal{G}_i \times_2 C_i, \forall i \tag{9}$$

It has a closed form solution as

$$\mathcal{G}_i = Q_i(W) \times_2 C_i \tag{10}$$

2. Update W:

$$\min_W \lambda \sum_{i=1}^{K} \|Q_i(W) - \mathcal{G}_i\|_F^2 + \mu \left\| H_{(3)} - W + \frac{V_1}{2\mu} \right\|_F^2 \tag{11}$$

It has a closed form solution as

$$W = \frac{1}{\lambda + \mu} \left(\mu H_{(3)} + \lambda G + \frac{V_1}{2} \right) \tag{12}$$

where G is an intermediate variable by rearranging FBPs from all $\mathcal{G}_i$ to the corresponding positions in HSI and averaging on overlapped portions.

3. Update Y:

$$\min_Y \|M_{(3)} - RY\|_F^2 + \mu \left\| H_{(3)} - Y + \frac{V_2}{2\mu} \right\|_F^2 \tag{13}$$

It has a closed form solution as

$$Y = \left(R^T R + \mu I \right)^{-1} \left(R^T M_{(3)} + \mu H_{(3)} + \frac{V_2}{2} \right) \tag{14}$$

4. Update $H_{(3)}$:

$$\min_{H_{(3)}} \|L_{(3)} - H_{(3)} S\|_F^2 + \mu \left\| H_{(3)} - W + \frac{V_1}{2\mu} \right\|_F^2 + \mu \left\| H_{(3)} - Y + \frac{V_2}{2\mu} \right\|_F^2 \tag{15}$$

It has a closed form solution as

$$H_{(3)} = \left(L_{(3)} S + \mu W + \mu Y - \frac{V_1}{2} - \frac{V_2}{2} \right) \left(SS^T + 2\mu I \right)^{-1} \tag{16}$$

Here we use conjugate gradient (CG) algorithm [23] to effectively solve the above equation.

5. Update Lagrangian Multipliers V_1 and V_2:

$$\begin{cases} V_1 = V_1 + 2\mu \left(H_{(3)} - W \right) \\ V_2 = V_2 + 2\mu \left(H_{(3)} - Y \right) \end{cases} \tag{17}$$

The proposed SPSP method is summarized in **Algorithm 1**. The sub-problems with respect to A, C_i, $\mathcal{G}_i$, W, Y and $H_{(3)}$ have the closed form solution. Then the optimization problem in (8) can be readily solved with a global optimal solution.

Algorithm 1: The proposed SPSP model

Input: LR-HSI $\mathcal{L}$, HR-MSI $\mathcal{M}$.

Initialize: V_1, V_2, μ, η, λ, $H_{(3)}$, W and Y.

   I.   Spectral subspace learning via Eq. (4);

   II.   Self-projected smooth prior on spectral subspace:

         (a). Divide FBPs of $\mathcal{P}$ into K groups;

         (b). Learning C_i via Eq. (6);

   III.   HSI super-resolution via SPSP:

While not converged do

     (1) Update $\mathcal{G}_i$ by Eq. (10);

     (2) Update W via Eq. (12);

     (3) Update Y via Eq. (14);

     (4) Update $H_{(3)}$ via Eq. (16) using CG;

     (5) Update Lagrangian multipliers V_1 and V_2 via Eq. (17);

End while

Output: HR-HSI $\mathcal{H}$.

---

3 Experiment and Results

3.1 Experimental Results on Synthetic Data

Two different public hyperspectral datasets, i.e., the University of Pavia [24] dataset and the CAVE [25] dataset, are chosen to evaluate the performances of the proposed method. Specifically, HSI in University of Pavia dataset is clipped as 256×256 pixels in each band are served as reference images. Then the HSIs in two datasets are normalized into the range of [0, 1]. The LR-HSIs are simulated by down-sampling the HR-HSI with averaging over disjoint 8×8 blocks. The corresponding HR-MSIs are generated by an IKONOS-like spectral reflectance response filter [26] and the response of the Nikon D700 camera, respectively. To be specific, we set $\lambda = 0.001$, $\mu = 0.0004$, and $\eta = 100$ in all experiments. To verify the effectiveness of the proposed method, the proposed SPSP method are compared with four state-of-the-art methods, including GSOMP (Generalization of Simultaneous Orthogonal Matching Pursuit) [13], BSR (Bayesian Sparse Representation) [9], CSU (Coupled Spectral Unmixing) [6], CMS (Clustering Manifold Structure) [8]. For visual comparison, parts of the reconstructed results are illustrated in Fig. 2. It can be observed that the compared methods lack of spatial details and produce some spectral distortions to a certain extent, while the proposed SPSP method provides better reconstructed results when compared to other methods. Furthermore, two quantitative metrics, namely peak signal-to-noise ratio (PSNR) and spectral angle mapper (SAM), are employed to evaluate the spatial and spectral reconstruction quality, respectively. As shown in Table 1, with the best values marked in bold for clarity, the proposed method still outperforms the other methods in terms of spatial and spectral information preservation.

Table 1. The average of PSNR/SAM results of compared methods on hyperspectral data

Methods	BSR	GSOMP	CSU	CMS	SPSP
chart&stuffed_toy	39.4/8.3	39.4/13.7	41.7/8.3	42.9/5.4	**44.1/4.8**
fake&real_lemons	42.4/7.4	43.0/27.4	47.0/4.2	47.3/3.4	**50.6/3.1**
fake&real_peppers	37.1/14.2	39.7/21.5	43.5/6.8	42.5/5.1	**46.8/4.6**
pompoms_ms	34.6/9.8	36.2/9.8	42.2/3.4	41.6/3.3	**44.2/2.7**
superballs_ms	37.7/11.4	38.7/27.2	43.3/8.2	44.8/6.6	**47.6/6.4**
stuffed_toys_ms	36.0/13.4	34.5/21.7	37.1/14.8	39.6/6.4	**44.5/5.2**
University of Pavia	42.5/2.1	40.0/2.5	41.9/1.9	36.7/5.2	**43.6/1.7**

3.2   Experimental Results on Real Data

In order to further validate the effectiveness of the proposed method, the SPSP method is subsequently applied to real HSI super-resolution. The real LR-HSI is collected by the Hyperion sensor onboard of Earth Observing-1 satellite, which has 220 spectral bands in the wavelength regions of 0.4−2.5 m and the spatial resolution of 30 m. 89 bands are preserved after removing the bands which suffer from noise. An area of size $150 \times 150 \times 89$ is used in the subsequent experiment. The real HR-MSI is captured by the Sentinel-2 satellite, which has 13 spectral bands. Four bands with the spatial resolution of 10 m are utilized for the data fusion. The spatial down-sampling factor is 3, and the used Sentinel-2 MSI is of size $450 \times 450 \times 4$. The spectral response matrix

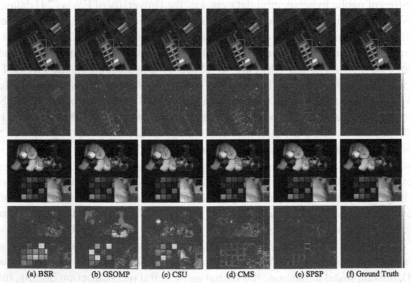

(a) BSR   (b) GSOMP   (c) CSU   (d) CMS   (e) SPSP   (f) Ground Truth

Fig. 2. Visual comparison of reconstructed results using selected HSIs in two datasets. The first and third rows show the reconstructed results of 14-th band; the second and last rows are the corresponding reconstruction error maps of the test HSIs.

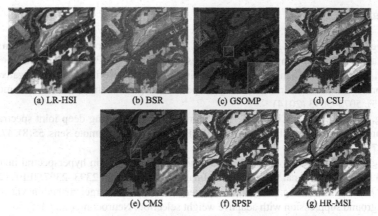

(a) LR-HSI   (b) BSR   (c) GSOMP   (d) CSU

(e) CMS   (f) SPSP   (g) HR-MSI

Fig. 3. The reconstructed HR-HSI at 10-th band by different methods. (a) Hyperion LR-HSI. (b) BSR. (c) GSOMP. (d) CSU. (e) CMS. (f) SPSP. (g) Sentinel-2 HR-MSI.

is estimated using the method in [27]. To be specific, we set $\lambda = 0.001$, $\mu = 0.0004$, and $\eta = 100$. Figure 3(b)-(f) show the reconstructed HR-HSI at 10-th band by different methods. As shown in the figure, all the methods can improve the spatial resolution of the Hyperion LR-HSI to different extent. The methods of GSOMP and CMS have relatively large spectral distortions compared to the others. The spatial details can be well reconstructed by the proposed method SPSP and CSU. Of particular note is the method CMS fails to fully preserve spatial and spectral information, mainly because the non-local self-similarities are directly exploited in the HR-MSI and the spectral dimension of HR-MSI is inconsistent with the intrinsic dimension of the desired HR-HSI. Therefore, the above experiments demonstrate that the proposed SPSP can perform well in the super-resolution of real hyperspectral data.

4 Conclusion

In this paper, a novel SPSP is proposed for the task of HSI super-resolution. The SPSP is adaptively learned to model non-local self-similarities on the intrinsic dimension of the desired HR-HSI by assuming that each group of clustered patches lies in its smooth subspace individually, namely smooth dependences of similar FBPs within each group of HSI. This work represents the first effort to exploit the non-local self-similarities on its spectral intrinsic dimension of desired HSI. In this way, spectral correlations and non-local self-similarities of HSI are incorporated into a unified paradigm to exploit spectral and spatial information simultaneously. As thus, the well learned SPSP is incorporated into the objective function solved by ADMM. Experimental results on synthetic and real hyperspectral data demonstrate the superiority of the proposed method.

Acknowledgements. This work was supported in part by the National Natural Science Foundation of China under Grant 61371152 and Grant 61771391, in part by the Shenzhen Municipal Science and Technology Innovation Committee under Grant JCYJ20170815162956949 and JCYJ20180306171146740.

References

1. Fauvel, M., Tarabalka, Y., Benediktsson, J.A., Chanussot, J., Tilton, J.C.: Advances in spectral-spatial classification of hyperspectral images. Proc. IEEE **101**(3), 652–675 (2013)
2. Wen, J., Zhao, Y., Zhang, X., Yan, W., Lin, W.: Local discriminant non-negative matrix factorization feature extraction for hyperspectral image classification. Int. J. Remote Sens. **35**(13), 5073–5093 (2014)
3. Yang, J., Zhao, Y.-Q., Chan, J.C.-W.: Learning and transferring deep joint spectral-spatial features for hyperspectral classification. IEEE Trans. Geosci. Remote Sens. **55**(8), 4729–4742 (2017)
4. Liang, J., Zhou, J., Bai, X., Qian, Y.: Salient object detection in hyperspectral imagery. In: 2013 IEEE International Conference on Image Processing, pp. 2393–2397. IEEE (2013)
5. Wu, K., Xu, G., Zhang, Y., Du, B.: Hyperspectral image target detection via integrated background suppression with adaptive weight selection. Neurocomputing **315**, 59–67 (2018)
6. Lanaras, C., Baltsavias, E., Schindler, K.: Hyperspectral super-resolution by coupled spectral unmixing. In: Proceedings of the IEEE International Conference on Computer Vision, pp. 3586–3594 (2015)
7. Yi, C., Zhao, Y.-Q., Chan, J.C.-W.: Hyperspectral image super-resolution based on spatial and spectral correlation fusion. IEEE Trans. Geosci. Remote Sens. **56**(7), 4165–4177 (2018)
8. Zhang, L., Wei, W., Bai, C., Gao, Y., Zhang, Y.: Exploiting clustering manifold structure for hyperspectral imagery super-resolution. IEEE Trans. Image Process. **27**(12), 5969–5982 (2018)
9. Akhtar, N., Shafait, F., Mian, A.: Bayesian sparse representation for hyperspectral image super resolution. In: Proceedings of the IEEE Conference on Computer Vision and Pattern Recognition, pp. 3631–3640 (2015)
10. Yi, C., Zhao, Y.-Q., Yang, J., Chan, J.C.-W., Kong, S.G.: Joint hyperspectral superresolution and unmixing with interactive feedback. IEEE Trans. Geosci. Remote Sens. **55**(7), 3823–3834 (2017)
11. Yokoya, N., Yairi, T., Iwasaki, A.: Coupled nonnegative matrix factorization unmixing for hyperspectral and multispectral data fusion. IEEE Trans. Geosci. Remote Sens. **50**(2), 528–537 (2012)
12. Zhao, Y., Yang, J., Chan, J.C.-W.: Hyperspectral imagery super-resolution by spatial-spectral joint nonlocal similarity. IEEE J. Sel. Top. Appl. Earth Observations Remote Sens. **7**(6), 2671–2679 (2014)
13. Akhtar, N., Shafait, F., Mian, A.: Sparse spatio-spectral representation for hyperspectral image super-resolution. In: Fleet, D., Pajdla, T., Schiele, B., Tuytelaars, T. (eds.) ECCV 2014. LNCS, vol. 8695, pp. 63–78. Springer, Cham (2014). https://doi.org/10.1007/978-3-319-10584-0_5
14. Dian, R., Fang, L., Li, S.: Hyperspectral image super-resolution via non-local sparse tensor factorization. In: Proceedings of the IEEE Conference on Computer Vision and Pattern Recognition, pp. 5344–5353 (2017)
15. Xu, Y., Wu, Z., Chanussot, J., Wei, Z.: Nonlocal patch tensor sparse representation for hyperspectral image super-resolution. IEEE Trans. Image Process. **28**(6), 3034–3047 (2019)
16. Dian, R., Li, S.: Hyperspectral image super-resolution via subspace-based low tensor multi-rank regularization. IEEE Trans. Image Process. **28**(10), 5135–5146 (2019)
17. Dian, R., Li, S., Fang, L.: Learning a low tensor-train rank representation for hyperspectral image super-resolution. IEEE Trans. Neural Netw. Learn. Syst. **30**(9), 2672–2683 (2019)
18. Bioucas-Dias, J.M., Nascimento, J.M.: Hyperspectral subspace identification. IEEE Trans. Geosci. Remote Sens. **46**(8), 2435–2445 (2008)
19. He, W., Yao, Q., Li, C., Yokoya, N., Zhao, Q.: Non-local meets global: an integrated paradigm for hyperspectral denoising. arXiv preprint arXiv:1812.04243 (2018)

20. Kolda, T.G., Bader, B.W.: Tensor decompositions and applications. SIAM Rev. **51**(3), 455–500 (2009)
21. Hu, H., Lin, Z., Feng, J., Zhou, J.: Smooth representation clustering. In: Proceedings of the IEEE Conference on Computer Vision and Pattern Recognition, pp. 3834–3841 (2014)
22. Boyd, S., Parikh, N., Chu, E., Peleato, B., Eckstein, J., et al.: Distributed optimization and statistical learning via the alternating direction method of multipliers. Found. Trends R Mach. Learn. **3**(1), 1–122 (2011)
23. Golub, G.H., Van Loan, C.F.: Matrix Computations, vol. 3. JHU press, Baltimore (2012)
24. Dell'Acqua, F., Gamba, P., Ferrari, A., Palmason, J.A., Benediktsson, J.A., Arnason, K.: Exploiting spectral and spatial information in hyperspectral urban data with high resolution. IEEE Geosci. Remote Sens. Lett. **1**(4), 322–326 (2004)
25. Yasuma, F., Mitsunaga, T., Iso, D., Nayar, S.: Generalized assorted pixel camera: post-capture control of resolution, dynamic range and spectrum. Technical report, Department of Computer Science, Columbia University CUCS-061-08, November 2008
26. Wei, Q., Dobigeon, N., Tourneret, J.-Y.: Fast fusion of multi-band images based on solving a Sylvester equation. IEEE Trans. Image Process. **24**(11), 4109–4121 (2015)
27. Simoes, M., Bioucas-Dias, J., Almeida, L.B., Chanussot, J.: A convex formulation for hyperspectral image superresolution via subspace-based regularization. IEEE Trans. Geosci. Remote Sens. **53**(6), 3373–3388 (2015)

3D Point Cloud Segmentation for Complex Structure Based on PointSIFT

Zeyuan Li, Jianzong Wang[✉], Xiaoyang Qu, and Jing Xiao

Ping An Technology (Shenzhen) Co., Ltd., Shenzhen 518063, China
{lizeyuan950,wangjianzong347,quxiaoyang343,
xiaojing661}@pingan.com.cn

Abstract. To solve the problem of the three-dimensional point clouds segmentation of complex structure, a new segmentation algorithm by discrete unit sampling module (DUSM) based on PointSIFT is proposed. Due to the inherent disorder and density difference of 3D point cloud, as well as the redundant surrounding noise points, which make some limitations on the representation and segmentation of complex components. Therefore, based on the PointSIFT input parameters and point cloud collection method, this paper improves the algorithm and completes the point cloud segmentation of complex structure. In this paper, PointSIFT that the point cloud segmentation network is improved appropriately. Secondly, point clouds data that selecting hull structure as a complex structure is derived from the CAD and annotation, which forms the training and validation data sets. Finally, the improved algorithm was adopted to complete the training and verification for the point cloud segmentation network model. The accuracy rate is 81.7% by verification. Experimental results show that the improved point cloud segmentation network model can be applied to the segmentation of complex structure, and has a good generalization effect.

Keywords: 3D point cloud · PointSIFT · DUSM · Complex structure · Point cloud segmentation

1 Introduction

In recent years, with the increasing use of point cloud-based intelligent applications such as autonomous driving, medical diagnosis, augmented reality and mixed reality, research on deep learning in 3D point cloud segmentation technology has become particularly urgent and important, especially for complex components Point cloud segmentation [1] research is still in the blank stage. According to the common point cloud task scene division, including object classification, target recognition, semantic segmentation, and instance segmentation. Among them, the point cloud already has a basic model that deals with image scenes in two-dimensional space and the segmentation effect is remarkable [2, 3], but in the three-dimensional spatial relationship, the point cloud is unstructured and disordered, and there is translational rotation The invariance of the category, as well as the interference of environmental noise. In addition, due to the imbalance of

Y. Peng et al. (Eds.): PRCV 2020, LNCS 12305, pp. 660–670, 2020.
https://doi.org/10.1007/978-3-030-60633-6_55

the sampling and division of the point cloud in the model itself, the random selection of the center point cannot guarantee that the point cloud can cover the target field, that is, it cannot better characterize and characterize the surface of complex components. These problems will make 3D point cloud segmentation research more challenging, and also put forward higher requirements for the robustness of the model calculation without the interference of point cloud noise. In this regard, the literature shows that three-dimensional point cloud data can be represented by a set of images captured from different viewpoints [4], it can also be simulated into a regular grid or voxel form [5, 6], to solve the problem of point cloud disorder, but these expressions exist before input to the neural network because of local information loss and object self-occlusion, etc., which will cause quantization loss and reduce accuracy. Compared with the former, the point cloud itself provides more complete data information and is built directly on its neural network, which also shows good results in the end-to-end segmentation task.

In response to the above problems, this paper based on PointSIFT neural network [7] from the point cloud parameters and sampling methods to improve the point cloud segmentation algorithm in two aspects, different from the traditional random center sampling method, this article will be based on the sampling point cloud increment, the 3D cells are divided for quantitative discrimination. Taking the hull segmented point cloud data set as an example, a fast and accurate logical division is realized for the point cloud of complex large-scale components to ensure that the target object is well characterized and input into the design network. To improve the recognition accuracy of point cloud segmentation, the specific network structure is shown in Fig. 1. It shows how sampling point clouds in the discrete unit format can continuously achieve feature abstraction and propagation in the design network. The major contributions of this paper are summarized as below.

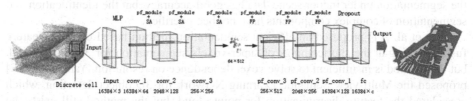

Fig. 1. Network structure diagram base on PointSIFT

1) An improved algorithm for point cloud segmentation of complex components is proposed. The point cloud is sampled in a unit format. While ensuring data integrity, it can also be completely embedded in the feature vector, thereby achieving the increase in the accuracy of point cloud segmentation recognition.
2) A large number of comparative experiments were carried out by applying the actual measured hull segmentation point cloud data set to verify the rationality of the proposed method, thus proving that the method has certain guiding significance for practical application.

Section 2 of this paper introduces the relevant technologies of 3D point cloud segmentation; Sect. 3 defines the research problems of this paper and makes the corresponding network structure Design; Sect. 4 introduces the point cloud data collection and annotation process; Sect. 5 verifies the accuracy and segmentation efficiency of the algorithm; finally summarizes the full text and proposes the next step.

2 Related Work

Deep learning of 3D data is currently a hot research field. In this paper, 3D data segmentation methods such as voxels, multi-views, and point clouds are studied. Due to the intuitive nature of point cloud data, it is easy to expand to high-dimensional data. Feature information to better characterize the surface of complex components. In this paper, the point cloud segmentation is studied in detail.

At present, a series of work has focused on deep learning methods for directly inputting point cloud data for processing.

Specifically, Shi et al. [8] pointed out the characteristics and applicable scenarios of various point cloud data segmentation algorithms, extending the 3D point Cloud concept; Charles et al. [9, 10] proposed the PointNet network structure, which pioneered the direct input of point clouds, and transformed the disordered point cloud into a regular form of expression through the T-Net network architecture. The network MLP outputs the category of each point. However, PointNet only extracts global point cloud features, and does not capture local structural information at multiple scales; in order to solve this problem, the team proposed the PointNet++ network, which extends the PointNet feature extraction block and adds a layered structure. In order to deal with local features, good segmentation results were obtained; on this basis, Lu et al. [7] proposed PointSIFT neural network module based on SIFT operator direction coding and size perception characteristics, which can be automatically optimized during the training process, and the segmentation under indoor scene data has good accuracy, but the identification and segmentation of complex components have not been verified.

Ren et al. [11] proposed an efficient segmentation method based on multi-feature fusion, aiming at the shortcomings of super-voxels segmentation for 3D point cloud, but the method is insufficient to solve voxel dependence on resolution. Yang et al. [12] proposed the Multi-view Semantic Learning Network for 3D object detection, which considered the feature discrimination for point cloud, but the method still exists the problem of information loss in multi-view representation. Jaritz et al. [13] proposed an MVPNet network framework, which is the integration of dense feature information of two-dimensional images and three-dimensional point cloud data. Although the sparseness of three-dimensional point cloud data is improved, there is no departure from the way of extracting features from two-dimensional images. Pham et al. [14] established a multi-task point-by-point network to represent the point cloud of the same instance. At the same time, combined with a multi-valued conditional random field model, the point cloud was segmented and verified by joint optimization. Wu et al. [15] extended the dynamic filter to a new convolution operation and constructed the PointConv deep convolution network, which performs operations on the 3D point cloud, and has a better result while reducing the occupation of video memory. However, for the sampling enhancement problem of point cloud segmentation of complex components, the research work is not mentioned, and further research is still needed.

3   3D Point Cloud Segmentation Network Structure Design

3.1   Problem Description

Three-dimensional point clouds are unstructured data. Under the influence of point cloud acquisition equipment, the distribution of point clouds used to describe the same object in the coordinate system is different. Assuming that there is a matrix composed of N point clouds, then there are $N!$ distributions. This way of distribution makes the point clouds have no fixed order when entering the network. Even so, the different point cloud matrices represent the same object, which requires that the features extracted from the point cloud by the network must be consistent.

Besides, the point clouds have the category invariance of rigid transformation. When the target object rotates and translates in the three-dimensional space, the point cloud coordinates will change correspondingly, but the object type is not affected. In this regard, this requires that the models can extract the same features of the same objects under different point cloud coordinates. In addition, due to the difference in the viewing angle of the target component from close to far, the observed point cloud densities will be different. Therefore, to better characterize the surface of complex components, the enhancement of the point cloud samples is required.

3.2   Objective Function

Defining the objective function of the network based on the above description. Firstly, assuming that

$$\mathbf{x} = (N, D) \in R^n, \tag{1}$$

where R^n denotes the discrete metric space, $N \subseteq R^n$ refer to the number of point clouds, D represents the feature dimension of each point, and the density of N in the discrete metric space is non-uniform. In order to obtain geometric information without loss from the disordered point cloud, a symmetric function g needs to be constructed that maps each point with three-dimensional information to a redundant high-dimensional space. Here, taking x and its features as input, and then using transformation function f to label and segment each point in $\mathbf{x}$. Based on the above assumptions, the unordered point cloud data sets $\mathbf{X} = \{\mathbf{x}_1, \mathbf{x}_2, \dots, \mathbf{x}_n\} \in R^n$ are defined.

The transformation function can be expressed as

$$f(\mathbf{x}_1, \mathbf{x}_2, \dots, \mathbf{x}_n) \approx g(h(\mathbf{x}_1), \dots, h(\mathbf{x}_n)), \tag{2}$$

where $f{:}2^{(R^N)} \to R$, $h{:}R^N \to R^K$, $g{:}R^{(K_1)} \times \dots \times R^{(K_n)} \to R$ denotes a symmetric function that is realized by maximum pooling, and $h(\bullet)$ denotes the multi-layer perceptron MLP which used for feature extraction. Then the set of single-valued functions are fed to the maximum pooling layer under the high-dimensional space g. Finally, the network further learn the point cloud information to obtain the attributes of the point cloud collection through γ, which can be expressed as

$$f(\mathbf{x}_1, \mathbf{x}_2, \dots, \mathbf{x}_n) = \gamma(\max_{i=1,2,\dots n} \{h(\mathbf{x}_i)\}), \tag{3}$$

where $\gamma(\bullet)$ and $h(\bullet)$ belong to the MLP network structure.

3.3 Model Structure

In order to make the three-dimensional point better describe the morphological characteristics of the target objects, it is necessary to introduce the neural network architecture,
which is used to process the sampling point set in the metric space and effectively extract
the features of input point set. Inspired by the feature descriptor SIFT, the parameterized deep learning module is nested in our model. Through the key attributes of scale
perception and direction coding, the proposed network can realize the scale-invariant
information coding of the 3D point cloud in different directions and complete the point
cloud segmentation.

Specifically, the SIFT feature descriptor considers two basic features of morphological expression. One of the basic features is direction coding that assigns directions to
each point after obtaining the matching feature point position, the other is scale perception that selects the most suitable size for feature extraction in the target image. Different
from the artificially designed SIFT, PointSIFT is a neural network module, which can
realize self-optimization according to the pre-training process. The basic module of
PointSIFT is the OE unit (orientation encoding unit), which can convolve in 8 directions
and extract features.

In order to better obtain the characteristic information of the point cloud, this paper
stacks information from different directions based on PointSIFT. First, dividing the three-
dimensional space into eight subspaces based on the center point p_n, and every subspace
contains eight different directions information. For center point p_n and the corresponding
$n \times d$ dimensional feature f_n, finding the nearest neighbor k_n of p_n to gain the characteristics of the neighbor points of the subspace. If there is no target point within the search
radius within a certain subspace, the available features f_n are used to represent eight subspaces. At the same time, in order for convolution to perceive direction information, the
convolutional operations are conducted on X, Y, Z axis, separately. The features of the
searched nearest neighbors k_n are coded counting into the tensor $N \subseteq R^{2 \times 2 \times 2 \times d}$. Here,
$R^{a \times b \times c}$ denotes three dimensions space corresponding to X, Y, Z. The convolutional
operations can be expressed as

$$
\begin{aligned}
N_1 &= g[Conv_x(A_x, N)] \in R^{2 \times 2 \times 1 \times d} \\
N_2 &= g[Conv_y(A_y, N)] \in R^{2 \times 1 \times 1 \times d}, \\
N_3 &= g[Conv_z(A_z, N)] \in R^{1 \times 1 \times 1 \times d}
\end{aligned}
\tag{4}
$$

where A_x, A_y, A_z represents the convolution weight to be optimized,
$Conv_x, Conv_y, Conv_z$ denote the convolution process along the X, Y, Z axis direction.

After three convolution stacks, each p_n point will be converted into one d dimension
vector, which will contain p_0 shape information of nearby neighborhoods. It can be seen
that by stacking multiple directional coding units through convolution, the directional
coding units of different convolutional layers can perceive the scale information in each
direction, and then connect the directional coding units of the previous layers through
shortcuts to extract the final feature information of the invariable scale, thus solving the
problem of point cloud disorder and invariance.

The network architecture process mainly includes two major stages of downsampling
and upsampling. The initial input of the network is the randomly sampled coordinates

and RGB information of 8192 points. In the downsampling stage, through using the SA (Set Abstraction) module in the PointNet++ network, the multi-layer perceptron MLP converts the input data into 64-dimensional feature vectors after three consecutive downsampling processes. In the upsampling stage, the FP (feature propagation) module in the PointNet++ network is used to complete feature aggregation through three upsampling processes aligned with downsampling. In the process of stage execution, this paper intersects the PointSIFT module between adjacent SA modules and FP modules, and finally passes all point cloud information through a fully connected layer to realize the realization of each point cloud with semantic tags prediction.

At the same time, considering that point cloud segmentation involves multiple classification problems, this paper designs a neural network with two different neurons at the output layer of the proposed network. The proposed network adopts the softmax activation function shown in Eq. (5) to classify and the cross-entropy loss function shown in Eq. (6) to optimize the network.

$$
\begin{bmatrix} a_1 \\ a_2 \end{bmatrix} \rightarrow \begin{bmatrix} \frac{e^{-1}}{e^{a_1}+e^{a_2}} \\ \frac{e^{a_2}}{e^{a_1}+e^{a_2}} \end{bmatrix}, \tag{5}
$$

where $\begin{bmatrix} a_1 \\ a_2 \end{bmatrix}$ represents the vector before the activation function of the last layer of the network, $\begin{bmatrix} \frac{e^{a_1}}{e^{a_1}+e^{a_2}} \\ \frac{e^{a_2}}{e^{a_1}+e^{a_2}} \end{bmatrix}$ represents the output vector of the last layer of the network after the activation function, and each item in the vector can represent each probability of category. If $\begin{bmatrix} \frac{e^{a_1}}{e^{a_1}+e^{a_2}} \\ \frac{e^{a_2}}{e^{a_1}+e^{a_2}} \end{bmatrix}$ can be expressed as $\begin{bmatrix} \hat{y}_1 \\ \hat{y}_2 \end{bmatrix}$, the loss function of one sample can be expressed as Eq. (6).

$$
loss = -\sum_{i=1}^{2} y_i \log \hat{y}_i, \tag{6}
$$

where y_i denotes the one-hot value of the $i-th$ component. For example, if the sample label is 0, the label's on-hot vector is $\begin{bmatrix} 1 \\ 0 \end{bmatrix}$, if its label is 1, the label's one-hot vector is $\begin{bmatrix} 0 \\ 1 \end{bmatrix}$, and $\hat{y}_i$ represents the $i-th$ component of the output vector in the last layer.

In addition, the complex components are different from the indoor scene data set oriented by the PointSIFT network structure. In order to ensure the validity of the training data and maintain the characteristics of the large-scale shape components, the input network point cloud parameters of each sample are increased from 8192 to 16384. Besides, the different scales of the point set are transferred to the same scale as the original data set to improve the generalization ability of the model. In addition, to avoid the extraction of sparse point clouds in the sampling of point clouds characterizing complex components and eliminate the invisible problem of low point cloud density, this paper designed a discrete unit sampling module (DUSM). Thus, the network enable us to perceive the local neighborhood point set density and adaptively select the target center sampling point. Specifically, randomly select a point as the center point for each sample in the data set, and select point cloud data within ±0.01 range around it, and divide all point cloud data discretely into 20 * 20 * 20 spatial cells to obtain all point clouds in the cell, as shown in Fig. 2. If the volume of the point cloud in the selected range can account for more than 1% of the volume of all cells (including the point cloud), the sampling center point is retained, and then 16384 points from each cell are randomly selected and included in the model for training and verification.

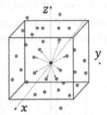

Fig. 2. Discrete unit grid diagram

4   3D Point Cloud Segmentation Model Experiment

4.1   Point Cloud Data Collection and Annotation

The construction of the data set has important significance for the iterative solution of the model and the final effect. The larger the data set, the more types and the more balanced, the faster the iterative solution of the model, and the better the final model effect. Therefore, it is necessary to construct a set of point cloud data sets of complex components. Point clouds data set that contains a complex component of the hull block is selected and established in this paper [16].

At this stage, there are usually two ways to obtain point cloud data sets of complex components: one is to use mature laser scanning technology to complete the point cloud data collection on site, and the other is to output the point cloud through the 3D model design drawing. Considering the factors such as the difficulty of collecting work on site by laser scanners and the high cost, this article adopts the second method. Use Tribon modeling software to model the hull segment, export the dxf file, and then manually mark it, specifying label 0 as a strong component (red), such as a flat plate; label 1 as a normal component (green), such as longitudinal bone, analysis the marked files to obtain point cloud data, as shown in Fig. 3.

In order to increase the data volume of the training model and accelerate the speed of model optimization and convergence, this paper enhances the point cloud data of hull segmentation through geometric transformation of arbitrary angles. Finally, this paper prepares 1000 marked point cloud data of complex components, and sets it as the training set and the test set according to the 7:3 ratio, which are used to iteratively solve the point cloud model and verify the effect.

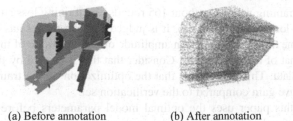

(a) Before annotation   (b) After annotation

Fig. 3. Sketch map of complex structure before and after annotation (Color figure online)

4.2 Model Training

At the beginning of model training, the parameters of the network model are initialized to random values. Therefore, in order to realize the point cloud segmentation of the complex components of the model, this paper will iteratively update the neural network parameters according to the loss function.

The training parameters of the model in this paper adopt an initial learning rate of 0.001, corresponding to a change rate of 0.01, a batch size of 8, and a training network of Epoch of 250. The test set is set to be evaluated every 5 rounds.

4.3 Point Cloud Segmentation Experiment Results

After iterative calculations of multiple rounds of training, the learning rate represented by Eq. (7) will decrease gradually, and this article selects the best training curve from several iteration calculations, as shown in Fig. 4.

$$lr = lr \times 0.68^{\frac{n}{100000}}, \tag{7}$$

where, 0.68 represents the decay rate of the learning rate, lr indicates the learning rate, n indicates the current number of training rounds on the network.

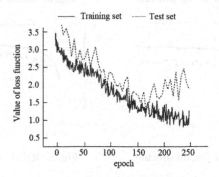

Fig. 4. Curve of loss function value

As the number of training rounds increases, the loss function values of the training set and the verification set oscillate at a high frequency and continue to decline. When the number of training rounds is about 165 rounds, the training loss still decreases, and the verification loss starts to pick up. It is judged that the model has overfitted at this time. At the same time, the oscillation amplitude of the loss value of the verification set is larger than that of the training set. Consider that this is caused by the small amount of verification data. This also means that the optimization on the training set may not produce a positive gain compared to the verification set.

Therefore, this paper uses the optimal model parameters before overfitting, and evaluates the optimal results of the model on the test set, through the two indicators of accuracy and recall [17].

$$\begin{cases} P = \frac{TP}{TP+FP} \times 100\% \\ R = \frac{TP}{TP+FN} \times 100\% \end{cases}, \tag{8}$$

where, P is the accuracy rate, R is the recall rate, TP is a true example, FP is a false positive example, FN is a false counterexample.

According to Eq. (8), the accuracy of the improved point cloud model on the test set is 81.7%. Among them, the accuracy rate of the component predicted to be label 1 and label 2 is 85.1% and 71.3%, the recall rate is 78.2% and 80.1%, as shown in Figs. 5, 6, 7, 8 shown. It can be seen that the recognition accuracy of the model performs better in the test set.

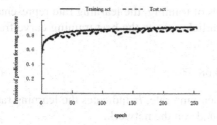

Fig. 5. Precision for strong structure

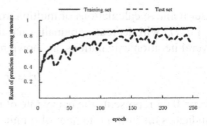

Fig. 6. Recall for strong structure

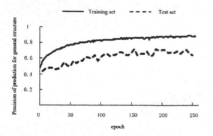

Fig. 7. Precision for general structure

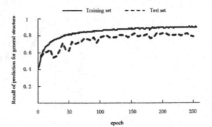

Fig. 8. Recall for general structure

Further, under the data set that a complex component of hull block, some point cloud segmentation models of related work are used for comparative experiments to obtain multiple accuracy indicators, which also shows that the improved algorithm has a certain degree of point cloud segmentation. The improvement is shown in Table 1.

Table 1. Accuracy, precision and recall rate under complex component data set.

Method	Accuracy/%	Precision rate/%	Recall rate/%
PointNet [9]	76.6	73.6	76.2
PointNet++ [10]	78.3	75.4	77.1
PointConv [15]	79.5	76.1	**82.3**
PointSIFT [7]	80.2	77.5	80.1
PointSIFT + DUSM (Ours)	**81.7**	**78.2**	79.1

On the other hand, the verification results are visualized, as shown in Fig. 9. Compared with the effect of manual segmentation (Fig. 9(a)), it can be seen that the model has been able to classify complex components, especially for large-area flat structures and horizontal and longitudinal stiffeners (Fig. 9(b)), and it is also more robust to blind spots caused by viewing angles. However, it can be seen from Fig. 9(b) and Fig. 9(c), there will be a certain error in the segmentation, which is more likely to occur at the connection between the strong structure and the general structure, because of it is not much different from that of general component, resulting in misidentification.

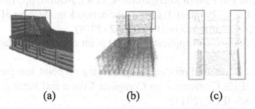

(a)   (b)   (c)

Fig. 9. Visualization of production effect for model test set

5 Conclusion

Aiming at the problem of 3D point cloud segmentation of complex components, this paper proposes an improved PointSIFT-based 3D point cloud segmentation algorithm, a point cloud method of discrete unit sampling, and the improvement of point cloud parameters to solve the inherent disorder of 3D point cloud at the same time that the rotation and the translation are not deformed, the enhanced sampling of the point cloud is realized to improve the density and effectiveness of the model into the parameter point cloud. Furthermore, the complex component data set of hull block is selected, and the improved algorithm model is compared with PointNet, PointNet++, PointConv, and PointSIFT. The accuracy of point cloud recognition for complex components is improved respectively by 5.1%, 3.4%, 2.2%, and 1.5%. Experimental results show that the improved point cloud segmentation network model can be applied to the segmentation of dense morphological structures of complex components, which improves the network's ability to recognize fine targets and generalize complex scenes. In addition, there are still some problems for improvement in the research process, mainly to further refine the sampling granularity at the intersection of the components, and to better eliminate the problem of cross misidentification by framing the area and performing local enhancements. Meanwhile, the optimization of training data labels can also solve the problem.

References

1. Zhang, X., Liu, J., Shi, Z., et al.: Review of deep learning based semantic segmentation. Laser Optoelectron. Prog. **56**(15) (2016)
2. Chen, L., Papandreou, G., Kokkinos, I., et al.: DeepLab: semantic image segmentation with deep convolutional nets, atrous convolution, and fully connected CRFs. IEEE Trans. Pattern Anal. Mach. Intell. **40**(4), 834–848 (2016)
3. He, K., Georgia, G., Piotr, D., et al.: Mask R-CNN. IEEE Trans. Pattern Anal. Mach. Intell. **42**(2), 386–397 (2018)
4. Hang, S., Maji, S., Kalogerakis, E., et al: Multi-view convolutional neural networks for 3D shape recognition. arXiv preprint arXiv:1505.00880 (2015)
5. Zhi, S., Liu, Y., Li, X., et al.: LightNet: a lightweight 3D convolutional neural network for real-time 3D object recognition. In: Eurographics Workshop on 3D Object Retrieval, Lyon, pp. 009–016 (2017)
6. Wu, Z., et al.: 3D ShapeNets: a deep representation for volumetric shapes. In: IEEE Conference on Computer Vision and Pattern Recognition (CVPR), Boston, pp. 1912–1920. IEEE (2015)
7. Jiang, M., Wu, Y., Lu, C., PointSIFT: a SIFT-like network module for 3D point cloud semantic segmentation. arXiv preprint arXiv:1807.00652 (2018)
8. Shi, Y., Chu, Z.: Principle and application of 3D point cloud segmentation. Sci. Technol. Inf. (24) (2016)
9. Qi, C., Su, H., Mo, K., et al.: PointNet: deep learning on point sets for 3D classification and segmentation. In: IEEE Conference on Computer Vision and Pattern Recognition (CVPR), Honolulu, pp. 77–85. IEEE (2017)
10. Qi, C., Yi, L., Su, H., et al.: PointNet++: deep hierarchical feature learning on point sets in a metric space. In: Proceedings of the Conference on Neural Information Processing Systems, Long Beach, pp. 77–85. NIPS (2017)
11. Ren, X., Wang, W., Xu, S.: An innovative segmentation method with multi-feature fusion for 3D point cloud. J. Intell. Fuzzy Syst. **38**(1), 345–353 (2020)
12. Yang, Y., Chen, F., Wu, F., Zeng, D., et al.: Multi-view semantic learning network for point cloud based 3D object detection. Neurocomputing **397**, 477–485 (2020)
13. Jaritz, M., Gu, J., Su, H.: Multi-view PointNet for 3D scene understanding. arXiv preprint arXiv:1909.13603 (2019)
14. Pham, Q., Duc, T., et al.: JSIS3D: joint semantic instance segmentation of 3D point clouds with multi-task pointwise networks and multi-value conditional random fields. In: 2019 IEEE Conference on Computer Vision and Pattern Recognition (CVPR), Long Beach, pp. 8827–8836. IEEE (2019)
15. Wu, W., Qi, Z., Li, F., PointConv: deep convolutional networks on 3D point clouds. In: 2019 IEEE Conference on Computer Vision and Pattern Recognition (CVPR), Long Beac, pp. 9621–9630. IEEE (2019)
16. Cheng, S., Wang, J., Liu, Y., Zhang, X.: Intelligent recognition of block erection surface based on PointNet++. Mar. Eng. **41**(12), 138–141 (2019)
17. Yang, B., Dong, Z., Liu, Y., et al.: Computing multiple aggregation levels and contextual features for road facilities recognition using mobile laser scanning data. ISPRS J. Photogramm. Remote Sens. **126**, 180–194 (2017)

Completely Blind Image Quality Assessment with Visual Saliency Modulated Multi-feature Collaboration

Min Zhang[1](✉), Wenjing Hou[2], Xiaomin Xu[1], Lei Zhang[2], and Jun Feng[2]

[1] School of Mathematics, Northwest University, Xi'an, China
dr.zhangmin@nwu.edu.cn
[2] School of Information Science and Technology, Northwest University, Xi'an, China

Abstract. It has been proved that opinion-unaware (OU) blind image quality assessment (BIQA) methods are of good generalization capability and practical usage. However, so far, no OU-BIQA methods have demonstrated a very high prediction accuracy. In this paper, we proposed a novel OU-BIQA method which incorporated a collection of quality-aware statistical features that are highly related to the distortions of image local structure, the hue, the contrast and the natural scene statistics (NSS). Then, to take the benefit from the psychophysical characteristics of the human visual system (HVS), a new fitting model from the visual saliency modulated multivariate gaussian distribution is proposed to fit these features to predict the perceptual image quality by comparing with a standard model learned from natural image patches. The experimental results show that the proposed method has excellent performance and good generalization capacity.

Keywords: Image quality assessment · Visual attention · Image saliency · Opinion unaware · Multivariate gaussian

1 Introduction

Various multimedia technologies have made tremendous progress, and perceived quality evaluation of visual media has drawn significant attention. However, precise evaluation of image quality by large groups of human subjects is an inconvenient, time-consuming task. Therefore, devising models to evaluate human-perceived quality automatically is challenging.

No-reference/Blind (NR/Blind) Image Quality Assessment (IQA) has gained considerable importance, aiming to predict image quality without any prior knowledge of the original image and distortion types. Most existing BIQA methods are opinion-aware (OA), which means these metrics were trained using a regression model applied to a labeled dataset [1]. Algorithms like [2–6] are OA-BIQA methods. However, these methods rely on samples with manual subjective labels, whose generalization capability are not satisfying.

Current progress in OU-BIQA algorithms is limited. Mittal et al. [1] proposed an OU-BIQA based on constructing a collection of 'quality aware' features from the spatial

© Springer Nature Switzerland AG 2020
Y. Peng et al. (Eds.): PRCV 2020, LNCS 12305, pp. 671–680, 2020.
https://doi.org/10.1007/978-3-030-60633-6_56

domain natural scene statistics (NSS) and fitting them to a multivariate Gaussian (MVG) model. The quality of a test image is evaluated by the distance between its parameters, which describes the MVG model and that of the MVG model learned from a corpus of pristine images. Lin Zhang et al. [7] proposed an OU-BIQA method which enriched the 'quality-aware' feature set with four more natural image statistics features, including spatial NSS, gradient, statistics of log-Gabor filter responses and the color derived from multiple cues to learn a multivariate Gaussian (MVG) model of pristine images. Min Zhang el al. [8] proposed a novel OU-BIQA method based on a simple and successful local structure descriptor which is called generalized local binary pattern (GLBP) to reflect the statistics of image local structures.

From above, the statistical features from the low-level vision of human visual system (HVS) play an important role in these methods. The characteristics of middle to high level vision, such as attention mechanism, have not been taken into consideration yet. It is known to us, the HVS is highly complex. The visual attention mechanism of the HVS indicates that human observers tend to concentrate more on salient areas than on other areas in static and dynamic scenes automatically. In effect, attention operates to filter out irrelevant information from nearby distracters. The degradation of images may bias of the sensitivity of visual features perception.

In this paper, a bag of quality-aware features including multi-order NSS, local structure features, image color statistics and contrast features are extracted for quality assessment. Meanwhile; going beyond the feature extraction and integration, the top-down influence of attention mechanism is incorporated to enhances perceptual information processing of difference features. The saliency computational model is employed to mimic the attention mechanism and used to modulate the visual quality prediction distance. The experimental results demonstrate that the performance of the proposed method is comparable with the current state-of-the-art OU-BIQA methods.

The rest of the paper is organized as follows. Section 2 introduces the general framework of the proposed method including visual feature descriptors and the proposed fitting model. Section 3 presents the experimental results, and Sect. 4 concludes the paper.

2   Method

2.1   The General Framework of the Proposed Method

The general framework of the proposed method is shown in Fig. 1. In the training stage, we extract the visual features from the natural scene statistics, local structure, color and local contrast from a collection of high-sharpness image patches from a pristine image set. Then the multivariate Gaussian distribution model is utilized to fit the feature vector to form the standard visual quality model. In the testing stage, the distorted image is divided into patches of the same size. And we extract quality-aware features from these distorted image patches to form a quality-aware feature vector. Then we propose a saliency modulated multivariate gaussian (S-MVG) model to fit the feature statistical distribution of the distorted images. Finally, the perceptual quality of the distorted image is evaluated as the distance between MVG distribution of the feature statistics from the pristine image set and the saliency modulated MVG distribution from the distorted image. In this work, the image is partitioned into patches with the size of $P \times P$, and the pristine image dataset and high-sharpness image patches selection strategy are those given in [1].

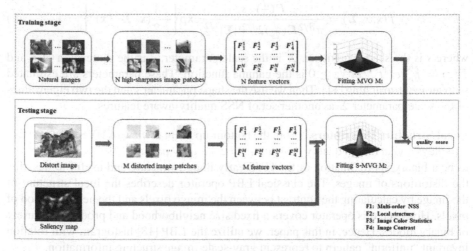

Fig. 1. The general framework of the proposed method.

2.2 Visual Feature Descriptors

In this section, we enumerate the different visual features to formulate the final form of the proposed method.

Multi-order Natural Scene Statistical Features. It has been proved that NSS are potent indicators of the severity of image distortion [1–6]. For example, the statistics of Mean Subtracted Contrast Normalized (MSCN) coefficients have been employed in some IQA metrics [1, 3, 6, 7]. MSCN is computed as:

$$\hat{I}(i,j) = \frac{I(i,j) - \mu(i,j)}{\sigma(i,j) + 1} \tag{1}$$

where $I(i,j)$ and $\hat{I}(i,j)$ denote the original and MSCN coefficients of a gray-scale image at the position (i,j), $\mu(i,j)$ is the local mean and $\sigma(i,j)$ is the standard deviation [1].

As pointed out in [1, 3], image distortion degree can be captured by the distribution of the products of pairs of adjacent MSCN coefficients computed along the vertical, horizontal and diagonal direction $\hat{I}(i,j)\hat{I}(i+1,j)$, $\hat{I}(i,j)\hat{I}(i,j+1)$, $\hat{I}(i,j)\hat{I}(i+1,j+1)$ and $\hat{I}(i,j)\hat{I}(i+1,j-1)$. These products of neighboring coefficients are well-modeled as follow a zero-mode Asymmetric Generalized Gaussian Distribution (AGGD) [10]. The parameters $(\gamma, \beta_l, \beta_r, \eta)$ of AGGD are extracted as the quality-aware features.

Furthermore; the second-order NSS are also included [11, 12]. We utilize the Kotz-type distribution [12] to fit the two-dimensional histograms of the pairs of adjacent

MSCN coefficients $(\hat{I}(i,j), \hat{I}(i+1,j))$, $(\hat{I}(i,j), \hat{I}(i,j+1))$, $(\hat{I}(i,j), \hat{I}(i+1,j+1))$ and $(\hat{I}(i,j), \hat{I}(i+1,j-1))$. The Kotz-type distribution is defined as:

$$f(\mathbf{x}; s, \boldsymbol{\Sigma}) = \frac{\Gamma(\frac{d}{2}) \cdot s}{\pi^{d/2} \Gamma(\frac{d}{2s}) 2^{d/2s} |\boldsymbol{\Sigma}|^{1/2}} \exp\left\{-\frac{1}{2}(\mathbf{x}^T \boldsymbol{\Sigma}^{-1} \mathbf{x})^s\right\} \tag{2}$$

where s is the shape parameter, $\boldsymbol{\Sigma}$ is the scale parameter, d is the dimension of $\mathbf{x}$ and $\Gamma(z) = \int_0^\infty e^{-t} t^{z-1} dt \ \forall z \geq 0$ is the gamma function. These parameters are estimated with moment-matching [11]. Then we use the shape parameter s and the two eigenvalues of the scale parameter $\boldsymbol{\Sigma}$ as another set of NSS quality-aware features.

Local Structure Descriptors. From the bottom-up point of view of HVS, image structure plays a vital role in quality evaluation. Local binary pattern (LBP) has been proved to be a binary approximation of the HVS early image structure and useful to describe the distortions of images. The classical LBP operator describes the local structure of the image by calculating the contrast between the image pixels and the neighborhood of pixels. However, this operator covers a fixed size neighborhood and produces a variety of binary modes. Hence, in this paper, we utilize the LBP [18] histogram with rotation invariant "uniform" pattern to represent gray-scale image structure information.

Meanwhile; we utilize the steerable Gaussian filters to obtain the image gradient components [19]. Then, we use the shape and variance parameters computed by fitting the histograms of the gradient components to form another set of local structure features [7].

Image Color Statistics and Contrast Descriptors. To capture color distortion, we extract the MSCN coefficients' arithmetic mean and variance from the chrominance component $\sqrt{(a*)^2 + (b*)^2}$ in CIELAB space, and the MSCN arithmetic mean and skewness from the yellow component as a set of color-aware features.

Additionally, we form the hue component by converting the image into an HSV (Hue, Saturation and Value) color space. The HSV is sensitive to the difference in hue values, and when the difference is higher, then the visual quality of the image decreases. In this paper, we have modeled the perception process. First, we calculate the histogram of the hue, denoted $H_h = (H_1, H_2, \ldots, H_m)$, which is in ascending order, that is $H_1 < H_2 < \ldots H_m$, where m is the number of the unique values. Then we find the difference between adjacent values in H_h, denoted $D = (d_1, d_2, \ldots d_{m-1})$, such that $d_i = H_{i+1} - H_i$. Consequently, we obtain the average difference to measure the hue distortion as:

$$\hat{D} = \frac{\sum_{i=1}^{m-1} d_i}{m-1} \tag{3}$$

Moreover, contrast distortion cannot be ignored. In this paper, we propose a simple but effective feature to evaluate the contrast distortion degree of the image patch. Then we compute the local feature for each patch:

$$contrast = \max(\hat{I}) - \min(\hat{I}) \tag{4}$$

that is, contrast is the difference between the maximum MCSN and minimum MSCN of the gray-scale image patch.

2.3 The Proposed Saliency Modulated Fitting Model

From the perspective of the visual attention mechanism, human observers concentrate more on salient areas than another irrespective areas, therefore, visual attention models are promising to boost the quality prediction accuracy [9].

In this paper, we propose to utilize a saliency modulated multivariate gaussian distribution (S-MVG) model to fit the features computed from distorted image patches. The representation of MVG is:

$$f(x_1, x_2, \ldots, x_n) = \frac{1}{(2\pi)^{n/2} \left| \hat{\Sigma} \right|^{1/2}} \exp\left(-\frac{1}{2}(x - \hat{\delta})^T \hat{\Sigma}^{-1} (x - \hat{\delta}) \right) \quad (5)$$

where $(x_1, x_2, \ldots, x_n)$ is the feature vector, $\hat{\delta}$ and $\hat{\Sigma}$ are the mean and covariance matrices of the MVG model.

Previous studies have shown that applying the visual attention mechanism of HVS on IQA models can have a striking effect. Therefore, in this paper, the saliency of each image patch is taken into consideration in the estimation of MVG parameters. $\hat{\delta}$ and $\hat{\Sigma}$ are calculated as:

$$\hat{\delta} = \sum_{i=1}^{M} \omega_i \times (x_1^i, x_2^i, \ldots, x_n^i) \quad (6)$$

$$\hat{\Sigma} = \begin{bmatrix} \hat{\Sigma}_{11} & \hat{\Sigma}_{1n} \\ & \cdots & \\ \hat{\Sigma}_{n1} & \hat{\Sigma}_{nn} \end{bmatrix} \quad (7)$$

where $\hat{\Sigma}_{p,q} = \sum_{i=1}^{M} \omega_i (x_p - \hat{\delta}_p)(x_q - \hat{\delta}_q)$ is the covariance between pth-dimension feature x_p and qth-dimension feature x_q. Further, the patch saliency ω_i is calculated as the mean of the saliency corresponding to each pixel of the i-th image patch to represent its quality importance in the total test image. That is,

$$\omega_i = \sum_{j=1}^{P \times P} \omega_{ij} \quad (8)$$

where j is a pixel of the i-th image, and P is the image patch size. ω_i is normalized to [0,1]. In this paper, the saliency detection method which is called SDSR is employed [20].

It should be noted that when estimating the parameters of MVG in training stage, the saliency for each patch is the same, i.e., it is set to 1 divided by the number of high-sharpness image patches.

Subsequently, to predict the image quality, We calculate the distance between the MVG model learned from the pristine high-sharpness image patches and the MVG model learned from the distorted image, the formula is expressed as follows [1]:

$$Q = \sqrt{\left((\hat{\delta} - \hat{\delta}_s)^T \left(\frac{\hat{\Sigma} + \hat{\Sigma}_s}{2} \right) (\hat{\delta} - \hat{\delta}_s) \right)} \tag{9}$$

Where the $(\hat{\delta}, \hat{\Sigma})$ are the parameters of MVG from pristine high-sharpness patches, and the $(\hat{\delta}_s, \hat{\Sigma}_s)$ are the parameters of saliency modulated MVG from distorted image.

3 Experiments and Results

3.1 Experimental Setup

In this paper, we extract multi-order NSS features, local structure features, color features, and image contrast from image patches to express the image distortion. Except for the color features, all the other features are obtained at two scales to capture multiscale behavior. In our implementation, we set image patch size P to 52, and the threshold of high-sharpness natural image patches to 0.76. The parameters related to LBP are set as follows: the number of neighbors is 4, and the radius is 1. In addition, we tuned the orientation and scale parameters of the steerable Gaussian filters as $\{0, \frac{\pi}{4}, \frac{\pi}{2}, \frac{3\pi}{4}\}$ and 1 respectively. This yields a total of 91 features.

Two benchmark databases are used to evaluate the performance of the proposed method, LIVE r2 [21] and LIVE Challenge [22]. The LIVE r2 database contains 779 distorted images generated by 29 reference images, including JP2K compression (JP2K), JPEG compression (JPEG), White Noise (WN), Gaussian blur (GBLUR) and Fast fading (FF), i.e., five different distortion types. It is worth noting that the LIVE Challenge dataset contains 1162 images captured using mobile devices subjected to authentic real-world distortions. Therefore, it is far more challenging to estimate quality for the distorted images from LIVE Challenge.

We compare the proposed model with state-of-the-art opinion-aware methods, including BRISQUE [3], BLIINDS2 [5], DIIVINE [2], and FRIQUEE [6], as well as three state-of-the-art opinion-unaware methods: NIQE [1], IL-NIQE [7] and the method in [8]. To evaluate the performance of the proposed method, the Spearman's Rank Ordered Correlation Coefficient (SROCC) and the Pearson's Linear Correlation Coefficient (PLCC) are applied between the subjective DMOS/MOS from human opinion and the predicted score. Before computing PLCC, the predicted scores are passed through a logistic nonlinearity [21] to map to DMOS/MOS space.

To validate the performance of the method, we first evaluate the correlation between objective judgments of the proposed method and human visual perception on the LIVE r2 database. Since the OA-BIQA approaches that we compare to require a training procedure, the LIVE r2 database is divided randomly into two subsets: 80% for training and 20% for testing, such that no overlap occurs between training and testing subsets. The random train-test procedure was repeated for 1000 iterations, and we report the median results to remove bias.

Since OU-BIQA models do not require a training process, we report the results on the whole dataset. Furthermore, we also test the performance on the LIVE Challenge database to validate the generalization capability of the proposed model, where all OA-BIQA quality prediction models are trained on the entire LIVE database.

3.2 Performance Evaluation and Comparison

Table 1 and Table 2 show the SROCC and PLCC performance of different distortion types on the LIVE r2 database, respectively. Table 3 shows the overall prediction accuracy of the state-of-the-art NR-IQA models on the LIVE r2 and LIVE Challenge databases. The highest correlation values are highlighted in bold.

Table 1. Evaluation using SROCC on each distortion type in the LIVE r2 database.

Model	Type	JP2K	JPEG	WN	GBLUR	FF
BRISQUE	Opinion aware	0.9139	**0.9647**	0.9786	0.9511-	0.8768
BLIINDS2		**0.9285**	0.9422	0.9691	0.9231	**0.8893**
DIIVINE		0.913	0.910	0.984	0.921	0.863
FRIQUEE		0.9114	0.9480	**0.9900**	**0.9731**	0.8821
NIQE	Opinion unaware	0.9187	0.9400	0.9716	0.9326	0.8632
IL-NIQE		0.8944	0.9410	0.9809	0.9154	0.8329
Method in [8]		0.911	0.963	0.974	0.945	0.864
Proposed		**0.9418**	**0.9680**	**0.9814**	**0.9450**	**0.8780**

Table 2. Evaluation using PLCC on each distortion type in the LIVE r2 database.

Model	Type	JP2K	JPEG	WN	GBLUR	FF
BRISQUE	Opinion aware	0.9229	**0.9734**	0.9851	0.9506	0.9030
BLIINDS2		**0.9348**	0.9676	0.9799	0.9381	0.8955
DIIVINE		0.922	0.921	0.988	0.923	0.888
FRIQUEE		0.9277	0.9654	**0.9963**	**0.9824**	**0.9158**
NIQE	Opinion unaware	0.9265	0.9507	0.9772	0.9444	0.8816
IL-NIQE		0.9051	0.9590	**0.9863**	0.9314	0.8640
Method in [8]		0.920	0.966	0.978	**0.957**	0.878
Proposed		**0.9517**	**0.9730**	0.9860	0.9523	**0.8989**

From Tables 1, 2 and 3, it is evident that the proposed method outperforms all OU-BIQA models and comparable with other OA-BIQA methods on both the two databases. Additionally, the proposed method outperforms most of the OA-BIQA models on the

Table 3. Overall evaluation results on the LIVE r2 and LIVE Challenge databases

Model	Type	LIVE r2		LIVE challenge	
		PLCC	SROCC	PLCC	SROCC
BRISQUE	Opinion aware	0.9424	0.9395	0.3296	0.2650
BLIINDS2		0.9302	0.9306	0.1791	0.1259
DIIVINE		0.917	0.916	0.3667	0.3328
FRIQUEE		**0.9576**	**0.9507**	**0.6289**	**0.6303**
NIQE	Opinion unaware	0.9033	0.9055	0.4935	0.4493
IL-NIQE		0.9017	0.8970	0.5021	0.4392
Method in [8]		0.907	0.908	–	–
proposed		**0.9336**	**0.9333**	**0.5703**	**0.5140**

LIVE challenge dataset, demonstrating that the proposed method has better generalization capacity. Meanwhile it should be noted that the LIVE challenge database contains no computer-synthesized distortion – unlike LIVE r2 – but only authentic and real-world images, which makes it more realistic.

3.3 Performance Sensitivity

Here we evaluated the performance of different type of feature including spatial NSS, local structure, color statistics and contrast features on two representative databases, respectively. SROCC is used as the performance metric. The results are listed in Table 4.

Table 4. SROCC performance evaluation of each feature type.

Feature type	LIVE	LIVE challenge
Multi-order NSS	0.8685	0.5081
Local structure	0.8606	0.4623
Color statistics	0.6219	0.1043
Image contrast	0.5431	0.4795

From Table 4, it can be seen that the performance of each type of visual feature independently deteriorates significantly. However, the collaboration of multi-feature yields the best performance.

To validate the effectiveness of the S-MVG model in this work, we also test different features by utilizing the MVG the weighting factor from the saliency. The results are shown in Table 5, where we can see that the introduction of the saliency weighting factor can further boost the performance of the method substantially.

Table 5. Performance comparison of saliency modulated MVG.

Fitting model	LIVE		LIVE challenge	
	PLCC	SROCC	PLCC	SROCC
MVG	0.9307	0.9293	0.5579	0.5027
S-MVG	0.9336	0.9333	0.5703	0.5140

4 Conclusion

In this paper, we proposed a novel OU-BIQA method. The new method extracts four types of quality-aware features from a collection of high-sharpness patches of the pristine image set, then a MVG model is learned from these feature vectors as a standard visual quality model. For a test image, we proposed a saliency modulated MVG (S-MVG) model to fit the feature vectors and then computed the distance between its pairwised parameters which describes each S-MVG models and that of each S-MVG models learned from a corpus of pristine images. Extensive experiments showed that the proposed method exhibits excellent performance and generalization capacity. It outperforms three OU-BIQA methods and competes well with the OA-BIQA methods.

Acknowledgement. This research work is financially supported by National Natural Science Foundation of China (No. 61701404) and partially supported by Major Program of National Natural Science Foundation of China (No. 81727802), Natural Science Foundation of Shaanxi Province of China (No. 2020JM-438), the Key Research and Development Program in Shaanxi Province of China (No. 2017ZDCXL-GY-03-01-01).

References

1. Mittal, A., Soundararajan, R., Bovik, A.C.: Making a completely blind image quality analyzer. IEEE Sig. Process. Lett. **22**(3), 209–212 (2013)
2. Moorthy, A.K., Bovik, A.C.: Blind image quality assessment: from natural scene statistics to perceptual quality. IEEE Trans. Image Process. **20**(12), 3350–3364 (2012)
3. Mittal, A., Moorthy, A.K., Bovik, A.C.: No-reference image quality assessment in the spatial domain. IEEE Trans. Image Process. Publ. IEEE Sig. Process. Soc. **21**(12), 4695 (2012)
4. Zhang, M., Muramatsu, C., Zhou, X., et al.: Blind image quality assessment using the joint statistics of generalized local binary pattern. IEEE Sig. Process. Lett. **22**(2), 207–210 (2015)
5. Saad, M.A., Bovik, A.C., Charrier, C.: Blind image quality assessment: a natural scene statistics approach in the DCT domain. IEEE Trans. Image Process. **21**(8) (2012)
6. Deepti, G., Bovik, A.C.: Perceptual quality prediction on authentically distorted images using a bag of features approach. J. Vis. **17**(1), 32 (2017)
7. Zhang, L., Zhang, L., Bovik, A.C.: A feature-enriched completely blind image quality evaluator. IEEE Trans. Image Process. **24**(8), 2579–2591 (2015)
8. Zhang, M., Li, Y., Chen, Y.: Completely blind image quality assessment using latent quality factor from image local structure representation. In: International Conference on Acoustics, Speech, and Signal Processing (2019)

9. Farias, M.C.Q., Akamine, W.Y.L.: On performance of image quality metrics enhanced with visual attention computational models. Electron. Lett. **48**(11), 631–633 (2012)
10. Lasmar, N.E., Stitou, Y., Berthoumieu, Y.: Multiscale skewed heavy tailed model for texture analysis. In: IEEE International Conference on Image Processing. IEEE (2010)
11. Gupta, P., Bampis, C.G., Glover, J.L., Paulter, N.G., Bovik, A.C.: Multivariate statistical approach to image quality tasks. Imaging **4**, 117 (2018)
12. Gómez, E., Gomez-Viilegas, M.A., Marín, J.M.: A multivariate generalization of the power exponential family of distributions. Commun. Stat. **27**(3), 12 (1998)
13. Field, D.J.: Relations between the statistics of natural images and the response properities of cortical cells. J. Opt. Soc. Amer. **4**(12), 2379–2394 (1987)
14. Lyu, S.: Dependency reduction with divisive normalization: justification and effectiveness. Neural Comput. **23**(11), 2942–2973 (2011)
15. Su, C.C., Cormack, L.K., Bovik, A.C.: Closed-form correlation model of oriented bandpass natural images. IEEE Sig. Process. Lett. **22**(1), 21–25 (2014)
16. Pascal, F., Bombrun, L., Tourneret, J.Y., et al.: Parameter estimation for multivariate generalized Gaussian distributions. IEEE Trans. Sig. Process. **61**(23), 5960–5971 (2013)
17. Su, C.C., Cormack, L.K., Bovik, A.C.: oriented correlation models of distorted natural images with application to natural stereopair quality evaluation. IEEE Trans. Image Process. **24**(5), 1685–1699 (2015)
18. Ojala, T., Pietikainen, M., Maenpaa, T.: Multiresolution gray-scale and rotation invariant texture classification with local binary patterns, pp. 971–987. IEEE (2002)
19. Sinno, Z., Caramanis, C., Bovik, A.C.: Towards a closed form second-order natural scene statistics model. IEEE Trans. Image Process. **27**(7), 3194–3209 (2018)
20. Seo, H.J., Milanfar, P.: Static and space-time visual saliency detection by self-resemblance. J. Vis. **9**(12), 15 (2009)
21. Sheikh, H.R., Sabir, M.F., Bovik, A.C.: A statistical evaluation of recent full reference image quality assessment algorithms. IEEE Trans. Image Process. **15**(11), 3440–3451 (2006)
22. Ghadiyaram, D., Bovik, A.C.: LIVE in the wild image quality challenge database (2015). http://live.ece.utexas.edu/research/ChallengeDB/index.htm

Blood Flow Velocity Detection of Nailfold Microcirculation Based on Spatiotemporal Analysis

Zhenkai Lin[1,2,3], Fei Zheng[4], Jianpei Ding[5], and Jinping Li[1,2,3]($\boxtimes$)

[1] School of Information Science and Engineering,
University of Jinan, Jinan 250022, Shandong, China
ise_lijp@ujn.edu.cn
[2] Shandong Provincial Key Laboratory of Network Based Intelligent
Computing, (University of Jinan), Jinan 250022, Shandong, China
[3] Shandong College and University Key Laboratory of Information Processing
and Cognitive Computing in 13th Five-Year, Jinan 250022, Shandong, China
[4] Jinan Shengquan Group Share Holding Co. Ltd, Jinan 250000, China
[5] Shandong ShenTu Intelligent Science and Technology Co. Ltd, Jinan 250022, China

Abstract. Nailfold microcirculation can reflect the state of health. It is an important research topic to detect the change of microcirculation blood flow velocity. In order to save cost, people usually choose the low-cost micro camera to collect samples. However, the microvascular video collected by such camera is often disturbed by various noises, resulting in poor contrast and clarity of the image. Due to the large number of microvessels, it is time-consuming and laborious to detect the blood flow rate manually, which is inefficient and difficult to accurately detect the subtle changes of the blood flow velocity. At present, the previous research of blood flow velocity detection is mainly focused on the clearer microcirculation video. It is difficult to find any related research of blood flow velocity detection for noise image. For the low-quality nailfold microcirculation video collected by the low-cost microscope camera, we propose an automatic detection method of blood flow velocity based on projection analysis of spatiotemporal image. The method is as follows: firstly, video preprocessing, correlation matching are used to remove jitter and image blur is eliminated by deconvolution, the row mean is used to eliminate reflective area by analyzing the characteristics of noise distribution; secondly, we use cumulative background modeling to segment blood vessels and propose an automatic detection algorithm of blood vessel centerline; thirdly, the direction of binary spatiotemporal image is detected by using rotation projection, and then the blood flow velocity is calculated. The experimental results show that the proposed method can detect the blood flow velocity of microcirculation automatically and efficiently. Meanwhile, the average correlation coefficient between the proposed method and the manual measurement standard value is 0.935.

Keywords: Nailfold microcirculation · Blood flow velocity detection · Rotational projection · Spatiotemporal image

© Springer Nature Switzerland AG 2020
Y. Peng et al. (Eds.): PRCV 2020, LNCS 12305, pp. 681–695, 2020.
https://doi.org/10.1007/978-3-030-60633-6_57

1 Introduction

Blood flow velocity can reflect the working state of microcirculation system dynamically [1, 2]. Therefore, it is of great significance for human health to detect the velocity changes of microcirculation blood flow [3, 4].

The nailfold microcirculation samples are easy to be collected and observed. The parameter can be obtained by the observation and statistics of microcirculation video by human eyes. There are many microvessels in the observation field (see Fig. 1). Obviously, it is difficult to obtain the change value of blood velocity accurately.

Fig. 1. Nailfold microcirculation image

The application environment of precision medical equipment is severely restricted. In order to facilitate the popularization and application, people put forward the microcirculation parameter detection method based on video analysis [8–13]. Up to now, to the utmost of our knowledge, it is not easy to find any related researches about such automatic detection method. The related works that can be investigated mainly focus on clear video and image. There are two main research methods, one is semi quantitative detection with reference [8–10], the other is automatic detection without reference [11–13]. The former calculates the velocity of blood flow according to the relative change information by selecting the moving reference point and setting the corresponding reference conditions; the latter directly extracts the effective information of blood flow by analyzing the characteristics of video.

In the study of blood flow velocity detection with reference, in order to improve the detection accuracy in multiple scenes, people have made in-depth research from the use of external reference materials to the selection of internal reference quantity. Chen proposed a method to measure the blood flow velocity of organism by using fluorescent markers [8]. The intensity of the fluorescent agent is easily affected by many factors. Liu detected the blood velocity by using the flying point method [9]. The method needs preset reference points, and the detection process is complex. Sheng proposed a spatial correlation method to measure the flow velocity [10]. This method can only detect the vessels with uniform cells and slow-moving speed.

In the research of automatic detection without reference, the related work is analyzed from spatial domain to frequency domain, and the corresponding algorithm of blood flow velocity detection is proposed. Chen put forward the spatial domain enhancement method of ST image based on multi-scale directional filter [11]. In the process of spatial domain enhancement, it was easy to lose the effective signal of blood flow. Lv proposed a method to detect blood flow velocity by analyzing the spectrum of ST image [12]. The method could not form an ideal straight line passing through the center of the spectrum, and the spectrum analysis results were greatly affected. Wang proposed a method to detect blood flow velocity with multiple tracking lines, and extracted multiple tracking lines according to the shape of blood vessels [13]. Many tracking lines extracted from the original image and the generated spatiotemporal image were greatly disturbed by noise.

The following problems exist in the blood flow velocity detection:

(1) The microcirculation samples collected are often interfered by various noises. However, there are few related works that can be investigated and the existing work has not focused on microcirculation video disturbed by noise;
(2) The detection methods with reference need artificial intervention. There is no standard reference quantity, which is suitable for specific types of blood vessels;
(3) Some effective information will be lost during the detection process, when the blood flow state changes in the video are used to detect the blood flow velocity. The detection results need to be further improved for noisy image.

We used common microscope camera to collect nailfold microcirculation samples. An efficient blood flow detection method is proposed by using rotation projection to analyze the motion trajectory of the binary spatiotemporal image.

There are two innovations:

(1) In view of the nail microcirculation video collected by low-cost ordinary micro camera, we propose an effective detection method. At present, there are few researches on the automatic detection of microcirculation blood flow. The proposed method provides a new research idea for the further calculation of microcirculation parameters.
(2) The proposed method can segment the microvascular accurately. Binary spatiotemporal image can reduce the loss of information during the processing. The detection of the blood flow velocity is converted into the calculation of the trajectory of the two-dimensional ST image, and the detection result is more objective and accurate.

The rest of paper is organized as follows: in Sect. 2, the proposed efficient and accurate blood velocity automatic detection algorithm is presented; in Sect. 3, experimental results are shown; in Sect. 4, conclusion of the method and possible future work are pointed out.

2 Proposed Method

The sample is collected by a microscope camera with a magnification of 250 times, a resolution of 1280 × 720. The samples are shown in Table 1, including various types of nailfold microcirculation videos under different ages, genders, pressures and states (Fig. 2).

An automatic detection algorithm of blood flow velocity is proposed for the nailfold microcirculation video collected by ordinary micro camera (Fig. 3).

Table 1. Video sample

Sample	Men	Women	Age(18~56)
Normal pressure	46	44	90
Compressed finger	46	44	90
Heated hands	40	40	80
Graphene contrast	40	40	80

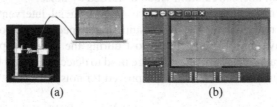

(a)   (b)

Fig. 2. Sample collection. (a) collection equipment (b) schematic diagram of collection interface

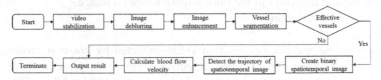

Fig. 3. Flow chart of blood flow velocity detection algorithm

2.1 Microcirculation Image Preprocessing

The distribution of microvessel varies in depth and form. Human breath and pulse beat will cause video jitter and image blur. There are two problems need to be solved:

(1) We need to obtain stable microcirculation video.
(2) It is necessary to eliminate noise and improve contrast in complex background.

Video Stabilization. The video jitter caused by human breath, pulse beat and other factors is more obvious under the microscope camera [14]. It is difficult for us to observe the blood vessels. Therefore, it is necessary to de jitter the original video.

The video jitter is removed by correlation matching. The first frame of the video image is taken as the initial frame standard image, and each frame in the video is matched with the selected initial frame standard image. Select a certain middle region in the image need to be matched, and calculate the similarity between the region and the initial frame standard image. The similarity calculation formula is shown in (1):

$$R(i,j) = \frac{\sum_{m=1}^{M} \sum_{n=1}^{N} \left[S_{i,j}(m,n) * T(m,n) \right]}{\sqrt{\sum_{m=1}^{M} \sum_{n=1}^{N} \left[S_{i,j}(m,n) \right]^2} * \sqrt{\sum_{m=1}^{M} \sum_{n=1}^{N} \left[T(m,n) \right]^2}} \tag{1}$$

Where $R(i,j)$ is the cross-correlation coefficient, S is the standard image of the initial frame, $S_{i,j}$ is the subgraph in the standard image, and T is the subgraph in the image to be modified. The pixel coordinate information is used to return the correction parameters, that is, the translation distance between the corrected image and the original frame standard image.

Image Deblurring. The blur of microcirculation image is mainly caused by the relative motion between camera and finger. In a short time, it is difficult to establish an accurate mathematical model for this kind of relative motion [15, 16]. It is assumed that there is a uniform motion between the camera and the finger in the microvascular video acquisition.

The key point of motion deblurring is to obtain the fuzzy parameters, i.e. the length and angle of relative motion. The deconvolution kernel is designed according to the image offset parameters, and the motion blur in the image is eliminated by deconvolution.

Image Enhancement. Wavelet transform is used to enhance the image. The experimental results of various enhancement algorithms are shown in Fig. 4.

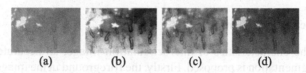

(a)　　　　(b)　　　　(c)　　　　(d)

Fig. 4. Image enhancement results. (a) original image (b) histogram equalization (HE) [17] (c) contrast limited adaptive histogram equalization (CLAHE) [18] (d) wavelet decomposition

The contrast of G-channel image is relatively high, which is conducive to further segmentation of blood vessels (see Fig. 5).

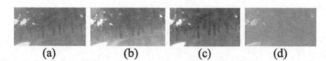

(a)　　　　(b)　　　　(c)　　　　(d)

Fig. 5. Channel separation. (a) gray image (b) B-channel (c) G-channel (d) R-channel (Color figure online)

G-channel image is accumulated to obtain mean image, which can eliminate noise. At the same time, it can fill the discontinuous area in the blood vessel and improve the contrast between the blood vessel and the background (Fig. 6).

The method of row average light homogenization is proposed. The results are shown in Fig. 7. The calculation formula is shown in (2), where h is the height of the image, w is the width of the image, and T_i is the gray mean value of each row of image.

$$
\begin{cases}
T_i = \sum_{i=0}^{i=h}\sum_{j=0}^{j=w} f(x_i, y_j)/w \\
f(x_i, y_j) = \begin{cases} T_i, f(x_i, y_j) \geq T_i \\ f(x_i, y_j), f(x_i, y_j) < T_i \end{cases}
\end{cases}
\tag{2}
$$

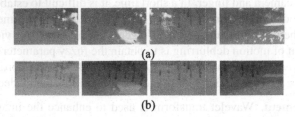

Fig. 6. Single frame image and mean image. (a) single frame image (b) mean image

Fig. 7. Results of light removal. (a) original image (b) light eliminate

2.2 Microvascular Segmentation

According to the characteristics of microcirculation image, an accurate algorithm of blood vessel segmentation is proposed. Firstly, the foreground in the image is segmented by local window mean binarization [19], then the noise and abnormal blood vessels are removed by contour information.

A custom morphological processing structure element is designed for the blood vessel contour. The pixel of mark position (▲) is the origin of the structure element (Figs. 8 and 9).

Fig. 8. Custom structure elements

2.3 Blood Flow Velocity Detection

The velocity of blood flow in microcirculation was measured by tracking the movement of white particles (Fig. 10).

Extraction of Blood Vessel Centerline. In this paper, the automatic detection method of blood vessel centerline is designed. Firstly, the contour of the blood vessel is refined; secondly, the centerline of the blood vessel is detected by BFS to remove the bifurcations; finally, the centerline of a single blood vessel is obtained (Fig. 11).

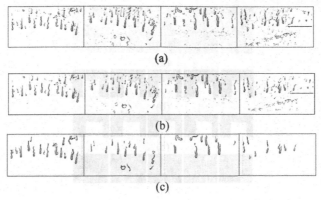

Fig. 9. Microcirculation image segmentation. (a) binarization of local window mean (b) morphological processing (c) contour information denoising

Fig. 10. Blood flow sequence

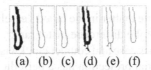

(a) (b) (c) (d) (e) (f)

Fig. 11. Blood vessel centerline detection. (a, d) vascular contour (b, e) refinement (c, f) BFS

Calculate the Gravity Center of Projection Histogram. The difference binary image is obtained by the binarization of the adaptive threshold. As shown in Fig. 12, projections are made in the normal direction along the tangent line of the vessel centerline. Suppose that the coordinates of P1 point are (x_1, y_1), the coordinates of P2 point are (x_2, y_2), the length of blood vessel centerline is defined as CL, and the projection calculation method is shown in (3) (Fig. 13):

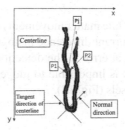

Fig. 12. Schematic diagram of vascular projection calculation

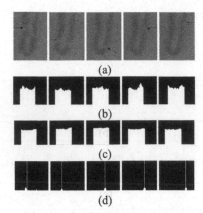

(a)

(b)

(c)

(d)

Fig. 13. Projection histogram. (a) continuous video frame (b) single frame image projection histogram (c) cumulative projection histogram (d) differential projection histogram

$$V_{P_j} = \frac{\sum_{i=x_1}^{x_2} f(x_i, \frac{y_2-y_1}{x_2-x_1} x_i)}{\sqrt{(x_2-x_1)^2 + (y_2-y_1)^2}}, x_1 \leq x_i \leq x_2, 0 < j \leq CL \tag{3}$$

The conclusion is as follows:

(1) The movement of the white matter in the blood vessel is consistent with the change of the center of gravity of the difference histogram;

(2) When the blood flows clockwise and counter clockwise, the movement of the center of gravity of the differential projection has a periodic rule (Fig. 14).

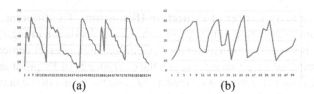

(a)   (b)

Fig. 14. Curve of gravity center movement. (a) counter clockwise flow (b) clockwise flow

When there are continuous white matter movements in the blood vessels, the position of the center of gravity of the current difference histogram cannot replace the position of the plasma, resulting in a great error in the detection of blood flow velocity. When multiple plasmas are moving, it is impossible to judge the current inflow and outflow state of plasma in the blood vessels (Fig. 15).

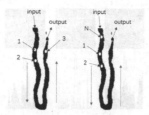

Fig. 15. Schematic diagram of multi plasma continuous movement

Calculate the Trajectory of Spatiotemporal Image. The difference binary image is used as the column vector of spatiotemporal image (ST image), and the abscissa is the number of video frames. ST image combines the continuous sequence of video with the spatial distribution, which can more intuitively observe the blood movement information in the video.

The plasma position information is transformed into the coordinate information in the corresponding frame column vector to generate a complete ST image (Fig. 16).

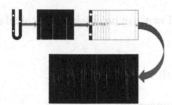

Fig. 16. ST image of vessel

As shown in Fig. 17, we project the motion trajectory W of vessel ST image in the S direction, rotate the value of the projection angle θ from 0 to π. The projection set is defined as H_w. The calculation is shown in (4).

$$\begin{cases} p(t, \theta) = \sum_{s=0}^{Q} W([t \cdot \cos\theta - s \cdot \sin\theta], [t \cdot \sin\theta + s \cdot \cos\theta]) \\ Q = [M \cdot \cos\theta + N \cdot \sin\theta] \\ 0 \leq t \leq [M \cdot \sin\theta + N \cdot \cos\theta] \end{cases} \quad (4)$$

Where M and N are the height and width of the image respectively, and the operator [] indicates rounding down.

Fig. 17. Correspondence between projection coordinates (t, s) and original coordinates (x, y)

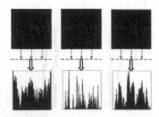

Fig. 18. Projection of motion trajectory

As can be seen from Fig. 18, the results of 180-degree rotation projection of ST image mainly show that there are obvious differences in the distribution of peaks in the projection results.

(1) The variance of each histogram in the projection set H_w is calculated and put into the set H_σ;
(2) The largest element is selected from the variance set H_σ, and then the corresponding projection histogram is found from the set H_w. Finally, the corresponding projection angle θ can be found;
(3) The velocity of blood flow v can be obtained by calculating the cotangent of angle θ, i.e. $v = cot\ \theta$.

3   Experimental Results and Analysis

3.1   Results of Image Enhancement

Peak signal-to-noise ratio (PSNR) is used as the objective quality evaluation index of video de jitter. The larger PSNR value is, the better the effect of video de jitter is.

$$PSNR(F_1, F_0) = 10 \bullet \log^{\frac{255^2}{MSE(F_1, F_0)}} \tag{5}$$

Where F_0 is the reference frame, F_1 is the current frame, and MSE is the mean square deviation of pixels between F_0 and F_1.

The comparison experiment results of video stabilization are shown in Table 2.

Table 2. Experimental results of video de jitter

Sample	Original video (PSNR/dB)	De jitter video (PSNR/dB)	Time(s)
SIFT	18.73	24.97	0.076
Proposed method	18.73	25.26	0.058

DeblurGAN network does not need to build a mathematical model of the fuzzy process manually, but it needs to build a clear-fuzzy image pair [20]. However, in practical applications, it is impossible to obtain clear microvascular video images. DeblurGAN network can eliminate image blur, but it also introduces a lot of noise. So, it is not suitable for microvascular image. Deconvolution deblurring will produce ringing effect. However, the ringing effect is effectively suppressed by using extension processing. This method is more suitable for deblurring microcirculation images.

The experimental results of image deblurring are shown in Fig. 19.

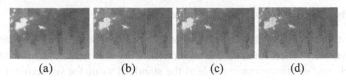

(a)   (b)   (c)   (d)

Fig. 19. Image deblur. (a) fuzzy image (b) deconvolution (c) eliminating ringing effect (d) DeblurGAN

3.2 Results of Image Segmentation

Combined with the characteristics of microcirculation image, we make full use of the continuity of video to segment blood vessels in complex background. We manually label different blood vessels. 100 images were labeled manually.768single vessels were labeled in total. The results of artificial labeled blood samples are shown in Fig. 20(a):

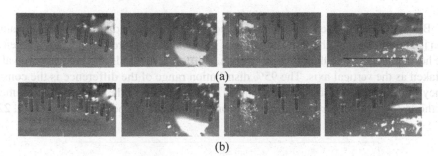

(a)

(b)

Fig. 20. Results of blood vessel detection. (a) manual label (b) segmentation results

In 100 images, 706 single blood vessels were segmented by the detection algorithm. The coincidence rate between the segmentation results and the number of manual labels was 91.93%. The calculation formula is shown in (6), where P_c is the average degree of coincidence, R_i is the vascular contour area marked artificially, and R_j is the vascular contour area segmented by the algorithm (Table 3).

$$P_c = \frac{\sum_{i,j=1}^{N} (R_i - R_j)^2}{N}, N = 706 \tag{6}$$

Table 3. Experimental results of blood vessel segmentation

Sample (100)	Time(s)	Number of vessels (consistency)	Vessel contour (coincidence)
Manual label	/	768	/
Proposed method	4	706 (91.93%)	90.02%

3.3 Results of Blood Flow Velocity Calculation

At present, there is no strictly defined standard that can be used as standard value of blood flow velocity. Therefore, we use manual marking method to calculate the tilt angle of each track, and use the average angle as the standard value for subsequent evaluation. The result of manual marking of binary ST image (see Fig. 21(b)).

(a)   (b)

Fig. 21. Results of manual marking of track direction in spatiotemporal image. (a) original binary spatiotemporal image (b) manual marking

Bland-Altman image is a common method for quantitative data consistency evaluation [21]. In this paper, the standard value of the manual measurement results is taken as the horizontal axis, and the difference between different methods and the standard value is taken as the vertical axis. The 95% distribution range of the difference is the consistency limit. If the consistency limit value is clinically acceptable, the two measurement methods can be considered to have good consistency and can be used instead (Fig. 22).

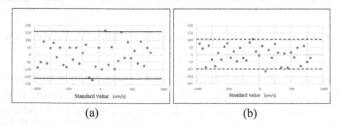

(a)   (b)

Fig. 22. Consistency experiment. (a) Bland Altman analysis results of projection histogram. (consistency limit value: −171 um/s~167 um/s). (b) Bland Altman analysis results of trajectory calculation of ST image (consistency limit value: −113 um/s~108 um/s).

The results show that the blood flow velocity detection method based on ST image motion trajectory calculation has higher consistency. In order to test the accuracy of this method, 40 video samples of nailfold microvascular were selected from the collected video samples. We track the number of video frames from the beginning to the end of the white material in a single blood vessel by the human eye, and then calculate the current blood flow velocity in the blood vessel by combining the length information of the blood vessel. The blood flow velocity of 40 blood vessels was calculated and the standard value database was established. The average correlation coefficient between the two detection methods and the standard value can be obtained by correlation calculation. The calculation method is shown in (7), where r is the correlation coefficient, X_i is the blood flow velocity calculated by the algorithm, Y_j is the standard value of manual measurement, and N_m is the number of tested samples (Table 4) (Fig. 23).

$$r = \frac{\sum_{i,j=1}^{N_m} \frac{Cov(X_i,Y_j)}{\sqrt{Var|X_i|Var|Y_j|}}}{N_m}, N_m = 40 \tag{7}$$

In practical application, the proposed method can replace the results of manual detection and meet the needs of actual detection.

Table 4. Experimental results of blood flow velocity detection

Number of blood vessels	Average correlation coefficient (histogram projection)	Average correlation coefficient (ST image)
40	0.85	0.935

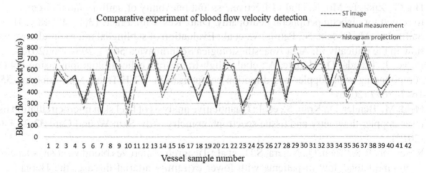

Fig. 23. Comparative experiment of blood flow velocity detection.

4 Conclusion

For the nailfold microcirculation video collected by ordinary micro camera, we propose an automatic detection method of microvascular blood flow velocity. Firstly, the image disturbed by noise is preprocessed to improve the image visualization effect; secondly, microvasculature was segmented by using cumulative background modeling; finally, a single vessel binary ST image is generated. The velocity of blood flow was measured by calculating the trajectories of binary space-time images of blood vessels with rotating projection. The detection method is easy to implement and low cost. The experimental results show that the proposed method can accurately detect the blood flow velocity in microcirculation in the collected samples, and the average correlation coefficient with the results of artificial measurement is 0.935. In the collected video samples, there is no obvious movement of white granular material in microvessels. In the future research, we need to research such vessels in depth. The accuracy of the whole blood flow detection algorithm will be improved by extracting more accurate and detailed blood flow change features.

Acknowledgement. Supported by The National Natural Science Foundation of China (61701192); Shandong Provincial Key Research and Development Project (2017CXGC0810); Shandong Education Science Plan "Special Subject for Scientific Research of Educational Admission Examination" (ZK1337212B008); Shandong University Science and Technology Program "Research on Navigation Technology of Wheeled Robot Based on Binocular Vision" (J18KA371).

References

1. Gurfinkel', I.I., Sasonko, M.L., Talov, N.A.: Correction of blood microcirculation parameters and endothelial function in chronic venous insufficiency of lower limbs. Angiol. Vasc. Surg. **23**(2), 89–98 (2017)
2. Kulikov, D., Glazkov, A., Dreval, A., et al.: Approaches to improve the predictive value of laser doppler flowmetry in detection of microcirculation disorders in diabetes mellitus. Clin. Hemorheol. Microcirc. **70**(11), 1–7 (2018)
3. Hu, G., Zhai, F., Mo, F., et al.: Effectiveness and feasibility of nailfold microcirculation test to screen for diabetic peripheral neuropathy. Diab. Res. Clin. Pract. **13**(1), 42–48 (2017)
4. Sallisalmi, M., Oksala, N., Pettilä, V., et al.: Evaluation of sublingual microcirculatory blood flow in the critically ill. Acta Anaesthesiologica Scandinavica **56**(3), 298–306 (2012)
5. Poree, J., Posada, D., Hodzic, A., et al.: High-frame-rate echocardiography using coherent compounding with doppler-based motion-compensation. IEEE Trans. Med. Imaging **35**(7), 1647–1657 (2016)
6. Held, M., Bender, D., Krau, S., et al.: Quantitative analysis of heel skin microcirculation using laser doppler flowmetry and tissue spectrophotometry. Adv. Skin Wound Care **32**(2), 88–92 (2019)
7. Rosenson, R.S., Chen, Q., Najera, S.D., et al.: Ticagrelor improves blood viscosity-dependent microcirculatory flow in patients with lower extremity arterial disease: the Hema-kinesis clinical trial. Cardiovasc. Diabetol. **18**(1), 77–85 (2019)
8. Chen, H.F.: Quantitative measurement system of microcirculation velocity. Zhejiang University (2002)
9. Liu, Y., Yang, J., Sun, K., et al.: Determination of erythrocyte flow velocity by dynamic grey scale measurement using off-line image analysis. Clin. Hemorheol. Microcirc. **43**(3), 263–267 (2009)
10. Sheng, Y.M.: Research on dynamic measurement technology and application of micro image characteristic parameters of microcirculation and myocardial cells. Peking Union Medical College (2011)
11. Chen, Y.: Measurement and recognition of human microcirculation parameters based on image analysis. Nanjing University of Aeronautics and Astronautics (2012)

12. Lv, F.: Development of microcirculation blood flow analysis system based on video image. Southeast University (2015)
13. Wang, Y.: Research on microcirculation parameter measurement algorithm based on video image. Southeast University (2017)
14. Dong, J., Liu, H.: Video stabilization for strict real-time applications. IEEE Trans. Circ. Syst. Video Technol. **27**(4), 716–724 (2017)
15. Pan, J.: Motion blur estimation: theories, algorithms and applications. Dalian University of Technology (2017)
16. Wang, J., Shi, G., Zhang, L., Han, B.: Methods to suppress edge ringing in motion blurred images. Electron. Measur. Technol. **36**(05), 62–67 (2013)
17. Liu, S.: Study on medical image enhancement based on wavelet transform fusion algorithm. J. Med. Imaging Health Inform. **7**(2), 388–392 (2017)
18. Magudeeswaran, V., Singh, J.F.: Contrast limited fuzzy adaptive histogram equalization for enhancement of brain images. Int. J. Imaging Syst. Technol. **27**(1), 98–103 (2017)
19. Zhang, X., Liu, Y., Zhou, J., et al.: Study on QR code image binaryzation under non-uniform lighting conditions based on grayscale morphology. IEEE Computer Society (2012)
20. Kupyn, O., Budzan, V., Mykhailych, M., et al.: DeblurGAN: blind motion deblurring using conditional adversarial networks. In: IEEE Conference on Computer Vision and Pattern Recognition (CVPR) (2017)
21. Misyura, M., Sukhai, M.A., Kulasignam, V., et al.: Improving validation methods for molecular diagnostics: application of Bland-Altman, deming and simple linear regression analyses in assay comparison and evaluation for next-generation sequencing. J. Clin. Pathol. **71**(2), 117–127 (2018)

Background Cleaning and Direction Weight in Salient Object Detection

Xiaodong Wang and Xiaoming Huang[⊠]

Computer School, Beijing Information Science and Technology University,
Beijing, China
wangxiaodong@mail.bistu.edu.cn, huangxm18@bistu.edu.cn

Abstract. Background prior is widely-used knowledge in salient object detection, and geodesic distance is based on background prior knowledge. To measure the difference between a pixel and image boundary, geodesic distance depends on a path which has the lowest cost to access the image boundary, and the cost is the value of distance. Areas with low geodesic distances are considered to be backgrounds, and the high are foregrounds. However, when some background areas are surrounded by areas that differ greatly in their colors, this best path to the boundary also needs a large cost and contributes to the high saliency of these areas. To address this problem, we propose a background cleaning (BC) method. For more ideal results in datasets, this BC method should be adaptively used. RBD (robust background detection) is a successful method in salient object detection. It is based on the global contrast information, but global contrast is easily affected by noise. In previous work, MDC (minimum directional contrast) was proposed. And we find that the directional contrast information can improve RBD method, so we use the minimum contrast information of four directions. Experimental results on two public datasets show that our method performs better than RBD, MDC and some widely-used methods.

Keywords: Salient object detection · Directional contrast · Background prior · Geodesic distance

1 Introduction

Salient object detection is inspired by the selectivity of the human visual system. It is used to select important targets in the image and use the saliency map [9] to represent the saliency of the image. Image saliency object detection has been applied in image classification [13,14], object detection [3,5], object segmentation [18] and other fields.

This work was supported by National Nature Science Foundation (NNSF: 61171118, 61673234, U1636124), Scientific Research Project of Beijing Educational Committee (KM202011232014).

© Springer Nature Switzerland AG 2020
Y. Peng et al. (Eds.): PRCV 2020, LNCS 12305, pp. 696–706, 2020.
https://doi.org/10.1007/978-3-030-60633-6_58

Many theories and methods have been produced in the field of salient object detection. Among them, the methods based on background prior [15, 20] and contrast [6, 12, 23] are more widely used. Geodesic distance (GD) is defined in [20], and minimum barrier distance (MBD) is defined in [15], both of which are used to measure the difference between two points. To use background prior, these two distances depend on a path which causes the lowest cost to access the image boundary, and the cost is the value of distance. The cost of GD is the sum of color differences upon the path, and the MBD is defined as max difference upon the path. Areas with low distance are considered to be backgrounds, and with high are foregrounds.

Literature [23] proposed a method RBD based on background prior and global contrast information. In [23], a measure called boundary connectivity is to define the similarity between pixel and image boundary. Areas with high boundary connectivity are considered to be backgrounds, and with low are foregrounds. This measure is computed and based on geodesic distance. Geodesic distance can define the saliency of background areas well in general situation. But when some background areas are surrounded by areas that differ greatly in their color, the best path to image boundary causes high cost. So the saliency of this area is high. The results in [20] and [23] both show this problem.

Fig. 1. Motivation of proposed method. (a) input image. (b) ground truth. (c) GD [20] saliency. (d) RBD [23] saliency. (e) BC saliency. (Color figure online)

For the areas marked by the red cross in Fig. 1(a), these areas are background areas in the ground truth Fig. 1(b). The result of GD [20] is shown in Fig. 1(c). The red cross areas show little saliency. The result of RBD [23] is shown in Fig. 1(d), and these red cross areas are salient regions. Because the best geodesic distance paths from these areas to the boundary all cause high cost. For areas marked by the red cross in Fig. 1(a), their colors are yellow and similar color is on the image boundary, so these areas are backgrounds in ground truth. GD needs a shortest path to reach the boundary, but the area path always leads to high cost. Because these areas are surrounded by the body of horses and some dark colors marked by blue cross. These shortest paths need to pass some areas to achieve the lowest cost. However, these paths obviously sum up high color differences, which lead to higher saliency of these areas. To address this problem, we propose one background cleaning (BC) method that can effectively solve it. After the BC process, the result of RBD is shown in Fig. 1(e). These areas show very little saliency.

In our example, these red cross-marked areas are actually background areas because their similar areas are on the image boundary. These shortest geodesic paths always have high cost because they have to across through some greatly different colors in one path. In general, the area surrounded by areas that differ greatly in its color, is the foreground because it actually gets a high saliency with the geodesic distance. And the area would not be similar with boundary in color. If the area is similar with boundary in color, it has a high probability of being the background, although the geodesic distance is high.

Fig. 2. Background cleaning method. (a) input image. (b) superpixels. (c) background cleaning. (d) saliency map.

Our proposal to clean areas similar to the boundary is called background cleaning (BC). As we said previously, if the area is similar to boundary in color and it has high geodesic distance, it has a high probability of being the background. But we define all the saliency of pixels by the geodesic distance, so, the area has to show low geodesic distance.

For the image in Fig. 2(a). Firstly, the image is segmented using SLIC [2] as shown in Fig. 2(b). The result after BC is shown in Fig. 2(c). All cleared pixels are in blue mask. The advantage of this method is that after a certain area is cleared, its adjacent areas will be adjacent to each other. In this image, these areas marked by red cross can be adjacent to the similar area in the boundary or have lower-cost path to boundary. Figure 2(d) shows our final saliency map. These areas show very little saliency.

RBD [23] is based on geodesic distance and global contrast information. We proposed MDC (Minimum directional contrast) in [6]. A region has high saliency, so there is higher contrast in all directions. We think global contrast is easily affected by noise. So, we calculated the contrasts of different directions and the minimum contrast was selected. It is a good measure of the saliency of this area and it shows great performance in [6]. We think minimum directional contrast is better than global contrast. So, we improve the RBD by using the minimum directional information.

We compared the improved method with some widely-used methods using background prior and contrast on public datasets. Experimental results proved that our method performs better than these methods.

The main contributions of this work include:

1) We propose a background cleaning method that cleans areas similar to the image boundary, which can reduce the saliency of challenging background areas.

2) RBD method uses the global information, and to be more robust, we replace the global information with minimum directional information.

The remainder of this paper is organized as follows. Section 2 is a review of related saliency detection methods. Then, we proposed the methods and improvements in Sect. 3 and experiment in Sect. 4. And Sect. 5 is the conclusion.

2   Related Work

With the efforts of researchers, many theories and methods have emerged for image saliency detection. These methods can be divided into bottom-up methods and top-down methods. The bottom-up approach relies on the underlying visual characteristics such as contrast, spatial distribution, and contours. The top-down method is a method commonly used in recent years, mainly by manually labeling truth values, and learning saliency models from training through machine learning or deep learning. The bottom-up method requires feature extraction, and feature extraction requires some prior knowledge. Prior knowledge generally includes location prior, background prior, spatial distribution prior, and contour prior. For human's visual system, contrast is one of the most important and intuitive features.

Contrast information can be divided into global contrast and local contrast. Global contrast considers the difference between the pixel/area and the entire image, while local contrast considers the difference between the pixel/area and the surrounding area. [4] proposed a pixel-level global contrast method HC, which defined the saliency of each pixel as the color distance from all other pixels. At the same time, [4] proposed a region-based saliency algorithm RC, which divided the input image into several regions, and then calculated the contrast of the two regions and weighted it with spatial distance. The local contrast method mainly considers the contrast in a small area. Literature [7] proposed a saliency detection method based on multi-scale contrast. Literature [12] combined the local and global contrast and estimated the saliency by using high-dimensional Gaussian filters.

We mentioned that the foreground of the image is usually surrounded by the surrounding background. The contrast between the foreground and the surrounding areas is high, and the contrast in each direction will be high, so we proposed the minimum directional contrast (MDC) in [6]. And the computational complexity is O (1) by integral graph operation [17].

Background prior is also useful prior knowledge. It considers that the boundary of an image is mostly a background area. Therefore, the difference between an area and image boundary can be used to measure saliency. Literature [20] proposed a method to measure the difference between a region/pixel and a boundary, which is called geodesic distance. Calculating the distance should be performed on all the paths from the area to the boundary, and then calculating the sum of the differences between adjacent pixels on the path is to be done, finally, the value of geodesic distance is the calculated minimum. In [15], the maximum difference on the path is calculated, and the MBD distance is proposed. The value of MBD takes the calculated minimum value. These two methods are commonly used in background prior. In [23], it is found that the background in the image is usually highly connected to the image boundary, while the foreground and the image boundary are lowly connected. He defined the boundary connectivity, which measures the connectivity of region to image boundary. This method shows better performance over other traditional methods.

In recent years, deep learning has been well applied in computer vision. Convolutional neural networks have the advantage of expressing high-level semantic information and have also been applied to saliency detection. Literature [19] used a convolutional neural network to predict the saliency of each region in local and global contexts, and then optimized the saliency of the target from a global perspective. Literature [22] used a variety of context-based convolutional neural networks to consider both global and local contexts and combined the two to predict saliency. Literature [10] extracted the global contrast by multiple scale estimated maps, and it generated saliency map by combining contrast and segment information. Recently, the saliency detection method based on deep learning has been performing better than traditional methods.

3   Proposed Method

In this section, we will introduce our background cleaning method. Our method can reduce the saliency of challenging background areas. And we will show the proposed way using minimum directional information in RBD method.

3.1   Background Cleaning

Background prior is a very common method in saliency detection, which measure the connectivity between pixels/regions and the background. The geodesic distance in [20] is widely used because of its advantages. In calculation of geodesic distance, first is to calculate the sum of the differences between adjacent pixels on all paths from the area to the image boundary. The minimum sum is the geodesic distance. Suppose I is a multi-channel image, the connection path between pixel i and pixel j in the image is $\pi = \{\pi(0), \pi(1), \ldots, \pi(k)\}$. Geodesic distance between pixel i and pixel j is defined as following:

$$gd(i,j) = \min_{\pi} \sum_{i=1}^{k} |\pi(i-1) - \pi(i)| \tag{1}$$

Using the background prior, the pixels at the boundary are taken as the elements of the set of seed S, and the geodesic saliency of pixel i is defined as following:

$$S(i) = \min_{j \in S} gd(i,j) \tag{2}$$

Simply using the above method to calculate the geodesic distance can easily lead to bad results in the final saliency. A background region surrounded by other colors background regions will have high saliency. Because the geodesic path to the boundary must cross other backgrounds with great color difference, which results in a large geodesic distance. Our BC method is to get accurate saliency of the challenging area.

For the calculation of geodesic distance, the adjacency matrix generated by image pixels is generally taken as 1 adjacent and 0 otherwise. Johnson's algorithm [8] is used to solve the shortest path problem. The result of using background cleaning is to generate an updated adjacency matrix. Using Johnson's algorithm on this matrix makes the geodesic distance calculation more reasonable. Figure 3 shows the pseudo code used for background cleaning. Firstly, we segment the image into superpixels as shown in Fig. 4(a). We use the superpixels that touch the boundary as the seed set S and get the original adjacency matrix. As shown in Fig. 4(b), all seeds are marked blue.

Algorithm 1(Background Cleaning)

Input: all superpixels SP, original adjcMatrix   //original adjacency matrix
Output: adjcMatrix   //adjacency matrix after background cleaning
seed superpixel S = all superpixels that touch the boundary
while (success to get one supixel r from seed S)
    for each superpixel t which is neighbor to r
      if ColDis(t, r) < threshold
        append superpixel t to seed S
        for p1,p2 which is neighbor to t
        adjcMatrix(p1, p2)=1

Fig. 3. Pseudo code of background cleaning.

The *ColDis* matrix records the LAB space differences of all superpixels. For each seed point, we calculate the color distance of the superpixels adjacent to it, and we use the biggest color difference value in the image multiplied by a coefficient to control the similarity threshold. If the distance is less than the threshold, we add the superpixel to the seed collection S. For each seed point, some adjacent and similar superpixels will be added to the seed set S. Figure 4(c) shows that the added superpixels are marked blue. The superpixels taken from the seed set cannot enter the seed set S again. The algorithm will stop when the seed set S is empty. The result of the background cleaning is shown in Fig. 4(d). The superpixels that can be cleared have already been cleared and they are marked blue.

After execution of Algorithm 1, the updated adjacency matrix is obtained. The advantage of this method is that after a certain area is cleared, its adjacent areas will be adjacent to each other. In this way, that challenging area

(a)   (b)   (c)   (d)

Fig. 4. The flowchart of background cleaning. (Color figure online)

connects similar areas on image boundary, and its geodesic distance shows low value. Later, Johnson's algorithm is used to calculate the geodesic distance of all superpixels in the image. Of course, the superpixels that are cleared here are background by default, and their geodesic distance is 0. After our method, the saliency defined by geodesic distance is more robust.

3.2 Minimum Directional Contrast Information

The method in [23] is very robust in the calculation of background prior and is one of the best methods to use background prior. We mentioned in [6] that the foreground is generally surrounded by the background. The contrast between the foreground and the surrounding areas is high, and the contrast in each direction will be high. So, we proposed the minimum directional contrast. The RBD method uses the global contrast information. We think global contrast is easily affected by noise, while the minimum directional contrast is more robust. So, we replace the global contrast with minimum directional contrast. We define the minimum contrast as follow:

$$MC(p) = \min_{TL,TR,BL,BR} \sum d(p,p_i)\, w_{sp}(p,p_i) \tag{3}$$

$d(p,p_i)$ denotes the color distance.

$$w_{sp}(p,p_i) = \exp\left(-\frac{d_{sp}^2(p,p_i)}{\sigma_{sp}^2}\right) \tag{4}$$

$d_{sp}^2(p,p_i)$ denotes the Euclidean distance. Here we expand in four directions (top left, top right, bottom left and bottom right) and use color and position weighting to get the contrast. The minimum directional contrast will be selected.

In RBD, the boundary connectivity is used to weight the global contrast, and this weight succeeds in background prior. We define the minimum directional contrast previously, and we also use the boundary connectivity for weight. We suggest the following definition:

$$wMC(p) = \min_{TL,TR,BL,BR} \sum d(p,p_i)\, w_{sp}(p,p_i)\, w_i^{bg} \tag{5}$$

w_i^{bg} is defined as following.

$$w_i^{bg} = 1 - \exp\left(-\frac{Bndcon^2(p_i)}{\sigma_{sp}^2}\right) \tag{6}$$

Because the minimum directional contrast is more robust, we calculate the minimum value of four directions when weighting position and color. In this way, the contrast information will be more robust. We also use the boundary connectivity for weighting. We only change the weighted contrast information in RBD, and then we follow the principled-optimized framework to define the saliency.

4   Experiment

We evaluate our method using minimum directional contrast with some widely used methods on public datasets PASCAL-S [11], DUTOMRON [21]. We use the PR curve, which converts the saliency map into a binary map with a threshold of 0–255 and compares it with the true map. The curve is the average of the entire data set. The PR curve considers that the saliency of the object is higher than the saliency of the background. F-measure is an index that comprehensively considers the accuracy rate and the recall rate. Therefore we also calculated the F-measure and Mean Absolute Error (MAE) between the saliency map and the ground truth.

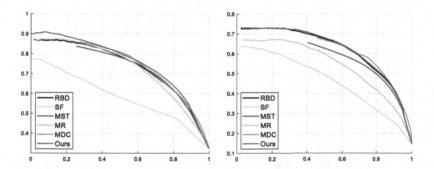

Fig. 5. PR curve. Left is on PASCAL-S, right is on DUTOMRON.

We compared our method with MR [21], MST [16], RBD [23], SF [12], MDC [6]. Figure 5 shows the PR curve comparison of our method to other methods on two public datasets. Our method MRBD shows better performance on two datasets. On the dataset PASCAL-S, our method MRBD is advantageous over other methods. And on the dataset DUTOMRON, our method performs little better than other methods. In Table 1, we evaluate the F-measure and MAE on all methods. The result is that our method MRBD shows notable F-measure and MAE on PASCAL-S, and functions better on DUTOMRON. In this paper, we proposed a BC method. This method is to reduce the saliency of challenging background areas, because it helps these area find a shortest path after background cleaning. If we combine the BC and minimum directional contrast information, it will show weakness in the experimental result on dataset

Table 1. F-measure and MAE comparison.

Methods/datasets	PASCAL-S		DUTOMRON	
	F-measure	MAE	F-measure	MAE
RBD	0.650	0.247	0.586	0.143
SF	0.520	0.309	0.453	0.144
MST	0.662	0.226	0.552	0.156
MR	0.636	0.266	0.503	0.177
MDC	0.658	0.221	0.585	0.144
Ours	0.664	0.235	0.585	0.144

ASD [1]. As the Fig. 6 has shown, this combination is little worse than original RBD method. So, the BC shows some advantages in some situations and weaknesses on the public datasets. For the most ideal results in datasets, this BC method should be adaptively used together with minimum directional contrast. However, only using the minimum directional contrast can greatly improve the RBD method.

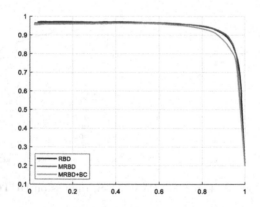

Fig. 6. BC combination result.

5   Conclusion

Background prior is widely-used knowledge in salient object detection, and geodesic distance is based on background prior knowledge. When measuring the difference between a pixel and image boundary, geodesic distance depends on a path which has the lowest cost to access image boundary. When some background areas are surrounded by areas that differ greatly in their colors, the best path to the boundary also causes a large cost and contributes to high saliency of these areas. Under this problem, we propose a background cleaning (BC) method. This method can reduce the saliency of these challenging background areas, and we've discussed its advantages and weaknesses in our experiments. For more ideal results in datasets, this BC method should be adaptively used. RBD (robust background detection) is a successful method in salient object detection, and it is based on the global contrast information, but global contrast is easily affected by noise. In previous work, MDC (minimum directional contrast) was proposed. And we find that the directional contrast information can improve RBD method, so we use the minimum contrast information of four directions. Experimental results on two public datasets showed that our method performs better than RBD, MDC and some widely-used methods.

References

1. Achanta, R., Hemami, S., Estrada, F., Susstrunk, S.: Frequency-tuned salient region detection. In: 2009 IEEE Conference on Computer Vision and Pattern Recognition, pp. 1597–1604. IEEE (2009)
2. Achanta, R., Shaji, A., Smith, K., Lucchi, A., Fua, P., Süsstrunk, S.: SLIC superpixels compared to state-of-the-art superpixel methods. IEEE Trans. Pattern Anal. Mach. Intell. 34(11), 2274–2282 (2012)
3. Alexe, B., Deselaers, T., Ferrari, V.: Measuring the objectness of image windows. IEEE Trans. Pattern Anal. Mach. Intell. 34(11), 2189–2202 (2012)
4. Cheng, M.M., Mitra, N.J., Huang, X., Torr, P.H., Hu, S.M.: Global contrast based salient region detection. IEEE Trans. Pattern Anal. Mach. Intell. 37(3), 569–582 (2014)
5. Frintrop, S., García, G.M., Cremers, A.B.: A cognitive approach for object discovery. In: 2014 22nd International Conference on Pattern Recognition, pp. 2329–2334. IEEE (2014)
6. Huang, X., Zhang, Y.J.: 300-FPS salient object detection via minimum directional contrast. IEEE Trans. Image Process. 26(9), 4243–4254 (2017)
7. Itti, L., Koch, C., Niebur, E.: A model of saliency-based visual attention for rapid scene analysis. IEEE Trans. Pattern Anal. Mach. Intell. 20(11), 1254–1259 (1998)
8. Johnson, D.B.: Efficient algorithms for shortest paths in sparse networks. J. ACM (JACM) 24(1), 1–13 (1977)
9. Koch, C., Ullman, S.: Shifts in selective visual attention: towards the underlying neural circuitry. In: Vaina, L.M. (ed.) Matters of Intelligence. SYLI, vol. 188, pp. 115–141. Springer, Dordrecht (1987). https://doi.org/10.1007/978-94-009-3833-5_5
10. Li, G., Yu, Y.: Deep contrast learning for salient object detection. In: Proceedings of the IEEE Conference on Computer Vision and Pattern Recognition, pp. 478–487 (2016)
11. Li, Y., Hou, X., Koch, C., Rehg, J.M., Yuille, A.L.: The secrets of salient object segmentation. In: Proceedings of the IEEE Conference on Computer Vision and Pattern Recognition, pp. 280–287 (2014)
12. Perazzi, F., Krähenbühl, P., Pritch, Y., Hornung, A.: Saliency filters: contrast based filtering for salient region detection. In: 2012 IEEE Conference on Computer Vision and Pattern Recognition, pp. 733–740. IEEE (2012)
13. Sharma, G., Jurie, F., Schmid, C.: Discriminative spatial saliency for image classification. In: 2012 IEEE Conference on Computer Vision and Pattern Recognition, pp. 3506–3513. IEEE (2012)

14. Siagian, C., Itti, L.: Rapid biologically-inspired scene classification using features shared with visual attention. IEEE Trans. Pattern Anal. Mach. Intell. **29**(2), 300–312 (2007)

15. Strand, R., Ciesielski, K.C., Malmberg, F., Saha, P.K.: The minimum barrier distance. Comput. Vis. Image Underst. **117**(4), 429–437 (2013)

16. Tu, W.C., He, S., Yang, Q., Chien, S.Y.: Real-time salient object detection with a minimum spanning tree. In: Proceedings of the IEEE Conference on Computer Vision and Pattern Recognition, pp. 2334–2342 (2016)

17. Viola, P., Jones, M.: Rapid object detection using a boosted cascade of simple features. In: Proceedings of the 2001 IEEE Computer Society Conference on Computer Vision and Pattern Recognition, CVPR 2001, vol. 1, p. I-I. IEEE (2001)

18. Wang, L., Xue, J., Zheng, N., Hua, G.: Automatic salient object extraction with contextual cue. In: 2011 International Conference on Computer Vision, pp. 105–112. IEEE (2011)

19. Wang, L., Lu, H., Ruan, X., Yang, M.H.: Deep networks for saliency detection via local estimation and global search. In: Proceedings of the IEEE Conference on Computer Vision and Pattern Recognition, pp. 3183–3192 (2015)

20. Wei, Y., Wen, F., Zhu, W., Sun, J.: Geodesic saliency using background priors. In: Fitzgibbon, A., Lazebnik, S., Perona, P., Sato, Y., Schmid, C. (eds.) ECCV 2012. LNCS, vol. 7574, pp. 29–42. Springer, Heidelberg (2012). https://doi.org/10.1007/978-3-642-33712-3_3

21. Yang, C., Zhang, L., Lu, H., Ruan, X., Yang, M.H.: Saliency detection via graph-based manifold ranking. In: Proceedings of the IEEE Conference on Computer Vision and Pattern Recognition, pp. 3166–3173 (2013)

22. Zhao, R., Ouyang, W., Li, H., Wang, X.: Saliency detection by multi-context deep learning. In: Proceedings of the IEEE Conference on Computer Vision and Pattern Recognition, pp. 1265–1274 (2015)

23. Zhu, W., Liang, S., Wei, Y., Sun, J.: Saliency optimization from robust background detection. In: Proceedings of the IEEE Conference on Computer Vision and Pattern Recognition, pp. 2814–2821 (2014)

A Robust Automatic Method for Removing Projective Distortion of Photovoltaic Modules from Close Shot Images

Yu Shen, Xinyi Chen, Jinxia Zhang[✉], Liping Xie, Kanjian Zhang,
and Haikun Wei[✉]

Key Laboratory of Measurement and Control of CSE, Ministry of Education,
School of Automation, Southeast University, Nanjing 210096, People's Republic of China
`{jinxiazhang,hkwei}@seu.edu.cn`

Abstract. Partial shading and hot spots may cause power loss and sometimes irreversible damage of photovoltaic (PV) modules. In order to evaluate the power generation of PV modules, it is necessary to calculate the area of shading and hot spots. The perspective distortion makes region closer to the camera have more pixels, which results in misestimate of shading area and hot spots. To solve this problem, a robust automatic method for removing projective distortion of photovoltaic modules from close shot images is proposed in this paper. Firstly, Images are converted to gray scale, and their edges are detected by Canny algorithm. Then, lines are detected by Hough transform, vanishing points are found by intersecting lines, and the line at infinity is specified by vanishing points. Next, the projective transformation matrix is decomposed into an affine transformation matrix and a simple projective transformation matrix, which has only two degrees of freedom. Affine rectification is derived from computing this simple projective transformation matrix, and finally, right angles are recovered by computing shear transformation matrix. The close shot visible and infrared images are collected for experiments. The results show that the proposed method performs better on rectification of PV modules from close shot images.

Keywords: Projective distortion · Photovoltaic · Affine rectification

1 Introduction

The field of photovoltaics has developed rapidly in recent years. In the operation of photovoltaic power station, partial shading [1, 2] and hot spots [3, 4] usually occur, and reduces the PV module output power. In order to evaluate the power loss caused by shading and hot spots of different types, different locations and different areas, it is feasible to take close shot images of photovoltaic (PV) modules, then combine computer vision [5, 6], pattern recognition [7], and machine learning [8] methods, to recognize the photovoltaic modules, shading and hot spots in the image. Object detection [9, 10], semantic segmentation [11–14], and salient analysis [15, 16] offer fruitful methods, but problems of projective distortion still remain. Projective distortion makes parallel lines

© Springer Nature Switzerland AG 2020
Y. Peng et al. (Eds.): PRCV 2020, LNCS 12305, pp. 707–719, 2020.
https://doi.org/10.1007/978-3-030-60633-6_59

in the real world imaged as converging lines and region closer to the camera occupy more pixels than those farther away.

In order to estimate the area of shading and hot spots, projective distortion of photovoltaic modules should be removed from the images. A common approach is to recognize a rectangular object in the real world and find its four corners in the image. It is possible to compute the transformation matrix and "undo" the projective distortion by the inverse transformation [17]. This method depends on the corners and the corners are usually need to be located manually. An automatic way to find corners can be realized by detecting and segmenting the rectangular object, which may cause deviation of corners and errors of transformation matrix. TILT [18] extracts a class of "low-rank textures" in a 3D scene, and automatically rectifies the texture in a 3D scene from user-specified windows in 2D images. TPS [19] rectifies objects by training a spatial transformer network to find the position of fiducial points. GCO [20] adopts a frequency approach for affine rectification of planar homogeneous texture, which manifests as regular repeating scene elements along a plane. Some works rectify images based on morphological analysis or registration techniques [21–23]. These methods are universal for various scenes and objects, but there is a paucity of work specially designed for PV modules rectification. Some existing methods may rectify inaccurately when the corners or edges of PV modules are covered.

The motivation of this paper is to propose a robust automatic method for removing projective distortion of photovoltaic modules from close shot images. The contribution of our work is: 1) creating a database for close-shot visible and Infrared (IR) images; 2) proposing a rectification method that makes full use of texture information of PV modules. Compared to existing methods, the proposed approach is more robust and performs well even if the corners or edges of PV modules are covered. Besides, the proposed method does not rely on a large amount of training samples, thus can be used for distributed PV stations with a small number of samples.

Firstly, 100 visible images and 600 infrared images are collected. The images are taken from close shot, thereby contain sufficient texture information of PV modules. Secondly, images are converted to gray scale and edges are detected by Canny algorithm [24–26]. Thirdly, lines are detected by Hough transform [27], vanishing points are specified, and the line at infinity is found. Next, the projective transformation matrix is decomposed into an affine transformation matrix and a simple projective transformation matrix, which has only two degrees of freedom. Then, affine rectification is conducted by computing this simple projective transformation matrix. Finally, right angles are recovered by computing shear transformation [28] matrix.

The rest parts of this paper are organized as follows: in Sect. 2, a robust automatic method for removing projective distortion of photovoltaic modules is proposed; the experiments and results are shown in Sect. 3 and conclusions are drawn in Sect. 4.

2   Method

In this section, a robust automatic method for removing projective distortion of photovoltaic modules from close shot images is described in detail. The framework is shown as

Fig. 1. The projective transformation is decomposed as a simple projective transformation and an affine transformation. The affine transformation can be further decomposed as a shear transformation and other three transformations (translation, rotation and scale).

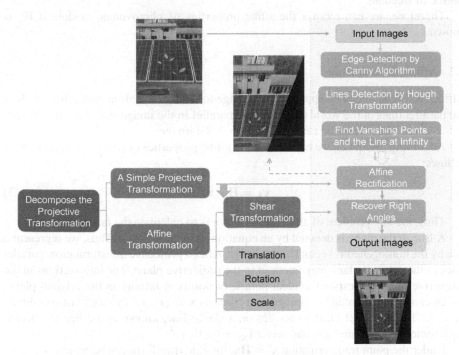

Fig. 1. The framework for removing projective distortion of photovoltaic modules

2.1 Decomposition of the Projective Transformation

A projective transformation can be represented by a homogeneous matrix $\mathbf{H} : \mathbf{x}' = \mathbf{Hx}$. The point is specified by a homogeneous vector $\mathbf{x} = (x_1, x_2, x_3)^T$, and the projective transformation is shown more detailedly as follows:

$$\begin{pmatrix} x_1' \\ x_2' \\ x_3' \end{pmatrix} = \begin{bmatrix} h_{11} & h_{12} & h_{13} \\ h_{21} & h_{22} & h_{23} \\ h_{31} & h_{32} & h_{33} \end{bmatrix} \begin{pmatrix} x_1 \\ x_2 \\ x_3 \end{pmatrix} \tag{1}$$

Where $\mathbf{H}$ is a non-singular 3-order square matrix. The projective transformation will not be changed if each element of $\mathbf{H}$ is multiplied by a non-zero factor, because the matrix $\mathbf{H}$ is homogeneous. That is, $\mathbf{H}$ has eight degrees of freedom, and h_{33} can be set as 1, thus a projective transformation can be decomposed as follows:

$$\mathbf{H} = \mathbf{H_P H_A} = \begin{bmatrix} 1 & 0 & 0 \\ 0 & 1 & 0 \\ v_1 & v_2 & 1 \end{bmatrix} \begin{bmatrix} a_{11} & a_{12} & a_{13} \\ a_{21} & a_{22} & a_{23} \\ 0 & 0 & 1 \end{bmatrix} \tag{2}$$

Where $\mathbf{H}_A$ is a affine transformation, which has six degrees of freedom. After an affine transformation, the parallel lines are mapped to parallel lines and ratio of areas remains unchanged. While $\mathbf{H}_P$ is another projective transformation, which has only two degree of freedom.

Therefore, we can recover the affine properties of photovoltaic module if $\mathbf{H}_p$ is known.

2.2   Affine Rectification

Affine rectification can be applied to the image to recover the affine properties, such as the parallel lines in the world plane are still parallel in the image, and the ratio of areas in the world plane keep the same as the ratio in the image.

According to Eq. (2), we can recover the affine properties of photovoltaic module as follows:

$$\mathbf{x}_A = \mathbf{H}_P^{-1}\mathbf{x}_P \tag{3}$$

Thus, the key problem of affine rectification is to calculate the matrix $\mathbf{H}_P$.

A line in the plane is denoted by an equation $ax + by + c = 0$. Thus, we represent a line by the homogeneous vector $\mathbf{l} = (a, b, c)^T$. After a projective transformation, parallel lines in the original plane may intersect in the projective plane. The intersections in the projective plane correspond to ideal points, or points at infinity in the original plane, which can be represented by homogeneous vectors $\mathbf{x} = (x_1, x_2, 0)^T$ with last coordinate $x_3 = 0$. The set of all ideal points lies on a single line, known as the line at infinity, represented by the homogeneous vector $\mathbf{l}_\infty = (0, 0, 1)^T$.

Under the point transformation $\mathbf{x}' = \mathbf{H}\mathbf{x}$, the line transform can be represented as:

$$\mathbf{l}' = \mathbf{H}^{-T}\mathbf{l} \tag{4}$$

It can be proved that after the affine transformation $\mathbf{H}_A$, the line at infinity remains the line at infinity. While after the projective transformation $\mathbf{H}_P$, the line at infinity may correspond to a line represented as a homogeneous vector $\mathbf{l}'_\infty = (l_1, l_2, 1)^T$:

$$\mathbf{l}'_\infty = \mathbf{H}_P^{-T}\mathbf{l}_\infty \tag{5}$$

That is:

$$\begin{bmatrix} 1 & 0 & v_1 \\ 0 & 1 & v_2 \\ 0 & 0 & 1 \end{bmatrix} \begin{pmatrix} l_1 \\ l_2 \\ 1 \end{pmatrix} = \begin{pmatrix} 0 \\ 0 \\ 1 \end{pmatrix} \tag{6}$$

v_1 and v_2 in matrix $\mathbf{H}_P$ can be derived as follows:

$$\begin{cases} v_1 = -l_1 \\ v_2 = -l_2 \end{cases} \tag{7}$$

Therefore, we can get the matrix $\mathbf{H}_P$ if the line $\mathbf{l}'_\infty$ has been found in the image.

2.3 How to Find the Line at Infinity?

Firstly, RGB images are converted to grayscale. Secondly, edges are found by Canny edge detection algorithm. Thirdly, lines are detected by Hough transformation. (see Fig. 6(b))

Hough transformation converts image space to parameter space. In the image space, each pair of x and y corresponds to a point, and a line is denoted by a pair of k and b:

$$y = kx + b \tag{8}$$

In the parameter space, each pair of k and b corresponds to a point, and a line is denoted by a pair of x and y:

$$b = -xk + y \tag{9}$$

That means lines and points are dual in the image space and parameter space. Thus, if many points lie on a line in the parameter space, it may represent that lines intersect at a vanishing point. The location of this vanishing point can be calculated by Eq. (9). (see Fig. 7(a))

Parallel lines in the photovoltaic module has two directions (see Fig. 4 and Fig. 5), and we can recognize the line at infinity in the image in different conditions: 1) If parallel lines of each direction intersect respectively at a point in the image after projective transformation, the line at infinity in the image is simply the line through these two points (see Fig. 2(a)). 2) If parallel lines of one direction intersect at a point, and parallel lines of the other direction keep parallel, the line at infinity in the image can be considered as the line through this point and parallels with these parallel lines in the image (see Fig. 2(b)).

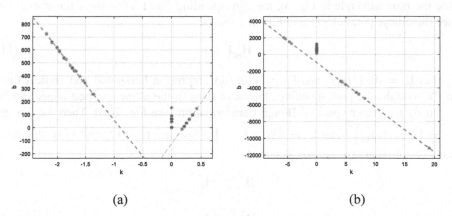

(a)   (b)

Fig. 2. Points in parameter space after Hough transformation (a) two lines in parameter space correspond to two vanishing points in image space; (b) a single line in parameter space corresponds to a single vanishing point in image space.

According to Sect. 3.2, recognizing the line at infinity in the image $\mathbf{l}'_\infty$ make it possible to recover affine properties. (see Fig. 6(c))

2.4   Recovery of Right Angles

Affine rectification is enough for area estimation of shading and hot spots. However, considering that the location of shading and hot spots also influence the power generation loss of PV modules, we can do one more step to recover the right angles of PV modules.
Right angles can be recovered by calculating shear transformation matrix $\mathbf{H}_{sh}$:

$$\mathbf{H}_{sh} = \begin{bmatrix} 1 & sh_y & 0 \\ sh_x & 1 & 0 \\ 0 & 0 & 1 \end{bmatrix} \tag{10}$$

sh_x denotes the shear factor along the x axis, and sh_y represents the shear factor along the y axis. (see Fig. 3)

(a)   (b)

Fig. 3. Shear transformation (a) shear transformation along the x axis; (b) shear transformation along the y axis.

According to Eq. (4), suppose $\mathbf{l}$ is the line in the plane without shear transformation (see the blue rectangle in Fig. 3), the corresponding line $\mathbf{l}'$ after shear transformation (see the orange parallelogram in Fig. 3) satisfies:

$$\mathbf{H}_{sh}^T \mathbf{l}' = \mathbf{l} \tag{11}$$

Let $\mathbf{l}_a = (0, 1, l_{3a})^T$ and $\mathbf{l}_b = (1, 0, l_{3b})^T$ represent horizontal and vertical lines in PV module, the corresponding slope k of lines in the affine rectified image can be found by parameter space of Hough transformation (see Fig. 7(b)). Then we can get $\mathbf{l}_a' = \left(k_a, -1, l_{3a}'\right)^T$ and $\mathbf{l}_b' = \left(k_b, -1, l_{3b}'\right)^T$. Solving Eq. (12) and (13), the shear transformation matrix can be calculated.

$$\mathbf{H}_{sh}^T \mathbf{l}_a' = \mathbf{l}_a \tag{12}$$

$$\mathbf{H}_{sh}^T \mathbf{l}_b' = \mathbf{l}_b \tag{13}$$

It can be proved that the shear transformation matrix is invertible. Thus, the right angles can be recovered (see Fig. 6(d)) as follows:

$$\mathbf{x}' = \mathbf{H}_{sh}^{-1} \mathbf{x} \tag{14}$$

3   Experiments

Data collection is described in Sect. 3.1. Experiments for removing projective transformation distortion of PV modules are shown in Sect. 3.2. In order to demonstrate the application value of our work, we also do one more step to estimate the area of shading and hot spots, shown in Sect. 3.3.

3.1   Data Collection

We collect 100 close-shot visible images of photovoltaic modules in our photovoltaic experimental system. A cellphone is used to get visible images (1440 × 1080 pixels) on sunny days, as shown in Fig. 4. The left image corresponds to a normal PV module, and other images show PV module with shading of leaves.

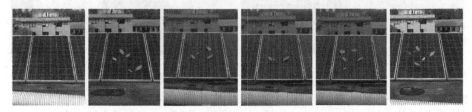

Fig. 4. Visible images of photovoltaic modules

600 infrared images (320 × 240 pixels) are provided by an enterprise specializing in manufacturing photovoltaic products. FLIR infrared camera was used to obtain the images on sunny days. Infrared images of photovoltaic modules are shown as Fig. 5. The left column of images corresponds to normal PV module and other images show hot-spotted PV module.

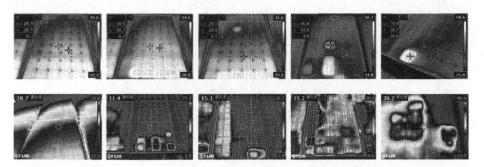

Fig. 5. Infrared images of photovoltaic modules

3.2 Removing Perspective Distortion of PV Modules

The results of the proposed method are shown in steps as follows: For an original image (see Fig. 6(a)), firstly it was converted to grayscale. Edges are found by Canny edge detection algorithm and lines are detected by Hough transformation. (see Fig. 6(b)) Affine properties are recovered as shown in Fig. 6(c) and right angles are recovered as shown in Fig. 6(d).

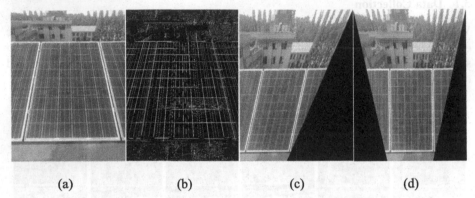

(a)   (b)   (c)   (d)

Fig. 6. Removing projective distortion of photovoltaic modules (a) original image; (b) lines detection by Hough transformation; (c) affine rectification; (d) recovery of right angles.

Figure 7(a) shows how the vanishing point is specified. The green line in parameter space corresponds to a point in the image space, and this point is a vanishing point.

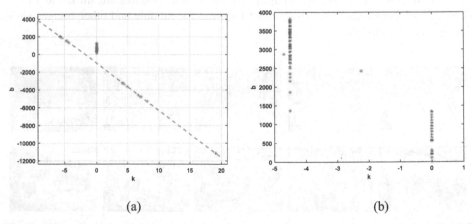

(a)   (b)

Fig. 7. Points in parameter space after Hough transformation (a) points in parameter space of original image; (b) points in parameter space of affine rectified image. (Color figure online)

Figure 7(b) shows the slopes of lines in Fig. 6(c). We can see clearly that there are two main directions of lines.

Results for more images are shown as follows: Fig. 8 displays the rectified visible images of PV modules corresponding to Fig. 4, Fig. 9 shows the rectified IR images of PV modules corresponding to Fig. 5.

Fig. 8. The rectified visible images of PV modules

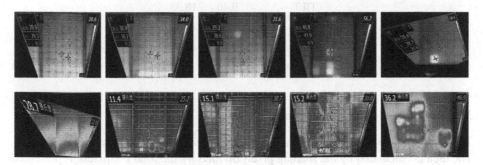

Fig. 9. The rectified IR images of PV modules

For a quantitative evaluation, we adopt the following metric:

$$error_1 = \frac{1}{N} \sum_{i=1}^{N} \sqrt{\left(\hat{a}_i - a_i\right)^2 + \left(\hat{b}_i - b_i\right)^2} \tag{15}$$

$$error_2 = \frac{1}{4N} \sum_{i=1}^{N} \sum_{c=1}^{4} \left|angle_{ic} - 90°\right| \tag{16}$$

Where $error_1$ is used to estimate the performance of affine rectification and $error_2$ is used for right angle estimation. N is the number of test samples: $N = 30$ for visible images and $N = 50$ for IR images. In Eq. (15), $(a, b, 1)^T$ represents the line at infinity homogeneously. For each test case, a pair of vanishing points was manually annotated, giving the vanishing line. Thus, a_i, b_i are the ground truth parameters and $\hat{a}_i$, $\hat{b}_i$ is the calculated by affine rectification algorithm. In Eq. (16), $angle_{ic}(c = 1, 2, 3, 4)$ are the angles of corners of the PV module after rectification for sample i.

We adopted TILT [18], TPS [19], and GCO [20] algorithms for comparison (Tables 1 and 2):

We can see clearly that the proposed method performs better than the other three methods for PV module rectification of visible images and IR images.

Table 1. The performance of visible images PV module rectification

Method/Metric	$error_1$	$error_2$
TILT	0.42	17.72
TPS	0.29	12.21
GCO	0.17	10.17
Proposed	0.06	3.32

Table 2. The performance of IR images PV module rectification

Method/Metric	$error_1$	$error_2$
TILT	0.46	18.33
TPS	0.32	14.27
GCO	0.16	10.68
Proposed	0.08	3.96

3.3   Anomalous Area Estimation

The area of shading in visible images are calculated, the results is shown in Table 3. The area is calculated by the shading pixels divided by the total PV pixels, represented in percentage. The ground truth is defined by manually specifying 4 corners of the PV modules.

Table 3. The area of shading in some visible images

	Image1	Image 2	Image 3	Image 4	Image 5	Error
Ground truth	3.15%	3.92%	2.66%	5.03%	4.70%	
Original	4.23%	2.86%	2.08%	4.00%	3.70%	16.47%
TILT	3.72%	3.88%	2.41%	4.67%	4.51%	6.33%
TPS	3.12%	3.85%	2.59%	4.85%	4.77%	2.09%
GCO	3.23%	3.81%	2.56%	4.88%	4.73%	6.16%
proposed	3.18%	3.96%	2.60%	4.97%	4.69%	1.04%

The area of hot spots in IR images of mono-crystalline silicon are calculated, the results is shown in Table 4.

Table 4. The area of hot spots in some IR images

	Image1	Image 2	Image 3	Image 4	Error
Ground truth	8.33%	5.00%	6.67%	1.67%	
Original	17.73%	9.17%	10.73%	2.39%	75.06%
TILT	8.15%	4.86%	6.79%	1.57%	3.19%
TPS	8.19%	4.88%	6.59%	1.76%	2.67%
GCO	8.21%	4.92%	6.82%	1.75%	3.47%
proposed	8.25%	4.96%	6.71%	1.72%	1.34%

4   Conclusion

A robust automatic method for removing projective distortion of photovoltaic modules from close shot images is proposed in this paper. Firstly, images are converted to gray scale, and edges are detected by Canny algorithm. Then, lines are detected by Hough transform, vanishing points are found by intersecting lines, and the line at infinity is recognized by vanishing points. Next, the projective transformation matrix is decomposed into an affine transformation matrix and a simple projective transformation matrix, which has only two degrees of freedom. Finally, affine rectification is achieved by computing this simple projective transformation matrix, and right angles are recovered by computing shear transformation matrix. The close-shot visible and infrared images are collected for experiments. The results show that the proposed method performs better than the other three methods for rectifying the PV modules in close-shot images.

Acknowledgement. This work was supported by the National Key Research and Development Program of China (Grant No. 2018YFB1500800), and by the National Natural Science Foundation of China (Grant No. 61773118, Grant No. 61703100, Grant No. 61973083, Grant No. 61802059), and by the Natural Science Foundation of Jiangsu (Grant No. BK20170692, Grant No. BK20180365), and by the Zhishan Young Scholar Program of Southeast University and the Fundamental Research Funds for the Central Universities (Grant No. 2242020R40119).

References

1. Bingol, O., Ozkaya, B.: Analysis and comparison of different PV array configurations under partial shading conditions. Solar Energy **160**, 336–343 (2018)
2. Zhu, L., Li, Q., Chen, M., Cao, K., Sun, Y.: A simplified mathematical model for power output predicting of building integrated photovoltaic under partial shading conditions. Energy Convers. Manage. **180**, 831–843 (2019)
3. Niazi, K.A.K., Akhtar, W., Khan, H.A., Yang, Y., Athar, S.: Hotspot diagnosis for solar photovoltaic modules using a naive bayes classifier. Sol. Energy **190**, 34–43 (2019)
4. Dunderdale, C., Brettenny, W., Clohessy, C., Dyk, E.E.: Photovoltaic defect classification through thermal infrared imaging using a machine learning approach. Prog. Photovoltaics Res. Appl. **28**(3), 177–188 (2019)

5. Forsyth, D., Ponce, J.: Computer Vision: A Modern Approach (2002)
6. Szeliski, R.: Computer Vision: Algorithms and Applications. Springer, London (2011). https://doi.org/10.1007/978-1-84882-935-0
7. Bishop, C.M.: Pattern Recognition and Machine Learning (Information Science and Statistics). Springer, New York (2006)
8. Li, H.: Statistical Learning Methods. Tsinghua University Press, Beijing (2012)
9. Ren, S., He, K., Girshick, R., Sun, J.: Faster R-CNN: towards real-time object detection with region proposal networks. IEEE Trans. Pattern Anal. Mach. Intell. **39**(6), 1137–1149 (2017)
10. Redmon, J., Divvala, S., Girshick, R., Farhadi, A.: You only look once: unified, real-time object detection (2015)
11. Kaiming, H., Georgia, G., Piotr, D., Ross, G.: Mask R-CNN. IEEE Trans. Pattern Anal. Mach. Intell. (2018)
12. Huang, Z., Huang, L., Gong, Y., Huang, C., Wang, X.: Mask scoring r-cnn (2019)
13. Chen, L.-C., et al.: Encoder-decoder with atrous separable convolution for semantic image segmentation. In: ECCV (2018)
14. Badrinarayanan, V., Kendall, A., Cipolla, R.: SegNet: a deep convolutional encoder-decoder architecture for scene segmentation. IEEE Trans. Pattern Anal. Mach. Intell. **39**(12), 2481–2495 (2017)
15. Zhang, J., Fang, S., Ehinger, K.A., Wei, H., Yang, J.: Hypergraph optimization for salient region detection based on foreground and background queries. IEEE Access **6**, 26729–26741 (2018)
16. Zhang, J., Ehinger, K.A., Wei, H., Zhang, K., Yang, J.: A novel graph-based optimization framework for salient object detection. Pattern Recogn. **64**(C), 39–50 (2016)
17. Hartley, R., Zisserman, A.: Multiple View Geometry in Computer Vision (2000)
18. Zhang, Z., Ganesh, A., Liang, X., Ma, Y.: TILT: transform invariant low-rank textures. Int. J. Comput. Vision **99**(1), 1–24 (2012)
19. Shi, B., Wang, X., Lyu, P., Yao, C., Bai, X.: Robust scene text recognition with automatic rectification. In: Computer Vision and Pattern Recognition (2016)
20. Ahmad, S., Cheong, L.-F.: Robust detection and affine rectification of planar homogeneous texture for scene understanding. Int. J. Comput. Vision **126**(8), 822–854 (2018). https://doi.org/10.1007/s11263-018-1078-2
21. Ahmad, S., Cheong, L.-F.: Facilitating and exploring planar homogeneous texture for indoor scene understanding. In: Leibe, B., Matas, J., Sebe, N., Welling, M. (eds.) ECCV 2016. LNCS, vol. 9906, pp. 35–51. Springer, Cham (2016). https://doi.org/10.1007/978-3-319-46475-6_3
22. Pritts, J., Chum, O., Matas, J.: Detection, rectification and segmentation of coplanar repeated patterns. In: Computer Vision and Pattern Recognition (2014)
23. Shi, B., Yang, M., Wang, X., Lyu, P., Yao, C., Bai, X.: ASTER: an attentional scene text recognizer with flexible rectification. IEEE Trans. Pattern Anal. Mach. Intell. **41**(9), 2035–2048 (2019)
24. Canny, J.: A computational approach to edge detection. IEEE Trans. Pattern Anal. Mach. Intell. PAMI **8**(6), 679–698 (1986)
25. Lim, J.S.: Two-Dimensional Signal and Image Processing, pp. 478–488. Prentice Hall, Englewood Cliffs (1990)
26. Parker, J.R.: Algorithms for Image Processing and Computer Vision, pp. 23–29. Wiley, New York (1997)
27. Duda, R.O., Hart, P.E.: Use of the hough transformation to detect lines and curves in pictures. Comm. ACM **15**, 11–15 (1972)
28. Zhuang, X.: Digital affine shear transforms: fast realization and applications in image/video processing. SIAM J. Imaging Sci. **9**(3), 1437–1466 (2016)

29. Xie, L., Tao, D., Wei, H.: Early expression detection via online multi-instance learning with nonlinear extension. IEEE Trans. Neural Netw. Learn. Syst. (TNNLS) **30**(5), 1486–1496 (2019)
30. Xie, L., Guo, W., Wei, H., Tang, Y., Tao, D.: Efficient unsupervised dimension reduction for streaming multi-view data. IEEE Trans. Cybern. https://doi.org/10.1109/tcyb.2020.2996684

Subset Ratio Dynamic Selection for Consistency Enhancement Evaluation

Kaixun Wang, Hao Liu$^{(\boxtimes)}$, Gang Shen, and Tingting Shi

College of Information Science and Technology, Donghua University, Shanghai 201620, China
liuhao@dhu.edu.cn

Abstract. Due to poor imaging conditions, large-scale underwater images need the consistency enhancement. According to the subset-guided consistency enhancement evaluation criterion, the existing subset selection methods need too much candidate samples from a whole imageset without any adaptation on data content. Therefore, this paper proposes a subset ratio dynamic selection method for consistency enhancement evaluation. The proposed method firstly divides the candidate samples into several sampling subsets. Based on a non-return sampling strategy, the consistency enhancement degree of an enhancement algorithm is obtained for each sampling subset. By using the student-t distribution under a certain confidence level, the proposed method can adaptively determine the subset ratio for a whole imageset, and the candidate subset is used to predict the consistency enhancement degree of the enhancement algorithm on the whole imageset. Experimental results show that as compared with the existing subset selection methods, the proposed method can reduce the subset ratio in all cases, and correctly judge the consistency performance of each enhancement algorithm. With similar evaluation error, the subset ratio of the proposed method can decrease by 2%~20% over that of the subset fixed ratio method, and decrease by 3%~13% over that of the subset gradual addition method, and thus the complexity is reduced for subset-guided consistency enhancement evaluation.

Keywords: Underwater image · Candidate subset · Dynamic selection · Confidence level · Consistency enhancement

1 Introduction

In recent years, underwater imaging technology has attracted considerable attention in image processing and computer vision [1]. But large-scale underwater images are often low quality and non-reference, so they need the consistency enhancement. Image enhancement refers to highlighting useful information and weakening useless information according to a specific requirement. The image enhancement aims to make the enhanced image more suitable for the visual characteristics of human eyes or machine vision [2]. Some representative underwater image enhancement algorithms can be used in challenging situations such as dynamic or unknown scenes. For example, the contrast enhancement algorithm enables the probability density function of the grayscale level of the image to meet the approximate uniform distribution, thereby this algorithm

© Springer Nature Switzerland AG 2020
Y. Peng et al. (Eds.): PRCV 2020, LNCS 12305, pp. 720–733, 2020.
https://doi.org/10.1007/978-3-030-60633-6_60

increases the dynamic range of the image and improves its contrast [3]. The contrast limited adaptive histogram equalization (CLAHE) [4] algorithm is an improvement on the traditional histogram equalization algorithm, and it can overcome noises by limiting image contrast. Voronin *et al.* [5] proposed a hybrid equalization (HE) algorithm by combining multiple physical model-based enhancement methods. The color channel compensation (3C) algorithm proposed by Ancuti *et al.* [6] is a kind of underwater image enhancement algorithm which has been representative in recent years. In addition, underwater image acquisition often faces unknown scenarios. Due to complex underwater environment and lighting conditions, underwater images can be degraded by wavelength dependence absorption, forward scattering and backward scattering [7]. Though Blasinski *et al.* [8] provided an open-source underwater image simulation tool, there still exists a gap between the synthetic underwater images with reference and real-world underwater images without reference. According to Shannon's information theory, the Entropy metric is widely used to represent the information uncertainty and complexity of an image [9]. In addition, the underwater color image quality evaluation (UCIQE) is served as a no-reference metric [10], where the measures of chrominance variation, average saturation, and luminance contrast of an underwater image is linearly combined. For a single image, its quality score is provide by an enhancement algorithm [11]. With the advent of big-data era, the amount of image data is increasing rapidly, and the enhancement task is gradually facing various imagesets. For an imageset, the average value of all image scores is usually selected to judge the enhancement algorithm [12–14]. Further, Liu *et al.* [15] proposed a subset-guided consistent enhancement evaluation criterion for the non-reference underwater imageset, and the criterion performs the data-driven performance evaluation on any underwater imageset, which can effectively find the optimal consistency enhancement algorithm for each imageset.

Due to a large number of images and rich enhancement algorithms, the candidate subset from the original imageset have to be extracted as a sampling test to reduce the complexity of the application system. The consistent enhancement assessment criterion utilizes a subset fixed ratio (SFR) method [15]. The SFR method selects a certain ratio of a candidate subset from the original imageset, where the consistency of an enhancement algorithm is evaluated quantitatively by calculating the consistency enhancement of the candidate subset. The subset ratio obtained by the SFR method is very empirical. Due to the large step between adjacent subset ratios, the sampling data often insufficient or overflow. So the SFR method cannot be adapted to the imageset during sample selection. In order to reduce the randomness of sample selection, Ancuti *et al.* [16] proposed a subset gradual addition (SGA) method to select a subset, where the subset selection is finished in a small step-by-step increment, and some samples from the large imageset are taken repeatedly to obtain multiple small-ratio subset information. The SGA method adjusts the subset ratio more finely than the existing SFR method. Though the SGA method utilizes a smaller step to reduce the random of sampling results, it still lacks the self-adaptability of the data content and difficultly selects the appropriate subset ratio for any imageset.

Therefore, according to the subset-guided consistent enhancement evaluation criterion, this paper proposes a dynamic selection method for the subset ratio during consistent enhancement evaluation. The proposed method utilizes a more precise procedure to divide the candidate subset into multiple sampling subsets. In order to gradually obtain the information of these sampling subsets and provide a comprehensive judgment, the proposed method adopts a novel sampling strategy to adaptively select a subset ratio for different original imagesets.

2 Consistently Enhancement Evaluation Criterion

To reduce system complexity, the consistent enhancement evaluation criterion selects a certain subset ratio from the original imageset to evaluate the subset, then it judges the consistency performance of an enhancement algorithm according to the evaluation results of the subset, and finally it applies the enhancement algorithm to the original imageset. Figure 1 is a flowchart of consistent enhancement evaluation with the subset selection. Firstly, a certain percentage of the subset images is selected from the original imageset, and the image quality metric Q is used to evaluate the quality score of all subset images (the initial subset $I_1, I_2, \cdots, I_n$). The initial quality score α_r of the initial image I_r is obtained as a parameter in the consistent enhancement evaluation criterion; r $(r = 1, 2, \cdots, n)$ is the image label, and n is the image quantity contained in the initial subset. Secondly, an image enhancement algorithm E is used to enhance the image quality of all initial images in the subset, the enhanced imageset is obtained as $I'_1, I'_2, \cdots, I'_n$, and the metric Q is used again to evaluate the quality of all the enhanced images I'_r, and the enhanced quality score β_r is obtained. Then the difference QSD_r is calculated between the quality scores before and after enhancement. If the value is positive, it means under the metric Q, the image enhancement algorithm E improves the quality of the initial image I_r; Otherwise, it reduces the quality of the initial image I_r, and then it calculates the mean value U of QSD_r. The equation is as follows:

$$U = \frac{1}{n} \sum_{r=1}^{n} QSD_r \tag{1}$$

Then, it calculates the standard deviation S of QSD_r:

$$S = \frac{1}{n} \sqrt{\sum_{r=1}^{n} (QSD_r - U)^2} \tag{2}$$

The proposed criterion determines the parameter λ according to the requirements of an application scenario, and substitute the parameter into the following equation:

$$\varepsilon = \lambda \cdot S \tag{3}$$

The parameter ε gives a confidence interval $[U - \varepsilon, U + \varepsilon]$, and this confidence interval is used to filter all the values of QSD_r, so as to keep the values of QSD_r within the confidence interval as valid test data. The valid data is sorted in ascending order by the QSD_r subscript number, i.e.,$QSD_1, QSD_2, \ldots, QSD_m$, where j $(j = 1, 2, \ldots, m)$ is the new image subscript, and m $(m \leq n)$ is the amount of valid data. Then the valid data will be compared with zero. If all valid data are greater than zero, it means that under the image quality metric Q, the image enhancement algorithm E is a consistency enhancement algorithm; Otherwise, the image enhancement algorithm E is a non-consistent enhancement algorithm, and the original image quality will deteriorate. When image enhancement algorithm E is a consistency enhancement algorithm, by the above effective test data, the minimum quality score of QSD_j (QSD_{min}) is determined to indicate the worst case scenario:

$$QSD_{min} = min\{QSD_1, QSD_2, \ldots, QSD_m\} \qquad (4)$$

Then, the mean quality score (QSD_{ave}) of QSD_j is also calculated as the average case of valid data:

$$QSD_{ave} = ave\{QSD_1, QSD_2, \ldots, QSD_m\} \qquad (5)$$

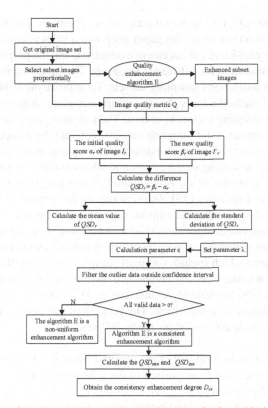

Fig. 1. Flowchart of consistent enhancement evaluation criterion with the subset selection

Finally, under the image quality metric Q, the criterion obtains the consistency enhancement degree D_{ce} of the image enhancement algorithm E for the imageset:

$$D_{ce} = (1 - \mu) \cdot QSD_{ave} + \mu \cdot QSD_{min} \qquad (6)$$

where a weighting factor $\mu \in [0, 1]$ indicates the relative importance of the QSD_{min} component which depends on a specific application scenario. The D_{ce} value of each enhancement algorithm can be obtained through the above evaluation process, and the highest D_{ce} value is used to select the best consistency enhancement algorithm for the original imageset. In the SFR method, the candidate subset is randomly selected from a large imageset. In order to reduce the problem of selecting a subset of excessive ratio because of the large step between the selected subset ratios, the SGA method adopts a step-by-step increment approach to select the candidate subset through domain knowledges, and it uses smaller steps to repeatedly select a portion of a large imageset so as to obtain multiple small subsets of information.

3 Subset Ratio Dynamic Selection

The SGA method can reduce the subset ratio, but it still cannot determine the proportion of the candidate subset by imageset adaptation. Therefore, based on the consistent enhancement evaluation criterion, this paper proposes a subset-ratio dynamic selection (DSCE) method for consistent enhancement evaluation, and the proposed method adaptively selects the subset ratio for a different imageset. The proposed method first divides the candidate subset into several sampling subsets. To improve the reliability during sampling, the proposed method adopts the non-return sampling strategy [18] for multiple sampling to get the sampling subset X_i in turn. Then an underwater image enhancement algorithm is used to enhance the sampling subsets X_i image-by-image. The consistency enhancement D_{ce} value of the image enhancement algorithm for each sampling subsets is recorded as D_i according to the consistent enhancement evaluation criterion. The mean and standard deviation of D_i is calculated. According to the central limit theorem of probability and statistics [19], D_i conforms to the normal distribution. Further, the proposed method can predict the consistency enhancement degree D_{ce} value of any imageset by an enhancement algorithm. Based on less sampling subset information, the student-t distribution is used to estimate the mean of a dataset with a normal distribution and unknown variance which includes a small sample ($n < 30$) [19]. In this paper, the student-t distribution under a certain confidence level is used to dynamically determine the subset ratio, and the consistency enhancement D_{ce} value of the original imageset under the image enhancement algorithm. The process of subset ratio dynamic selection is shown in Fig. 2, whose detailed steps are as follows:

Step 1: The candidate subset ratio R of the original imageset is further equally divided into the subsampling ratio of P proportion of the subsampling ratio f.

Step 2: A subset of R proportions are given according to the non-return sampling strategy, and the sampling subset X_i is obtained in turn, where i is the subset label, $i = 1, 2, \ldots, n$, where n is the final total number of extracted subsets. k is the image label in the sampling subset X_i, $k = 1, 2, \ldots, m$, where m is the total number of images in the sampling subset X_i.

Step 3: The underwater image enhancement algorithm E used to enhance the sampling subsets X_i, and the consistency enhancement D_{ce} value of image enhancement algorithm E is calculated according to the consistent enhancement evaluation criterion for each sampling subset X_i, which is recorded as D_i.

Step 4: The mean value M of the consistency enhancement D_i is calculated for all subsets. The equation is as follows:

$$M = \frac{1}{n} \sum_{i=1}^{n} D_i \tag{7}$$

Then the standard deviation S of D_i is calculated as follows:

$$S = \sqrt{\frac{1}{n} \sum_{i-1}^{n} (D_i - M)^2} \tag{8}$$

Step 5: According to the requirements of the application scenario, the student-t distribution at a certain confidence level is selected, and the significance level is a. Typically, when the confidence level is 95%, $a = 0.05$, then the proposed method calculates the fluctuation range δ of D_i:

$$\delta = t_{0.5a}(n-1) \frac{S}{\sqrt{n}} \tag{9}$$

In the above formula, n is the number of sampling subsets, and γ is the threshold, which are determined according to different imagesets. The smaller the value, the higher the accuracy of the sampling results. If δ is greater than or equal to γ, go to Step 2 to continue sampling; Otherwise, go to Step 6.

Step 6: When δ is less than γ, the sampling process is terminated, and the X_i of multiple sampling subsets is obtained. The images of the resulting samples are combined into a total sample imageset. The proposed method calculates the consistency enhancement D_{ce} value of the image enhancement algorithm E for total sample imageset according to consistent enhancement evaluation criterion, which is denoted as D_{ce_all}. Finally, at a certain confidence level, the proposed method concludes that the consistency enhancement degree of the underwater image enhancement algorithm E towards the original imageset is D_{ce_all}.

4 Experimental Results

4.1 Parameter Setting

Experimental simulation is carried out to verify the performance of the proposed method. The experiments are performed on a PC with MATLAB R2018a, which is configured as Intel i5-4210 M CPU (2.60 GHz), 4.0 GB memory, and Windows 10 operating system. The proposed method requires some parameter settings: underwater image enhancement algorithm E, image quality metric Q, threshold γ, significance level a. Based on the consideration of the experimental complexity, the parameter values in the experiment are

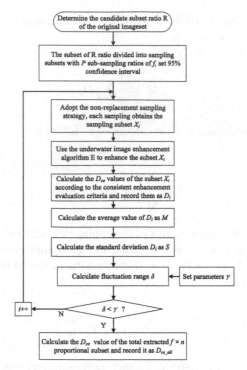

Fig. 2. Process of subset ratio dynamic selection

selected as follows: the student-t distribution under the 95% confidence level which the significance level $a = 0.05$, and the threshold γ is set according to different imagesets. Because the goal of the proposed method is to evaluate the enhancement performance of an enhancement algorithm towards the imageset, and the HE and 3C algorithms are often regarded as the good consistency enhancement algorithm for the original imageset. The popular UCIQE metric is selected to evaluate the quality of each image, where standard parameters are $\lambda = 2$, $\mu = 0.5$.

This experiment selects 1000 underwater images from a large underwater imageset as the original imageset **A** [17]. According to the consistency enhancement evaluation criterion, the consistency performance of the HE algorithm is evaluated on the original imageset **A**. The consistency enhancement degree D_{ce} of the imageset **A** is 0.0881, then the imageset **A** is tested by the SFR and SGA methods. The consistency enhancement degree D_{ce} for each candidate subset are calculated in turn according to the consistent enhancement criterion, whose error value φ with original imageset is calculated.

4.2 Experimental Results and Analysis

In order to eliminate the accidental results of the experiments, the proposed method conducts an experimental analysis on three underwater imagesets. Imageset **A** has 1000 images; Imageset **B** from another large underwater imageset by selecting 1000 underwater images [20]; Imageset **C** is the collection of imageset **A** and imageset **B**, which has

2000 underwater images. According to the experimental steps of the consistent enhancement evaluation criterion, the consistency enhancement degree D_{ce} of the HE algorithm are respectively obtained on the original imageset **A**, imageset **B** and imageset **C**, which is recorded in Table 1.

Table 1. The D_{ce} value of each original imageset

Imageset	Imageset A	Imageset B	Imageset C
D_{ce}	0.0881	0.0887	0.0884

In the following, different subset selection methods will be carried out on imageset **A**, imageset **B**, and imageset **C**, where the advantages of the proposed method are discussed in detail. The experiments are as follows:

Experiment 1: The imageset **A** is tested by the DSCE method, where $\gamma = 0.015$. First, the candidate subsets are divided into 10, 20, 25, 40 and 50 copies, and each consists of 10 images, and a subsample ratio of 1% step is obtained. Then, according to the steps of the DSCE method to dynamically sample the candidate subsets. The experimental results are shown in Table 2.

Table 2. Experimental results of the DSCE method on imageset **A**

Subset ratio	D_{ce} value	Error value φ
8%	0.0947	0.0066
19%	0.0927	0.0046
22%	0.0914	0.0033
36%	0.0889	0.0008
42%	0.0884	0.0003

Figure 3 gives the experimental results of three selection methods. The proposed DSCE method can effectively reduce the ratio of subset and the error value φ. Compared with the SFR method, the advantages of the DSCE method are gradually significant, when the subset ratio is increased from 25% to 50%, the DSCE method can reduce the proportion of subsets by up to 8%. Compared with the SGA method, when the subset ratio is between 5% and 25%, the DSCE method is more obvious for reducing the error value φ, so it improves the robustness of subset guidance. When the subset ratio is between 25% and 50%, the DSCE method can steadily reduce the subset ratio in a smaller range, by up to 4%. In general, the DSCE method can be more flexible during the selection of a candidate subset from the imageset.

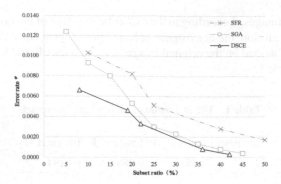

Fig. 3. Comparison of three subset selection methods on imageset **A**

Experiment 2: This experiment performs three subset selection methods on the imageset **B**, and it takes $\gamma = 0.015$. The experimental results are shown in Table 3. This shows that compared with the SFR and SGA methods, the DSCE method can dynamically determine the subset ratio while maintaining a certain accuracy, so it effectively reduces the subset ratio, and at the same time robustly determines the consistent performance of the enhancement algorithm.

For the imageset **B**, Fig. 4 compares the experimental results of three selection methods. Compared with the existing methods, the DSCE method can reduce the error value for different subset ratios. The subset ratio can be reduced by up to 7% as compared to the SFR method. The subset ratio can be reduced by up to 6% as compared to the SGA method. Experiment 2 further verified that the DSCE method is also suitable for different imagesets. Compared with the other subset selection methods, the proposed method can dynamically select a subset ratio for a different imageset.

Experiment 3: In order to verify the robustness of the DSCE method on the large imageset, this experiment repeats the above experiment procedure on the imageset **C**, where 20 images are taken at each step during the dynamic sampling and it takes the $\gamma = 0.008$. The experimental results are shown in Table 4.

Further, Fig. 5 selects some experimental results for a visual comparison. With the increase in the number of images, the DSCE method has more advantages during the subset-guided consistency enhancement evaluation. When the subset ratio is between 5% and 15%, the DSCE method can effectively reduce the subset error value as compared to the SFR method and the SGA method, it makes the conclusion of subset guidance more robust. When the subset ratio is between 15% and 25%, the SGA method does not improve the existing SFR method significantly, while the DSCE method performs better, it can reduces the subset ratio by up to 13%. When the subset ratio is between 25% and 50%, the DSCE method provides more excellence in reducing system complexity by reducing the number of subset images by up to 20%, and it greatly improves the performance of the system. The DCSE method is more competitive than the SFR method and the SGA method, and it can reduce the system complexity more effectively for the subset ratio selection of different imagesets.

Table 3. Experimental results of three subset selection methods on imageset **B**

Selection method	Subset ratio	D_{ce} value	Error value φ
SFR	10%	0.0967	0.0080
	20%	0.0951	0.0064
	25%	0.0933	0.0046
	40%	0.0911	0.0024
	50%	0.0896	0.0009
SGA	5%	0.0979	0.0092
	10%	0.0945	0.0058
	15%	0.0932	0.0045
	20%	0.0929	0.0042
	25%	0.0924	0.0037
	30%	0.0913	0.0026
	35%	0.0908	0.0021
	40%	0.0895	0.0008
	45%	0.0892	0.0005
DSCE	**7%**	0.0922	0.0035
	13%	0.0916	0.0029
	21%	0.0909	0.0022
	34%	0.0881	0.0006
	43%	0.0885	0.0002

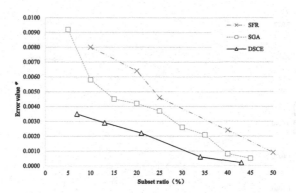

Fig. 4. Comparison of three subset selection methods on imageset **B**

Experiment 4: This experiment uses the 3C enhancement algorithm proposed by Ancuti *et al.* [6] to experiment on imageset **C**, and verifies whether the proposed method can be robust to guide other enhancement algorithms. First of all, the 3C algorithm is

Table 4. Experimental results of three subset selection methods on Imageset **C** (HE algorithm)

Selection method	Subset ratio	D_{ce} value	Error value φ
SFR	10%	0.0970	0.0086
	20%	0.0938	0.0054
	25%	0.0909	0.0025
	40%	0.0893	0.0009
	50%	0.0892	0.0008
SGA	5%	0.0981	0.0097
	10%	0.0955	0.0071
	15%	0.0947	0.0063
	20%	0.0924	0.0040
	25%	0.0901	0.0017
	30%	0.0893	0.0009
	35%	0.0892	0.0008
	40%	0.0888	0.0004
	45%	0.0886	0.0002
DSCE	7%	0.0915	0.0031
	18%	0.0904	0.0020
	21%	0.0874	0.0010
	30%	0.0879	0.0005
	37%	0.0885	0.0001

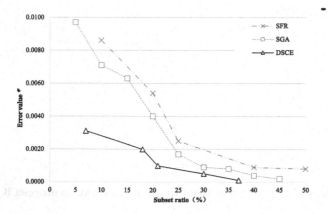

Fig. 5. Comparison of three subset selection methods on imageset **C**

used to enhance the imagist **C**, and the consistent performance of the 3C algorithm is evaluated on the imageset **C**. The consistency enhancement degree D_{ce} of imageset **C** under the 3C algorithm is 0.0898. Finally, the experimental steps of Experiment 3 are repeated. The experimental results of the imageset **C** under the 3C algorithm are shown in Table 5. Figure 6 selects some of the experimental results for visual comparison. For typical enhancement algorithms, the DSCE method is equally reliable. The 3C algorithm is more stable than the HE algorithm in image enhancement. When the imageset is evaluated consistently by the DSCE method, the consistent performance of each enhancement algorithm can be correctly judged by a smaller subset ratio, and the system complexity can be effectively reduced.

Table 5. Experimental results of three subset selection methods on imageset **C** (3C algorithm)

Selection method	Subset ratio	D_{ce} value	Error value φ
SFR	10%	0.0810	0.0088
	20%	0.0950	0.0052
	25%	0.0855	0.0043
	40%	0.0917	0.0019
	50%	0.0904	0.0006
SGA	5%	0.1004	0.0106
	10%	0.0967	0.0069
	15%	0.0951	0.0053
	20%	0.0856	0.0042
	25%	0.0925	0.0027
	30%	0.0920	0.0022
	35%	0.0913	0.0015
	40%	0.0889	0.0009
	45%	0.0893	0.0005
DSCE	6%	0.0839	0.0059
	14%	0.0940	0.0042
	19%	0.0875	0.0023
	31%	0.0907	0.0009
	36%	0.0901	0.0003

In summary, when an imageset is evaluated consistently, the candidate subset of the proposed method can be adapted to reduce the complexity of the system. When the subset ratio is large, the subset ratio can be reduced by 2% to 20% compared to the existing SFR method. When the subset proportion is smaller, the subset ratio can be reduced by 3% to 13% as compared to the SGA method.

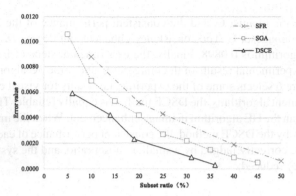

Fig. 6. Comparison of three subset selection methods on imageset **C**

5 Conclusion

In this paper, a subset ratio dynamic selection method is proposed for the subset-guided consistent enhancement evaluation criterion. Compared with the existing selection methods, the proposed method can adaptively select the subset ratio for different imagesets. The consistent performance of each enhancement algorithm can be correctly judged by a smaller subset ratio while maintaining a certain accuracy. The proposed method can effectively reduce the complexity of consistent enhancement evaluation on any large-scale imageset.

Acknowledgments. This work is supported by the Natural Science Foundation of Shanghai (18ZR1400300).

References

1. Jaffe, J.: Underwater optical imaging: the past, the present, and the prospects. IEEE J. Oceanic Eng. **40**(3), 683–700 (2015)
2. Xu, Y., Wen, J., Fei, L., Zhang, Z.: Review of video and image defogging algorithms and related studies on image restoration and enhancement. IEEE Access **4**, 165–188 (2016)
3. Lu, H., Li, Y., Zhang, L., Serikawa, S.: Contrast enhancement for images in turbid water. J. Opt. Soc. Am. A – Opt. Image Sci. Vis. **32**(5), 886–893 (2015)
4. Hitam, M., Awalludin, E., Yussof, W., Bachok, Z.: Mixture contrast limited adaptive histogram equalization for underwater image enhancement. In: International Conference on Computer and Application Technology, Sousse, pp. 1–5 (2013)
5. Voronin, V., Semenishchev, E., Tokareva, S., Zelenskiy, A., Agaian, S.: Underwater image enhancement algorithm based on logarithmic transform histogram matching with spatial equalization. In: 14th IEEE International Conference on Signal Processing, Beijing, pp. 434–438 (2018)
6. Ancuti, C.O., Vleeschouwer, C., Bekaert, P.: Color balance and fusion for underwater image enhancement. IEEE Trans. Image Process. **27**(1), 379–393 (2018)
7. Akkaynak, D., Treibitz, T.: A revised underwater image formation model. In: IEEE International Conference on Computer Vision and Pattern Recognition (CVPR), Salt Lake City, pp. 6723–6732 (2017)

8. Blasinski, H., Lian, T., Farrell, E.: Underwater image systems simulation. In: International Conference on Image Application Optical, San Francisco, p. ITh3E.3 (2017)

9. Tan, S., Wang, S., Zhang, X., Wang, S., Wang, S., Ma, S., Gao, W.: Visual information evaluation with entropy of primitive. IEEE Access **6**, 31750–31758 (2018)

10. Yang, M., Sowmya, A.: An underwater color image quality evaluation metric. IEEE Trans. Image Process. **24**(12), 6062–6071 (2015)

11. Liu, R., Fan, X., Zhu, M., Hou, M., Luo, Z.: Real-world underwater enhancement: challenges, benchmarks, and solutions under natural light. IEEE Trans. Circ. Syst. Video Technol. (2020)

12. Yang, P., Cosman, P.: Underwater image restoration based on image blurriness and light absorption. IEEE Trans. Image Process. **26**(4), 1579–1594 (2017)

13. Li, C., Guo, J., Cong, R.: Underwater image enhancement by dehazing with minimum information loss and histogram distribution prior. IEEE Trans. Image Process. **25**(12), 5664–5677 (2016)

14. Li, C., Guo, J., Chen, S.: Underwater image restoration based on minimum information loss principle and optical properties of underwater imaging. In: International Conference on Image Processing, Phoenix, pp. 1993–1997 (2016)

15. Liu, H., Wei, D., Li, D.: Subset-guided consistency enhancement assessment criterion for an imageset without reference. IEEE Access **7**, 83024–83033 (2019)

16. Ancuti, C.O., Vleeschouwer, C., Sbert, M.: Color channel compensation (3C): a fundamental pre-processing step for image enhancement. IEEE Trans. Image Process. **29**, 2653–2665 (2020)

17. Jian, M., Qi, Q., Dong, J., Yin, Y., Zhang, W., Lam, K.M.: The OUC-vision large-ratio underwater image database. In: IEEE International Conference on Multimedia and Expo (ICME), Hong Kong, pp. 1297–1302 (2017)

18. Hossfeld, T., Heegaard, P., Varela, M., Skorin-Kapov, L.: Confidence interval estimators for MOS values. arXiv: 1806.01126 (2018)

19. Wunderlich, A., Noo, F., Gallas, B., Heilbrun, M.: Exact confidence intervals for channelized hotelling observer performance in image quality studies. IEEE Trans. Med. Imaging **34**(2), 453–464 (2015)

20. Li, C., Guo, C.: An underwater image enhancement benchmark dataset and beyond. IEEE Trans. Image Process. **29**, 4376–4389 (2020)

An Attention Enhanced Graph Convolutional Network for Semantic Segmentation

Ao Chen and Yue Zhou(✉)

Shanghai Jiao Tong University, Shanghai, China
{llykevin,zhouyue}@sjtu.edu.cn

Abstract. Global modeling and integrating over feature relations are beneficial for semantic segmentation. Previous researches excel at modeling features through convolution operations but are inefficient at associating features between distant regions. Consequently, key information within one semantic category can be neglected if the features between distant regions are not bridged. To solve this issue, we propose a novel Attention Enhanced Graph Convolutional Network to explore relational information and preserve details in semantic segmentation tasks. We first build a graph over feature maps, where each node is represented by features in spatial and channel dimensions. Then the interdependent semantic information is obtained through applying the GCN to globally bridge co-occurrence features regardless of their distance. Finally, we introduce the self-attention mechanism in spatial and channel dimensions respectively to project the semantic-aware information into feature maps so as to capture the discriminative information. We examine the merit of our model on three challenging scene segmentation datasets: Cityscapes, PASCAL VOC2012 and COCO Stuff. Experimental results show that our model boosts the performance over other state-of-the-arts. Code will be available.

Keywords: Graph convolutional network · Self-attention mechanism · Semantic segmentation

1 Introduction

Semantic segmentation is a fundamental and challenging problem, with the intention to segment a scene image into separate image regions associated with semantic categories including various objects (e.g. person, bus and bicycle). The study of this task has a wide application including automatic driving, sensing and image processing.

However, some complicated situations remain to be refined. Objects can be affected by illumination and occlusion so that their features are hard to be integrated. Some works try to improve the segmentation accuracy through

© Springer Nature Switzerland AG 2020
Y. Peng et al. (Eds.): PRCV 2020, LNCS 12305, pp. 734–745, 2020.
https://doi.org/10.1007/978-3-030-60633-6_61

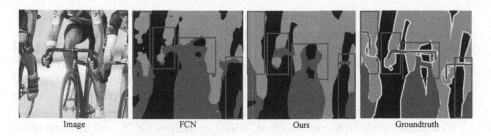

Image   FCN   Ours   Groundtruth

Fig. 1. Illustration about the difference between the results of Fully Convolutional Networks (FCN) and our approach. We apply GCN in our method to bridge co-occurrence features so as to preserve details in semantic segmentation tasks.

multi-scale feature fusion. [1] and [2] aggregate features through combining feature maps generated by different convolutions and pooling operations. Besides, Deeplabv3(+Xception) [3] focuses on the encoder-decoder structures to fuse multi-scale features. Although these researches have made progress, they tend to loss key information when dealing with complicated situation. Therefore, it is necessary to enhance the ability of associating features of same semantic category to preserve key information.

Employing recurrent neural networks to exploit relational features is another type of methods to improve semantic segmentation accuracy. The method guided by LSTM networks [4] is proposed to capture complex spatial dependencies. DAG-RNN [5] employs a recurrent neural network to capture contextual dependencies over local features. Although these methods grasp local semantic concepts within regions, the efficiency remain to be improved.

To improve the performance under the complicated circumstance, we propose a novel Attention Enhanced Graph Convolutional Network to extract key information and enhance effectiveness for semantic segmentation, as illustrated in Fig. 2. We first build a graph over feature maps, where each node is represented by features in spatial and channel dimensions. Then the interdependent semantic information is obtained through applying the GCN to globally bridge co-occurrence features [6] regardless of their distance. Finally, we introduce the self-attention mechanism in spatial and channel dimensions respectively to project the semantic-aware information into feature maps so as to capture the discriminative information.

For the features in spatial dimension, the GCN finds out the semantic category of each spatial region and the spatial feature map is updated with a weighted sum. The weights are decided by the semantic similarities between the corresponding two regions. Features of the same semantic category retain a high weight and they would be alienated if they belong to separate semantic categories. Any two regions with similar semantic features contribute mutual improvement regardless of their distance in spatial dimension. In the channel dimension, our approach characterizes each channel map by its own feature map. The GCN instructs which channels focus on a specific semantic category and we update each channel map with a weighted sum of all channel maps. Channels of the same semantic category will retain high weights in our channel module. After that, we introduce the self-attention mechanism in spatial and channel dimensions respectively to project the semantic-aware information into feature maps so as to capture the discriminative information. Finally, the outputs of these two attention enhanced GCN modules are fused to further enhance the feature representations in both spatial and channel dimensions.

Compared with [7,8], our method, which combines GCN and self-attention mechanism, can overcome environmental factors and help preserve key information so as to enhance the performance in semantic segmentation tasks.

Our main contributions can be summarized as follows:

- We apply the GCN to explore relational information and bridge co-occurrence features in semantic segmentation tasks.
- We introduce the self-attention mechanism in spatial and channel dimensions respectively. The self-attention mechanism projects the semantic-aware information into feature maps so as to preserve the discriminative information.
- Experimental results indicate that our Attention Enhanced Graph Convolutional Network attains new state-of-the-art results on three publicly achieved datasets including Cityscapes dataset, PASCAL VOC2012 dataset and COCO Stuff dataset.

2   Related Works

Semantic Segmentation. Fully Convolutional Networks (FCNs) [7] guided methods have made great progress in semantic segmentation. Several model variants are proposed to enhance feature aggregation. Deeplabv3 [1] adopts spatial pyramid pooling to embed contextual information, which consists of parallel dilated convolutions with different dilated rates. PSPNet [2] designs a pyramid pooling module to collect the effective contextual prior, containing information of different scales. DANet [8] fuses channel features to enhance global representation. DeepLabv3+(Xception-JFT) [3] utilizes the encoder-decoder structures to fuse mid-level and high-level semantic features to obtain different scale contexts.

Previous researches have made progress in multi-scale feature fusion. However, the loss of co-occurrence features caused by the diversity of scenes remains to be solved in semantic segmentation tasks.

Graph Convolutional Networks. Graph Convolutional Networks (GCN) are capable of exploiting relationships between nodes [9]. The basic idea of GCN is updating the feature of one node with another, according to the adjacent matrix of this graph. The work [10] utilizes GCN to explore inherent similarity structure among input images. [11] adopts GCN to propagate relation layer-wisely and combine the rich structural information and region features. [12] applies the GCN to achieve more robust network embedding for fully utilizing multi-relation information. In this work, we make full use of the merits of GCN, to explore relational information and globally bridge co-occurrence features regardless of their distance so that we can obtain interdependent semantic information.

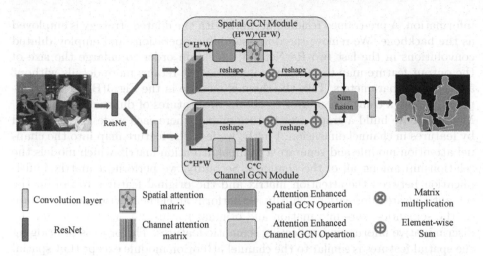

Fig. 2. An overview of the attention enhanced graph convolutional network. (Best Viewed in color)

Self-attention Mechanism. Attention modules can model long-range dependencies and have been applied in many tasks. In particular, [13] proposes the self-attention mechanism to capture global dependencies of inputs. DANet [8] applies the self-attention mechanism and receives progress in scene segmentation tasks. However, the accuracy partly rely on the quality of input information.

In our work, we introduce the self-attention enhanced GCN module in spatial and channel dimensions respectively to project the semantic-aware information into feature maps so as to capture the discriminative information. This can ensure that the discriminative information can help enhance the accuracy of semantic segmentation.

3   Methodology

In this section, we first present a general framework of our network and then introduce the two modules which associate feature information in spatial and channel dimensions respectively. Then we describe how to aggregate them together for further refinement.

3.1   Overview

Given an image of semantic segmentation, the accuracy can be affected by scales, illumination and occlusion of objects. To address this issue, we explore relational information and globally bridge features regardless of their distance. Our method can adaptively aggregate co-occurrence feature information, thus enhancing details representation in semantic segmentation tasks.

As illustrated in Fig. 2, we design two types of attention enhanced GCN modules to bridge co-occurrence features in order to obtain interdependent semantic information. A pre-trained residual network with the dilated strategy is employed as the backbone. We remove the down-sampling operations and employ dilated convolutions in the last two ResNet [14] blocks in order to enlarge the size of the output feature map to 1/8 of the input size. It retains more details without adding extra parameters. Take the channel module in the Fig. 3(B) as an example, we apply a convolution layer to obtain the features of dimension reduction. Meanwhile we build a graph over feature maps, where each node is represented by features in channel dimension. Firstly, we feed the feature map into the channel attention module and generate a channel attention matrix which models the relationship among all of the channels. Secondly, we perform a matrix multiplication between the attention matrix and the original features to obtain the semantic-aware information. Thirdly, we perform an element-wise sum operation on the semantic-aware information and original features in order to obtain the discriminative information, reflecting semantic features. The process of bridging the spatial features is similar to the channel attention module except that spatial matrix is calculated in spatial dimension. Finally, we aggregate the outputs from the two dimension modules to obtain better pixel-level feature representations in semantic segmentation tasks.

3.2 Attention Enhanced Spatial GCN Module

Feature representations are essential for scene understanding which could be obtained through capturing relational feature information. However, [2] points out that local features generated by traditional FCNs could lead to misclassification of objects. In order to build rich spatial feature relationships, we introduce an attention enhanced spatial GCN module. This GCN operation encodes a more precise spatial feature information and the self-attention mechanism helps further enhance feature representation capability. Next, we elaborate the process to aggregate spatial feature information.

As illustrated in Fig. 3, given a feature map $F_A \in R^{C \times H \times W}$, we first feed it into a convolution layer to generate two new features F_B and F_C, respectively, where $\{F_B, F_C\} \in R^{C \times H \times W}$. Then we reshape F_B to $R^{C \times N}$, where $N = H \times W$ is the number of pixels and send it into the spatial operation.

We build up the spatial operation to bridge connections between spatial regions and perform reasoning GCN to aggregate features with semantic relationships. For an image, we aim to represent it with a set of features $F_B = \{f_B^1, ..., f_B^k\} \in R^D$, such that each feature f_B^i encodes a region in this image. We measure the pairwise affinity between image regions in an embedding space to construct their relationship using Eq. 1:

$$R(f_B^i, f_B^j) = \varphi(f_B^i)^T \phi(f_B^j), \tag{1}$$

where $\varphi(f_B^i) = W_\varphi f_B^i$ and $\phi(f_B^j) = W_\phi f_B^j$ are two embedings. The weight parameters W_ϕ and W_φ can be learned through back propagation.

Then a fully-connected relationship graph $G_r = (V, E)$, where V is the set of detected regions and edge set E is described by the affinity matrix R. R is

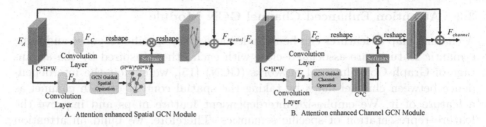

Fig. 3. The details of spatial attention module and channel attention module are illustrated in (A) and (B).

obtained by calculating the affinity edge of each pair of regions using Eq. 1. That means there will be an edge with high affinity score connecting two image regions if they have strong semantic relationships and are highly correlated.

We apply the Graph Convolutional Networks (GCN) [15] to perform reasoning on this graph. Response of each node is computed based on its neighbors defined by the graph relations. We add residual connections to the original GCN as follows:

$$V_s = W_R(RF_BW_G) + F_B, \qquad (2)$$

where W_G is the weight matrix of the GCN layer with dimensions of $D \times D$. W_R is the weight of matrix of residual structure. R is the affinity matrix with shape of $k \times k$. We follow the routine to normalize the affinity matrix R. The output V_s is the relationship enhanced spatial information for image region nodes. After that we apply a softmax layer to calculate the spatial attention map $S \in R^{N \times N}$:

$$s_{ji} = \frac{exp(V_{si} \cdot V_{sj})}{\sum_{i=1}^{N} exp(V_{si} \cdot V_{sj})}, \qquad (3)$$

where s_{ji} measures the impact of ith position on j^{th} position. The more similar the feature representations between the two spatial regions, the greater the correlation between them is. Meanwhile, we feed feature F_A into a convolution layer to generate a new feature map $F_C \in R^{C \times H \times W}$ and perform a matrix multiplication between F_C and the transpose of S and reshape the result to $R^{C \times H \times W}$. Finally, we multiply it by a scale paramater α and perform an element-wise sum operation with the features F_A to obtain the final output $F_{spatial} \in R^{C \times H \times W}$ as follows:

$$F_{spatial}^j = \alpha \sum_{i=1}^{N} (s_{ji}F_{Ci}) + F_{Aj}, \qquad (4)$$

where α is initialized to zero and gradually learns to assign more weight [16]. It can be inferred from Eq. 4 that the resulting feature $F_{spatial}$ at each position is a weighted sum of the features across all positions and original features. Therefore, it has a global contextual view and selectively aggregates contexts according to the spatial attention map. The similar semantic features achieve mutual results so as to improve semantic consistency.

3.3   Attention Enhanced Channel GCN Module

A channel map of features can be regarded as a class-specific response and same semantic features are associated close with each other. Inspired by the advantage of Graph Convolutional networks (GCN) [15], we exploit the interdependence between channel maps by taking the spatial context of each channel as a feature of it. We emphasize interdependent feature maps and improve the feature representation of specific semantics. Therefore, we build an attention enhanced channel GCN module to explicitly model interdependence between channels. The structure of channel GCN module is illustrated in Fig. 3(B). Similar with the spatial GCN module, we feed the local feature $F_A \in R^{C \times H \times W}$ into a convolutional layer to generate two new F_B and F_C, respectively, where $\{F_B, F_C\} \in R^{C \times H \times W}$. In channel dimension, we represent channel maps with a set of features $F_B = \{f_B^1, ..., f_B^k\}$, $f_B^i \in R^{1 \times H \times W}$, and we measure pairwise affinity between channel maps using Eq. 1. The relationship enhanced feature V_c for image channel dimension follows the Eq. 2. Different from the process of spatial module, we take channel feature maps as the nodes and the spatial contexts as the features in $G_r = (V, E)$. The interdependence between channel maps is exploited according to the spatial contexts. After that we apply a softmax layer to calculate the channel attention map $X \in R^{C \times C}$:

$$x_{ji} = \frac{exp(V_{ci} \cdot V_{cj})}{\sum_{i=1}^{C} exp(V_{ci} \cdot V_{cj})}, \tag{5}$$

where x_{ji} measures the impact of ith channel on j^{th} channel. In addition, we perform a matrix multiplication between the tranpose of X and F_C and reshape their result to $R^{C \times H \times W}$. Then we multiply it by a scale parameter β and perform an element-wise sum operation with the feature F_A as residual to obtain the final output $F_{channel} \in R^{C \times H \times W}$:

$$F_{channel}^j = \beta \sum_{i=1}^{C} (x_{ji} F_{Ci}) + F_{Aj}, \tag{6}$$

where β gradually learns a weight from zero. The Eq. 6 shows that the final feature of each channel is a weighted sum of the features of all channels and origin features, which models the global channel features between channel maps. It helps boost feature discriminability.

Noted that we do not employ convolution layers to embed features before computing relationships of two channels, since it can maintain relationship between different channel maps. In addition, different from recent work [16] which explores channel relationships by a global pooling or a encoding layer, we exploit all corresponding feature information to model channel correlations.

3.4   Model Optimization

In the standard training process, the network is learned with a per-pixel multinomial logistic loss (for given input image and ground truth labels) and validate with the standard metric of mean pixel intersection over union. We aggregate the features from these two attention enhanced GCN modules. Specifically, the outputs of two modules are transformed by a convolution layer and perform an element-wise sum to accomplish feature fusion. At last, a convolution layer is followed to aggregate the final prediction map. Our modules are simple and can be directly inserted in the existing FCN pipeline. They do not increase too many parameters and strengthen feature representations effectively.

4 Experiments

To evaluate the proposed method, we carry out comprehensive experiments on PASCAL VOC2012 [17], Cityscape dataset [18] and COCO Stuff dataset [19]. Experimental results demonstrate that our method achieves state-of-the-art performance on these datasets. In the following subsections, we first introduce the datasets and implementation details. Then we carry out a series of ablation experiments and visualization results on PASCAL VOC2012 dataset. Finally, we report our results on Cityscapes and COCO Stuff.

4.1 Implementation Details

We implement our method based on Pytorch. Following [16], a poly learning rate policy is employed, where the initial learning rate is modified after each iteration. The base learning rate is set to 0.001 for PASCAL VOC2012 dateset. Momentum and weight decay coefficients are set to 0.9 and 0.0001 respectively. We train our model with Synchronized BN [16]. Batchsize are set to 16 for PASCAL VOC2012 and 8 for other datesets. We set training time to 240 epochs for PASCAL VOC2012 and 200 epochs for other datasets when adopting multi-scale augmentation. Following [20], multi-loss is adopted on the end of the network when both two attention modules are used. For data augmentation, we apply random cropping (cropsize 384) and random left-right flipping during training in the ablation study for PASCAL VOC2012 dateset.

Table 1. Ablation study on PASCAL VOC2012 *test* set. *ASM* represents Attention enhanced Spatial GCN Module, *ACM* represents Attention enhanced Channel GCN Module.

Method	BaseNet	ASM	ACM	MIoU%	Method	BaseNet	ASM	ACM	MIoU%
FCN [7]	Res50			75.7	FCN [7]	Res101			76.9
Ours	Res50	✓		86.5	Ours	Res101	✓		88.9
Ours	Res50		✓	88.6	Ours	Res101		✓	90.2
Ours	Res50	✓	✓	89.9	Ours	Res101	✓	✓	92.1

4.2   Ablation Study and Results

We employ our model on top of the dilation network to capture global dependencies for better scene understanding. We verify the performance of each component by conducting some ablation experiments.

As shown in Table 1, the attention enhanced modules improve the performance remarkably. Compared with the baseline FCN (ResNet-50), employing the attention enhanced spatial module yields a result of 86.5% in Mean IoU, which brings 14.3% improvement. Meanwhile, employing attention enhanced channel module individually outperforms the baseline by 17%. The performance further improves to 89.6% when we integrate the two modules together. Furthermore, when we adopt a deeper pre-trained network (ResNet-101), the network with two modules significantly improves the segmentation performance over the baseline by 21.1%. Results show that our model brings great benefit to semantic segmentation.

Table 2 summarizes the experimental results on different attention arranging methods. From the results, we can find that arranging modules in parallel performs better than doing it in sequence. In addition, the spatial-first order performs slightly better than the channel-first order. Note that all the arranging methods outperform those using only the channel or spatial module independently.

Table 2. Performance on PASCAL VOC2012 *test* set through combining channel attention module and spatial attention module. Using both attention modules is critical while the best-combining strategy further improves the accuracy.

Method	MIoU%
ResNet101 + ACM + ASM	90.7
ResNet101 + ASM + ACM	91.6
ResNet101 + ACM & ASM in parallel	**92.1**

4.3   Comparisons with State-of-the-art Methods

Results on PASCAL VOC 2012 Dataset. Our method is compared with existing state-of-the-art methods on PASCAL VOC 2012 Dataset. Different from the previous methods, our approach associate key information explicitly, and we achieve a better performance as shown in Table 3. Our method exceeds DeepLabv3+(Xception-JFT) by 1.01%. When we adopt a deeper network ResNet-101 the model further achieves a Mean IoU of 92.1% and outperforms all the previous works, as shown in Table 3.

Visualization results in Fig. 4 indicate that our approach improve performance significantly, especially the key details on images. Taking the first image in the first row as an example, we preserve the details better in the bottom left corner of the chair than DeeplabV3(+Xception) does. Besides, on the bottom row, the results of our framework reveals the details of the front hoof of the horse as well as the tail.

Table 3. Performance on PASCAL VOC 2012 *test* set.

Method	MIoU%
FCN [7]	62.2
DeepLab-v2 [20]	71.6
Res101+PSPNet [2]	85.4
DeepLab-v3+(Xception) [3]	89.0
Res50+AEGCN(Ours)	**89.9**
Res101+AEGCN(Ours)	**92.1**

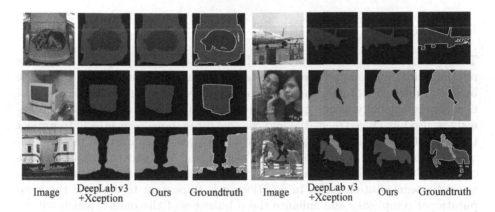

Image   DeepLab v3 +Xception   Ours   Groundtruth   Image   DeepLab v3 +Xception   Ours   Groundtruth

Fig. 4. Visualization results on the PASCAL VOC 2012 *test* set: input images, segmentation results with DeeplabV3(+Xception), segmentation results with our framework and groudtruths.

Results on Cityscape Dataset. In this subsection, we carry out experiments on the Cityscape dataset to further evaluate the effectiveness of our method. Quantitative results of Cityscape dataset are shown in Table 4. The baseline (Dilated FCN-50) yields Mean IoU 70.3%. With a deeper pre-trained network ResNet101, our model results achieve a Mean IoU of 82.3%. Among previous works, they adopt a deeper model (ResNet152) to improve their segmentation results. Table 4 indicates that our method outperforms previous methods by a large margin.

Results on COCO Stuff Dataset. Experiments on the COCO Stuff dataset are conducted to verify the generalization of our proposed network. Among the compared methods, Ding [22] adopts a gating mechanism in the decoder stage for improving inconspicuous objects and background stuff segmentation. Comparisons with previous state-of-the-art methods are reported in Table 4. Results show that our model achieves 39.7% in Mean IoU, which outperforms these methods.

Table 4. Performance on COCO Stuff *test* set and Cityscape *test* set.

COCO Stuff		Cityscape	
Method	MIoU%	Method	MIoU%
DeepLab-v2 [20]	26.9	Dilated FCN [7]	70.3
Dilated FCN [7]	31.9	DeepLab-v2 [20]	70.4
Res101+Refinenet [21]	33.6	PSPNet [2]	78.4
DANet [8]	39.7	DANet [8]	81.5
Res101+AEGCN(Ours)	**39.7**	Res101+AEGCN(Ours)	**82.3**

5   Conclusion

The Attention Enhanced Graph Convolutional Network is proposed to explore relational information and preserve details in semantic segmentation tasks. We build a graph over feature maps and obtain the interdependent semantic information through globally bridging co-occurrence features. The self-attention mechanism is also introduced in spatial and channel dimensions respectively to project the semantic-aware information into feature maps so as to capture the discriminative information. The visualization results and the experimental statistics on three datasets indicate that our model achieves outstanding performance in semantic segmentation tasks. In addition, it is important to decrease the computational complexity and enhance the robustness of the model, which will be further studied in future work.

References

1. Chen, L.C., Papandreou, G., Schroff, F., Adam, H.: Rethinking atrous convolution for semantic image segmentation. arXiv preprint arXiv:1706.05587 (2017)
2. Zhao, H., Shi, J., Qi, X., Wang, X., Jia, J.: Pyramid scene parsing network. In: Proceedings of the IEEE Conference on Computer Vision and Pattern Recognition, pp. 2881–2890 (2017)
3. Chen, L.C., Zhu, Y., Papandreou, G., Schroff, F., Adam, H.: Encoder-decoder with atrous separable convolution for semantic image segmentation. In: Proceedings of the European Conference on Computer Vision (ECCV), pp. 801–818 (2018)
4. Byeon, W., Breuel, T.M., Raue, F., Liwicki, M.: Scene labeling with LSTM recurrent neural networks. In: Proceedings of the IEEE Conference on Computer Vision and Pattern Recognition, pp. 3547–3555 (2015)
5. Shuai, B., Zuo, Z., Wang, B., Wang, G.: Scene segmentation with DAG-recurrent neural networks. IEEE Trans. Pattern Anal. Mach. Intell. **40**(6), 1480–1493 (2017)
6. Seo, Y., Defferrard, M., Vandergheynst, P., Bresson, X.: Structured sequence modeling with graph convolutional recurrent networks. In: Cheng, L., Leung, A.C.S., Ozawa, S. (eds.) ICONIP 2018. LNCS, vol. 11301, pp. 362–373. Springer, Cham (2018). https://doi.org/10.1007/978-3-030-04167-0_33
7. Long, J., Shelhamer, E., Darrell, T.: Fully convolutional networks for semantic segmentation. In: Proceedings of the IEEE Conference on Computer Vision and Pattern Recognition, pp. 3431–3440 (2015)

8. Fu, J., et al.: Dual attention network for scene segmentation. In: Proceedings of the IEEE Conference on Computer Vision and Pattern Recognition, pp. 3146–3154 (2019)
9. Park, J., Lee, M., Chang, H.J., Lee, K., Choi, J.Y.: Symmetric graph convolutional autoencoder for unsupervised graph representation learning. In: Proceedings of the IEEE International Conference on Computer Vision, pp. 6519–6528 (2019)
10. Xu, R., Li, C., Yan, J., Deng, C., Liu, X.: Graph convolutional network hashing for cross-modal retrieval. In: Proceedings of the Twenty-Eighth International Joint Conference on Artificial Intelligence, IJCAI, pp. 10–16 (2019)
11. Wu, Y., Lian, D., Jin, S., Chen, E.: Graph convolutional networks on user mobility heterogeneous graphs for social relationship inference. In: Proceedings of the 28th International Joint Conference on Artificial Intelligence, pp. 3898–3904. AAAI Press (2019)
12. Ye, R., Li, X., Fang, Y., Zang, H., Wang, M.: A vectorized relational graph convolutional network for multi-relational network alignment. In: Proceedings of the Twenty-Eighth International Joint Conference on Artificial Intelligence, IJCAI-19, pp. 4135–4141 (2019)
13. Li, K., Zhang, Y., Li, K., Li, Y., Fu, Y.: Visual semantic reasoning for image-text matching. In: Proceedings of the IEEE International Conference on Computer Vision, pp. 4654–4662 (2019)
14. He, K., Zhang, X., Ren, S., Sun, J.: Deep residual learning for image recognition. In: Proceedings of the IEEE Conference on Computer Vision and Pattern Recognition, pp. 770–778 (2016)
15. Kipf, T.N., Welling, M.: Semi-supervised classification with graph convolutional networks. arXiv preprint arXiv:1609.02907 (2016)
16. Zhang, H., Goodfellow, I., Metaxas, D., Odena, A.: Self-attention generative adversarial networks. arXiv preprint arXiv:1805.08318 (2018)
17. Everingham, M., Van Gool, L., Williams, C.K., Winn, J., Zisserman, A.: The pascal visual object classes (VOC) challenge. Int. J. Comput. Vision 88(2), 303–338 (2010)
18. Cordts, M., et al.: The cityscapes dataset for semantic urban scene understanding. In: Proceedings of the IEEE Conference on Computer Vision and Pattern Recognition, pp. 3213–3223 (2016)
19. Caesar, H., Uijlings, J., Ferrari, V.: Coco-stuff: thing and stuff classes in context. In: Proceedings of the IEEE Conference on Computer Vision and Pattern Recognition, pp. 1209–1218 (2018)
20. Chen, L.C., Papandreou, G., Kokkinos, I., Murphy, K., Yuille, A.L.: DeepLab: semantic image segmentation with deep convolutional nets, atrous convolution, and fully connected CRFs. IEEE Trans. Pattern Anal. Mach. Intell. 40(4), 834–848 (2017)
21. Lin, G., Milan, A., Shen, C., Reid, I.: RefineNet: multi-path refinement networks for high-resolution semantic segmentation. In: Proceedings of the IEEE Conference on Computer Vision and Pattern Recognition, pp. 1925–1934 (2017)
22. Ding, H., Jiang, X., Shuai, B., Qun Liu, A., Wang, G.: Context contrasted feature and gated multi-scale aggregation for scene segmentation. In: Proceedings of the IEEE Conference on Computer Vision and Pattern Recognition, pp. 2393–2402 (2018)

Variational Regularized Single Image Dehazing

Renjie He[1(✉)], Jiaqi Yang[2], Xintao Guo[1], and Zhongke Shi[1]

[1] Northwestern Polytechnical University, Xi'an, China
{davidhrj,zkeshi}@nwpu.edu.cn, guoxintao@mail.nwpu.edu.cn
[2] HuaZhong University of Science and Technology, Wuhan, China
850207131@qq.com

Abstract. Dehazing from a single hazy image has been a crucial task for both computer vision and computational photography applications. While various methods have been proposed in the past decades, estimation of the transmission map remains a challenging problem due to its ill-posed nature. Moreover, because of the inevitable noise generated during imaging process, visual artifacts could be extremely amplified in the recovered scene radiance in densely hazy regions. In this paper, a novel variational regularized single image dehazing (VRD) approach is proposed to accurately estimate the transmission map and suppress artifacts in the recovered haze-free image. Firstly, an initial transmission is coarsely estimated based on a modified haze-line model. After that, in order to preserve the local smoothness property and depth discontinuities in the transmission map, a novel non-local Total Generalized Variation regularization is introduced to refine the initial transmission. Finally, a transmission weighted non-local regularized optimization is proposed to recover a noise suppressed and texture preserved scene radiance. Compared with the state-of-the-art dehazing methods, both quantitative and qualitative experimental results on both real-world and synthetic hazy image datasets demonstrate that the proposed VRD method is capable of obtaining an accurate transmission map and a visually plausible dehazed image.

Keywords: Image dehazing · Transmission estimation · Variational regularized model · Non-local TGV

1 Introduction

Images obtained in outdoor environments are often degraded by atmospheric phenomena like haze, fog and smoke, which are mainly generated by the substantial presence of atmospheric particles scattering and absorbing the light. As a result, acquired outdoor scenery images are often of low visual quality, such as reduced contrast, limited visibility, and weak color fidelity. In addition,

Supported by NSFC (61671387, 61420106007, 61871325) and NSSX (2019JQ-572).

Y. Peng et al. (Eds.): PRCV 2020, LNCS 12305, pp. 746–757, 2020.
https://doi.org/10.1007/978-3-030-60633-6_62

this type of degradation is spatially variant as the level of scattering depends on the distance between scene objects and the imaging device. Consequentially, the performance of computer vision-based algorithms like detection, recognition, and surveillance may be severely limited under such hazy days. The goal of dehazing is to remove the influence of haze and recover a clear scene image, which is crucial for both computational photography and computer vision applications. It is also beneficial to obtain depth information of the scene during the dehazing process. Removing the influence of haze has been widely studied in the past decades. However, haze model is highly under-constrained, and removing haze from only a single image remains a challenging ill-posed task. A milestone in single image dehazing was introduced by He *et al.* [7] in 2009. Based on the observation that natural images are typically composed of colorful and dark objects, they proposed a novel Dark Channel Prior (DCP) assuming that at least one color channel should have a significantly small pixel value in a clear image. These approaches achieved satisfactory dehazing results in the early years In recent years, the performance of single image dehazing bas been improved continuously [2,5,11–15,23]. More recently, machine learning methods, especially the convolutional neural network (CNN) based methods, have been introduced to single image dehazing [1,8,9,17,19]. In addition, the latest residual-based deep CNN technique also has been brought into image dehazing field [10,20,21]. Although these learning-based methods can produce impressive dehazed results, they depend heavily on a large number of annotated training examples due to its data-driven property.

Accurately estimating the transmission map is crucial in the dehazing process, most of the state-of-the-art single image dehazing algorithms estimate transmission map based on the assumption that transmission is locally constant [2,4–7,15,16]. However, the piece-wise constant transmission map may lead to undesired halo artifacts in the dehazed results. In fact, since the value of a transmission map is solely related to scene depth, the transmission should be spatially smooth and only have jumps value at edges with discontinuing depth. In addition, while most dehazing approaches focused on estimating the transmission map, less attention has been paid to visual artifacts boosting in the dehazed results. Conventionally, with all the unknowns are estimated, the scene radiance is recovered directly according to the hazy imaging model. However, noise and other visual artifacts are often revealed and amplified in the dehazed results, especially in regions far away from the observer.

To address the problems mentioned above, a novel variational regularized single image dehazing (VRD) approach is proposed in this paper to accurately estimate the transmission map and suppress artifacts in the recovered haze-free image. The proposed approach follows the coarse-to-fine transmission map estimation framework. The main contributions of our paper are: **1)** A variational regularized framework is proposed to address the single image dehazing problem. **2)** A non-local TGV regularization model is proposed to estimate the transmission map, where the piece-wise smoothness property and depth structures are well preserved. **3)** A transmission weighted optimization is introduced to recover the scene radiance without noise amplification. Compared with the state-of-the-art dehazing methods, both quantitative and qualitative experimental results indicate that the proposed method is capable of obtaining an accurate transmission map and a visually plausible dehazed image.

The rest of the paper is organized as follows: Sect. 2 introduces the background of single image dehazing and existing limitations. Section 3 thoroughly describes the proposed method. Section 4 shows the experimental results and analysis. Finally, Sect. 5 concludes the paper.

2   Background

2.1   Haze Image Modeling

In hazy days, sensors not only receive the attenuated reflection from the scene objects, but also the additive light in the atmosphere. To describe the formation of the hazy imaging process, in computer vision, a hazy image can be mathematically modeled as the pixel-wise convex combination of the scene radiance and the airlight according to the Koschmieder's law as follows:

$$I^c(\mathbf{x}) = J^c(\mathbf{x}) t(\mathbf{x}) + A^c(1 - t(\mathbf{x})), \tag{1}$$

where $I(\mathbf{x})$ and $J(\mathbf{x})$ represent the observed hazy image and the haze-free scene radiance, respectively. A is the global atmospheric light, which is assumed to be spatially constant. $c \in \{R, G, B\}$, and $\mathbf{x}$ indicate the color channels and pixel index, respectively. $t(\mathbf{x}) \in [0, 1]$ is the depth dependent transmission map describing the portion of the light that is unscattered and observed by the camera.

In RGB color space, by denoting vectors $\mathbf{I}(\mathbf{x}) - \mathbf{A}$ as $\mathbf{I}_A(\mathbf{x})$ and $\mathbf{J}(\mathbf{x}) - \mathbf{A}$ as $\mathbf{J}_A(\mathbf{x})$, it can be inferred that $\mathbf{I}_A(\mathbf{x})$ and $\mathbf{J}_A(\mathbf{x})$ have the same orientation and the transmission $t(\mathbf{x})$ is the magnitude ratio between $\mathbf{I}_A(\mathbf{x})$ and $\mathbf{J}_A(\mathbf{x})$, $t(\mathbf{x})$ can thus be rewritten as follows:

$$t(\mathbf{x}) = \frac{\|\mathbf{I}_A(\mathbf{x})\|}{\|\mathbf{J}_A(\mathbf{x})\|}, \quad \hat{\mathbf{I}}_A(\mathbf{x}) = \hat{\mathbf{J}}_A(\mathbf{x}), \tag{2}$$

where $\|\mathbf{I}\|$ and $\hat{\mathbf{I}}$ indicate the magnitude and the orientation of vector $\mathbf{I}$, respectively.

3   Methodology

As schematically shown in Fig. 1, our method is composed of two major components for estimating the transmission map and recovering the scene radiance, respectively. The framework and the principle of each component will be presented in detail in the following subsections.

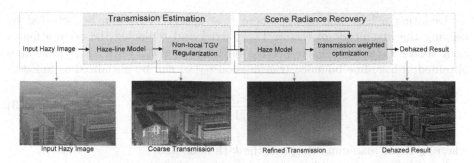

Fig. 1. The framework of our method.

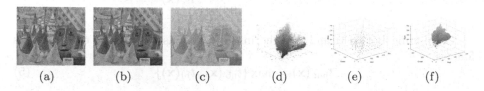

Fig. 2. Illustration of haze-line model. (a): original image, (b): quantized image, (c): hazy image of (b), (d)-(f): pixel distribution in RGB space of (a)-(c).

3.1 Transmission Map Estimation

Transmission Initialization. In our work, the haze-line model [2] is modified to give an initial estimation on the transmission t. According to the color quantization theory, the number of unique colors used in an image can be reduced to a small number without distorting visual similarity to the original input image [22]. Therefore, by clustering pixels to a given number of clusters, a haze-free image can be faithfully expressed using the cluster center, as shown in Fig. 2(b) and (e). Since pixels are distributed all over the image plane, pixels in the hazy image will form lines spanned by their original color of the quantized haze-free image and the airlight in RGB color space, as shown in Fig. 2(c) and (f). Pixels in the same haze-line should have a similar color in the haze-free image. Berman *et al.* [2] indicate these lines as haze-lines.

Theoretically, the maximum value of transmission map $t = 1$ is obtained when vectors $\mathbf{I}_A(\mathbf{x})$ and $\mathbf{J}_A(\mathbf{x})$ have the same magnitude. Since pixels on the same haze-line are distributed all over the whole image with different depth values, the transmission of the same haze-line can thus be obtained by setting the maximum magnitude of one detected haze-line as the initial estimation of $\|\mathbf{J}_A(\mathbf{x})\|$. Therefore, each haze-line can be detected by clustering the orientations of $\mathbf{I}_A(\mathbf{x})$ for every pixel in the observed hazy image to N clusters. Practically, because of the inevitable noise in the observed image, orientations may be incorrectly estimated for vectors $\mathbf{I}_A(\mathbf{x})$ with a small magnitude. Therefore, we propose to estimate the haze-lines using a patched hazy image based on the assumption that transmission and surface albedo are constant in a small local patch [4,5].

Denoting a haze-line as L_k, where $k = 1, \ldots, N$, L_k can be detected by keeping the magnitude of vectors $\mathbf{I}_A(\mathbf{x})$ fixed and replacing the orientation of each vector $\mathbf{I}_A(\mathbf{x})$ with its respective cluster center $\hat{\mathbf{I}}_k$. With all N haze-lines obtained, a haze-line bounded transmission $t_{\mathrm{hl}}(\mathbf{x})$ can be initialized as follows:

$$t_{\mathrm{hl}}(\mathbf{x}) = \frac{\|\mathbf{I}_A(\mathbf{x})\|}{\max\limits_{\mathbf{x} \in L_k} \{\|\mathbf{I}_A(\mathbf{x})\|\}}, \mathbf{x} \in L_k, \tag{3}$$

where l_k is the maximum length of the k-th haze-line.

In addition, a natural lower boundary of transmission t_{nb} can be inferred as follows:

$$t_{\mathrm{nb}}(\mathbf{x}) = 1 - \min_{c \in \{R,G,B\}} \frac{I^c(\mathbf{x})}{A^c} \tag{4}$$

Combining both (3) and (4), the initial transmission can be calculated as follows:

$$t_{\mathrm{init}}(\mathbf{x}) = \max \{t_{\mathrm{nb}}(\mathbf{x}), t_{\mathrm{hl}}(\mathbf{x})\} \tag{5}$$

As shown in Fig. 1, the initial coarse transmission estimated above contains too many textures and artifacts, which will result in halos and block artifacts in the dehazed result. Therefore, it is necessary to refine the initial transmission map.

Transmission Refinement. Generally, the refinement of the transmission map can be modeled as a regularized optimization. Various approaches have been proposed to refine the coarse transmission map based on the assumption that the transmission map is piece-wise smooth and only changes at depth jumps. Traditional Total Variation based regularizations generate piece-wise constant images, which may produce inaccurate depth discontinuities and lead to staircasing artifacts in the dehazed results. Non-local Total Variation based methods show better performance on edge preserving. However, the staircasing artifacts still exist. To encourage the piece-wise smoothness property, TGV based regularization is employed in [3]. Unfortunately, the depth discontinuities can not be well preserved using TGV.

In order to preserve the depth discontinuities as well as the piece-wise smoothness of the transmission map, a second-order non-local TGV regularization is introduced in our method. Specifically, the refined transmission map can be obtained by minimizing the following energy function:

$$t_{\mathrm{opt}}(\mathbf{x}) = \underset{t}{\mathrm{argmin}} \left\| \frac{t(\mathbf{x}) - t_{\mathrm{init}}(\mathbf{x})}{\sigma(\mathbf{x})} \right\|_1 + \mathrm{NLTGV}_\alpha^2(t(\mathbf{x})), \tag{6}$$

where the first term is the data term embodying the assumption that the refined transmission $t_{\mathrm{opt}}(\mathbf{x})$ and the initial transmission $t_{\mathrm{init}}(\mathbf{x})$ are close to each other, and σ is the standard deviation of $t_{\mathrm{init}}(\mathbf{x})$ in each haze-line. Instead of utilizing the commonly used squared Euclidean norm, the $\ell1$-norm is employed in our method to ensure robustness to outliers caused by noise and ambiguity between image color and depth.

The second term is a non-local TGV regularization term indicating the assumptions or priors on the transmission, which is defined as follows:

$$
\begin{aligned}
\text{NLTGV}_\alpha^2 \left(t(\mathbf{x})\right) = \min_\omega \sum_{\mathbf{x}} \sum_{\mathbf{y} \in \Omega(\mathbf{x})} \alpha_0\left(\mathbf{x}, \mathbf{y}\right) \left|\omega^1\left(\mathbf{x}\right) - \omega^1\left(\mathbf{y}\right)\right| \\
+ \sum_{\mathbf{x}} \sum_{\mathbf{y} \in \Omega(\mathbf{x})} \alpha_0\left(\mathbf{x}, \mathbf{y}\right) \left|\omega^2\left(\mathbf{x}\right) - \omega^2\left(\mathbf{y}\right)\right| \\
+ \sum_{\mathbf{x}} \sum_{\mathbf{y} \in \Omega(\mathbf{x})} \alpha_1\left(\mathbf{x}, \mathbf{y}\right) \left|t\left(\mathbf{x}\right) - t\left(\mathbf{y}\right) + \left\langle \omega\left(\mathbf{x}\right), \mathbf{x} - \mathbf{y}\right\rangle\right|
\end{aligned}
\tag{7}
$$

where $\omega(\mathbf{x}) = \left(\omega^1, \omega^2\right)^{\mathrm{T}}$ is a auxiliary vector with low variation. $\langle \centerdot, \centerdot \rangle$ represents the inner product, measuring the total deviation of t. α_1 and α_0 are two weighting functions.

By reformulating (6) to a convex-concave saddle point problem with the Legendre-Fenchel transform, the primal-dual formulation of (6) can thus be written as follows:

$$
\begin{aligned}
\min_{t,\omega} \max_{p \in P, q \in Q} \left\| \frac{t(\mathbf{x}) - t_{\text{init}}(\mathbf{x})}{\sigma(\mathbf{x})} \right\|_1 \\
+ \sum \sum \left(t(\mathbf{x}) - t(\mathbf{y}) + \langle \mathbf{y} - \mathbf{x}, \omega(\mathbf{x})\rangle\right) p_{\mathbf{xy}} \\
+ \langle \omega(\mathbf{x}) - \omega(\mathbf{y}), q_{\mathbf{xy}}\rangle,
\end{aligned}
\tag{8}
$$

where p and q are dual variables with feasible solution sets P and Q, respectively, which are defined as follows:

$$
\begin{aligned}
P = \left\{ p \in \mathbb{R}^{2MN}, \|p\|_\infty \leq \alpha_1\left(\mathbf{x}, \mathbf{y}\right)\right\} \\
Q = \left\{ q \in \mathbb{R}^{4MN}, \|q\|_\infty \leq \alpha_0\left(\mathbf{x}, \mathbf{y}\right)\right\}
\end{aligned}
\tag{9}
$$

The optimization of (8) can be derived iteratively.

The first row of Fig. 4 demonstrates a comparison on refined transmission by DCP [7], CL [5], TGV [3], NL [2] and the proposed VRD, respectively. It is observed that the proposed VRD is capable of obtaining a smooth transmission while preserving depth discontinuity.

3.2   Scene Radiance Restoration

It is evident that recovering J directly will lead to artifacts boosting, such as noise magnification, color distortion and false edges emergence in the dehazed results. Based on the observation that visual artifacts are depth-related, where the noise amplification becomes severe in regions with small transmission values, a transmission weighted regularization is proposed in our approach to suppress

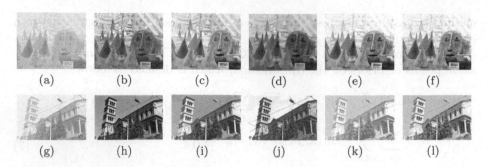

Fig. 3. Comparison on synthetic hazy images *Cones* and *Church*. (a): synthetic hazy image *Cones*, (g): synthetic hazy image *Church*, (b)(h): results by DCP [7], (c)(i): results by CL [5], (d)(j): results by TGV [3], (e)(k): results by NL [2], (f)(l): results by our VRD.

the amplified noise in the recovered scene radiance. Accordingly, the objective function is constructed as follows:

$$J = \underset{J}{\arg\min} \|t\,(J - J_0)\,\|_2^2 + \lambda_n \sum_{\mathbf{x}} \sum_{\mathbf{y} \in \Omega(\mathbf{x})} \omega_J\,(\mathbf{x}, \mathbf{y})\,|J\,(\mathbf{x}) - J\,(\mathbf{y})|. \tag{10}$$

The first term is a ℓ-2 norm based data fidelity term. $J_0\,(\mathbf{x})$ is the initial value of the scene radiance. The fidelity between J and J_0 is weighted by the value of transmission t based on the assumption that for pixels with large t, J_0 at corresponding region should be trusted, while for pixels with small t, the corresponding J_0 is not reliable due to the magnified noise. The second term is a non-local total variation regularization term to preserve local smoothness and suppress noise, where λ_n is the regularization parameter, $\omega_J\,(\mathbf{x}, \mathbf{y})$ is a weighting function, and $\Omega\,(\mathbf{x})$ denotes a patch centred at pixel $\mathbf{x}$. For pixels with similar transmission value and similar color, the value of the weighting function $\omega_J\,(\mathbf{x}, \mathbf{y})$ should be large to encourage spatial smoothness. While for pixels with large depth variance or color difference, $\omega_J\,(\mathbf{x}, \mathbf{y})$ should be small. Therefore, $\omega_J\,(\mathbf{x}, \mathbf{y})$ is defined as follows:

$$\omega_J\,(\mathbf{x}, \mathbf{y}) = \exp\left(-\left(\frac{|t(\mathbf{x}) - t(\mathbf{y})|^2}{2\sigma_t^2} + \frac{\sum_c \sum_{\mathbf{y}} (\mathbf{J}_0(\mathbf{x}) - \mathbf{J}_0(\mathbf{y}))^2}{2\sigma_J^2}\right)\right), \tag{11}$$

where σ_t and σ_J denote the standard deviation of t and $\mathbf{J}_0$.

By introducing an auxiliary variable v_y, (10) can be reformulated as follows:

$$J = \underset{J,v}{\arg\min} \|t\,(J - J_0)\,\|_2^2 + \sum_{\mathbf{y} \in \Omega(\mathbf{x})} \|\omega \circ v_y\|_1 + \beta \sum_{\mathbf{y} \in \Omega} \|v_j - \nabla J(\mathbf{x})\|_2^2. \tag{12}$$

The solution of (12) is obtained by:

$$v^{k+1}(\mathbf{x}) = \frac{\nabla J^k(\mathbf{x})}{|\nabla J^k(\mathbf{x})|} \max\left(|\nabla J^k(\mathbf{x})| - \frac{1}{\beta}, 0\right) \tag{13}$$

4   Experimental Results

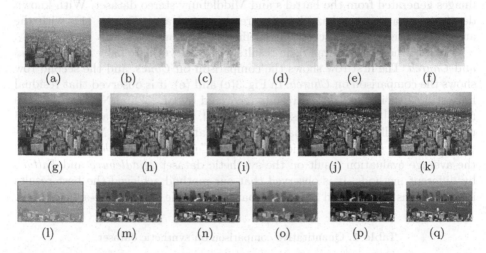

(a)   (b)   (c)   (d)   (e)   (f)

(g)   (h)   (i)   (j)   (k)

(l)   (m)   (n)   (o)   (p)   (q)

Fig. 4. Comparison on *Manhattan* image. (a): original hazy image, (g): result by DCP [7], (h): result by CL [5], (i): result by TGV [3], (j): result by NL [2], (k): result by our VRD, (b)–(f): estimated transmission of (g)–(k), (l): local enlargement of (a), (m)–(q): local enlargements of (g)–(k).

Comprehensive experiments are conducted in this section to demonstrate the performance of the proposed method. The proposed algorithm is implemented using MATLAB 2018a on PC with i7-7700K CPU and 16 GB RAM. In addition, the proposed method is evaluated on datasets containing both natural and synthetic images. Particularly, the O-HAZE dataset and real-world benchmark images used by various methods are employed for evaluation. The O-HAZE dataset contains 45 different outdoor scenes depicting the same visual content recorded in haze-free and man-made hazy conditions. Fattal's dataset and images from the Middlebury stereo Datasets are selected to build the synthetic data. The Fattal's dataset contains 11 pairs of haze-free images on natural scenes and the corresponding hazy images generated using true depth maps.

The performance of the proposed VRD method is compared with representative classic and state-of-the-art single image dehazing methods, including the DCP [7], CL [5], NL [2] and TGV [3], where the parameters are set as defined in the references. In the implementation, the regularization parameters for transmission refinement are set as $\alpha_0 = 0.5$, $\alpha_0 = 0.01$. The regularization parameters for scene radiance restoration are set as $\lambda_n = 0.01$, $\lambda_e = 0.005$ and the patch size is 5×5. For simplicity, airlight A is obtained directly by the robust airlight estimation method [18].

4.1   Experiments on Synthetic Images

We first compare the dehazing results of different methods on the synthetic images generated from the Fattal's and Middlebury stereo dataset. With known depth information, the hazy images are generated according to (1), with the atmospheric light **A** and scattering coefficient β set manually.

Figure 3 shows two comparison results with the synthetic hazy images *Cones* and *Church*. The first row shows the comparison on *Cones*, and the second row shows the comparison on *Church*. In Fig. 3(c) and (e), it is observed that residual haze still exists in the dehazed results recovered by CL [5] and NL [2]. Also, the result by TGV [3] in Fig. 3(d) is over-smoothed. Both DCP [7] and our method recovered visually plausible dehazing results. Table 1 shows the MSE and structural similarity (SSIM) evaluation results on *Cones* and *Church*, together with the average evaluation result on the synthetic dataset *Middlebury* and *Fattal's* by different methods. It is observed that our method achieved the best results in both terms of MSE and SSIM, comparing with the state-of-the-art methods.

Table 1. Quantitative comparison on synthetic dataset.

		DCP [7]	CL [5]	TGV [3]	NL [2]	VRD
Cones	MSE	0.0095	0.0103	0.0125	0.0091	**0.0067**
	SSIM	0.9281	0.9722	0.9701	0.9837	**0.9886**
Avg. on *Middlebury*	MSE	0.0137	0.0122	0.0098	0.0085	**0.0081**
	SSIM	0.9733	0.9881	0.9830	0.9923	**0.9967**
Church	MSE	0.0081	0.0097	0.0088	0.0080	**0.0076**
	SSIM	0.9534	0.9680	0.9712	0.9868	**0.9890**
Avg. on Fattal's	MSE	0.0126	0.0113	0.0101	0.0084	**0.0079**
	SSIM	0.9680	0.9702	0.9725	0.9855	**0.9879**

4.2   Experiments on Real-World Images

The proposed VRD is also evaluated on real-world natural datasets both qualitatively and quantitatively. For images in O-HAZE dataset with ground-truth haze-free images, the PSNR and SSIM metrics are employed to quantitatively evaluate the performance of each method. For real-world hazy images without ground-truth, a no-reference haze density metric (HDM) is applied for evaluation. The HDM includes three evaluators, namely e, Σ and $\bar{r}$, which indicate the rate of newly appeared edges, the percentage of pure white or pure black pixels and the mean ratio of edge gradients, respectively.

Figure 4 shows one comparison result on the test image *Manhattan* in the benchmark dataset, to demonstrate the effectiveness on noise suppression of the proposed VRD. The original hazy image and transmissions estimated with different methods are shown in the first row of Fig. 4. The dehazing results and the enlarged region of the red and blue rectangles are demonstrated in the second and the third rows, respectively. It is observed that texture details and object boundaries are preserved in the transmission maps estimated by DCP, CL and NL, texture edges and sharp edges effectively Both TGV and our VRD are capable of generating a smooth transmission map. However, our VRD can preserve depth discontinued edges It is observed that all five model-based methods are capable of removing haze from the original image. However, magnified noise and degradation of original information are observed in the sky regions in results of DCP (Fig. 4(g)) and NL (Fig. 4(j)). In addition, in Fig. 4(h) and (j), the results by CL [5] and NL [2] are over-saturated, which lead to unnatural appearance in the results. Also, in Fig. 4(i), distant objects are over-smoothed in the dehazed result by TGV [3]. As observed in Fig. 4(k), the proposed method is capable of recovering a visually plausible dehazing result without introducing magnified noise and artifacts in the recovered haze-free image.

Figure 5 shows a comparison result on the test image *Playground* in O-HAZE dataset. Additionally, a deep learning based method DehazeNet [1] is also employed for comparison. All the methods are capable of enhancing the contrast of the scene. However, color distortion is noticed in the result by DCP and CL due to the invalid prior in bright regions and inaccurate airlight. TGV has limitation on removing dense haze and the contrast of the scene remains low in the result. NL and the proposed VRD are capable of removing most of the haze and increasing the scene contrast significantly. In addition, it is also noticed that DehazeNet can achieve a satisfying fidelity on color. However, haze remains obvious in the dehazed result. Table 2 shows the HDM evaluation results on *Manhattan* and the average evaluation result on the real-world dataset by different methods. It is shown that the proposed VRD achieved the best performance for all indicators. Table 3 shows the MSE and SSIM evaluation results on *Playground* and the average evaluation result on the O-HAZE dataset by different methods. It is observed that our method achieved better results in both terms of PSNR and SSIM, comparing with the state-of-the-art methods.

Table 2. Quantitative comparison on real-world benchmark dataset.

		DCP [7]	CL [5]	TGV [3]	NL [2]	VRD
Manhattan	e	1.53	1.84	1.39	1.40	**1.86**
	Σ	**0**	**0**	**0**	0.03	**0**
	$\bar{r}$	4.80	5.05	4.88	3.95	**5.08**
Avg. on real-world	e	0.75	0.81	0.79	0.76	**0.82**
	Σ	0.01	0.01	0.01	**0**	**0**
	$\bar{r}$	2.19	2.12	1.97	2.20	**2.35**

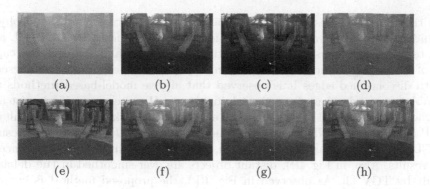

Fig. 5. Comparison on *Playground* image. (a): hazy image, (b): result by DCP [7], (c): result by CL [5], (d): result by TGV [3], (e): haze-free image, (f): result by NL [2], (g): result by DehazeNet, (h): result by our VRD.

Table 3. Quantitative comparison on O-HAZE dataset.

		DCP [7]	CL [5]	TGV [3]	NL [2]	VRD
Playground	PSNR	0.7020	0.6840	0.7270	<u>0.7600</u>	**0.7650**
	SSIM	16.8650	16.5730	16.3010	**17.2310**	<u>17.0500</u>
Avg.on O-HAZE	PSNR	0.7320	0.7050	0.7300	<u>0.7500</u>	**0.7580**
	SSIM	16.5860	15.6350	16.0500	<u>16.6120</u>	**16.8610**

5 Conclusion

In this paper, we present a novel variational regularized single image dehazing method (VRD) to accurately estimate the transmission map and suppress artifacts in the recovered haze-free image. For transmission estimation, we proposed a non-local Total Generalized Variation regularization to refine a coarse transmission map by preserving the local smoothness property and depth discontinuities in the transmission map. For scene radiance recovery, we introduced a transmission weighted non-local regularized optimization to recover a noise suppressed and texture preserved scene radiance. We have conducted extensive experiments on both real-world and synthetic datasets. In the comparisons with the state-of-the-art, both the quantitative and qualitative evaluations indicate the superiority of our proposed VRD.

References

1. Cai, B., Xu, X., Jia, K., Qing, C., Tao, D.: DehazeNet: an end-to-end system for single image haze removal. IEEE Trans. Image Process. **25**(11), 5187–5198 (2016)
2. Berman, D., Treibitz, T., Avidan, S.: Non-local image dehazing. In: Proceedings of IEEE Conference on Computer Vision and Pattern Recognition, pp. 1674–1682 (2016)

3. Chen, C., Do, M., Wang, J.: Robust image and video dehazing with visual artifact suppression via gradient residual minimization. In: Proceedings of European Conference on Computer Vision, pp. 576–591 (2016)
4. Fattal, R.: Single image dehazing. ACM Trans. Graphics **27**(3), 1–9 (2008)
5. Fattal, R.: Dehazing using color-lines. ACM Trans. Graphics **34**(1), 1–14 (2014)
6. Gibson, K., Nguyen, T.: An analysis of single image defogging methods using a color ellipsoid framework. Eurasip J. Image Video Process. **2013**(1), 37 (2013)
7. He, K., Sun, J., Tang, X.: Single image haze removal using dark channel prior. In: Proceedings of IEEE Conference on Computer Vision and Pattern Recognition, pp. 1956–1963 (2009)
8. Kim, G., Kwon, J.: Robust pixel-wise dehazing algorithm based on advanced haze-relevant features. In: Proceedings of the British Machine Vision Conference, pp. 79.1–79.12 (2017)
9. Li, B., Peng, X., Wang, Z., Xu, J., Feng, D.: AOD-Net: all-in-one dehazing network. In: Proceedings of IEEE Conference on Computer Vision and Pattern Recognition, pp. 4770–4778 (2017)
10. Li, J., Li, G., Fan, H.: Image dehazing using residual-based deep CNN. IEEE Access **6**, 26831–26842 (2018)
11. Li, Z., Zheng, J.: Edge-preserving decomposition-based single image haze removal. IEEE Trans. Image Process. **24**(12), 5432–5441 (2015)
12. Li, Z., Zheng, J.: Single image de-hazing using globally guided image filtering. IEEE Trans. Image Process. **27**(1), 442–450 (2018)
13. Li, Z., Zheng, J., Zhu, Z., Yao, W., Wu, S.: Weighted guided image filtering. IEEE Trans. Image Process. **24**(1), 120–129 (2015)
14. Liu, Y., Shang, J., Pan, L., Wang, A., Wang, M.: A unified variational model for single image dehazing. IEEE Access **7**, 15722–15736 (2019)
15. Meng, G., Wang, Y., Duan, J., Xiang, S., Pan, C.: Efficient image dehazing with boundary constraint and contextual regularization. In: Proceedings of IEEE International Conference on Computer Vision, pp. 617–624 (2013)
16. Nishino, K., Kratz, L., Lombardi, S.: Bayesian defogging. Int. J. Comput. Vision **98**(3), 263–278 (2012)
17. Ren, W., Liu, S., Zhang, H., Pan, J., Cao, X., Yang, M.: Single image dehazing via multi-scale convolutional neural networks. In: Proceedings of European Conference on Computer Vision, pp. 154–169 (2016)
18. Sulami, M., Glatzer, I., Fattal, R., Werman, M.: Automatic recovery of the atmospheric light in hazy images. In: Proceedings of IEEE International Conference on Computational Photography, pp. 1–11 (2014)
19. Tang, K., Yang, J., Wang, J.: Investigating haze-relevant features in a learning framework for image dehazing. In: Proceedings of IEEE Conference on Computer Vision and Pattern Recognition, pp. 2995–3002 (2014)
20. Wang, C., Li, Z., Wu, J., Fan, H., Xiao, G., Zhang, H.: Deep residual haze network for image dehazing and deraining. IEEE Access **8**, 9488–9500 (2020)
21. Single image dehazing based on learning of haze layers. Neurocomputing (2020). https://doi.org/10.1016/j.neucom.2020.01.007
22. Zhang, Q., Xiao, C., Sun, H., Tang, F.: Palette-based image recoloring using color decomposition optimization. IEEE Trans. Image Process. **26**(4), 1952–1964 (2017)
23. Zhu, Q., Mai, J., Shao, L.: A fast single image haze removal algorithm using color attenuation prior. IEEE Trans. Image Process. **24**(11), 3522–3533 (2015)

Author Index

Printed in the United States
By Bookmasters